Designing New Heterogeneous Catalysts

ondon, United Kingdom

–6 April 2016

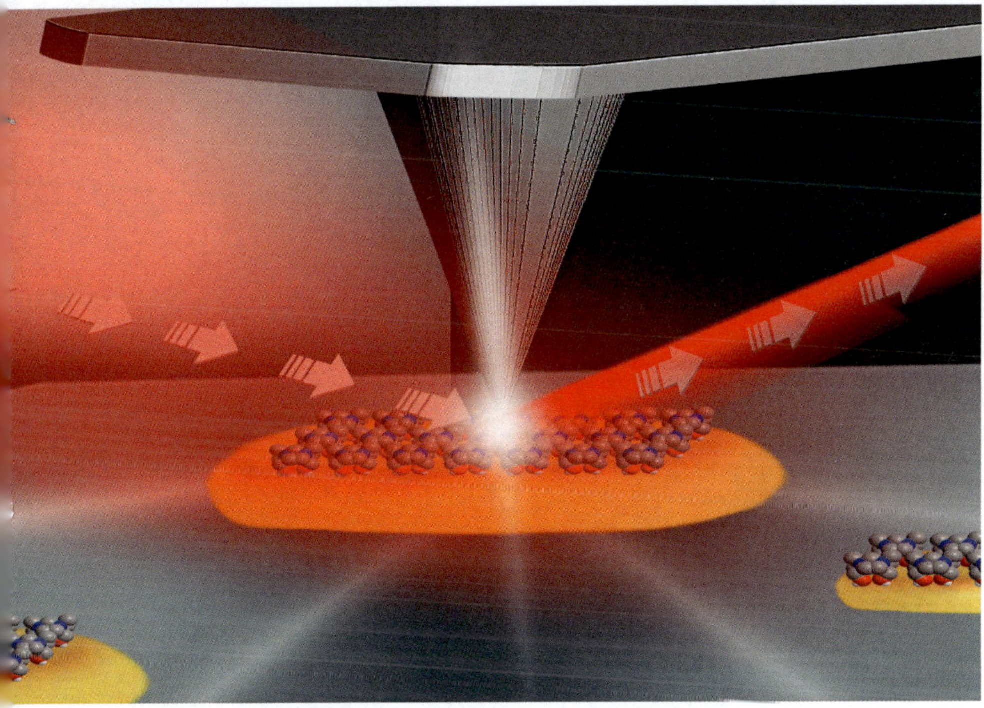

FARADAY DISCUSSIONS

Volume 188, 2016

ROYAL SOCIETY OF CHEMISTRY

The Faraday Division of the Royal Society of Chemistry, previously the Faraday Society, founded in 1903 to promote the study of sciences lying between Chemistry, Physics and Biology.

Faraday Discussions (Print ISSN 1359-6640, Electronic ISSN 1364-5498) is published 10 times a year by the Royal Society of Chemistry, Thomas Graham House, Science Park, Milton Road, Cambridge, UK CB4 0WF.

Volume 188 ISBN-13: 978-1-78262-475-2

2016 annual subscription price: print+electronic £929 US $1635; electronic only £883, US $1553. Customers in Canada will be subject to a surcharge to cover GST. Customers in the EU subscribing to the electronic version only will be charged VAT. All orders, with cheques made payable to the Royal Society of Chemistry, should be sent to RSC Order Department, Royal Society of Chemistry, Thomas Graham House, Science Park, Milton Road, Cambridge, CB4 0WF, UK.

Tel +44 (0) 1223 432398; E-mail www.orders@rsc.org

If you take an institutional subscription to any RSC journal you are entitled to free, site-wide web access to that journal. You can arrange access via Internet Protocol (IP) address at www.rsc.org/ip. Customers should make payments by cheque in sterling payable on a UK clearing bank or in US dollars payable on a US clearing bank.

Printed in the UK

MIX
Paper from responsible sources
FSC www.fsc.org FSC® C013604

Faraday Discussions are unique international discussion meetings that focus on rapidly developing areas of chemistry and its interfaces with other scientific disciplines.

)esigning New Heterogeneous :atalysts

araday Discussions
'ww.rsc.org/faraday_d

General Discussion on Designing New Heterogeneous Catalysts was held in London, UK on the n, 5th and 6th of April 2016.

:ONTENTS

;N 1359-6640; ISBN 978-1-78262-475-2

Cover
See Levratovsky and Gross, *Faraday Discuss.*, 2016, **188**, 345–353.

High spatial resolution tip-enhanced IR spectroscopy of N-heterocyclic carbine molecules that were anchored to the surface of catalytically active Pt nanoparticles.

Image reproduced by permission of Elad Gross, *Faraday Discuss.*, 2016, **188**, 345–353.

ITRODUCTORY LECTURE

APERS AND DISCUSSIONS

ABSTRACT BOOK SPONSOR

INDIUM CORPORATION®

EXHIBITOR

SPECS™

CONCLUDING REMARKS

ADDITIONAL INFORMATION

Faraday Discussions

PAPER

Spiers Memorial Lecture

Heterogeneous catalysis: understanding the fundamentals for catalyst design

Avelino Corma

Received 1st April 2016, Accepted 27th April 2016
DOI: 10.1039/c6fd00066e

Taking the chemoselective hydrogenation of substituted nitroaromatics as a base case, it will be shown that it is possible to design improved and new catalysts by attacking the problem in a multidisciplinary way. By combining molecular modeling with *in situ* operando spectroscopy, and with micro-kinetic and isotopic studies, it is possible to determine how and where on the catalysts the reactant molecules interact. Then, materials synthesis methods can be applied to prepare catalysts with the desired surface active sites and their selective interaction with the reactants.

Introduction

The final objective in heterogeneous catalysis is to reach catalysts by design. This could be done by increasing our arsenal of fundamental knowledge on the molecular interactions at the solid–gas or solid–liquid interface, and the synthesis techniques required for the preparation of the material that will act as a catalyst.[1] The development of new spectroscopic techniques able to work under reaction conditions should bring key knowledge on surface homogeneity and surface dynamics, especially when interacting with reactants. Indeed, the determination of active sites and reaction mechanism under dynamic conditions, *i.e.* under realistic working conditions, can strongly help to establish a structure–reactivity selectivity relationship. This will be strongly helped by operando spectroscopy for investigating the dynamic structure of a solid catalyst under real working conditions, and the simultaneous measurement of activity and selectivity. It is clear that extrapolation of the results obtained by following the operando methodology is only possible if the reactor catalyst activation and catalytic reactor conditions are analogous to those used under realistic operation. In this respect, several reviews have been published.[2–4] It is then clear that for in depth knowledge of the catalytic phenomena, a combination of material synthesis, multitask surface characterization techniques, molecular simulation,

Instituto de Tecnología Química (UPV-CSIC), Universitat Politècnica de València Consejo Superior de Investigaciones Científicas, Valencia, 46022, Spain. E-mail: acorma@itq.upv.es

and micro-kinetic studies will be required. From these combined efforts, not only will our scientific knowledge increase, but we will be able to design better solid catalysts for improving existing processes and for opening new reaction avenues. In our studies, after selecting the reaction to be catalyzed, the reaction mechanism is considered first in order to postulate the potential active sites. We will show that this could be achieved by combining micro-kinetic and mechanistic studies with molecular modeling and *in situ* spectroscopy studies, preferentially under working conditions. Finally, through the materials synthesis techniques, catalysts with well defined single or multisites will be generated. Following this strategy, it should be possible not only to explain and understand how a given catalyst works, but to design the next catalyst generation.

Results and discussion

In our studies in heterogeneous catalysis, regardless of whether we work with zeolites, structured inorganic materials, hybrid organic–inorganic materials, metal nanoparticles or multifunctional catalysts, we attempt to select the reaction to be studied not only because of its intrinsic chemical interest, but also with the objective to apply the knowledge acquired to solve an industrially relevant problem. In other words, we have always attempted to combine the generation of scientific knowledge and the potential application. For achieving this, the methodology given in Fig. 1 has been applied and will be illustrated here for the development of chemoselective hydrogenation catalysts based on uni and bimetallic nanoparticles and clusters.

The reaction

Amines are important building blocks in inorganic synthesis and are widely used industrially. They can be obtained by the hydrogenation of $-NO_2$ groups into the corresponding amines. This, in principle, is not a difficult process and, for instance, aniline is successfully produced industrially from nitrobenzene with H_2 and different metal catalysts. However, the nitro group could be difficult to selectively reduce with H_2 when other reducible groups are present in the same

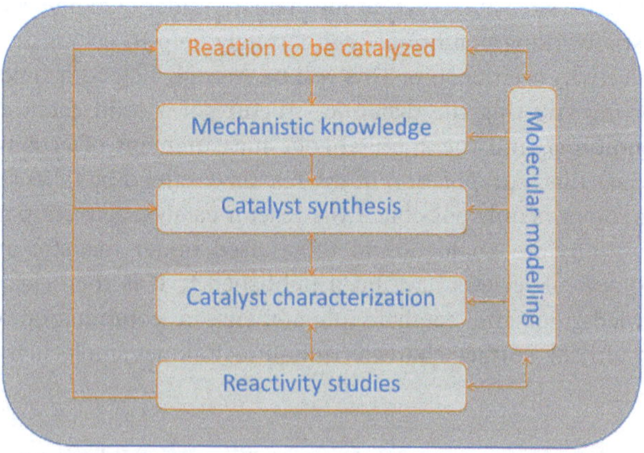

Fig. 1 Methodology followed for heterogeneous catalytic studies.

molecule and have to be retained in the final product. This will be the case when the nitro compounds have to be hydrogenated in the presence of olefin, alkyne, carbonyl or nitrile groups. To tackle this problem, and owing to its industrial interest, researchers from Ciba-Geigy[5] have carried out the chemoselective hydrogenation of $-NO_2$ in the presence of olefins and alkyne groups, with supported Pt catalysts. Since supported Pt was not selective, those authors modify the metal with PbO or H_3PO_4 and added vanadium and iron salts into the solution to avoid the accumulation of partially reduced nitro compounds. In fact, as presented by Haber,[6] the reaction mechanism for the reduction of nitro compounds follows the sequence presented below:

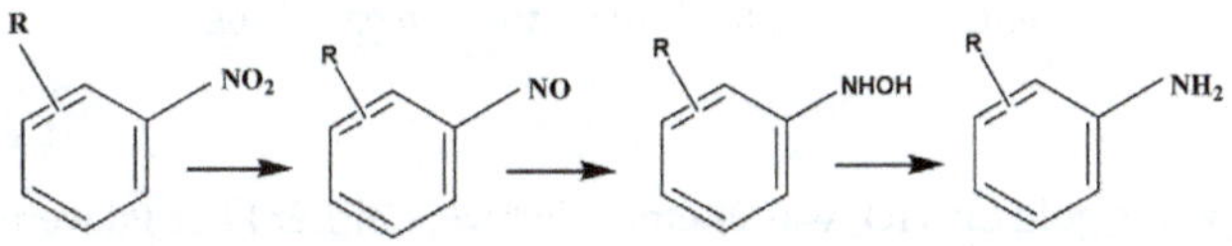

Then, with the Pt catalyst described above, the potentially explosive hydroxylamine compound accumulated and vanadium and iron were required to decrease its concentration. While the results obtained are interesting, it is clear that the above catalytic system was not optimum from an environmental point of view and a more elegant catalyst should be developed.

The catalyst

Chemical steps in a catalytic reaction involve reactant adsorption and activation, followed by the surface reaction. In the case of chemo, regio and enantioselective hydrogenations, the adsorption step can be determining for the final selectivity. In the case of substituted nitrobenzene, and more specifically nitrostyrene, a potentially chemoselective catalyst could be achieved with a solid that adsorbs the nitrostyrene selectively through the nitro group instead of adsorbing the molecule of nitrostyrene flat on the surface (Fig. 2).

Since non-selective catalysts such as Pt, Pd or Ni adsorbed nitrostyrene flat on the surface, and gold is a good material for adsorbing nitrogen oxides while being a poor olefin hydrogenation catalyst, we thought about using nanoparticulated gold for carrying out the selective reduction of nitrostyrene. Results in Table 1

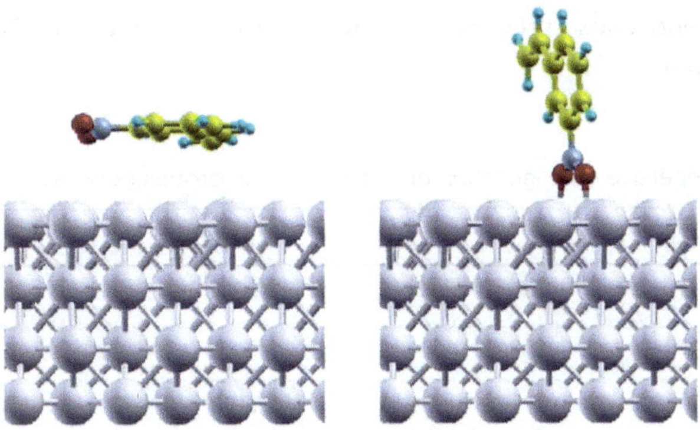

Fig. 2 Adsorption modes of nitrostyrene on the metal catalyst's surface.

Table 1 Nitrostyrene reduction on supported metal catalysts

Substrate	Catalysts	% metal (mol)	T (°C)	P (bar)	Time (hours)	Conversion (%)	Selectivity[5]
3-Nitrostyrene	Au/TiO_2	0.23	120	9	6.00	98.5	95.9
	Pd/C	0.11	120	9	0.03	99.0	0.0
	Pt/C	0.12	120	9	0.03	96.7	2.9
	$Au–Pt/TiO_2$	0.31	120	9	0.02	93.0	1.8
4-Nitrobenzaldehyde	Au/TiO_2	1.14	100	10	1.25	99.0	96.8
	Pd/C	0.35	100	10	0.03	92.0	29.8
	Pt/C	0.38	100	10	0.03	96.7	68.6

show[7] that while gold on TiO_2 was chemoselective, neither Pt or Pd were. Then the question was whether the selectivity of a gold catalyst was because of the selective adsorption of nitrostyrene through the nitro group or due to differences in the intrinsic rate of hydrogenation of the nitro and olefinic groups on the gold catalyst.

Understanding the interactions at the solid–liquid, solid–gas interface

To understand the mechanism of action of the catalyst, styrene, nitrobenzene and a 50% mixture of nitrobenzene and styrene were reacted on the Au/TiO_2 catalyst. The results presented in Table 2 clearly demonstrate the key role played by the reactant selective adsorption. Indeed, the higher intrinsic hydrogenation activity of Au/TiO_2 for nitrobenzene cannot explain the very high chemoselectivity observed in Table 1. Meanwhile, when the 50% mixture was reacted, results in Table 2 show a very strong decrease in the rate of hydrogenation of the olefinic group, while there is no effect on the hydrogenation of the nitro group.

The question then was: where and how does specific adsorption through the nitro group occur on the catalyst? To answer the question, molecular modeling of nitrostyrene adsorption on well defined Au metal faces, surface unsaturated atoms and gold nanoparticles was carried out, using periodic DFT calculations with VASP code PW 91 functional – PAW pseudopotential, Γ point, a cut-off of 400–500 eV and Au clusters and NPs in a 20 × 20 × 20 Å box.[8] The results showed no adsorption of nitrostyrene on extended Au (111) and Au (100) surfaces and some preferential adsorption of C=C on defective surfaces and on isolated Au_{38} NPs (see Fig. 3).

Table 2 Competitive hydrogenation of substituted nitroaromatics on Au/TiO_2 catalyst

Feeding (% mol)		TOF (h^{-1})	
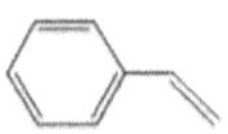	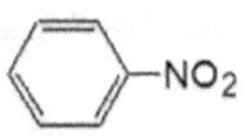	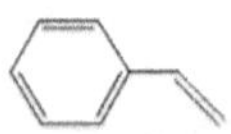	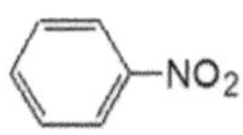
8.5	0	160	0
0	8.5	0	360
4.25	4.25	16	405

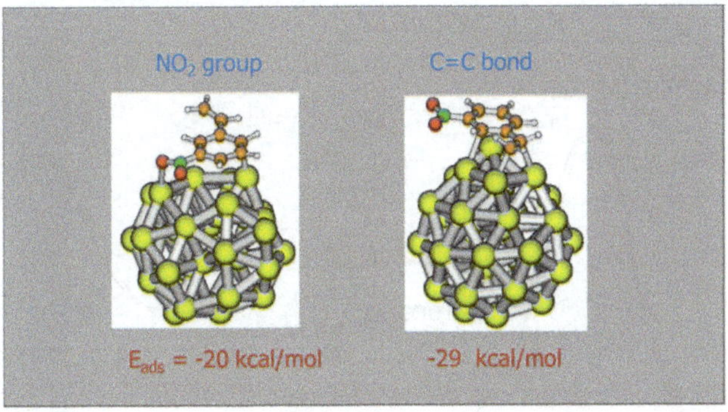

Fig. 3 DFT results of nitrostyrene adsorbed on gold nanoparticles.

The answer from molecular modeling was that specific adsorption on gold could not be responsible for the observed chemoselectivity. On this basis, an experimental study was carried out, and IR spectroscopy of adsorbed nitrobenzene, styrene and nitrobenzene + styrene on Au/SiO_2 and AuTi/O_2 showed that the key adsorption of nitrostyrene was not taking place on gold but on TiO_2. Moreover, the adsorbed nitrostyrene gave two adsorption bands, one at 1526 and another at 1520 cm^{-1} that can be associated with $\gamma(NO_2)$–TiO_2 and $\gamma(NO_2)$–Au–TiO_2 (gold–TiO_2 interface) (Fig. 4).

Then, when the reaction was performed *in situ*, the IR clearly showed that the nitrostyrene adsorbed at the interface was reacting, while that adsorbed on Ti was a spectator (Fig. 5).

At this point, combining kinetics, *in situ* spectroscopy and molecular modeling, we know that there is indeed a preferential adsorption of nitrostyrene

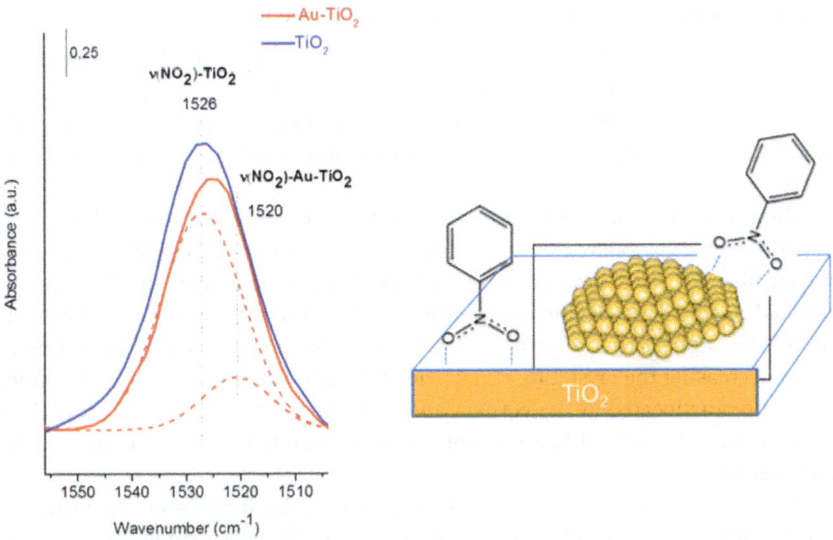

Fig. 4 Nitrobenzene adsorption on the Au/TiO_2 catalyst followed by IR spectroscopy.

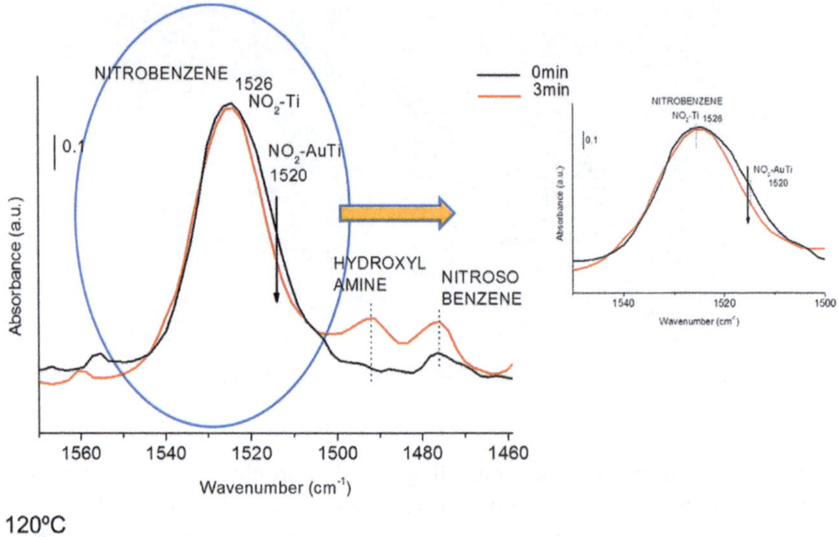

Fig. 5 Nitrobenzene hydrogenation on the Au/TiO_2 catalyst and IR spectroscopy.

via the NO_2 group at the gold–titania interface. Furthermore, one could assume in a first approximation that H_2 will be dissociated on gold. With the knowledge already acquired and with the objective of designing a better catalyst, one should find what the reaction controlling step is and which are the catalytic active sites involved. To do that, different surface kinetic models were built assuming that the controlling steps were nitrostyrene adsorption, H_2 dissociation, and the surface reaction between the dissociated H_2 and activated nitrostyrene on the same or in different sites. Kinetic experiments indicated that H_2 dissociation was the controlling step, and this was confirmed by H_2/D_2 isotopic exchange.

Improving the catalysts through fundamental knowledge

It was now clear that if H_2 dissociation is the controlling step, the design of an improved catalyst should increase the rate of H_2 dissociation. This could be achieved by: (a) increasing the number of active sites responsible for such activation, (b) increasing the intrinsic H_2 dissociation activity by site modification, and (c) by increasing both.

The dissociation of H_2 was studied theoretically on different gold surfaces and supported gold nanoparticles. It was found that the unsaturated gold atoms were responsible for such dissociation.[9] This is in agreement with the conclusion reached previously by Hammer and Norskov.[10] Taking the above into account, it would then appear that more active sites for H_2 dissociation should be achieved when decreasing the crystallite size of the metal nanoparticles, since the fraction of gold atoms located at the corners and borders will be larger. Indeed, higher catalyst activity was obtained for Au/TiO_2 catalysts with smaller gold nanoparticles.

The second way to increase catalyst activity is by increasing the intrinsic activity of the H_2 dissociation sites, without losing the specific adsorption properties. Along that direction, it is clear that Pt metal is more active than gold for H_2

dissociation but it is not chemoselective. However, we could accomplish both conditions, *i.e.* faster H_2 dissociation and selective nitrostyrene adsorption, by introducing small amounts of Pt on the gold catalyst. The Pt should be at the level of a few atoms per each gold nanoparticle to avoid a dominant role of Pt. Indeed, when this was done, no activation energy for H_2 dissociation was observed, which is in contrast to the existence of an activation energy (though low) for H_2 dissociation on pure gold nanoparticles (Fig. 6). Moreover, nitrostyrene was still adsorbed through the nitro groups at the interface of gold–TiO_2 in the bimetallic particles (Fig. 7).[11]

Catalytic results confirm the above calculations and the catalytic activity of Au/TiO_2 strongly increased, while preserving selectivity when 100 ppm of Pt is introduced in the catalyst (Fig. 8).

Transforming a non-selective catalyst into a selective catalyst from the knowledge acquired

If a selective catalyst requires the adsorption of nitrostyrene at the metal/TiO_2 interface, avoiding the adsorption of nitrostyrene flat on the catalyst surface, this could be achieved if one prepares Pt and Ni catalysts with very small metal domains, *i.e.* only exposing low surface coordinated sites. This could occur if one decorates Pt or Ni crystallites with titania by calcination of Pt or Ni supported on TiO_2 at 450 °C.[12] When this is done, good activities and selectivities are achieved as shown by the results in Table 3.[13]

Recently, decoration of Pt nanoparticles by ZnO in Pt/ZnO catalysts have shown a similar increase in catalyst selectivity.[14] High activity and selectivity for hydrogenation of nitrostyrene derivatives have also been obtained by preparing silver–cerium oxide core–shell nanoparticles.[15]

As a result of all the above, it is possible to achieve a wide scope of substituted anilines synthesized by hydrogenation of the corresponding nitro compounds

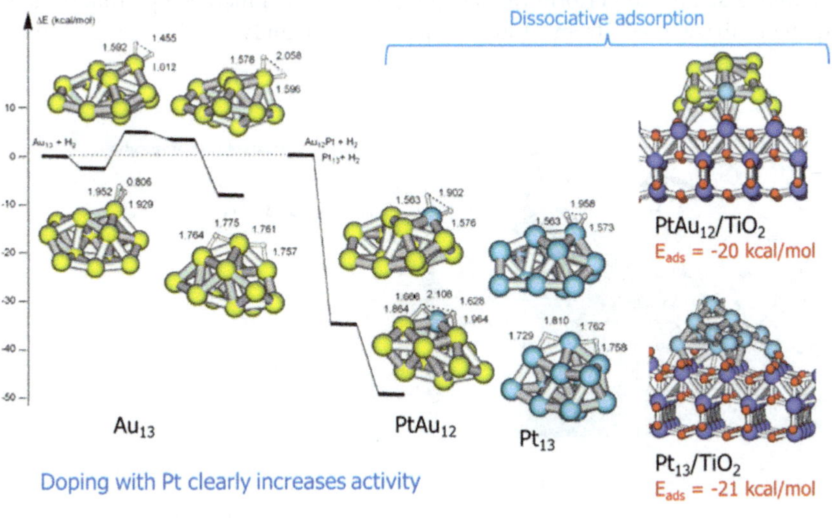

Fig. 6 Reaction coordinates for the hydrogenation of nitrostyrene: the role of doping gold with Pt.

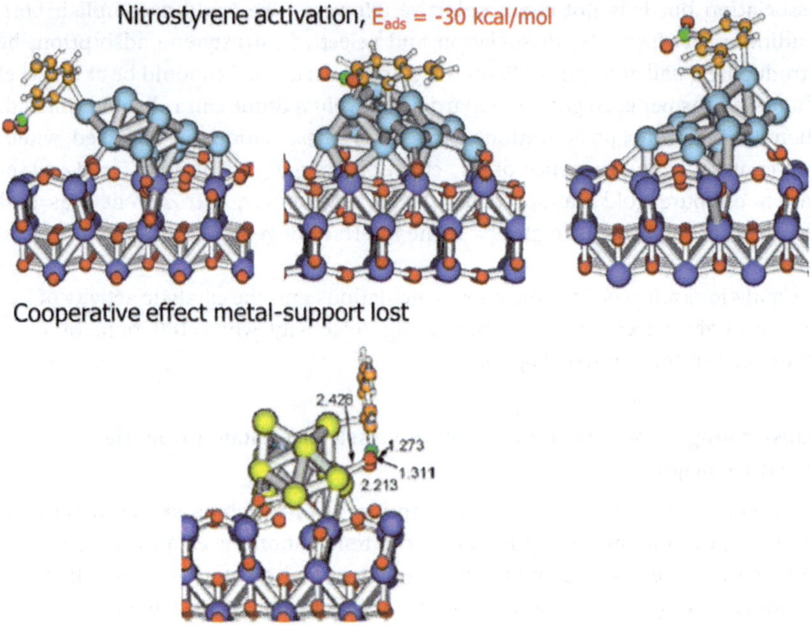

Fig. 7 Hydrogenation of styrene by doping gold with Pt.

with metal nanoparticles supported on TiO_2. Moreover, by having such chemoselectivity, one can perform one pot cascade reactions starting with nitro compounds and reacting the amines formed with molecules bearing alkyne, carbonyl, alkene, nitrile groups, *etc.* (Fig. 9).

In these ways, imines, secondary amines, nitrones, and β-amino carbonyls can be produced in high conversions and selectivities.[16–21]

By combining the selective metals with other acid and base catalytic functions, we were also able to perform the one pot synthesis of 3-methyl 2-pyrrolidones,[22] 1,5-benzodiazepines from nitroaromatic compounds, quinoxalines from

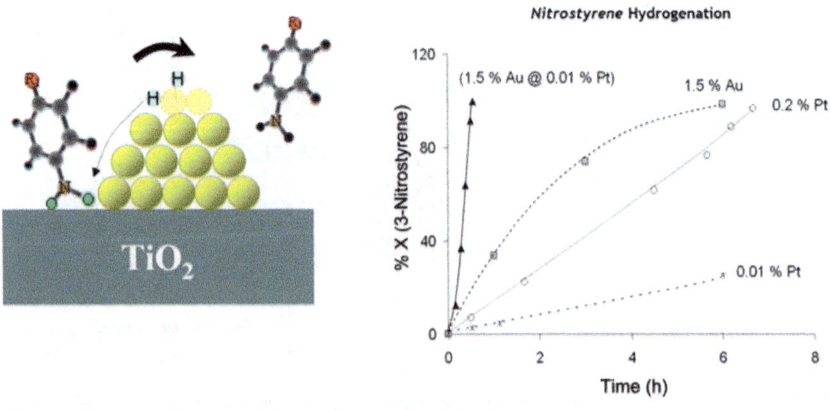

Fig. 8 Hydrogenation of nitrostyrene on Au@Pt catalyst.

Table 3 Hydrogenation of 3-nitrostyrene on different metals with their surface decorated with TiO_2. (a) 3-Aminostyrene; (b) nitroethylbenzene; (c) 3 ethylaniline; (d) hydroxylamine derivatives

Catalytic system	Reduction temperature (°C)	% X	% Sa	% Sb	% Sc	% Sd
1.5% Au/TiO_2	—	98.5	95.9	0.0	3.9	0.2
0.2% Pt/TiO_2	450	95.1	93.1	0.1	6.8	n.d.
5% Ni/TiO_2	450	93.0	90.2	0.2	7.3	2.3

dinitroaromatics and glycols,[23] and 2′-aminochalcones from 2-nitroacetophenones and benzaldehydes.

Chemoselective hydrogenations with non-noble metal catalysts

Considering the cost and limited availability of noble metals, it is important to prepare non-noble metal catalysts for chemoselective hydrogenation reactions. In fact, modified non-noble metals can present comparable properties to noble metals in photocatalysis, electrocatalysis and heterogeneous catalysis.[24–27]

Non-noble metals such as CoO_x and Fe_2O_3 on N doped carbon materials have been successfully used for the hydrogenation of nitroarenes.[28,29] These catalysts give good chemoselectivity, though they are required to work at 50 bar of H_2 instead of the 10 bar required by gold. Furthermore, the role of the layered carbon component on the final catalytic performance is still unknown. It may protect metal NPs, but it may also block accessibility of substrates to the metal NPs. One hypothesis on the role of carbon in the above catalyst[30–32] is that the hydrogenation may occur on the surface of the carbon layers.

In our case, we have prepared cobalt–carbon nanoparticles (Co@CNPs) by reduction of Co–EDTA complexes in H_2. The morphology of Co@NPs consists of nanodispersed metal NPs ranging from 20 to 150 nm (Fig. 10). The CoNPs are covered by a few carbon layers, and HRTEM images indicate that the CoNPs are

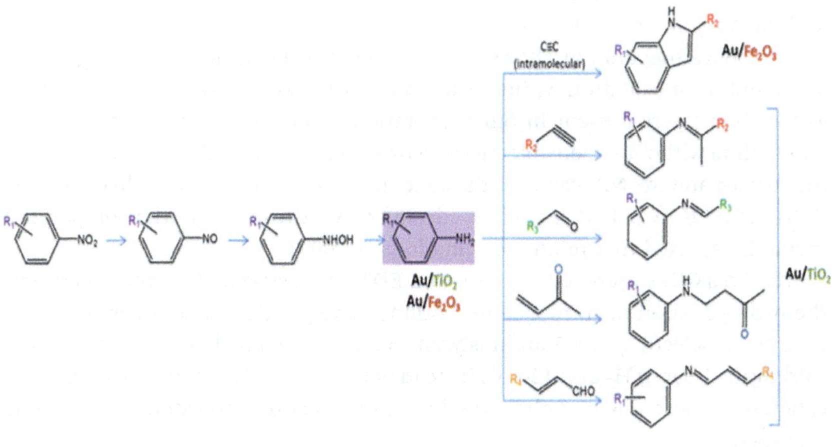

Fig. 9 Cascade reactions starting with nitroanilines.

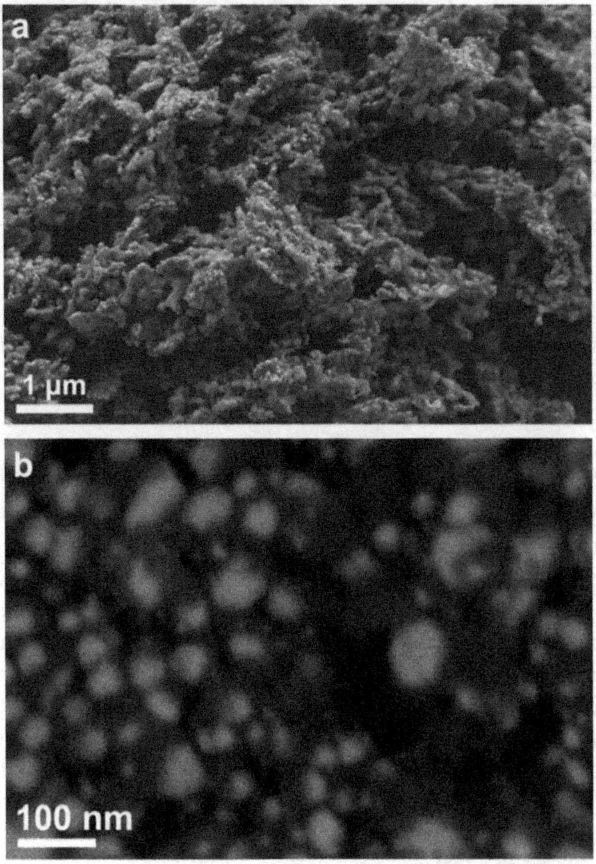

Fig. 10 Field-emission scanning electron microscopy (FESEM) images of Co@C NPs obtained by SE2 detector (a) and ESB detector (b). In (a), Co NPs are surrounded by foam-like carbon and those Co NPs with particle size ranging from 20 to 100 nm can be seen in (b).

formed by a core–shell structure with Co metal at the core, and CoO_x at the shell and the outside carbon layers.

The above preparation allows having stable CoNPs while avoiding agglomeration and deep oxidation without the necessity of capping with organic ligands. The carbon layers present in our preparation allow the CoNPs to be stored for a long time without observable morphological changes. Oxidation occurs only at the surface and the NPs can be reduced to metallic Co under very mild conditions (7 bar H_2, 120 °C). This is a clear advantage with respect to other preparation methods reported that require calcinations at 800 °C.

The Co@CNPs prepared from Co and EDTA are active and chemoselective for the hydrogenation of nitrostyrene, working at very mild reaction conditions. In fact, 93% selectivity to 3-aminostyrene can be obtained at 95% conversion working at 7 bar of H_2 and 120 °C, instead of the 50 bar H_2 required with previous catalysts. No leaching was observed during the reaction and catalyst recyclability was good.

We have extended this type of catalyst to bimetallic materials, the preparation method avoids metal segregation, and excellent metal dispersion can be achieved.

The reactant interactions at the interphase were studied using *in situ* IR spectroscopy, and the results obtained indicate that metallic Co is responsible for the preferential adsorption of the nitro group, while the olefinic group will be preferentially adsorbed in the presence of cobalt oxide.

This work opens the possibility to prepare stable multiatomic catalyst nanoparticles and to control the adsorption properties towards different reactants.

Conclusions

We have attempted to show that improving the knowledge of the molecular interactions at the catalyst surface is the key to understanding the catalytic phenomena. That includes the nature of the solid surface under the reaction conditions, the nature of the active sites at work, and their interaction (adsorption–reactivity) with reactants. The study of such complex systems requires the combination of different techniques that make the structures and interactions "visible". These include HRTEM in the presence of the reactants and *in situ* spectroscopic techniques working at reaction conditions. Molecular modeling allows us to understand the molecular interactions, helps the imagination and directs the prediction. Last but not least, microkinetic studies, together with spectroscopy work for the detection of intermediates, help to delve into the reaction mechanism. By following that multitask methodology, it should be possible to go from understanding to prediction and to achieve improved or even completely new types of catalysts.

Acknowledgements

This work was funded by the Spanish Government (Severo Ochoa program SEV-2012-0267). The support of the European Union by (ERC-AdG-2014-671093 – SynCatMatch) is also acknowledged.

References

1 J. M. Thomas, *Microporous Mesoporous Mater.*, 2011, **146**, 3.
2 B. M. Weckhuysen, *Phys. Chem. Chem. Phys.*, 2003, **5**, 4351.
3 M. A. Bañares, *Catal. Today*, 2005, **100**, 71.
4 Y. Zhang, D. Fu, X. Xu, Y. Sheng, J. Xu and Y. Han, *Curr. Opin. Chem. Eng.*, 2016, **12**, 1.
5 H. U. Blaser, H. Steiner and M. Studer, *ChemCatChem*, 2009, **1**, 210.
6 F. Haber, *Z. Elektrochem.*, 1898, **22**, 506.
7 A. Corma and P. Serna, *Science*, 2006, **313**, 332.
8 M. Boronat, P. Concepcion, A. Corma, S. Gonzales, F. Illas and P. Serna, *J. Am. Chem. Soc.*, 2007, **129**, 16230.
9 A. Corma, M. Boronat and I. Illas, *Chem. Commun.*, 2007, 3371.
10 B. Hammer and J. K. Norskov, *Nature*, 1995, **376**, 238.
11 M. Boronat and A. Corma, *Langmuir*, 2010, **26**, 16607.
12 S. J. Tauster, S. C. Fung and R. Garten, *J. Am. Chem. Soc.*, 1978, **100**, 170.
13 P. Serna, M. Boronat and A. Corma, *Top. Catal.*, 2011, **54**, 439.

14 A. Yarulin, C. Berguerand, A. O. Alonso, I. Yuranov and L. Kiwi-Minsker, *Catal. Today*, 2015, **256**, 241.
15 T. Mitsudome, Y. Mikami, M. Matoba, T. Mizugaki, K. Jitsukawa and K. Kaneda, *Angew. Chem., Int. Ed.*, 2012, **51**, 136.
16 L. L. Santos, P. Serna and A. Corma, *Chem.–Eur. J.*, 2009, **15**, 8196.
17 L. Cisneros, P. Serna and A. Corma, *Angew. Chem.*, 2014, **126**, 9460.
18 X. Zhang and A. Corma, *Angew. Chem., Int. Ed.*, 2008, **47**, 4358.
19 Y. Yamane, X. Liu, A. Hamasaki, T. Ishida, M. Haruta, T. Yokoyama and M. Tokunaga, *Org. Lett.*, 2009, **11**, 5162.
20 O. Peng, Y. Zhang, F. Shi and Y. Deng, *Chem. Commun.*, 2011, **47**, 6476.
21 M. J. Climent, A. Corma, S. Iborra and L. Marti, *ACS Catal.*, 2015, **5**, 157.
22 D. Vidal, M. J. Climent, A. Corma and S. Iborra, *ACS Catal.*, 2015, **5**, 5812.
23 M. J. Climent, A. Corma, S. Iborra and S. Martinez-Silvestre, *J. Catal.*, 2012, **292**, 118.
24 J. Ran, J. Zhang, J. Yu, M. Jaroniecc and S. Zhang Qiao, *Chem. Soc. Rev.*, 2014, **43**, 7787.
25 H. R. Byon, J. Suntivich and Y. Shao-Horn, *Chem. Mater.*, 2011, **23**, 3421.
26 R. V. Jagadeesh, H. Junge and M. Beller, *Nat. Commun.*, 2014, **5**, 4123.
27 B. Su, Z.-C. Cao and Z.-J. Shi, *Acc. Chem. Res.*, 2015, **48**, 886.
28 F. A. Westerhans, R. V. Jagadeesh, G. Wienhöfer, M. M. Pohl, J. Radnik, A. E. Surkus, J. Rabeah, K. Junge, H. Junge, M. Nielsen, A. Brückner and M. Beller, *Nat. Chem.*, 2013, **5**, 537.
29 R. V. Jagadeesh, A. E. Surkus, H. Junge, M. M. Pohl, J. Radnik, J. Rabeah, H. Huan, V. Schünemann, A. Brückner and M. Beller, *Science*, 2013, **342**, 1073.
30 F. A. Westerhans, R. V. Jagadeesh, G. Wienhöfer, M. M. Pohl, J. Radnik, A. E. Surkus, J. Rabeah, K. Junge, H. Junge, M. Nielsen, A. Brückner and M. Beller, *Nat. Chem.*, 2013, **5**, 537.
31 R. V. Jagadeesh, A. E. Surkus, H. Junge, M. M. Pohl, J. Radnik, J. Rabeah, H. Huan, V. Schünemann, A. Brückner and M. Beller, *Science*, 2013, **342**, 1073.
32 Z. Wei, J. Wang, S. J. Mao, D. F. Su, H. Y. Jin, Y. H. Wang, F. Xu, H. R. Li and Y. Wang, *ACS Catal.*, 2015, **5**, 4783.

Faraday Discussions

PAPER

Using degrees of rate control to improve selective *n*-butane oxidation over model MOF-encapsulated catalysts: sterically-constrained $Ag_3Pd(111)$

Sean T. Dix,[a] Joseph K. Scott,[a] Rachel B. Getman*[a] and Charles T. Campbell*[b]

Received 29th November 2015, Accepted 21st December 2015

DOI: 10.1039/c5fd00198f

Metal nanoparticles encapsulated within metal organic frameworks (MOFs) offer steric restrictions near the catalytic metal that can improve selectivity, much like in enzymes. A microkinetic model is developed for the regio-selective oxidation of *n*-butane to 1-butanol with O_2 over a model for MOF-encapsulated bimetallic nanoparticles. The model consists of a $Ag_3Pd(111)$ surface decorated with a 2-atom-thick ring of (immobile) helium atoms which creates an artificial pore of similar size to that in common MOFs, which sterically constrains the adsorbed reaction intermediates. The kinetic parameters are based on energies calculated using density functional theory (DFT). The microkinetic model was analysed at 423 K to determine the dominant pathways and which species (adsorbed intermediates and transition states in the reaction mechanism) have energies that most sensitively affect the reaction rates to the different products, using degree-of-rate-control (DRC) analysis. This analysis revealed that activation of the C–H bond is assisted by adsorbed oxygen atoms, O*. Unfortunately, O* also abstracts H from adsorbed 1-butanol and butoxy as well, leading to butanal as the only significant product. This suggested to (1) add water to produce more OH*, thus inhibiting these undesired steps which produce OH*, and (2) eliminate most of the O_2 pressure to reduce the O* coverage, thus also inhibiting these steps. Combined with increasing butane pressure, this dramatically improved the 1-butanol selectivity (from 0 to 95%) and the rate (to 2 molecules per site per s). Moreover, 40% less O_2 was consumed per oxygen atom in the products. Under these conditions, a terminal H in butane is directly eliminated to the Pd site, and the resulting adsorbed butyl combines with OH* to give the desired 1-butanol. These results demonstrate that DRC analysis provides a powerful approach for optimizing catalytic process conditions, and that highly selectivity oxidation can sometimes be achieved by using a mixture of O_2 and H_2O as the oxidant. This was further demonstrated by DRC analysis of a second

[a]*Department of Chemical and Biomolecular Engineering, Clemson University, Clemson, SC 29634, USA. E-mail: rgetman@clemson.edu*

[b]*Department of Chemistry, University of Washington, Seattle, WA 98195-1700, USA. E-mail: charliec@uw.edu*

microkinetic model based on a related but hypothetical catalyst, where the activation energies for two of the steps were modified.

1. Introduction

The steric environment surrounding the metal site in metallo-enzymes is often thought to impart selectivity in its catalytic products. It has been shown that this same type of effect can be achieved in synthetic catalyst structures where catalytic nanoparticles (NPs) are encapsulated by metal–organic frameworks (MOFs), known as NP@MOF systems.[1–5] For example, Stephenson *et al.*[5] achieved 95% selectivity in the hydrogenation of 1,3-hexene to 3-hexene by encapsulating ~3 nm Pt particles in the MOF ZIF-8. Without ZIF-8, *n*-hexane is formed instead with 80% selectivity.

With the goal of using DFT calculations to help develop NP@MOF catalysts, but aiming to circumvent the unknown details of the NP/MOF interface, Gomez-Gualdron *et al.*[6] introduced a simplified approach where steric constraints near the catalyst surface are provided by a "surrogate" pore formed by a cylinder of chemically inert helium atoms. They showed that this steric constraint on a Pd catalyst may facilitate the C–H activation of *n*-butane due to the different relative destabilization of *n*-butane and 1-butyl, relevant to the selective oxidation of *n*-butane to 1-butanol.

In a concurrent study,[7] we used this "surrogate pore" approach together with DFT to investigate whether this steric constraint is sufficient to achieve regio-selective oxidation of *n*-butane to 1-butanol on Ag_3Pd bimetallic nanoparticles. Specifically, we studied the (111) facet of a Ag_3Pd alloy, where the MOF-mimetic pore is modelled by two rings of 9 He atoms each, stacked on top of each other and frozen in space in the geometry shown on the left in Fig. 1.[7] Except for the replacement of 3/4 of the Pd atoms with Ag, this is the same structure as that used by Gomez-Gualdron *et al.*[6] A butanol molecule is included in the right image in Fig. 1 to demonstrate that only its terminal C atoms have access to the metal atoms in this catalyst, due to steric constraints imposed by the (frozen) He_{18} ring. This allowed oxygen addition only to the terminal carbons.

We use that same model here to develop a microkinetic model to predict the reaction rates to the various potential products under a specific set of reaction conditions chosen to approximately simulate experimental conditions for gas phase butane oxidation. The energies of its adsorbed intermediates and transition states were taken from the DFT calculations reported in that concurrent study.[7] We also analyse this new microkinetic model to determine the degree of rate control (DRC) of every elementary step's transition state and every adsorbed intermediate, with respect to the rate of production of every possible product and the rate of consumption of the most expensive reactant, butane. The DRC values (X_i) of each adsorbed intermediate and transition state for every elementary step are defined as outlined in ref. 8:

$$X_i = \left(\frac{\partial \ln r}{\partial \left(\frac{-G_i^0}{RT} \right)} \right)_{G_{j \neq i}^0} \quad (1)$$

Fig. 1 (a) Top view of adsorbed *n*-butanol on the He ring-encapsulated $Ag_3Pd(111)$ surface. (b) Side view of the same system shown in panel (a) but with the He atoms removed to view the surface sites better. Ag = grey, Pd = yellow, C = black, H = off-white, O = red, He rings = white.

where r is the net rate of production of one product (*e.g.* butanol, butanal, or 1-butene oxide gas), or the rate of consumption of one reactant (*e.g.* butane), G_i^0 is the standard-state free energy of species i (*e.g.* one adsorbed intermediate or transition state), R is the gas constant, and T is the absolute temperature. In practice, these partial derivatives are calculated numerically, with the change in G_i^0 easily implemented by simply changing the electronic energy of species i by a tiny amount. The DRC values are nearly zero for most species, but by finding the species with large values, we are able to define which species' energies need to be tuned in what directions by further material modifications in order to achieve a more effective catalyst in terms of activity and selectivity, which should be useful in supplying ideas for catalyst optimization. A fast computational method for using such DRC values from such a reference catalyst as studied here to search for better catalyst materials has recently been developed.[9] We have discovered here, through this DRC analysis, how to tune the reactant feed to achieve much better selectivity to butanol (95%) and a faster rate with much less consumption of O_2.

2. Methods

We studied the selective butane oxidation, with butanol as the desired product, over a MOF-encapsulated $Ag_3Pd(111)$ catalyst model at 423 K, initially with O_2 as the only oxidant (at its pressure in air) and 1898 kPa C_4H_{10} (g). The $Ag_3Pd(111)$ catalyst surface was modelled by cutting a 4-layer, 4 metal atom × 4 metal atom (111) surface from the bulk $L1_2$ structure of Ag_3Pd. Steric constraints imposed by the MOF are provided by the surrogate pore approach, which has been introduced, described, and validated in detail by Gomez-Gualdron *et al.*[6] The surrogate pore limits the way C_4 species can interact with the catalyst surface, while imparting only weak dispersion interactions. It is composed of chemically inert helium atoms (Fig. 1), the positions of which remain fixed during the simulations. The helium atoms form two rings, each composed of nine helium atoms.

The reaction network is comprised of 20 reactions, each in forward and reverse, including the desired formation of 1-butanol and competing undesired formation of 1-butanal and 1-butene oxide, which represent the first steps in the over-dehydrogenation of the primary and secondary carbons, respectively.[7] The elementary steps are listed in Table 1 along with each step's reaction energy and activation energy in the forward direction as written. These energies were all calculated using density functional theory (DFT) following the methods outlined below and in a concurrent publication,[7] using large unit cells so that they represent the low-coverage limit. The energies listed here were all taken from that publication[7] with the following exceptions. Reactions (2), (3), (7), (8), (9), and (11) were reported in that publication at higher total or local coverages where adsorbate–adsorbate interactions affected the energies. Table 1 lists the equivalent energies for the low-coverage limit for each separate adsorbate, which we used in this present study (for the interested reader, the first-nearest-neighbour interaction energies for some of these reactions are reported in ref. 7 and in ref. 10). The energies for reactions (14)–(19) were not included in that concurrent publication, so these are new values here.

The use of low-coverage energies is justified as follows. The first two sets of reaction conditions studied below resulted in high O* coverages, and these were higher than they should have been due to our use of low-coverage energies. However, even using the proper energies, these O* coverages would have been too high to get a good catalyst performance (see below). By the third set of reaction conditions, which are the most important, we managed to get a very low O*

Table 1 Elementary reaction steps considered in the microkinetic model and the DFT-derived reaction energy and activation barrier for each, in units of eV. The * refers to a metal surface site

Step #	Reaction	ΔE_{rxn}	ΔE_{act}
1	$O_2 + * = O_2*$	−0.52	0
2	$O_2* + * = 2O*$	−1.05	0.66
3	$O_2* + C_4H_{10}* = C_4H_9OH* + O*$	−2.86	1.12
4	$C_4H_{10}* + O* = C_4H_9O* + H*$	−0.18	1.14
5	$C_4H_{10}* + O* = C_4H_9* + OH*$	−0.34	0.92
6	$C_4H_{10}* + O* = C_4H_9OH* + *$	−1.10	1.22
7	$C_4H_9O* + O* = C_4H_8Oep* + OH*$	−0.17	0.94
8	$C_4H_9O* + O* = C_4H_8Oal* + OH*$	−0.85	0.39
9	$C_4H_9O* + H* = C_4H_9OH* + *$	−0.76	0.43
10	$C_4H_9* + OH* = C_4H_9OH* + *$	−1.36	0.76
11	$C_4H_9O* + OH* = C_4H_9OH* + O*$	+0.05	0.14
12	$C_4H_{10} + * = C_4H_{10}*$	−0.47	0
13	$C_4H_9OH* = C_4H_9OH + *$	+0.79	0.79
14[a]	$C_4H_8Oal* = C_4H_8Oal + *$	+0.70	0.70
15[b]	$C_4H_8Oep* = C_4H_8Oep + *$	+0.49	0.49
16	$2H* = H_2\ (g) + 2*$	+0.20	0.20
17	$O* + H* = OH* + *$	−1.50	0.72
18	$H_2O + * = H_2O*$	−0.44	0
19	$H_2O* + O* = 2OH*$	−0.15	0.12
20	$C_4H_9* + H* = C_4H_{10}* + *$	−0.66	0.11

[a] "al" means 1-butanal. [b] "ep" means 1-butene oxide.

coverage and excellent performance. In the last set of conditions, the coverage of *n*-butane is high, but this is not expected to influence the energies to any significant amount since, as we report in a concurrent publication,[7] *n*-butane exhibits only weak neighbour interactions.

All DFT calculations were performed with the VASP code,[11,12] which utilizes periodic boundary conditions and plane wave basis sets. Valence electron exchange and correlation were calculated with the PBE functional,[13,14] while core electrons were simulated with PAW pseudopotentials[15,16] to an energy cut-off of 400 eV. The D2 method of Grimme[17] was employed in all calculations to improve the modelling of dispersion interactions. Gamma-centred *k*-point meshes of $7 \times 7 \times 1$ were used to sample the first Brillouin zones. Transition states (TSs) were located with a combination of the Climbing Image Nudged Elastic Band (CI-NEB)[18,19] and dimer methods.[20] Other details of the DFT calculations are given in ref. 7.

We supplied the temperature and inlet flow rates of all reactants and products, and solved for the steady state concentrations in the limits of low conversion using all 20 reactions presented in Table 1 in the equation solver under six different sets of conditions (temperature, and partial pressures of gas phase species). The equation solver was the ode15s function, as implemented in MATLAB.

We model here a CSTR (continuous stirred tank reactor). The conversion of butane is kept below 1% in all six sets of conditions. The CSTR is operated differentially (meaning that the percent change in partial pressure for all gas phase species is <1%) in the first three sets of conditions, but in the fourth, fifth, and sixth sets of conditions, differential operation is forgone in favour of changing one partial pressure at a time in order to demonstrate its effect on the results. To track the rates of butane adsorption, butanol desorption, butanal desorption, and 1-butene oxide desorption, we used units of moles per second for the rates. This way, the rates to products can be directly compared with the rates of the surface reactions. To obtain the rates of the surface reactions, we assumed a value for the number of catalyst sites per reactor volume of 1.1×10^{25} sites per m^3.

The degrees of rate control of each adsorbed intermediate and the transition state for every elementary step were determined numerically using the definition of the DRC,[8] except for those reaction steps that have no real barrier (*i.e.* where the potential energy only drops from reactants to products). In those cases, it is impossible to incrementally decrease the energy of the transition state, as required by that exact definition, without also decreasing that for the reactant. For such "downhill-only" steps, we instead incrementally increased the transition state energy (to be slightly larger than the reactant), and changed the sign on the resulting change in rate needed in the definition of this transition state's DRC. This gives a value for the DRC that has the same essential meaning as intended by the original definition in ref. 8, but broadens the applicability to such usual steps (*i.e.* "downhill only"). The absolute magnitude of all incremental energy changes used to calculate the DRC values was 10^{-6} eV for the first three sets of conditions and 10^{-8} eV for the fourth, fifth, and sixth sets. The DRC values reported in Tables 9, 10, and 12 below do not sum to exactly 1 for each column, as is normally expected. This is because the residence time was somewhat too long to achieve true differential reactor conditions. This does not affect the qualitative results.

3. Results

3.1. Native catalyst reaction energetics

We studied the reaction kinetics under steady-state conditions at $T = 423$ K and initially used inlet partial pressures of 21.3 kPa O_2 (g), 1898 kPa C_4H_{10} (g), 9.83 × 10^{-9} kPa C_4H_9OH (g), 22.3 kPa C_4H_8O (al, g), 2.93 × 10^{-5} kPa C_4H_8O (ep, g), 1.05 × 10^{-9} kPa H_2 (g), and 1.8 kPa H_2O. Here, "al" indicates that the C_4H_8O species is 1-butanal, and "ep" indicates that it is 1-butene oxide. These conditions are meant to represent a length element very early in a flow reactor, but far enough along that we can calculate steady-state conversions with small changes in all concentrations. The residence time used here was 1 ns.

The TOFs for the individual steps and their DRC values for making butanal (the only significant product besides water) and for the consumption of butane are shown in Table 2 (the steps with large DRC values are the same for both). These DRC values were calculated by perturbing the electronic energy of each transition state and calculating the new net rates, and then using the definition of DRC in eqn (1). The ratio of the forward TOF to the net TOF for any step, column 3, reveals if that step is fast to equilibrium: when the absolute magnitude of this ratio is ≫1, it is fast to equilibrium, whereas when it is 1, the reaction proceeds

Table 2 Steady-state net turnover frequencies (TOFs), ratio of the forward TOF to net TOF, and degrees of rate control (X_i) for each elementary step obtained by solving the microkinetic model at $T = 423$ K and inlet partial pressures of 21.3 kPa O_2 (g), 1898 kPa C_4H_{10} (g), 9.83 × 10^{-9} kPa C_4H_9OH (g), 22.3 kPa C_4H_8O (al, g), 2.93 × 10^{-5} kPa C_4H_8O (ep, g), 1.05 × 10^{-9} kPa H_2 (g), and 1.8 kPa H_2O. The TOFs are defined as molecules per site per s, where the molecules per s for each species is divided by its stoichiometric coefficient, so that it is the same value for every reactant and product of that elementary step. Negative net TOFs indicate that this step goes mainly in the reverse direction

Reaction #	Net TOF (molecules per site per s)	Forward TOF/net TOF	$X_{butanal}$	X_{butane}
1	6.60	2.61×10^6	-5.28×10^{-7}	0
2	6.60	1	**−0.955**	**−0.955**
3	4×10^{-6}	1	4.40×10^{-7}	1.13×10^{-5}
4	0.02	1	5.90×10^{-6}	2.25×10^{-5}
5	6.6	1	**1.96**	**1.96**
6	0.002	1	5.21×10^{-4}	5.19×10^{-4}
7	2×10^{-6}	1	-5.28×10^{-7}	2.25×10^{-5}
8	6.60	1	8.47×10^{-5}	5.64×10^{-5}
9	2×10^{-10}	1	-1.76×10^{-7}	2.25×10^{-5}
10	6.60	1	0.002	0.002
11	−6.60	−13	6.25×10^{-6}	1.13×10^{-5}
12	6.60	1.73×10^8	1.76×10^{-7}	-1.13×10^{-5}
13	−0.006	−0.07	0.001	3.95×10^{-4}
14	6.60	1.82×10^6	3.52×10^{-7}	-1.13×10^{-5}
15	2×10^{-6}	8.59×10^6	8.80×10^{-8}	-1.13×10^{-5}
16	-5×10^{-5}	−0.04	-2.53×10^{-5}	-3.38×10^{-5}
17	1×10^{-4}	1	3.79×10^{-5}	6.76×10^{-5}
18	−6.60	-2.91×10^5	1.76×10^{-7}	-3.38×10^{-5}
19	−6.60	-7.87×10^6	0	0
20	0.02	191	-2.64×10^{-7}	1.13×10^{-5}

only in one direction. The TOF to 1-butanol (BuOH) was found to be negative, *i.e.* BuOH was consumed through the reverse of reaction (11), even at the ultralow initial concentration used in the feed. Both butanal and water had TOFs of 6.6 molecules per site per s, and everything else was negligible. The only surface intermediates with significant coverages were O* (0.971 ML), OH* (0.013 ML) and butanal (0.006 ML). About 0.7% of the sites are unoccupied.

As seen, the steps with large net TOFs (of absolute magnitude 6.6) prove that the dominant reaction pathway (neglecting simple adsorption and desorption steps) under the conditions is as listed below, where "=" indicates that the step is fast to equilibrium:

$$(2)\ O_2{}^* + {}^* \rightarrow 2O^*$$

$$(5)\ C_4H_{10}{}^* + O^* \rightarrow C_4H_9{}^* + OH^*$$

$$(10)\ C_4H_9{}^* + OH^* \rightarrow C_4H_9OH^* + {}^*.$$

The BuOH* product of step (10) then immediately undergoes the reverse of step (11):

$$(-11)\ C_4H_9OH^* + O^* = C_4H_9O^* + OH^*.$$

This $C_4H_9O^*$ product then reacts immediately with another O* *via*

$$(8)\ C_4H_9O^* + O^* \rightarrow C_4H_8Oal^* + OH^*.$$

The OH* from this step combines with the OH* from step (−11) *via*

$$(-19)\ 2OH^* = H_2O^* + O^*,$$

thus regenerating the O* used in step (8).

The net reaction is thus:

$$C_4H_{10} + O_2 \rightarrow C_4H_8Oal + H_2O.$$

Importantly, the $C_4H_9O^*$ intermediate needed to form butanal in step (8) is produced from adsorbed butanol *via* the reverse of reaction (11).

It is clear that the reason that BuOH is not produced is that every BuOH* produced is immediately consumed by the reverse of step (11) before it can desorb. Indeed, that step even consumes some of the low-concentration BuOH included in the feed, so the TOF to BuOH is negative. Since one of every two O* produced is used in this step to consume BuOH*, the DRC of step (2) to BuOH (not shown in Table 1) is ~ -1. One could decrease this by modifying the catalyst to slow down step (2) (increase the transition state (TS) energy for $O_2{}^*$ dissociation), or reduce the O_2 pressure. However, these would decrease the net rate too, since step (5) also requires O* and is integral in the production of 1-BuOH. The ratio of the forward TOF to the net TOF for step (11) is −13, so it is fast to equilibrium. One could thus push this undesired reverse reaction equilibrium, step (11), in the desired forward direction by increasing its reactant OH* coverage, for example by modifying the catalyst to stabilize the TS for step (19), or simply by increasing the H_2O pressure.

3.1.1. Increased water pressure in the feed. We thus increased the H_2O pressure to its saturation value, so that the reaction conditions were: 423 K with inlet partial pressures of 21.3 kPa O_2 (g), 1898 kPa C_4H_{10} (g), 9.83×10^{-9} kPa C_4H_9OH (g), 22.3 kPa C_4H_8O (al, g), 2.93×10^{-5} kPa C_4H_8O (ep, g), 1.05×10^{-9} kPa H_2 (g), and 486.2 kPa H_2O. Solving the microkinetic model under these conditions, the TOF to butanol was found to indeed become positive at 6.6×10^{-5} molecules per site per s, but with a selectivity, defined as the amount of 1-butanol formed divided by the amount of butane reacted, of only 0.0015%. The only other significant products were 1-butanal, still at nearly 100% selectivity, and H_2O (g). Table 3 shows the corresponding DRC values for all the transition states (*i.e.* for each elementary step). Also shown are the net steady-state reaction rates for each step, and the ratio of the forward rate for that step to its net rate.

From these results, the dominant reaction pathway is still identical to that listed above when less water was present, and the net reaction is still strongly dominated by the same process:

$$C_4H_{10} + O_2 \rightarrow C_4H_8Oal + H_2O.$$

Table 3 Steady-state net turnover frequencies (TOFs), ratio of the forward TOF to net TOF, and degrees of rate control (X_i) for each elementary step for the conditions: $T = 423$ K and inlet partial pressures of 21.3 kPa O_2 (g), 1898 kPa C_4H_{10} (g), 9.83×10^{-9} kPa C_4H_9OH (g), 22.3 kPa C_4H_8O (al, g), 2.93×10^{-5} kPa C_4H_8O (ep, g), 1.05×10^{-9} kPa H_2 (g), and 486.2 kPa H_2O

Reaction #	Net TOF (molecules per site per s)	Forward TOF/ net TOF	$X_{butanol}$	$X_{butanal}$	$X_{butene\ oxide}$	X_{butane}
1	4.53	3.16×10^6	-1×10^{-5}	0	9×10^{-5}	0
2	4.53	1	**−39.7**	**−0.79**	**−0.79**	**−0.79**
3	2.68×10^{-6}	1	5×10^{-5}	-1×10^{-5}	1×10^{-5}	0.003
4	0.0108	1	-5×10^{-4}	0	1×10^{-4}	0
5	4.53	1	**117**	**1.79**	**1.79**	**1.79**
6	0.0012	1	0.03	4×10^{-4}	6×10^{-4}	0.003
7	1.27×10^{-6}	1	-2×10^{-5}	-1×10^{-5}	**1.00**	0.001
8	4.53	1	**−77.5**	0.002	**−1.00**	0.001
9	2.82×10^{-9}	1	8×10^{-8}	0	1×10^{-4}	0
10	4.52	1	0.04	8×10^{-4}	8×10^{-4}	0.003
11	−4.52	−217	**−0.36**	0	9×10^{-5}	0.001
12	4.53	2.09×10^8	6×10^{-7}	0	2×10^{-4}	0
13	6.65×10^{-5}	78	**0.99**	-1×10^{-5}	2×10^{-4}	0.001
14	4.53	2.20×10^6	-6×10^{-9}	1×10^{-5}	1×10^{-4}	0.001
15	1.27×10^{-6}	1.04×10^7	-8×10^{-10}	-1×10^{-5}	0	0.001
16	4.15×10^{-6}	9.13	2×10^{-4}	-1×10^{-5}	4×10^{-5}	0.003
17	0.0015	1	0.04	6×10^{-4}	6×10^{-4}	0.001
18	−4.53	-9.61×10^7	-4×10^{-7}	-1×10^{-5}	1×10^{-4}	0
19	−4.53	-2.15×10^9	-2×10^{-8}	-1×10^{-5}	1×10^{-4}	0
20	0.0094	220	-2×10^{-4}	-1×10^{-5}	1×10^{-4}	0

Now, however, a small amount of BuOH is produced. Because the competitive pathways that lead to butanal are so much faster than those to BuOH, elementary steps that accelerate or slow them (relative to the steps that make BuOH) show DRC values to BuOH with huge magnitudes (*i.e.* steps (5) and (8)) (changes in their rates by a few % can accelerate BuOH production by much larger factors).

Table 4 shows the corresponding steady-state coverages of adsorbed intermediates under these conditions. The most abundant surface intermediate is O*, which occupies 80% of the available surface sites. The only other adsorbed intermediate with a non-tiny coverage is OH* (0.18 ML). The free site coverage is essentially equal to 0. The corresponding DRC values for all adsorbates are also shown in Table 4. These DRC values are approximately proportional to their coverages in Table 3, with the proportionality constant being negative, as expected[8] (−1 in the case of butanol, and −2 in the cases of butane, butanal, and 1-butene oxide). This shows that the usual situation occurs here: when adsorbed intermediates that have significant coverages are stabilized in energy, their coverages increase and this poisons the rate controlling step(s).[8]

To produce 1-butanol more selectively over 1-butanal, we focus on the steps in this dominant path that exhibit DRC values to either of these two products with both large magnitude and disparity in their DRC values to butanol *versus* butanal. The steps that exhibit disparity in Table 3 are steps 2, 5, 8, 11, and 13. Based on the definition of the DRC, one should be able to improve the relative rate of butanol production by modifying the catalyst to increase the transition state energy for steps 2, 8, or 11 (since they have negative DRC values to BuOH), or decreasing the transition state energy for step (5). However, this DRC analysis suggests a simpler way to improve the selectivity. It is clear that the reason that BuOH is not produced nearly as fast as butanal is that almost every BuOH* produced is immediately consumed by the reverse of step (11) before it can desorb. Since that step requires O*, one could suppress its rate and thus

Table 4 Steady state coverages and DRC values of the adsorbed intermediates for the same reactions and conditions as Table 3

Intermediate	θ (ML)	X_{butanol}	X_{butanal}	$X_{\text{butene oxide}}$	X_{butane}
O*	**0.805**	**−0.802**	**−1.61**	**−1.61**	**−1.61**
O_2*	0.0001	−0.0001	−0.0002	−0.0002	−0.0016
C_4H_{10}*	0.0011	−0.0011	−0.0021	−0.0021	0
C_4H_9*	6×10^{-5}	-6×10^{-5}	−0.0001	-4×10^{-5}	0.0033
OH*	**0.183**	**−0.182**	**−0.365**	**−0.365**	**−0.364**
C_4H_9O*	5×10^{-10}	-2×10^{-9}	-1×10^{-5}	4×10^{-5}	0.0016
H*	2×10^{-9}	-1×10^{-9}	-1×10^{-5}	9×10^{-5}	0.0033
C_4H_8Oep*[a]	2×10^{-11}	-1×10^{-9}	0	4×10^{-5}	−0.0016
C_4H_8Oal*[b]	0.0050	−0.0050	−0.0100	−0.0100	−0.0066
H_2O*	3×10^{-11}	-5×10^{-9}	-1×10^{-5}	0.0001	0.0033
C_4H_9OH*	0.0007	−0.0007	−0.0014	−0.0012	0
*	0.0060	—	—	—	—

[a] "al" means 1-butanal. [b] "ep" means 1-butene oxide.

achieve higher selectivity to BuOH by decreasing the O* coverage, simply by using less O_2 in the feed (consistent with the large negative DRC value to BuOH for step (2)). Since O* is consumed by step (5), and step (5) has a huge positive DRC value to BuOH, one could alternatively decrease the O* coverage by adding more butane to the feed, thus accelerating step (5). This has the advantage over decreasing the O_2 pressure of increasing the TOF rather than decreasing it. We show next that decreasing the O_2 pressure and increasing that for butane indeed lead to the desired results.

3.1.2. Decreased O_2 pressure and increased butane pressure in the feed. Thus, we ran the microkinetic model at $T = 423$ K but decreased the O_2 pressure by a factor of 10^5 and doubled the pressure of butane to its saturation vapour pressure, *i.e.* with inlet partial pressures of 2.13×10^{-4} kPa O_2 (g), 3796 kPa C_4H_{10} (g), 1.49×10^{-5} kPa C_4H_9OH (g), 2.23×10^{-5} kPa C_4H_8O (al, g), 2.93×10^{-5} kPa, C_4H_8O (ep, g), 1.05×10^{-5} kPa H_2 (g), and 486.2 kPa H_2O (g). Under these conditions, the TOF to butanol increased by 30 000-fold (to 1.8 molecules per site per s), and the selectivity to butanol increased by many orders of magnitude to 95%. The selectivity to butanal dropped to 5%. Table 5 shows the corresponding TOFs, TOF ratios and DRC values for each elementary step.

Table 5 Steady state turnover frequencies (TOFs), ratio of the forward TOF to net TOF, and degrees of rate control (X_i) for all elementary steps under reaction conditions of: $T =$ 423 K and inlet partial pressures of 2.13×10^{-4} kPa O_2 (g), 3796 kPa C_4H_{10} (g), 1.49×10^{-5} kPa C_4H_9OH (g), 2.23×10^{-5} kPa C_4H_8O (al, g), 2.93×10^{-5} kPa, C_4H_8O (ep, g), 1.05×10^{-5} kPa H_2 (g), and 486.2 kPa H_2O (g)

Reaction #	Net TOF (molecules per site per s)	Forward TOF/ net TOF	X_{butanol}	X_{butanal}	$X_{\text{butene oxide}}$	X_{butane}
1	0.562	2.83×10^5	2×10^{-5}	0.0008	0	−0.005
2	0.562	1	**0.670**	**2.36**	**2.45**	**0.770**
3	7×10^{-7}	1	8×10^{-7}	0.002	−0.003	0.005
4	0.0003	1	9×10^{-5}	1×10^{-6}	−0.003	−0.005
5	0.144	1	−0.044	**−0.151**	**−0.165**	−0.040
6	4×10^{-5}	1	-1×10^{-6}	-8×10^{-5}	−0.005	0.010
7	3×10^{-8}	1.05	-3×10^{-8}	-6×10^{-8}	**1.00**	0
8	0.103	1	**−0.119**	**0.781**	**−0.235**	−0.065
9	0.0033	1	−0.001	−0.004	−0.008	−0.005
10	1.89	1	**0.493**	**−1.99**	**−2.07**	**0.362**
11	−0.106	-1.85×10^5	-7×10^{-7}	4×10^{-6}	−0.003	0.010
12	1.89	1.11×10^{11}	0	-5×10^{-9}	−0.003	0
13	1.79	410	0.0003	−0.002	−0.005	0.005
14	0.103	1.07×10^5	1×10^{-10}	-2×10^{-8}	−0.003	0.010
15	3×10^{-8}	5.34×10^{10}	0	0	−0.005	0.005
16	0.872	2403	0.0001	−0.0004	−0.003	−0.010
17	0.0015	1	−0.0005	−0.002	−0.003	0
18	0.768	6.31×10^{10}	-2×10^{-11}	5×10^{-9}	0	−0.005
19	0.768	2.02×10^8	1×10^{-9}	-7×10^{-8}	−0.008	0.005
20	−1.75	−2920	2×10^{-5}	-7×10^{-5}	0	0

The dominant reaction pathway (neglecting simple adsorption and desorption steps) is now:

$$(2)\ O_2{}^* + {}^* \rightarrow 2O^*$$

$$(10)\ C_4H_9{}^* + OH^* \rightarrow C_4H_9OH^* + {}^*$$

$$(16)\ 2H^* = H_2 + {}^*$$

$$(19)\ H_2O^* + O^* = 2OH^*$$

$$(-20)\ C_4H_{10}{}^* + {}^* = C_4H_9{}^* + H^*.$$

The net reaction is approximately:

$$2O_2 + 3H_2O + 7C_4H_{10} \rightarrow 7C_4H_9OH + 3H_2.$$

Note from Table 5 that ~40% of the O in the BuOH product now comes from H_2O. This could save O_2 feed costs by ~40%, beyond the dramatic improvement that results from the increased selectivity to BuOH relative to butanal. More importantly, the latter also saves butane costs dramatically.

There are also important but 4- to 10-fold slower contributions from these steps:

$$(5)\ C_4H_{10}{}^* + O^* \rightarrow C_4H_9{}^* + OH^*$$

$$(8)\ C_4H_9O^* + O^* \rightarrow C_4H_8Oal^* + OH^*$$

$$(-11)\ C_4H_9OH^* + O^* = C_4H_9O^* + OH^*.$$

Steps (−11) and (8) combine to give the minor butanal product.

The reaction conditions were not further optimized. For example, it seems that it should be possible to get faster rates while maintaining a similar selectivity by increasing the O_2 pressure some. Tuning of the catalyst material to slow down step (8) and/or speed up step (10) could further improve the rate and selectivity toward butanol.

Table 6 shows the corresponding steady-state coverages of adsorbed intermediates under these conditions. The most abundant surface intermediate is now butane*, which occupies 24% of the available surface sites. H_2O has a coverage of 7.5%. All other adsorbed intermediates have tiny coverages, and the free site coverage is high, 66%. The DRC values of the adsorbed surface intermediates equal these values times −2.

Table 7 summarizes the reaction conditions and resulting dramatic improvements in the rate and selectivity to 1-butanol.

3.2. Modified catalyst material

To further demonstrate the power of DRC analysis, we also studied a related catalyst material, where a few of the adsorbate and transition state energies were modified. This corresponds to a hypothetical material at this stage. The energies that were changed are summarized in Table 8. The main difference is that the activation energy for step (4) was decreased by ~0.8 eV.

Table 6 Steady-state coverages of surface intermediates under the same reaction conditions as in Table 5

Intermediate	θ (ML)
O^*	1×10^{-4}
O_2^*	1×10^{-7}
$C_4H_{10}^*$	**0.238**
$C_4H_9^*$	0.0002
OH^*	0.023
$C_4H_9O^*$	8×10^{-8}
H^*	1×10^{-5}
$C_4H_8Oep^{*a}$	2×10^{-9}
$C_4H_8Oal^{*b}$	6×10^{-6}
$C_4H_9OH^*$	4×10^{-6}
H_2O^*	0.075
*	**0.664**

[a] "al" means 1-butanal. [b] "ep" means 1-butene oxide.

Table 7 Summary of how changes in the reaction conditions were used to markedly improve the butanol production rate, selectivity toward butanol, and amount of O in the products from O_2 (rather than H_2O). In all cases, the conditions were: $T = 423$ K, $P_{C_4H_9OH} = 1.49 \times 10^{-5}$ kPa, $P_{C_4H_8O,ep} = 2.93 \times 10^{-5}$ kPa, $P_{H_2} = 1.05 \times 10^{-5}$ kPa

Conditions	TOF to 1-butanol (molecules per site per s)	TOF for butane consumption (molecules per site per s^{-1})	1-Butanol selectivity	O from O_2/(total O in product)
$P_{H_2O,0} = 1.8$ kPa, $P_{O_2,0} = 21.3$ kPa, $P_{C_4H_{10},0} = 1858$ kPa, $P_{C_4H_8O,al} = 22.3$ kPa	−0.0058	6.6	1-Butanol is consumed	100%
$P_{H_2O,0} = 486.2$ kPa, $P_{O_2,0} = 21.3$ kPa, $P_{C_4H_{10},0} = 1858$ kPa, $P_{C_4H_8O,al} = 22.3$ kPa	6.6×10^{-5}	4.5	1.5×10^{-3}%	100%
$P_{H_2O,0} = 486.2$ kPa, $P_{O_2,0} = 2.1 \times 10^{-4}$ kPa, $P_{C_4H_{10},0} = 3796$ kPa, $P_{C_4H_8O,al} = 2.23 \times 10^{-5}$ kPa	1.8	1.9	95%	60%

Table 8 List of reaction steps whose energies were changed and their new values used in this hypothetical "modified catalyst" material

Reaction #	Reaction	Actual ΔE_{rxn}	Modified ΔE_{rxn}	Actual ΔE_{act}	Modified ΔE_{act}
4	$C_4H_{10}^* + O^* = C_4H_9O^* + H^*$	−0.18	−0.03	1.14	0.35
5	$C_4H_{10}^* + O^* = C_4H_9^* + OH^*$	−0.34	−0.12	0.92	1.19

We studied the reaction kinetics on this modified catalyst under steady-state conditions at a lower temperature ($T = 373$ K), since activation of butane is so much faster now. We initially used inlet partial pressures of 21.3 kPa O_2 (g),

42.6 kPa C_4H_{10} (g), 0.010 kPa C_4H_9OH (g), 0.010 kPa C_4H_8O (al, g), 0.010 kPa H_2 (g), and 1.5 kPa H_2O (g).

Table 9 gives the TOFs, TOF ratios and DRC values. As with the unmodified catalyst, step (11) still goes so fast in reverse that it consumes most of the BuOH* as it is produced. Step (9) now dominates production of 1-BuOH*, but it is now in direct competition for the same butoxy used in step (8) to make butanal. The net production of butanal is now ~40-fold faster than that of 1-BuOH, so the DRC value to 1-BuOH for step (8) is large and negative (~−4), since every increase in its rate decreases the rate to 1-BuOH by ~40 times that same amount (*i.e.* the DRC to butanol for step (8) is roughly this factor of 40 times its DRC to butanal). For a similar reason, there is a large negative DRC value to 1-BuOH for step (2) of −1, since step (2) produces the O* reactant needed in step (8), which, in turn, robs the butoxy needed to make 1-BuOH*. Similarly, there is a large positive DRC value to 1-BuOH for step (4) of +3, since this step produces both the C_4H_9O* and the H* needed to make 1-BuOH* in step (9) in this competition with step (8) to butanal, and, more importantly, it consumes the O* that drives step (11) so fast in the reverse direction to destroy most of the 1-BuOH* just produced. The surface coverage is relatively low under these conditions.

Table 9 Steady state net turnover frequencies (TOFs) on the "modified catalyst", ratio of the forward TOF to net TOF, and degrees of rate control (X_i) for the reaction conditions: T = 373 K and inlet partial pressures of 21.3 kPa O_2 (g), 42.6 kPa C_4H_{10} (g), 0.010 kPa C_4H_9OH (g), 0.010 kPa C_4H_8O (al, g), 0.010 kPa H_2 (g), and 1.5 kPa H_2O (g)

Reaction #	Net TOF (molecule site per s)	Forward TOF/net TOF	$X_{butanol}$	$X_{butanal}$	$X_{butene\ oxide}$	X_{butane}
1	44 845	3.98×10^4	0.00508	−0.00451	−0.00451	−0.00424
2	44 845	1.00	**−0.971**	**1.01**	**1.01**	**0.964**
3	8.40×10^{-4}	1.00	0.00343	−0.00305	−0.00305	−0.00290
4	53 755	2.71	**2.9983**	−0.0993	−0.0993	−0.0311
5	6.51×10^{-7}	1.00	0.00583	−0.00518	−0.00518	−0.00490
6	2.56×10^{-7}	1.00	0	0	0	0
7	1.95×10^{-3}	1.00	−0.00302	0.00268	**1.00**	0.00246
8	52 570	1.00	**−4.05**	**0.198**	**−0.802**	0.105
9	19 443	1.00	**0.881**	−0.0990	−0.0990	−0.0774
10	6.52×10^{-7}	1.00	0	0	0	0
11	−18259	-1.01×10^5	0.00779	−0.00698	−0.00698	−0.00673
12	53 755	4.91×10^4	0.00222	−0.00182	−0.00182	−0.00171
13	1184	522	0.0164	-9.17×10^{-4}	-9.17×10^{-4}	-5.12×10^{-4}
14	52 570	69.3	0.00352	−0.00313	−0.00313	−0.00302
15	1.95×10^{-3}	58.7	0.00650	−0.00577	−0.00577	−0.00559
16	17 156	378	0.00946	−0.00349	−0.00349	−0.00322
17	6.97×10^{-2}	1.00	-7.73×10^{-4}	6.88×10^{-4}	6.88×10^{-4}	6.83×10^{-4}
18	−35414	-4.84×10^3	0.00411	−0.00438	−0.00438	−0.00422
19	−35414	−42.0	−0.010498	9.02×10^{-4}	9.02×10^{-4}	6.59×10^{-4}

3.2.1. Adding H_2O to the feed, modified catalyst. It is clear from the DRC values in Table 9 that adding more water to reverse the net direction of steps (18) and (19), so that they go forward in net, should make the selectivity and rate to 1-BuOH go up. Table 10 shows the results of adding the maximum amount of water possible, its saturation vapour pressure, or 1 bar at this temperature of 373 K. Now step (11) has a major positive contribution to 1-BuOH production, rather than going in reverse to consume it, as in Table 9. This dramatically increases the selectivity for 1-butanol from 2.2% up to 70%, and the TOF to 1-butanol increases 30-fold over the low-water-pressure conditions of Table 9. Moreover, O_2 was consumed in making only 85% of the oxygen in products, with the remainder coming from the added H_2O. At the lower water pressure in Table 9, 100% of this oxygen came from O_2. This lower O_2 consumption could represent cost and energy savings beyond that from the dramatically increased selectivity and TOF. The surface has a relatively high coverage of adsorbed butanol under these conditions, ~0.25 ML. The OH* coverage increased 6-fold, from 0.0125 ML to 0.0734 ML.

Table 11 shows the corresponding DRC values and coverages of the adsorbed intermediates.

3.2.2. Increasing the butane pressure, modified catalyst. It is clear from Table 10 that step (4) offers an opportunity to further improve the selectivity to 1-BuOH over butanal, since its DRC value is +0.4 for 1-BuOH but −0.5 for butanal. Modifying the catalyst to decrease its activation barrier would lead to improved

Table 10 Steady state net turnover frequencies (TOFs) on the "modified catalyst", ratio of the forward TOF to net TOF, and degrees of rate control (X_i). In this case, 101 kPa of H_2O was included in the feed, otherwise the conditions were the same as in Table 9[a]

Reaction #	Net TOF (molecules per site per s)	Forward TOF/net TOF	X_{butanol}	X_{butanal}	$X_{\text{butene oxide}}$	X_{butane}
1	21 444	5.75×10^4	-2.16×10^{-5}	-5.99×10^{-5}	-5.97×10^{-5}	6.52×10^{-5}
2	21 444	1.00	0.293	1.21	1.21	0.568
3	4.04×10^{-4}	1.00	0.00188	0.00568	0.00569	0.00308
4	50 247	1.52	0.385	−0.502	−0.502	0.120
5	3.42×10^{-7}	1.00	4.01×10^{-4}	1.22×10^{-3}	1.22×10^{-3}	6.91×10^{-4}
6	1.34×10^{-7}	1.00	0	0	0	0
7	5.55×10^{-4}	1.00	0.00157	0.00477	1.00	0.00256
8	14 887	1.00	−0.298	0.519	−0.481	−0.054
9	5553	1.00	−0.00852	−0.0826	−0.0826	−0.0306
10	3.42×10^{-7}	1.00	0	0	0	0
11	29 707	1.37×10^5	0.00123	0.00372	0.00372	0.00197
12	50 247	3.64×10^4	0.00200	0.00599	0.00599	0.00321
13	35 260	51.2	0.0153	0.00264	0.00264	0.0116
14	14 987	66.8	0.00213	0.00644	0.00644	0.00344
15	5.55×10^{-4}	41.0	-3.68×10^{-4}	−0.00112	−0.00112	-5.09×10^{-4}
16	22 347	167	−0.00104	−0.00561	−0.00561	−0.00237
17	4.00×10^{-2}	1.00	4.67×10^{-4}	0.00142	0.00142	7.70×10^{-4}
18	7360	1.06×10^6	0.00116	0.00353	0.00353	0.00190
19	7360	7.01×10^4	8.30×10^{-4}	0.00250	0.00250	0.00137

[a] Since step (20) had a negligible contribution on this modified catalyst under the conditions studied here, we have omitted it completely from all analyses of this modified catalyst.

selectivity. A far simpler method to increase the rate of step (4) is to increase the butane coverage, since adsorbed butane is its reactant but adsorbed butane does not participate in any other kinetically relevant steps. To achieve this, we increased the partial pressure of butane up to its saturation value at 373 K, *i.e.* 1580 kPa, and recalculated the rates and DRC values under the same reaction conditions as in Tables 10 and 11. Indeed, the selectivity increased to 100%. The results are shown in

Table 11 Coverages and degrees of rate control for the surface intermediates on the "modified catalyst" under the conditions of Table 10

Intermediate	θ (ML)	$X_{butanol}$	$X_{butanal}$	$X_{butene\ oxide}$	X_{butane}
O*	6.45×10^{-4}	-9.60×10^{-5}	−0.000291	−0.000291	-9.13×10^{-5}
O_2*	0.0864	−0.0652	−0.198	−0.198	−0.105
C_4H_{10}*	0.015	−0.00987	−0.0299	−0.0299	−0.0157
C_4H_9*	2.03×10^{-10}	0	0	0	0
OH*	0.0734	−0.0539	−0.163	−0.163	−0.0865
C_4H_9O*	0.0102	−0.00856	−0.0260	−0.0260	−0.0137
H*	0.000830	0.00253	0.00767	0.00767	0.00411
C_4H_8Oep*	2.79×10^{-13}	0.00231	0.00701	0.00701	0.00380
C_4H_8Oal*	0.00844	−0.00636	−0.0193	−0.0193	−0.0102
C_4H_9OH*	0.245	−0.184	−0.558	−0.558	−0.296
H_2O*	0.0785	−0.060	−0.181	−0.181	−0.096
*	0.483	—	—	—	—

Table 12 Steady state turnover frequencies (TOFs) on the "modified catalyst", ratio of the forward TOF to net TOF, and degrees of rate control (X_i) for the elementary steps obtained by solving the microkinetic model using the reactions listed in Tables 1 and 8. In this case, 101 kPa of H_2O was included in the feed, and butane was fed at its saturation vapor pressure of 1580 kPa. All other conditions are the same as in Tables 10 and 11

Reaction #	TOF (molecules per site per s)	Forward TOF/ net TOF	$X_{butanol}$	$X_{butanal}$	$X_{butene\ oxide}$	X_{butane}
1	9718	8.54×10^4	−0.00121	−0.00227	−0.00226	−0.00219
2	9718	1.00	0.753	1.89	1.89	0.753
3	6.82×10^{-3}	1.00	0.00870	0.0163	0.0163	0.00766
4	38 332	1.06	0.0618	−1.34	−1.34	0.0580
5	1.81×10^{-7}	1.00	0.00542	0.0102	0.0102	0.00438
6	7.11×10^{-8}	1.00	0	0	0	0
7	1.51×10^{-6}	1.00	0.00628	0.0118	1.01	0.00602
8	41	1.00	0.000552	1.00	0.00110	0
9	456	1.00	−0.00946	−0.0237	−0.0237	−0.0131
10	1.81×10^{-7}	1.00	0.00224	0.00419	0.00420	0.00164
11	37 835	1.65×10^3	0.0125	0.0228	0.0228	0.0109
12	38 332	1.20×10^6	0.00259	0.00486	0.00486	0.00109
13	38 292	34.2	−0.00877	−0.0478	−0.0478	−0.00985
14	41	6.41×10^3	0.000865	0.00162	0.00162	−0.00219
15	1.51×10^{-6}	27.9	0.00456	0.00854	0.00854	0.00274
16	18 938	80.2	−0.00153	−0.00334	−0.00334	−0.00383
17	5.39×10^{-4}	1.00	−0.00987	−0.0185	−0.0185	−0.0120
18	18 897	2.79×10^5	4.40×10^{-5}	7.86×10^{-5}	8.05×10^{-5}	−0.00219
19	18 897	38.8	0.00196	0.00228	0.00229	0.00109

Table 13 Coverages and degrees of rate control for surface intermediates on the "modified catalyst" under the conditions of Table 12

Intermediate	θ (ML)	X_{butanol}	X_{butanal}	$X_{\text{butene oxide}}$	X_{butane}
O*	1.36×10^{-5}	0.00270	0.00506	0.00506	0.00055
O_2*	0.0582	−0.0918	−0.172	−0.172	−0.0941
C_4H_{10}*	0.374	−0.577	−1.08	−1.08	−0.578
C_4H_9*	9.04×10^{-10}	0.00401	0.00752	0.00752	0.00274
OH*	0.00874	−0.0213	−0.0400	−0.0400	−0.0235
C_4H_9O*	0.00131	−0.00680	−0.0127	−0.0127	−0.00985
H*	0.000530	0.00783	0.0147	0.0147	0.00711
C_4H_8Oep*	0	0.00609	0.0114	0.0114	0.00602
C_4H_8Oal*	0.00220	−0.00449	−0.00842	−0.00842	−0.00547
C_4H_9OH*	0.177	−0.276	−0.517	−0.517	−0.277
H_2O*	0.0528	−0.0844	−0.158	−0.158	−0.0854
*	0.325	—	—	—	—

Tables 12 and 13. The TOF to butanol also increased slightly, and the fraction of the oxygen in the products that comes from O_2 decreased by 34%, to 51%, another major improvement. Almost exactly half the oxygen in the products is coming from O_2 and half from H_2O. Thus, the net reaction here is essentially:

$$2C_4H_{10} + H_2O + \tfrac{1}{2}O_2 \rightarrow 2C_4H_9OH + H_2,$$

which is still thermodynamically favorable ($\Delta G^0 = -40$ kJ mol^{-1} at 373 K based on standard thermodynamic data). This addition of butane increased the butane coverage from 0.01 ML to 0.37 ML and correspondingly decreased the fraction of empty sites from 47% to 32%, which resulted in a similar relative decrease in the butanol coverage (from 0.25 ML to 0.18 ML). The coverage of OH* decreased 8-fold, to 0.01 ML.

The use of DRC values to improve the performance of this hypothetical modified catalyst are summarized in Table 14. The DRC values when very little water was added to the feed revealed that the rate of production of adsorbed OH, required in the step to make adsorbed butanol, was dominated by the dehydrogenation of butoxy to make butanal plus OH. This severely limits the selectivity.

Table 14 Summary of how changes in the reaction conditions for the "modified catalyst" were used to markedly improve the butanol production rate, selectivity toward butanol, and amount of O in the products from O_2 (rather than H_2O). In all cases, the conditions were: $T = 373$ K, $P_{O_2,0} = 21.3$ kPa, $P_{C_4H_9OH,0} = P_{C_4H_8O,0} = P_{H_2,0} = 0.01$ kPa

Conditions	TOF to 1-butanol (molecules per site per s)	1-Butanol selectivity	O from O_2/(total O in product)
$P_{H_2O,0} = 1.5$ kPa, $P_{C_4H_{10},0} = 42.6$ kPa	923	2%	100%
$P_{H_2O,0} = 101$ kPa, $P_{C_4H_{10},0} = 42.6$ kPa	35 200	70%	85%
$P_{H_2O,0} = 101$ kPa, $P_{C_4H_{10},0} = 1580$ kPa	38 300	100%	51%

This suggested that water should be added to the feed to make more OH, which indeed increased the rate to butanol and its selectivity increased dramatically. Moreover, less O_2 was consumed per oxygen atom in the products. The degree of rate control for oxygen adatom addition to adsorbed butane to make butoxy plus H was positive for making 1-BuOH but negative for butanal, suggesting to increase the butane pressure. This indeed increased the selectivity to 1-BuOH to 100%, increased its rate and reduced O_2 consumption.

4. Conclusions

A $Ag_3Pd(111)$ surface decorated with a 2-atom-thick ring of (immobile) helium atoms which creates an artificial pore was used to model the sterically-constrained access to metal sites offered by a MOF-encapsulated PdAg nanoparticle catalyst. Using energies and kinetic parameters derived from DFT, a microkinetic model was developed to predict reaction rates and selectivities of this model catalyst for the regio-selective oxidation of *n*-butane to 1-butanol with O_2. The degrees of rate control of all species (adsorbed intermediates and transition states) were analysed to determine the rate controlling species and provide ideas for improvement. Table 7 shows that this TOF analysis allowed us to find conditions whereby the TOF to 1-butanol could be increased from 0 to ~2 molecules per site per s and the selectivity from 0% to 95%. On a hypothetical modified catalyst where the activation energy for step (4) was decreased by ~80 kJ mol^{-1}, Table 14 shows that the TOF could be increased from ~1000 to ~40 000 molecules per site per s and the selectivity increased from 2% to 100% by using DRC analysis to select better reaction conditions. These results demonstrate that analysis of DRC values provides a powerful approach to optimizing catalytic process conditions. They also show that highly selective oxidation can sometimes be achieved by using a mixture of O_2 and H_2O as the oxidant, since these couple on the surface to make OH* *via* $\frac{1}{2}O_2 + H_2O + 2^* \rightarrow 2\ OH^*$.

5. Note added in proof

We recently discovered that the reaction energies and barriers initially studied for steps (4) and (5) in Table 1 were incorrect. The data in the version of the manuscript presented to the delegates before the meeting were later found to be inaccurate. Therefore, the energies and data presented above in this final version of the manuscript have been revised accordingly, and are now correct.

Acknowledgements

This research was primarily supported as part of the Inorganometallic Catalyst Design Center (ICDC), an Energy Frontier Research Center funded by the U. S. Department of Energy (DOE), Office of Science, Basic Energy Sciences (BES), under Award # DE-SC0012702. Undergraduate STD, who performed part of the DFT calculations and helped construct the microkinetic modelling code used in this work, acknowledges support from the National Science Foundation (Award DMR-1334928) and the Clemson University Calhoun Honors College. Simulations were performed on the Palmetto Supercomputer Cluster, which is maintained by the Cyberinfrastructure Technology Integration Group at Clemson University.

References

1 G. Lu, S. Li, Z. Guo, O. K. Farha, B. G. Hauser, X. Qi, Y. Wang, X. Wang, S. Han, X. Liu, J. S. DuChene, H. Zhang, Q. Zhang, X. Chen, J. Ma, S. C. J. Loo, W. D. Wei, Y. Yang, J. T. Hupp and F. Huo, *Nat. Chem.*, 2012, **4**, 310–316.

2 S. L. James, *Chem. Soc. Rev.*, 2003, **32**, 276–288.

3 H. Furukawa, K. E. Cordova, M. O'Keeffe and O. M. Yaghi, *Science*, 2013, **341**, 1230444.

4 H. Li, M. Eddaoudi, M. O'Keeffe and O. M. Yaghi, *Nature*, 1999, **402**, 276–279.

5 C. J. Stephenson, J. T. Hupp and O. K. Farha, *Inorg. Chem. Front.*, 2015, **2**, 448–452.

6 D. A. Gomez-Gualdron, S. T. Dix, R. B. Getman and R. Q. Snurr, *Phys. Chem. Chem. Phys.*, 2015, **17**, 27596–27608.

7 S. T. Dix, D. A. Gomez-Gualdron and R. B. Getman, *Surf. Sci.*, 2016, DOI: 10.1016/j.susc.2016.05.004.

8 C. Stegelmann, A. Andreasen and C. T. Campbell, *J. Am. Chem. Soc.*, 2009, **131**, 8077–8082.

9 C. A. Wolcott, A. J. Medford, F. Studt and C. T. Campbell, *J. Catal.*, 2015, **330**, 197–207.

10 We calculated the first-nearest-neighbour interaction energies between H_2O^* and O^* to be −0.25 eV, and between two OH^*s to be +0.13 eV, but these were not included in the energy for reaction (19) used here as listed in Table 1.

11 G. Kresse and J. Furthmuller, *Comput. Mater. Sci.*, 1996, **6**, 15–50.

12 G. Kresse and J. Furthmuller, *Phys. Rev. B: Condens. Matter Mater. Phys.*, 1996, **54**, 11169–11186.

13 J. P. Perdew, K. Burke and M. Ernzerhof, *Phys. Rev. Lett.*, 1996, **77**, 3865–3868.

14 J. P. Perdew, K. Burke and M. Ernzerhof, *Phys. Rev. Lett.*, 1997, **78**, 1396.

15 P. E. Blochl, *Phys. Rev. B: Condens. Matter Mater. Phys.*, 1994, **50**, 17953–17979.

16 G. Kresse and D. Joubert, *Phys. Rev. B: Condens. Matter Mater. Phys.*, 1999, **59**, 1758–1775.

17 S. Grimme, *J. Comput. Chem.*, 2006, **27**, 1787–1799.

18 G. Henkelman and H. Jonsson, *J. Chem. Phys.*, 2000, **113**, 9901–9904.

19 G. Henkelman and H. Jonsson, *J. Chem. Phys.*, 2000, **113**, 9978–9985.

20 G. Henkelman and H. Jonsson, *J. Chem. Phys.*, 1999, **111**, 7010–7022.

Faraday Discussions

PAPER

The cluster beam route to model catalysts and beyond

Peter R. Ellis,[a] Christopher M. Brown,[a] Peter T. Bishop,[a] Jinlong Yin,[b] Kevin Cooke,[b] William D. Terry,[c] Jian Liu,[c] Feng Yin[c] and Richard E. Palmer*[c]

Received 18th November 2015, Accepted 19th January 2016
DOI: 10.1039/c5fd00178a

The generation of beams of atomic clusters in the gas phase and their subsequent deposition (in vacuum) onto suitable catalyst supports, possibly after an intermediate mass filtering step, represents a new and attractive approach for the preparation of model catalyst particles. Compared with the colloidal route to the production of pre-formed catalytic nanoparticles, the nanocluster beam approach offers several advantages: the clusters produced in the beam have no ligands, their size can be selected to arbitrarily high precision by the mass filter, and metal particles containing challenging combinations of metals can be readily produced. However, until now the cluster approach has been held back by the extremely low rates of metal particle production, of the order of 1 microgram per hour. This is more than sufficient for surface science studies but several orders of magnitude below what is desirable even for research-level reaction studies under realistic conditions. In this paper we describe solutions to this scaling problem, specifically, the development of two new generations of cluster beam sources, which suggest that cluster beam yields of grams per hour may ultimately be feasible. Moreover, we illustrate the effectiveness of model catalysts prepared by cluster beam deposition onto agitated powders in the selective hydrogenation of 1-pentyne (a gas phase reaction) and 3-hexyn-1-ol (a liquid phase reaction). Our results for elemental Pd and binary PdSn and PdTi cluster catalysts demonstrate favourable combinations of yield and selectivity compared with reference materials synthesised by conventional methods.

1 Introduction

Catalysis has always been "nanotechnology", in the sense that catalyst particles have nanometre dimensions, but the increased level of material control which

[a]Johnson Matthey Technology Centre, Blount's Court, Sonning Common, Reading RG4 9NH, UK. E-mail: r.e.palmer@bham.ac.uk
[b]Teer Coatings Ltd., Miba Coating Group, West Stone House, West Stone, Berry Hill Industrial Estate, Droitwich, Worcestershire, WR9 9AS, UK
[c]Nanoscale Physics Research Laboratory, School of Physics and Astronomy, University of Birmingham, Birmingham B15 2TT, UK

nanotechnology now offers is an obvious attraction for those who seek to understand and improve the function of catalysts. Even within the bounds of nanotechnology, there are many ways to create "designer" catalyst particles, including the adoption of micro/nanolithographic approaches[1] and colloidal routes.[2] A relatively new player in this scene is the cluster beam route to model catalyst production,[3] in which atomic clusters (*i.e.* nanoparticles) are generated in the gas phase (typically) and deposited onto the catalyst support as a beam in high vacuum, sometimes with a mass-selection step prior to the deposition stage. Both heterogeneous and electro/photo-catalysis are the subjects of active current research.

The potential advantages of the cluster beam approach are several: (i) the size of the catalyst particle can be selected even to single atom precision;[4] (ii) the (direct) interaction between the metal cluster and the support can sometimes be tuned by the energy of the impacting particle;[5] (iii) immobilised clusters can show robust behaviour against sintering at elevated temperatures and pressures;[6,7] (iv) the "metal-to-metal" processing produces limited effluent and avoids the cost of ligand molecules; (v) binary and ternary nanoclusters can be prepared in addition to elemental clusters.[8–10]

In this paper we will report recent results which address two of the main challenges to the widespread adoption of the cluster beam route. We will show that the cluster beam approach is not confined to ultra high vacuum surface science experiments but instead enables model catalyst studies under realistic reaction conditions and, furthermore, that clusters can be deposited onto industrial catalyst powders and not just planar supports. We will also show that the constraints imposed by the limited flux characteristics of conventional cluster beam sources can now be overcome.

The field of cluster physics can be traced back at least 30 years to the time when "magic numbers" (corresponding to closed electronic or atomic shells) were discovered through use of the cluster beam sources invented at the time.[11] The deposition of such clusters onto surfaces opens up the prospect of generating model heterogeneous catalysts. A modern mass-selected cluster source designed for deposition onto solid surfaces in vacuum is likely to generate a cluster beam current up to about 1 nA, or a flux of about 10^{10} clusters per second.[12] Taking as an example a gold cluster with 100 atoms, Au_{100}, this flux equates to a production rate of (only) about 1 microgram per hour. As a result, catalysis studies to date have generally been limited to the surface science approach in ultra high vacuum or have required the use of specially designed microreactors.

Catalysts are of course ubiquitous in manufacturing, from fuels, bulk chemicals and polymers to pharmaceuticals. The nature of the catalyst varies enormously, depending on the reactor in which it is used, the economics of the process, and the nature of the reaction catalysed. In all cases, however, one can argue that the most important parameter in a catalyst's performance is selectivity. Whilst other factors such as activity and lifetime are undeniably important, it is selectivity which determines how efficiently a feedstock or reagent is used and hence the economic viability of a plant or process using the catalyst. The selectivity depends on the sum of the reactions taking place over all active sites on the catalyst, including those on the catalyst support and interfacial metal-support sites as well as those on the metal itself. In this context, the use of pre-formed nanoparticles in heterogeneous catalysts is particularly attractive.[13]

If all the metal-based active sites in the catalyst are the same, then they will perform the same catalytic transformation at the same rate, inherently leading to high selectivity materials. However, despite significant progress in the synthesis of nanoparticles, with high levels of control of their shape, size and composition (core–shell, homogenous alloy, *etc.*), the adoption of nanoparticle-based catalysts is not as yet widespread. In large part this is because pre-formed nanoparticles typically require stabilisers such as polymers or ligands which interfere with the catalytic activity, for example, by blocking the active site with a donor atom such as sulphur or nitrogen. Attempts to remove these ligands[14] are not always successful. For example, calcination can lead to residues (*e.g.* carbon) remaining on the nanoparticle, rendering it inactive. Washing and centrifuging methods are successful in some cases, but are laborious and tedious, especially when performed at scale. In some syntheses simple washing with hot water is successful[15] but still such materials are not commonly used in industry.

The synthesis of nanoparticles without the use of stabilisers is therefore attractive. This is one of the key features of cluster beam methods, alongside the option of mass selection to control particle size further and control of the cluster–support interaction. Moreover, methods in which a metal is processed directly to metal-containing catalysts, without the manufacture of salts, are economically attractive, as they avoid the energy and reagent costs of making metal salts, as well as the associated waste products. Drying, calcining and hydrogen reduction of catalysts are all energy intensive and there are economic and environmental drivers to avoid such processes. Another interesting aspect of catalyst synthesis using cluster beam deposition is the opportunity to make new materials which cannot easily be produced by conventional methods such as impregnation, deposition or precipitation. This could be because water-soluble salts of a particular element are not readily available, or contain catalyst poisons such as sulphur.

In this paper we describe methods for making catalysts using cluster beam techniques, including magnetron sputtering plus gas condensation and a new high flux method (matrix assembly), and introduce some materials which are made readily with this technology but are not easily accessed *via* wet chemical processing. The paper is organised as follows. In Section 2 we describe the development of the cluster beam method to enable the production of model catalysts (including alloys) by cluster deposition, in vacuum, onto powder supports. In Section 3 we report model catalyst studies of elemental (Pd) and binary clusters (also based on Pd) focused on selective hydrogenation reactions (both liquid and vapour phase). Finally, in Section 4, we discuss the prospects for further scale-up of cluster beam production to enable more facile model catalyst studies and, possibly, to move towards a small-scale manufacturing level.

2 Preparation and characterisation of model catalysts

2.1 Size-controlled cluster deposition on powders

To evaluate the performance of cluster-based catalysts and compare them with catalysts made using conventional wet chemistry techniques, it is helpful to test the catalysts in reactors relevant to industry and thus obtain meaningful results under conditions which are close to the real industrial end-use environment. The minimum amount of catalyst required by a catalyst evaluation reactor is typically at the level of a few hundred mg, loaded with a few percent of metal (here clusters), which means a few mg of clusters must be prepared on conventional particulate support materials within a reasonably short time. This production rate for gas aggregated nano-clusters is a huge challenge for conventional cluster-beam apparatus.

The original cluster apparatus at Teer Coatings consisted of a magnetron sputtering and gas aggregation source[16] and a Time-of-Flight mass filter.[17] The configuration was very similar to the Birmingham system,[12,18] but with the addition of a high volume deposition system for multiple planar substrates. The system was a joint development between Teer Coatings Ltd, the University of Birmingham, and Inanovate (UK) Ltd. With this system, only the mass-selected clusters could be collected, and the typical production rate was limited to approximately 1 μg $hour^{-1}$ (based on a cluster beam current of up to approximately 1 nA). To make a few mg of clusters as required by a representative reactor, thousands of hours' deposition time would be needed, which is obviously unrealistic. In addition, there was no means to deposit clusters directly onto powders, the support format of choice for many industrial catalysts.

To address these issues, the experimental apparatus has been redesigned and reconstructed. Fig. 1 shows the schematic of the modified system used for the current work.

The system consists of four sections: (1) a magnetron sputtering section; (2) an ion optics section; (3) a mass filter section; and (4) a powder deposition section. Within the magnetron sputtering section metal atoms are sputtered out of the magnetron target, and these atoms are subsequently cooled down and condensed to form clusters of various sizes. Because of the nature of plasma sputtering, a good proportion of the clusters produced are ionised. After they leave the magnetron sputtering chamber *via* a small nozzle (5 mm in diameter), the clusters of positive charge are accelerated and steered by the ion optical electrostatic lenses which sit in the second vacuum chamber. In the next step, the ion beam is focused into the

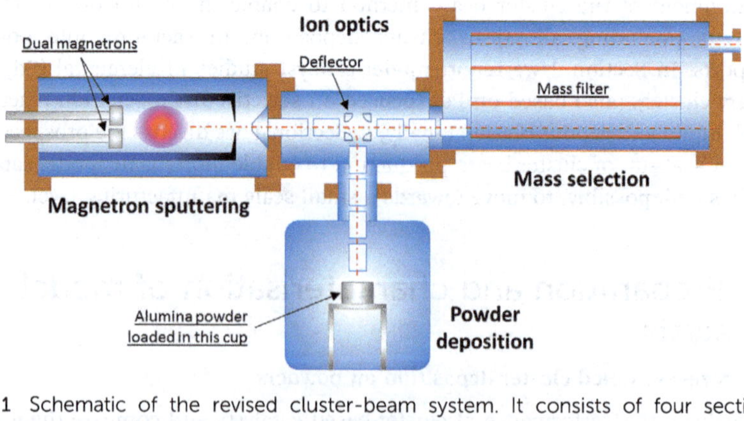

Fig. 1 Schematic of the revised cluster-beam system. It consists of four sections: magnetron sputtering, ion optics, mass selection and powder deposition. Note that the mass filter is only used for cluster size monitoring, not for deposition. The clusters are instead deposited directly onto powders in the chamber at the bottom of the figure.

third vacuum chamber for mass selection. By using the Birmingham Time-of-Flight mass filter,[17] the mass distribution of clusters can be monitored in real time. The neutral fraction present in the cluster beam is rejected by the mass filter.

The cluster size distribution is influenced by the aggregation length, sputtering gas (Ar) flow rate, condensation-assisting gas (He) flow rate, and temperature and pressure inside the condensation chamber. While tuning these parameters, the cluster size distribution is closely monitored using the TOF mass filter. When the desired cluster size distribution within the sampled ion beam is achieved, the high voltages applied to the deflector in the centre of the ion optics chamber are switched to deflection mode, so that the positively ionised fraction of the cluster beam is bent vertically downwards. The beam thus enters the powder deposition chamber, eventually depositing onto the particulate support material (gamma-alumina powder, HP14-150, Sasol) loaded inside a cup, which is constantly agitated to maximise exposure of all the particles. A high voltage can be applied to the cup in order to control the impact energy of a cluster landing on the support. This is so-called 'size-controlled' cluster deposition. Although the 'size-controlled' clusters deposited on the powder substrate have a much broader size distribution than those of the fully 'size-selected' clusters which are collected after mass selection, the throughput (*i.e.* material flux) of the former is about 100 times that of the latter, *i.e.* ~100 μg h^{-1} *versus* 1 μg h^{-1}. This new approach has made it possible to produce one catalyst sample in about 10 hours – rather than in 1000 hours!

To produce alloy clusters, specifically Pd/Ti and Pd/Sn, two magnetrons were mounted in the magnetron sputtering chamber. The sputtering power on each magnetron was individually controlled, to influence the average metal ratio in the resulting alloy clusters. For all the cluster samples, the DC magnetron power was chosen between 5 W and 10 W, except that about 20 W power was used for Ti, which has a much lower sputtering yield than the other metals.

The vacuum system was first pumped down to a base pressure of 2.0×10^{-6} mbar with the combination of a rotary pump and a turbo-molecular pump. The walls of the cluster condensation chamber were cooled with liquid nitrogen to enhance cluster growth, taking about two hours to reach base temperature. For sputtering, the argon gas flow rate was 80 sccm, and a 20 sccm flow of helium gas (for condensation) was also admitted into the condensation chamber. An aggregation length of 24 cm between the sputtering target(s) and the exit nozzle was used for all the cluster samples.

Fig. 2 shows an example of the mass spectra observed for Pd/Sn and Pd/Ti clusters. The Pd/Sn alloy cluster distribution has a peak at a mass of ~500k amu; the corresponding mass is 750k amu for Pd/Ti clusters. It is estimated that both types of clusters have a similar diameter, around 5 nm, assuming quasi-spherical structures. Approximately the same particle size was measured using TEM analysis (see below) of the supported catalysts produced. Of course, from the mass spectra alone one cannot determine the composition of the cluster, specifically, the ratio between the two metals.

2.2 Preparation of conventional catalyst materials as references

To assess the suitability of the cluster beam materials as catalysts, reference catalysts are required. However, PdSn and especially PdTi catalysts are difficult to make using conventional routes such as impregnation or deposition–

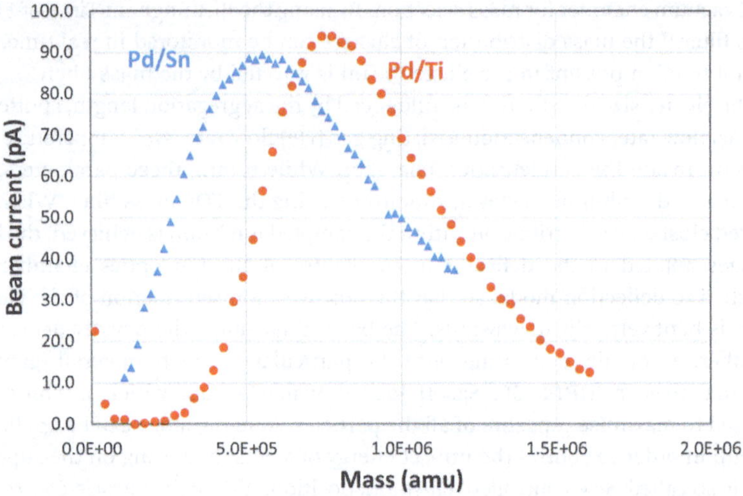

Fig. 2 Mass spectra of Pd/Sn and Pd/Ti clusters. The clusters' size distribution varies with magnetron sputtering conditions, two examples are shown here. It's estimated that both types of clusters shown in this figure have a diameter around 5 nm.

precipitation. For example, the lack of a readily-available water-soluble titanium precursor means that aqueous-based methods, desirable from environmental and economic standpoints, are not possible. Thus, the reference materials made are representative formulations but were synthesised by chemical routes which would be difficult to operate at any meaningful manufacturing scale.

Palladium–tin catalysts were prepared by co-impregnation of palladium acetate with tin(II) acetate. Although this is not a typical synthesis method for palladium–tin catalysts, it avoids the use of corrosive halides. In this synthesis, the appropriate amounts of palladium acetate and tin acetate were dissolved in THF (tetrahydrofuran). The volume of the solution was calculated to match the pore volume of the support. Alumina powder (HP14-150, Sasol) was added to the solution with stirring. The resulting wet solid was air dried, then dried at 100 °C overnight, and finally reduced in flowing hydrogen at 250 °C for two hours.

Palladium–titanium catalysts were prepared by co-impregnation of palladium acetate and titanium tetraisopropoxide in THF. This is a system which would be somewhat challenging for manufacturing, with THF chosen as a solvent for both metal precursors. The need for an organic solvent arises from the lack of simple water-soluble titanium precursors. The appropriate amounts of palladium acetate and titanium tetraisopropoxide were dissolved in THF, with the solution volume again being chosen to match the support pore volume. Alumina powder (HP14-150, Sasol) was added to the solution in one portion with stirring. The resulting wet solid was air dried, then dried at 100 °C overnight, and then reduced in flowing hydrogen at 250 °C for two hours.

2.3 Catalyst characterisation

The metal content of the materials synthesised is reported in Table 1. The cluster beam materials contain significantly less metal than those prepared using wet chemistry.

Transmission electron microscopy (TEM) coupled with energy-dispersive X-ray spectroscopy (EDX) has been used extensively to understand the local structure of the catalysts. Fig. 3 shows representative images of the Pd–Sn and Pd–Ti catalysts. TEM shows that the nanoparticles are better dispersed through the support in the case of the "conventional" (impregnated) sample PdSn–I2, compared with the cluster beam material PdSn–C2, where the clusters appear to be aggregated. For the Pd–Ti catalysts, the cluster beam sample PdTi–C3 presents more discrete particles than the impregnated material PdTi–I2.

X-ray photoelectron spectroscopy (XPS) was used to gain information about the chemical composition of the catalyst's surface. Although the spectra were inevitably dominated by the alumina support, it was still possible to measure the palladium, tin and titanium in the samples. Table 2 shows the surface elemental compositions as measured using XPS. The chemical environments of Pd, Sn and Ti are similar when comparing the impregnated and cluster beam materials. Palladium is present as a metal whilst both tin and titanium are found as oxidic species. While both tin and titanium are assembled as metal nanoparticles in the cluster beam technique, they will be oxidised as a result of exposure to air when the materials are removed from the apparatus.

3 Catalyst performance

The catalysts were evaluated in the selective hydrogenation of alkynes to alkenes. Two reactions were selected: the vapour-phase hydrogenation of 1-pentyne and the liquid phase hydrogenation of 3-hexyn-1-ol. The aim of these reactions is to hydrogenate the alkyne to the corresponding alkene, but without further hydrogenation to the alkane, or isomerisation of the double bond. The reaction schemes are shown in Fig. 4. A wide range of metals are active for these hydrogenation reactions, such as nickel, cobalt, copper, platinum, ruthenium, rhodium and palladium.[19] Of these, palladium is typically chosen for fine chemicals applications because of its superior performance despite its higher cost.

Table 1 Metal content of the catalysts measured using ICP-ES

Catalyst	Synthesis method	Palladium content/wt%	Tin content/wt% (Pd/Sn molar ratio)	Titanium content/wt% (Pd/Ti molar ratio)
PdSn–I1	Impregnation	0.96	3.27 (0.33)	—
PdSn–I2	Impregnation	0.94	1.01 (1.0)	—
PdSn–I3	Impregnation	1.07	0.36 (3.3)	—
PdSn–C1	Clusterbeam	0.09	0.06 (1.7)	—
PdSn–C2	Clusterbeam	0.15	0.12 (1.4)	—
PdSn–C3	Clusterbeam	0.09	0.05 (2.0)	—
PdTi–I1	Impregnation	0.88	—	1.05 (0.38)
PdTi–I2	Impregnation	0.89	—	0.61 (0.66)
PdTi–I3	Impregnation	1.10	—	0.15 (3.3)
PdTi–C1	Clusterbeam	0.053	—	0.11 (0.22)
PdTi–C2	Clusterbeam	0.041	—	0.0075 (2.5)
PdTi–C3	Clusterbeam	0.09	—	0.01 (4.1)
PdTi–C4	Clusterbeam	0.0085	—	0.008 (0.48)

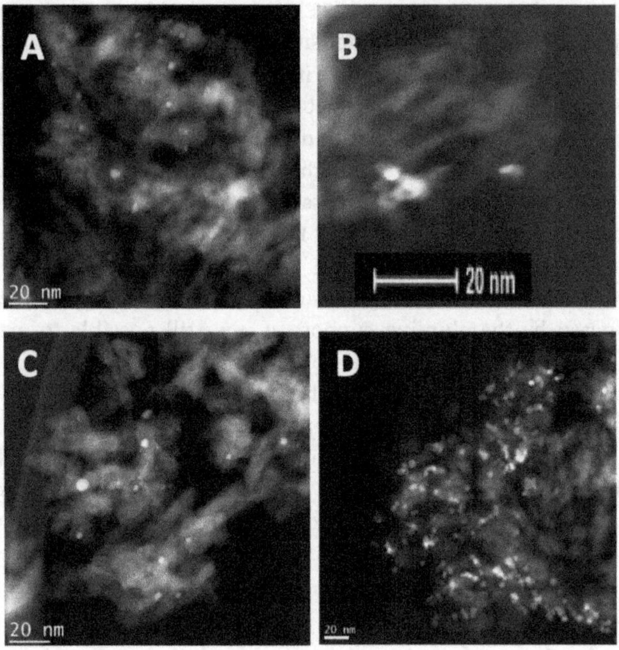

Fig. 3 TEM analysis of (A) PdSn–I2, (B) PdSn–C2, (C) PdTi–I2, and (D) PdTi–C3.

The selective hydrogenation of alkynes to the corresponding alkene is of interest to the bulk as well as fine chemicals sectors. In the bulk chemical industry it is applied to the purification of ethylene streams by removal of acetylene impurities.[20] These impurities are typically present at very low levels. The catalyst is required to hydrogenate the acetylene selectively to ethylene without over-hydrogenation to ethane. Also it must not hydrogenate the ethylene which makes up the main part of the gas mixture. In the fine chemicals industry, selective hydrogenation is used in many different ways to effect functional group inter-conversions in complex molecules. In this context, very high selectivity is again required along with tolerance of other functional groups.[21] In this work, we use the selective hydrogenation of 1-pentyne and 3-hexyn-1-ol as model reactions for both types of scenario. Additionally, the selective hydrogenation product of 3-hexyn-1-ol, *cis*-3-hexen-1-ol, is a fragrance compound used in the perfumery industry.

Table 2 Representative XPS analysis of the catalysts

		Binding energy/eV			Surface atomic ratios	
Catalyst	Synthesis method	Pd 3d	Sn 3d	Ti 2p	Pd/Sn	Pd/Ti
PdSn–I2	Impregnation	335.6	487.3	—	0.38	—
PdSn–C2	Clusterbeam	335.2	486.5	—	0.77	—
PdTi–I2	Impregnation	335.2	—	458.6	—	0.37
PdTi–C3	Clusterbeam	335	—	458	—	3.3

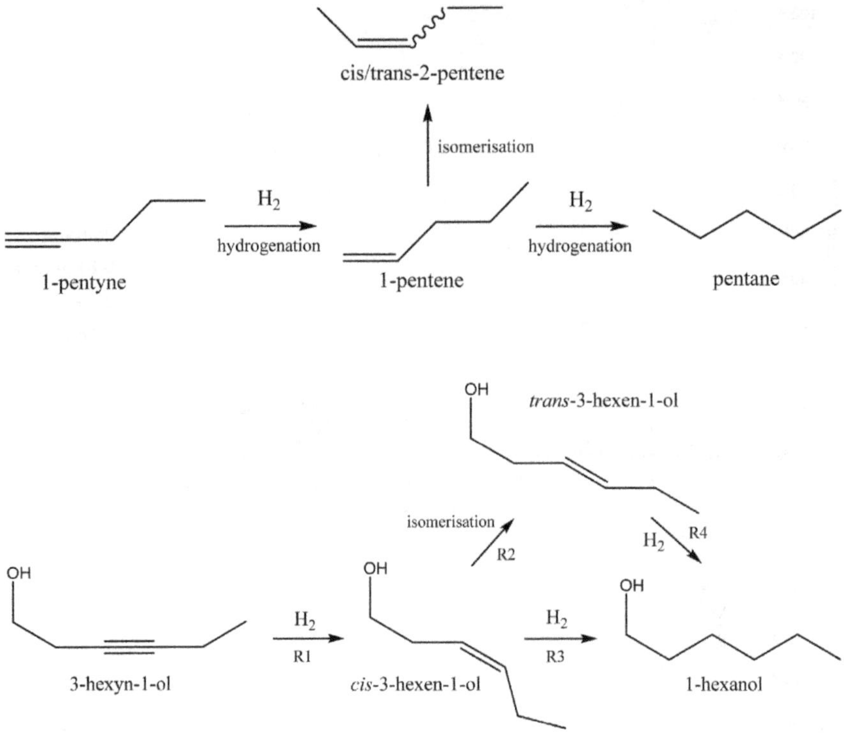

Fig. 4 Reaction scheme for 1-pentyne hydrogenation (top) and 3-hexyn-1-ol hydrogenation (bottom).

3.1 Results for 1-pentyne selective hydrogenation

The selective hydrogenation of pentyne was performed using a fixed bed reactor. 10 mg of catalyst was held between two plugs of quartz wool and placed into a furnace. The gas feed consisted of 40% H_2/He at 250 ml min^{-1} and 1 M pentyne solution in *n*-hexane, with 0.5 M iso-hexane as an internal standard, at 0.06 ml min^{-1}. Once flushing was complete, the catalyst was heated to 250 °C at 2 °C min^{-1}. Analysis was performed at different temperatures using an online GC.

Typical data acquired from two catalysts prepared by cluster beam deposition, PdTi/γ-Al_2O_3 and Pd/γ-Al_2O_3, are shown in Fig. 5. Both catalysts show good activity for the reaction, with complete conversion of the 1-pentyne feedstock. The monometallic palladium catalyst shows some selectivity to 1-pentene, but the levels of pentane (from overhydrogenation) and the *cis* and *trans* 2-pentene (from isomerisation) are also high. These latter reactions are thought likely to occur on the reactive sites on the alumina support.[22] The palladium–titanium catalyst behaves similarly. It is less active, complete conversion of the 1-pentyne feedstock requires a higher reaction temperature. Interestingly, the side-reactions to the 2-pentene isomers and to pentane are suppressed, giving a higher selectivity to 1-pentene. This could be caused by a number of factors: an electronic interaction between palladium and titanium; the presence of titania in the palladium particles leading to a better palladium dispersion; a direct interaction of the reactants with titania; or a bimetallic palladium–titania active site for the reaction.

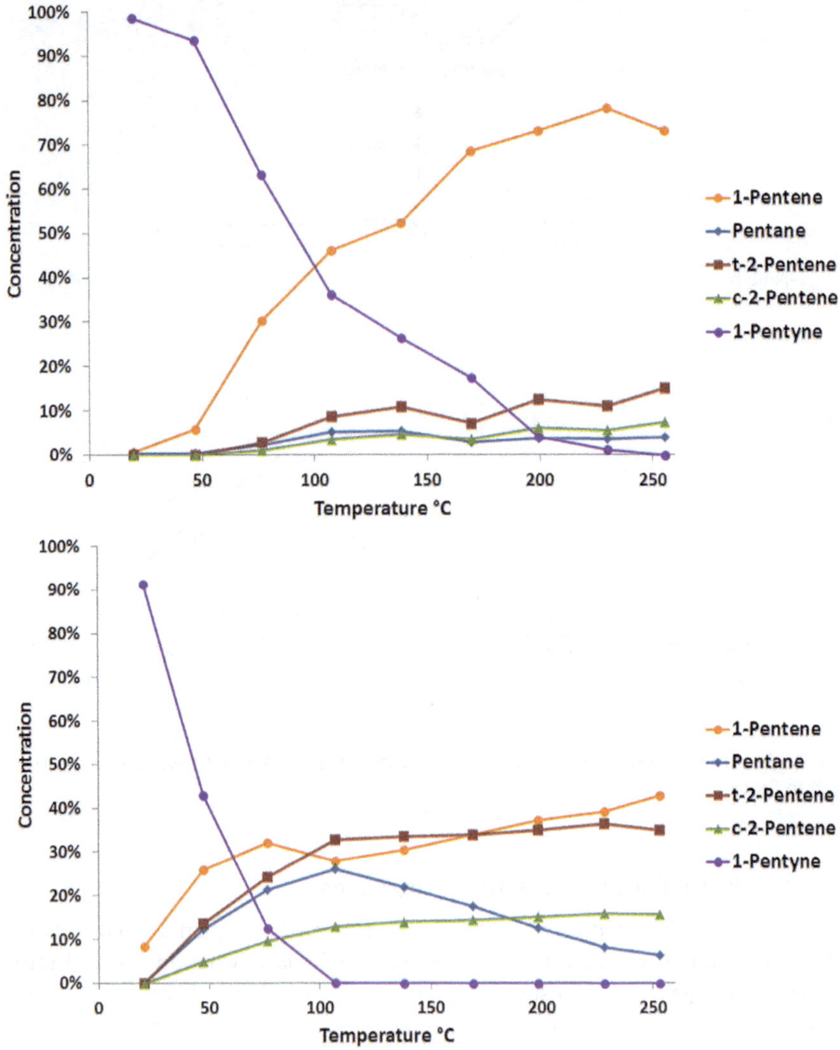

Fig. 5 Data for the catalytic 1-pentyne hydrogenation with PdTi/γ-Al_2O_3 (top) and Pd/γ-Al_2O_3 (bottom) prepared using the clusterbeam method. The charts show the outlet concentration of each compound at different reaction temperatures.

Fig. 6 compares the performance of a range of catalysts produced *via* cluster beam and conventional methods. It can be seen that the palladium–tin catalysts do not offer good performance in 1-pentyne hydrogenation. The materials prepared by impregnation are only selective at a low conversion, while those prepared using the clusterbeam method are poorly active, even if they show some selectivity. It seems that tin is a poison for the palladium catalyst. Unlike the Lindlar palladium–lead catalyst,[23] the poisoning does not lead to extra product selectivity. The palladium–titanium catalysts, on the other hand, are more promising. When prepared by impregnation, the catalysts exhibit very high activity but low selectivity. The materials prepared *via* cluster beam methods, on

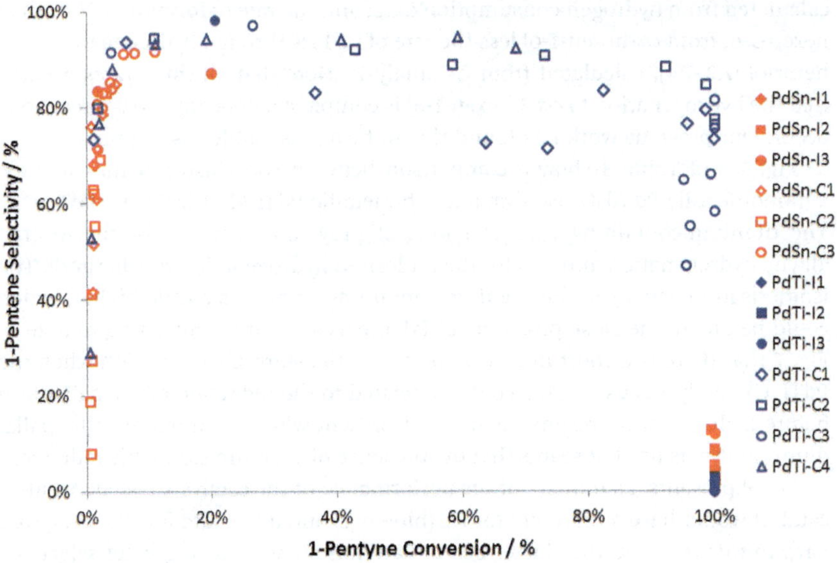

Fig. 6 Summary of 1-pentyne hydrogenation performance for PdSn and PdTi catalysts. Note that the data arises from experiments performed at a range of temperatures from room temperature to 250 °C.

the other hand, show excellent selectivity (85–90%) to 1-pentene at high 1-pentyne conversion (over 95%). The best catalysts (top right corner of Fig. 6, which means good conversion and good selectivity) are cluster beam materials. Whilst the low loading of the cluster beam samples could be a factor (this is currently under investigation) – and certainly makes catalyst characterisation more difficult – it is likely that the catalyst structure arising from the cluster beam synthesis method also impacts positively on the performance.

3.2 Results for 3-hexyn-1-ol selective hydrogenation

Selective hydrogenation of 3-hexyn-1-ol was carried out using a Chemscan reactor. This consists of eight small autoclaves (8 ml volume), which can be run in parallel, with monitoring of hydrogen uptake. Each autoclave was filled with catalyst (24.5 mg) and 0.5 M 3-hexyn-1-ol solution in ethanol (5 ml) with 0.5 M 1,4-dioxane as the internal standard. The autoclave was pressurised to 3 bar with hydrogen, and the reaction temperature was 30 °C. The reaction time was 90 minutes, which was generally enough time to allow complete conversion to 3-hexen-1-ol. The reactions were analysed by hydrogen uptake rate and also by offline GC analysis at the end of the reaction.

Four main reactions occur during the catalysis: the desired hydrogenation of 3-hexyn-1-ol to *cis*-3-hexen-1-ol (R1), the isomerisation of the *cis* isomer to *trans*-3-hexen-1-ol (R2), and the hydrogenation of the *cis*-(R3) and *trans*-(R4) alkenes to 1-hexanol. This is summarised pictorially in Fig. 4. Three apparent reaction rates were derived from the experimental data: the rate of hydrogenation of 3-hexyn-1-ol to *cis*-3-hexen-1-ol (R1), calculated from the hydrogen uptake data; the rate of 1-hexanol formation from both *cis*- and *trans*-3-hexen-1-ol (R3 + R4), again calculated from hydrogen consumption data; and the rate of formation of *trans*-3-hexen-1-ol from *cis*-hexen-1-ol less the rate of its loss through hydrogenation to 1-hexanol (R2–R4), calculated from GC analysis. Note that all three rates assume that the hydrogenation to *cis*-3-hexen-1-ol is complete before any further reactions occur. Our previous work has found this to be a reasonable assumption.

Fig. 7 and Table 3 show a comparison between two cluster beam catalysts: a monometallic Pd/Al_2O_3 catalyst and a bimetallic PdTi/Al_2O_3 catalyst (PdTi–C1). The titanium-containing catalyst shows slightly more activity for the desired alkyne hydrogenation and less for the undesired hydrogenation to 1-hexanol. The isomerisation activity of the titanium-containing catalyst is a little higher, which could be due to the close proximity of Pd and Ti sites. It is interesting to note in Fig. 7 that there is a short induction period at the start of the reaction when the PdTi–C1 catalyst is used. This could be related to the reduction of the palladium before hydrogenation begins. As it is not present when the monometallic palladium catalyst is used, it seems that the presence of titanium slows this reduction.

Fig. 8 presents a summary of the performance of the catalysts tested. An ideal catalyst would have a high rate for R1 (blue bar) and a low rate for (R3 + R4) (red bar). In this reaction, the cluster beam materials show a much greater selectivity for 3-hexen-1-ol than the impregnated materials. For the PdSn materials, the catalysts prepared *via* impregnation are more active than those prepared *via* the cluster beam route. However, the cluster beam materials are more selective, and for the most selective cluster beam material the overhydrogenation reaction is almost completely eliminated. Tin can act as a selective poison for palladium in selective hydrogenation in a similar manner to the Lindlar catalyst.[5] This

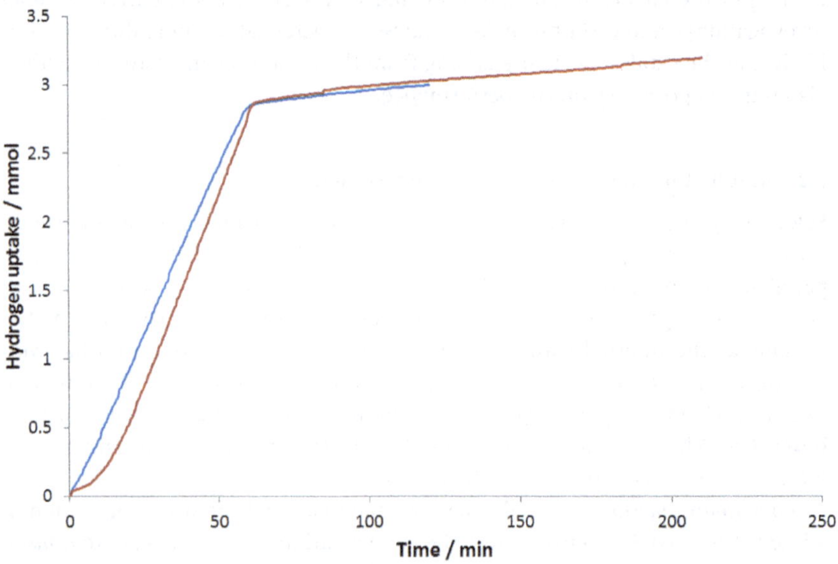

Fig. 7 Hydrogen uptake data in the selective 3-hexyn-1-ol hydrogenation using Pd/Al_2O_3 (blue) and PdTi/Al_2O_3 (red) catalysts prepared *via* the clusterbeam technique. The steeper slope between 0 and 60 minutes is the hydrogenation of 3-hexyn-1-ol (R1), whilst the shallower slope above 60 minutes is the hydrogenation of *cis*- and *trans*-3-hexen-1-ol (R3 + R4).

Table 3 Reaction rates determined for Pd/Al_2O_3 and $PdTi/Al_2O_3$ prepared *via* the clusterbeam method. For definitions, please see the text.

Catalyst	R1/ μmol min^{-1} g_{cat}^{-1}	R3 + R4/ μmol min^{-1} g_{cat}^{-1}	R2–R4/ μmol min^{-1} g_{cat}^{-1}
Pd/Al_2O_3	2041	102	135
$PdTi/Al_2O_3$	2286	69	147

advantage of the cluster beam materials over the conventional materials is even more pronounced in the case of the PdTi catalysts. All three of the impregnated materials are very poorly selective to the alkene, with the rate of alkene hydrogenation being similar to the rate of alkyne hydrogenation. The cluster beam materials offer much better selectivity to 3-hexen-1-ol regardless of their Pd/Ti ratio. Good catalytic performance in this reaction has previously been reported for monometallic palladium nanoparticles, albeit at lower hydrogen pressure, which will tend to reduce the amount of overhydrogenation observed.[24] However, the cluster beam technique allows PdSn and PdTi catalysts to be made which exhibit superior performance, especially the PdTi clusters.

3.3 Discussion of selective hydrogenation results

It is clearly of interest to understand the origin of the good performance (especially selectivity) observed in alkyne hydrogenation by the cluster beam materials. When a catalyst is prepared *via* impregnation, the metal salt contacts the support,

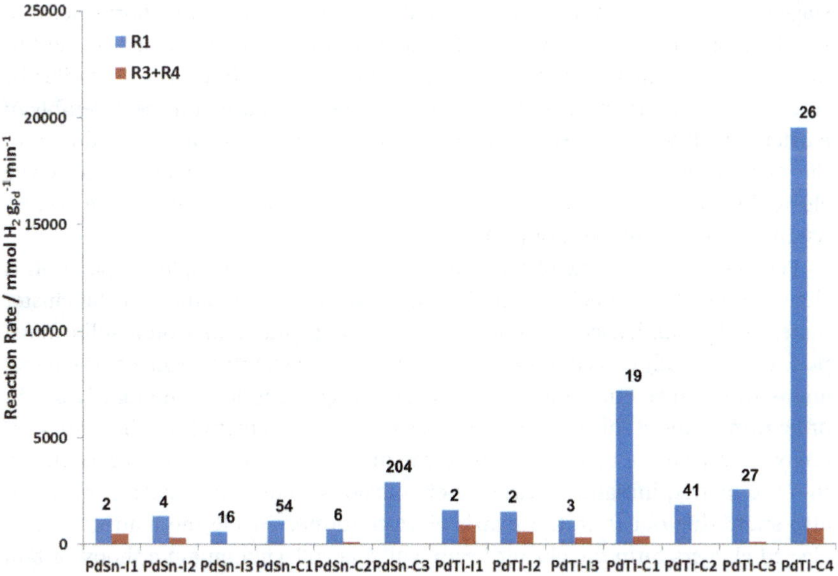

Fig. 8 Summary of the performance of PdSn and PdTi catalysts in 3-hexyn-1-ol hydrogenation. The chart shows the desired hydrogenation rate (blue) and the undesired overhydrogenation (red). R1, R3 and R4 are defined in the text. The numbers in the figure show the ratio R1/(R3 + R4) which is a measure of the selectivity of the catalysts.

which allows ion exchange to occur. The nature and number of these ion-exchange sites will depend on the nature of the metal and the support, but they will be different from the main nanoparticle sites. Having a second reaction site of uncontrolled structure is likely to be detrimental to selectivity. When catalysts are synthesised *via* the cluster beam method, and thus the nanoparticles are synthesised away from the support, the metal only contacts the support in a metallic nanoparticle form. This means that ion exchange cannot occur. A similar phenomenon can be observed when using nanoparticles synthesised *via* solution routes, but one key benefit of the cluster beam method is the absence of stabilising ligands, so the nanoparticles are deposited cleanly onto the support.

In summary, we have shown that novel catalyst compositions can be readily synthesised using the cluster beam method, including those which are difficult to access by conventional "wet chemistry" approaches. The method makes active catalysts for a range of industrially relevant catalytic processes, as illustrated here by two selective hydrogenation reactions, and in some cases creates materials which outperform those synthesised following conventional methods.

4 Outlook: further scale-up of cluster beam production with the matrix assembly cluster source (MACS)

The catalysis results presented in Section 3 demonstrate the promise of cluster beam materials for catalyst research but such experiments would be greatly assisted by further scale up of the rate of cluster beam production. The preparation of the catalysts evaluated in Section 3 depended on the modification of the cluster beam source (Section 2): extraction of clusters before the mass selection stage resulted in a hundred times more flux. In this section, we report briefly on the demonstration of a new kind of cluster beam source, developed in the Birmingham lab, which offers the prospect of a further rise in cluster beam flux by several more orders of magnitude. The new source is based on the assembly of clusters inside a condensed inert matrix, assisted by ion beam bombardment of the matrix. The ion beam also sputters clusters out of the matrix to make the cluster beam which is deposited on the support. The source is termed the "Matrix Assembly Cluster Source", or (MACS).[25]

Fig. 9(a) is a schematic of the MACS instrument; the principle of operation is shown in Fig. 9(b), which depicts inert gas (Ar) atoms and atoms of the cluster material (Ag) condensed together to form a composite matrix on a solid metal plate cryogenically cooled to <25 K in a high vacuum chamber. Some of the metal atoms will diffuse and form small clusters by aggregation in the matrix but in order to grow these clusters to a larger size and to extract them from the matrix, an (argon) ion beam is used to bombard the matrix. The ion impact injects energy into the matrix, initiating a cascade of collisions leading to cluster growth and also sputtering out from the matrix clusters formed inside the matrix.[26] These ejected clusters form the cluster beam which is collected on amorphous carbon TEM grids for analysis by High Angle Annular Dark Field Scanning Transmission Electron Microscopy (HAADF STEM).

We refer to the geometry shown in Fig. 9 as "reflection mode", in contrast to the "transmission mode" (this employs a semi-transparent grid as the matrix

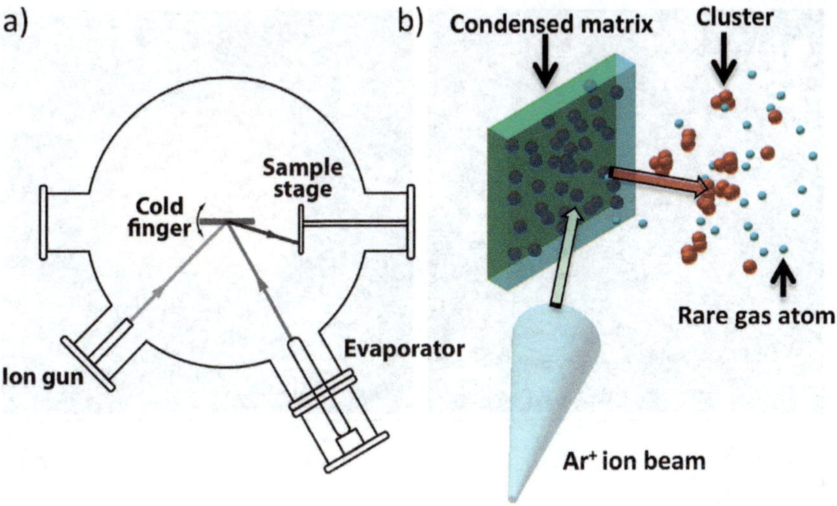

Fig. 9 (a) Scheme of the Matrix Assembly Cluster Source (MACS) employed in the work, showing the ion gun (Omicron ISE-5), the cryogenically cooled matrix, the sample collection stage and the evaporator (Createc effusion cell). (b) Schematic of the MACS process.

support).[25] The matrix employed in the present work was prepared with a concentration of 1.6% Ag atoms in an argon matrix. The Ar gas was dosed through a leak valve, with an Ar pressure of 5×10^{-6} Torr, and Ag was evaporated at the same time at a deposition rate of 0.05 Å s^{-1} for 200 s. The matrix was subsequently sputtered with an Ar^+ ion beam (2.5 keV, 6 μA Ar^+) for 30 seconds.

Fig. 10 shows the HAADF STEM images of the Ag clusters produced with the MACS and deposited onto TEM grids. Fig. 10(a) shows an array of clusters and Fig. 10(b) a close-up view of one of them. The intensities of the clusters in these images allow measurement of their size distribution *via* the atom counting method.[27–30] We used the average intensity of single atoms as a reference. The cluster size distribution of Fig. 10(c) has a peak between 100 and 150 silver atoms. The half width at half maximum is about 100% of the peak cluster size, which is only about ±33% in terms of diameter (as more commonly quoted in catalyst work). Thus, although no mass filtering stage is employed in the present work, the size distribution is quite narrow in comparison with conventional methods of catalyst particle synthesis.

We obtained an equivalent cluster beam current of ~28 nA for a 3 μA ion beam current incident on the matrix, giving a cluster yield per incident ion of close to 1%. The "equivalent cluster beam current", which pretends that each cluster bears a single charge, is chosen to compare the flux with the output of conventional cluster sources (Section 2), but the clusters we collect on the TEM grids may have any, or no, charge. To demonstrate that scale up (and sustained operation) of the MACS method is feasible, we investigated the generation of clusters with a higher ion beam current of 30 μA over a period of 4.5 minutes in a different vacuum chamber. In this case, the Ag concentration in the matrix was 1.8%. Clusters were deposited into a carousel able to present 21 glass slides (75 mm × 25 mm), or equivalent surfaces, to the beam. The maximum equivalent beam

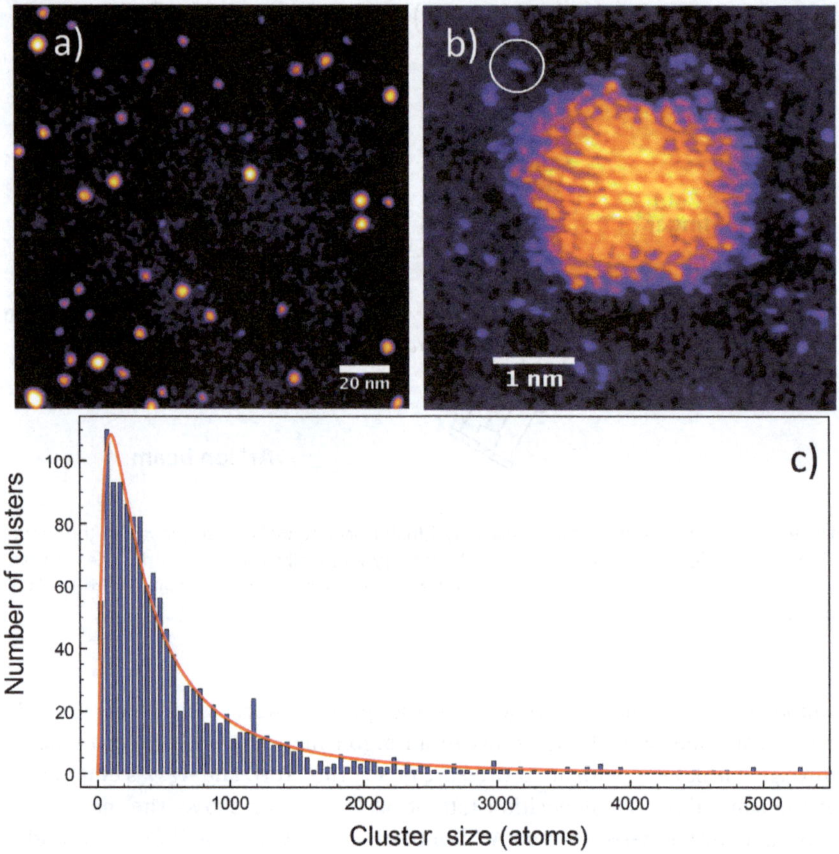

Fig. 10 (a) HAADF STEM image of Ag clusters created in the MACS. (b) A high-resolution image of one such Ag cluster, the circled region shows a single Ag atom, as used for mass calibration. (c) Size distribution of the clusters produced, calculated from the integrated HAADF STEM intensities.

current achieved was 210 ± 36 nA and a relatively stable flux was obtained at the 100nA level.

In both demonstration experiments the efficiency (number of clusters per incident atomic ion) approaches 1%. Since large area commercial ion guns are available with currents of more than 1 A, it should be possible to produce several more orders of magnitude of the cluster flux in MACS. Ultimately, a cluster current of 10 mA (1% of 1 A) would be equivalent to about 10 g of clusters per hour, or 1 kg of catalyst (loaded at 1% metal) per hour, sufficient for small batch pharmaceutical work. The demonstration experiments reported here are sufficient to demonstrate that the principle of the MACS is valid and indeed already exceeds the flux reported in Section 2.

5 General conclusions

The research reported here demonstrates some of the advantages of the cluster beam approach in the study of model catalysts under realistic reaction conditions and is made possible by innovations in cluster beam source design, which enable the scale-up of the cluster beam flux and the direct deposition of clusters onto powder supports. We have demonstrated favourable combinations of efficiency and selectivity for binary metal clusters in both liquid and gas phase hydrogenation experiments, when compared with catalysts prepared *via* conventional routes. Moreover, the new MACS technology offers at least the prospect of several more orders of magnitude of cluster beam flux, to the point where small scale manufacturing of commercial catalysts may become feasible in the foreseeable future. No doubt, however, much further innovation will be required to reach that point.

Acknowledgements

We thank Innovate UK (formerly the Technology Strategy Board), EPSRC and the Leverhulme Trust for financial support of this project. The STEM employed in this work was provided by the Birmingham Science City project funded by AWM and the ERDF. We are grateful to the following colleagues at Johnson Matthey for their analytical work: Laura Stead and Matthew Rose (ICP); Tugce Erlap Erden and Richard Smith (XPS); Gregory Goodlet, Dogan Ozkaya and Winson Kuo (TEM). The MACS concept is the subject of the following patent applications: Palmer, R. E., (2011) *Provisional GB Patent Application GB1113168.7;* (2012) *PCT Patent Application PCT/GB2012/051854;* (2014) *Europe Patent Application EP12753793.4;* (2014) *China Patent Application CN201280048410.2;* (2014) *Japan Patent Application JP2014-523387;* (2014) *U.S. Patent Application US14/236386.*

References

1 J. Grunes, J. Zhu, E. A. Anderson and G. A. Somorjai, *J. Phys. Chem. B*, 2002, **106**(44), 11463–11468.
2 I. Lee, R. Morales, M. A. Albiter and F. Zaera, *Proc. Natl. Acad. Sci. U. S. A.*, 2008, **105**(40), 15241–15246.
3 E. C. Tyo and S. Vajda, *Nat. Nanotechnol.*, 2015, **10**, 577–588.
4 M. Arenz, U. Landman and U. Heiz, *ChemPhysChem*, 2006, **7**(9), 1871–1879.
5 R. E. Palmer, S. Pratontep and H.-G. Boyen, *Nat. Mater.*, 2003, **2**(7), 443–448.
6 F. Yin, S. Lee, A. Abedela, S. Vadja and R. E. Palmer, *J. Chem. Phys.*, 2011, **134**, 141101.
7 F. Yin, C. Xirouchaki, Q. Guo and R. E. Palmer, *Adv. Mater.*, 2005, **17**(6), 731–734.
8 F. Yin, Z. W. Wang and R. E. Palmer, *J. Am. Chem. Soc.*, 2011, **133**(27), 10325–10327.
9 P. Hernandez-Fernandez, F. Masini, D. N. McCarthy, C. E. Strebel, D. Friebel, D. Deianan, P. Malacrida, A. Nierhoff, A. Bodin, A. M. Wise, J. H. Nielsen, T. W. Hansen, A. Nilisson, I. E. L. Stephens and I. Chorkendorff, *Nat. Chem.*, 2014, **6**(8), 732–738.
10 C. Holse, C. F. Elkjaer, A. Nierhoff, J. Sehested, I. Chorkendorff, S. Helveg and J. H. Nielsen, *J. Phys. Chem. C*, 2015, **119**(5), 2804–2812.
11 W. A. de Heer, *Rev. Mod. Phys.*, 1993, **65**(3), 611–676.
12 S. Pratontep, S. J. Carroll, C. Xirouchaki, M. Streun and R. E. Palmer, *Rev. Sci. Instrum.*, 2005, **76**(4), 045103–045109.

13 A. T. Bell, *Science*, 2003, **299**(5613), 1688–1691.
14 C.-J. Jia and F. Schuth, *Phys. Chem. Chem. Phys.*, 2011, **13**, 2457–2487.
15 J. A. Lopez-Sanchez, N. Dimitratos, C. Hammond, G. L. Brett, L. Kesavan, S. White, P. Miedziak, R. Tiruvalam, R. L. Jenkinsm, A. F. Carley, S. Knight, C. J. Kiely and G. J. Hutchings, *Nat. Chem.*, 2011, **3**, 551–556.
16 H. Haberland, M. Mall, M. Moseler, Y. Qiang, T. Reiners and Y. Thurner, *J. Vac. Sci. Technol., A*, 1994, **12**(5), 2925–2930.
17 B. von Issendorff and R. E. Palmer, *Rev. Sci. Instrum.*, 1999, **70**(12), 4497–4501.
18 I. M. Goldby, B. von Issendorff, L. Kuipers and R. E. Palmer, *Rev. Sci. Instrum.*, 1997, **68**(9), 3327–3334.
19 M. W. Tew, M. Nachtegaal, M. Janousch, T. Huthwelker and J. A. van Bokhoven, *Phys. Chem. Chem. Phys.*, 2012, **14**, 5761–5768.
20 F.-M. McKenna, R. P. K. Wells and J. A. Anderson, *Chem. Commun.*, 2011, **47**, 2351–2353.
21 A. F. Kyle, P. Jakubec, D. M. Cockfield, E. Cleator, J. Skidmore and D. J. Dixon, *Chem. Commun.*, 2011, **47**, 10037–10039.
22 J. Medema, *J. Catal.*, 1975, **37**, 91–100.
23 A. K. Ghosh and K. Krishnan, *Tetrahedron Lett.*, 1998, **39**(9), 947–948.
24 J. C. A. A. Roelofs and P. H. Berben, *Chem. Commun.*, 2004, 970–971.
25 R. E. Palmer, L. Cao and F. Yin, *Rev. Sci. Instrum.*, 2016, **87**, 046103, DOI: 10.1063/1.4947229.
26 R. Smith, *Atomic and ion collisions in solids and at surfaces: theory, simulation and applications*, Cambridge University Press, 2005.
27 N. Young, Z. Li, Y. Chen, S. Palomba, M. Di Vece and R. E. Palmer, *Phys. Rev. Lett.*, 2008, **101**, 246103, DOI: 10.1103/PhysRevLett.101.246103.
28 Z. W. Wang, Z. Y. Li, S. J. Park, A. Abdela, D. Tang and R. E. Palmer, *Phys. Rev. B: Condens. Matter Mater. Phys.*, 2011, **84**, 073408, DOI: 10.1103/PhysRevB.84.073408.
29 Z. W. Wang, O. Toikkanen, F. Yin, Z. Y. Li, B. M. Quinn and R. E. Palmer, *J. Am. Chem. Soc.*, 2010, **132**, 2854–2855.
30 N. Jian, C. Stapelfeldt, K. J. Hu, M. Fröba and R. E. Palmer, *Nanoscale*, 2015, **7**(3), 885–888.

Faraday Discussions

PAPER

Active site densities, oxygen activation and adsorbed reactive oxygen in alcohol activation on npAu catalysts

Lu-Cun Wang,[a] C. M. Friend,[ab] Rebecca Fushimi†[cd] and Robert J. Madix[b]

Received 11th November 2015, Accepted 14th December 2015

DOI: 10.1039/c5fd00161g

The activation of molecular O_2 as well as the reactivity of adsorbed oxygen species is of central importance in aerobic selective oxidation chemistry on Au-based catalysts. Herein, we address the issue of O_2 activation on unsupported nanoporous gold (npAu) catalysts by applying a transient pressure technique, a temporal analysis of products (TAP) reactor, to measure the saturation coverage of atomic oxygen, its collisional dissociation probability, the activation barrier for O_2 dissociation, and the facility with which adsorbed O species activate methanol, the initial step in the catalytic cycle of esterification. The results from these experiments indicate that molecular O_2 dissociation is associated with surface silver, that the density of reactive sites is quite low, that adsorbed oxygen atoms do not spill over from the sites of activation onto the surrounding surface, and that methanol reacts quite facilely with the adsorbed oxygen atoms. In addition, the O species from O_2 dissociation exhibits reactivity for the selective oxidation of methanol but not for CO. The TAP experiments also revealed that the surface of the npAu catalyst is saturated with adsorbed O under steady state reaction conditions, at least for the pulse reaction.

1. Introduction

Metallic gold is used as an exceptional catalyst for a wide range of reactions, in particular the oxygen-assisted chemical transformations, such as CO oxidation and esterification of alcohols.[1-3] Gold nanoparticles dispersed on reducible oxides are the most commonly used gold catalysts for various catalytic applications.[1,4] This type of catalyst, however, is susceptible to deactivation resulting from the growth of gold particle size under working conditions.[4] Recently, a new type of

[a]*Department of Chemistry and Chemical Biology, Harvard University, Cambridge, MA, 02138, USA. E-mail: rmadix@seas.harvard.edu; Fax: +617 496 9489; Tel: +617 496 9489*

[b]*School of Engineering and Applied Sciences, Harvard University, Cambridge, MA, 02138, USA*

[c]*Parks College of Engineering, Aviation and Technology, Saint Louis University, Saint Louis, MO, 63103, USA*

[d]*The Langmuir Research Institute, Saint Louis, MO, 63108*

† Present address: Idaho National Laboratory, Biological and Chemical Processing Department, Idaho Falls, ID, 83415, USA

unsupported, nanoporous gold (npAu) catalyst with ligament sizes between 20–100 nm has been demonstrated to be highly efficient for these key oxidation processes.[3,5–10] In addition to its excellent activity, the npAu material is also highly selective for ester formation.[3,7,8,10] More importantly, these materials are very robust, showing negligible losses of activity or selectivity even for months under the reaction conditions.[8]

It is well established that on model gold surfaces the adsorbed *atomic* oxygen, usually formed *via* ozone decomposition[11] or electron-induced dissociation of NO_2,[12] directly activates and drives the catalytic cycles of alcohol coupling reactions.[3,7,8,10,13] Since molecular O_2 is used under practical catalytic conditions, understanding the mechanism of O_2 activation processes on the surfaces of 'real-world' gold catalysts is critical to develop new materials with improved catalytic performances.

Molecular O_2 is readily dissociated on npAu materials even at low temperature; the nature and density of the reactive centers for this process, however, still remain elusive. The npAu employed in the catalytic studies contains a small amount of Ag (1–3%) that results from its synthesis from an Ag/Au bulk alloy.[3,5,6] Taking into account the fact that O_2 can readily dissociate on Ag surfaces, the Ag(110) surface in particular,[14] it is reasonable to speculate that the residual Ag is involved in the dissociation of molecular O_2 on npAu catalysts, as proposed on the basis of numerous experimental and theoretical studies.[15–19] On the other hand, low-coordinated Au sites, such as edge, step, or kink sites, have also been suggested to be responsible for O_2 activation and catalytic activity.[6,20,21] This viewpoint is suggested by transmission electron microscopy (TEM) studies[22] that reveal a high incidence of step and kink sites on the material. However, recent studies with extremely pure npAu synthesized from an intermetallic AuZr alloy (bulk residual Zr content ~0.03%) exhibit very low activity toward CO oxidation,[23] which indicates that Ag impurities play a key role in the dissociation of O_2 on npAu.

As a transient reaction technique, the temporal analysis of products (TAP) reactor has been successfully applied in numerous studies of the mechanism and dynamics of adsorption and reaction processes on catalyst surfaces.[24] The high detection sensitivity and time resolution render this technique an ideal tool to study the dynamic and kinetic behaviors of catalytic reactions as well as to investigate the nature and reactivity of the intermediate species, such as adsorbed oxygen, in both qualitative and quantitative ways.

In this work, using a TAP reactor, we address the issue of O_2 activation on npAu catalysts by measuring the saturation coverage of atomic oxygen on npAu, its surface-averaged initial collisional dissociation probability, the activation barrier for O_2 dissociation, and the facility with which the adsorbed O species activates methanol, the initial step in the catalytic cycle of esterification. The results from these TAP experiments support the hypothesis that molecular oxygen dissociation is associated with surface silver, that the density of the reactive sites is quite low, that the adsorbed oxygen atoms do not spill over from the sites of activation onto the surrounding surface, and that methanol reacts quite facilely with the adsorbed oxygen atoms.

2. Experimental

2.1 Synthesis of hollow npAu microspheres

Hollow npAu microspheres were synthesized according to a previously reported method.[8] Briefly, 10 μm cross-linked polystyrene–divinylbenzene (PS–DVB) microspheres were coated with Ag and Au in a ratio of 85 : 15 *via* electroless deposition in aqueous solution. The AgAu-coated microspheres were subsequently calcined at 450 °C for one hour in air (39 mL min^{-1}) in a flow reactor to alloy the metals and to remove the PS–DVB core and any residual organic stabilizing agents from the synthesis procedure. The annealed hollow AgAu microspheres were dealloyed in concentrated nitric acid until the Ag composition was 1%. The BET surface area was determined to be 2 m^2 g^{-1} after activation (see below for the detailed procedure).

2.2 TAP experiments

The pulse experiments were carried out in a commercial TAP-2 reactor (Fig. 1).[24] In a typical pulse response experiment, a narrow pulse of reactant gas (*ca.* 10^{-8} mol per pulse, pulse width 0.135 ms) was injected into a packed-bed microreactor using a high-speed, magnetically controlled pulse valve. The microreactor was a quartz tubular reactor with a length of 38 mm and a diameter of 6.35 mm, housing an internal thermocouple. The microreactor sat on top of a high-throughput vacuum chamber (10^{-8} to 10^{-9} Torr) containing an RGA200 mass spectrometer, which was used to detect the effluent from the reactor. The npAu catalyst was sandwiched in a thin-zone configuration in the middle of the microreactor packed among inert quartz particles with diameters between 250 and 300 μm. Pulse sizes were chosen to ensure Knudsen flow throughout the reactor.

The diffusion–reaction processes in the TAP pulse experiment, taking O_2 dissociation on the npAu catalyst for example, can be described using a model based on the following partial differential equation:[24]

$$\varepsilon_b \frac{\partial C_{O_2}}{\partial t} = D_{k,\mathrm{eff}} \frac{\partial^2 C_{O_2}}{\partial x^2} - k C_{z,\mathrm{total}} C_{O_2} f(\theta_O) \tag{1}$$

where $f(\theta)$ is an unspecified coverage dependence of the dissociation probability of O_2. The initial condition is

$$0 \le x \le L, \quad t = 0, \quad C_{O_2} = \delta_z \frac{N_{O_2}}{\varepsilon_b A} \tag{2}$$

and the boundary conditions are

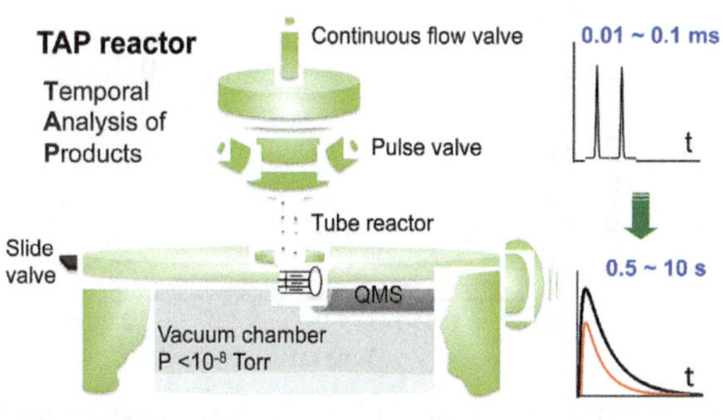

Fig. 1 Scheme of the TAP-2 reactor.

$$x = 0, \quad \frac{\partial C_{O_2}}{\partial z} = 0 \tag{3}$$

$$x = L, \ C_{O_2} = 0 \tag{4}$$

In the case of a very low reaction probability, the diffusion–reaction processes can be approximated as the diffusion-only case (Fig. 2):

$$\varepsilon_b \frac{\partial C_{O_2}}{\partial t} = D_{k,\mathrm{eff}} \frac{\partial^2 C_{O_2}}{\partial x^2} \tag{5}$$

For a standard diffusional response curve, the dimensionless peak position τ_p is a constant,[24] namely,

$$\tau_p = \frac{t_p D_e}{\varepsilon L^2} = \frac{1}{6} \tag{6}$$

where t_p is the maximum of the experimentally derived response curves (0.077 s), D_e the effective Knudsen diffusivity of the gas in the packed bed (26.2 cm^2 s^{-1}), ε the fractional voidage (0.84) of the packed bed, and L the length of the reactor (3.8 cm). Accordingly, the Knudsen diffusion characteristics of the species throughout the reactor, such as methanol and O_2, can be directly determined from the corresponding experimental pulse responses (Fig. 2). The reaction rate constants in eqn (1) can be determined either by solving the equation numerically or from moment analysis with presumed reaction kinetics.[24]

Prior to all the TAP experiments, the pulse size was calibrated using argon gas. To do this, a calibrated volume (V) filled with Ar at a given pressure was connected to the pulse valve. The pressure drop (ΔP) after a certain number of pulses (n) was used to calculate the average pulse size (number of molecules per pulse, N) as follows:

$$N = \frac{\Delta P V}{RTn} \tag{7}$$

The detection sensitivity factor, λ, for Ar could then be determined from the measured pulse intensity (I, the integrated area under the pulse curve of mass 40)

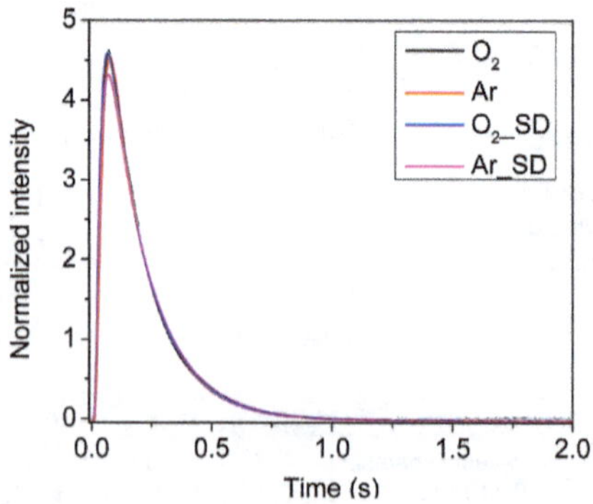

Fig. 2 Experimental and simulated standard diffusion (SD) pulse response curves of O_2 and Ar on the npAu catalyst at 423 K.

and the corresponding pulse size (N, the number of molecules in each pulse) as follows:

$$\lambda = \frac{N}{I} \tag{8}$$

The detection sensitivity factors for reactants and products, including methanol, O_2, and CO_2, were determined by pulsing a mixture with Ar with known composition. With a mixture of O_2/Ar (ratio η), the sensitivity factor for O_2 (λ_{O_2}), can be determined using:

$$\lambda_{O_2} = \frac{N_{O_2}}{I_{O_2}} = \frac{\eta N_{Ar}}{I_{O_2}} = \frac{\eta \lambda_{Ar} I_{Ar}}{I_{O_2}} \tag{9}$$

The amount of stable adsorbed active oxygen formed on the npAu catalyst by O_2 dissociation was determined by titration with a sequence of methanol pulses. Prior to the pulse experiments, the catalyst was activated by ozone treatment *via* a leak valve (10^{-6} Torr for 1 min) at 423 K and completely reduced with CH_3OH/Ar pulses. Note that under the pulse reaction conditions in the TAP reactor, formaldehyde was detected as the predominant product (selectivity > 98%) in the reaction of methanol with the adsorbed oxygen. The O_2 uptake was quantified either from the total amount of formaldehyde formation ($m/z = 30$, parent ion) from the calibrated methanol pulses or directly from the O_2 uptake during O_2 exposure (see Fig. 3). The packing material was determined to be inert in these experiments under identical conditions.

Since the $m/z = 30$ fragment is not unique to formaldehyde, the overlapping contributions from the cracking of methanol were subtracted according to the relative ion yields obtained in a blank experiment with methanol pulsing over the inert packing materials, *i.e.*, SiO_2. After subtracting the fragment contributions from methanol, the mass intensity was then converted into the number of formaldehyde molecules from each pulse using the detection sensitivity factor of methanol relative to Ar, assuming that these molecules have comparable mass spectrometer detection factors, which proves to be the case for CO_2 and Ar. No formaldehyde was produced upon exposure to methanol pulses on the pristine sample, *i.e.* without prior exposure to O_2.

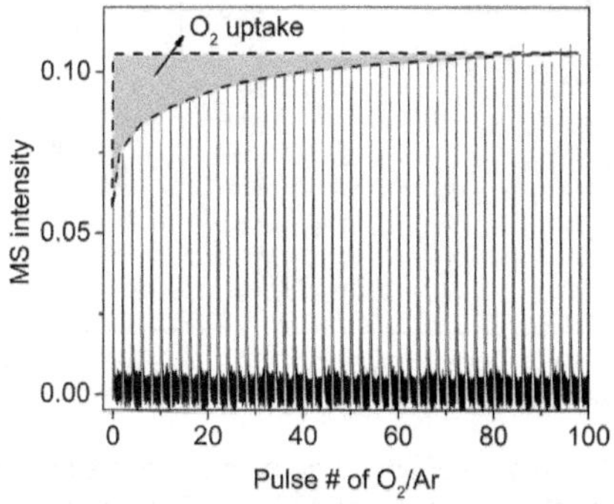

Fig. 3 Sequence of multiple O_2/Ar pulses (pulse separation = 2 s) dosed on activated npAu at 423 K. The grey area indicates the total O_2 uptake during the pulsing.

3. Results and discussion

3.1 Capacity of the npAu catalyst in O_2 dissociation

We first examined the activation of O_2 on the npAu catalyst by exposing the catalyst at 423 K sequentially to 100 O_2 pulses and 100 CH_3OH/Ar pulses. The oxygen uptake from O_2 exposure, determined either by the O_2 response pulse exiting the reactor or by the total amount of formaldehyde formed during the subsequent methanol titration, increases rapidly as the O_2 dosage increases and then approaches saturation (Fig. 4a).

Based on the amount of npAu (27 mg), the BET surface area (2 $m^2\ g^{-1}$) after activation, and the amount of formaldehyde formed at the saturation point (3.1×10^{15} molecules), the surface concentration of adsorbed O at saturation coverage resulting from the O_2 pulsing was calculated to be 6.2×10^{16} O atoms per m^2 or 4.4×10^{-3} ML, assuming a surface atomic density of 1.4×10^{19} atoms per m^2 for the Au(111) surface. It is possible that O could spill over onto the surface (the saturation density on gold being near 1.0 M)[25] but this apparently does not occur at these temperatures.

From the points in the initial linear portion of the uptake *vs.* exposure curve (Fig. 4a) the initial dissociative sticking probability of O_2 at the lowest initial oxygen coverage on the activated npAu catalyst at 423 K was computed to be $\sim 1 \times 10^{-7}$ at 423 K. This value is significantly lower than that on single-crystal Ag surfaces[26,27] and on Ag powder[28] or even on supported Ag catalysts.[29]

By methanol titration following O_2 exposure at different temperatures (473 K to 423 K), the apparent activation energy (E_a) for O_2 dissociation on the npAu catalyst was determined to be 5.0 ± 0.4 kcal mol^{-1} (Fig. 4b). This result is comparable to that reported on Ag single-crystal surfaces[27] but significantly below that estimated for single-crystal Au surfaces.[25,30,31] Experimentally, neither stepped single crystals nor polycrystalline Au surfaces dissociate molecular oxygen even at a high temperature or under high-pressure conditions.[14,25,31]

3.2 Stability of O_a from O_2 dissociation

By varying the delay time between the O_2 pulse and the subsequent methanol pulse, we further investigated the stability of the active oxygen species formed

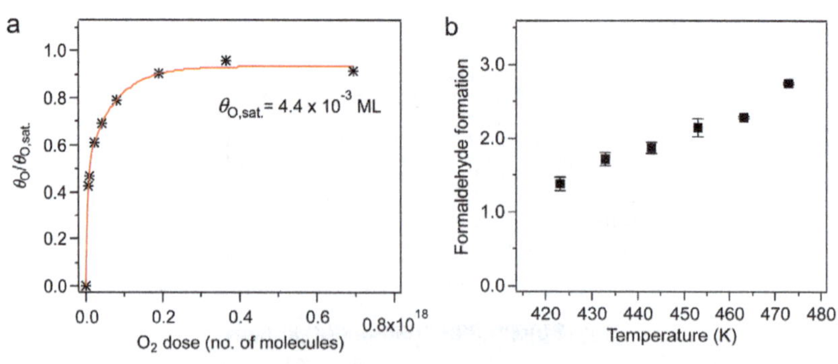

Fig. 4 (a) The fractional coverage of adsorbed oxygen atoms on the npAu catalyst *versus* the O_2 dose from sequential pulses of O_2 at 423 K. (b) Formaldehyde formation during the titration of active oxygen from O_2 dissociation on the npAu catalyst at different temperatures (single pulse of O_2 followed by titration with 100 pulses of CH_3OH/Ar).

from O_2 dissociation on the npAu catalyst. The results showed that the amount of formaldehyde formation decreased rapidly with the delay time below 0.5 s and then kept almost constant as the delay time was further extended up to 2 s (Fig. 5). The higher formaldehyde formation when the delay time is below 0.5 s may be explained by the enhanced total O_2 uptake due to the overlap between the O_2 and methanol pulses (see Fig. 2). Since methanol can readily react with adsorbed O, it is reasonable to deduce that in the presence of excess methanol, the active sites (adsorbed O) can be rapidly regenerated, resulting in a higher total O_2 uptake at a shorter delay time. With an extended delay time between the O_2 and methanol pulses, more O_2 will leave the catalyst bed before the methanol pulse arrives at the catalyst surface, leading to less O_2 dissociation and thus decreasing formaldehyde formation.

Note that adsorbed atomic O is quite stable on both Ag and Au single crystal surfaces, desorbing by atom recombination near 550 K.[14,27,30] It is possible, however, that with increased time between pulses the adsorbed O could change reactivity by conversion to another binding state. Since there was little variation in the amount of various products at a delay time longer than 0.5 s (Fig. 5), it appears that the active oxygen species from O_2 dissociation on the npAu surface at 423 K does not undergo a transformation into another form with different reactivity.

3.3 Reactivity of active adsorbed O with CO and methanol

To determine the reactivity of CO and methanol with the active adsorbed oxygen species, sequential pulsing of O_2 and CO/CH_3OH was conducted. Following an initial 100 pulses of O_2, the introduction of CO pulses resulted in the formation of only a very small amount of CO_2 at the beginning of the pulse cycle (Fig. 6a), but there was no measurable consumption of CO in parallel with the CO_2 formation. In addition, negligible O_2 uptake was detected when O_2 was re-introduced after CO pulsing (Fig. 6b). These results clearly indicate that the CO_2 formation was not from the reaction of CO with the active adsorbed oxygen. We attribute the CO_2 to displacement from other regions of the reactor. When methanol is introduced after the CO pulses, however, initially almost all the methanol is converted with concomitant formation of formaldehyde, and the reaction abates after approximately 100 methanol pulses (Fig. 6a).

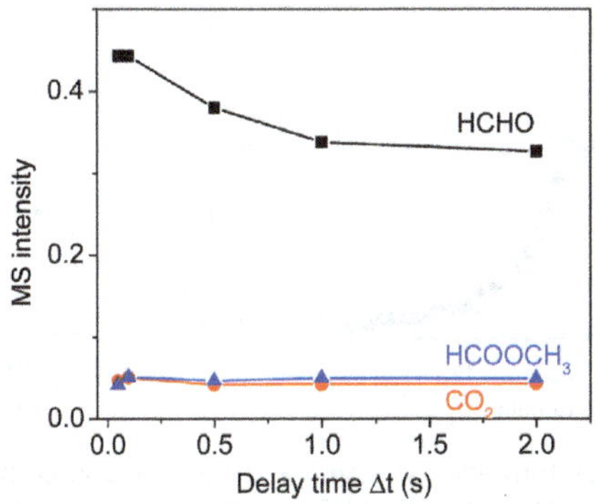

Fig. 5 The formation of various products including formaldehyde, methyl formate, and CO_2 as a function of the delay time between 1 pulse of CH_3OH/Ar and 1 pulse of O_2.

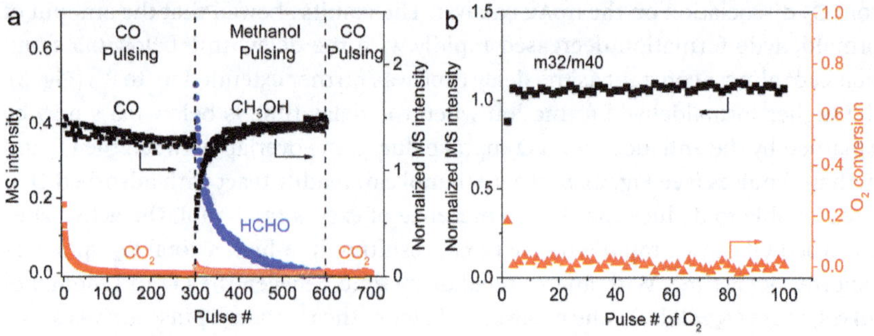

Fig. 6 (a) Sequential pulsing of CO and methanol over npAu after 100 pulses of O_2. (b) Normalized O_2 pulse intensity and the corresponding O_2 conversion in each pulse during O_2 pulsing after exposure to CO pulses.

It is clear that methanol can be catalytically activated on the npAu catalyst when pulsed simultaneously with O_2 (see Fig. 7a), producing formaldehyde as the predominant product under the pulse conditions. Surprisingly, when O_2 pulses were interrupted at the steady state, there was still a notable amount of formaldehyde formation, which, nevertheless, underwent an exponential decay with continuous methanol pulses. This observation indicates that there is a certain amount of oxygen species ($\sim 3 \times 10^{-3}$ ML) on the catalyst surface under the steady-state methanol oxidation reaction in the TAP reactor, which might also be the case for the reaction in the flow reactor under steady state conditions.

Assuming first-order kinetics for the reaction of methanol with adsorbed O (second stage in Fig. 7a), the apparent kinetic constant k_{app} can be calculated using the following expression:[32,33]

$$k_{app} = \frac{X}{1 - X}\frac{1}{\tau_{residence}} \tag{10}$$

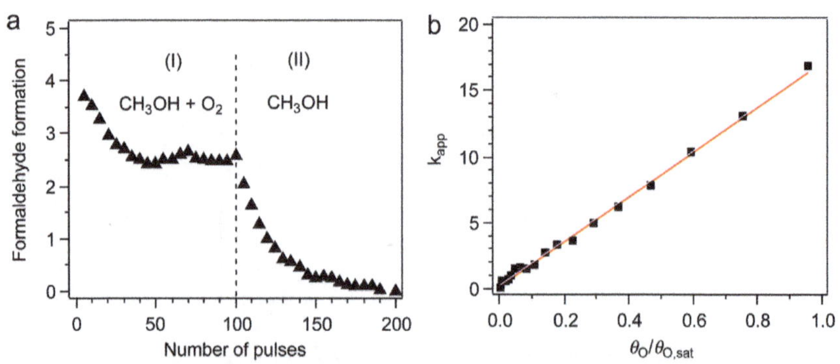

Fig. 7 (a) Formaldehyde formation during the pulse reaction of methanol with O_2 on the activated npAu catalyst at 423 K. (b) The apparent rate constant as a function of the fractional O coverage during titration after simultaneous pulsing of O_2 and CH_3OH at 423 K.

$$\tau_{res} = \varepsilon\left(\frac{L_{cat}L_{in2}}{D}\right) \tag{11}$$

where X is the conversion of methanol in each pulse, τ_{res} (s) is the mean residence time of methanol gas in the catalytic zone and is the same for each pulse, L_{cat} is the length of the catalyst zone, L_{in2} is the length of the second inert zone, and D is the diffusivity in the catalyst zone.

On the other hand, k_{app} can be related to the surface coverage and the intrinsic constant by:

$$k_{app} = \frac{m_{cat}}{V_G}C_{z,total}k\theta_O \tag{12}$$

where m_{cat} (g) is the catalyst mass, V_G (m^3) is the void volume in the catalyst zone, $C_{z,total}$ is the total active sites for O_2 dissociation, k ($m^3\ mol^{-1}\ s^{-1}$) is the intrinsic rate constant, and θ_O is the fractional coverage of the adsorbed oxygen.

Based on the data shown in Fig. 7a, $C_{z,total}$ can be determined from the total amount of formaldehyde formation, and X and θ_O for each pulse of methanol can be calculated from the formaldehyde formation and the corresponding methanol dose in each pulse. τ_{res} is calculated to be 0.006078 s. The k_{app} derived from eqn (10) is then plotted as a function of θ_O (see Fig. 7b) and the slope is determined to be 17 s^{-1}. Combining the m_{cat} (27 mg), V_G (0.041 cm^3), and $C_{z,total}$ (1.564×10^{-7} mol $g_{cat}{}^{-1}$), the intrinsic rate constant k was calculated to be 165 $m^3\ mol^{-1}\ s^{-1}$, which, in turn, yields a reaction probability of $\sim 1.3 \times 10^{-8}$ of methanol with an active surface O atom. It is notable that this reaction probability is nearly an order of magnitude smaller than that for activation of O_2 on the bare surface ($\sim 10^{-7}$), leading to the conclusion that the active sites should be nearly saturated with O in the steady state. In fact, the quantification result indeed showed that the total amount of the remnant adsorbed O species at the steady-state pulse reaction, *i.e.*, after 100 simultaneous pulses of methanol and O_2 (Fig. 7a), is close to the saturation coverage of O as determined above (see Fig. 4a).

4. Conclusions

In this work, we have determined the saturation coverage of atomic oxygen on np(Ag)Au microspheres ($\sim 4 \times 10^{-3}$ ML), the initial collisional dissociation probability of O_2 with the npAu surface ($\sim 1 \times 10^{-7}$), and the apparent activation energy of O_2 dissociation on the npAu catalyst (5.0 kcal mol^{-1}). These results suggest that the active site density for O_2 dissociation on the npAu catalyst is very low and that some form of AgAu bimetallic site is responsible for O_2 activation. The oxygen species formed by exposure to O_2 pulses under TAP conditions is only active for methanol and gives partial oxidation products. The kinetic study of O_2 activation suggests that the surface of the npAu catalyst is saturated with adsorbed O under steady state reaction conditions, at least for the pulse reaction in the TAP reactor.

Acknowledgements

This work was supported as part of the Integrated Mesoscale Architectures for Sustainable Catalysis, an Energy Frontier Research Center funded by the U.S.

Department of Energy, Office of Science, Basic Energy Sciences under Award # DE-SC0012573.

Notes and references

1 M. Haruta, T. Kobayashi, H. Sano and N. Yamada, *Chem. Lett.*, 1987, **16**, 405–408.

2 E. Taarning, K. Egeblad, R. Madsen and C. H. Christensen, *Catal. Lett.*, 2007, 35–40.

3 A. Wittstock, V. Zielasek, J. Biener, C. M. Friend and M. Bäumer, *Science*, 2010, **327**, 319–322.

4 G. C. Bond and D. T. Thompson, *Catal. Rev.: Sci. Eng.*, 1999, **41**, 319–388.

5 V. Zielasek, B. Jürgens, C. Schulz, J. Biener, M. M. Biener, A. V. Hamza and M. Bäumer, *Angew. Chem., Int. Ed.*, 2006, **45**, 8241–8244.

6 C. Xu, J. Su, X. Xu, P. Liu, H. Zhao, F. Tian and Y. Ding, *J. Am. Chem. Soc.*, 2007, **129**, 42–43.

7 K. M. Kosuda, A. Wittstock, C. M. Friend and M. Bäumer, *Angew. Chem., Int. Ed.*, 2012, **51**, 1698–1701.

8 M. L. Personick, B. Zugic, M. M. Biener, J. Biener, R. J. Madix and C. M. Friend, *ACS Catal.*, 2015, **5**, 4237–4241.

9 L. C. Wang, K. J. Stowers, B. Zugic, M. M. Biener, J. Biener, C. M. Friend and R. J. Madix, *Catal. Sci. Technol.*, 2015, **5**, 1299–1306.

10 L. C. Wang, K. J. Stowers, B. Zugic, M. L. Personick, M. M. Biener, J. Biener, C. M. Friend and R. J. Madix, *J. Catal.*, 2015, **329**, 78–86.

11 B. K. Min, A. R. Alemozafar, D. Pinnaduwage, X. Deng and C. M. Friend, *J. Phys. Chem. B*, 2006, **110**, 19833–19838.

12 B. K. Min, X. Deng, D. Pinnaduwage, R. Schalek and C. M. Friend, *Phys. Rev. B: Condens. Matter Mater. Phys.*, 2005, **72**, 121410.

13 B. Xu, R. J. Madix and C. M. Friend, *Acc. Chem. Res.*, 2014, **47**, 761–772.

14 A. G. Sault, R. J. Madix and C. T. Campbell, *Surf. Sci.*, 1986, **169**, 347–356.

15 A. Wittstock, B. Neumann, A. Schaefer, K. Dumbuya, C. Kübel, M. M. Biener, V. Zielasek, H.-P. Steinrück, J. M. Gottfried, J. Biener, A. Hamza and M. Bäumer, *J. Phys. Chem. C*, 2009, **113**, 5593–5600.

16 J. L. C. Fajín, M. N. D. S. Cordeiro and J. R. B. Gomes, *Chem. Commun.*, 2011, **47**, 8403–8405.

17 L. V. Moskaleva, S. Röhe, A. Wittstock, V. Zielasek, T. Klüner, K. M. Neyman and M. Bäumer, *Phys. Chem. Chem. Phys.*, 2011, **13**, 4529–4539.

18 L. C. Wang, Y. Zhong, D. Widmann, J. Weissmüller and R. J. Behm, *ChemCatChem*, 2012, **4**, 251–259.

19 L. C. Wang, Y. Zhong, H. Jin, D. Widmann, J. Weissmüller and R. J. Behm, *Beilstein J. Nanotechnol.*, 2013, **4**, 111–128.

20 C. Xu, X. Xu, J. Su and Y. Ding, *J. Catal.*, 2007, **252**, 243–248.

21 N. Lopez, T. Janssens, B. S. Clausen, Y. Xu, M. Mavrikakis, T. Bligaard and J. K. Nørskov, *J. Catal.*, 2004, **223**, 232–235.

22 T. Fujita, P. Guan, K. McKenna, X. Lang, A. Hirata, L. Zhang, T. Tokunaga, S. Arai, Y. Yamamoto, N. Tanaka, Y. Ishikawa, N. Asao, Y. Yamamoto, J. Erlebacher and M. Chen, *Nat. Mater.*, 2012, **11**, 775–780.

23 T. Déronzier, F. Morfin, L. Massin, M. Lomello and J.-L. Rousset, *Chem. Mater.*, 2011, **23**, 5287–5289.

24 J. T. Gleaves, G. S. Yablonskii and P. Phanawadee, *Appl. Catal., A*, 1997, **160**, 55–88.

25 J. Kim, E. Samano and B. E. Koel, *Surf. Sci.*, 2006, **600**, 4622–4632.

26 M. Bowker, M. A. Barteau and R. J. Madix, *Surf. Sci.*, 1980, **92**, 528–548.

27 C. T. Campbell, *Surf. Sci.*, 1985, **157**, 43–60.

28 M. Bowker, P. Pudney and G. Roberts, *J. Chem. Soc., Faraday Trans.*, 1989, **85**, 2635–2640.

29 M. Dean and M. Bowker, *Appl. Surf. Sci.*, 1988, **35**, 27–40.

30 J. M. Gottfried, K. J. Schmidt and S. Schroeder, *Surf. Sci.*, 2003, **525**, 184–196.

31 P. Legare, L. Hilaire, M. Sotto and G. Maire, *Surf. Sci.*, 1980, **91**, 175–186.

32 E. A. Redekop, G. S. Yablonsky, D. Constales, P. A. Ramachandran, C. Pherigo and J. T. Gleaves, *Chem. Eng. Sci.*, 2011, **66**, 6441–6452.

33 G. S. Yablonsky, M. Olea and G. B. Marin, *J. Catal.*, 2003, **216**, 120–134.

Faraday Discussions

PAPER

Photo-spectroscopy of mixtures of catalyst particles reveals their age and type†

M. M. Kerssens,[a] A. Wilbers,[b] J. Kramer,[b] P. de Peinder,[c] G. Mesu,[d] B. J. Nelissen,[e] E. T. C. Vogt[ae] and B. M. Weckhuysen*[a]

Received 3rd December 2015, Accepted 12th January 2016

DOI: 10.1039/c5fd00210a

Within a fluid catalytic cracking (FCC) unit, a mixture of catalyst particles that consist of either zeolite Y (FCC-Y) or ZSM-5 (FCC-ZSM-5) is used in order to boost the propylene yield when processing crude oil fractions. Mixtures of differently aged FCC-Y and FCC-ZSM-5 particles circulating in the FCC unit, the so-called equilibrium catalyst (Ecat), are routinely studied to monitor the overall efficiency of the FCC process. In this study, the age of individual catalyst particles is evaluated based upon photographs after selective staining with substituted styrene molecules. The observed color changes are linked to physical properties, such as the micropore volume and catalytic cracking activity data. Furthermore, it has been possible to determine the relative amount of FCC-Y and FCC-ZSM-5 in an artificial series of physical mixtures as well as in an Ecat sample with unknown composition. As a result, a new practical tool is introduced in the field of zeolite catalysis to evaluate FCC catalyst performances on the basis of photo-spectroscopic measurements with an off-the-shelf digital single lens reflex (DSLR) photo-camera with a macro lens. The results also demonstrate that there is an interesting time and cost trade-off between single catalyst particle studies, as performed with *e.g.* UV-vis, synchrotron-based IR and fluorescence micro-spectroscopy, and many catalyst particle photo-spectroscopy studies, making use of a relatively simple DSLR photo-camera. The latter approach offers clear prospects for the quality control of *e.g.* FCC catalyst manufacturing plants.

Introduction

Fluid catalytic cracking (FCC) is a major chemical process used in the oil refining industry to convert crude oil fractions into smaller, more useful molecules such as

[a]*Inorganic Chemistry and Catalysis, Debye Institute for Nanomaterials Science, Utrecht University, Universiteitslaan 99, 3584 CG Utrecht, The Netherlands. E-mail: b.m.weckhuysen@uu.nl*

[b]*DSM Resolve, Urmonderbaan 22, Geleen, The Netherlands*

[c]*Vibspec, Haaftenlaan 28, 4006 XL Tiel, The Netherlands*

[d]*Albemarle Corporation, 13000 Bay Park Road, Pasadena, TX 77507, USA*

[e]*Albemarle Catalysts Company BV, Nieuwendammerkade 1-3, 1022 AB Amsterdam, The Netherlands*

† Electronic supplementary information (ESI) available: FCC laboratory deactivation methods, N_2 physisorption measurements, catalytic activity testing and photo-spectroscopy data analysis. See DOI: 10.1039/c5fd00210a

gasoline and propylene.[1,2] FCC catalysts are multi-component hierarchically structured particles of about 50–150 μm containing a complex mixture of clay, silica, alumina, and zeolite. The choice of the zeolite material depends on the application in mind, although zeolite Y (FCC-Y) and ZSM-5 (FCC-ZSM-5) are the main workhorses in FCC catalysis. The addition of FCC-ZSM-5 to the catalyst inventory of an FCC process is known to enhance the propylene yield.

In a real-life crude oil refinery plant, FCC catalyst particles are poisoned by metals, such as nickel, iron and vanadium, from the crude oil feedstock as well as by coke formation, which both affect their short- and long-term activity.[3–6] The coke on the catalyst particles is removed in a regenerator unit before the catalyst is returned to the riser reactor. This cycle between the riser reactor and regenerator unit is repeated continuously. Part of the catalyst inventory in an FCC unit is replaced by fresh catalyst material on a daily basis. As a result, the so-called equilibrium catalyst (Ecat) has a range of ages and activities. Since with an increasing amount of FCC-ZSM-5 the propylene yield plateaus, there is less FCC-ZSM-5 than FCC-Y in Ecat mixtures when using a ZSM-5 additive.[7]

Several studies showed that the deactivation of FCC-Y and FCC-ZSM-5 differs.[5,8] For example, dealumination resulting from steam treatment in the regenerator unit will mainly cause a decrease in the number of acid sites in zeolite ZSM-5, while it will also have an effect on the acid strength in zeolite Y.[8] Laboratory deactivation methods, such as steam (ST), two step cyclic (CD), and Mitchell-steam (MI) deactivation, are commonly used to simulate the influence of hydrothermal conditions, coke formation and/or metal poisoning, although none of those laboratory deactivation methods yield the same results as a genuine Ecat material.

For example, it was found that after applying a selective staining reaction with a Brønsted acid selective probe, the fluorescence intensity of the different laboratory deactivated samples shows sensitivity towards the specific deactivation method applied. More specifically, the following fluorescence intensity decrease was noted for two series of laboratory deactivated FCC catalyst particles (FCC1 and FCC2): fresh > ST > CD > MI.[9,10] An Ecat sample was found to display Brønsted acidity in between those of the CD and MI catalyst samples.

In a similar manner, synchrotron-based IR microscopy in combination with pyridine as a probe molecule for assessing Brønsted and Lewis acidity could be used to determine the degree of FCC catalyst deactivation at the single catalyst particle level, leading to the following decreasing order in acidity: fresh > ST > CD > MI.[11] Also, an Ecat sample was found to display an overall acidity in between those of the CD and MI catalyst samples.

The ability to distinguish FCC-Y from FCC-ZSM-5 within Ecat samples would be interesting to combine with a separation technique as this would allow studying them independently and the obtained knowledge could allow more deliberate designs of FCC catalyst particles. In order to achieve this ambitious goal, first a simple identification should be possible for a single particle, preferably based upon a simple characterization tool. It has been shown that zeolite ZSM-5 crystals can be mapped in great detail by making use of the optical absorption of polarized light.[12] Recently, our group used the anisotropy of zeolite ZSM-5 crystals to determine the dispersion, aggregate size and amount of ZSM-5 within a single FCC particle.[13]

Inspired by the work of Schwaebel *et al.*,[14] we have developed in this work a versatile photo-spectroscopy method to determine both the age and type of an individual FCC catalyst particle simply on the basis of its colour after applying a selective staining procedure. To determine the specific colour of a large number of FCC catalyst particles we have used an off-the-shelf digital single lens reflex (DSLR) photo-camera with a macro lens. An outline of the approach taken is given in Fig. 1. In the work described, the amount of FCC-ZSM-5 in mixtures differing in structure (FCC-Y *vs.* FCC-ZSM-5) as well as in an Ecat sample is quantified based upon the photographs taken after staining a mixture of FCC catalyst particles. Furthermore, the differences in Brønsted acid strength in Ecat–Y samples, obtained by different methods of aging, are evaluated by colour changes after catalyst staining, and related to catalytic cracking data as well as to physico-chemical properties, such as the micropore volume. As a result, a simple but powerful analytical method emerges, which can be used to evaluate catalytic performances based on specific colour changes.

Results and discussion

In this work, we discuss the versatility of photo-spectroscopy for catalyst characterization studies by taking FCC catalyst particles as a showcase system. As modern off-the-shelf digital reflex photo-cameras are equipped with very sensitive CCD detectors, it is possible to determine very accurately the colour of the FCC catalyst particles after selective staining with a probe molecule. The advantage of this approach is that a large number of FCC catalyst particles ($\sim 10^4$) can be analysed in "one photographic shot", as compared to previous characterization studies in which only a handful of FCC catalyst particles (<10) could be analysed

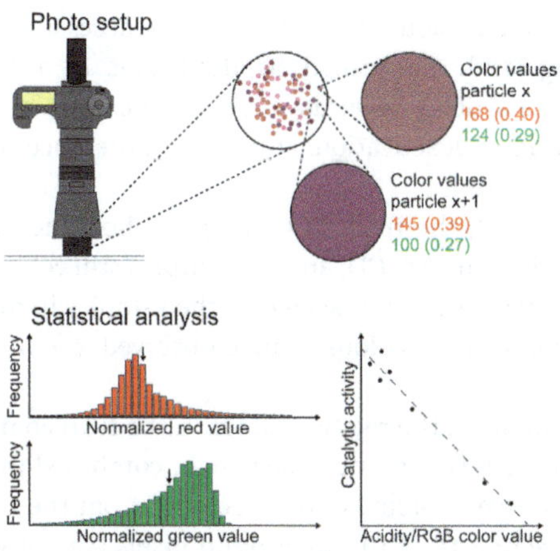

Fig. 1 Outline of the developed photo-spectroscopy method to determine the age and type of individual catalyst particles on the basis of their color. In this specific case, the photographs are taken from mixtures of deactivated FCC-Y catalyst particles after staining with 4-methoxystyrene as a Brønsted acidity probe. The RGB values of each FCC-Y catalyst particle within the mixtures are then analyzed, normalized, and plotted in a histogram. The average values are plotted *versus* bulk properties, such as the catalytic cracking activity, to establish if a relation or correlation exists.

with various micro-spectroscopic techniques, including UV-vis, synchrotron-based IR, and (confocal) fluorescence microscopy.[9–11] The principles of the photo-spectroscopy method will now be illustrated for two examples: one focusing on the determination of the age of the FCC catalyst particles, while the second example demonstrates the potential to discriminate between FCC catalyst particles containing either zeolite Y or ZSM-5.

Determining the age of catalyst particles

In the first series of experiments we have focused on a set of 11 distinct FCC catalyst samples. More specifically, we have made use of FCC density separated equilibrium catalysts (Ecat, 3 samples in total), and two sets of laboratory deactivated FCC-Y (8 samples in total). For the latter FCC catalyst samples we use the sample codes FCC1 and FCC2, which is identical to that of Buurmans *et al.*[9] It is important to remark that these 11 FCC catalyst materials have been described in detail in previous publications from our research group, including their physico-chemical and catalytic characterization.[9–11,15]

Due to severe deactivation after the laboratory deactivation procedures (*i.e.*, ST, CD and MI), the colour change resulting from a 4-fluorostyrene probe reaction is limited and differences between the FCC-Y cannot be observed. Therefore, the FCC catalyst samples were stained with 4-methoxystyrene instead, which is known to react with all Brønsted acid sites of the zeolite material, as opposed to 4-fluorostyrene, which only reacts with the strongest Brønsted acid sites of a zeolite material.[12]

As shown in Fig. 2, distinct colour differences were observed for the FCC1 and FCC2 samples across the different deactivation techniques, indicating that the Brønsted acid sites, although strong enough to catalyze the selective 4-methoxystyrene oligomerization reaction, do not have sufficient strength to catalyze the 4-fluorostyrene reaction. This is in agreement with previous studies.[10] It is often thought that the observed colour changes are due to accessibility differences, but a Nile-blue staining study performed by Buurmans *et al.*[9] showed that the accessibility of FCC particles towards larger fluorescent molecules does not change significantly upon deactivation, and colour differences are thus not solely due to accessibility differences.

As illustrated in Fig. 2, the fresh FCC samples show the most severe colour change towards purple/pink. ST, CD, and MI samples still colour pink and purple upon reaction, but a gradient can be seen over the samples in the order: ST > CD > MI. The more pronounced the colour change observed, the higher the acidity of the FCC particle.

For quantitative analysis, the relative values in the RGB channels (RGB stands for the intensities of Red, Green and Blue which, combined in different intensities, result in the observed colour) were calculated from the photographs taken (Fig. 2) for an average of 3.16×10^6 foreground pixels (*i.e.*, pixels part of an FCC particle) per sample (Table S3†). This roughly translates to ~23 000 catalyst particles as opposed to the six catalyst particles studied by Buurmans *et al.*[9] The average red values are fairly similar for the ST, CD, and MI samples, while the values found for the fresh catalyst samples are higher and show a broader distribution. The average green values also show the highest distribution for the fresh catalyst samples. Average green values for series FCC1 and FCC2 show

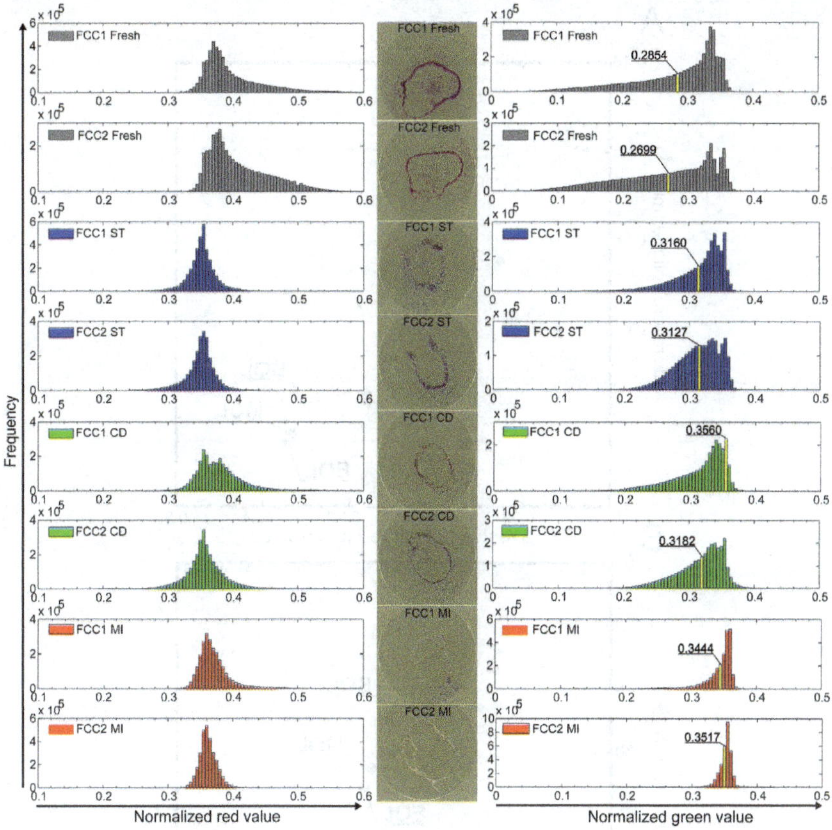

Fig. 2 Histograms of the normalized red and green values for the FCC1 and FCC2 foreground pixels of photographic images of fresh (gray), steamed (blue), cyclic deactivated (green), and Mitchell impregnated (red) samples after a 4-methoxystyrene probe reaction displayed with an example of the images taken.

a fresh < ST ≈ CD < MI trend and obvious differences can be observed (Fig. 2). The tail at lower values disappears across the samples, with the tail being the longest for the fresh and the shortest for the MI samples. The highest frequency in the green distribution of the fresh catalyst samples is ~0.34 and for the MI samples ~0.36, and across the catalyst samples the relative intensities of these bands seem to change.

When plotting the average green value *versus* bulk properties (*i.e.*, catalytic cracking activity and micropore volume)[9] the same inverse proportional trend is found, as illustrated in Fig. 3. As the average green value is inversely proportional to the activity and bulk properties related to activity, it can be used to estimate the activity. The same experiments were performed on beginning of life (BOL), middle of life (MOL) and end of life (EOL) density separated Ecat samples (Fig. 3 and Table S3†). The catalytic activity and micropore volume of the density separated Ecats are between MI and CD, as are the average normalized green values.

It is evident that the experiments presented in this study take much less time than the confocal fluorescence microscopy measurements conducted at the single catalyst particle level. Furthermore, the statistics are more robust as a much larger

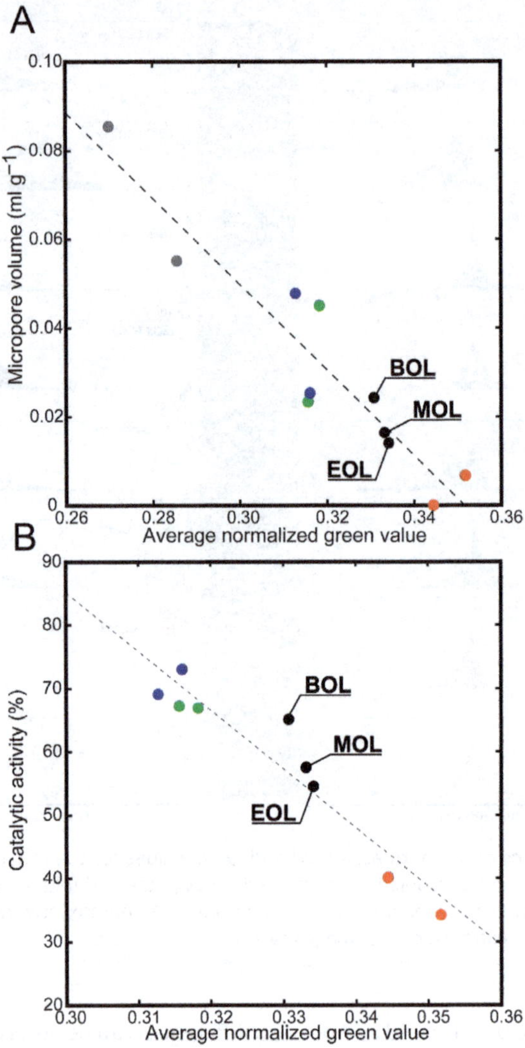

Fig. 3 Plots illustrating the average normalized green values *versus* bulk properties for FCC catalyst particles. (A) Micropore volumes obtained from nitrogen physisorption measurements; (B) activity data obtained for a catalyst-to-oil ratio of 3 when performing a catalytic cracking test. The color coding is as follows: fresh in gray, steamed in blue, cyclic deactivated in green, Mitchell impregnated in red, and density separated Ecat in black. Numerical data for these plots can be found in the ESI† of this article.

amount of catalyst particles is analyzed. Previous confocal fluorescence microscopy studies were based upon the analysis of six particles or 150 domains per sample,[9] in contrast to the thousands of FCC catalyst particles which can be captured in one single photograph with the method used in this work.

Determining the type of zeolite in a particle

In the second series of experiments, both FCC-ZSM-5 and FCC-Y have been analyzed in mixtures using 4-fluorostyrene as a selective staining probe. Upon reaction, colour differences appear between FCC-ZSM-5 (pink) and FCC-Y (orange), which results from the zeolite type and is caused by the different reaction products that are formed. The result is shown in Fig. 4. For example, FCC-Y will produce an absorption band at around 520 nm due to the formation of mainly cyclic dimers, while FCC-ZSM-5 will give rise to several absorption bands due to the formation of linear conjugated dimers and higher oligomers. In Fig. 4, sample FCC-ZSM-5 transmits ~70% of the blue light, and 60% and 95% of the green and red light, respectively. This results in an overall pink colour (panel B). Instead, sample FCC-Y transmits ~50% of the blue, 60% of the green, and 70% of the red light, resulting in an orange/brown colour.

As the colour is related to the type of zeolite, the colours can be used to estimate the amount of FCC-ZSM-5 and FCC-Y particles by quantifying the amount of particles with a specific colour. This can be accomplished with a micro-spectroscopy set-up, as shown in previous work,[9,10] but also – as done in this study – with a digital single lens reflex (DSLR) photo-camera with a macro lens. For this purpose, one can make use of the amount of catalyst particles (Table S4, Fig. S1 and S2†) and the area of the catalyst particles (Table S5†). The findings were then plotted *vs.* the wt% of the constructed samples (Fig. 5 and S4†). The equation of the calibration lines can be used to translate the values found for the observed amount of FCC-ZSM-5 to a weight percentage, which was used for an Ecat with unknown composition (Table S7, Fig. S3†).

It is thought that the area method (the amount of pixels depicting a certain type of catalyst) is more robust than the number of particles because it is less sensitive to agglomerated catalyst particles. The *R*-squared value of the area calibration line is higher than that of the number of catalyst particles

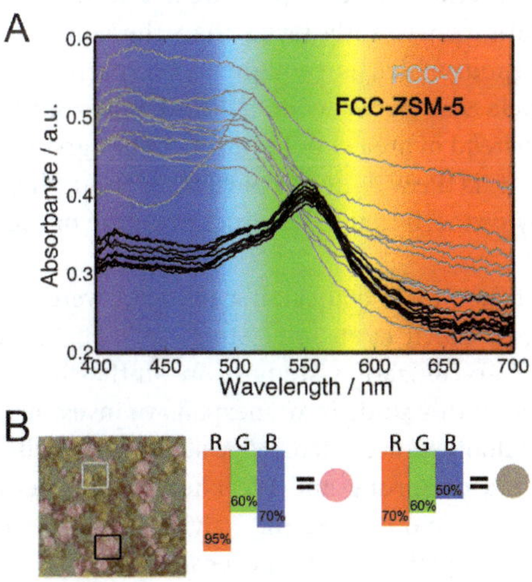

Fig. 4 Plots illustrating the relationships between zeolite type, color and specific absorption maxima. (A) Absorption spectra of samples FCC-ZSM-5 (black) and FCC-Y (gray); (B) photograph of a mixture of samples FCC-ZSM-5 (pink) and FCC-Y (orange). The transmission percentages for red, green, and blue are displayed with the resulting color, explaining the spectroscopic base on which a distinction can be made between FCC-ZSM-5 and FCC-Y.

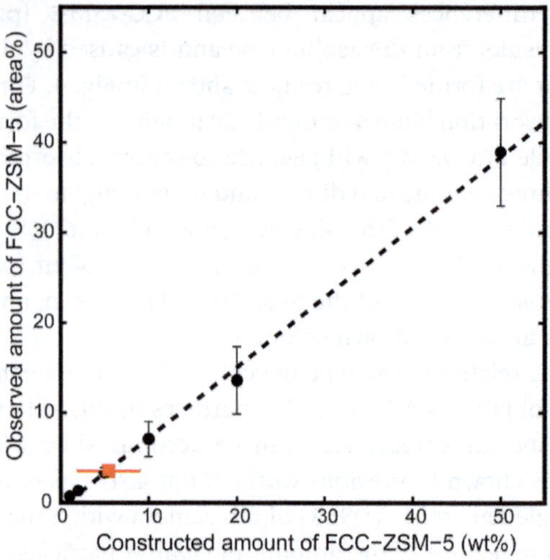

Fig. 5 Graphical representation of the observed amount of FCC-ZSM-5 *versus* the constructed amount for all data based upon the area of FCC catalyst particles (closed circles, $y = 0.7804x - 0.5919$). The Ecat sample is displayed with a red square. Numerical data can be found in the ESI† of this article.

calibration line and the spread in the wt% values of samples taken from the same Ecat is smaller (lower standard deviation), also indicating that using the area is a more robust method. Based upon the area of the catalyst particles, values for the fraction of the FCC-ZSM-5 sample were found between 0.6 and 9.8%, which translated to 1.6 and 13.3 wt% (Table S7†). An average of 5.5 wt% was found for the method based upon area. The observed amount of the FCC-ZSM-5 sample is consistently lower than the constructed amount when using the area method, which is thought to be (partly) caused by the slight difference in particle size between FCC-Y and FCC-ZSM-5. A method based upon the volume instead of area should improve the agreement of the observed and constructed wt% as the weight of a particle is (assuming equal density over the sphere) directly proportional to the linear volume of the sphere and not to the cross sectional area.

On average, slightly over 25 000 catalyst particles were analyzed from only 8 photographic images, which is an interesting trade-off between single catalyst particle and bulk characterization studies. An off-the-shelf DSLR camera with a macro lens as used in this study is an inexpensive investment compared to, for example, a confocal fluorescence microscope setup with multi-laser excitation, as employed in the work of Sprung and Weckhuysen. Furthermore, the data are acquired quicker and more efficient, giving way to measurements on manufacturing sites for quality control purposes.

Experimental

Equilibrium catalysts (Ecat), density separated Ecats, laboratory deactivated FCC-Y (FCC1 and FCC2, see ESI, Section 1†) and fresh FCC-Y and FCC-ZSM-5 were provided by Albemarle Catalysts. FCC-Y was heated to 600 °C in 3 h and then calcined for 5 h. FCC-ZSM-5 and the set of Ecats were used as received.

Photographs of the samples were taken with an EOS 6D camera body (Canon) equipped with an EF 100 mm IS USM macro lens (Canon) and EF 25II lens extension tube (Canon). A Copylizer eVision exe.cutive HF camera stand with lighting unit RB 4.36 HF (both Kaiser Fototechnik) was used to illuminate the sample under 45° angles and secure the camera at the desired distance to the sample.

Mixtures of 1, 2, 5, 10, 20 and 50 wt% FCC-ZSM-5 in a FCC-Y sample were prepared and finely sprinkled upon a borosilicate microscope window heated to 30 °C, after which 10 μl of 4-fluorostyrene (99%, Sigma Aldrich) was added to impregnate the sample. After evaporation the sample was rinsed twice with mesitylene (97%, Sigma) to remove the remaining probe molecules from the outer surface. The temperature was then increased to 180 °C (50 °C min^{-1}), after which a color change (pink for FCC-ZSM-5 and orange for FCC-Y) was observed and a photograph was taken for further analysis. This procedure was performed 8 times for all samples under study (Fig. S1†).

For the laboratory deactivated samples measurements were performed by adding 4-methoxystyrene (97% Sigma Aldrich) directly at 180 °C, without the washing steps, as this would save a significant amount of time in future applications. Here, again, the procedure was performed 8 times for all samples under study. It is important to state that the photographs taken with the additional washing steps and temperature ramp showed a similar trend in color. Photographs were color calibrated using the SpyderCheckr tool and loaded into Photoshop where the images were cropped and the background and rim were removed. Images were then loaded into Matlab and the RGB values for each pixel in the foreground were normalized: R_norm = R/(R + G + B), G_norm = G/(R + G + B) and B_norm = B/(R + G + B). As the sum of the normalized values equals 1 for each pixel, any combination of the two values is sufficient to describe the color, and only R and G are displayed.[14]

Conclusions

4-Fluorostyrene was used as a selective probe molecule for the quantification of the number of FCC-Y and FCC-ZSM-5 catalyst particles in mixtures based upon their respective colors. An off-the-shelf DSLR camera with a macro lens was used, which yields clear prospects for automation. Calibration lines were constructed, based upon the number and area of the FCC catalyst particles, and used to calculate the amount of FCC-ZSM-5 in an equilibrium FCC catalyst. The area method is believed to be more robust and gave a value of 5.5 wt% FCC-ZSM-5 for the Ecat catalyst material under study.

For the analysis of the acid strength, and related age, of a series of FCC-Y catalyst particles, 4-methoxystyrene has been used as a selective probe molecule. It was found that the normalized green value, as obtained from the photo-spectroscopy data, is inversely proportional to the catalytic activity and micropore volume of the series of FCC catalysts under study. This correlation demonstrates that the age of FCC catalyst particles can be simply determined on the basis of their colour.

Both showcase examples demonstrate that photo-spectroscopy is an interesting trade-off method between single catalyst particle and bulk characterization studies as measurements are relatively fast and a large amount of catalyst particles are captured in a single photograph.

Acknowledgements

This work was supported by the Dutch National Science Foundation (NWO-TASC project 053.23.101), while the FCC samples were provided by Albemarle Catalysts. Dr F. Meirer (Utrecht University) is acknowledged for his help with the artwork in Fig. 4.

Notes and references

1 (*a*) W. Vermeiren and J. P. Gilson, *Top. Catal.*, 2009, **52**, 1131; (*b*) A. Corma, *Chem. Rev.*, 1995, **95**, 559; (*c*) *Zeolites and Catalysis: Synthesis, Reactions, and Applications*, ed. J. Cejka, A. Corma and S. Zones, Wiley, Weinheim, 2010; (*d*) *Handbook of Heterogeneous Catalysis*, ed. G. Ertl, H. Knozinger, F. Schuth and J. Weitkamp, 2nd edn, Wiley-VCH, 2008.

2 (*a*) E. T. C. Vogt and B. M. Weckhuysen, *Chem. Soc. Rev.*, 2015, **44**, 7342; (*b*) E. T. C. Vogt, G. T. Whiting, A. Dutta Chowdhury and B. M. Weckhuysen, *Adv. Catal.*, 2015, **58**, 143.

3 (*a*) J. Ruiz-Martínez, A. M. Beale, U. Deka, M. G. O'Brien, P. D. Quinn, J. F. W. Mosselmans and B. M. Weckhuysen, *Angew. Chem., Int. Ed.*, 2013, **52**, 5983; (*b*) S. Kalirai, U. Boesenberg, G. Falkenberg, F. Meirer and B. M. Weckhuysen, *ChemCatChem*, 2015, **7**, 3674; (*c*) F. Meirer, S. Kalirai, J. N. Weker, Y. Liu, J. C. Andrews and B. M. Weckhuysen, *Chem. Commun.*, 2015, **51**, 8097; (*d*) F. Meirer, D. T. Morris, S. Kalirai, Y. J. Liu, J. C. Andrews and B. M. Weckhuysen, *J. Am. Chem. Soc.*, 2015, **137**, 102.

4 S. Haitao, D. Zhijian, Z. Yuxia and T. Huiping, *Catal. Commun.*, 2011, **16**, 70.

5 J. Ruiz-Martínez, I. L. C. Buurmans, W. V. Knowles, D. van der Beek, J. A. Bergwerff, E. T. C. Vogt and B. M. Weckhuysen, *Appl. Catal., A*, 2012, **419–420**, 84.

6 H. S. Cerqueira, G. Caeiro, L. Costa and F. Ramôa Ribeiro, *J. Mol. Catal. A: Chem.*, 2008, **1–2**, 292.

7 T. Amano, J. Wilcox, C. Pouwels and T. Matsuura, *Petrol. Tech. Q.*, 2012, **17**, 17–27.

8 E. Rautiainen, R. Pimenta, M. Ludvig and C. Pouwels, *Catal. Today*, 2009, **140**, 179.

9 I. L. C. Buurmans, J. Ruiz-Martínez, W. V. Knowles, D. van der Beek, J. A. Bergwerff, E. T. C. Vogt and B. M. Weckhuysen, *Nat. Chem.*, 2011, **3**, 862.

10 I. L. C. Buurmans, J. Ruiz-Martínez, S. L. van Leeuwen, D. van der Beek, J. A. Bergwerff, W. V. Knowles, E. T. C. Vogt and B. M. Weckhuysen, *Chem.–Eur. J.*, 2012, **18**, 1094.

11 I. L. C. Buurmans, F. Soulimani, J. Ruiz-Martinez, H. E. Van der Bij and B. M. Weckhuysen, *Microporous Mesoporous Mater.*, 2013, **166**, 86.

12 (*a*) M. H. F. Kox, E. Stavitski and B. M. Weckhuysen, *Angew. Chem., Int. Ed.*, 2007, **46**, 3652; (*b*) E. Stavitski, M. H. F. Kox and B. M. Weckhuysen, *Chem.–*

Eur. J., 2007, **13**, 7057; (*c*) M. H. F. Kox, E. Stavitski, J. C. Groen, J. Pérez-Ramírez, F. Kapteijn and B. M. Weckhuysen, *Chem.–Eur. J.*, 2008, **14**, 1718.
13 C. Sprung and B. M. Weckhuysen, *Chem.–Eur. J.*, 2014, **20**, 3667.
14 T. Schwaebel, S. Menning and U. H. F. Bunz, *Chem. Sci.*, 2014, **5**, 1422.
15 F. Meirer, S. Kalirai, D. Morris, S. Soparawalla, Y. Liu, G. Mesu, J. C. Andrews and B. M. Weckhuysen, *Sci. Adv.*, 2015, **1**, e1400199.

Faraday Discussions

PAPER

New crystalline complex metal oxides created by unit-synthesis and their catalysis based on porous and redox properties†

Zhenxin Zhang,[a] Satoshi Ishikawa,[a] Yuta Tsuboi,[a] Masahiro Sadakane,[b] Toru Murayama[c] and Wataru Ueda*[a]

Received 13th January 2016, Accepted 28th January 2016
DOI: 10.1039/c6fd00006a

The development of new complex metal oxides having structural complexity suitable for solid-state catalysis is of great importance in fundamental catalysis research and practical applications. However, examples of these materials are rare. Herein, we report two types of crystalline complex metal oxides with new structures and their catalytic properties. The first one is an all-inorganic ε-Keggin polyoxometalate-based material with intrinsic microporosity. The framework of the material is formed by the assembly of ε-Keggin polyoxomolybdate units with metal ion linkers in a diamondoid topology. The micropores of the material can be opened without change of the structures, and the material adsorbs small molecules. This material has both redox properties and acidity and can be applied to O_2 adsorption, selective oxidation of methacrolein, and hydrolysis of cellobiose. The other material is a crystalline metal oxide based on molecular nanowires. The hexagonal POM units stack along the *c* axis to form prismatic clusters as molecular wires. The molecular wires further assemble in a hexagonal fashion to form the crystals, and NH_4^+ and water are present in between the molecular wires. The material is active as an acid catalyst for cellobiose conversion.

Introduction

Transition metal oxides, mainly composed of metal–oxygen octahedra, show interesting properties such as multi-electron redox properties, structural stability, acidity, and elemental diversity and can be applied to many fields such as

[a]*Department of Material and Life Chemistry, Faculty of Engineering, Kanagawa University, Rokkakubashi, Kanagawa-ku, Yokohama-shi, Kanagawa, 221-8686, Japan. E-mail: uedaw@kanagawa-u.ac.jp*
[b]*Department of Applied Chemistry, Graduate School of Engineering, Hiroshima University, 1-4-1 Kagamiyama, Higashi Hiroshima 739-8527, Japan*
[c]*Department of Applied Chemistry, Graduate School of Urban Environmental Sciences, Tokyo Metropolitan University, 1-1 Minami-osawa, Hachioji, Tokyo 192-0397, Japan*

† Electronic supplementary information (ESI) available. See DOI: 10.1039/c6fd00006a

catalysis, photocatalysis, materials science, magnetism, biology, and medicine. The design and development of new transition metal oxides with unique structural and functional features, such as porosity and nano-structure, are of great interest and particularly important for catalysis.[1]

The assembly of early-transition metal–oxygen units forms inorganic molecular metal–oxygen clusters known as polyoxometalates (POMs).[2–5] POMs are well-defined building blocks for the construction of POM-based materials,[6,7] such as covalent POM-based organic–inorganic hybrid frameworks,[8,9] POM-based complex metal oxides,[10–14] and ionic porous organic–inorganic materials.[15–19] Previous research shows that using POMs as sub-units is an effective approach to obtain transition metal oxides with new structures. Actually, we have recently succeeded in synthesizing two new kinds of structural materials of transition metal-oxides using POM units. One is an all-inorganic crystalline ε-Keggin POM-based framework with intrinsic microporosity[20–24] and the other is a crystalline metal oxide based on molecular nanowires.[25] Their structural features are very rare in the field of metal oxides.

The all-inorganic crystalline ε-Keggin POM-based microporous materials are formed by the connection of ε-Keggin POM units with transition metal ions as linkers in a diamond-like topology (Fig. 1a–c). There are three sites in the framework: surrounding sites, central sites, and linker sites (Fig. 1a). The surrounding sites are mainly occupied by Mo ions, and the Mo ions can be substituted by V ions. The central and linker sites are occupied by other metal ions such as V, Bi, Zn, and Mn. The frameworks of the materials are anionic, and there are cations in the bulk to neutralize the negative charge of the framework. The chemical composition is readily tuned in each site mentioned above. Cages and channels form the micropores of the materials. Ten ε-Keggin POM units and the linker metal ions surround the cage. The cages are tetrahedrally connected with the channels to form a periodical 3D pore system as FAU-type zeolites do, and the tunnel of the pore was not straight but in a zig-zag fashion. The micropore can be opened using calcination that removes water and NH_4^+ in the micropore without collapsing the structures. The materials have interesting properties and applications such as gas adsorption,[20,23] gas separation,[23] ion-exchange,[20,21] and redox properties.[21] The ε-Keggin POM-based microporous material, Mo–V–Bi oxide with a proton as the cation, showed catalytic activity for the etherification of benzyl alcohol.[20,22]

Another example of the newly developed material is the crystalline metal oxides derived from molecular nanowires (Fig. 1d–f). The structures of the molecular wires are formed by the assembly of the hexagonal POM units, $[XMo_6O_{21}]^{2-}$ (X = Se or Te), along the c axis, which further pack in a hexagonal symmetry to form crystals with NH_4^+ and water in between the molecular wires. The molecular wires in the crystal are separable and isolable. In fact, ultrathin molecular wire-based materials can be formed and act as acid catalysts for esterification.[25]

In our previous research, 4 iso-structural all-inorganic crystalline ε-Keggin POM-based microporous materials have been reported, which show different properties and have been applied to different fields.[20–23] Herein, we extended our research and introduced two more iso-structural materials of ε-Keggin POM-based materials, denoted as Na–Mo–Mn oxide and NH_4–Mo–Fe oxide. The new compositional materials were characterized using powder X-ray diffraction (XRD),

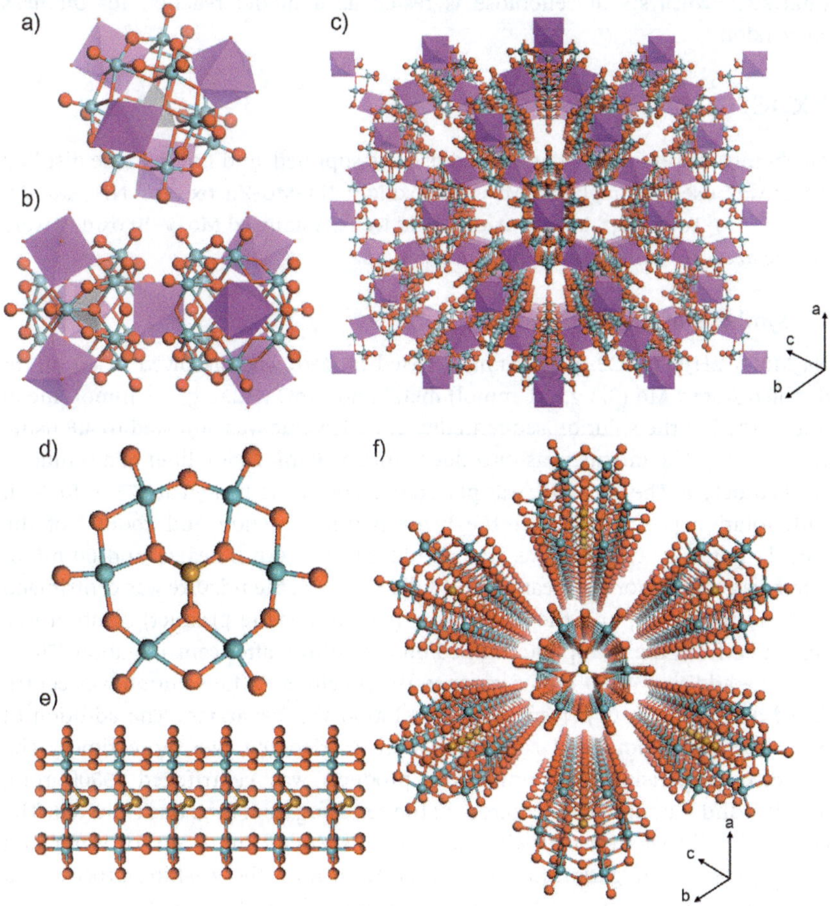

Fig. 1 Structural representation of (a) an ε-Keggin core with four metal ions in the linker site, (b) the connection of ε-Keggin POM units with linker ions, (c) the framework of porous crystalline ε-Keggin POM-based materials, blue sphere: surrounding site, central grey tetrahedron: central site, purple octahedron: linker site, and red sphere: O, (d) a hexagonal POM unit, (e) growing the hexagonal POM unit to form a molecular wire, and (f) the crystal structure formed by a molecular wire, blue sphere: Mo, yellow sphere: Te or Se, and red sphere: O.

Fourier transform infrared (FT-IR) spectroscopy, scanning electron microscopy (SEM), high resolution transmission electron microscopy (HR-TEM), X-ray photoelectron spectroscopy (XPS), and elemental analysis.

NH_4–Mo–Fe oxide is found to adsorb O_2 in its micropores at room temperature. The unique O_2 adsorption is probably on the basis of the formation of peroxo species in the Fe site in the material as indicated in the FT-IR spectra. Mo–V–Bi oxide is found to be a new catalyst member, active and selective for methacrolein oxidation to methacrylic acid using O_2 as an oxidant. Since heat-treatment removed NH_3 from the materials to leave acid sites, the all-inorganic crystalline ε-Keggin POM-based microporous materials and the hexagonal POM-based molecular nanowires can be used as solid acid catalysts. Using these catalysts, hydrolysis of cellobiose is tested as a model reaction for biomass conversion.

Experimental

All chemicals were reagent grade and used as supplied, and house made distilled water was used throughout. Mo–V–Bi oxide,[22] Na–Mo–Zn oxide,[23] NH_4–Mo–Zn oxide,[23] Mo–Te oxide,[25] Mo–Se oxide,[25] and well-crystallized Mo–V–Bi oxide[20] were synthesized according to our previous papers.

1. Synthesis of Na–Mo–Mn oxide

$Na_2MoO_4 \cdot 2H_2O$ (2.823 g, 11.7 mmol based on Mo) was dissolved in 40 mL of distilled water. Mo (0.2 g, 2.1 mmol) metal and MnO (0.235 g, 3.3 mmol) metal were added to the solution sequentially. The pH value was adjusted to 4.8 using H_2SO_4 (1 M). The mixture was introduced into a 50 mL Teflon liner of a stainless-steel autoclave. The autoclave was placed in an oven and heated at 175 °C for 24 h with rotation (~1 rpm). After the hydrothermal reaction and cooling of the autoclave, the crude solid was transferred into a 100 mL beaker and 60 mL of water was added. For purification and solid recovery, the mixture was centrifuged (1700 rpm, 2 min), and the suspension (containing the product) solution was separated from the precipitate formed on the bottom after centrifugation. Then, after the addition of 60 mL of water to the precipitate, the solution was centrifuged and the new upper suspension solution was separated. The addition of water, centrifugation, and separation were carried out two more times. The collected suspension (containing the product) was centrifuged (3500 rpm, 30 min), and the solid on the bottom of the centrifugation tube was collected. The collected solid was washed with water by dispersion in 10 mL of water and subsequent centrifugation (3500 rpm, 30 min). After the washing process had been carried out two more times, the obtained solid was dried at 60 °C overnight. Then, 0.94 g of Na–Mo–Mn oxide (yield: 46% based on Mo) was obtained. Elemental analysis: calcd for $Na_2Mn_{2.6}Mo_{12}O_{47}H_{24.8}$: Mn, 6.75; Mo, 54.39; Na, 2.17; H, 1.17, found: Mn, 6.65; Mo, 54.84; Na, 2.29; H, 1.08.

2. Synthesis of NH_4–Mo–Fe oxide

$(NH_4)_6Mo_7O_{24} \cdot 4H_2O$ (2.060 g, 11.7 mmol based on Mo) was dissolved in 40 mL of distilled water. Mo (0.2 g, 2.1 mmol) metal and Fe_3O_4 (0.258 g, 1.1 mmol) were added to the mixture sequentially (pH was 5.1). The mixture was introduced into a 50 mL Teflon liner of a stainless-steel autoclave. The autoclave was placed in an oven and heated at 175 °C for 48 h with rotation (~1 rpm). The purification process was the same to that of Na–Mo–Mn oxide. Then, 0.87 g of NH_4–Mo–Fe oxide (yield: 44% based on Mo) was obtained. Elemental analysis: calcd for $N_2Fe_{2.6}Mo_{12}O_{43}H_{23.4}$: Fe, 7.10; Mo, 56.29; N, 1.37; H, 1.14, found: Fe, 7.14; Mo, 56.46; N, 1.29; H, 1.00.

3. Synthesis of Mo–V–Bi oxide

Mo–V–Bi oxide was synthesized in nano-size *via* the following method.[22] $Bi(NO_3)_3 \cdot 5H_2O$ (0.68 g, 1.40 mmol) was dissolved in a solution (1.7 mL) of glycerol and water with a volume ratio of 1 : 1. $(NH_4)_6Mo_7O_{24} \cdot 4H_2O$ (1.471 g, 8.33 mmol based on Mo) was dissolved in 20 mL of water. $VOSO_4 \cdot 5H_2O$ (0.5365 g, 2.08 mmol) was dissolved in 20 mL of water. After the solids had been completely dissolved, the solution of $VOSO_4 \cdot 5H_2O$ was rapidly poured into the solution of $(NH_4)_6Mo_7O_{24} \cdot 4H_2O$. After stirring at room temperature for 3 min, $Bi(NO_3)_3$ solution was added. Then, the mixture was stirred for 7 min. The pH of the precursor was adjusted to 3.7 with 28% of ammonia aqueous solution. After the mixture had been purged with N_2 for 10 min, the mixture was refluxed for 48 h in an oil bath with a setting temperature of 110 °C. The black solid was collected by filtration, washed with 20 mL of water 3 times, and dried at 353 K overnight. Then, 1.28 g of Mo–V–Bi oxide (yield: 56% based on Mo) was obtained.

4. Catalytic oxidation of methacrolein

The catalytic oxidation of methacrolein (MAL) to methacrylic acid (MA) was conducted over the Mo–V–Bi oxide using a fixed bed stainless tubular reactor at atmospheric pressure. Prior to the catalytic reaction, the catalyst was heat-treated at 350 °C under 40.5 mL min^{-1} of N_2 flow for 2 h (designated as Mo–V–Bi–NC350). After the heat treatment, the catalyst (0.5 g) was mixed with 2.5 g of silica sand introduced into the stainless tubular reactor. Then, the reactor was heated gradually from room temperature to 280 °C at a rate of 10 °C min^{-1}. When the temperature reached 280 °C, a reaction gas with the composition of $MAL/O_2/(N_2 + He)/H_2O = 1.2/8.0/71.0/18.5$ mL min^{-1} was fed in and stabilized for 30 min before the start of the gas analysis. MAL was supplied by N_2 bubbling with a flow rate of 20.6 mL min^{-1} to a MAL liquid at 0 °C. Water was fed by He bubbling with a flow rate of 30.5 mL min^{-1} to a bottle of hot water (80 °C). The flow rates of MAL and water were determined based on gas chromatographic analysis. The catalytic reaction was carried out at setting temperatures of 280, 270, 260, 250 °C with the reaction temperature decreasing stepwise, and the real reaction temperature was monitored by the thermal-couple inserted in the middle of the catalyst zone. The reactants and the products were analyzed with three online gas chromatographs (Molecular Sieve 13X for O_2, N_2, and CO with a TCD detector, Gascropack54 for H_2O and CO_2 with a TCD detector, and Porapak Q for MAL, CH_3COOH, MA with an FID detector). Blank runs showed that no reaction took place without the catalyst under the experimental conditions in this study. Carbon balance was always *ca.* 97–100% and selectivity was calculated based on the sum of the products.

5. Cellobiose conversion

In a typical reaction, the catalyst (0.01 g), cellobiose (0.05 g), and H_2O (0.5 mL) were added into a 10 mL-test tube. The tube was sealed using a cap with a Teflon liner. The tube was packed in a 300 mL autoclave with a Teflon liner, followed by the addition of 20 mL of water in the autoclave to ensure the pressure-balance of the test tube. The autoclave was heated at 160 °C in an oven for 2 h. After the reaction, the solution in the tube was analyzed using high performance liquid chromatography (HPLC). The products were identified by comparison of the retention time with standard chemicals. The yield was calculated based on the carbon number.

Characterization

Nitrogen (at −196 °C) and oxygen (at 25 °C) adsorption–desorption isotherms were obtained with a BELSORP MAX (BEL, Japan) sorption analyzer. BET surface area was calculated *via* the BET method and the external surface area was calculated by the *t*-plot method using the adsorption branch. The samples were evacuated at 200 °C for 2.5 h before the measurement. The powder X-ray diffraction (XRD) pattern was obtained on a RINT2200 (Rigaku, Japan) with Cu Kα radiation (tube voltage: 40 kV, tube current: 20 mA). SEM images were obtained with a HD-2000 (HITACHI, Japan). HR-TEM images were taken with a 200 kV TEM (JEOL, JEM-2100F, Japan). FT-IR was carried out on a PARAGON 1000, Perkin Elmer. Temperature-programmed desorption mass spectrometry (TPD-MS) measurements were carried out from 30 °C to 600 °C at a heating rate of 10 °C min^{-1} under helium (flow rate: 50 mL min^{-1}). Samples were set up in between two layers of quartz wool. A TPD apparatus (BEL, Japan) equipped with a quadrupole mass spectrometer (M-100QA, Anelva) was used to detect NH_3 ($m/z = 16$) and H_2O ($m/z = 18$). For the TPD-MS measurements of the materials after heat treatment, the samples were heated at 200 °C under high vacuum for 2.5 h in the TPD instrument before the measurements. XPS was performed on a JPS-9010MC (JEOL, Japan). The spectrometer energies were calibrated using the C 1s peak at 284.8 eV. Elemental compositions were determined using an inductive coupling plasma (ICP-AES) method (ICPE-9000, Shimadzu, Japan). CHN elemental composition was determined at the Instrumental Analysis Division, Equipment Management Center, Creative Research Institution, Hokkaido University.

1. Rietveld analysis

Rietveld refinement was performed with the Materials Studio v6.1.0 package (Accelrys Software Inc.). The initial structures of Na–Mo–Mn oxide and NH_4–Mo–Fe oxide were refined by powder XRD Rietveld refinement.[26] The lattice parameters and pattern parameters of the materials were refined by Pawley refinement first. The initial structures were from Mo–V–Bi oxide.[20] Then, isotropical temperature factors were given for every atom in the initial structures. Rietveld analysis was started with the initial models of the materials and lattice parameters and pattern parameters from the Pawley refinement. Every atom position was refined. Occupancy of the atoms in the framework was fixed without further refinement and occupancies of atoms in micropores were refined with consideration of the elemental analysis results. Finally, the pattern parameters were refined again to obtain the lowest R_{wp} value. The crystallographic data are provided in Table S1–S3.†

Results and discussion

1. Synthesis of ε-Keggin POM-based porous metal oxides

The crystalline ε-Keggin polyoxomolybdate-based porous metal oxides were synthesized *via* the hydrothermal method. Three kinds of precursor compounds, the Mo source, the reducing agent, and the third metal ion, were necessary for the materials. As shown in Table 1, different metal ions were used as the precursors for the materials, which yielded 6 iso-structural materials with different chemical

Table 1 Comparison of the synthetic conditions for the POM-based microporous materials

	Mo source[a]	Reductant	The third metal ion	Additional process[b]	Hydrothermal conditions	Reference
Mo–V–Bi oxide	AHM	$VOSO_4$	$Bi(OH)_3$	—	175 °C, 48 h	20 and 22
Na–Mo–Zn oxide	Na_2MoO_4	Mo	$ZnCl_2$	pH ~ 4.8	175 °C, 24 h, rotation	21 and 23
NH_4–Mo–Zn oxide	AHM	Mo	$ZnCl_2$	pH ~ 4.8	175 °C, 24 h, rotation	23
Na–Mo–Mn oxide	Na_2MoO_4	Mo	MnO	pH ~ 4.8	175 °C, 24 h, rotation	This work
NH_4–Mo–Mn oxide	AHM	Mo	MnO	pH ~ 4.8	175 °C, 24 h	21
NH_4–Mo–Fe oxide	AHM	Mo	Fe_3O_4	—	175 °C, 48 h, rotation	This work

[a] AHM: ammonium heptamolybdate. [b] In some cases, pH was adjusted by the addition of 1 M H_2SO_4.

compositions. The Mo sources, $(NH_4)_6Mo_7O_{24}\cdot 4H_2O$ (AHM) and $Na_2MoO_4\cdot 2H_2O$, with reducing agents produced ε-Keggin POMs in aqueous solution, which is well-known in POM chemistry.[27] The produced ε-Keggin POMs were further connected by metal ions to form the materials.

2. Structures of crystalline POM-based microporous materials

The ε-Keggin POM-based microporous materials can incorporate different transition metal ions. The chemical composition can be tuned in each site in the materials. The iso-structural materials were synthesized. As shown in Fig. 2A, the powder XRD patterns of the two new materials, Na–Mo–Mn oxide and NH_4–Mo–Fe oxide, were similar to those of the previously reported materials. The space group of the materials was $Fd\bar{3}m$ with slightly different lattice parameters (Table 2). The XRD patterns demonstrated that the basic structures of the materials were similar. FT-IR spectra of the materials were also similar, indicating that the POM molecular structures of the materials were similar (Fig. 2B).

The elemental analysis exhibited Na : Mn : Mo = 2 : 2.6 : 12 for Na–Mo–Mn oxide and N : Fe : Mo = 2 : 2.6 : 12 for NH_4–Mo–Fe oxide, respectively. The proposed structures of Na–Mo–Mn oxide and NH_4–Mo–Fe oxide were on the basis of the ε-Keggin POM units ($Mn_{0.6}Mo_{12}O_{40}$ or $Fe_{0.6}Mo_{12}O_{40}$) where the central sites were partially occupied with the Mn or Fe ions and the surrounding sites were with the Mo ions. The POM units ($Mn_{0.6}Mo_{12}O_{40}$ or $Fe_{0.6}Mo_{12}O_{40}$) were linked by Mn and Fe, respectively. Similar structure models of NH_4–Mo–Zn oxide and NH_4–Mo–Mn oxide were reported.[21] The frameworks of the materials were negatively charged, so that there were cations, Na^+ (Na–Mo–Mn oxide), NH_4^+ (NH_4–Mo–Fe oxide), and water in the void spaces. The simulated XRD patterns using the Rietveld refinement were almost identical to the experimental patterns, indicating that the proposed structures were correct (Fig. S1†). There was no additional peak being observed in comparison to the experimental pattern with the simulated pattern, indicating that the powder samples were pure.

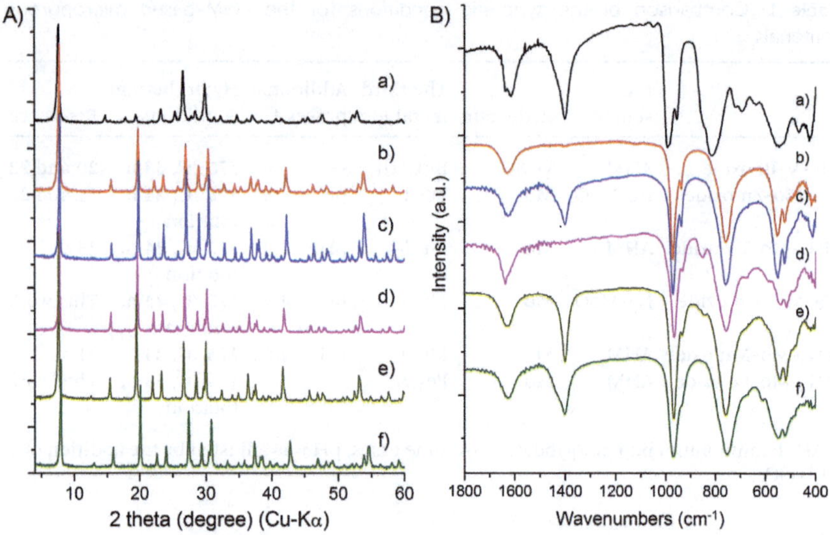

Fig. 2 (A) XRD patterns of (a) Mo–V–Bi oxide, (b) Na–Mo–Zn oxide, (c) NH_4–Mo–Zn oxide, (d) Na–Mo–Mn oxide, (e) NH_4–Mo–Mn oxide, and (f) NH_4–Mo–Fe oxide and (B) FT-IR spectra of (a) Mo–V–Bi oxide, (b) Na–Mo–Zn oxide, (c) NH_4–Mo–Zn oxide, (d) Na–Mo–Mn oxide, (e) NH_4–Mo–Mn oxide, and (f) NH_4–Mo–Fe oxide.

The oxidation states of the metal elements in Na–Mo–Mn oxide and NH_4–Mo–Fe oxide were studied using XPS (Fig. S2† and Table 2). The surrounding 12 Mo ions were mostly reduced to Mo^V. In Na–Mo–Mn oxide, Mn^{II} was present (Table 2). In the case of NH_4–Mo–Fe oxide, the ratio of Fe^{III}/Fe^{II} was 0.4/2.2. Considering the results of the elemental analysis and the oxidation state analysis, the detailed chemical formulae of the materials can be estimated (Table 2).

The SEM images (Fig. 3a and c) of Na–Mo–Mn oxide and NH_4–Mo–Fe oxide show that the materials are polyhedron-shaped crystallites with a particle size of

Table 2 The chemical properties of Na–Mo–Mn oxide and NH_4–Mo–Fe oxide

	Na–Mo–Mn oxide	NH_4–Mo–Fe oxide
Crystal system	Cubic	Cubic
Space group	$Fd\bar{3}m$	$Fd\bar{3}m$
Lattice parameter	19.7047	19.4012
Surrounding site	$Mo^{VI}/Mo^{V} = 2/10$	$Mo^{VI}/Mo^{V} = 3/9$
Central site	Mn^{II}	$Fe^{III}/Fe^{II} = 0.4/2.2$
Linker site	Mn^{II}	$Fe^{III}/Fe^{II} = 0.4/2.2$
Cation	Na^+	NH_4^+
Chemical formula	$Na_2H_{10.8}[Mn^{II}_{2.6}Mo^{VI}_2Mo^{V}_{10}O_{40}]\cdot 7H_2O$	$(NH_4)_2H_{9.4}[Fe^{III}_{0.4}Fe^{II}_{2.2}Mo^{V}_9Mo^{VI}_3O_{40}]\cdot 3H_2O$
BET surface area ($m^2\ g^{-1}$)	22	20
Pore volume ($cm^3\ g^{-1}$)	0.0051	0.0058

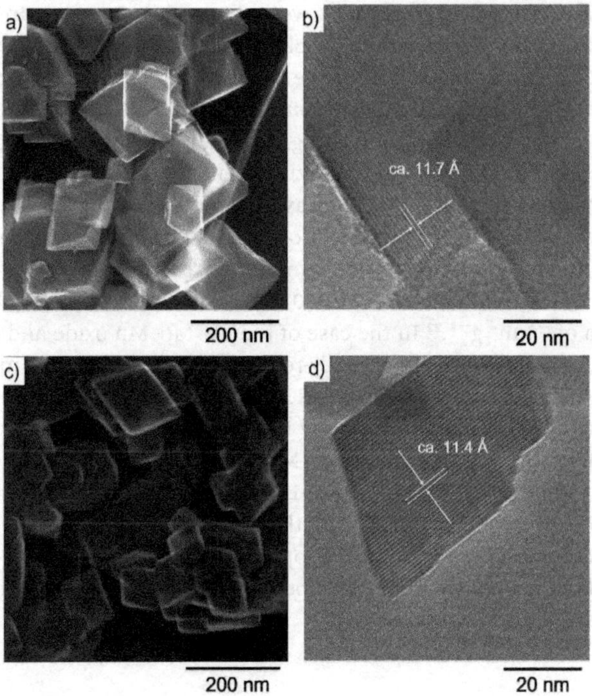

Fig. 3 (a) SEM image and (b) HR-TEM image of Na–Mo–Mn oxide and (c) SEM image and (d) HR-TEM image of NH_4–Mo–Fe oxide.

100–200 nm in diameter. The HR-TEM images further confirmed the structures of the materials (Fig. 3b and d) and showed the clear lattice images of the materials, illustrating that the materials are well-ordered sub-micrometer-sized single crystals. Layers were observed in the HR-TEM images, which corresponded to the (111) plane of the materials. The distances of the (111) plane obtained from the HR-TEM images, *ca.* 11.7 Å for Na–Mo–Mn oxide and *ca.* 11.4 Å for NH_4–Mo–Fe oxide, were in good agreement with the results from the crystal structures of the materials.

3. Microporosity of the materials

The presence of water and NH_4^+ in the POM-based materials was confirmed by FT-IR analysis (Fig. 2B). NH_4–Mo–Fe oxide showed peak maxima at *ca.* 1630 cm^{-1} and 1400 cm^{-1}, which corresponded to water and NH_4^+, respectively. In Na–Mo–Mn oxide, the peak at 1630 cm^{-1} that corresponded to water was observed. The TPD-MS analysis showed that water and NH_4^+ in the materials desorbed during a temperature increase (Fig. S3†). The mass numbers of $m/z = 16$ and $m/z = 18$ were attributed to the signals of desorbed NH_3 and water. Water desorbed from the materials from 30 °C to 600 °C in both cases. NH_4–Mo–Fe oxide showed a peak at 320–380 °C for NH_3 desorption in the TPD profile ($m/z = 16$).

The structure of Mo–V–Bi oxide remained after calcination at 350 °C for 2 h, denoted as Mo–V–Bi–NC350.[20] On the other hand, Na–Mo–Mn oxide and NH_4–Mo–Fe oxide were stable after calcination at 200 °C for 2 h under a N_2 flow

(Fig. S4†), and the calcined samples were denoted as Na–Mo–Mn–NC200 and NH_4–Mo–Fe–NC200. The remaining molecules in the materials after calcination were investigated by comparison of the TPD-MS profiles before and after calcination (Fig. S3†). The peaks of water ($m/z = 18$) and NH_3 ($m/z = 16$) decreased remarkably after calcination, indicating that the guest molecules in the micropores were partly removed by the heat treatment.

Nitrogen adsorption–desorption measurements were conducted for Mo–V–Bi–NC350, Na–Mo–Mn–NC200 and NH_4–Mo–Fe–NC200 (Fig. 4). In the case of Mo–V–Bi oxide, all guest molecules were removed by calcination at 350 °C without structure collapse, and Mo–V–Bi–NC350 was a microporous material with a BET surface area of 75 $m^2\ g^{-1}$.[22] In the case of the Na–Mo–Mn oxide and NH_4–Mo–Fe oxide the NH_3 and water were partially removed. Nevertheless, the micropores of Na–Mo–Mn–NC200 and NH_4–Mo–Fe–NC200 were also opened (Fig. 4) and the surface areas were calculated using the BET method to be 22 $m^2\ g^{-1}$ and 20 $m^2\ g^{-1}$, for Na–Mo–Mn–NC200 and NH_4–Mo–Fe–NC200, respectively, which were apparently smaller than that of Mo–V–Bi–NC350. The high BET surface area of Mo–V–Bi–NC350 mainly resulted from the highly opened micropores due to the high calcination temperature. The less opened micropores of Na–Mo–Mn–NC200 and NH_4–Mo–Fe–NC200 might be ascribed to the remaining NH_4^+ and Na^+, which blocked the micropores.

4. Redox properties and O_2 adsorption

The ε-Keggin POM-based microporous materials show redox properties in heating conditions under an H_2 or O_2 atmosphere, because the Mo ions in the materials are a mixture of Mo^V and Mo^{VI}, which can be oxidized to Mo^{VI} and reduced to Mo^V, respectively.[21]

An interesting O_2 adsorption phenomenon of NH_4–Mo–Fe oxide was observed. The materials used for O_2 adsorption were calcined at 200 °C for 2.5 h under high

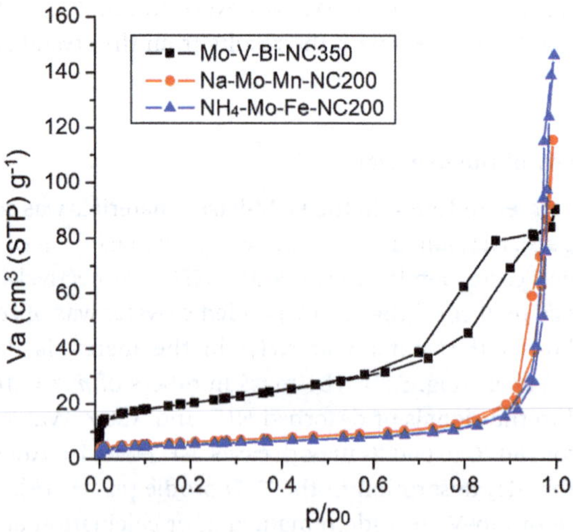

Fig. 4 The N_2 adsorption–desorption isotherms of Mo–V–Bi–NC350, Na–Mo–Mn–NC200, and NH_4–Mo–Fe–NC200.

vacuum before the experiments were conducted, denoted as NH_4–Mo–Fe–NC200. Calcined Na–Mo–Zn oxide and NH_4–Mo–Zn oxide were denoted as Na–Mo–Zn–NC200 and NH_4–Mo–Zn–NC200. NH_4–Mo–Fe–NC200 showed an interesting O_2 adsorption property at room temperature. The adsorption–desorption of O_2 isotherms in Fig. 5a and b shows that the O_2 adsorption in NH_4–Mo–Fe–NC200 increased quickly in the initial adsorption step when the O_2 pressure was low. When the adsorbed amount reached the step, the O_2 uptake in the material almost stopped even when O_2 pressure was still increasing. Most of the adsorbed O_2 in NH_4–Mo–Fe–NC200 could not desorb from the materials. For other isostructural materials of NH_4–Mo–Fe oxide, O_2 adsorption behavior was completely different from that of NH_4–Mo–Fe oxide, although the surface areas (Table S4†) and basic structures were similar. The result indicated that the unique O_2 adsorption was dependent on the chemical composition, and only the Fe containing material was active for O_2 adsorption. The Fe site in the material played an important role in O_2 adsorption.

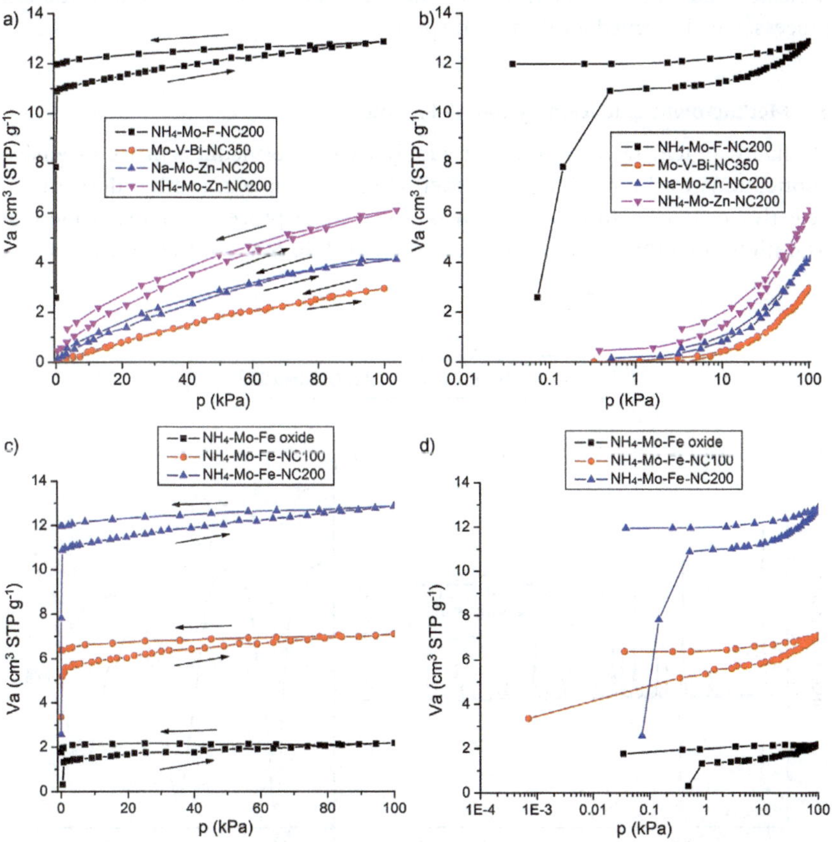

Fig. 5 O_2 adsorption isotherms of different ε-Keggin POM-based microporous materials under (a) normal pressure and (b) low pressure at 25 °C. O_2 adsorption isotherms of NH_4–Mo–Fe oxide with different pre-treatment temperatures, at (c) normal pressure and (d) low pressure at 25 °C.

The pre-treatment temperature affected the amount of adsorbed O_2 in NH_4–Mo–Fe oxide. As shown in Fig. 5c and d, as the pre-treatment temperature increased, the total amount of adsorbed O_2 increased. A high pre-treatment temperature removed more water and NH_3 from the material and opened the micropores of the material (Fig. S3†). The result indicated that the O_2 adsorption occurred in the micropores of NH_4–Mo–Fe–NC200.

After O_2 adsorption, NH_4–Mo–Fe oxide was characterized. XRD patterns (Fig. 6a) show that the basic structure of the material did not change, because all the characteristic diffraction peaks remained. The XRD peaks shifted to a higher 2 theta value, indicating that the lattice parameter decreased after O_2 adsorption.

Adsorption of O_2 in NH_4–Mo–Fe–NC200 at room temperature gave rise to a different FT-IR spectrum compared with that of the as-synthesized material in Fig. 6b, which can be explained in terms of the binding state change in the material. A new IR band emerged at *ca.* 797 cm^{-1}, corresponding to a ν(O–O) vibrational mode, which indicated the formation of a peroxo species, and it might coordinate to the Fe site (Fe–O–O).[28,29] The band at 970 cm^{-1} corresponding to Mo=O shifted to a higher wavenumber (*ca.* 978 cm^{-1}), which demonstrated that the bond length of Mo=O shortened due to dehydration. The FT-IR spectra indicated that the O_2 adsorption in the material was not a simple oxidation process, which formed an Fe peroxo species.

5. Methacrolein oxidation by Mo–V–Bi oxide

Methacrylic acid (MA) becomes a widely used intermediate in its extensive applications to the production of methyl methacrylate and other derivatives. Selective oxidation to produce MA is a two-stage process. Firstly, isobutene oxidation of methacrolein (MAL), and then MAL to MA. For the second step of

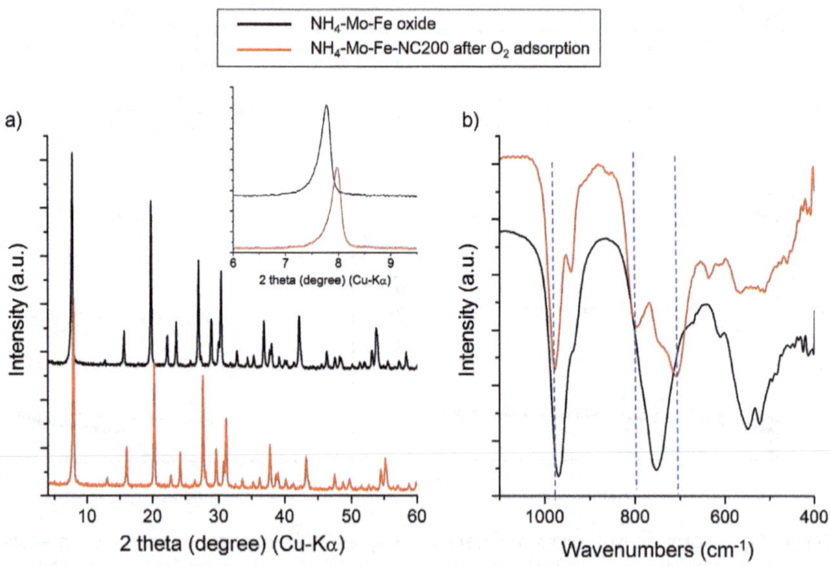

Fig. 6 (a) XRD patterns, inset figure: enlarged XRD patterns and (b) FT-IR spectra of NH_4–Mo–Fe oxide before and after O_2 adsorption.

the reaction, both oxidation ability and solid acidity are essential, so that Keggin type POM system has been widely used. However, the catalytic performance of the POM catalysts is not satisfactory, although commercialized, and still needs improvement. Therefore, new types of oxide catalysts have been long desired for the reaction. Based on our recent discovery of a new ε-Keggin POM-based porous material and considering that Mo–V–Bi oxide has constituents necessary for the selective oxidation of MAL, we applied this material as a catalyst to the reaction.

We first tested the Mo–V–Bi oxide with crystal particles in nanosize (20–50 nm in diameter[22]) for the MAL oxidation after calcination at 350 °C in N_2 stream (designated as Mo–V–Bi–N350(nano)). Conversions of MAL and O_2 and selectivity to MA, acetic acid (AcOH), CO, and CO_2 are shown in Fig. 7. Very interestingly, the reaction appreciably occurred over the catalyst even at 228 °C and yielded MA with 60% selectivity, which is comparable to that of a normal Keggin POM catalyst. The MAL conversion increased with increasing reaction temperature and was 28.1% at 270 °C. The MA selectivity gradually decreased as the reaction temperature increased. The AcOH selectivity, on the other hand, lineally increased with the reaction temperature and was 21.2% and 31.4% at 220 °C and 270 °C, respectively. The CO and CO_2 selectivity also increased with increasing reaction temperature. After the catalytic reaction, no changes in the XRD patterns and FT-IR spectra were observed (Fig. S5†), indicating that the basic crystal structure of the catalyst was maintained during the reaction. Evidently, the Mo–V–Bi oxide can be a new member of the oxidation catalysts for the selective oxidation of MAL.

The pore size of Mo–V–Bi oxide is 3.4 nm,[20] which is much smaller than the size of MAL, for which the molecular size was calculated to be 4.9 nm in minimum as calculated using the DMol^3 program, and thus the reaction would take place on the external surface of the material. To test that the catalytic activity was dependent on the external surface area of the materials, a well-crystallized Mo–V–Bi–NC350 with a large particle size (1 μm in diameter) was synthesized (Fig. S5Ac†)[20] and used for comparison. The activity of Mo–V–Bi–NC350 (large) was poor as expected. MAL conversion at 228 °C was *ca.* 5%, which was half of that of Mo–V–Bi–NC350(nano). The selectivity to MA was a little affected by the size of the catalyst particles.

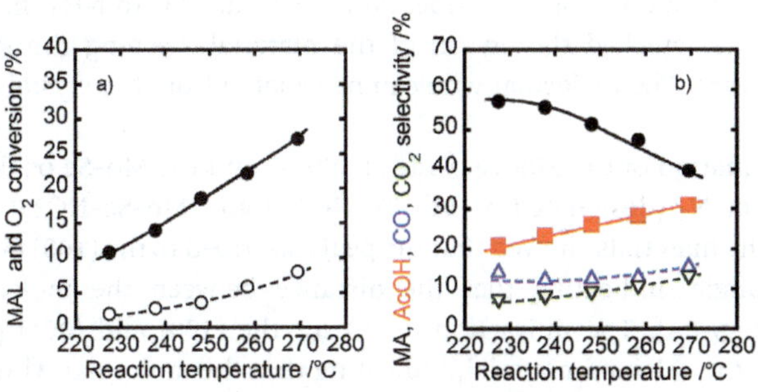

Fig. 7 MAL oxidation catalyzed by Mo–V–Bi–NC350(nano) with O_2 as an oxidant. The conversion of MAL (closed circle) and O_2 (open circle) in (a) and the selectivity (mol %) to MA (closed circle), AcOH (closed square), CO (triangle), and CO_2 (reverse triangle) in (b) as a function of the reaction temperature.

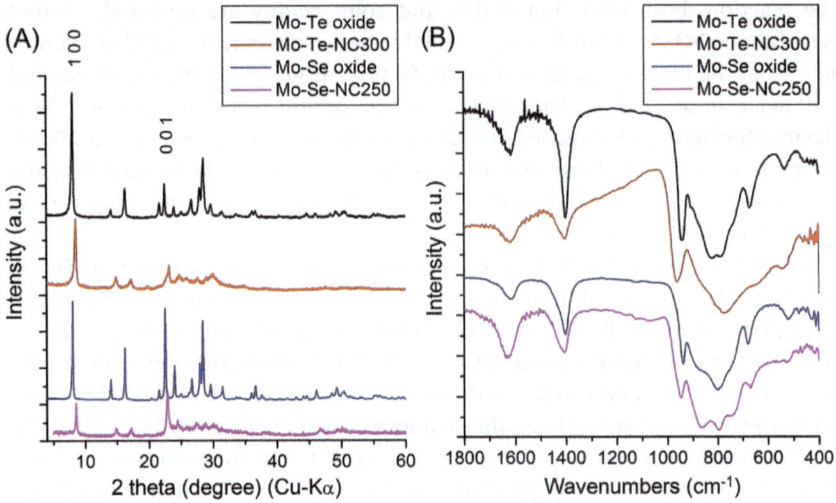

Fig. 8 (A) XRD patterns of (a) Mo–Te oxide and (b) Mo–Te–NC300 and (B) FT-IR spectra of (a) Mo–Te oxide and (b) Mo–Te–NC300.

6. Synthesis and characterization of crystalline metal oxides based on a molecular wire

Using the VI A group metal ions, such as Te and Se, a new POM-based molecular wire structure was prepared. The hydrothermal synthesis of $(NH_4)_6Mo_7O_{24}\cdot 4H_2O$ with Te^{IV} and Se^{IV} ions yielded novel metal oxides, Mo–Te oxide and Mo–Se oxide. The XRD patterns and FT-IR spectra are shown in Fig. 8. Six MoO_6 units surround one Te or Se ion in the *a*–*b* plane, which form a unit of $[XMo_6O_{21}]^{2-}$ (X = Te or Se). The hexagonal units stack along the *c* axis to form prismatic clusters as molecular wires. The molecular wires assembled in parallel in a hexagonal fashion to form the crystalline material with NH_4^+ and water, which existed in between the molecular wires and interacted with them (Fig. 1d–f).

The existence of water and NH_4^+ in Mo–Te oxide and Mo–Se oxide was confirmed by FT-IR (Fig. 8B), which exhibited bands at *ca.* 1630 cm^{-1} for water and at *ca.* 1400 cm^{-1} for NH_4^+. Water and NH_4^+ stabilized the crystal structures of the Mo–Te oxide and the Mo–Se oxide. Proton-exchange with NH_4^+ in the Mo–Te oxide effectively cracked the crystal of the material, keeping the structure of a molecular wire. The molecular wire can be isolated from the crystal and can be observed.[25]

After calcination of the Mo–Te oxide at 300 °C and the Mo–Se oxide at 250 °C for 2 h under N_2 (designated as Mo–Te–NC300 and Mo–Se–NC250), the XRD patterns of the materials showed that the peaks ascribed to the (100) plane shifted to a high angle, indicating that the distance between the molecular wires decreased (Fig. 8A). The diffraction peaks for the (100) and (001) planes still remained in the calcined material, indicating that the molecular wires were still in the structures after calcination. A comparison of the FT-IR spectra before and after calcination indicated that the molecular structure of the Mo–Te oxide was stable after calcination at 300 °C. After calcination, the IR peak for NH_4^+

(1400 cm^{-1}) decreased remarkably, and NH_3 was removed from the material to leave protons as acid sites.

7. Catalytic hydrolysis of cellobiose

Biomass utilization is of great importance, because of increasing attention to the consumption of energy resources for sustainable society and development. Cellulose is a big class of biomass, and conversion of cellulose to useful chemicals is one of the most important biomass conversion routes. The first step of cellulose conversion is hydrolysis of the glycosidic bond to form hexose that further transforms to other chemicals under certain reaction conditions. Brønsted acids are active for cellulose hydrolysis.[30–32] Cellobiose, formed by two glucose molecules, is an ideal model substrate for the investigation of glycosidic bond hydrolysis, and the novel complex metal oxides mentioned above are expected to be active for the reaction.

After calcination, NH_3 desorbed from the porous POMs and the transition metal oxide molecular wires, and protons emerged in the materials which may act as acid sites. The materials were used as catalysts for the hydrolysis of cellobiose (Table 3). The identified products in the resulting solution are glucose, mannose, formic acid (FA), AcOH, and 5-hydroxymethylfurfural (HMF). Glucose was a direct product formed by the hydrolysis of the glycosidic bond of cellobiose. Mannose was formed by the isomerization of glucose in aqueous solution. HMF was formed by further dehydration of hexoses. The fragment molecules such as FA and AA are formed by C–C bond cleavage. The undetectable products might be ascribed to humin, which commonly forms in biomass conversion.[33,34] Mo–V–Bi–NC350 showed the best performance among the ε-Keggin POM-based porous materials (Table 3, entries 1–3). For Mo–Te–NC300 and Mo–Se–NC250, the conversion of cellobiose and the yield of glucose and mannose were higher than all the ε-Keggin POM-based porous materials (Table 3, entries 4 and 5), which might be due to well-dispersed catalysts in the reaction media. The activity was low when no catalyst was used (Table 3, entry 6). After the reaction, the XRD patterns and FT-IR spectra of the porous POMs were identical, indicating that the materials were stable (Fig. S6†). The nanowires were too small to be recovered.

Table 3 Hydrolysis of cellobiose catalyzed by different materials[a]

Entry	Catalyst	Conversion (%)	Yield based on carbon (%)					
			Glucose	Mannose	FA	AcOH	HMF	Other
1	Mo–V–Bi–NC350	50.7	14.9	12.9	2.8	0.5	1.3	18.3
2	NH_4–Mo–Mn–NC200	37.6	4.1	4.5	0.6	0.4	0.7	27.3
3	NH_4–Mo–Fe–NC200	50.4	10	10.3	0.9	0.3	1.0	27.9
4	Mo–Te–NC300	74.4	16.1	16.6	1.8	0.3	1.6	38.0
5	Mo–Se–NC250	90.0	18.1	20.4	1.9	0.4	2.3	46.9
6	—	7.4	6.8	0	0	0	0.6	0

[a] Reaction conditions: hydrolysis of cellobiose catalyzed by different materials, reaction conditions: cellobiose (0.05 g), catalyst (0.01 g), water (0.5 mL), reaction time (2 h), and reaction temperature (160 °C).

Conclusion

Assembly of Mo POM units with octahedra of other elements forms POM unit-structured materials. By this assembling methodology, two kinds of novel complex Mo oxides are constructed, all-inorganic ε-Keggin POM-based 3D microporous materials and molecular wire-based crystalline metal oxides. The results strongly indicate that the use of POM units is an effective approach for the design and development of novel transition metal oxide catalysts.

In this research, two new all-inorganic ε-Keggin POM-based 3D framework materials, Na–Mo–Mn oxide and NH_4–Mo–Fe oxide, were synthesized and characterized. Structural analysis clearly demonstrates that ε-Keggin POMs are connected with metal ions (Mn and Fe) to form $\{H_{10.8}[Mn^{II}_{2.6}Mo^{VI}_2Mo^{V}_{10}O_{40}]\}^{2-}$ and $\{H_{9.4}[Fe_{2.6}Mo^{V}_9Mo^{VI}_3O_{40}]\}^{2-}$. These two materials are iso-structural to the reported Mo–V–Bi oxide. Furthermore, a unique O_2 adsorption phenomenon is observed in NH_4–Mo–Fe oxide at 25 °C. The adsorption of O_2 in the Fe site may form Fe peroxo species. Mo–V–Bi oxide shows activity for MAL oxidation to produce MA with O_2. The novel materials, ε-Keggin POM-based porous materials and the molecular wire-based metal oxides, are effective solid acids for the hydrolysis of cellobiose in aqueous solution. The present research provides a new approach using POM as units to construct new transition metal oxides for catalysis.

Acknowledgements

This work was financially supported by a Grant-in-Aid for Scientific Research (A) (grant No. 2324-6135) from the Ministry of Education, Culture, Sports, Science, and Technology, Japan.

References

1 M. Sadakane and W. Ueda, Ordered Porous Crystalline Transition Metal Oxides, in *Porous Materials*, ed. D. W. Bruce, D. O'Hare, and R. I. Walton, Wiley, 2011, vol. 11, pp. 147–216.
2 Y. Kamiya, M. Sadakane, and W. Ueda, Heteropoly Compounds, in *Comprehensive Inorganic Chemistry II*, ed. J. Reedijk and K. Poeppelmeier, Oxford, Elsevier, 2013, vol. 7, pp. 185–204.
3 M. T. Pope, *Heteropoly and Isopoly Oxometalates*, Springer-Verlag, Berlin, 1983.
4 Special thematic issue on polyoxometalates, *Chem. Soc. Rev.*, ed. L. Cronin and A. Müller, 2012, vol. 41, pp. 7325–7648.
5 Special thematic issue on polyoxometalates, *Chem. Rev.*, ed. C. L. Hill, 1998, vol. 98, pp. 1–390.
6 H. N. Miras, L. Vilà-Nadal and L. Cronin, *Chem. Soc. Rev.*, 2014, **43**, 5679–5699.
7 D.-L. Long, R. Tsunashima and L. Cronin, *Angew. Chem., Int. Ed.*, 2010, **49**, 1736–1758.
8 B. Nohra, H. El Moll, L. M. Rodriguez Albelo, P. Mialane, J. Marrot, C. Mellot-Draznieks, M. O'Keeffe, R. N. Biboum, J. Lemaire, B. Keita, L. Nadjo and A. Dolbecq, *J. Am. Chem. Soc.*, 2011, **133**, 13363–13374.
9 L. M. Rodriguez-Albelo, A. Rabdel Ruiz-Salvador, A. Sampieri, D. W. Lewis, A. Gomez, B. Nohra, P. Mialane, J. Marrot, F. Secheresse, C. Mellot-

Draznieks, R. N. Biboum, B. Keita, L. Nadjo and A. Dolbecq, *J. Am. Chem. Soc.*, 2009, **131**, 16078–16087.
10 M. Sadakane, K. Kodato, T. Kuranishi, Y. Nodasaka, K. Sugawara, N. Sakaguchi, T. Nagai, Y. Matsui and W. Ueda, *Angew. Chem., Int. Ed.*, 2008, **47**, 2493–2496.
11 M. Sadakane, K. Yamagata, K. Kodato, K. Endo, K. Toriumi, Y. Ozawa, T. Ozeki, T. Nagai, Y. Matsui, N. Sakaguchi, W. D. Pyrz, D. J. Buttrey, D. A. Blom, T. Vogt and W. Ueda, *Angew. Chem., Int. Ed.*, 2009, **48**, 3782–3786.
12 M. Sadakane, N. Watanabe, T. Katou, Y. Nodasaka and W. Ueda, *Angew. Chem., Int. Ed.*, 2007, **46**, 1493–1496.
13 S. G. Mitchell, C. Streb, H. N. Miras, T. Boyd, D.-L. Long and L. Cronin, *Nat. Chem.*, 2010, **2**, 308–312.
14 S. G. Mitchell, T. Boyd, H. N. Miras, D.-L. Long and L. Cronin, *Inorg. Chem.*, 2011, **50**, 136–143.
15 S. Uchida, R. Eguchi and N. Mizuno, *Angew. Chem., Int. Ed.*, 2010, **49**, 9930–9934.
16 S. Uchida, M. Hashimoto and N. Mizuno, *Angew. Chem., Int. Ed.*, 2002, **41**, 2814–2817.
17 S. Uchida, S. Hikichi, T. Akatsuka, T. Tanaka, R. Kawamoto, A. Lesbani, Y. Nakagawa, K. Uehara and N. Mizuno, *Chem. Mater.*, 2007, **19**, 4694–4701.
18 S. Uchida and N. Mizuno, *J. Am. Chem. Soc.*, 2004, **126**, 1602–1603.
19 S. Uchida and N. Mizuno, *Coord. Chem. Rev.*, 2007, **251**, 2537–2546.
20 Z. Zhang, M. Sadakane, T. Murayama, S. Izumi, N. Yasuda, N. Sakaguchi and W. Ueda, *Inorg. Chem.*, 2014, **53**, 903–911.
21 Z. Zhang, M. Sadakane, T. Murayama, N. Sakaguchi and W. Ueda, *Inorg. Chem.*, 2014, **53**, 7309–7318.
22 Z. Zhang, M. Sadakane, T. Murayama and W. Ueda, *Dalton Trans.*, 2014, **43**, 13584–13590.
23 Z. Zhang, M. Sadakane, S. Noro, T. Murayama, T. Kamachi, K. Yoshizawa and W. Ueda, *J. Mater. Chem. A*, 2015, **3**, 746–755.
24 Z. Zhang, M. Sadakane and W. Ueda, Assembly of Epsilon-Keggin-Type Polyoxometalates to form Crystalline Microporous Metal Oxides in Trends, in *Polyoxometalates Research*, ed. L. Ruhlmann and D. Schaming, Nova Science Publishers, 2015.
25 Z. Zhang, T. Murayama, M. Sadakane, H. Ariga, N. Yasuda, N. Sakaguchi, K. Asakura and W. Ueda, *Nat. Commun.*, 2015, **6**, 7731.
26 R. A. Young, *The Rietveld method*, Oxford University Press, Oxford, 1995.
27 P. Mialane, A. Dolbecq, L. Lisnard, A. Mallard, J. Marrot and F. Secheresse, *Angew. Chem., Int. Ed.*, 2002, **41**, 2398–2401.
28 E. D. Bloch, L. J. Murray, W. L. Queen, S. Chavan, S. N. Maximo, J. P. Bigi, R. Krishna, V. K. Peterson, F. Grandjean, O. G. J. Long, B. Smit, S. Bordiga, C. M. Brown and R. Long, *J. Am. Chem. Soc.*, 2011, **133**, 14814–14822.
29 D. Barats, G. Leitus, R. Popovitz-Biro, L. J. W. Shimon and R. Neumann, *Angew. Chem., Int. Ed.*, 2008, **47**, 9908–9912.
30 K. Shimizu, H. Furukawa, N. Kobayashi, Y. Itaya and A. Satsuma, *Green Chem.*, 2009, **11**, 1627–1632.
31 S. Suganuma, K. Nakajima, M. Kitano and D. Yamaguchi, *J. Am. Chem. Soc.*, 2008, **130**, 12787–12793.
32 Y.-B. Huang and Y. Fu, *Green Chem.*, 2013, **15**, 1095–1111.

33 R. Weingarten, W. C. Conner and G. W. Huber, *Energy Environ. Sci.*, 2012, **5**, 7559–7574.

34 R. Weingarten, Y. T. Kim, G. a. Tompsett, A. Fernández, K. S. Han, E. W. Hagaman, W. C. Conner, J. a. Dumesic and G. W. Huber, *J. Catal.*, 2013, **304**, 123–134.

Faraday Discussions

PAPER

Hydrothermal synthesis of bi-functional nanostructured manganese tungstate catalysts for selective oxidation†

Xuan Li,[ab] Thomas Lunkenbein,[a] Jutta Kröhnert,[a] Verena Pfeifer,[ac] Frank Girgsdies,[a] Frank Rosowski,[bd] Robert Schlögl[a] and Annette Trunschke*[a]

Received 23rd November 2015, Accepted 8th December 2015
DOI: 10.1039/c5fd00191a

The mechanism of C–H activation in selective oxidation reactions of short-chain alkane molecules over transition metal oxides is critically affected by the balance of acid–base and redox sites at the surface of the catalyst. Using the example of manganese tungstate we discuss how the relative abundance of these sites can be controlled *via* synthetic techniques. Phase-pure catalysts composed of the thermodynamic stable monoclinic $MnWO_4$ phase have been prepared using hydrothermal synthesis. Variation of the initial pH value resulted in rod-shaped nano-crystalline $MnWO_4$ catalysts composed of particles with varying aspect ratio. The synthesis products have been analysed using transmission electron microscopy, X-ray diffraction, infrared, and photoelectron spectroscopy. *In situ* Raman spectroscopy was used to investigate the dissolution–re-crystallization processes occurring under hydrothermal conditions. Ethanol oxidation was applied to probe the surface functionalities in terms of acid–base and redox properties. Changes in the aspect ratio of the primary catalyst particles are reflected in the product distribution induced by altering the fraction of acid–base and redox sites exposed at the surface of the catalysts in agreement with the proposed mechanism of particle growth by re-crystallization during ageing under hydrothermal conditions.

Introduction

Metal oxides are widely used as heterogeneous catalysts in the synthesis of chemicals, and in energy conversion and storage applications.[1–4] High

[a]*Department of Inorganic Chemistry, Fritz-Haber-Institut der Max-Planck-Gesellschaft, Faradayweg 4-6, 14195, Berlin, Germany. E-mail: trunschke@fhi-berlin.mpg.de*
[b]*UniCat-BASF Joint Lab, Technische Universität Berlin, 10623, Berlin, Germany*
[c]*Helmholtz-Zentrum Berlin für Materialien und Energie GmbH, Elektronenspeicherring BESSY II, Albert-Einstein-Str. 15, 12489 Berlin, Germany*
[d]*BASF SE, Process Research and Chemical Engineering, Heterogeneous Catalysis, Carl-Bosch-Straße 38, 67056, Ludwigshafen, Germany*

† Electronic supplementary information (ESI) available: SEM, TEM, XRD, FTIR, DRIFTS, XPS. See DOI: 10.1039/c5fd00191a

performance, selectivity, and stability of oxides in heterogeneous catalysis are bound to the homogeneity of the solid.[5] In complex reactions, such as selective oxidation of hydrocarbons, multi-functionality is necessarily required and can be achieved either chemically or by nano-structuring.[6,7]

Solvothermal techniques have been efficiently used in the preparation of metastable phases or oxides with particular morphological properties.[8–13] However, the underlying preparation strategies are often based on experience and parameter variation. Targeted design of metal oxide catalysts with predictable functionalities needs a deeper understanding of the chemistry in precursor solutions and during nucleation, growth, and ageing.

In situ Raman spectroscopy has proven beneficial in the investigation of condensation reactions of transition metal oxide species occurring at elevated temperature and pressure inside an autoclave.[14,15] Our recent study concerning the speciation of molybdates in aqueous media revealed that the molecular structure of the oxo-anions in the temperature range between 100 and 200 °C differs significantly from that at room temperature at comparable concentration and pH values.[15] Implicit understanding of the known condensation chemistry of molybdates under ambient conditions is, therefore, not beneficial in terms of knowledge-based planning of the hydrothermal synthesis of molybdenum oxide-based catalysts. Likewise, systematic *in situ* spectroscopic studies of the synthesis of complex metastable mixed oxide phases, like the so-called M1 phase of MoVTeNb oxide, resulted in the development of modular preparation techniques that yield the desired phase more effectively in shorter synthesis time with improved catalytic properties.[14]

Herein we present *in situ* Raman spectroscopic and transmission electron microscopy studies that provide insight into the mechanism of particle growth and re-crystallization during the synthesis of the thermodynamic stable monoclinic manganese tungstate phase in wolframite-type structure (ICSD 67906).[16] Nano-structuring of manganese tungstate features a feasible strategy to control bi-functional properties of the mixed metal oxide with implications on the selectivity in oxidation reactions applied to upgrade alkanes. We demonstrate the viability of such a strategy by studying rod-shaped nano-crystalline $MnWO_4$ catalysts consisting of particles with varying aspect ratio (AR). The particle morphology was controlled *via* the concentration of OH^- ions under hydrothermal synthesis conditions. Oxidation of ethanol was chosen to probe the surface chemistry of the synthesized materials in a catalytic reaction. The substrate molecule may undergo multiple pathways depending on the nature of the active sites at the catalyst surface. Generally, the alcohol is oxidized to the aldehyde at redox active centres whereas dehydration of ethanol to ethylene or the formation of diethyl ether indicate surface sites that facilitate acid/base reactions.

Experimental

Hydrothermal synthesis

The hydrothermal synthesis of $MnWO_4$ was performed in an analytical autoclave HPM-PT-040 (Premex Reactor GmbH) described before,[15] adopting a synthesis method that has been reported previously.[16] In the first step, a 0.2 M aqueous solution of $Mn(NO_3)_2$ ($Mn(NO_3)_2\cdot4H_2O$, 98%, Roth) was added to a 0.2 M aqueous solution of Na_2WO_4 ($Na_2WO_4\cdot2H_2O$, 99%, Sigma Aldrich) while stirring, leading to a mixed solution of pH = 6.7. Subsequently, the pH of the mixed solution was adjusted to 6.3, 8.0, 9.1, and 9.9 by adding appropriate amounts of 0.1 M HNO_3 (64–66%, Sigma Aldrich) or 0.1 M NaOH (98%, Alfa Aesar). The mixtures were transferred to the autoclave and the temperature was raised from 20 °C to 180 °C at a rate of 5 °C min^{-1}. The synthesis temperature was kept at 180 °C for 12 h. During hydrothermal synthesis the pH was recorded using a pH probe (ZrO_2 probe Model A2 and Ag/AgCl reference electrode, both with a 1/2" outer tubing made from Hastelloy C-276; Corr Instruments). The pH probes were calibrated by use of four buffer solutions at the given reaction temperatures prior to the experiments. At the same time the Raman spectra of the synthesis gels were recorded using a Raman probe (RAMAN RXN1, immersion optic 1/4" OD (HC-276); Kaiser Optical Systems). After cooling down the gel at a rate of 5 °C min^{-1}, the products of the hydrothermal synthesis were filtered by centrifugation and washed twice with de-ionized water (MilliPore®). In the final step, the solids were dried in a muffle furnace in air at 80 °C for 12 h. Depending on the pH value before hydrothermal treatment, yellowish to brownish solids were collected. The solids were annealed in argon (flow rate 50 mL min^{-1}) at 400 °C (heating rate 5 °C min^{-1}) for 2 h using a rotary tube furnace (XERION) resulting in five phase-pure $MnWO_4$ catalysts with different mean aspect ratios (AR) characterized by the identification numbers 19 112 (AR = 1.5), 19 113 (AR = 1.7), 19 114 (AR = 3.2), 19 251 (AR = 3.9), and 19 116 (AR = 5.1). The catalysts are called ARx.x, where x.x corresponds to the mean aspect ratio of the particles in the materials after thermal activation as determined by analysis of transmission electron microscopy images.

Characterization of catalysts

Transmission electron microscopy (TEM) studies were conducted using a Philips CM200 FEG transmission electron microscope operating at 200 kV. High resolution TEM (HRTEM) and high resolution high angle annular dark field scanning transmission electron microscopy (HAADF-STEM) were performed using a Cs corrected FEI TITAN 80–300 operated at 300 kV. TEM samples were prepared by drop deposition from ethanolic suspensions onto lacey-carbon coated Cu grids, and by cross section preparation.

Field emission scanning electron microscopy (FESEM) was carried out with a Hitachi S4800 instrument operating at 5 kV.

The X-ray diffraction (XRD) measurements were performed in Bragg–Brentano geometry on a Bruker AXS D8 Advance theta/theta diffractometer, using Ni filtered Cu Kα radiation and a position sensitive LynxEye silicon strip detector. The XRD data were evaluated by whole powder pattern fitting according to the Rietveld method as implemented in the TOPAS software [version 4.2, copyright 1999–2009 Bruker AXS]. During the routine fitting, which uses an isotropic peak width model (*i.e.* the diffraction profile widths are described as a smooth function of the diffraction angle, independent of *hkl*), systematic peak profile mismatches of varying degrees were observed. With the anisotropic crystallite shape observed by electron microscopy in mind, we developed an appropriate anisotropic (*i.e.*, *hkl* dependent) peak width model. An efficient model was obtained by modifying the phenomenological model published by Stephens.[17] Due to the macro language implemented in TOPAS, user defined peak models can be implemented easily.

The original Stephens model, which was derived to describe anisotropic strain broadening, did not work well with our data. Since we expected anisotropic crystallite size to be the predominant peak broadening factor in our case, we replaced the angular dependent term tan(θ) (representing strain) of the original Stephens model with a 1/cos(θ) (*i.e.*, size) term, while retaining the *hkl* dependent expression. In addition to a good overall fit, this modified model allowed us to obtain the nominal crystallite size for different crystal directions. It should be noted that such values represent volume weighted average lengths of unit cell columns, L_{Vol}–IB. This includes averaging over parallel columns of different lengths within crystallites (shape dependent), as well as averaging over different crystallites of possibly different sizes (size distribution dependent). Thus, the reported L_{Vol}–IB values cannot be directly compared to the physical dimensions of discrete crystallites as, *e.g.*, observed by electron microscopy. Nevertheless, the XRD derived dimensions may be considered to represent the (volume weighted) average crystallite morphology.

The specific surface area was measured in a volumetric N_2 adsorption device (Autosorb-6-B, Quantachrome) at the temperature of liquid nitrogen. The sample was degassed under dynamic vacuum at a temperature of 300 °C for 2 h prior to adsorption. The relative N_2 pressure was varied and 11 data points were measured. The linear range of the adsorption isotherm ($p/p_0 = 0.05$–0.3) was considered to calculate the specific surface area according to the BET method.

Ethanol oxidation was performed at atmospheric pressure in a feed composed of 30.2 mL of N_2, 2.8 mL of O_2 and 1.2 mL of CH_4. The feed was passed through a saturator at 15 °C to achieve a concentration of 4 vol% ethanol. A laboratory quartz U-tube fixed bed reactor (4 mm inner diameter, 6 mm outer diameter, 26 cm length) was used, which contained 200 mg of the catalyst previously pressed and sieved to a particle size of 250–355 μm. Gas analysis was performed using online gas chromatography (GC 6890A, Agilent) equipped with two channels. A combination of two capillary columns (GS-Carbonplot and Plot Mole Sieve 5A) in connection with a thermal conductivity detector (TCD) was used to analyse the permanent gases CO_2, O_2, N_2 and CO. A combination of two capillary columns (HP-FFAP and HP Plot Q) connected to a flame ionization detector (FID) was applied to analyse alkanes, olefins and oxygenates.

Transmission Fourier transform infrared spectroscopy (FTIRS) measurements were carried out using a Varian 670 spectrometer equipped with a MCT detector. The spectra were recorded at a resolution of 4 cm^{-1} accumulating 512 scans. Self-supported wafers (area weight of 23–29 mg cm^{-2}) were transferred into an IR cell that was connected to a vacuum system, in which residual pressures of *ca.* 1×10^{-6} mbar can be employed. Prior to the adsorption of gases, the catalysts were heated under dynamic vacuum at 300 °C for 1 h. A reference spectrum of the solid was taken after cooling down to 40 °C. Then, the sample chamber was charged with 7 mbar equilibrium pressure of NH_3 at 40 °C. After 30 min, spectra were recorded applying the spectrum of the pretreated $MnWO_4$ as background. The Brønsted acid site density was calculated applying the extinction coefficient 16 cm μmol^{-1}.[18]

Near-ambient-pressure X-ray photoelectron spectroscopy (NAP-XPS) was conducted at the end station of the ISISS beam line at BESSY II/HZB (Helmholtz-Zentrum Berlin, Germany). Details of the setup have been published previously.[19] For the XPS measurements, Mn 2p, O 1s, and W 4f core level spectra were collected at constant kinetic energies (KE) of the photoelectrons of 150 eV and 750 eV, resulting in inelastic mean free paths (IMFP) of the excited photoelectrons of ~0.6 nm (150 eV, denominated as surface) and ~1.6 nm (750 eV, denominated as deep). The experiments were performed at 300 °C and a total pressure of 0.25 mbar in a 1/1.1 O_2/He mixture with a total gas flow of 4.2 sccm.

Results and discussion

Phase formation and particle growth under hydrothermal conditions

The preparation of monoclinic manganese tungstate composed of the wolframite-type structure (ICSD 67906) was performed by hydrothermal synthesis for 12 hours at 180 °C in aqueous medium. The formation of the solid within the autoclave was monitored using *in situ* Raman spectroscopy. Fig. 1 shows the spectra (Fig. 1a) taken during the synthesis of a slightly acidulated synthesis gel (pH of the starting mixture 6.3, Fig. 1b) and a mixture in which the pH has been adjusted to 9.1 (Fig. 1c) by addition of NaOH solution (Fig. 1d). Reaction between manganese nitrate and sodium tungstate occurs immediately after mixing the precursor solutions at room temperature as is evident from the Raman spectra. The Raman spectrum of the Mn precursor solution shows only the nitrate peak at 1048 cm^{-1} and no peaks due to Mn–O stretching vibrations, since divalent manganese ions are present under these conditions as an octahedral aquo complex. In the Raman spectrum of the sodium tungstate solution peaks at 933, 837, and 324 cm^{-1} occur due to the presence of dissolved $[WO_4]^{2-}$ ions. The peaks belong to the four Raman active fundamental vibrational modes (A, +E, +2F_2) of an undistorted tetrahedron (T_d symmetry).[20]

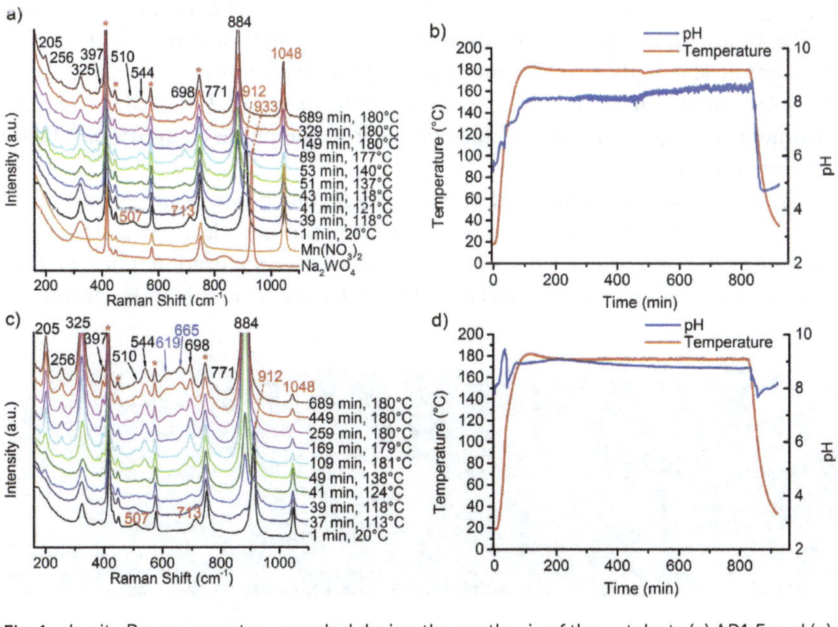

Fig. 1 *In situ* Raman spectra recorded during the synthesis of the catalysts (a) AR1.5 and (c) AR3.9 and the corresponding profiles of temperature and pH during synthesis of (b) AR1.5 and (d) AR3.9; the symbol * in the Raman spectra indicates the bands of the sapphire window of the Raman probe.

By mixing the two precursor solutions, the peak at 837 cm^{-1} disappears immediately, peaks at 507 and 713 cm^{-1} emerge, and the peak at 933 cm^{-1} is shifted to 912 cm^{-1} suggesting reaction between the manganese nitrate and sodium tungstate solution at room temperature. The bands are attributed to the formation of a partially crystalline product of unknown crystal structure that exhibits a layered-like morphology (Fig. S1†), indicated also by a peak in the XRD patterns at small angles that suggests regular stacking with a d spacing of 0.8 nm. The intermediate contains Mn and W in a molar ratio close to 1. During heating, bands at 205, 256, 325, 397, 510, 544, 698, 771, and 884 cm^{-1} indicate phase formation of $MnWO_4$ above 100 °C that is completed at 120 °C.[21,22] It has to be noted at this point that the Raman spectra measured by using the immersion probe originate from contributions of dispersed nanoparticles as well as from molecular species dissolved in the mother liquor. The phase formation is additionally indicated by a sharp drop of the pH in the temperature range from 110–120 °C. The pH at 180 °C rises above 8 during the synthesis of all catalysts. The average value increases with increasing starting pH and amounts to 8.1 in the synthesis of the catalysts AR1.5 and AR1.7, to 9.0 in the synthesis of the catalysts AR3.2 and AR3.9, and to 9.3 in the synthesis of catalyst AR5.1. The regulation of the initial pH involves changes in the ionic strength of the mother liquor from 1.20 mol L^{-1} for catalyst AR1.5 to 1.22 mol L^{-1} for catalyst AR5.1. The interpretation of the evolution of the pH values during hydrothermal synthesis is not straightforward, since the measured value results from superimposing condensation and hydrolysis reactions. All peaks observed at 180 °C, except weak bands at 619 and 665 cm^{-1}, belong to the normal modes of crystalline $MnWO_4$.[21,22] The extra peaks are tentatively attributed to manganese hydroxide species as will be discussed below. Irrespective of the initial pH value, crystallization occurs at 120 °C (Fig. 1) implying that the crystallization process is determined by thermodynamics, since the monoclinic wolframite-type $P2/c$ structure is the thermodynamic stable phase under the applied synthesis conditions.[16] According to the XRD, all synthesis products are phase-pure materials (Fig. S2†). Thermal treatment in inert gas at 400 °C does not change the phase composition of the products (Fig. S3†).

The morphology of nano-sized $MnWO_4$ can be tuned kinetically by adjusting the chemical potential during synthesis within the autoclave. The structural alterations are induced by different dissolution–recrystallization rates of the involved manganese tungstate species. These rates are predominantly controlled

Fig. 2 Electron microscopy images of the as-synthesized (top row) and thermally treated (bottom row) nanostructured $MnWO_4$ materials AR1.5 (a) and (f), AR1.7 (b) and (g), AR3.2 (c) and (h), AR3.9 (d) and (i), and AR5.1 (e) and (j); uncoloured TEM images are presented in the ESI (Fig. S4†).

Table 1 Specific surface area, results of shape analysis based on TEM, and crystallite size calculated from anisotropic fitting in Rietveld refinement of the XRD patterns

Catalyst	Catalyst ID[a]	pH[b]	Surface area of hydrothermal product[c] (m^2 g^{-1})	Surface area of catalyst[a] (m^2 g^{-1})	Number of particles[d]	Mean particle length[d] (nm)	Mean particle diameter[d] (nm)	Mean aspect ratio[d]	$D_{<100>}$[e] (nm)	$D_{<010>}$[e] (nm)	$D_{<001>}$[e] (nm)
AR1.5	19 112	6.3	27.7	25.9	132	53.8	35.7	1.5	27.1	27.3	41.4
AR1.7	19 113	6.7	26.1	22.9	124	59.0	35.1	1.7	27.2	26.3	46.8
AR3.2	19 114	8.0	23.3	22.1	93	104	32.6	3.2	28.5	25.6	70.3
AR3.9	19 251	9.1	25.7	24.0	125	119	30.8	3.9	27.2	22.1	65.3
AR5.1	19 116	9.9	31.0	28.7	131	122	24.4	5.1	24.0	17.3	55.4

[a] Required to identify different batches of catalyst synthesis, catalyst after thermal activation at 400 °C for 2 hours in argon. [b] pH value adjusted before the hydrothermal synthesis. [c] After drying at 80 °C. [d] Number average values analysed based on TEM images as shown in Fig. 2, S4 and S5. [e] Based on XRD of the catalysts (Fig. S3).

by the starting pH value of the synthesis gel. Thus, the shape of the primary $MnWO_4$ nanoparticles in the hydrothermal product crucially depends on the starting pH and varies from cube-like particles (starting pH = 6.3) to anisotropic nano-rods (starting pH = 9.9) (Fig. 2). These changes are reflected in the aspect ratio of the $MnWO_4$ nanoparticles, which can be obtained by measuring the length and width of the nanoparticles from the TEM images (Fig. 2, S4, and S5,† Table 1). Before catalytic testing, the as-synthesised materials have to be thermally treated above the reaction temperature of the catalytic reaction. Annealing in argon at 400 °C has no significant influence on the shape and size of the nanoparticles (Fig. 2), but the specific surface areas are slightly reduced (Table 1). Electron microscopy indicates a defect-rich structure and preferential growth of the rods along the <001> axis. Normally, crystallization and particle growth are induced by minimization of the free energy of the system. At the given chemical potential this observation suggests that the {001} surface is a high-energy surface. Thus, growing along <001> avoids its exposure to the environment. In addition, the fraction of surfaces with lower energies increases significantly, which improves the stabilization of the nanoparticle. In fact, analysis of the HAADF-STEM images of AR1.5 and AR5.1 allows for the allocation of specific facets of the particles, including the {100}, {010} and {110} crystal planes as illustrated in Fig. 3.

The dissolution–recrystallization processes occurring during ageing under hydrothermal conditions in the autoclave at 180 °C are illustrated in Fig. 4. The dissolution of $[WO_4]^{2-}$ ions and the hydrolysis of $[Mn(H_2O)_6]^{2+}$ species by deprotonation of water ligands is facilitated in a basic environment in the presence of

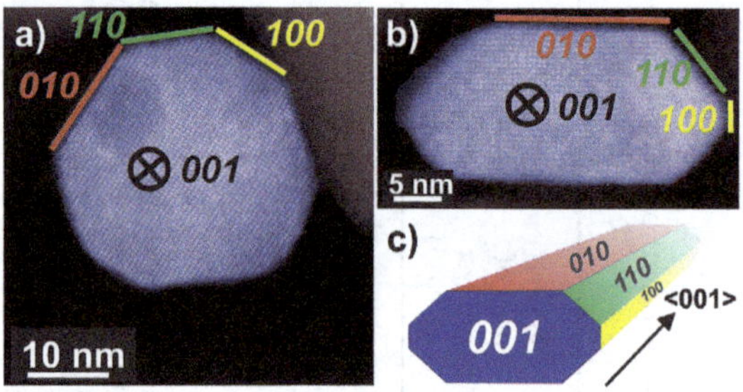

Fig. 3 HAADF-STEM images of $MnWO_4$ nanoparticles viewed along <001> with different aspect ratios: (a) AR1.5, (b) AR5.1 and (c) perspective model for a typical faceted nanoparticle. The original uncoloured images including Fast Fourier transform analysis are given in the ESI, Fig. S6.†

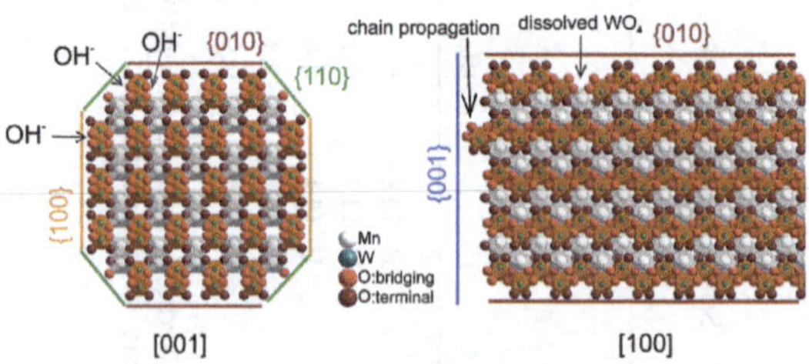

Fig. 4 Schematic representation of the proposed anisotropic mechanism of particle growth.

a high concentration of OH^- ions. Anisotropic growth is imaginable when the local structure of the WO_6 chains at different planes at the surface of the $MnWO_4$ crystals is taken into consideration. At the {010} crystal planes, edge sharing –O–W–O– bonds are exposed that can be easily attacked by OH^- ions leading to dissolution of $[WO_4]^{2-}$ species. Hydrolysis is hindered at other planes, for example at {100} and {110}, due to geometric reasons. Only half of the –O–W–O– bridges in a WO_6 octahedron are exposed at these planes. Therefore, the WO_4 units cannot be dissolved. Free $[WO_4]^{2-}$ species could condense with W–O(H) groups at the high-energy {001} basal planes propagating the tungsten oxide zigzag chains by forming new edge sharing O–W–O bonds (Fig. 4). Divalent manganese ions may interact with negatively charged tungstate chains and are in this way incorporated into the structure. With increasing OH^- concentration, the dissolution–recrystallization process is enhanced and nanoparticles with increased aspect ratio are formed as evidenced by particle size analysis (Fig. S5,† Table 1). The statistical analysis based on the TEM images is in agreement with the average particle diameter found by XRD. With increasing initial pH value, the (010) peaks in the XRD patterns (Fig. S3†) become more broadened, providing additional support for smaller crystallite sizes along the <010> direction with increasing pH (Table 1). During dissolution–recrystallization, hydrolysis of $[Mn(H_2O)_6]^{2+}$ species leads to the formation of $Mn(OH)_x$ species at the surface of the catalysts as indicated by an increasing intensity of the two bands at 619 and 665 cm^{-1} in the Raman spectra recorded during hydrothermal synthesis that are tentatively attributed to surface manganese oxide–hydroxide species.[23] The appearance of the corresponding bands is more distinct in reaction mixtures that contain a higher concentration of OH^- ions (compare Fig. 1a and c). However, the concentration of dissolved tungsten oxide and manganese species is apparently rather low, since no peaks due to such species in solution are detectable by Raman spectroscopy using the immersion probe.

Ethanol oxidation

The side products ethylene and diethyl ether in the oxidation of ethanol to acetaldehyde reveal the presence of extra acid/base functionalities at the catalyst

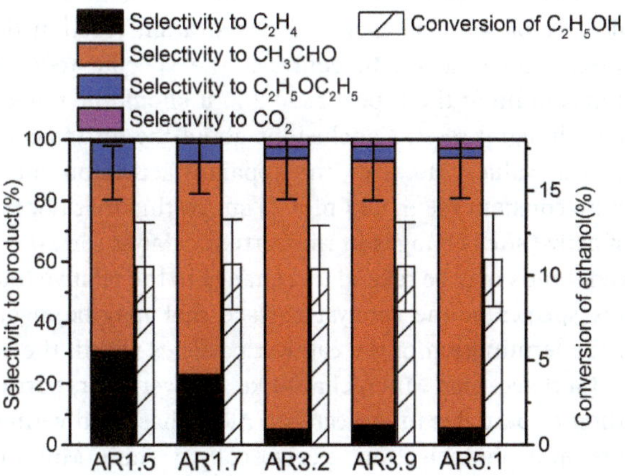

Fig. 5 Selectivity in ethanol oxidation over the nanostructured $MnWO_4$ catalysts at 10% ethanol conversion.

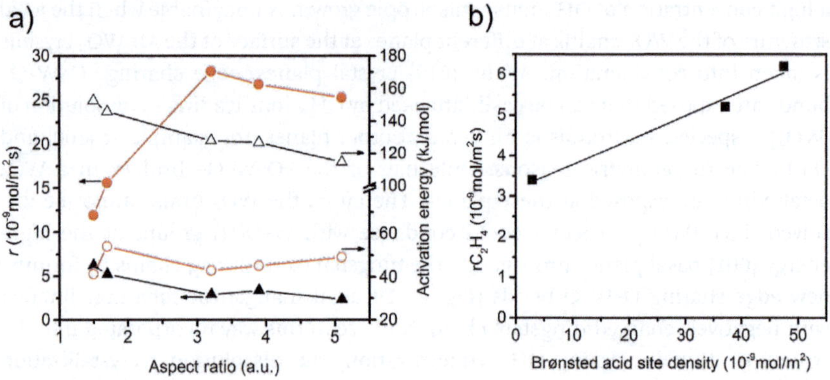

Fig. 6 (a) Specific reaction rates measured at $T = 310$ °C and normalized to surface area, and apparent activation energy as a function of the aspect ratio; red solid circle: acetaldehyde formation rate; red open circle: apparent activation energy of acetaldehyde formation; black solid triangle: ethylene formation rate; black open triangle: apparent activation energy of ethylene formation; (b) rate of formation of ethylene as a function of Brønsted acid site density at the catalyst surface determined by ammonia adsorption and specific surface area measurements.

surface of an oxidation catalyst. Ethanol oxidation has been performed over all catalysts within the reaction temperature range from 280 to 310 °C achieving conversion of ethanol from 5 to 15% (Fig. S7†). Under these conditions acetaldehyde as well as ethylene are formed over all catalysts, but the selectivity is different for the catalysts with different aspect ratios of their primary particles. In Fig. 5 the product distribution is compared at 10% ethanol conversion. The ethylene selectivity is quite high over the catalyst AR1.5, but decreases as the aspect ratio increases.

The selectivity to acetaldehyde decreases with increasing temperature and conversion, whereas ethylene selectivity rises (Fig. S7†). At constant temperature the selectivity to acetaldehyde decreases in general slightly with time on stream, whereas ethylene selectivity is quite constant or even increases (Fig. S7†) suggesting no structural relations between redox and acid–base sites.

The specific rate of the acid-catalysed dehydration reaction decreases with increasing aspect ratio (Fig. 6). In contrast, the specific redox reaction rate exhibits a flat maximum at the aspect ratio 3.2. It should be noted at this point that the rates are integral rates. Conclusions about the intrinsic activity are not possible from these values. However, the apparent activation energy of ethanol oxidation is quite constant (48 ± 6 kJ mol^{-1}) suggesting that not the nature, but the number of active sites changes in the current catalyst series.

The observed trends may be related to a change in the relative fraction of acid–base and redox species at the catalyst surface that may be, again, related to differences in the termination of the catalyst particles within the current series. The presence of a defect rich MnO_x chain-like 2D over-layer, particularly, at the {010} termination of particles in the catalyst AR5.1 has been verified using electron microscopy and photoelectron spectroscopy.[24] These MnO_x chains might bear oxygen defects, which are believed to provide the active sites in the activation of propane. The same manganese oxide sites might be relevant for the oxidation of ethanol in the present experiments as well. Due to the preferential growth of the particles along the <001> direction, the relative abundance of {010} planes that host these manganese oxide sites for oxidation catalysis may increase with increasing aspect ratio (Fig. 3). Therefore, an aspect ratio above three is reflected in increased integral rates of the oxidation reaction compared to the integral rates of the acid-catalysed dehydration reaction over the corresponding catalysts. The sites responsible for dehydration may be, in contrast, preferentially located at the two ends of the rods, which are terminated by {001} crystal planes. According to structural considerations, Brønsted acid sites might be preferentially located at these planes (Fig. S8†). Therefore, the two catalysts AR1.5 and AR1.7 exhibit enhanced acid–base functionalities, which is in agreement with the shape (Fig. 2 and 3a) and the proposed dissolution–recrystallization mechanism (Fig. 4 and S8†).

Experimentally, the acidity of the catalysts has been studied using infrared spectroscopy. After heating to the reaction temperature of ethanol oxidation (300 °C) under vacuum, three bands located at 3470, 3400, and 3366 cm^{-1} are observed in the O–H stretching region on top of a broad feature (Fig. 7a). The latter is attributed to hydroxy groups that undergo hydrogen bonding or strongly adsorbed water molecules. A residual band is observed at 1642 cm^{-1} that might be due to the bending mode of molecular water, but discrimination of this band from the overtones and combination vibrations of $MnWO_4$ is difficult. Based on the very low frequency and the unusual narrow bandwidth, the three sharp bands at 3470, 3400, and 3366 cm^{-1} are tentatively assigned to hydroxy groups that form well-ordered hydrogen bonding networks at the surface. Further investigations are necessary to clarify the origin of these bands. To investigate the acidity of the hydroxy groups, NH_3 was adsorbed. Unfortunately, the bands of the adsorbed ammonia in the N–H stretching vibration region between 3400 and 3000 cm^{-1} overlap with the hydroxy bands due to the low frequency of the latter (Fig. 7a). Therefore, the consumption of particular OH species in the reaction with ammonia cannot be monitored. But weak acidity is confirmed by the formation of a weak but distinct band at 1439 cm^{-1} due to the asymmetric deformation vibration of ammonium ions formed by reaction of ammonia with Brønsted acid sites at the surface of the catalysts AR1.5, AR1.7, and AR3.9 (Fig. S9†). No clear indications regarding redox processes of adsorbed ammonia molecules have been observed in the spectra. The concentration of acid sites at the surface of catalyst AR5.1 was below the detection limit. The small number of acid sites (in the range from 0 for AR5.1 to 1.2 μmol g^{-1} for AR1.5) suggests that most of the OH groups observed in the infrared spectra of the catalysts after dehydroxylation at 300 °C under vacuum (Fig. 7a) are nonacidic in nature. Essentially all the OH groups observed at the surface of catalyst 5.1 are not able to protonate ammonia. Therefore, the band at 3366 cm^{-1} that represents the dominant hydroxy species at the surface of catalyst AR5.1 is tentatively assigned to Mn–OH groups. A sharp band below 3500 cm^{-1} has been attributed to Mn–OH in the layered Mn^{2+} hydroxide $Mn_2(OH)_2SO_4$.[25] The surface concentration of acid sites correlates with the formation rate of ethylene (Fig. 6b), demonstrating the involvement of these sites in the dehydration of ethanol. The non-zero intercept of the fitting line in Fig. 6b suggests that Lewis acid sites may be involved in the dehydration of ethanol as well. Pathways leading to ethylene from ethanol adsorbed on Ce^{4+} have been discussed.[26] Similar reaction routes might also be possible at coordinatively

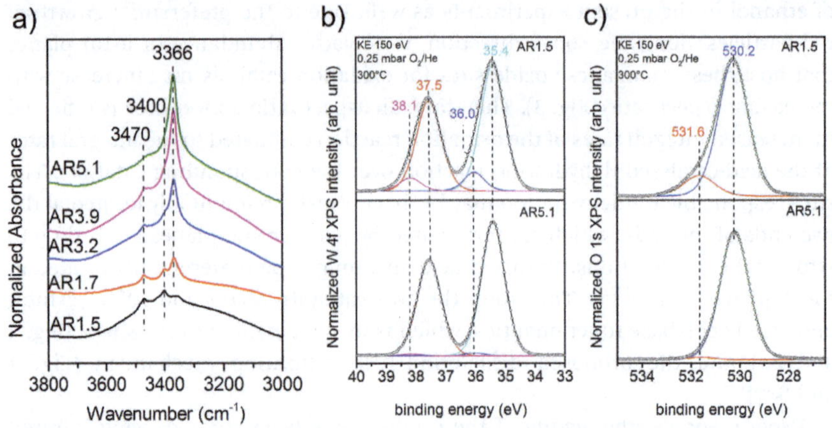

Fig. 7 (a) Infrared spectra in the region of OH stretching vibrations after thermal treatment of the catalyst in the infrared cell under vacuum at 300 °C; the measurement was performed under vacuum at 40 °C; W 4f spectra (b), and O 1s spectra (c) of the catalysts AR1.5 and AR5.1 measured using synchrotron-based near ambient pressure X-ray photoemission spectroscopy (NAP-XPS) at an inelastic mean free path (IMFP) of *ca.* 0.6 nm in 0.25 mbar in O_2/He at a total gas flow of 4.2 sccm at 300 °C.

unsaturated manganese or tungsten sites that are present at the surface of the catalysts as it becomes evident from ammonia adsorption (Fig. S9†). The peak at 1591 cm^{-1} may arise from the asymmetric bending vibration of ammonia molecules adsorbed at Lewis acid sites.

A high concentration of hydroxy groups at the surface of the catalyst AR1.5 was also confirmed by NAP-XPS measured in the presence of oxygen at 300 °C. The O 1s core level spectra of the catalysts AR1.5 and AR5.1 are shown in Fig. 7c. The main peak at 530.2 eV is assigned to lattice oxygen, while the component at 531.6 eV is attributed to hydroxy groups. Fig. 7c shows that catalyst AR1.5 comprises a higher concentration of hydroxy groups than catalyst AR5.1. A detailed analysis of the W 4f core level reveals that also the W $4f_{7/2}$ and W $4f_{5/2}$ doublet (Fig. 7b) can be deconvoluted into two contributions. The low binding energy doublet at 35.4 and 37.5 eV is in good agreement with the binding energy values reported in the literature for metal tungstates.[27,28] The high binding energy doublet at 36.0 and 38.1 eV is more pronounced in the AR1.5 catalyst than in the AR5.1 catalyst, in line with the higher hydroxyl concentration observed in the O 1s spectrum of the AR1.5 catalyst. Furthermore, the contribution of the high binding energy W 4f doublet as well as the hydroxide component in the O 1s spectrum is decreasing with increasing probing depth (compare Fig. 7b and c, S10a and b†). Therefore, we tentatively assign the high binding energy doublet in the W 4f spectrum to modified tungsten atoms bearing hydroxy groups at the surface of the catalyst. Thus, the XPS measurements indicate that more W–OH groups are present at the surface of the catalyst containing primary particles with a low aspect ratio (AR1.5). These results are complementary to the characterization of the catalysts using ammonia adsorption. The acid sites probed using FTIRS of adsorbed ammonium ions are attributed to W–OH groups at the surface of the catalyst, which is also plausible in terms of the acid–base chemistry of tungsten compared to manganese. Mn–OH groups are apparently not resolved by XPS, neither in the O 1s spectra (Fig. 7c), nor in the Mn 2p spectra (Fig. S11†). The discrepancy between the FTIRS and XPS concerning the presence of hydroxy groups at the surface of the two catalysts might be due to the fact that the Mn–OH contributions are enveloped by M − O contributions in the O 1s (Fig. 7c) and Mn 2p (Fig. S11†) spectra, or due to differences in the pretreatment procedures before IR (vacuum, 300 °C) and XPS (0.25 mbar O_2 at 300 °C) measurements, respectively.

In summary, the variation of the initial pH in the hydrothermal synthesis of nanostructured $MnWO_4$ affects the dissolution–recrystallization kinetics of the freshly formed particles during ageing at 180 °C. Although these dissolution and recrystallization processes are responsible for anisotropic particle growth, more important is that ageing provides the basis for the formation of the catalytically active sites. With increasing pH under hydrothermal conditions the fraction of acid W–OH groups at the surface of the final catalyst decreases and, consequently, acid–base catalysed side reactions during ethanol oxidation are increasingly suppressed, which is reflected in enhanced selectivity to acetaldehyde with increasing starting pH (Fig. 5).

Selectivity is the major issue in oxidation catalysis.[6,29] Activation of C–H bonds in saturated hydrocarbons is challenging because the formed reaction products and intermediates easily undergo consecutive and parallel reactions. Minimization of rates of undesired pathways requires co-ordinated design of bulk electronic properties and surface dynamics of oxidation catalysts.[30] C–H bond activation may involve multiple mechanisms including carbenium or carbonium intermediates and homolytic splitting of C–H bonds at metal oxide surface functional groups under formation of radical species.[31,32] Model calculations, generally based on small cluster models, favour the homolytic pathway over transition metal oxide catalysts.[33] However, Lewis acid sites in terms of coordinatively unsaturated metal cations and Brønsted acid sites may be present at the surface of a transition metal oxide under reaction conditions as well. Brønsted acid sites are particularly expected since water is an unavoidable coproduct in oxidation reactions and, as demonstrated in the current example, the dehydroxylation temperature is often above the reaction temperature.

Whereas acidity at the surface of nanostructured $MnWO_4$ catalysts is attributed to the presence of W–OH groups and coordinatively unsaturated metal cations, the origin of the redox activity is not that straightforward. During particle growth under hydrothermal conditions the formation of defects may occur. In the course of dissolution and recrystallization, chemical defects in terms of cationic vacancies or OH^- groups at anionic positions are generated that may be related to structural defects.[34] Establishing relations between defect chemistry and redox activity requires further kinetic studies and more comprehensive characterization of the catalysts, which is currently under way.

Conclusions

Nanostructured $MnWO_4$ catalysts have been prepared using hydrothermal synthesis. The aspect ratio of the primary $MnWO_4$ particles increases with increasing pH during ageing at 180 °C under hydrothermal conditions. Electron microscopy revealed that the particles grow along the <001> axis. A mechanism of particle growth by dissolution re-crystallization is proposed that leads to an increase in the fraction of terminating {010} planes with increasing aspect ratio.

At the same time, the fraction of {001} planes, which are terminated by W–OH groups, is decreasing. These changes are reflected in the selectivity patterns of the ethanol oxidation reaction that probes both, redox and acid–base sites. Metastable structures are required to achieve catalytic activity over $MnWO_4$. Solid-state synthesis provides theses structures only on a limited scale. The implementation of chemical defects as a basis for the catalytic activity of $MnWO_4$ succeeded *via* variation of the chemical potential under mild hydrothermal conditions.

Acknowledgements

This work was conducted in the framework of the BasCat collaboration between BASF SE, TU Berlin, FHI, and the cluster of excellence "Unified Concepts in Catalysis" (UniCat http://www.unicat.tu-berlin.de). X.L. acknowledges the Berlin International Graduate School of Natural Sciences and Engineering (BIG NSE) as part of UniCat for financial support. The authors thank Maike Hashagen and Jasmin Allan for technical assistance. We thank the HZB staff for their continual support of the electron spectroscopy activities of the FHI at BESSY II.

Notes and references

1 I. E. Wachs and K. Routray, *ACS Catal.*, 2012, **2**, 1235–1246.
2 E. W. McFarland and H. Metiu, *Chem. Rev.*, 2013, **113**, 4391–4427.
3 W. Ueda, *J. Jpn. Pet. Inst.*, 2013, **56**, 122–132.
4 R. Schlögl, *Angew. Chem., Int. Ed.*, 2015, **54**, 3465–3520.
5 A. Trunschke, in *Chemical Energy Storage*, ed. R. Schlögl, Walter de Gruyter GmbH, Berlin/Boston, 2013, ch. 4.2, pp. 277–301.
6 R. Schlögl, *Top. Catal.*, 2011, **54**, 627–638.
7 K. Amakawa, Y. V. Kolen'ko, A. Villa, M. E. Schuster, L.-I. Csepei, G. Weinberg, S. Wrabetz, R. Naumann d'Alnoncourt, F. Girgsdies, L. Prati, R. Schlögl and A. Trunschke, *ACS Catal.*, 2013, **3**, 1103–1113.
8 M. Rajamathi and R. Seshadri, *Curr. Opin. Solid State Mater. Sci.*, 2002, **6**, 337–345.
9 C. S. Cundy and P. A. Cox, *Microporous Mesoporous Mater.*, 2005, **82**, 1–78.
10 M. K. Devaraju and I. Honma, *Adv. Energy Mater.*, 2012, **2**, 284–297.
11 C. Perego and R. Millini, *Chem. Soc. Rev.*, 2013, **42**, 3956–3976.
12 G. R. Patzke, Y. Zhou, R. Kontic and F. Conrad, *Angew. Chem., Int. Ed.*, 2011, **50**, 826–859.
13 W. Shi, S. Song and H. Zhang, *Chem. Soc. Rev.*, 2013, **42**, 5714–5743.
14 M. Sanchez Sanchez, F. Girgsdies, M. Jastak, P. Kube, R. Schlögl and A. Trunschke, *Angew. Chem., Int. Ed.*, 2012, **51**, 7194–7197.
15 J. Noack, F. Rosowski, R. Schlögl and A. Trunschke, *Z. Anorg. Allg. Chem.*, 2014, **640**, 2730–2736.
16 S. H. Yu, B. Liu, M. S. Mo, J. H. Huang, X. M. Liu and Y. T. Qian, *Adv. Funct. Mater.*, 2003, **13**, 639–647.
17 P. Stephens, *J. Appl. Crystallogr.*, 1999, **32**, 281–289.
18 A. A. Budneva, E. A. Paukshtis and A. A. Davydov, *React. Kinet. Catal. Lett.*, 1987, **34**, 63–67.
19 A. KnopGericke, E. Kleimenov, M. Hävecker, R. Blume, D. Teschner, S. Zafeiratos, R. Schlögl, V. I. Bukhtiyarov, V. V. Kaichev, I. P. Prosvirin,

A. I. Nizovskii, H. Bluhm, A. Barinov, P. Dudin and M. Kiskinova, in *Advances in Catalysis*, ed. C. G. Bruce and K. Helmut, Academic Press, 2009, vol. 52, pp. 213–272.

20 F. Gonzalez-Vilchez and W. P. Griffith, *J. Chem. Soc., Dalton Trans.*, 1972, 1416–1421, DOI: 10.1039/dt9720001416.

21 L. H. Hoang, N. T. M. Hien, W. S. Choi, Y. S. Lee, K. Taniguchi, T. Arima, S. Yoon, X. B. Chen and I.-S. Yang, *J. Raman Spectrosc.*, 2010, **41**, 1005–1010.

22 M. N. Iliev, M. M. Gospodinov and A. P. Litvinchuk, *Phys. Rev. B: Condens. Matter Mater. Phys*, 2009, **80**, 212302.

23 F. Kapteijn, A. D. Vanlangeveld, J. A. Moulijn, A. Andreini, M. A. Vuurman, A. M. Turek, J. M. Jehng and I. E. Wachs, *J. Catal.*, 1994, **150**, 94–104.

24 X. Li, T. Lunkenbein, V. Pfeifer, M. Jastak, P. K. Nielsen, F. Girgsdies, A. Knop-Gericke, F. Rosowski, R. Schlögl and A. Trunschke, *Angew. Chem., Int. Ed.*, 2016, **55**, 4092–4096.

25 M. B. Salah, S. Vilminot, T. Mhiri and M. Kurmoo, *Eur. J. Inorg. Chem.*, 2004, 2272–2276.

26 M. Li, Z. Wu and S. H. Overbury, *J. Catal.*, 2013, **306**, 164–176.

27 S. Muthamizh, R. Suresh, K. Giribabu, R. Manigandan, S. Praveen Kumar, S. Munusamy and V. Narayanan, *J. Alloys Compd.*, 2015, **619**, 601–609.

28 T.-D. Nguyen, D. Mrabet, T.-T.-D. Vu, C.-T. Dinh and T.-O. Do, *CrystEngComm*, 2011, **13**, 1450–1460.

29 J.-M. M. Millet, *Top. Catal.*, 2006, **38**, 83–92.

30 M. Eichelbaum, M. Hävecker, C. Heine, A. M. Wernbacher, F. Rosowski, A. Trunschke and R. Schlögl, *Angew. Chem., Int. Ed.*, 2015, **54**, 2922–2926.

31 C. Coperet, *Chem. Rev.*, 2009, **110**, 656–680.

32 R. A. van Santen, I. Tranca and E. J. M. Hensen, *Catal. Today*, 2015, **244**, 63–84.

33 X. Rozanska, R. Fortrie and J. Sauer, *J. Am. Chem. Soc.*, 2014, **136**, 7751–7761.

34 Y. Chabre and J. Pannetier, *Prog. Solid State Chem.*, 1995, **23**, 1–130.

Faraday Discussions

PAPER

Design and stabilisation of a high area iron molybdate surface for the selective oxidation of methanol to formaldehyde

Stephanie Chapman,[ab] Catherine Brookes,[bc] Michael Bowker,[bc] Emma K. Gibson[bd] and Peter P. Wells*[bd]

Received 4th November 2015, Accepted 8th December 2015
DOI: 10.1039/c5fd00153f

The performance of Mo-enriched, bulk ferric molybdate, employed commercially for the industrially important reaction of the selective oxidation of methanol to formaldehyde, is limited by a low surface area, typically 5–8 $m^2 g^{-1}$. Recent advances in the understanding of the iron molybdate catalyst have focused on the study of MoO_x@Fe_2O_3 (MoO_x shell, Fe_2O_3 core) systems, where only a few overlayers of Mo are present on the surface. This method of preparing MoO_x@Fe_2O_3 catalysts was shown to support an iron molybdate surface of higher surface area than the industrially-favoured bulk phase. In this research, a MoO_x@Fe_2O_3 catalyst of even higher surface area was stabilised by modifying a haematite support containing 5 wt% Al dopant. The addition of Al was an important factor for stabilising the haematite surface area and resulted in an iron molybdate surface area of ~35 $m^2 g^{-1}$, around a 5 fold increase on the bulk catalyst. XPS confirmed Mo surface-enrichment, whilst Mo XANES resolved an amorphous MoO_x surface monolayer supported on a sublayer of $Fe_2(MoO_4)_3$ that became increasingly extensive with initial Mo surface loading. The high surface area MoO_x@Fe_2O_3 catalyst proved amenable to bulk characterisation techniques; contributions from $Fe_2(MoO_4)_3$ were detectable by Raman, XAFS, ATR-IR and XRD spectroscopies. The temperature-programmed pulsed flow reaction of methanol showed that this novel, high surface area catalyst (3ML-HSA) outperformed the undoped analogue (3ML-ISA), and a peak yield of 94% formaldehyde was obtained at ~40 °C below that for the bulk $Fe_2(MoO_4)_3$ phase. This work demonstrates how core–shell, multi-component oxides offer new routes for improving catalytic performance and understanding catalytic activity.

[a] *University of Southampton, Southampton, SO17 1BJ, UK*
[b] *UK Catalysis Hub, Research Complex at Harwell, RAL, Oxford, OX11 0FA, UK. E-mail: peter.wells@rc-harwell.ac.uk*
[c] *Cardiff University, Cardiff, CF10 3XQ, UK*
[d] *University College London, London, WC1H 0AJ, UK*

Introduction

Mixed-metal oxides are employed extensively as selective-oxidation catalysts and, in this capacity, iron and bismuth molybdates are a paragon. Whilst bismuth molybdate is a well-known catalyst for the oxidation and ammoxidation of olefins,[1,2] iron molybdate is employed extensively for the selective oxidation of methanol to formaldehyde. The latter is a reaction of particular industrial importance, given that global demand for formaldehyde exceeds 30 million tonnes pa.[3]

Despite their industrial significance, the application of iron and bismuth molybdates is hindered by low surface areas (typically < 6 m^2 g^{-1}),[4,5] which limit the catalytic performance. Recent catalysis literature is largely devoted to the study of nanoparticle (NP) catalysts. Such systems can afford high mass activities and often exhibit exceptional properties compared to those materials prepared by conventional routes. Since nearly 50% of industrial formaldehyde production uses bulk-phase ferric molybdate,[6] the development of nanoscale systems as a means of catalyst optimisation is hugely relevant.

Bowker,[7] Brookes,[8,9] and others[10,11] have investigated haematite-supported molybdena, MoO_x@Fe_2O_3, a shell@core system that comprises layers of molybdena deposited on haematite nanoparticles by thermal spreading,[10] or incipient wetness impregnation.[7–9,11] The shell@core catalyst was designed to isolate the catalytic contribution of surface molybdenum from that in the ferric molybdate bulk.[8] By studying the model system, the catalyst activity has since been attributed to a well-dispersed layer of amorphous or low-dimensional crystalline molybdena (MoO_x) at the surface.[12] Notably, Brookes *et al.* reported improved catalytic activity using the MoO_x@Fe_2O_3 catalyst, *versus* the ferric molybdate bulk phase, which was attributed to a 3–4-fold greater surface area in the former.[9] It is thus proposed that by supporting the monolayer of catalytic molybdena at higher surface areas, an enhanced performance can be achieved.

Whilst dispersion of molybdena on inherently high surface area supports, such as alumina, titania and silica, has been documented,[13] these materials favour inhomogeneous growth and distribution of molybdena. An exposed support reduces formaldehyde selectivity by promoting the formation of carbon monoxide and carbon dioxide by-products.[14] In contrast, XAFS studies have shown that a solid state reaction between surface MoO_3 crystallites and bulk haematite generates a subsurface layer of catalytic $Fe_2(MoO_4)_3$.[15] The interaction between the molybdena and ferric molybdate subsurface promotes high selectivity for formaldehyde; hence haematite is the favoured catalyst support.

In addition to applications as a support, nanoparticulate haematite finds widespread use in catalysis, including the oxidation of carbon monoxide,[16] photocatalytic splitting of water,[17] biomass conversion,[18] and catalytic combustion.[19] Although an inherently high energy of crystallisation of iron oxides affords surface areas in excess of 100 m^2 g^{-1},[20] high-temperature crystallisation of haematite leads to sintering and surface area losses.[21]

For the purpose of methanol oxidation, it is desirable to preserve the chemical properties of haematite and yet maintain higher surface areas to improve the catalyst performance. Herein, aluminium-doped haematite has been used as a support for a molybdena catalyst. The support has been found to retain the characteristics of hematite but a greater resistance to sintering has sustained a higher surface area of catalytic molybdena. Although the influence of the Al-dopant on the structure of iron oxide has been discussed extensively,[22–24] to our knowledge, the application of such materials in the context of a catalyst support material is a novel application.

Experimental

Catalyst preparation

The bulk materials were synthesised by co-precipitation. Dilute HNO_3 (70%, Sigma Aldrich) at pH 2 was maintained at 60 °C under stirring, whilst an aqueous solution of iron(III) nitrate nonahydrate (Sigma Aldrich) was added in a dropwise manner. The solution was then evaporated to dryness under stirring at 90 °C. The brick-red solid was dried at 120 °C for 24 hours before calcination in air at 500 °C for 24 hours. A 5 wt% aluminous haematite support was prepared by an analogous route, except that the aqueous iron nitrate solution was added dropwise to a stirred solution of aluminium(III) nitrate nonahydrate (Sigma Aldrich) at 60 °C.

The shell@core catalysts were prepared by incipient wetness impregnation of haematite.[7,9] An aqueous solution of ammonium heptamolybdate tetrahydrate (Sigma Aldrich) was prepared at a concentration to deliver the desired number of molybdena monolayer equivalents at the point of incipient wetness.

Characterization

Brunauer–Emmett–Teller (BET) surface area measurements were performed under nitrogen at 77 K using the Quantachrome quadrasorb SI gas sorption system. Samples were prepared by outgassing at 250 °C for 2 hours and then backfilled with nitrogen prior to analysis. Scanning electron microscopy energy dispersive X-ray (SEM-EDX) analysis was performed using the JEOL 6610LV apparatus, with solid samples adhered to a conductive carbon tab. EDX analysis was performed using AZtec software (Oxford Instruments plc) with the elemental composition determined under 'Point and ID' acquisition mode. Transmission electron microscopy (TEM)-EDX and scanning transmission electron microscopy (STEM) used the JEOL JEM 2100 EM system by dispersing the sample in ethanol before transfer to a lacey carbon-coated Cu grid. Raman spectroscopy was performed using the Renishaw inVia Raman microscope with laser irradiation at 830 nm. Data acquisition was performed at 1% maximal laser power, with four accumulations, each with 10 seconds exposure time. The Rigaku Miniflex goniometer was used to obtain powder X-ray diffraction (XRD) spectra, with analyses performed under continuous scan mode using Cu K_α radiation ($\lambda = 1.5406$ Å). Rietveld refinement was performed using CelRef software. X-ray absorption fine structure (XAFS) spectra were obtained on beamline B18 at the Diamond Light Source, UK. A QEXAFS setup was used for the measurements, performed using a fast-scanning Si(311) double crystal monochromator. Samples were diluted with cellulose and pelletised prior to analysis, which comprised six scans at the Mo K edge (20 000 eV), each of 1 minute ($K_{max} = 15$ Å^{-1}). *In situ* diffuse reflectance infrared Fourier transform spectroscopy (DRIFTS) was performed in a Harrick high temperature DRIFTS cell fitted with ZnSe windows. The cell was attached to the Praying Mantis Optics and spectra collected with an Agilent Carey 680 FTIR spectrometer, taking 64 scans with a resolution of 4 cm^{-1} using the DTGS detector. X-ray photoelectron spectroscopy (XPS) spectra were collected using the ESCALAB 220 spectrometer equipped with Al and Mg K_α sources. Al K_α (1486.6 eV) irradiation was used to prevent Fe Auger peaks overlapping with the Fe $2p_{1/2}$ and Fe $2p_{3/2}$ peaks.

Catalyst testing

The catalyst reactivity was studied using the CATLAB combined microreactor–mass spectrometer setup (Hiden Ltd., Warrington, UK). For the temperature-programmed desorption (TPD) analysis, injections of methanol (1 μL) were dosed on to the catalyst at 40 °C, under He (30 mL min^{-1}), until real-time mass spectroscopic monitoring showed saturation (~8 injections). A temperature ramp of 8 °C min^{-1} was applied to 400 °C, which was then sustained for 10 minutes. The temperature-programmed pulsed flow reaction (TPPFR) involved injection of methanol (1 μL) at 2 minute intervals under a flow of 10% O_2/He (30 mL min^{-1}), during which time the temperature was ramped by 8 °C min^{-1} to 400 °C. In all analyses, the outlet gas was continuously monitored by mass spectrometry. Mass spectroscopic data has been corrected for spectrometer-induced processes. The formaldehyde signal (mass 30) is corrected for the contributions of methanol cracking (masses 28 and 29) and oxidation to CO_2 (mass 44) on the mass spectrometer filament. The CO_2 signal is adjusted for methanol (mass 31) and formaldehyde combustion on the filament. The CO (mass 28) signal is corrected for methanol and formaldehyde fragmentation and oxidation, and CO_2 cracking.

Results and discussion

In order to establish that the doped support retained the properties of the host compound, a commercially-sourced haematite nanopowder (<50 nm particle size (BET), Sigma Aldrich) was used as a reference. Gas sorption studies revealed that there was negligible decrease in the surface area (22 $m^2\ g^{-1}$ to 21 $m^2\ g^{-1}$) of the commercial haematite following a typical calcination protocol. However, by introducing 5 wt% Al dopant, the post-calcination surface area was doubled, to 46 $m^2\ g^{-1}$ (from 111 $m^2\ g^{-1}$ as-synthesised).

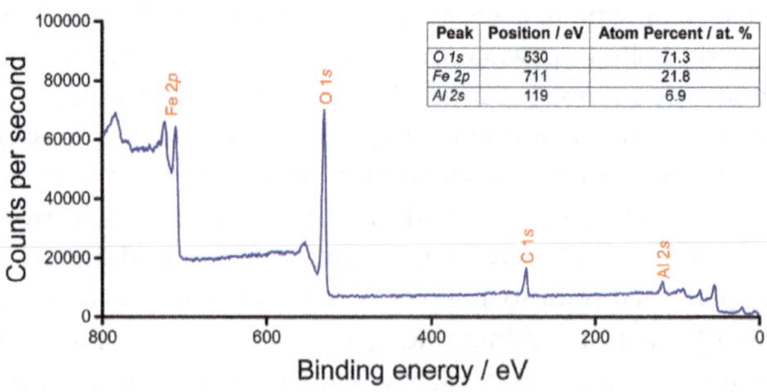

Peak	Position / eV	Atom Percent / at. %
O 1s	530	71.3
Fe 2p	711	21.8
Al 2s	119	6.9

Fig. 1 Widescan XPS spectrum of $Al_xFe_{(2-x)}O_3$ and table, inset, of the associated surface composition.

The nature of the aluminium incorporation in $Al_xFe_{(2-x)}O_3$ was investigated by XPS, and this was consistent with isomorphous substitution of Al into the bulk (Fig. 1). The XPS spectrum did not indicate any preferential segregation of Al. This contrasts with ferric molybdate systems (Mo/Fe < 1.5), for which segregation is apparent by a surface Mo loading 2–3 times the theoretical bulk phase.[25] In addition to the Al^{3+} peak at 119 eV, Fe $2p_{3/2}$ and $2p_{1/2}$ peaks at 725 eV and 711 eV, respectively, are consistent with high-spin Fe^{3+} of haematite. Lattice oxygen at a binding energy of 530 eV was identified.[26]

XRD analysis of $Al_xFe_{(2-x)}O_3$ was similar to that of haematite except that the peaks for $Al_xFe_{(2-x)}O_3$ were displaced to a higher angle than in commercial α-Fe_2O_3 (Fig. 2). Under Bragg's Law, this implied a reduction in the *d*-spacing on introducing aluminium.[26]

Changes in lattice parameters were quantified by Rietveld refinement in the space group $R\bar{3}cH$ (Table 1), which indicated reductions along both the *a*- and *c*-axes. Suppressed unit cell growth in aluminous haematite has been documented previously.[23,27] Where α-Al_2O_3 and α-Fe_2O_3 share the corundum crystal structure, the formation of a solid solution of trivalent cations is relatively facile. In affecting isomorphous substitution of Fe^{3+} ($r = 0.645$ Å) by the smaller Al^{3+} ion ($r = 0.535$ Å),[28] atomic displacement and altered electrostatic interactions introduce lattice strain. This is manifested as smaller crystallites and an enhanced surface area in Al-doped haematites.[22,26] Raman spectroscopy confirmed the reduced particle dimensions of $Al_xFe_{(2-x)}O_3$ *versus* commercial haematite through a blue shift in the excitation bands originating from quantum confinement effects (Fig. 3).[29] Additionally, where the substitution of Fe^{3+} with smaller Al^{3+} centres decreases atomic separation and concomitantly increases bond force constants, Raman bands have been shifted to a higher energy.[23,30] Another consequence of the reduced grain size is peak broadening, since phonon confinement necessitates an increase in photon momentum distribution. Some loss of crystallinity on the introduction of Al (supported by a less intense XRD pattern) may have contributed to the broadened Raman bands.[23,30–32]

TPD analysis was used to compare the reaction profile of Al-doped haematite with that of phase-pure haematite. Mass spectrometric measurements revealed high-temperature CO_2 and H_2 production, which was attributed to methanol decomposition *via* the formate intermediate. Water was desorbed in two peaks, the first coincident with methanol (an aqueous solution), and the second high-temperature desorption was accompanied by CO_2 as a product of combustion.

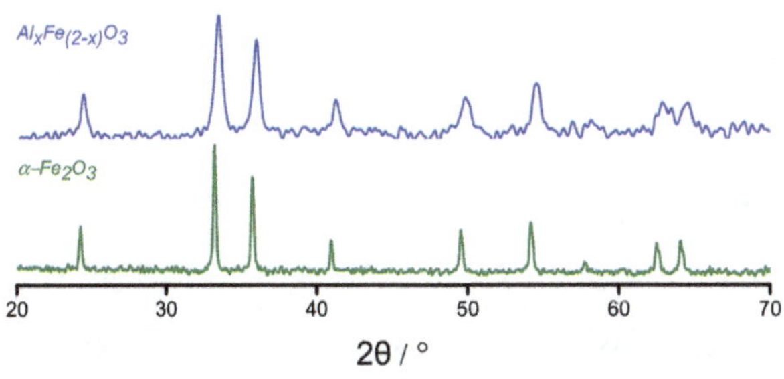

Fig. 2 XRD powder patterns of $Al_xFe_{(2-x)}O_3$ and a phase-pure haematite reference sample.

Table 1 Optimised lattice dimensions, determined by Rietveld refinement in the space group $R\bar{3}cH$, and Scherrer crystallite sizes for the haematite supports

Haematite support	BET multipoint surface area/$m^2\ g^{-1}$	Optimised lattice parameters/Å		Optimised cell volume/$Å^3$	Scherrer crystallite size/nm	$E_a\ (CO_2)$/ kJ mol^{-1}
		a	c			
Theoretical[34]	—	5.0382	13.7721	302.7	—	—
Commercial	21	5.0273	13.7245	300.4	37	182
$Al_xFe_{(2-x)}O_3$	45	4.9933	13.6266	294.2	14	186

The combustion of methanol under oxidative conditions is typical of haematite catalysis.[6] However, methanol TPD reveals a notable distinction between the behaviour of $Al_xFe_{(2-x)}O_3$ and that of commercial haematite (Fig. 4). For the Al-doped sample, CO_2 evolution is shifted to a higher temperature, which is consistent with the smaller particle size, noted previously. The decreased grain size leads to a concomitant increase in the surface free energy due to more acute curvature of the surface and a prevalence of low-coordinate sites. Where the adsorption of the bidentate formate intermediate satisfies surface valencies, mutual stabilisation yields stronger surface–adsorbate interactions that compel a larger energy input to initiate the reaction. Based on first order CO_2 production,[26] the Redhead equation has been used to approximate the activation energy of formate decomposition from the peak desorption temperature of CO_2 (Table 1).[33]

Characterisation of the shell@core catalysts

The commercial haematite (ISA) and the Al-doped haematite (HSA) support materials were coated with molybdena to produce shell@core catalysts. Methanol TPD performed on the theoretical one monolayer (ML) catalysts showed high-temperature CO_2 production, indicative of exposed iron clusters.[8,25] Since 1 ML molybdena did not provide full surface coverage, catalysts of theoretical 3 ML coverage were studied.

Relative to the haematite supports, α-MoO_3 has an exceptionally low surface area (<1 $m^2\ g^{-1}$). Whilst our own work shows that the surface area of the shell@core catalyst is diminished with increasing monolayer coverage (Table 2),[9]

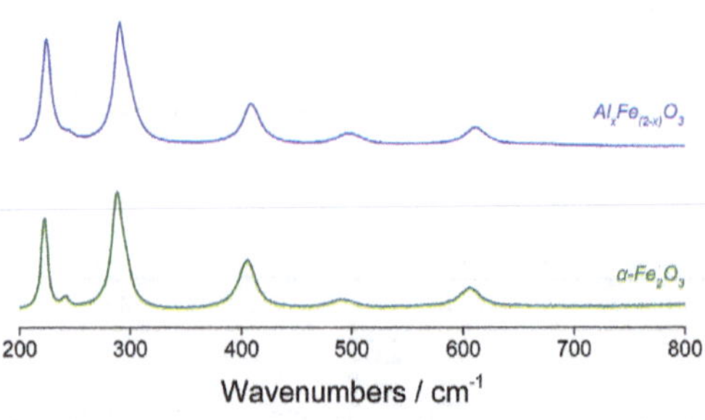

Fig. 3 Raman spectra of Al-doped and phase-pure haematite.

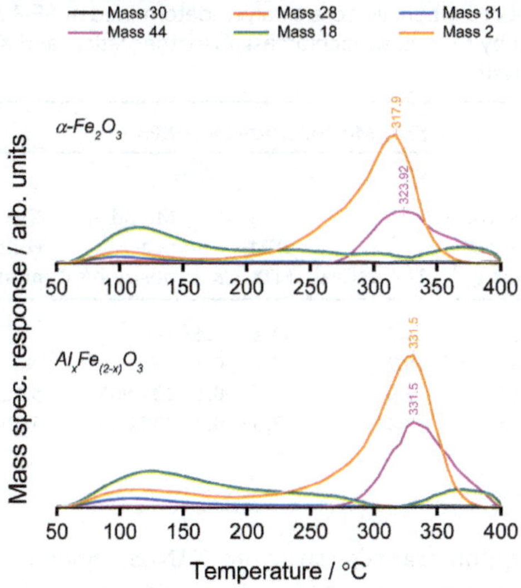

Fig. 4 TPD profiles of the haematite supports under linear temperature ramp, following methanol adsorption. The masses displayed correspond to HCHO (black), CO (red), CH_3OH (blue), CO_2 (purple), H_2O (green) and H_2 (orange).

these systems still offer a higher surface area than bulk $Fe_2(MoO_4)_3$; one of their principal advantages.

The Mo content of 3 ML systems, determined by SEM-EDX analysis, correlated with their theoretical values (Table 3). XPS analyses provided evidence of surface segregation, with Mo loading in vast excess of a theoretical bulk phase (Table 3). If XPS was completely surface-specific, the 3 ML catalysts should show the same atom percent Mo, corresponding to the MoO_x surface monolayer. However, XPS does, to some extent, penetrate the surface (~1 nm depth).[6] At a higher surface area, conversion of haematite to $Fe_2(MoO_4)_3$ is more extensive and hence there is a greater concentration of Mo at the surface. Electron binding energies from XPS affirmed that Al, Fe and Mo cations were in their maximum oxidation states of +3, +3 and +6, respectively.[35]

The Mo surface-segregation, evidenced by XPS, is well-documented in the literature.[8,9,11,36] Therefore, it is reasonable to treat Mo as being localised at the surface. On tuning the energy of the X-ray radiation to the molybdenum K-edge (20 000 eV),[37] XAFS can be used to analyse the surface structure of the shell@core catalysts.

Table 2 Evidence of surface area losses with increasing molybdena loading on aluminium-doped haematite

Theoretical number of MoO_x-monolayers on the $Al_xFe_{(2-x)}O_3$ support/ML	Surface area/$m^2\ g^{-1}$
0	46
1	45
3	35
6	30

Table 3 Mo content of 3 ML shell@core catalysts determined by SEM-EDX (theoretical Mo content determined by relative surface areas). Also, theoretical and XPS Mo content of 3 ML shell@core catalysts

		SEM Mo content/wt%			XPS		
Material	Surface area/ $m^2\ g^{-1}$	Theoretical	SEM-EDX	σ	Mo 3d peak position/eV	XPS Mo content/ atom%	Mo content of theoretical bulk phase/ atom%
α-MoO_3	1	66.7	62.2	0.4	—	—	—
$Fe_2(MoO_4)_3$	3	48.7	46.1	0.4	—	—	—
3 ML MoO_x@ISA	15	3.0	2.8	0.1	232.065	5.12	1.1
3 ML MoO_x@HSA	35	7.3	7.6	0.2	232.743	5.29	2.0

The X-ray absorption near-edge structure (XANES) region was used to establish the coordination environment of Mo in the shell@core catalysts with 1, 3 and 6 ML surface coverages on the HSA support (Table 4). Two features are assigned in the XANES spectra (Fig. 5).[7,38] The first is a pre-edge peak at 20 010 eV, attributed to the dipole-forbidden, quadrupole-allowed 1s → 4d transition. The pre-edge absorption is weak for Mo in a distorted octahedral environment (as in molybdena) but increases in intensity with tetrahedral character (as in $Fe_2(MoO_4)_3$). The second feature is an intense white line at ~20 020 eV from the dipole-allowed 1s → 5p transition, characteristic of Mo in (distorted) octahedral geometry.

The structure of the Mo-XANES is typical for a MoO_x@Fe_2O_3 catalyst.[8,9] From initial inspection, the increasing intensity of the pre-edge peak, with simultaneous diminution of the K-edge, indicates that the proportion of Mo in tetrahedral geometry increases with the level of surface doping. It is possible to quantify the proportion of octahedral and tetrahedral Mo by performing a linear combination fit (LCF), against suitable references (Table 4).

For the 1 ML catalyst, all Mo is in the (pseudo-)octahedral geometry and comprises the amorphous MoO_x surface phase.[9] At loadings in excess of 1 ML, an octahedral component from the MoO_x monolayer remains but subsurface layers are increasingly converted to $Fe_2(MoO_4)_3$. For the 3 ML and 6 ML catalysts, a tetrahedral component to Mo geometry indicates dopant levels in excess of monolayer coverage, where surplus Mo has reacted with the haematite core to form $Fe_2(MoO_4)_3$.

Table 4 The ratio of Mo environments in calcined MoO_x@HSA catalysts determined by a linear combination fit against octahedral (pre-calcined catalyst) and tetrahedral ($Fe_2(MoO_4)_3$) references

Theoretical molybdena coverage/ML	Proportion of Mo in octahedral geometry	Proportion of Mo in tetrahedral geometry	*R*-Factor
1	1.000	0.000	0.0055310
3	0.605	0.395	0.0007580
6	0.275	0.725	0.0007589

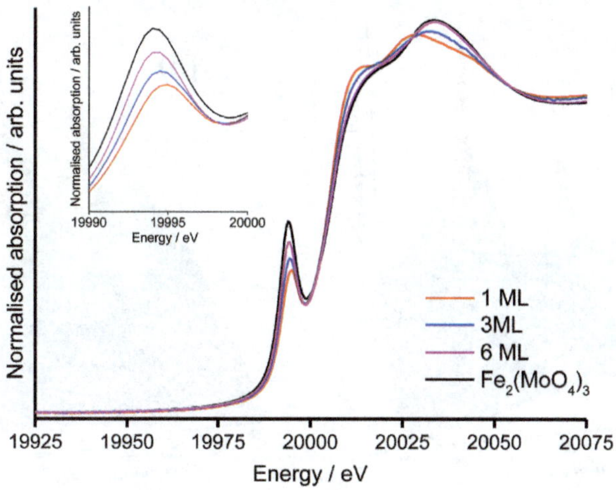

Fig. 5 Normalised Mo-XANES for 1, 3, 6 ML MoO_x@HSA catalysts, including an enlargement of the pre-edge region.

DRIFTS analysis of the shell@core catalysts

In situ diffuse reflectance infrared Fourier transform spectroscopy (DRIFTS) was used to monitor surface-bound intermediates during TPD of methanol. In studying the surface structure and Lewis acidity of the shell@core catalysts, methanol is a particularly informative probe, revealing the catalyst behaviour under reaction conditions.[39]

The DRIFTS study of the 3ML@ISA and 3ML@HSA catalysts is consistent with a dual adsorption pathway,[40] two methanolic surface species being detected (Fig. 6). The O–H group of methanol, adsorbed non-dissociatively, is identified by stretching vibrations at 3100–3500 cm^{-1} and a bending mode at ~1370 cm^{-1}. Also, a pair of intense bands at ~2950 and 2850 cm^{-1} correspond to the symmetric stretch and the first overtone of the symmetric bend of the methanol C–H bonds, respectively.[40] An analogous pair of bands in the C–H region (2930 and 2830 cm^{-1}) identifies methoxy formed by dissociative chemisorption of methanol.

The C–H region is particularly informative and combination bands possess unusual intensities due to Fermi resonance between the symmetric C–H stretch and the first overtone of the corresponding symmetric deformation.[40,41] Through the C–H band position, DRIFTS can reveal surface acid–base character. Over metal oxides of strong Lewis base character, surface oxygen or adsorbed hydroxyl may abstract the alcoholic proton from adsorbed CH_3OH. Thus, IR can distinguish the low frequency C–H stretch of anionic methoxide on a basic support (~2910/2800 cm^{-1}) from the intermediate bands of methoxy on moderately Lewis-acidic oxides (2930/2830 cm^{-1}) and the high frequency C–H vibrations of undissociated methanol on Lewis acidic surfaces (2960/2850 cm^{-1}).[42] In the case of the 3 ML catalysts, the strong Lewis acid character, particularly of the 3ML@ISA catalyst, is revealed by the greater intensity of the C–H stretches from adsorbed methanol, relative to methoxy. Overall, the C–H bands are less resolved in

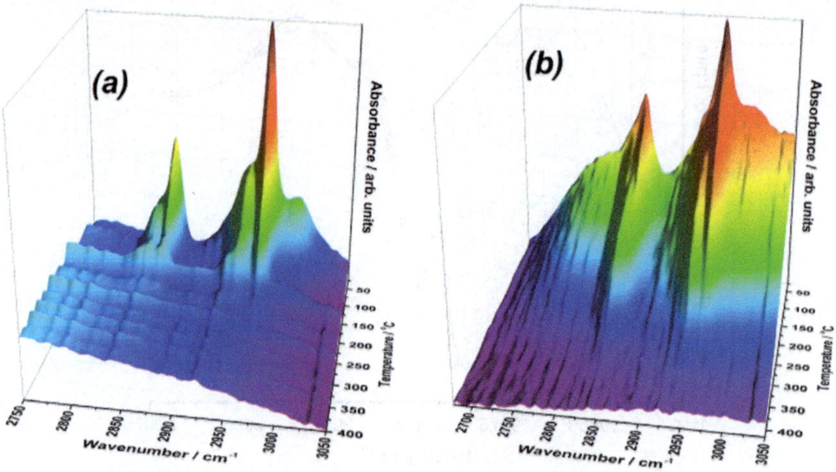

Fig. 6 3D maps of the C–H region of the IR spectra during TPD of methanol from the (a) 3ML@ISA and (b) 3ML@HSA catalysts.

3ML@HSA, which might indicate a greater range of adsorption sites, of varying affinity, for binding the methanolic adsorbate.

Furthermore, by tracking the C–H modes between 3000 and 2750 cm^{-1}, the reaction of surface-bound methanol can be monitored. On applying a temperature ramp, the intensity of the bands associated with surface methoxylation decline, presumably accompanying the reaction of methanol/methoxy to formaldehyde and CO, as evidenced by TPD. Whilst molecular methanol may be the dominant surface species at low temperature, associated C–H bands are lost at lower temperatures compared to those of methoxy, reflecting the stronger surface interactions of the latter.

Some distinctions can be made between the DRIFTS responses of the 3ML@ISA and 3ML@HSA materials. Firstly, bands associated with methanol and methoxy intermediates are absent at 255 °C from the ISA catalyst, whereas these bands are still detected at 315 °C for 3ML@HSA. This result supports microreactor TPD, for which formaldehyde (and CO) were stabilised on the HSA catalyst.

For the 3ML@ISA catalyst, the two sharp peaks between 3600 and 3100 cm^{-1} are typical for a non-bonded –OH group, but may be attributed to methanol and hydroxy species, each making consistent interactions with the surface, such that their O–H stretching frequency is well-defined and more intense. In contrast, on exposing 3ML@HSA to methanol, a broad O–H band is observed, indicating extensive intermolecular hydrogen bonding to the adsorbate. The C–H stretching region analysed by DRIFTS previously revealed (from broad C–H stretching bands) a range of binding interactions between 3ML@HSA and the methanolic adsorbate. Variations in the Lewis acid–base interactions between the surface and adsorbate may modify the O–H bond polarity and hence the strength of the hydrogen bond interactions. Since the O–H bond force constant is sensitive to modifications in the intermolecular hydrogen bonding capabilities, a broad range of O–H stretching frequencies is detected.

Temperature-programmed pulse flow reaction of methanol over bulk and shell@core catalysts

Under the temperature-programmed pulse flow reaction (TPPFR) of methanol, the bulk metal oxides show disparate activity (Fig. 7). For $Fe_2(MoO_4)_3$, conversion of methanol is initiated at a relatively low temperature (~100 °C) but only reaches 100% conversion (T_{100}) at 284 °C. Ferric molybdate sustains 100% formaldehyde selectivity (T_{sel}) to 330 °C, above which temperature CO is evolved.

In contrast, the Al-doped haematite operates a combustion pathway, converting methanol to CO_2 (with a trivial quantity of CO at low conversion).[43] The catalytic activity of the aluminous haematite is inferior to that of ferric molybdate, $Al_xFe_{(2-x)}O_3$ achieves 50% methanol conversion (T_{50}) at 20 °C above that for $Fe_2(MoO_4)_3$, which reflects the stabilisation of surface intermediates on the HSA support.

However, when the haematite support is doped with molybdena, the reactivity of the resultant shell@core catalyst is dominated by the surface, where large energetic barriers to oxygen insertion favour formaldehyde desorption.[25] Compared to the bulk phases, T_{50} is universally lower for the shell@core systems but decreases as the surface area, and hence access to active sites, is improved (Table 5).

The TPPFR indicated that, of all the shell@core systems, a higher formaldehyde yield is attained with the 3ML@HSA catalyst and the maximum formaldehyde yield (T_{max}) is also shifted to a lower temperature (Table 5). Given that, in TPD, the 3ML@HSA catalyst desorbs formaldehyde at the highest temperature of all the shell@core catalysts, the catalyst must therefore operate a high methanol conversion in the T_{sel} range.

Unlike the $Fe_2(MoO_4)_3$ catalyst, the 3ML@HSA catalyst does not achieve 100% formaldehyde yield but between 190 and 255 °C the HSA shell@core catalyst outperforms the other systems, selectively oxidising methanol to formaldehyde with high yield. Significantly, the HSA supported molybdena catalyst achieves its

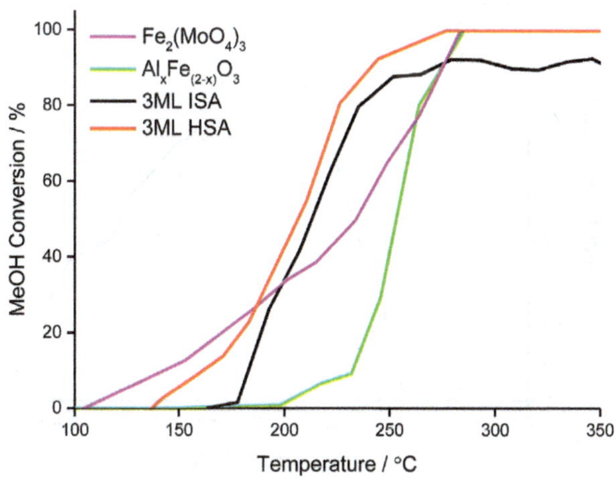

Fig. 7 Methanol conversion during the TPPFR over bulk and shell@core iron molybdate catalysts.

Table 5 The temperatures at which the onset (T_0), 50% (T_{50}) and 100% (T_{100}) methanol conversion are observed for the bulk and shell@core iron molybdate catalysts and the temperature range over which these exhibit 100% formaldehyde selectivity (T_{sel}). Also, the temperatures of the maximum formaldehyde yield (T_{max}) for the bulk and shell@core iron molybdate catalysts during the TPPFR of methanol

System	Surface area/m^2 g^{-1}	T_0/°C	T_{50}/°C	T_{100}/°C	T_{sel}/°C	Maximum formaldehyde yield/%	T_{max}/°C
$Al_xFe_{(2-x)}O_3$	46	173	253	285	—	0	—
$Fe_2(MoO_4)_3$	3	104	233	283	104–344	100	283–344
3ML@HSA	35	143	206	277	143–245	94	244
3ML@ISA	14	178	213	—	178–207	84	252

maximum formaldehyde yield (94%) at T_{max} ~ 40 °C below that for the industrial bulk phase (Fig. 8).

Conclusions

A haematite support with a surface area of 46 m^2 g^{-1} was successfully synthesised by the introduction of 5 wt% aluminium dopant. XPS analyses favoured a homogeneous distribution of Al and Fe in the haematite framework. Lattice strain induced by the isomorphous substitution of Fe^{3+} for the smaller Al^{3+} cation was found to constrain unit cell growth, particularly along the *c* dimension, and a reduction in particle size *versus* commercial haematite nanoparticles was evidenced by line broadening in the XRD and Raman spectra. Besides quantum confinement effects, the spectroscopic characteristics of $Al_xFe_{(2-x)}O_3$ were consistent with phase-pure haematite. Moreover, TPD of methanol revealed combustion products typical of haematite catalysis. A trend towards increased

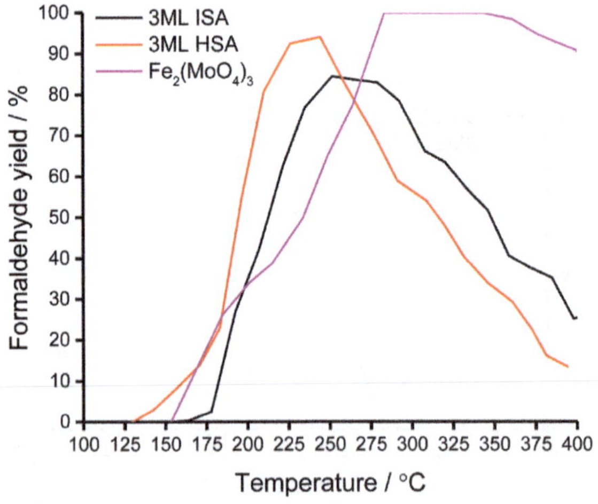

Fig. 8 Formaldehyde yield of the bulk and shell@core iron molybdate catalysts under the TPPFR of methanol.

activation energy of CO_2 production with a higher surface area support was attributed to stabilisation of the antecedent formate intermediate on smaller particles, of higher surface free energy.

In a novel preparation, $Al_xFe_{(2-x)}O_3$ was coated with molybdena to yield a shell@core catalyst of surface area 35 $m^2\ g^{-1}$. On confirming a Mo-rich exterior by XPS, surface analysis was performed by tuning XAFS to the Mo K-edge. All systems maintained a (*pseudo*)-octahedral Mo component of surface MoO_x, though an increase in tetrahedral Mo at higher surface dosing was consistent with a more extensive subsurface $Fe_2(MoO_4)_3$ formation. For the high surface area shell@core system, the sub-surface $Fe_2(MoO_4)_3$ phase was identified by XRD, Raman and IR spectroscopies.

In the TPD of methanol, the reactivity of the shell@core catalysts was directed by the molybdena surface. Formaldehyde and CO products were consistent with the methoxy intermediate detected by DRIFTS. The TPPFR of methanol over the shell@core catalyst revealed an increase in peak formaldehyde yield with catalyst surface area, accompanied by a shift of the maximal yield to a lower temperature. Between 190 and 255 °C, the formaldehyde yield over the 3ML@HSA catalyst exceeded that of the other systems. Significantly, the 3ML@HSA catalyst reached the maximum formaldehyde yield (94%) ~40 °C below that for the bulk phase. As such, further catalytic testing of the 3ML@HSA catalyst would focus on isothermal methanol oxidation between 190 and 255 °C to monitor formaldehyde yield over an extended reaction period. As is, the improved catalytic activity of the 3ML@HSA catalyst might be trialled for other reactions for which ferric molybdate catalysis is employed, such as propene,[44,45] methane[46] and toluene[47] oxidation.

Acknowledgements

The UK Catalysis Hub is kindly thanked for resources and support provided *via* our membership of the UK Catalysis Hub Consortium and funded by EPSRC (portfolio grants EP/K014706/1, EP/K014668/1, EP/K014854/1 and EP/K014714/1). EPSRC are thanked for supporting the activities of the UK Catalysis Hub at RCaH and the grant for funding Dr P. P. Wells (Catalytic Science in the Harwell Research Centre EP/I019693/1). Diamond Light Source and the Cardiff Catalysis Institute (CCI) are acknowledged for funding the studentship of Catherine Brookes. The authors wish to acknowledge the Diamond Light Source for provision of beam-time (SP10306). The RCaH are also acknowledged for use of facilities and support of their staff. Dave Morgan of the CCI is thanked for assisting with the XPS data acquisition.

References

1 W. J. Linn and A. W. Sleight, *J. Catal.*, 1976, **41**, 134–139.
2 J. L. Callahan, R. K. Grasselli, E. C. Milberger and H. A. Strecker, *Ind. Eng. Chem. Prod. Res. Dev.*, 1970, **9**, 134–142.
3 S. Kowatsch, *Formaldehyde*, Springer, Berlin Heidelberg, 2010.
4 G. W. Keulks, *J. Catal.*, 1970, **19**, 232–235.
5 U. Chowdhry, A. Ferretti, L. E. Firment, C. J. Machiels, F. Ohuchi, A. W. Sleight and R. H. Staley, *Appl. Surf. Sci.*, 1984, **19**, 360–372.

6 M. Bowker, R. Holroyd, A. Elliott, P. Morrall, A. Alouche, C. Entwistle and A. Toerncrona, *Catal. Lett.*, 2002, **83**, 165–176.

7 M. Bowker, C. Brookes, A. F. Carley, M. P. House, M. Kosif, G. Sankar, I. Wawata, P. P. Wells and P. Yaseneva, *Phys. Chem. Chem. Phys.*, 2013, **15**, 12056–12067.

8 C. Brookes, P. P. Wells, G. Cibin, N. Dimitratos, W. Jones, D. J. Morgan and M. Bowker, *ACS Catal.*, 2014, **4**, 243–250.

9 C. Brookes, P. P. Wells, N. Dimitratos, W. Jones, E. K. Gibson, D. J. Morgan, G. Cibin, C. Nicklin, D. Mora-Fonz, D. O. Scanlon, C. R. A. Catlow and M. Bowker, *J. Phys. Chem. C*, 2014, 26155–26161.

10 Y. Huang, L. Cong, J. Yu, P. Eloy and P. Ruiz, *J. Mol. Catal. A: Chem.*, 2009, **302**, 48–53.

11 K. Routray, W. Zhou, C. J. Kiely, W. Grünert and I. E. Wachs, *J. Catal.*, 2010, **275**, 84–98.

12 M. R. Sun-Kou, S. Mendioroz, J. L. G. Fierro, J. M. Palacios and A. Guerrero-Ruiz, *J. Mater. Sci.*, 1995, **30**, 496–503.

13 Y. Matsuoka, M. Niwa and Y. Murakami, *J. Phys. Chem.*, 1990, **94**, 1477–1482.

14 C. J. Machiels, W. H. Cheng, U. Chowdhry, W. E. Farneth, F. Hong, E. M. Mc Carron and A. W. Sleight, *Appl. Catal.*, 1986, **25**, 249–256.

15 C. Brookes, M. Bowker, E. K. Gibson, D. Gianolio, K. M. H. Mohammed, S. Parry, S. M. Rogers, I. P. Silverwood and P. P. Wells, *Catal. Sci. Technol.*, 2016, 722–730.

16 Y. Zheng, Y. Cheng, Y. Wang, F. Bao, L. Zhou, X. Wei, Y. Zhang and Q. Zheng, *J. Phys. Chem. B*, 2006, **110**, 3093–3097.

17 K. Sivula, F. le Formal and M. Grätzel, *ChemSusChem*, 2011, **4**, 432–449.

18 A. Khelfa, V. Sharypov, G. Finqueneisel and J. V. Weber, *J. Anal. Appl. Pyrolysis*, 2009, **84**, 84–88.

19 G. Picasso Escobar, A. Quintilla Beroy, M. P. Pina Iritia and J. Herguido Huerta, *Chem. Eng. J.*, 2004, **102**, 107–117.

20 R. M. Cornell and U. Schwertmann, *The Iron Oxides: Structure, Properties, Reactions, Occurrences and Uses*, Wiley, 2006.

21 M. Hermanek, R. Zboril, I. Medrik, J. Pechousek and C. Gregor, *J. Am. Chem. Soc.*, 2007, **129**, 10929–10936.

22 U. Schwertmann, R. Fitzpatrick, R. Taylor and D. Lewis, *Clays Clay Miner.*, 1979, **27**, 105–112.

23 A. Zoppi, C. Lofrumento, E. M. Castellucci and P. Sciau, *J. Raman Spectrosc.*, 2008, **39**, 40–46.

24 U. Schwertmann and R. M. Cornell, *Iron Oxides in the Laboratory*, Wiley, 2008.

25 M. P. House, A. F. Carley, R. Echeverria-Valda and M. Bowker, *J. Phys. Chem. C*, 2008, **112**, 4333–4341.

26 R. Li, Q. Li, S. Gao and J. K. Shang, *J. Am. Ceram. Soc.*, 2011, **94**, 584–591.

27 M. Wells, R. Gilkes and R. Anand, *Clay Miner.*, 1989, **24**, 513–530.

28 R. Shannon, *Acta Crystallogr., Sect. A: Cryst. Phys., Diffr., Theor. Gen. Crystallogr.*, 1976, **32**, 751–767.

29 L. Truffault, B. Choquenet, K. Konstantinov, T. Devers, C. Couteau and L. J. M. Coiffard, *J. Nanosci. Nanotechnol.*, 2011, **11**, 2413–2420.

30 H. Liu, T. Chen, X. Zou, C. Qing and R. L. Frost, *J. Raman Spectrosc.*, 2013, **44**, 1609–1614.

31 D. L. A. de Faria, S. Venâncio Silva and M. T. de Oliveira, *J. Raman Spectrosc.*, 1997, **28**, 873–878.
32 M. Hanesch, *Geophys. J. Int.*, 2009, **177**, 941–948.
33 P. A. Redhead, *Vacuum*, 1962, **12**, 203–211.
34 R. L. Blake, R. E. Hessevic, T. Zoltai and L. W. Finger, *Am. Mineral.*, 1966, **51**, 123.
35 A. P. V. Soares, M. F. Portela and A. Kiennemann, *Catal. Rev.: Sci. Eng.*, 2005, **47**, 125–174.
36 M. P. House, M. D. Shannon and M. Bowker, *Catal. Lett.*, 2008, **122**, 210–213.
37 W. McMaster, N. K. Del Grande, J. Mallett and J. Hubbell, *International Tables for Crystallography, Mathematical, Physical and Chemical Tables, Compilation of X-Ray cross sections*, Section III, Springer Science & Business Media, 1969.
38 A. M. Beale, S. D. M. Jacques, E. Sacaliuc-Parvalescu, M. G. O'Brien, P. Barnes and B. M. Weckhuysen, *Appl. Catal., A*, 2009, **363**, 143–152.
39 A. M. Turek, I. E. Wachs and E. DeCanio, *J. Phys. Chem.*, 1992, **96**, 5000–5007.
40 L. J. Burcham, L. E. Briand and I. E. Wachs, *Langmuir*, 2001, **17**, 6164–6174.
41 G. Busca, *Catal. Today*, 1996, **27**, 457–496.
42 L. J. Burcham, L. E. Briand and I. E. Wachs, *Langmuir*, 2001, **17**, 6175–6184.
43 M. P. House, Selective oxidation of methanol over iron molybdate catalysts, *PhD Thesis*, Cardiff University, 2007.
44 K. Manseri, H. Hentit, E. H. Elandaloussi, B. Benaichouba and M. S. Ouali, *Hyperfine Interact.*, 2010, **198**, 243–257.
45 B. Benaichouba, P. Bussiere and J. C. Vedrine, *Appl. Catal., A*, 1995, **130**, 31–45.
46 A. S. Chellappa and D. S. Viswanath, *Ind. Eng. Chem. Res.*, 1995, **34**, 1933–1940.
47 K. L. Madhok and K. P. Srivastava, *Proc. - Indian Acad. Sci., Chem. Sci.*, 1981, **90**, 527–535.

Faraday Discussions

DISCUSSIONS

Designing new catalysts: synthesis of new active structures: general discussion

Parasuraman Selvam, Joachim Sauer, Bruce Garrett, Charles Campbell, Rutger van Santen, Philip Davies, Amy L. Miller, Michael Bowker, Graham Hutchings, Daniel Wotton, Hans-Joachim Freund, Mark Howard, Mark Feaviour, Robbie Burch, Anna Maria Raspolli Galletti, Elad Gross, Francisco Ivars Barcelo, Andrzej Kotarba, Simon Kondrat, Bert Weckhuysen, Biju Majumdar, Albert Bruix, Nico Fischer, Bruce Gates, Jacob Moulijn, Alberto Roldan, Natasa Novak Tusar, Tomasz Jakubek, David Willock, Michael Craven, Kalidhasan Sethu, Richard Catlow, Robert Madix, Haresh Manyar, Cynthia Friend, Avelino Corma, Peter Wells, Wataru Ueda, Annette Trunschke and Richard Palmer

DOI: 10.1039/C6FD90015A

Philip Davies opened the discussion of the introductory lecture by Avelino Corma: The metal nanoparticles inside the zeolites are in a different environment from those outside. Is there any difference in their chemistry and their catalytic behaviour?

Avelino Corma answered: We were not able to determine the differences in reactivity, other than the accessibility of reactants with different sizes. It should be said that we did not use molecules specifically suited for showing potential electronic–entropic differences. I agree that this is an important point to be considered. What we clearly observed is that the clusters inside the channels were stable towards sintering.

Cynthia Friend asked: Have you considered the possible effect of ligands bound to your clusters? Have van der Waals' interactions been explicitly included?

Avelino Corma replied: The theoretical calculation of the interactions of nitrobenzene with the nanoparticles includes the interactions with the support and van der Waals' interactions. In the case of clusters, H_2 dissociation has been carried out on isolated Pt, Au and Au-Pt clusters and van der Waals' interactions were not considered.

Graham Hutchings continued: You rightly point out in your design strategy that the adsorption of the reactant is the key factor, and that adsorption occurs preferentially *via* the nitro group when reacting substituted nitrobenzenes. What happens if you add an additional nitro group? Will this inevitably lead to oligomers being formed?

Avelino Corma responded: If the molecule has two nitro groups to be reduced, instead of one, it is not a problem.

On the contrary, for the coupling of the corresponding amino groups derived from nitrobenzene to form the azo compounds, we did not see the formation of trimers or tetramers.

The consecutive steps are slow and the higher molecular weight products would be difficult to desorb at the lower reaction temperatures.

Bruce Gates remarked: A valuable goal with metal cluster catalysts is to control the cluster nuclearity. With a zeolite as a support, the clusters may be stabilized. What confers that stability? Is it a confinement effect associated with the structure of the zeolite? Are there other issues at play?

Avelino Corma answered: Indeed I believe that besides a confinement effect, which should be more important in small pore zeolites, the zeolite framework with the electric fields associated and interactions with the framework oxygen atoms can also stabilize charged metal clusters.

Parasuraman Selvam asked: Where do you draw a line to distinguish between the clusters and nanoparticles in terms of the size and shape? Are these clusters amorphous or crystalline? What do we understand about these aspects in terms of their activity towards a chosen reaction?

In general, the driving force for the adsorption is the coordinative unsaturation of the active sites. In this context, the nanoclusters/nanoparticles have definite sizes and shapes, *viz.*, the corners, edges and faces, and predominantly the sites at the faces are utilized for the reaction. Indeed, all three sites have a different coordination and the number of such sites varies with size. Are they not influencing the activity when the size becomes smaller and smaller?

Avelino Corma responded: We define the limit between the clusters and nanoparticles by considering that in the clusters, all atoms are directly accessible for potential interactions. This certainly implies subnanometric sites. In the case of the clusters that are normally formed by 3–15 atoms, it is difficult to consider crystallinity. Nevertheless, it is possible to find atoms with different coordination in the clusters, and they can show different reactivity. In the case of nanoparticles with sizes >1 nm, definitive surface atoms with different coordinations can show different reactivity, especially for reactions sensitive to the structure, according to Boudart's nomenclature.

Bruce Garrett remarked: You mentioned that the confinement of zeolite pores can affect the overall rates of a catalytic process by controlling diffusion. Sufficiently small pores can also affect the rates of elementary reactions. The confined space of the pore can change the entropy of activation (*e.g.*, the transition state structure can be more compact and better accommodated into the constrained space than the separated reactants, thus increasing the entropy). How small do the pores need to be to see these types of effects?

Avelino Corma replied: From the point of view of diffusion, this can become very slow (slower than the rate of the reaction) or may even be totally impeded when the size of the pore starts to match that of the molecule. The solution to increase the zeolite catalyst activity in those cases, where the diffusion through the micropore is rate controlling, is to decrease the zeolite crystallite size, *i.e.* to decrease pore length or, in other words, to increase the number of more accessible sites at or close to the external surface.

With respect to the transition state stabilization, the size of the Ts and that of the pore of the zeolite should allow to maximize the van der Waals' interactions to get the desired stabilization. I believe that what is determinant is the relative size and shape of the zeolite and the reaction transition state.

Robbie Burch addressed Robert Madix, Charles Campbell and Avelino Corma: In 1926, H. S. Taylor first suggested that different locations on a catalyst's surface could have different activities, and he introduced the expression 'active site' to identify the most important sites. At that time, an active site was treated as a very static entity but we now know that adsorption and desorption, and the whole catalytic process, is quite violent at the atomic level and so the 'active' surface may be in a state of flux during the catalytic action. With this in mind, how do you think we can best use fundamental, essentially static investigations, whether theoretical or experimental, into the nature of a catalytic action to design a new catalyst? In your opinion, is there any realistic prospect of designing a genuinely new catalyst from first principles or should we be satisfied if such fundamental investigations simply point us in new directions where subsequent 'trial and error' experiments may lead to a new catalyst?

Robert Madix answered: I completely agree that catalyst surfaces are dynamic in nature. As I have noted in previous comments during this meeting and from the responses to questions, there are numerous experiments reported in the literature that document the importance of this and suggest that theorists address the relevant timescales on which this may be important in actual catalytic reactions.

Charles Campbell responded: You raise a very important point. As you probably know, many DFT calculations allow the atoms in the topmost atomic layer(s) of the catalyst's surface to move upon the adsorption of reactants (or transition states). Thus, there is already quite a bit of restructuring of the catalyst as each step proceeds in many DFT microkinetic models. So if one is to define the "active site" in such advanced models, one must describe at a certain adsorbate coverage or in the absence of adsorbates. It is not a static entity. Still, I do not think such adsorbate-induced motions preclude the value of the "active-site" concept.

Avelino Corma replied: I believe that theoretical methods are considering the dynamics of surfaces (at least at the active site) better and better during catalytic processes. This, together with operando techniques and isotopic-microkinetic studies, direct the design of better catalysts. I believe that it is the combination of theoretical and experimental studies that gives the best information.

Biju Majumdar opened the discussion of the paper by Charles Campbell†: Steric constraints in the absorbed intermediates over metal nanoparticles explain the selectivity of the reaction. However, for a catalytic reaction, the catalytic conversion increases with an increasing number of exposed metal active sites for the incoming reactant material. So, if the active metal surface is sterically hindered, how can the reactant materials approach the active metal surface efficiently to have a good catalytic conversion?

Charles Campbell replied: You raise a good point. Certainly the rate per total area of the metal nanoparticles would usually go down when they are encapsulated in MOFs, like the materials we model here (DOI: 10.1039/C5FD00198F), simply due to the corresponding loss of exposed surface area. There are two higher-order effects that can mitigate this harmful effect. It is known that the buildup of decomposed adsorbates that can poison the surface is often inhibited when large ensembles of contiguous metal sites are not available. More importantly, for the reaction studied here, the selectivity is often more important than the total rate per metal atom for industrial applications. The limited access to metal atoms in this material, where the reactant butane must align within the pore of the MOF with only one carbon atom 'touching' the metal surface results in only that terminal carbon atom getting oxidized, thus giving 100% regio-selectivity to 1-butanol in the optimized conditions (according to DFT energies).

Albert Bruix continued: Using the He ring looks like an excellent idea to model the effect of the encapsulating MOF. However, how inert is the MOF edge (or the MOF-metal interface), especially with respect to He?

Also, encapsulation is used to increase the selectivity towards a particular product, but at the same time this drastically reduces the surface area available to reactants. Could you comment on the trade-off between the loss of surface area and gain in selectivity?

Charles Campbell answered: Thank you. Yes, using a frozen ring of helium atoms is a great simplification. Since He atoms are very inert, except for steric repulsions, they offer the advantage of allowing one to make a model wherein only the steric constraints of the pore size are isolated and evaluated, separately from all other types of interactions of the real pore with the adsorbed reaction intermediates, transition states and metal nanoparticles. Of course, this misses many effects that could be crucial in the real catalyst.

Your second question is the same as that from Biju Majumdar, please see my response to that above. The trade-off always comes down to economic details of the process, about which I am not an expert.

† We recently discovered that the reaction energies and barriers initially studied for steps (4) and (5) in Table 1 (DOI: 10.1039/C5FD00198F)were incorrect. The data in the version of the manuscript presented to the attendees at the meeting were later found to be inaccurate. Therefore, the data presented in the final version of the manuscript have been revised accordingly.

Jacob Moulijn remarked: In your modelling of spatial constraints you use a 'surrogate pore' built from two layers of He atoms. Clearly the method works well. Perhaps you could get additional insights from simulations involving surrogate pores with different properties. For instance, a blank experiment without a surrogate pore would be an extreme case. Also, you could define a pore that has a certain interaction with the molecule.

Charles Campbell responded: Thank you, those are excellent suggestions!

Avelino Corma added: Molecular sieves have the problem of the rate of diffusion, you have a good system.

Philip Davies asked: The window created by the helium cage is quite narrow and its positioning will therefore determine where the reactant can bind to the surface. Have you looked at how moving the window across the surface affects the predicted chemistry?

Charles Campbell answered: No. This has not been done. The He "pore" was just centrally located over the most active site (a Pd atom completely surrounded by Ag atoms).

Graham Hutchings enquired: In your mechanism you indicated that the surface hydroxyls are important. Have you considered the possibility of hydroperoxy species or even hydrogen peroxide being formed in this reaction? Pd is a very good catalyst for forming these species, so such species might be present.

Charles Campbell responded: Prof. Hutchings, as always, raises a very good point. Unfortunately, we have not considered –OOH species here yet.

Bruce Garrett said: Is there a difference between your approach to degrees of rate control and sensitivity analysis? Sensitivity analysis has a long history of being used to understand the parameters in kinetic mechanisms that influence the overall rate of a process the most (*e.g.*, for atmospheric models of reaction mechanisms). Sensitivities are usually formulated in terms of partial derivatives of the overall rate with respect to the rate constants. In your approach (DOI: 10.1039/C5FD00198F) you formulate degrees of rate control in terms of partial derivatives of the overall rate with respect to activation free energies and equilibrium free energies. At first glance it appears that these degrees of rate control should contain similar information to standard sensitivities, since the free energies are proportional to the log of the rate constants. Please explain the difference.

Charles Campbell replied: Thanks for the perceptive question, Bruce. You are correct that the partial derivative I defined as the degree of rate control (DRC) is closely related to the sensitivity analyses, where the derivative of the log of the rate is taken relative to each rate constant. Our derivation that shows the relationship between these partial derivatives is found in reference 8 in DOI: 10.1039/C5FD00198F. The major difference is in the subscripts on the partial derivative in my DRC, which ensures thermodynamic consistency by holding the equilibrium constant for each step constant, in addition to all other rate constants (equivalent to keeping all standard state free energies constant, except that for the transition state of interest).

Michael Bowker remarked: You suggest that adsorbate interactions are not important, do you really believe that? And do you think it is possible that we will get to the stage of total predictability of single step reaction kinetics?

Charles Campbell responded: Adsorbate–adsorbate interactions are usually large (easily up to 50 kJ mol^{-1}). Many DFT groups are doing great work trying to include these in their DFT-based microkinetic models. However, the helium-ring pores here separate the adsorbates from each other by a large distance when in different pores, so they interact strongly only with those within the same pore. This is generally only the reacting adsorbates themselves for the cases treated here, and those interactions are included in some steps.

Regarding the predictability of single-step reaction kinetics based on DFT, the energy errors in DFT still average at ~30 kJ mol^{-1} (see: a benchmark database for adsorption bond energies to transition metal surfaces and comparison to selected DFT functionals[1]). If this error appears in an Arrhenius activation energy, the rate would have an error of several orders of magnitude. So we are not there yet. However, I think and hope that DFT will be getting much more accurate soon, at least for metal surfaces.

1 J. Wellendorff, T. L. Silbaugh, D. Garcia Pintos, J. K. Nørskov, T. Bligaard, F. Studt and C. T. Campbell, *Surf. Sci.*, 2015, **640**, 36.

Rutger van Santen remarked: Is there information on the state of the alloy surface? Is it metallic, as assumed in the simulations, or actually an oxide layer? It is known for Pt and Pd that direct CH activation is enhanced on cationic transition metal atoms, compared to activation by the metal. Heterolytic activation of CH has a higher barrier. Possibly O insertion then occurs in two reaction steps and hydrogenation of O atoms may also play a role.

Charles Campbell replied: Thanks for the good question. The DFT model did not really allow the oxide to be formed. However, since it treats the thermodynamics of adsorbed oxygen with reasonable accuracy, and since the coverage by surface oxygen turns out to be very low at steady state in the microkinetic model derived from the DFT energetics (since all the oxygen is being rapidly consumed by reaction with butane), it certainly appears that oxides would not be formed in such a DFT model, even if it did allow oxide formation. I do not think any experiments have been done yet to determine this.

Alberto Roldan communicated: The selection of elementary reaction steps is a crucial process to develop reliable micro-kinetic studies. According to the manuscript (DOI: 10.1039/C5FD00198F), the inclusion of water has an effect on the reaction mechanism, nevertheless the step OH + H = H_2O is missed in elementary reactions, which can be important for the whole discussion as either H* or OH* are present on the surface. Considering this step may lead to the formation of H_2O instead of H_2, as suggested in the original mechanism. Have the authors verified that this elementary reaction is not needed for a well-founded description of the process?

Charles Campbell communicated in response: Good point! The elementary step $H_2O^* = OH^* + H^*$ was omitted from the set of elementary reactions we included, just to simplify the analysis. This is based on my knowledge of the literature for late transition metals like Pd, Pt, Ag and Cu, where it is always observed that the activation barrier for this step is very high. When oxygen is around on the surface (which is the case here since we have lots of oxygen gas in the reactant feed), water dissociates much more rapidly *via*: $H_2O^* + O^* = 2\ OH^*$, which has a very low barrier. We did include this reaction and its reverse (see Table 5, DOI: 10.1039/C5FD00198F).

Microscopic reversibility suggests that the reverse of this reaction will be much faster than the reverse of $H_2O^* = OH^* + H^*$. That being said, we cannot be completely sure of my logic above, so it would have been wiser for us to include this step also in the mechanism. I am pretty sure it would have little contribution to the rates.

Mark Howard opened the discussion of the paper by Richard Palmer: The comparison between cluster beam and impregnation prepared catalysts for 1-pentyne hydrogenation in Fig. 6 (DOI: 10.1039/C5FD00178A) has no data for impregnated catalysts between 20% and 100% conversion. The selectivities look similar up to 20% conversion. The low selectivity for impregnated catalysts at 100% conversion may be a consequence of running these reactions for too long with higher activity catalysts. Data for impregnated catalysts at intermediate conversions, *i.e.* between 20% and 100%, are needed to support the conclusion that the cluster beam prepared material is more selective.

During the liquid phase 3-hexyn-1-ol hydrogenation tests, is there any possibility of the metal clusters being washed off the support into the suspension? How have you tested the immobilisation?

Richard Palmer answered: The reason for the lack of data as a function of temperature, between 20% and 100% conversion, is that the conversion lights off at a particular temperature, something which is quite common in gas phase catalysis. We do not have evidence for the leaching of metal during the reaction at this stage, but these are early studies.

Hans-Joachim Freund asked: How softly can you land the clusters on the support (energy per atom) and is there information about how the various possible supports influence the landing?

Richard Palmer answered: We have identified three regimes of cluster–surface interactions as a function of impact energy per atom; soft landing[1], cluster pinning and cluster implantation. Of course the detailed behaviour will depend on the support. In the soft landing regime gold clusters do not wet the amorphous carbon (*i.e.* they are quasi spherical[2]) but they certainly wet MgO.

1 R. E. Palmer, S. Pratontep and H.-G. Boyen., *Nat. Mater.*, 2003, **2**, 443–448.

2 N. P. Young, Z. Y. Li Y. Chen, S. Palomba, M. Di Vece and R. E. Palmer, *Phys. Rev. Lett.*, 2008, **101**, 246103.

Jacob Moulijn said: I have a general question and comment. In your contribution you claim that you made steps toward a practical application. An important step is that you extended the use of the cluster beam route to particles. This step is very relevant. For the synthesis of practical catalysts it would be attractive to apply porous supports to land the clusters. However, I suppose that this would be impossible. Do you have ideas of how this could be done? Could additive methods similar to 3D-printing be used?

Richard Palmer replied: I think it may be possible to deposit clusters from the beam directly into macroporous materials, but not, say, zeolites. Certainly one can envisage additive approaches in which both the clusters and support are deposited to a template. Recently we achieved the deposition of a multilayer architecture based on repeating stacks of carbon support, clusters and polymer release layers.

Bruce Gates remarked: You stated that it is not practical to deposit catalysts within the pores of the supports with your method. But what do you get with non-porous powder supports? Do you make non-uniform materials with such supports, or can you make them uniform?

Richard Palmer responded: It might be possible to deposit clusters into large pores. We have some images showing good dispersion of the clusters over non-porous powder particles, but this analysis is only just beginning, and we are also checking how effectively the powder is agitated.

Avelino Corma commented: what is your hypothesis of the activity of impregnation? What about selectivity effects?

Richard Palmer responded: Our hypothesis is that one of the key differences between the impregnated and cluster beam materials is the nature of the active sites. In an impregnated catalyst, a metal salt (palladium nitrate in this case) comes into contact with the support, allowing ion exchange processes to take place. These ion-exchanged sites probably persist during processing and in some cases become highly active and unselective in catalysis. The cluster beam method pre-fabricates the clusters in the gas phase prior to their intact deposition, so contact between the metal salts and the support is avoided, giving rise to a different set of active sites which may well be more uniform. Such a difference would account for the enhanced selectivity of the cluster beam materials observed.

Graham Hutchings said: This is a really exciting way of making nanoparticles that can be supported and you are making headway in scaling up so that you can obtain mg quantities for catalyst evaluation. With bimetallic nanoparticles are you able to control the composition with particle size? Fig 10a (DOI: 10.1039/C5FD00178A) shows that there is a distribution of particle sizes for Ag clusters and presumably this will be the same for bimetallic clusters? If the composition could be controlled so that it is not a function of the particle size, that would be very valuable as many current methods give supported nanoparticles in which the composition of the bimetallic is a function of the particle size.[1]

1 G. J. Hutchings and C. J. Kiely, *Acc. Chem. Res.*, 2013, **46**, 1759.

Richard Palmer replied: One of the reasons for coming to this Faraday Discussion was to learn what catalysis scientists want, so we will now regard 'cluster composition independent of cluster size' as a key challenge for our future work with the new MACS. We recently made our first binary clusters this way (Au and Pd) but we do not know the details of the structure/composition yet. On the matter of size control, the best we have achieved to date in the MACS is $\pm 10\%$ control over the cluster diameter.[1]

1 R. E. Palmer, L. Cao and F. Yin, *Rev. Sci. Instrum.*, 2016, **87**, 046103.

Anna Maria Raspolli Galletti commented: The cluster beam route appears very interesting. In the perspective of its scale-up, do you have a preliminary evaluation of its sustainability (also from an economic point of view) with respect to other approaches, such as electric explosion of wires, metal vapour deposition or sonochemical decomposition? I am interested in this route's sustainability also from the energy consumption and environmental point of view.

Richard Palmer responded: From an environmental point of view the cluster beam route appears promising. There are no solvents and the material is contained within the vacuum chamber, one can call it 'metal to metal' processing. We estimate that the power consumption of the MACS system employed in the current work is of the order of 1kW. The cryogenic cooling is a significant contribution.

Andrzej Kotarba commented: Several important advantages of the cluster beam route for model catalyst preparation are pointed out in the paper (DOI: 10.1039/C5FD00178A). There is no doubt that we can control the composition, size and thus dispersion of the nanoparticles (NPs) on the support. Another important point is the preparation of the catalyst without additional chemicals like NP stabilizers. However, in practice we use the additives not only to stabilize NPs but also to improve the wetting of the support's surface with the active phase precursor, and control the nanoparticle formation (crystallization) to expose a specific, optimal crystallographic plane. Therefore, the question arises if there is any possibility to control the cluster orientation *via* modifying the cluster beam parameters?

Richard Palmer responded: It should be possible to chemically functionalize clusters produced by the MACS after they are formed. But as far as bare clusters are concerned, their orientation on the support would probably depend on the degree of epitaxy. The Leuven group carried out some classic early work on this[1,2] and we have recently identified a dependence of epitaxy of Au clusters on MgO as a function of cluster size.

1 N. Vandamme, E. Janssens, F. Vanhoutte, P. Lievens and C. Van Haesendonck, *J. Phys.: Condens. Matter*, 2003, **15**, S2983.

2 N. Vandamme, G. Verschoren, A. Depuydt, M. Cannaerts, W. Bouwen, P. Lievens, R. E. Silverans and C. Van Haesendonck, *Appl. Phys. A*, 2001, 72, S177.

Cynthia Friend asked: Have you investigated the effects of using a viscous liquid or polymer in order to dissipate the energy of your cluster source?

Richard Palmer answered: We have not used these systems to slow the clusters. In the magnetron source the sample is biased to control the deposition energy of the ionised clusters. We have not investigated the kinetic energy of the (apparently mainly neutral) clusters from the MACS yet. In relation to polymers, we have deposited clusters from both sources onto PVP films, which we have subsequently dissolved to form PVP-capped colloidal clusters; we have also implanted energetic clusters into polymers to make nanocomposites for sensors and we plan to write up these results shortly.

David Willock said: In solution chemistry the size of metal particles produced can be controlled by influencing the rate at which they precipitate. Is this also the case in the cooling step used in your methodology? For example could inert gases of heavy molecular species, such as SF_6, be used to increase the rate of cooling so that smaller particles are favoured?

Richard Palmer replied: At the present time, the parameters known to control the size of the clusters produced in the MACS are (i) metal concentration in the matrix, (ii) matrix temperature, and (iii) incident ion energy. The effect of the matrix atoms/molecules remains to be elaborated in detail.

Robert Madix added: What specific cluster sizes are being targeted and for what specific reactions?

Richard Palmer responded: The cluster beam route is well suited to clusters in the range of 1–10 nm. With the MACS we have achieved a diameter definition of ±10% in the best case. As far as the reactions are concerned, our focus has been on exploiting the homogeneity of the samples to achieve high selectivity, especially as they are relevant to pharmaceuticals and fine chemicals, *e.g.* selective hydrogenation[1], but of course cluster catalysts can be used for any reaction, in principle, it depends on the amount of catalyst required.

1 V. Habibpour, M. Y. Song, Z. W. Wang, J. Cookson, C. M. Brown, P. T. Bishop and R. E. Palmer, *J. Phys. Chem. C*, 2012, **116**, 26295.

Daniel Wotton remarked: Currently you are using a dual magnetron sputterer to sputter two different metals and are hoping that these clusters condense together to create an alloyed cluster. Is there a good chance that a lot of the clusters landing on your surface are simply of one metal and haven't condensed with the other?

Also, have you considered sputtering from an alloy target instead of two separate targets? Because then the sputtered material may already contain both metals and you may get nanoparticles that have consistently constant ratios of the metals.

Richard Palmer answered: Actually, even with a single mixed sputtering target, elemental clusters can be generated in addition to binary clusters.[1] However, in the present work (DOI: 10.1039/C5FD00178A) we found no evidence of elemental clusters in the TEM analysis. As far as the MACS is concerned, we can evaporate more than one metal in the matrix where the clusters are made. A program of work is needed to explore how consistent the resulting composition of the binary clusters is.

1 F. Yin, Z. W. Wang and R. E. Palmer, *J. Am. Chem. Soc.*, 2011, **133**, 10325.

Mark Howard added: In response to a previous question, cobalt catalysed Fischer Tropsch is an example of a catalytic system where the metal particle size is reported as being important for selectivity – I believe a size in the region of 10nm is preferred.

What is the range of metals and particle sizes accessible to the cluster beam method, whilst retaining good size selection?

Richard Palmer replied: Thank you for your comment on preferred sizes for Fischer Tropsch. With the magnetron sputtering gas condensation source we can reach cluster sizes of 10nm and above,[1] with precise size control but limited flux. In the case of the MACS we have concentrated to date on the sub-5 nm range[2] and only explored briefly the production of larger clusters.

1 N. P. Young, Z. Y. Li, Y. Chen, S. Palomba, M. Di Vece and R. E. Palmer, *Phys. Rev. Lett.*, 2008, **101**, 246103.
2 R. E. Palmer, L. Cao and F. Yin, *Rev. Sci. Instrum.*, 2016, **87**, 046103.

Nico Fischer said: If transition metal clusters are deposited in their reduced form onto the support of choice (for example the described alumina) how can these be transferred to a reactor? I am worried that their interaction with the support might not be strong enough to avoid sintering during the exothermic oxidation when exposed to air.

Richard Palmer replied: Indeed there is evidence that the sintering of size-selected clusters (under reaction conditions) can be driven by exothermic reactions.[1] However, it is possible to transfer clusters out of the cluster source under a controlled atmosphere of argon gas.

1 K. J. Hu, S. R. Plant, P. R. Ellis, C. M. Brown, P. T. Bishop and R. E. Palmer, *J. Am. Chem. Soc.*, 2015, **137**, 15161.

Amy L. Miller opened the discussion of the paper by Robert Madix: As the active sites are thought to be AuAg bimetallic sites and their concentration is low, would increasing the concentration of silver be beneficial? In addition, the dispersion of silver in the surface could be expected to have an effect on the active site concentration. Do you know if the silver is in small clusters on the surface or atomically dispersed among the gold atoms?

Robert Madix replied: We have recent high pressure XPS and TEM results to be published that show that during the activation procedure using ozone, silver is drawn to the surface and oxidized. In this process there is a local increase of the concentration of Ag at the surface. Reduction of this surface in the reactant mixture of methanol and dioxygen leads to a local dissolution of the Ag into the near surface region of the Au host material. This alloy surface produces the selective cross coupling of the alcohols. An increase in the bulk concentration of Ag from 1–3% up to 8% leads to loss of selectivity due to combustion. These higher Ag concentrations have not been studied extensively to identify the differences in the state of the material under operating conditions.

Amy L. Miller communicated: What is the lateral dispersion of silver on the surface? Is the silver thought to be in small clusters on the surface or atomically dispersed with gold? The lateral dispersion of silver at the surface could be expected to have an impact on the surface concentration of AuAg bimetallic sites.

Robert Madix communicated in reply: This is an excellent and important question. Recent TEM studies in our program show that in the activation process, silver is drawn to the surface and oxidized in the form of small particles on the surface of the nanoporous material. Upon exposure to the reactant mixture, these particles are reduced and the silver redistributed locally into the gold matrix. Theory suggests that the silver atoms prefer to disperse in the surface, forming a two-dimensional alloy. These studies should be published soon.

Jacob Moulijn remarked: My question regards the interpretation of the results given in Fig. 5 (DOI: 10.1039/C5FD00161G). In your experiments with varying delay time between the oxygen pulse and the methanol pulse you get information on the stability of the adsorbed oxygen species. For a delay time >1 s, the formaldehyde signal is constant, showing the high stability of the adsorbed O-species. At a delay time of 0.5 s the intensity is significantly higher. You attribute this to the presence of more adsorbed O_2, hence the lower delay time. Is this in agreement with your O_2 response curve showing that the signal is already close to zero at 0.5 s?

Robert Madix responded: Yes, these results are compatible. The simplest way to think of the origin of the rise in the formaldehyde signal at delay times of 0.5 s and below is that the dioxygen concentration has not reached zero in the reactor at these times. Thus, methanol removes the adsorbed atomic oxygen to form formaldehyde, and the atomic oxygen is rapidly replaced by the dissociation of ambient dioxygen, thus yielding more formaldehyde.

Bruce Gates remarked: Please tell us about the advantages of the TAP reactor as you used it. You have evidently found it beneficial in ways that have not been reported before, is that correct? What did you find when you used a conventional once-through flow reactor, and what are the trade-offs between the reactor types in your research?

Robert Madix answered: The advantage of the TAP reactor is that we can make measurements impossible with a standard flow reactor. First of all, the short time, calibrated pulses allow the determination of active site densities for the npAu catalyst – a quantity more difficult to derive from flow-through measurements. Secondly, reaction probabilities can be quantitatively determined for steps that most influence the reaction, namely the dissociation probability of molecular oxygen and the activation of methanol. We have limited kinetic data from the flow reactor configuration which we are analyzing. To date, we have not conducted a detailed comparison. As a more general answer to the question, the TAP reactor allows reactions to be addressed in the Knudsen flow regime on a relatively short timescale on the actual catalytic material, yielding quantities like reaction probabilities, site densities and transient kinetics, which affords a perspective on the reactions that is different from and complementary to the flow reactor approach.

Rutger van Santen commented: How does the TOF you measure for the oxidation by the Ag/Au alloy compare with that of the non alloyed Ag? Does the porous structure of your material contribute to low selectivity?

Robert Madix replied: Under similar reaction conditions pure Ag combusts the methanol. We do not have accurate information which allows us to make this quantitative comparison.

Parasuraman Selvam remarked: What kind of facets can you have in this nanoporous gold? Does it vary with pore size as well as cluster/particle size? Can we manipulate the surface structure of porous gold?

Robert Madix answered: Due to the radius of the curvature of the ligaments in the nanoporous gold, the surface consists primarily of (111) facets and steps. In principle, the step density can be varied by changing the ligament diameters, but we have not explored this in detail. Overall, however, the activity of the catalyst does not change significantly as the reaction proceeds, even over days of operation. Since we know that the material coarsens (ligament size increases) over that period of time, it appears that these coupling reactions are not very structure sensitive.

Jacob Moulijn opened the general discussion: Clusters play a large role in catalysis and cluster size is a fundamental property. A good example is Fischer Tropsch synthesis, where there exists usually a minimum cluster size for satisfactory catalyst performance. Is it in general true that a minimum cluster size is needed for optimal performance? Can isolated atoms be active and selective?

Wataru Ueda replied: It is generally accepted that there exits a cluster size most suitable for a certain reaction to occur in high catalytic performance. This is a general requisite for solid-state catalysis. However, it is hard to predict what the most suitable sizes and numbers of constituent elements for a given reaction are. Nevertheless, in catalytic selective oxidations for example, the number of electrons that are involved in the catalytic cycle of a reaction will determine a minimum number of elements in an oxide cluster. In addition, the oxide cluster may need more elements for providing functions, for example adsorption, oxygen insertion, activation of oxygen, and so forth. By taking these factors into account, one can propose a suitable size of the oxide cluster and kind of elements. But if these elements do not form oxide clusters that are isolated structurally but form normal oxides, they inevitably form other unnecessary catalytic functions that usually cause low catalytic performance. Of course an isolated single atom can complete one catalytic cycle in a certain case, but again the isolation of the atom is suitable for the catalytic performance compared to a non-isolated catalyst system.

More important is how to synthesize these catalytic materials. It should be emphasized that the development of new synthetic methodology for solid-state materials that can introduce cluster structures, as well as complexity of constituent elements, is indispensable for future design of solid-state catalysts.

Charles Campbell commented: In our paper (DOI: 10.1039/C5FD00198F), after we optimized the selective oxidation of butene, 50% of the oxygen in the product came from water, and 50% from oxygen gas. I was wondering whether any of you experts in selective oxidation, like Prof. Bowker, know if such a stoichiometric contribution from water as the oxidant has ever been used before in selective oxidation catalysis, to your knowledge? And if so, why not? Is it worth trying?

Michael Bowker responded: It is true that the presence of a little water can affect some processes beneficially in terms of a changed distribution of oxidising surface species, but to use it as a net reactant has thermodynamic implications since, as a reactant, it is further down the thermodynamic scale. Oxygen is defined as zero enthalpy and water is ~ -240 kJ mol^{-1}, so it is unlikely to provide a useable source of oxygen, unless other energy inputs (such as light) are used. Hence why steam reforming is so highly endothermic.

Joachim Sauer opened the discussion of the paper by Bert Weckhuysen: The catalyst particle you showed had a tiny bit of zeolite in it, is this typical?

Bert Weckhuysen replied: This is indeed the case. Fluid catalytic cracking or FCC catalyst particles consist of zeolite material, next to alumina, silica and clay material, the latter is called a binder or matrix material. The zeolite is literally embedded within this binder material. The amount of zeolite is relatively low and is typically about 20 wt%.

Bruce Gates asked: Your design is evidently based on the roles of the acidic and metal catalyst functions; would you please explain the basis for the design?

Bert Weckhuysen answered: The performance of an individual FCC catalyst particle is related to the number and strength of the acid sites; being it either Brønsted or Lewis acid sites. The photo-spectroscopy method developed only probes the Brønsted acid sites, and therefore only visualizes and quantifies indirectly the effect of Brønsted acidity on catalytic cracking performance. The catalyst performance is negatively impacted by the presence of the metal (ions), for example Ni metal nanoparticles, originating from the crude oil feed molecules, are responsible for the hydrogenation and dehydrogenation reactions. Metal ions, in part, will cancel acidity, hence lower the reactivity of the catalyst particles towards the staining reaction, and therefore will be less colored. The photo-spectroscopy method developed is able to discriminate between the catalyst particles containing a lot of Brønsted acid sites and no or not many metal poisons, and catalyst particles containing little or no Brønsted acid sites and are full of metal poisons.

Graham Hutchings remarked: How general can you make this methodology? It works very well for acid–base catalysts, but are there suitable probes for redox reactions?

Bert Weckhuysen replied: The method is general in the sense that it can be applied to a variety of acid catalysts and appearances. For example, it can also be applied to investigate extrudates or catalyst bodies of millimeter size loaded with zeolite material. It is also possible to have different probe reactions, *e.g.* furfuryl alcohol condensation, styrene oligomerization and thiophene oligomerization, are currently the most used for the purpose of acidity probing. With respect to the second part of the question, we are currently working on the use of this approach for redox reactions. There are already some reports in the literature in which *e.g.* photocatalytic applications can be monitored with a specific dye molecule, where a redox process can be assessed at the level of even a single molecule in a single catalyst particle. On the other hand, the dominant number of studies is still devoted to the spatial mapping of Brønsted acid sites.

Avelino Corma asked: Interactions between the matrix and the zeolite are already well known, it is known that alumina can migrate into the zeolite. Silica alumina can also be formed at the external surface of the zeolite. Can you distinguish between the acidity in the zeolite and that generated in the reaction? What is the correlation? When you have a combination of particles of ZSM-5, it is going to be very difficult; amorphous silica alumina in the periphery has also been investigated with respect to acidity.

Bert Weckhuysen replied: The article does not discuss in detail the binder effects in zeolite matrix systems nor the interaction of the amorphous silica or alumina with the zeolite fraction. Instead, it discusses the use of a powerful digital photo-camera to distinguish between fluid catalytic cracking (FCC) particles differing in their type (zeolite Y *vs.* zeolite ZSM-5) and the amount of zeolite material. For this we make use of a selective staining reaction, which turned out to be primarily catalyzed by the zeolite material, and separate experiments on the individual matrix components have shown that they only react very mildly (clay components) or not at all (silica components) with the styrene oligomerization reaction. In other words, we probe dominantly the zeolite fraction of the FCC catalyst particles.

Cynthia Friend continued: How specific is the binding of the dyes to particular sites and materials? Can you design dyes for other types of materials that would enable this optical imaging?

Bert Weckhuysen responded: That is indeed possible, although what I have shown today is the discrimination between different types of Brønsted acid sites in a catalyst material. Actually, we wish to have dyes which can discriminate between Brønsted acid sites, Lewis acid sites and redox active sites, and determine their relative amounts. Currently, we and others are working on the development of different probes varying in their reactivity, in their type of reactivity, as well as in their size. Ultimately, we wish to determine what active site is present in which pore. That is currently not possible, but I am optimistic that this might become available in the years to come. Signs of such development are selective probes for *e.g.* the characterization of photocatalytic materials.

Bruce Gates asked: Presumably there is an optimal balance among the different components in a cracking catalyst. Would it be beneficial in practice if the manufacturing methods gave more homogeneous materials than are evidently typical of industrial cracking catalysts?

Bert Weckhuysen answered: It is indeed so that there is a balance among the different components present; *i.e.*, there should be the right amount of zeolite and binder material, the latter consisting of clay, low and high crystalline alumina, as well as silica. These components give strength to the catalyst particle (important for resistance against attrition and abrasion) as well as additional cracking properties (*e.g.* amorphous silica-alumina) and resistance against metal poisons (*e.g.* alumina, as well as additional scavengers added to the catalyst material). It is the right balance between each of these components which will lead to the best performing and most robust catalyst material. Nobody really knows in my opinion whether these components should be homogeneously present or not. I believe that a certain degree of intraparticle heterogeneity will help to make the catalyst particle more robust and active. It is the challenge of future research to find out how much we have to control intra- and inter particle heterogeneities to make the best performing catalyst material possible.

Rutger van Santen said: A major criterion for good cracking catalyst performance is sustained stability by streaming. An important parameter that affects zeolite catalyst stability is the La distribution in the zeolite. This distribution should also be considered an important parameter for catalyst stability.

Bert Weckhuysen answered: This is fully true, and as we will show in the near future, it is possible to also map La in a fluid catalytic cracking particle in space at the level of a single particle. We can even demonstrate that the zeolite domains do not have a homogeneous distribution of Al within these zeolite domains; and that this is related to the presence of La.

Elad Gross asked: You have demonstrated that the acidic sites within a micron-sized particle can be identified by staining the particle with light-absorbing molecules. 1. Is this method mostly sensitive to the surface properties of the particle or can it also resolve its bulk properties? 2. Can you detect molecules that reside in the inner part of the particle without being masked by the light absorption properties of the particle?

Bert Weckhuysen responded: These are very good points. The method is in principle able to detect both the surface and bulk of the material. On the other hand, it is also clear that as it concerns light, it can only be penetrating into a material to a certain extent. Care has to be taken to make sure that we do not have too much dye or coloration, as the absorption of light is then too strong and you will only probe the upper parts of the material. This also points towards the need for a clear well defined protocol for catalyst staining. For your information, we have developed such a protocol which has been transferred to our industrial partner as an operator and should in principle be able to perform these experiments.

Michael Bowker remarked: Zeolites are very well defined materials, while the catalyst spheres themselves have to be very uniform in size, have great abrasion resistance, and also convey the strength of the particle. So what you looked at is not entirely heterogeneous, but the real active phase is very well defined in there. In general, the surfaces involved in catalysis are very well defined, in the sense that it is the job of the catalytic chemist to maximise these active sites in the material.

Bert Weckhuysen responded: The zeolite domains themselves embedded in the matrix of clay, alumina and silica are well-defined materials, but they drastically differ from domain to domain in the catalyst particle. In a previous study, with the aid of single molecule fluorescence microscopy, we have been able to demonstrate that each of these zeolite domains has different reactivity, most probably because not all of these domains have the same Si/Al ratio or amount of La present. The goal of a catalytic chemist is to stabilize these zeolite domains and also perfect the access to these domains in order to make sure that a high activity can be obtained and maintained.

Rutger van Santen commented: This deals with the question of the heterogeneity of heterogeneous catalysts. I propose that heterogeneous catalysts are intrinsically non homogeneous because symmetry is broken due to the presence of a surface and bulk. Intrinsic to the catalyst is therefore the presence of concentration gradients. Since the practical catalyst contains a distribution of different surfaces (it cannot be a sphere), catalytic transformations will happen at different sites. Because of possible reconstruction and also deactivation, this adds to contributions of inhomogeneity. In addition to this there is also a temporal component to heterogeneity. Surface diffusion may be slow compared to the reaction, which will lead to surface concentration gradients and may even lead to island formation.

Bert Weckhuysen responded: Many thanks for these insightful comments; I fully agree that heterogeneous catalysts are intrinsically non-homogeneous and that they should not be. The heterogeneities are present both in space and time; it is the spatiotemporal changes in the active sites and the concentration of reactants and reaction products, which all influence the overall performance of a solid catalyst. Understanding these spatiotemporal heterogeneities and steering them in the right direction will be key for making new or better catalyst materials.

Graham Hutchings opened the discussion of the paper by Wataru Ueda: Concerning the nature of the catalyst, is long range order really needed? Is the crystallinity of the material needed. If you made these POMs and co-crystallised them with another suitably inert matrix would it be possible to observe catalysis? Is it a single POM unit or an ensemble that is important for catalysis?

Wataru Ueda responded: We have experiences and facts that well-crystallized catalysts with a long range order of the cluster units show superior catalytic performance compared to ill-crystallized materials.[1] This is simply due to the lower content of active cluster units on the surface. In fact, the intrinsic activity of the cluster units appears the same for both cases, since the local structure is similar and the difference is only whether the material has long range order or not. However, if we precisely analyze the local structure for both cases, we will find small differences, for example bond length and bond angle, that may cause additional catalytic functions possibly affecting the durability and selectivity of a reaction. As a consequence, it can be said that the long range order is not necessary for catalytic performance unless the cluster units are created. But at the same time it has to be emphasized that the synthesis of crystalline materials is the best way to introduce an isolated cluster unit into solid-state materials, because at the moment we do not have synthetic methodology for isolating the cluster unit in an inert matrix, while keeping the structure of the cluster unit. For this purpose, we must have structure-tunable metal oxide ligands in the solid state.

It is known that a simple α-Keggin POM showed activity for the oxidation of mathacrolein with O_2. Since our porous material based on an ε-Keggin POM has various structurally different elemental positions thanks to the 3D structure, we can introduce a variety of catalytic elements while maintaining the whole structure. Thus we expect a synergistic effect of each element in the different sites (hetero-atom site, polyatom site, linker site, and ion-exchange sites) of the materials. By using the element positioning control methodology, we can clarify not only the catalytic function of the POM unit but also the other elements outside the POM.

1 T. Konya, T. Katou, T. Murayama, S. Ishikawa, M. Sadakane, D. Buttrey and W. Ueda, *Catal. Sci. Technol.*, 2013, **3**, 380.

Simon Kondrat continued: For your gas phase reaction of methacrolein with your porous Keggin POM materials, you acknowledge that the pores are too small for the substrate to enter and the reaction occurs on the surface.

My question is whether the long range order of your POM structure is required and whether a catalyst with a disordered arrangement of POM units would be active for this reaction? Can you prepare such a disordered matrix?

Wataru Ueda answered: The first part of the question is answered above.

For ε-Keggin POM based porous materials, so far, we cannot prepare a disordered sample. We tried to damage the crystal structure without collapsing the molecular structure of the POM, but it failed. The molecular structure of the POM is damaged when the whole crystal structure is collapsed.

Philip Davies remarked: Your materials look to have very interesting catalytic properties but how stable are they over longer timescales?

Wataru Ueda responded: The oxidation of methacrolein was carried out at 280 °C for 20 h using Mo–V–Bi oxide. Over the hours, the catalyst showed almost stable catalytic performance and after the reaction, the structure of the material was completely maintained.

Francisco Ivars Barcelo asked:

1) How do you keep the molecular wires isolated and stabilized in the aqueous media?

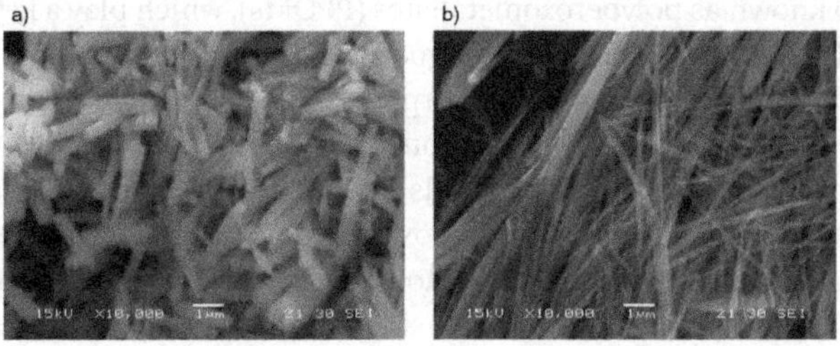

Fig. 1 SEM images of a) Mo–Te oxide prepared at 25 °C and b) Mo–Te oxide prepared at 100 °C.

2) What is the concentration threshold to avoid the assembly of these molecular wires in superstructures and keep them isolated in the aqueous media?

3) How do you prevent the stabilizer, if used, from interfering in the esterification reaction which these molecular wires are supposed to be used for?

4) Are you able to control the length of the molecular wires?

5) Is the length affecting the catalytic properties of these molecular wires?

6) Have you measured the wire length distribution you get with the synthesis of these molecular wires?

Wataru Ueda replied:

1) The single molecular wire of Mo–Te oxide can be isolated from the crystal of Mo–Te oxide in ethanol or water using ultrasonication for 1 h. The nanowire thus obtained is stable in the media.

2) Details for obtaining the nanowire are as follows: The Mo–Te oxide crystal (5 mg) was dispersed in 5 mL of ethanol. The mixture was sonicated for 1 h at room temperature, followed by centrifugation for 2 h (3500 rpm). Only *ca.* 2 wt% of the Mo–Te oxide crystal formed a nanowire. Therefore, we assume that 0.1 mg of Mo–Te oxide per 5 mL of ethanol is the suitable concentration for maintaining the dispersed nanowire.

3) No stabilizer was used at all.

4) The length of the molecular wire can be tuned in the corresponding crystal by changing the synthesis conditions or precursor compositions. For example, the crystal of Mo–Te oxide prepared at room temperature exhibited the length of *ca.* 3 μm (Fig. 1a). When the temperature increased to 100 °C, the crystal of Mo–Te oxide was formed at over 10 μm in length (Fig. 1b). The length of the isolated molecular wire will change according to the length of the crystals.

5) Intrinsically not but morphologically yes. The length might affect the activity of the material because a decrease in the length would decrease the size of the material and increase the exposed molecular wires.

6) Yes, we measured the length distribution of Mo–Te oxide. The distribution maximum was *ca.* 50 nm.

Haresh Manyar asked: The Keggin type polyoxometalate (POM) based materials are known to have good catalytic activity for oxidation reactions. As per the well-established Ishii–Venturello chemistry, Keggin type POM structures are shown to collapse and form smaller dinuclear and tetranuclear metal peroxo species, also known as polyperoxometallates (PPOMs), which play a key role in the oxidation of substrates. There is also good evidence of the formation of metal peroxo species in your paper upon adsorption of O_2, giving rise to a new IR band at 797 cm^{-1} indicating a ν(O–O) vibrational mode of peroxo species. What is the fate of Keggin type POM based materials in your studies? Do you observe structural changes in POM materials during oxidation or from the characterisation of the post-reaction or spent catalysts? How does it affect the reusability of the catalysts?

Wataru Ueda responded: Thank you for your question. We have only tested what will happen to the sample after O_2 adsorption and have not tested any catalytic reactions. The formed peroxo species should be active for oxidation. However, the catalyst has a harrow-sized channel, so hydrocarbon oxidation might be retarded.

Jacob Moulijn asked: The POM based materials you describe are excellent model catalysts and might be the basis for future industrial catalysts. Do they have potential for applications in the synthesis of novel membranes?

Wataru Ueda responded: The crystalline materials based on POM have potential applications in novel membranes. Tuning the preparation method can easily change the particle size and morphology of the materials. Furthermore, the surface properties might also be tuned using methods such as ion-exchange. All of the properties of the materials help obtaining new membranes based on the materials.

Mark Feaviour asked: You made the interesting observation that the NH_4–Mo–Fe system displayed a unique oxygen adsorption property compared to the other material produced here, in that it stored oxygen as an Fe peroxide species. Would this be expected to form under partial oxidation reaction conditions and what are the potential implications, for example would it favour total oxidation?

The high activity of the Mo–V–Bi system was linked to the high surface area but also shown to occur on the external surface area only since the MAL molecules are too large to access the pores. Would it be possible to use a soft template to introduce additional porosity to enhance catalyst performance?

Wataru Ueda replied: According to the isotherm of O_2 adsorption, almost all of the O_2 adsorbed in the materials at low O_2 partial pressure ($p(O_2)$ <1 kPa). Moreover, the oxygen adsorption can also occur in air. Therefore, we believe it would occur under partial oxidation reaction conditions.

The peroxo species formed after O_2 adsorption is expected to catalyze some oxidation reactions, including total oxidation, since it is reported that the peroxo species formed in POMs can catalyze the oxidation of organic compounds.[1]

1 R. Neumann and M. Dahan *J. Am. Chem. Soc.*, 1998, **120**, 11969.

Michael Craven communicated: The type of heteroatom present in heteropoly acids has a strong influence on the stability and electronic properties of the compound as a whole. The precursor Keggin polyoxometalates that you used to make your compounds contained either Se or Te as the heteroatom. How did the choice of heteroatom used in these Keggin precursors affect the properties of your final compound?

Wataru Ueda communicated in reply: The properties of the resulting materials can be caused by the heteroatoms, Te(IV) and Se(IV), even though the structures are the same between the Mo–Te oxide and the Mo–Se oxide. The difference between the analogues of Te(IV) and Se(IV) seems to be resulted from the ionic size and electronegativity of the heteroatoms. Thus, the local electronic structure around the heteroatoms will also be different and therefore we currently think that the electronic structure affects the protonic acid properties.

Andrzej Kotarba opened the discussion of the paper by Annette Trunschke: The pH of the solution for the synthesis was adjusted by NaOH addition. In hydrothermal conditions the Na^+ cations exhibit the tendency to diffuse into the bulk and form bronzes, especially when the WO_6 building blocks are formed in the solution. The residual Na^+ cations are very difficult to eliminate and they may affect the material reactivity to a large extent. In the applied conditions one can expect formation of Na_xWO_3. Have you noticed the formation of tungsten bronzes in the applied conditions?

Annette Trunschke answered: We agree that the sodium contamination may affect the catalytic performance of the synthesized catalysts. Chemical analysis discloses sodium content below 1 wt% in the bulk. According to XRD and HRTEM, the catalysts are phase pure and of high crystallinity. The catalysts remain phase-pure after catalysis. The formation of tungsten bronze phases has not been observed in either the fresh or used catalyst. However, we cannot exclude that bronze formation occurs locally. Moreover, the presence of sodium at the surface may influence acid–base properties and, consequently, reactivity. Therefore, we tried to substitute Na^+ with NH_4^+. Ammonium ions can also cause bronze formation, but the ions can be more easily removed. However, we did not succeed in the synthesis of nanostructured products when ammonium was used as the counter ion under the synthesis conditions applied so far.

Based on the elemental composition near the surface analysed by XPS, no relation between the reactivity and sodium content has been found. The Na/(Mn+W) ratio changes as follows: AR1.5 = 0.11; AR1.7 = 0.13; AR3.2 = 0.10; AR3.9 = 0.08; AR5.1 = 0.10.

Andrzej Kotarba continued: The catalysts were calcined under reducing conditions (Ar flow, 400 °C) and thus the surface oxygen vacancies were formed. However, the catalytic performance was investigated in oxidizing conditions (ethanol oxidation). Was the catalyst surface stable while passing from a reducing to an oxidizing atmosphere?

Annette Trunschke responded: In the oxidation of ethanol the catalysts show no or only slight deactivation with the exception of the most active catalyst (the catalyst composed of needles with a high aspect ratio of 5.1) that loses activity (2.5% over five hours) at all reaction temperatures applied. Interestingly, the latter catalyst deactivates slightly during the first 20 h under conditions of oxidative dehydrogenation of propane (ODP), but then shows an increase in the conversion with time on stream, reaching the steady state after 100 h at the same conversion as at the beginning of the reaction. All other catalysts deactivate slightly in ODP and reach the steady state after 100 h on stream. *In situ* UV-vis measurements suggest that the catalyst surface is stabilized in a partially oxidized state under the reaction conditions in oxidative dehydrogenation of propane. The steady state is reflected in a constant redox-state (degree of oxidation) of the catalyst surface.

Tomasz Jakubek remarked: During the synthesis of the manganese tungstate catalyst, one of the precursors contained sodium (sodium tungstate). Alkali metals are well known promoters in oxidation catalysis, typically increasing the catalytic activity of materials. Therefore, residual sodium could influence the model reactions used here. Were there any traces of sodium left in the materials after their synthesis? If so, is there a correlation between the concentration of sodium and the catalytic activity of the tested materials? Were other alkali precursors (*e.g.* potassium tungstate) tested to synthesize the same catalytic material?

Annette Trunschke answered: The catalysts contain sodium as an impurity, but no systematic correlation between sodium content and reactivity was found. Except ammonium, no other counter ions have been used. To apply different alkali precursors in the synthesis is an interesting suggestion, the effect of alkali on reactivity could be studied in the future.

Jacob Moulijn commented: In your systematic way of synthesizing unsupported manganese tungstates, you have got a series with different aspect ratios. The selectivity appears to change with the aspect ratio, and you attribute this phenomenon to changes in the relative abundance of acid–base and redox sites. Could other factors also play a role, such as the accessibility of sites, assuming that there might be some porosity? Is the rate of reaction dependent on the aspect ratio?

Annette Trunschke replied: The material exhibits no structural porosity. Nitrogen adsorption indicates a non- or macro-porous solid. The selectivity towards the oxidation product increases within the series of catalysts because the rate of oxidation increases with increasing aspect ratio. The aspect ratio is, however, just a proxy. By changing the aspect ratio with the help of synthesis means we increase the abundance of {010} planes at the surface of the catalyst. Selective dissolution of WO_6 species is possible during ageing under hydrothermal conditions, in particular at the latter planes, leaving behind a chemically modified surface that is enriched in manganese hydroxide species. The overall near surface concentration of Mn increases with increasing aspect ratio as evidenced by XPS. However, manganese hydroxide rows formed during the synthesis at the 010 surfaces represent only the precursor of the active sites. Active sites for C–H activation are supposed to be oxygen defects formed by dehydration of $Mn(OH)_x$ during the thermal activation of the catalyst.

Bruce Gates asked: Can you articulate any design principles that guided your catalyst preparation?

Annette Trunschke responded: Selective oxidation, in particular of alkanes, is a challenging class of heterogeneously catalysed reactions. The underlying complex reaction networks are generally composed of numerous consecutive and/or parallel reaction steps. Essentially, the formulation of a design concept requires detailed knowledge of the reaction network, which is, if at all, only fragmentarily known. In any case, the investigation of high-performance (high selectivity at high conversion) oxidation catalysts, such as vanadyl pyrophosphate for the selective oxidation of n-butane to maleic anhydride or the M1 phase of MoVTeNb oxide for the oxidation of propane to acrylic acid, reveal that the following design parameters are important:

1) Phase purity of the bulk is an important criterion in terms of catalyst stability. Impurities lead to continuing phase segregation processes with time on stream and, consequently, to catalyst deactivation.

2) The various reaction steps require different catalytic functions, such as the ability for oxygen dissociation/reduction, C–H activation, hydrogen abstraction and oxygen insertion. An important design principle is, therefore, the generation of multifunctionality, which may be achieved chemically (addition of various elements) and/or structurally.

3) Usually a high number of electrons have to be transferred in selective oxidation reactions. Successful oxidation catalysts, such as vanadyl pyrophosphate or the M1 phase of MoVTeNb oxide, are bulk catalysts with semiconducting properties that exhibit the formation of a thin layer at the surface with insulating or semiconducting properties (heterojunction) under reaction conditions.

4) Balanced oxygen activation is a prerequisite to ensure selectivity by avoiding total oxidation to carbon dioxide.

Based on these four criteria, our current design concept involves the synthesis of phase pure bulk catalysts with semiconducting properties. We are looking for phases that generate a thin surface layer, which is chemically and/or structurally different from the bulk. The layer may be formed during synthesis, activation or under reaction conditions. The thin, self-supported layer contains the redox active element(s) and perhaps other elements that may have stabilizing or spacer functions leading to geometric site isolation with impact on the electronic properties of the system as well. The specific termination creates localized electronic surface states that may be represented by $M^{n+}/M^{(n+1)+}$ redox couples. Since the Fermi energy of the bulk phase and surface phase usually differ, charge transfer occurs, resulting in the formation of a surface charge and an oppositely charged space charge layer in the sub-surface reflected in band bending in the surface region. The catalyst shows properties like solid-state sensors. Band bending depends on the occupation of the surface states, and, therefore, on the reaction conditions (oxidizing or reducing feed). The surface states are responsible for activation of the alkane molecule. At the same time, charge transfer to oxygen is limited due to the barrier in the space charge region. Accordingly, oxidation of the hydrocarbon molecule and reduction of oxygen can only occur sequentially or are spatially separated if the layer is not homogeneous. Such a situation avoids the total oxidation of reaction intermediates due to a low concentration of highly reactive electrophilic oxygen species at the location of hydrocarbon oxidation, implying site isolation induced by electronic properties.

The $MnWO_4$ catalyst is a phase-pure, semiconducting bulk catalyst that exhibits a surface layer that is chemically different from the bulk. Further studies of the electronic properties of the catalyst under reaction conditions are necessary to find out whether the design concept outlined above is indeed a more general concept in oxidation catalysis or if the concept applies only to specific bulk structures and oxidation reactions.

Francisco Ivars Barcelo remarked: Regarding the sodium traces found in your material, have you investigated whether that sodium is on the surface, in the bulk or in both locations?

Annette Trunschke responded: The sodium is located in the bulk and at the surface of the material. At the low pressure (0.2 mbar) and high temperature (400 °C) of near-ambient pressure XPS measurements, the sodium content at the surface increases with time (hours).

Michael Bowker enquired: The conversions reported here are very low, so what happens when you go to high conversion in terms of selectivity, is combustion then favoured? Also, relating to Fig. 5 (DOI: 10.1039/C5FD00191A), can you tell us what the situation would be for tungsten oxide of a similar surface area in terms of selectivity/conversion?

Annette Trunschke replied: The $MnWO_4$ catalysts have been studied in ethanol oxidation in the temperature range of 250–300 °C and in the meantime also in the oxidative dehydrogenation of propane between 360 and 400 °C. Conversion has been adjusted below or close to 10% to reduce falsification by heat and mass transport phenomena. Regarding ethanol oxidation, higher conversion (up to 20%) was also measured. With increasing ethanol conversion, only a slight reduction of the selectivity to acetaldehyde was observed at the comparatively low reaction temperatures. In oxidative dehydrogenation of propane the selectivity decreases with increasing conversion as is typically observed for many oxidation catalysts.

Tungsten oxide has not been tested in the oxidation of ethanol. High selectivity towards ethene as a product of acid-catalyzed dehydration of ethanol is expected based on literature concerning alcohol conversion over tungsten oxide, for example Tanner *et al.*[1] Tungsten oxide characterized by high specific surface area (71 m2/g) exhibits negligible conversion in oxidative dehydrogenation of propane and the major product formed was carbon dioxide.

1 Robert E. Tanner, Pornthep Meethunkij and Eric I. Altman, *J. Phys. Chem. B*, 2000, **104**, 12315.

Graham Hutchings said: What is the effect of water in the reaction? If you were to increase the concentration of water, would it affect the surface of the catalyst? Water is a product from C–H activation and perhaps it is water that is driving the formation of the active and selective surface?

Annette Trunschke answered: We varied the concentration of water in the feed in the oxidative dehydrogenation of propane to propene by adding auxiliary steam (5–25%). The product composition does not change after the addition of water, *i.e.* no oxygenates are formed. Furthermore, the formation rate of propene was not affected by steam, but the formation rate of CO_2 decreases with increasing steam concentration (>10%). Hence, the addition of steam improves the selectivity towards propylene. This indicates that water has an impact on either the nature of the catalyst surface, the coverage of the surface by reacting molecules, or the reaction mechanism. To explore the reason, spectroscopic studies of the working catalyst are necessary. We have not yet studied the impact of water addition on the surface properties of $MnWO_4$. However, *in situ* studies by UV-vis, XPS, and NEXAFS have been initiated in which the feed composition (oxygen : alkane ratio) has been varied. The degree of oxidation of the catalyst is affected, but noticeable changes in the chemical composition of the surface have not been observed under the reaction conditions explored so far.

In the particular case of nanostructured $MnWO_4$, an enrichment of manganese oxide species at the surface of the catalyst develops during the synthesis of the catalyst under hydrothermal conditions. Further changes under working conditions are limited, perhaps due to the dense crystal structure of the wolframite-type $MnWO_4$ phase. The metastable M1 phase of MoVTeNb oxide is another bulk catalyst that bears a thin layer at the surface. Segregation of vanadium and in particular tellurium occurs during the hydrothermal synthesis and activation of M1. But in contrast to $MnWO_4$, the surface composition of the selective oxidation catalyst M1 changes further under the reaction conditions dependent upon feed composition, in particular in the presence of steam. The bulk crystal structure of M1 remains intact at the same time. The reversible surface reconstruction experimentally observed under the working conditions may arise from the open crystal structure of M1 that comprises micropores, which are partially filled by Te and V oxo species. These species may be detached by hydrolysis and migrate to the surface, forming stable surface compounds there.

Simon Kondrat opened the discussion of the paper by Peter Wells: In Fig. 8 (DOI: 10.1039/C5FD00153F), you show that the 3 monolayer Mo catalysts produce good formaldehyde yield at very low temperatures. However, you observed a very narrow temperature window with respect to the formaldehyde yield. Why does the yield drop so quickly as the temperature is increased? Is this a structural effect, such as a change in the MoO_x surface speciation? Did you follow the XANES under the reaction conditions to understand this phenomenon?

Peter Wells answered: This is a good question and something we are still investigating. We have followed the XANES of other MoO_x/Fe_2O_3 catalysts and not observed any changes in Mo speciation during continuous formaldehyde production.

Mark Howard addressed Peter Wells and Michael Bowker: Commercial iron molybdate catalysts are believed to slowly lose molybdenum from the catalyst surface during industrial operation, which is continuously replenished by migration of excess molybdenum from the bulk. Being able to lower the operating temperature by 40 °C, as in the comparison between your 3ML@HSA catalyst and the bulk material, could be very significant for molybdenum loss. Do you have any insights into this from your work?

Michael Bowker responded: You are right, they do slowly lose Mo. This wasn't the case in our work (DOI: 10.1039/C5FD00153F), which takes place on a much shorter timescale. We have worked with a commercial organisation to elucidate the mechanism of this loss, which can be simulated under the right conditions of flow, temperature and gas phase composition.

Peter Wells responded: We have looked at the reduction–oxidation cycles for similar catalysts to mimic possible deactivation pathways and seen that the catalyst is robust and able to regenerate. However, to date, we have not performed the long term stability studies required to assess the extent of Mo loss. This is something we intend to do in the future.

Michael Bowker commented: Indeed, if we can reduce the temperature window for high yield, it could reduce the vapour phase molybdenum loss in the material and hence help lifetime in the plant.

Jacob Moulijn continued: I have a question about the sizes and morphology of the multiple Mo-oxide layers put on haematite. Are these layers homogeneous and closed, or does segregation play a role? Does the thickness scale with the number

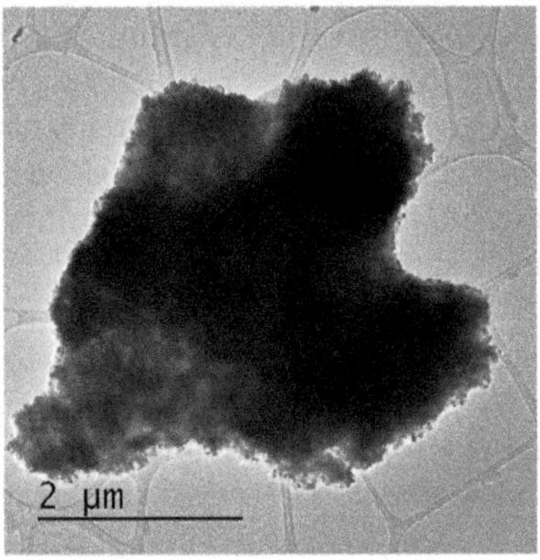

Fig. 2 TEM image of 3ML MoO_x@HSA Fe_2O_3.

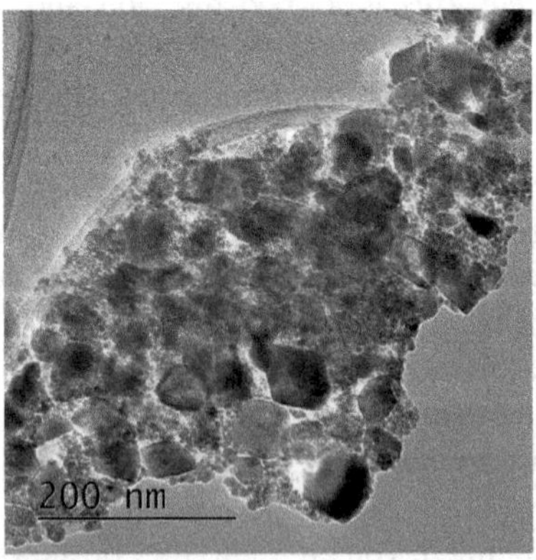

Fig. 3 TEM image of HSA Fe_2O_3.

of monolayers? You might get information from the surface area measurements, such as those reported in Table 2 (DOI: 10.1039/C5FD00153F).

Peter Wells responded: The indication from the XANES, which can be used to show the ratio of surface to subsurface Mo (through changes in Mo coordination geometry), is that the layers are relatively even. Significant clustering/heterogeneity in Mo would lead to a ratio of surface/sub-surface Mo far different to what we were expecting.

Graham Hutchings addressed Peter Wells and Michael Bowker: What is known about the mechanism of the loss of molybdenum? What is the mechanism by which the metals are moving through the lattice?

Peter Wells responded: The current view in the literature is that Mo is lost through volatilisation of MoO_3 after the reaction: $Fe_2(MoO_4)_3 + CH_3OH = 2FeMoO_4 + CH_2O + MoO_3 + H_2O$. The role of the excess Mo added in the industrial catalyst is to provide a reservoir of MoO_3 for the subsequent reoxidation: $2FeMoO_4 + \frac{1}{2}O_2 + MoO_3 = Fe_2(MoO_4)_3$

Michael Bowker answered: It is thought that the Mo is lost through gas volatilisation of a Mo (possibly oxy-hydroxide) species. Indeed we have done experiments using a glass reactor, which show Mo to appear at the cold spot at the end of the reactor when run under certain conditions which favour it.

Regarding movement through the lattice, Mo is mainly lost from the surface layers of the catalyst pellets, so that the very outside is red (iron oxide-like), but most of the core is still yellowish. It is also lost from the surface atomic layers of the material. During reduction events, the main loss is of the anionic species and so anion vacancies appear to have considerable mobility.

Natasa Novak Tusar remarked: Why was the addition of aluminium in the sample fixed? Have you also performed TEM investigations? Why you did not include them in the manuscript?

Peter Wells replied: We intend to look at varying the Al content in the future. TEM was performed but not included for brevity. We have uploaded a couple of images to this discussion post (Fig. 2 and Fig. 3).

Joachim Sauer opened the general discussion: When seeing the title of the conference 'Designing New Heterogeneous Catalysts' I had a problem with 'heterogeneous catalyst'. I assumed it was a negligence, and what was meant was a 'solid catalyst' (used in a heterogeneous process). Does the heterogeneity in the catalyst particle you have shown improve the catalytic properties?

Bert Weckhuysen responded: This is an interesting question and comment. Firstly about the comment, it is indeed so that in my opinion many or maybe most of the industrially used heterogeneous catalysts are heterogeneous in nature, and maybe should be better called solid catalysts. Something to be considered in the future when discussing the subfields in catalysis. The fluid catalytic cracking or FCC catalyst materials under study in our work (DOI: 10.1039/C5FD00210A) are prototype examples of a heterogeneous catalyst in the sense that they are full of heterogeneities. I believe that heterogeneities are often an advantage, especially when complex feedstock molecules, such as crude oil molecules containing different poisons, are entering the catalyst particle and converting. Heterogeneities in the composition/type of material in a 3D catalyst object may give strength, resistance against poisoning, and the question is whether we can get full control over these heterogeneities and if we can design them for the benefit of a specific catalytic reaction. In other words, in the case of FCC catalysts the heterogeneities are helpful, although I am also confident that there are most probably more heterogeneities present than strictly required/needed.

Graham Hutchings remarked: Just following on from the discussion concerning the surface and bulk structures of catalysts being different, it is important to remember a key example of this is the vanadium phosphate catalyst that is used for butane oxidation. In this case, there is a phosphorus-rich surface layer that is disordered, yet this was not observed until the mid 1990s, even though earlier XPS of the catalysts in the mid 1970s showed the surface was very different from the bulk, was phosphorus rich and had V^{5+} present. Yet all models were based on vanadyl pyrophosphate which was the crystalline bulk comprising V^{4+} and had a V:P ratio of unity. This is because trying to study such surfaces is still proving to be very difficult, especially under *in situ* conditions.

Annette Trunschke responded: It is difficult, indeed. We have studied the surface of the catalyst by synchrotron-based XPS in the presence of the reactants. The total pressure under these experimental conditions is in the mbar-range, which may be an issue as shown by pressure-dependent X-ray absorption spectroscopy of vanadyl pyrophosphate in the selective oxidation of n-butane. Studies concerning the pressure dependency of the electronic, surface, and catalytic properties of oxidation catalysts at work are currently performed in our laboratory.

Charles Campbell added: Just a comment about the earlier remark that in a good (industrial) catalyst there is a distribution of different types of sites. It is clear to me that the most selective catalyst should be very homogeneous in its sites, according to simple microkinetic modelling. It is well known that different sites (for example terrace sites *versus* step sites *versus* kinks on metal surfaces) each have different activation energies and reaction energies for each elementary step. Thus, one can only optimize reaction conditions for maximum selectivity for one type of site at a time. That is, the conditions that lead to the highest selectivity at terrace sites could be quite different from those for the highest selectivity on steps. With a mixture of sites, you will always be forced to work at conditions that are not optimum for some of the site types.

Richard Catlow said: I think it is still the case that except for a limited number of cases, we do not have detailed models for the surface structures of oxides, including those of importance in catalysis. This problem should, I consider, be a high priority for computational and experimental work in surface and catalytic science.

Michael Bowker answered: This is true. A future direction for surface science is to advance to more complex oxide surfaces, especially the fabrication of multi-cationic materials. In contrast to metals, we have few rules to indicate the surface composition of multi-cationic oxides, and so it would be a considerable advance to define such rules, both theoretically and experimentally.

Andrzej Kotarba remarked: Catalytic reactions are complex *per se*, since several consecutive processes (adsorption, diffusion, surface reactions, desorption) take place simultaneously at the catalyst surface. Therefore, the fundamental understanding of the intimate reaction mechanism, even for simple catalytic processes involving small molecules, is challenging and requires extensive investigation. That is the reason why we need well-defined model systems, which can be studied by surface science methods and corroborated with molecular modeling. At the same time the catalytic systems are designed to operate in specific industrial conditions, where they adjust their surfaces or structures to the process environment. There are many examples when not only the catalyst surface but the whole bulk structure transforms substantially during the on-line operation (see *e.g.* potassium promoted iron-oxide catalyst for styrene production). This directly implies the necessity to investigate the dynamic behavior of the catalyst. I think we all agree that at the moment the well-defined model catalyst approach is the only way to get insights into bulk–surface–activity relationships, molecular description of the reaction steps, energetic profile of the reaction pathway, *etc.* The development of an adequate model for complete catalyst function is not yet available and we are confined to deal with well-defined and dynamic catalyst models separately and then do our best to bridge the gap between them.

Cynthia Friend replied: The comment about the dynamic nature of catalyst materials is absolutely correct and is a challenge that needs to be met in the catalysis community. The development of new tools that can be used under operating conditions has already provided insight into some of these issues.

At the same time, model studies provide a detailed understanding of a mechanism that can be used in a predictive manner and that can be used to benchmark theory. The future challenge is to construct model systems that are closely related to working catalysts.

Faraday Discussions

PAPER

Methanol formation from CO_2 catalyzed by $Fe_3S_4\{111\}$: formate *versus* hydrocarboxyl pathways†

A. Roldan and N. H. de Leeuw*

Received 20th November 2015, Accepted 30th November 2015
DOI: 10.1039/c5fd00186b

Carbon capture and utilisation is one of the most promising techniques to minimize the impact of the increasing amount of carbon dioxide in the atmosphere. Recently, the mineral greigite was shown to be capable of catalysing CO_2 conversion, leading to useful small organic molecules. Here, we have carried out a systematic study of the adsorption and selective reduction of CO_2 on the $Fe_3S_4\{111\}$ surface. We have considered both formate and hydrocarboxyl key intermediates, leading to different reaction pathways *via* Eley–Rideal and Langmuir–Hinshelwood mechanisms, and we have built a kinetic model considering the wide range of intermediates in the reaction network. Our results show that the mechanism to produce formic acid takes place *via* formate intermediate mostly on Fe_A sites, while methanol is formed *via* hydrocarboxyl intermediates on Fe_B sites. From the kinetic model, we have derived a reaction constant comparison and determined the limiting step rates. The overall process takes place under very mild conditions, requiring only a small energy input that might come from a chemiosmotic potential.

Introduction

Our present dependence on fossil fuels means that, as our demand for energy inevitably increases, so do emissions of greenhouse gases, most notably carbon dioxide (CO_2). To avoid the obvious consequence of severe climate change,[1] the concentration of such greenhouse gases in the atmosphere must be stabilized.[2–5] Integrated carbon capture and subsequent sequestration is generally recognized as the most promising option to tackle CO_2 gases in the short to medium term. A long-term solution is its reduction and conversion to suitable compounds: CO_2 is a source of carbon and may be used as a feedstock for conversion to useful chemicals.[6] Reductive reactions involving CO_2 are a grand challenge because they require energy to generate reduced forms of CO_2, such as formate or CO.[7] In particular, production of CO is considered an important objective in the context

School of Chemistry, Cardiff University, Main Building, Park Place, CF10 3AT, Cardiff, UK. E-mail: roldanmartineza@cardiff.ac.uk; deLeeuwN@cardiff.ac.uk

† Electronic supplementary information (ESI) available. See DOI: 10.1039/c5fd00186b

of production of renewable carbon feedstock chemicals.[8] The main challenge in CO_2 reduction is increasing the energy efficiency of the activation process of the molecule, which is hindered primarily by high CO_2 reduction over-potentials,[9] which is also suggested to be the rate limiting step in the conversion process. In most cases, mild temperatures but high CO_2 pressures (>10 bar) and long reaction times are required.[10,11] Nevertheless, experiments on iron sulfide have shown that electroreduction under room conditions and small overpotentials is achievable.[12]

Reactions of interest for CO_2 conversion are classified into two types: reductive and non-reductive reactions, both requiring the activation of the C=O double bond, which is accompanied by charge transfer to the lowest unoccupied molecular orbital.[13] Solid state catalysts may act as a redox center, transferring electrons from its valence band to the molecular orbital while decreasing the O–C–O bond angle.[9,14] In this process, two problems are encountered: (i) the large thermodynamic stability of the molecule[15] and (ii) the appreciable kinetic energy barrier to activate CO_2.[16] Several possible reaction pathways have been proposed to transform CO_2 into one-carbon organic molecules, such as CH_4 or CH_3OH. These involve the reverse water gas-shift (rev-WGS) reaction, yielding CO followed by the subsequent hydrogenation *via* formyl intermediate (HCO) to methanol[17] or to a variety of long chain hydrocarbons *via* the Fischer–Tropsch process.[18] This reaction mechanism on Cu showed that, under ambient conditions, the reaction takes place through carboxyl intermediate.[19] Other possible pathways have shown formate intermediate (HCOO) with subsequent O–CO bond scission, yielding similar products.[11,12] Recent studies have claimed that CH_3OH is produced by the subsequent hydrogenation of HCOOH to H_3CO *via* C–O bond scission,[20] although formic acid might dissociate back to formate with a low energy barrier.[17] It has also been proposed that the hydrogenation of CO_2 towards CH_3OH takes place through dihydroxycarbene (COHOH), hydroxymethylidyne (COH) and hydroxymethylene (HCOH) intermediates.[20]

Based on such mechanistic information, we have analyzed, in detail, the CO_2 interaction and conversion on greigite (Fe_3S_4) surfaces. This magnetite (Fe_3O_4) isomorphic mineral is formed as an intermediate in the solid-state transformation of mackinawite into pyrite, playing a crucial role in the pyrite formation pathway.[21–26] Fe_3S_4 has been widely identified in marine soils and sedimentary rocks of up to a few million years old.[27,28] Greigite has also been associated as a catalyst in a number of key biochemical reactions associated with the "iron–sulfur world" hypothesis for the origin of life.[12,29–37] Inspired by the CO_2 conversion in this hypothesis and recent experimental results, we have explored the entire reaction network on the $Fe_3S_4\{111\}$ surface, mapping out the complete energy profile, and implemented it into a comprehensive kinetic model under particular initial conditions. This model is based on unrestricted elementary reaction steps derived rigorously from density functional theory (DFT) calculations, which have been shown to be instrumental in unravelling multiple aspects of heterogeneous catalysis.[17,20,38,39]

Computational details

Electronic structure calculations

Periodic plane-wave DFT calculations were carried out to study the CO_2 adsorption and its reactivity with adsorbed hydrogen on the greigite surface $Fe_3S_4\{111\}$.

All calculations were performed using the VASP software.[40,41] Ion–electron interactions were represented by the projector-augmented wave (PAW) method[42] and the generalized gradient approximation (GGA) with the Perdew–Wang 91 functional[43] within the spin interpolation formula of Vosko *et al.*[44] We also considered non-spherical contributions from the gradient corrections to the PAW spheres. All the calculations include the long-range dispersion correction approach by Grimme,[45] which is an improvement on pure DFT when considering large polarizable atoms.[46–51] We have used the global scaling factor parameter optimized for PBE ($s_6 = 0.75$). The Kohn–Sham valence states were expanded in a plane-waves basis set with a cut-off of 600 eV for the kinetic energy.[52] This high value for the cut-off energy ensured that no Pulay stresses occurred within the cell during relaxations.

The initial magnetic moment was described by high spin distributions in both types of Fe, octahedral and tetrahedral, with antiparallel orientation as reported previously.[53,54] For an accurate treatment of the electron correlation in the localized d-Fe orbital, we have used the U approximation[55,56] ($U_{\rm eff} = 1$ eV),[57–59] which improves the description of localized states in this type of system where standard LDA and GGA functionals fail.[60] The choice of the U parameter is rather empirical, a feature that also appears when using computationally expensive hybrid functionals since the amount of Fock exchange is system-dependent.[60–63] Calculations were described by a Monkhorst–Pack grid of $5 \times 5 \times 1$ K-points, which ensures electronic and ionic convergence.[64] The geometry of all stationary points was found with the conjugate-gradient algorithm and was considered converged when the force on each ion dropped below 0.03 eV Å^{-1}. The energy threshold defining self-consistency of the electron density was set to 10^{-5} eV. In order to improve the convergence of the Brillouin-zone integrations, the partial occupancies were determined using the tetrahedron method with Blöch corrections smearing, with a set width for all calculations of 0.02 eV. All total energies have been extrapolated to $K_{\rm B}T = 0$ eV.

Computational model

The greigite unit cell consists of eight Fe_3S_4 subunits with a cubic lattice parameter of ~9.8 Å,[65,66] which is close to the calculated parameter resulting from the cell optimization (9.671 Å). The inverse thio-spinel arrangement is reflected by the formula AB_2S_4, where there are two possible locations for the Fe ions: the tetrahedral sites (A), filled by Fe^{3+} ions, and the octahedral sites (B), where both Fe^{3+} and Fe^{2+} ions co-exist.[24,25,58,67–70] In agreement with experiments, the electronic structure shows a high-spin configuration for both Fe sites in antiparallel alignments, resulting in a ferromagnetic material.[24,25,59,67,68,70–72] The orbital spin-splitting in the valence region results in localized outermost 3d-electrons and in ordered magnetism.[69,73,74] Good agreement with experimental evidence has been obtained by using the same computational details as listed above.[53,54]

We prepared the $Fe_3S_4\{111\}$ surface as a slab model by cutting the bulk structure with the METADISE code,[75] which considers periodicity in the plane direction and provides atomic layer stacking, resulting in non-dipolar reconstructions.[76] The slab contains 56 atoms (24 Fe and 32 S) per unit cell, where the atoms are positioned in four layers of two Fe_3S_4 units each, exposing an area of 81.0 Å^2 and a thickness of sufficient size to relax the two uppermost layers (four Fe_3S_4 units) until energy convergence, keeping the bulk structure frozen at the bottom. We added a vacuum width of 12 Å between periodic slabs, big enough to avoid the interaction between images. Isolated molecules were placed in the center of a 15 × 16 × 17 Å^3 cell and optimized with the same criteria.

Every intermediate is linked to the next minimum along the reaction mechanism by a saddle point in the pathway of minimum energy across the potential surface. These saddle points are the reaction transition states (TS), which determine the kinetics of the process. We have determined the TS by means of either the dimer method[77,78] or Climbing Image Nudge Elastic Band (CI-NEB),[79,80] depending on how far the starting structure is from the products. For instance, CI-NEB was used when in the initial structure hydrogen is further than 2.5 Å from the hydrogenated centre. The identified TS were further confirmed by a vibrational frequency calculation, in which only one imaginary frequency is obtained, corresponding to the reaction coordinate. The dimer image is relaxed to both local minima, leading to the initial state and the final state.

All binding energies (E_B) are given with respect to the naked slab and the reactants in the gas phase, see eqn (1). In the discussion of the DFT results, we have referred to energy values where the zero point energy (ZPE) correction is neglected unless noted otherwise.

$$E_B = E_{Mol+Fe_3S_4\{111\}} - \left(E_{Fe_3S_4\{111\}} + \sum_i E^i_{Mol}\right) \quad (1)$$

$E_{Mol+Fe_3S_4\{111\}}$ corresponds to the total energy of the adsorbate(s) on the $Fe_3S_4\{111\}$ slab, $E_{Fe_3S_4\{111\}}$ is the energy of the $Fe_3S_4\{111\}$ slab and E_{Mol} is the energy of the isolated molecule(s) in vacuum. The activation energy (E_A) of a certain step is the energy required to surmount the potential barrier characteristic of the transition state. We have defined this energy barrier as the difference between the initial state and transition state energy for the forward process. The reaction energy (E_R) of each step is calculated as the total energy difference between the final state (product(s)) and the initial state (reactant(s)). We have discussed the feasibility of particular steps and the overall reaction trends (ΔE) as a function of the reference energy, *i.e.* clean surface and gas phase molecules, H_2 and CO_2.

Kinetic model

We have presented an extensive kinetic model[81–83] for the reduction of CO_2 on $Fe_3S_4\{111\}$ surfaces on the basis of 140 elementary steps, with the migration of the CO intermediate between the two different active sites, Fe_A and Fe_B. No assumptions regarding the mechanism or the rate-limiting step are made, all the processes are bi-directional, and all model parameters are rigorously derived from DFT calculations, which we have compared with experimental data collected under realistic conditions, see Fig. S1 in ESI.† In the frame of transition state theory,[84,85] this kinetic model may poorly describe the surface kinetics when local ordering is important and diffusion of adsorbents as H ad-atoms is slow. In these cases, computationally expensive approximations can be used, for instance kinetic Monte Carlo (kMC) is capable of accounting for the correct local structure.[86] Unfortunately, such an analysis remains completely impractical for reaction networks as complicated as the one described here. Lateral adsorbate–adsorbate interactions can also influence the stability of intermediates.[87]

Although we have included in the reaction network the effect of H ad-atoms near the catalytic center and the diffusion of CO_2 and CO molecules among Fe-sites, we have nevertheless considered the total surface H coverage (θ_H) to be low under typical conditions, and we have also included the H_2 evolution reaction as a function of θ_H. The maximum surface coverage was restricted to 1 ML, and multilayer adsorption was not considered. We have considered three different adsorption sites, a surface sulfur for the H ad-atom adsorption (eight per unit cell) and the metal centers, Fe_A or Fe_B (one site per cell).

Results and discussion

In order to develop a comprehensive model, firstly we have systematically studied the adsorption–desorption processes for the different reactants, followed by their co-adsorption and reaction leading to an extensive reaction network. Secondly, we have implemented these results in a micro-kinetics model to obtain a description of the process dynamics. Given the large amount of data, we have not described all the results in detail; the interested reader can find more relevant data in the ESI.† Finally, we have discussed the overall reaction trends and their implications in the CO_2 conversion process.

Density functional theory results

Greigite (Fe_3S_4) is an inverse spinel-structured material whose particles in hydrothermal synthesis expose {001} and {111} surfaces.[88] We have focused our attention on the Fe_3S_4{111} surface as this surface has shown conversion of carbonate species.[37] The Fe_3S_4{111} surface contains eight S atoms and three Fe atoms at the top layer of a unit cell, two from a tetrahedral bulk position (Fe_A) and one from an octahedral bulk position (Fe_B).

CO_2 adsorption and activation. Carbon dioxide interacts with the Fe centers as well as with a surface depleted in S through the oxygen lone pair electrons. The most stable configurations led to practically undisturbed molecules, see Table 1. The weak interactions with the surface, and the lack of charge transfer and molecule distortion, indicate physical adsorption on the naked surface, see Fig. 1.

Table 1 Summary of binding energies (E_B) and optimized geometry parameters for the different adsorbents as well as for the activated CO_2 on Fe_3S_4{111} (# denotes the activated state). The sorption sites are described as Fe_B, Fe_A and S vacancy (v_S). The distances (d), angles ($\angle$), and asymmetric stretching frequencies (ν_{C-O}) are given. The transferred charge (q) refers to the number of electrons transferred from the surface. Adsorbed structures are represented in the ESI, Fig. S3.1†

System	Site	E_B (eV)	d_{O-C} (Å)	$\angle_{O-C-O}$ (°)	ν_{C-O} (cm^{-1})	q (e$^-$)
CO_2	Fe_A	−0.02	1.176	178.3	2359	0.0
CO_2	Fe_B	−0.19	1.175	179.4	2363	0.0
CO_2	v_S	−0.20	1.175	179.4	2366	0.0
$CO_2^{\#}$	Fe_A	0.40	1.233	138.1	1879	0.4
$CO_2^{\#}$	Fe_B	0.19	1.207	157.4	2103	0.2
$CO_2^{\#}$ + 2H	Fe_A	−0.33	1.250	131.9	1700	0.8
$CO_2^{\#}$ + 2H	Fe_B	−0.31	1.213	153.9	2050	0.8

The activation of CO_2 from the physisorbed state is an endothermic process, but it is more likely to take place on Fe_B than on Fe_A, by 0.2 eV, perhaps due to the different ability of Fe_A and Fe_B to exchange electron density. The atomic charge analysis indicates that the transferred charge by Fe_A is twice that of Fe_B. We have noticed that, although most of the charge is transferred directly through the Fe site, its spin density remains unchanged. Fig. S3† represents the electron density depletion on the sulfur and appearance on both S–C and O lone pairs; rearrangement around the Fe atoms displays changes in the atomic orbital hybridization. Thus, the transferred charge is originally from the sulfur atoms near the metal centers. We have also considered the presence of atomic hydrogens on the CO_2 activation process. Accordingly to the adsorption energies, the presence of H ad-atoms on the surface stabilizes the activated species ($CO_2^{\#}$) on both metal centers owing to the hydrogen providing labile electrons on the sulfur, which transfers them to the CO_2 antibonding orbital. CO_2 activation produces notable changes in the molecular structure: the O–C distance elongates by 0.07 Å, the O–C–O angle bends (~130–160°), and the ν(C–O) shifts by 300 cm^{-1} on Fe_B and up to 660 cm^{-1} on Fe_A.

The positive E_B values of the activated CO_2 with respect to the energy reference indicate the preference of the molecule to desorb without modification before being activated. However, the presence of co-adsorbed hydrogens stabilizes $CO_2^{\#}$ and its transition state, as shown by the reaction energies in Table 2. The lack of H ad-atoms on the surface makes the activation of CO_2 thermodynamically unfavourable and high pressures are required to increase the CO_2 activity. On the other hand, the presence of electron-donating species, *i.e.* H ad-atoms, enhances the CO_2 activation by decreasing the energy barrier, especially on electron deficient centers such as Fe_B.

Formate and hydrocarboxyl intermediates. From CO_2 adsorption and activation in the presence of H ad-atoms, we have systematically studied its hydrogenation on both Fe_A and Fe_B, leading to formate and hydrocarboxyl intermediates *via* Langmuir–Hinshelwood (LH) reaction mechanism, where both reactants lie on the surface. The hydrogenation with the lowest activation energy is on the oxygen (**3c** from Table 3), leading to COOH, which binds to S and Fe_A by C and O, respectively, see Fig. 2. The product is practically in thermodynamic equilibrium with the reactants by an energy difference of only 0.02 eV. The most favourable process on Fe_B is reaction **3d**, which proceeds through the hydrogenation of the

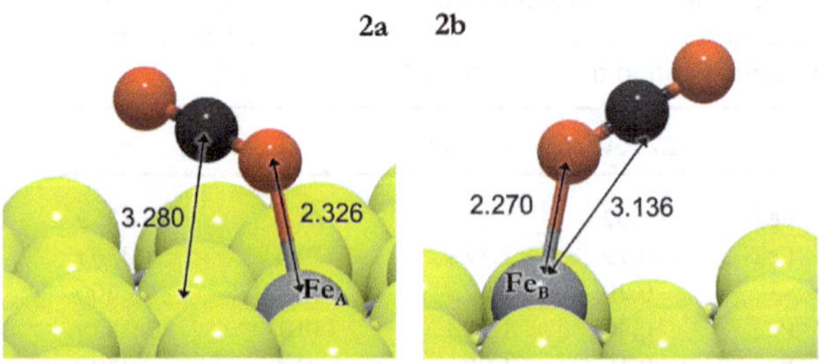

Fig. 1 Schematic representation of non-activated CO_2 on Fe_A (left) and on Fe_B (right). The light grey and yellow balls represent Fe and S, respectively, and the red and dark grey balls represent O and C, respectively.

Table 2 Elementary reaction energies for CO_2^* activation on $Fe_3S_4\{111\}$ (* denotes the adsorbed state and # denotes the activated state). E_A is the activation energy, E_R is the product reaction energy and ΔE indicates the product energy with respect to isolated molecules and the $Fe_3S_4\{111\}$ slab. The structures are represented in the ESI, Fig. S3.1†

Label	Reaction	Site	E_A (eV)	E_R (eV)	ΔE (eV)
2a	$CO_2^* \rightarrow CO_2^\#$	Fe_A	0.43	0.39	0.36
2b	$CO_2^* \rightarrow CO_2^\#$	Fe_B	0.41	0.38	0.20
2c	$CO_2^* + 2H^* \rightarrow CO_2^\# + 2H^*$	Fe_A	0.62	0.06	−0.33
2d	$CO_2^* + 2H^* \rightarrow CO_2^\# + 2H^*$	Fe_B	0.36	0.32	−0.30

oxygen coordinated to Fe, at −0.16 eV below co-adsorbed $CO_2^\#$ and two H ad-atoms.

We have also investigated the Eley–Rideal (ER) mechanism, describing the interaction between gas phase CO_2 molecules and pre-adsorbed H on $Fe_3S_4\{111\}$. We have brought the molecule from a vacuum to the Fe center, exposing independently carbon or oxygen to the H ad-atom, which produces formate or hydrocarboxyl intermediates on the surface. The reaction energies are summarized in Table 3 (**3e–3h**). HCOO is derived from reaction **3e** on Fe_A, which is the most suitable pathway for the formation of formic acid after surmounting a barrier of 0.3 eV above the energy reference. The generation of HCOO on Fe_B has an energy barrier of over 1 eV. Optimized HCOO binds to both Fe_A and Fe_B with an Fe–O distance of ~1.8 Å, see Fig. 2. The formate intermediate was also identified experimentally on metallic Cu upon exposure to CO_2 and H_2 gases (ref. 89 and references therein).

The barrier energies for COOH formation *via* the ER mechanism are 0.1 (**3g**) and 0.6 eV (**3h**) lower than those for **3e** and **3f** on Fe_A and Fe_B, respectively. The formation of COOH as an intermediate is not surprising as it has been identified as the key reaction intermediate in the water gas shift reaction.[19] Hydrocarboxyl has two distinguishable configurations depending on the direction of the hydroxyl group: *trans*-COOH and *cis*-COOH. The energy difference makes the *trans*-configuration the most stable by 0.4 eV.

Considering both mechanisms, *i.e.* LH and ER, the most favorable paths leading to HCOO intermediates are **3e** *via* ER and **3b** *via* LH on Fe_A and Fe_B,

Table 3 Elementary reaction energies for the $CO_2^\#$ hydrogenation *via* Langmuir–Hinshelwood (LH) and Eley–Rideal (ER) mechanisms on the $Fe_3S_4\{111\}$ surface. The structures are represented in the ESI, Fig. S3.2†

Label	Mechanism	Reaction	Site	E_A (eV)	E_R (eV)	ΔE (eV)
3a	LH	$CO_2^\# + 2H^* \rightarrow HCOO^* + H^*$	Fe_A	1.21	−0.28	−0.60
3b		$CO_2^\# + 2H^* \rightarrow HCOO^* + H^*$	Fe_B	1.15	−1.25	−1.56
3c		$CO_2^\# + 2H^* \rightarrow COOH^* + H^*$	Fe_A	0.10	0.02	−0.31
3d		$CO_2^\# + 2H^* \rightarrow COOH^* + H^*$	Fe_B	0.71	−0.16	−0.46
3e	ER	$CO_2 + 2H^* \rightarrow HCOO^* + H^*$	Fe_A	0.94	0.03	−0.60
3f		$CO_2 + 2H^* \rightarrow HCOO^* + H^*$	Fe_B	1.86	−0.92	−1.56
3g		$CO_2 + 2H^* \rightarrow COOH^* + H^*$	Fe_A	0.84	−0.17	−0.81
3h		$CO_2 + 2H^* \rightarrow COOH^* + H^*$	Fe_B	1.28	0.19	0.65

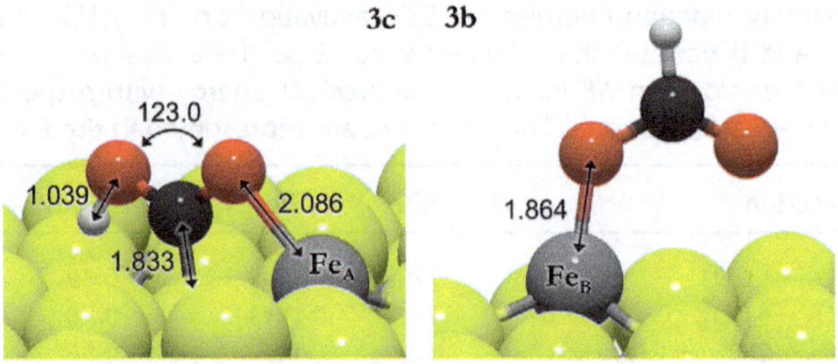

Fig. 2 Schematic representation of hydrocarboxyl (from **3c**) on Fe_A (left) and formate (from **3b**) on Fe_B (right). The light grey and yellow balls represent Fe and S, respectively, and the red and dark grey balls represent O and C, respectively.

respectively. In the transition state structure for the **3e** process, almost linear CO_2 hovers close to the H ad-atom at a distance (C–H) of 1.526 Å and the molecule starts bending with an angle of 145°. The activation energy for the process **3b** is 1.15 eV above that of the reactants and the C–H distance (1.692 Å) indicates an early transition state. The lowest activation energy for the formation of COOH is reaction **3c**, although the energy required for the adsorption of the reactants and products on Fe_A is higher than that for the ER reaction **3g**. The early saddle point for the hydrogenation of $CO_2^{\#}$ *via* **3c** places the H at 1.300 Å from the oxygen on Fe_A, while on Fe_B, the CO_2 in the **3d** transition state lies at a distance of 1.374 Å from the H. Upon COOH formation, the HO–C bond becomes weaker and elongates by 0.1 Å compared with CO_2. We have further investigated the COOH dissociation to CO and OH, and found that it is thermodynamically and kinetically likely ($E_A = 0.60$ eV; $E_R = -0.11$ eV), proceeding to the formation of H_2O + CO.

The small difference between the energy barriers (**2c** and **3e** < 0.05 eV) to HCOO and COOH intermediates indicates the low selectivity of the process on Fe_A, while a difference of 0.44 eV between the activation barriers of **3b** and **3d** indicates the selectivity for COOH intermediates on Fe_B. However, external conditions and the presence of dopants and promoters may influence the stability of intermediates, increasing the energy difference and therefore the selectivity.

Formic acid, formaldehyde and carbon monoxide. Formate intermediates remain adsorbed on the surface, allowing further changes such as C–O bond scission or additional hydrogenation. We have studied the breaking of the C–O bond but the process is thermodynamically and kinetically unfavourable ($E_R > 1.5$ eV; $E_A > 1.8$ eV). We have also studied the H migration process from formate to hydrocarboxyl but the process is kinetically unaffordable ($E_A \sim 3$ eV). A consecutive hydrogenation of the formate intermediate leads to compounds such as formic acid (HCOOH) and formaldehyde (H_2CO), see Table 4.

We have explored the energy profile for HCOOH formation, see Fig. 3, and found that, from HCOO, the transition states leading to hydrogenation of the uncoordinated oxygen (**4a,b**) are more favourable than those leading to hydrogenation of the oxygen coordinated to Fe (**4c,d**). We have also noticed that the process is thermodynamically and kinetically more likely on Fe_B than on Fe_A. Between the two sites, pathway **4a** is more feasible, leaving the HCOOH molecule adsorbed on Fe_A with a binding energy of −0.86 eV. The formation of HCOOH is

Table 4 Elementary reaction energies for the HCOO hydrogenation on the $Fe_3S_4\{111\}$ surface. Notice that we have differentiated O^1 and O^2 from the same CO_2 molecule. The structures are represented in the ESI, Fig. S3.3†

Label	Reaction	Site	E_A (eV)	E_R (eV)	ΔE (eV)
4a	$HCOO^* + H^* \rightarrow HCOO^1H^*$	Fe_A	0.08	−0.45	−1.05
4b	$HCOO^* + H^* \rightarrow HCOO^1H^*$	Fe_B	0.56	0.28	−1.28
4c	$HCOO^* + H^* \rightarrow HCOO^2H^*$	Fe_A	1.28	0.19	−0.42
4d	$HCOO^* + H^* \rightarrow HCOO^2H^*$	Fe_B	2.12	0.71	−0.84
4e	$HCOO^* + H^* \rightarrow H_2CO + O^*$	Fe_A	1.83	1.51	1.74
4f	$HCOO^* + H^* \rightarrow H_2COO^*$	Fe_B	2.12	1.30	−0.26

considerably more likely than the formation of formaldehyde (**4e** and **4f**), although this species has been identified on other catalysts as a potential intermediate in methanol production from CO_2.[17,90] Once HCOO is hydrogenated to form formaldehyde, its precursor (H_2COO) remains adsorbed on Fe_B, but on Fe_A, the C–O bond is completely broken and H_2CO is released from the surface, leaving an O ad-atom strongly adsorbed on Fe_A, thereby poisoning the active site. Formaldehyde is also a product of CO hydrogenation as described below.

Similarly to the formate intermediate, further hydrogenation of COOH leads to thermodynamically favorable HCOOH, but the activation energy ($E_A > 1.2$ eV) makes the reaction kinetically implausible (reactions **5a** and **5b**). As summarized in Table 5, dihydroxycarbene (HOCOH) is an alternative hydrogenation route as experimentally proposed in the methanol synthesis on oxide catalysts,[91] but its formation is highly inaccessible from a kinetic point of view (**5c** and **5d**). Nevertheless, we have considered HOCOH formation and dissociation to COH* + OH*, as suggested on Cu{111}.[17] However, on $Fe_3S_4\{111\}$, the reaction pathway is inhibited not only by the activation energy forming HOCOH but also by the unlikely scission of HO–C, which has $E_R > 0.8$ eV and $E_A > 1.4$ eV on both Fe sites. Under certain conditions, HOCOH may be observable experimentally but it is not part of the main reaction pathway on greigite. The hydrogenation of the OH group on the COOH intermediate takes place mostly on Fe_B and leads to the formation

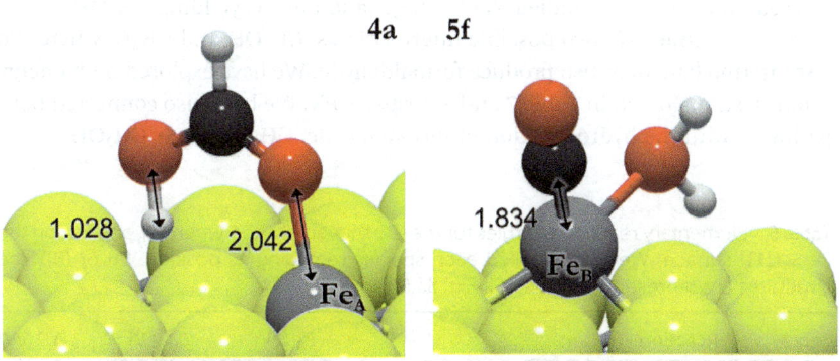

Fig. 3 Schematic representation of $HCOO^1H$ (from **4a**) on Fe_A (left) and $CO + H_2O$ (from **5f**) on Fe_B (right). The light grey and yellow balls represent Fe and S, respectively, and the red and dark grey balls represent O and C, respectively.

Table 5 Elementary reaction energies for the COOH hydrogenation on the $Fe_3S_4\{111\}$ surface. The structures are represented in the ESI, Fig. S3.4†

Label	Reaction	Site	E_A (eV)	E_R (eV)	ΔE (eV)
5a	COOH* + H* → HCOOH*	Fe_A	1.21	−0.74	−1.05
5b	COOH* + H* → HCOOH*	Fe_B	1.38	−0.24	−0.70
5c	COOH* + H* → HOCOH*	Fe_A	1.39	−0.13	−0.44
5d	COOH* + H* → HOCOH*	Fe_B	1.67	0.07	−0.40
5e	COOH* + H* → CO* + H_2O	Fe_A	1.84	0.07	−0.24
5f	COOH* + H* → CO* + H_2O*	Fe_B	0.73	−0.49	−0.95

of co-adsorbed H_2O and CO (reactions **5e** and **5f**). While the process is slightly endothermic on Fe_A with a large energy barrier above the reference of 1.6 eV, on Fe_B it is energetically favourable, see Fig. 3. The co-adsorbed H_2O stabilizes the system by 0.58 eV and, compared with HCOOH formation *via* reaction **3e**, its energy barrier is higher by less than 0.05 eV.

We have evaluated the movement of CO between both Fe_B and Fe_A from **5f** and **5e**, respectively, and also considered the presence of co-adsorbed H_2O (on Fe_B) or hydrogen ad-atoms, see Table 6. Before and after H_2O desorption from Fe_B, the migration of CO to Fe_A is endothermic and kinetically impractical ($E_A > 1.5$ eV), see Fig. 4. We have found that once water desorbs, the CO molecule shrinks the Fe_B–C bond by 0.08 Å and lies perpendicular to the surface at 1.754 Å. The intramolecular bond is 1.161 Å, which, together with a shift in its stretching frequency of 103 cm^{-1} with respect to the isolated molecule ($d_{(C\text{–}O)} = 1.142$ Å, $\nu_{(C\text{–}O)} = 2129$ cm^{-1}), indicates a weak C–O bond. The presence of co-adsorbed hydrogens decreases the energy barrier by up to 0.82 eV, making the process more likely, although $_{(A)}$CO is 0.16 eV less stable than $_{(B)}$CO. The effect of the hydrogens is similar to the CO_2 activation process, which is related to the availability of charge density. The stabilization of the CO adsorption on Fe_A by the H ad-atoms is observable in the Fe–C distance differences – the Fe_A–C bond is 0.18 Å larger than the Fe_B–C bond compared to 0.12 Å in the presence of hydrogens – and the asymmetric stretching frequencies vary by 23 and 11 cm^{-1}, respectively, without and with H ad-atoms. We have carried out the hydrogenation of CO on both sites, Fe_A and Fe_B, independently.

Methanol. We have studied the hydrogenation of CO yielding CH_3OH on the $Fe_3S_4\{111\}$ surface *via* two possible intermediates, CH_yOH and CH_xO, where the last intermediate may also produce formaldehyde. We have explored both energy profiles, summarized in Table 7 and 8, respectively. We have also connected both pathways with the hydrogenation of formaldehyde: CH_2O + H → CH_2OH.

Table 6 Elementary reaction energies for the CO movements between Fe_B and Fe_A on the $Fe_3S_4\{111\}$ surface. We have labelled each species regarding its position: (A) or (B). The structures are represented in the ESI, Fig. S3.5†

Label	Reaction	E_A (eV)	E_R (eV)	ΔE (eV)
6a	$_{(B)}$CO* + $_{(B)}H_2O$* → $_{(A)}$CO* + $_{(B)}H_2O$*	1.57	0.46	−0.49
6b	$_{(B)}$CO* → $_{(A)}$CO*	1.84	0.39	0.02
6c	$_{(B)}$CO* + 2H* → $_{(A)}$CO* + 2H*	0.82	0.16	−0.37

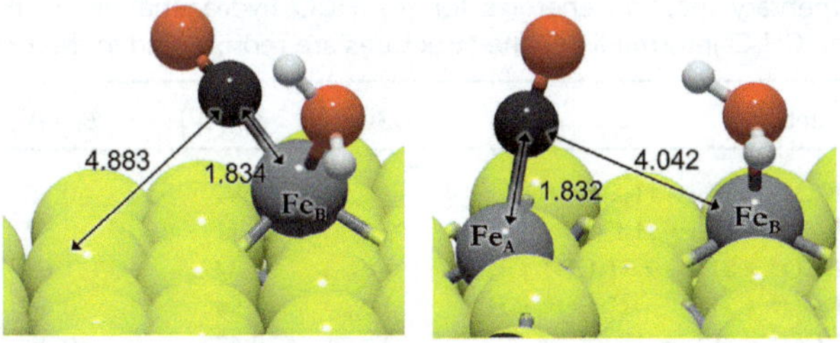

Fig. 4 Schematic representation of CO migration from Fe_B (left) to Fe_A (right) according to the process **6a**. The light grey and yellow balls represent Fe and S, respectively, and the red and dark grey balls represent O and C, respectively.

The carbon monoxide hydrogenation step is slightly unfavorable thermodynamically, $E_R(Fe_B) < 0.2$ eV, although the overall energy is lower than the reference energy (ΔE), indicating that it is a viable process. On Fe_A, the reaction **7a** is endothermic by 0.3 eV whereas it is also kinetically controlled as the transition state lies above the energy reference. The transition states for the formation of HCOH (**7c,d**) and H_2CO (**8a,b**) also lie above the reference energy on both Fe_A and Fe_B, see Fig. 5. These processes, thus, require energy from external sources, such as a natural chemiosmotic or applied potential to surmount the energy barrier. The over-potential needed for the formation of HCOH on Fe_A (**7c**) is the smallest one (0.5 eV), leading to a downhill conversion towards CH_3OH. The HCOH intermediate has also been detected during CO reduction on transition metal surfaces.[92] The presence of co-adsorbed hydrogens stabilises HCOH, whereas the production becomes thermodynamically unfavourable in pathway **8**. The formation of HCOH (**7c**) has an energy barrier 0.4 eV lower than that for the production of H_2CO (**8a**), indicating selectivity of this process on Fe_A. On the Fe_B adsorption site, the E_A values are bigger than those for Fe_A because the reactants start from a more stable position (by 0.30 eV), although the transition states for **7d** and **8b** are 0.1 and 0.2 eV more favourable than their counterparts on Fe_A.

Although CO reduction is likely to take place following process **7c**, the hydrogenation of HCOH (**7e**) has a slightly higher energy barrier than the **8c** process, by 0.1 eV, leading to competing processes. The reduction of H_2CO *via* the **9a** process is 0.2 eV lower than that *via* **7e**, allowing the hydrogenation of the oxy

Table 7 Elementary reaction energies for the CO hydrogenation on the $Fe_3S_4\{111\}$ surface *via* the CH_yOH intermediate. The structures are represented in the ESI, Fig. S3.6†

Label	Reaction	Site	E_A (eV)	E_R (eV)	ΔE (eV)
7a	$CO^* + 2H^* \rightarrow HCO^* + H^*$	Fe_A	0.52	0.31	−0.06
7b	$CO^* + 2H^* \rightarrow HCO^* + H^*$	Fe_B	0.37	0.17	−0.36
7c	$HCO^* + H^* \rightarrow HCOH^*$	Fe_A	0.55	0.26	0.20
7d	$HCO^* + H^* \rightarrow HCOH^*$	Fe_B	1.09	0.26	−0.11
7e	$HCOH^* + 2H^* \rightarrow H_2COH^* + H^*$	Fe_A	1.11	−0.81	−1.04
7f	$HCOH^* + 2H^* \rightarrow H_2COH^* + H^*$	Fe_B	0.86	−0.58	−1.08
7g	$H_2COH^* + H^* \rightarrow H_3COH^*$	Fe_A	1.17	−0.75	−1.79
7h	$H_2COH^* + H^* \rightarrow H_3COH^*$	Fe_B	0.85	−0.99	−2.07

Table 8 Elementary reaction energies for the HCO hydrogenation on the $Fe_3S_4\{111\}$ surface *via* the CH_xO intermediate. The structures are represented in the ESI, Fig. S3.7†

Label	Reaction	Site	E_A (eV)	E_R (eV)	ΔE (eV)
8a	$HCO^* + H^* \rightarrow H_2CO^*$	Fe_A	0.94	0.75	0.69
8b	$HCO^* + H^* \rightarrow H_2CO^*$	Fe_B	1.11	−0.57	−0.93
8c	$H_2CO^* + 2H^* \rightarrow H_3CO^* + H^*$	Fe_A	0.34	−1.68	−1.23
8d	$H_2CO^* + 2H^* \rightarrow H_3CO^* + H^*$	Fe_B	1.04	−0.58	−1.79
8e	$H_3CO^* + H^* \rightarrow H_3COH^*$	Fe_A	0.42	−0.98	−1.79
8f	$H_3CO^* + H^* \rightarrow H_3COH^*$	Fe_B	1.14	−0.28	−2.07

group and lying −1.04 eV below the energy reference (Table 9). However, the **7f**, **8d** and **9b** processes on Fe_B require less energy to surmount their transition states, leading to intermediates stabilised by up to 1.8 eV below the reference. Once methanol is formed through either the **7** or **8** pathway, it binds to Fe_A or Fe_B with an adsorption energy of −0.54 and −0.82 eV, respectively. Overall, the CH_3OH formation is more feasible on Fe_B where only **7d** and **8b** are limiting steps.

Reaction energy profile. We have expressed, in energy profiles, all the steps described above for the CO_2 reduction at both Fe_A and Fe_B sites on the $Fe_3S_4\{111\}$ surface, see Fig. 6 and 7, respectively. The activation of CO_2 is known to be an endothermic process, although the presence of atomic H on the surface stabilizes the activated molecule, making this a favourable process on Fe_B (**2d**). Upon CO_2 activation (**2**), the molecule may undergo hydrogenation on either C or O. Although reactions such as **3b** may be driven by favourable thermodynamics, the activation energy is higher than the reference energy, indicating that the molecule would desorb before being hydrogenated. Reaction **3c** is an exception as it has an activation energy of only 0.1 eV with a transition state 0.2 eV below the reference. Processes that do not require CO_2 pre-activation and proceed *via* the ER mechanism (**3e–h**) still require an energy input to surmount the activation energy and not all of them are thermodynamically driven. The difference in activation energies for the formation of HCOO and COOH is only 0.1 eV (**3e** *vs.* **3d**) and therefore the selectivity for specific products, *i.e.* HCOOH and CH_3OH, is very low. Many of the following processes studied are energetically inhibited under mild conditions and the pathways are limited to reactions **4a** on Fe_A, and **4b**, **4d** and **5f** on Fe_B.

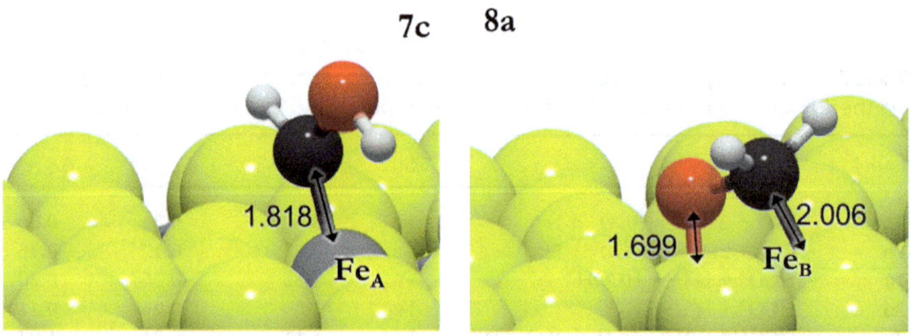

Fig. 5 Schematic representation of HCOH (from **7c**) on Fe_A (left) and H_2CO (from **8a**) on Fe_B (right). The light grey and yellow balls represent Fe and S, respectively, and the red and dark grey balls represent O and C, respectively.

Table 9 Elementary reaction energies for the H_2CO hydrogenation on the $Fe_3S_4\{111\}$ surface. The structures are represented in Fig. S3.8†

Label	Reaction	Site	E_A (eV)	E_R (eV)	ΔE (eV)
9a	$H_2CO^* + 2H^* \rightarrow H_2COH^* + H^*$	Fe_A	0.18	−1.49	−1.04
9b	$H_2CO^* + 2H^* \rightarrow H_2COH^* + H^*$	Fe_B	1.14	0.13	−1.08

The formate adsorbed on Fe_A, the product of reaction **3e**, is followed by a fast process producing HCOOH, which remains adsorbed, suggesting the possibility for secondary reactions, *e.g.* C–C formation or combining with NH_2 to give formamide.[12] This agrees with previous experiments, which observed large amounts of formate on metal surfaces upon exposure to CO_2 and H_2 gases.[89,93] Thus, the rate-limiting step for HCOOH formation is the first CO_2 hydrogenation with a TS energy 0.3 eV above the reference.

Recently identified as a key intermediate in the water gas shift reaction,[19,94] we have identified hydrocarboxyl as the first intermediate in the CO formation prior to methanol yield. The reverse water gas shift reaction takes place when oxygen from the CO_2 molecule reacts with H to produce CO and H_2O, *i.e.* processes **3d** and **5f**. This pathway on Fe_B led to carbon monoxide bonded to Fe_B with an adsorption energy of −1.1 eV, which is stabilized by the presence of H ad-atoms (by −0.17 eV). Although the formation of HCO is thermodynamically feasible, its hydrogenation has an energy barrier of at least 0.75 eV, processes (**7** and **8**), which are driven by very stable intermediates. The reduction of HCO is, hence, the rate-limiting step for CH_3OH formation from CO_2 on $Fe_3S_4\{111\}$. Once the limiting step is surmounted, the methoxy route (**8**) is the most favourable despite the fact that pathway 7 is also downhill energetically. Methoxy intermediates are very stable and remain on the surface for longer, which may lead to interactions with other molecules or intermediates, leading to *e.g.* longer chain hydrocarbons.[12]

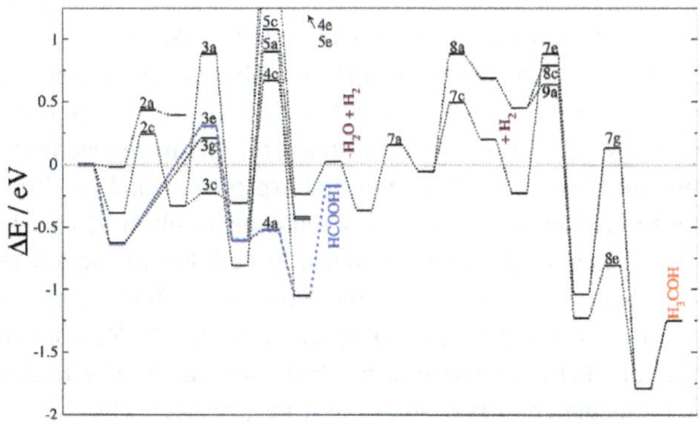

Fig. 6 Reaction energy profiles for the CO_2 hydrogenation process taking place on Fe_A. The blue line highlights the formate pathway. The labels corresponding to elementary steps are placed at the transition state. Processes such as water molecule release ($-H_2O$) and surface hydrogenation ($+H_2$) are also included. The horizontal brown line indicates the energy reference of the naked surface and isolated molecules.

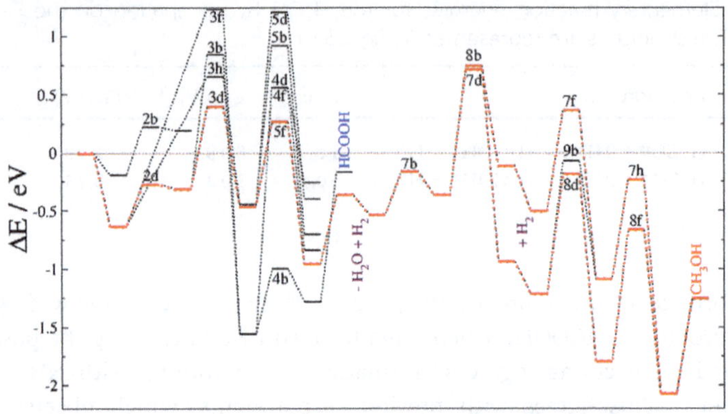

Fig. 7 Reaction energy profiles for the CO_2 hydrogenation process taking place on Fe_B. The blue line highlights the formate pathway. The labels corresponding to elementary steps are placed at the transition state. Processes such as water molecule release ($-H_2O$) and surface hydrogenation ($+H_2$) are also included. The horizontal brown line indicates the energy reference of the naked surface and isolated molecules.

Kinetic model results

We have applied a kinetic model, including the adsorption–desorption of the molecules and all the steps described above, in a consistent reaction network. We have characterized the desorption of CO_2, H_2O, H_2, HCOOH and CH_3OH individually and modelled the temperature programmed desorption (TPD) starting with full coverage and raising the temperature by 1 K s^{-1}. Molecules like CO_2, HCOOH and CH_3OH that have different adsorption sites, *i.e.* Fe_A and Fe_B, were equally distributed among them. For example, there are two configurations of HCOOH on Fe_A and two on Fe_B, and we set each configuration at an initial coverage of 0.25 ML. We have also considered secondary reactions like the exchange between CO_2 adsorption sites and the dissociation–association of the water molecule. The formation of H_2 is considered as a function of the H-coverage, for which we have investigated the associative desorption with different coverages, *i.e.* 0.25, 0.5, 0.75 and 1 ML, over the eight sulfur sites of an $Fe_3S_4\{111\}$ cell unit. The overlap of these individual TPDs led to Fig. 8.

Physisorbed CO_2 desorbs from the surface with a minimal thermal energy, in agreement with the weak interaction with either Fe_B or Fe_A. Separately, H adatoms remain scattered on the surface until the temperature reaches 500 K, when they recombine and evolve as H_2 giving a sharp peak, which is linked to a fast process. Water molecules desorb from Fe_B relatively easily at 135 K with a narrow signal; note that it is a single molecule per unit cell. Formic acid is desorbed in three signals instead of the four expected from the different adsorbed conformations due to the overlapping of two signals around 378 K, from reactions **4a** and **4d**. Methanol TPD also presents multiple signals, at 119 K and at 185 K, related to the molecule adsorbed on Fe_A and Fe_B, respectively.

We have derived the reaction constants for the forward and the reverse reactions, tabulated in Table S1.† The relationship between these constants ($K = K_{+1}/K_{-1}$) provides important information with respect to a particular step. For instance, although reaction **3f** has a high energy barrier compared with other CO_2

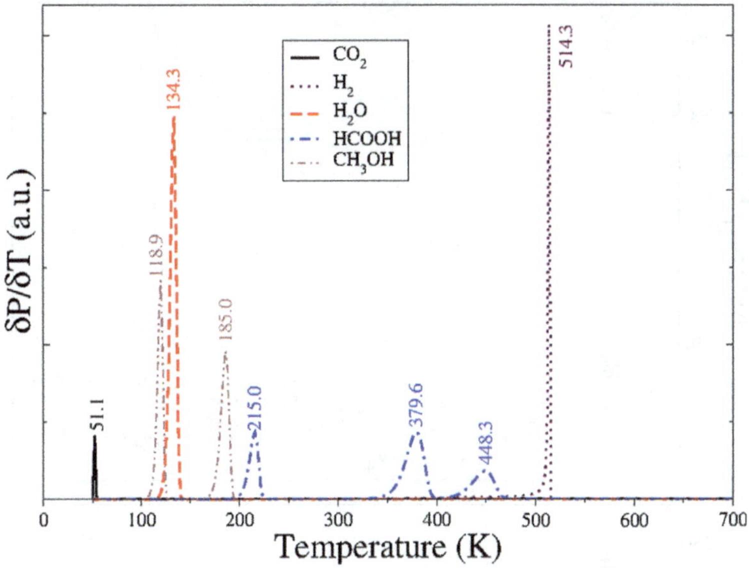

Fig. 8 Temperature programmed desorption of the molecules involved in the reduction mechanism of CO_2 on Fe_3S_4{111}. The temperature range considered is from 50 to 700 K, evaluated every 2 K each 2 s, leading to a heating rate of 1 K s^{-1}.

hydrogenation processes, the driving force of the HCOO formation makes K_{+1} bigger than K_{-1} and therefore the equilibrium constant (K) shifts towards formate. We have plotted the logarithm of K for every hydrogenation process in Fig. 9.

Although the activation of CO_2 on the surface is unlikely (reactions **2a–d**), the presence of H ad-atoms enhances its molecular adsorption on either Fe_A or Fe_B as the top left values of Fig. 9 indicate. However, the Eley–Rideal mechanism for the formate formation on Fe_A is preferred and this is followed by a second hydrogenation (**4a**), leading to formic acid, whose desorption, however, is the slowest step, thereby reducing its production. On Fe_B, the formation of formic acid is capped by the second hydrogenation (**4b**) with the reverse process being more favourable, which means that formic acid released from Fe_A would adsorb on Fe_B and remain as formate, limiting the site availability. The processes **5c**, **7a** and **8a** on Fe_A have similar consequences with respect to methanol formation. The methanol formation on Fe_B may take place *via* the highly favourable Eley–Rideal mechanism (**3h**) or through CO_2 activation and hydrocarboxyl formation (**2b** and **3d**, respectively). The only slightly reverse step is the hydrogenation of CO (**7b**), which limits the process *via* the H_2CO intermediate, mechanism **8**.

The mechanism showed HCOOH and CH_3OH as products of the CO_2 conversion with the reaction constants indicating that intermediate species like formate, carbon monoxide and CH_xO have a long enough residence time on the surface to allow them to react with other C-species, leading to *e.g.* acetic and pyruvic acid. Indeed these C_2 and C_3 products were detected experimentally after exposing Fe_3S_4 particles to a saturated aqueous solution of CO_2 under a mild reduction potential.[12]

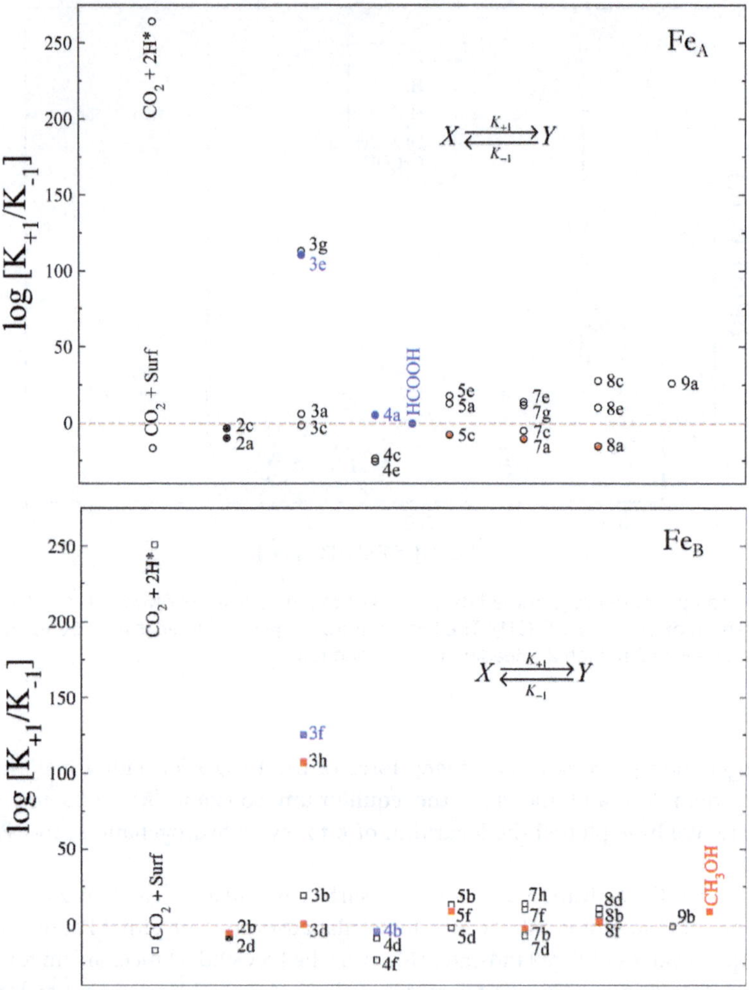

Fig. 9 Relationship between the equilibrium constants ($\log[K_{+1}/K_{-1}]$) at a temperature of 300 K for the CO_2 reduction towards HCOOH (blue) and CH_3OH (red) on Fe_A (circles) and Fe_B (squares) sites on the $Fe_3S_4\{111\}$ surface. The labels indicate the processes according to Fig. S3.† Formic acid and methanol desorption/adsorption constants are included on Fe_A and Fe_B, respectively. The solid symbols show the limiting step for each mechanism.

Conclusions

We have carried out an extensive density functional theory study to map the CO_2 reduction mechanism network on the mineral surface $Fe_3S_4\{111\}$. Among the multiple reaction pathways explored, we have discerned the most favourable by using arguments based on energy profiles and reaction constants. Our results show distinctive behaviour for the different adsorption sites: Fe_A and Fe_B. On Fe_A, the CO_2 conversion occurs through a formate intermediate *via* Eley–Rideal mechanism, yielding HCOOH. On Fe_B, the formation of the hydrocarboxyl intermediate is favoured, which dissociates and yields CO on the surface plus a water molecule. Further hydrogenations of CO are limited by the formation of the HCO intermediate; despite being energetically favourable compared with the energy reference, the reaction constant towards HCO is smaller than that of the reverse processes. Following reductions are driven by the stability of their products. Methanol, in contrast to formic acid, desorbs from the surface easily, releasing an active site for further CO_2 conversion. Although the production of formic acid or methanol is determined by the adsorption site, CO_2 and intermediates are able to move from one site to another, as shown in step **6.**

Thus, $Fe_3S_4\{111\}$ catalyzes the CO_2 transformation by hydrogen, yielding mostly formic acid and methanol, where the rate-limiting step is the CO_2 activation *via* Langmuir–Hinshelwood mechanism, and indicating that the production of methanol is through the formation of the HCO intermediate. The selectivity towards particular products increases, *e.g.* by blocking selectively an Fe site, where capping Fe_A, for instance, leads to the reduction of CO_2 into methanol.

Acknowledgements

We acknowledge the Engineering & Physical Sciences Research Council (grant EP/H046313, EP/K001329 and EP/K016288) for funding. This work made use of the HECToR and ARCHER facilities, the UK's national high-performance computing service, which is provided by UoE HPCx Ltd at the University of Edinburgh, Cray Inc. and NAG Ltd, funded by the Office of Science and Technology through EPSRC's High End Computing Programme and provided *via* our membership of the HPC Materials Chemistry Consortium (EPSRC grants EP/D504872, EP/F067496 and EP/L000202). The authors also acknowledge the use of the UCL Legion High Performance Computing Facility and the Advanced Research Computing @ Cardiff (ARCCA) at Cardiff University, and associated support services, in the completion of this work. All data created during this research is openly available from the University of Cardiff Research Portal at http://dx.doi.org/10.17035/d.2016.0008385635.

References

1 M. C. Urban, *Science*, 2015, **348**, 571–573.
2 J. Rogelj, J. Nabel, C. Chen, W. Hare, K. Markmann, M. Meinshausen, M. Schaeffer, K. Macey and N. Hohne, *Nature*, 2010, **464**, 1126–1128.
3 H. Arakawa, M. Aresta, J. N. Armor, M. A. Barteau, E. J. Beckman, A. T. Bell, J. E. Bercaw, C. Creutz, E. Dinjus, D. A. Dixon, K. Domen, D. L. DuBois, J. Eckert, E. Fujita, D. H. Gibson, W. A. Goddard, D. W. Goodman, J. Keller, G. J. Kubas, H. H. Kung, J. E. Lyons, L. E. Manzer, T. J. Marks, K. Morokuma, K. M. Nicholas, R. Periana, L. Que, J. Rostrup-Nielson, W. M. Sachtler, L. D. Schmidt, A. Sen, G. A. Somorjai, P. C. Stair, B. R. Stults and W. Tumas, *Chem. Rev.*, 2001, **101**, 953–996.
4 L. Plasseraud, *ChemSusChem*, 2010, **3**, 631–632.
5 M. Meinshausen, N. Meinshausen, W. Hare, S. C. B. Raper, K. Frieler, R. Knutti, D. J. Frame and M. R. Allen, *Nature*, 2009, **458**, 1158–1162.
6 M. Aresta and A. Dibenedetto, *Dalton Trans.*, 2007, 2975–2992.
7 S. C. Roy, O. K. Varghese, M. Paulose and C. A. Grimes, *ACS Nano*, 2010, **4**, 1259–1278.
8 G. Centi and S. Perathoner, *Catal. Today*, 2009, **148**, 191–205.

9 H. J. Freund and M. W. Roberts, *Surf. Sci. Rep.*, 1996, **25**, 225–273.
10 M. R. Kember, A. Buchard and C. K. Williams, *Chem. Commun.*, 2011, **47**, 141–163.
11 N. S. Spinner, J. A. Vega and W. E. Mustain, *Catal. Sci. Technol.*, 2012, **2**, 19–28.
12 A. Roldan, N. Hollingsworth, A. Roffey, H. U. Islam, J. B. Goodall, C. R. Catlow, J. A. Darr, W. Bras, G. Sankar, K. B. Holt, G. Hogarth and N. H. de Leeuw, *Chem. Commun.*, 2015, **51**, 7501–7504.
13 H.-J. Freund and M. W. Roberts, *Surf. Sci. Rep.*, 1996, **25**, 225–273.
14 J. P. Pradiser and C. M. Pradier, *Carbon Dioxide Chemistry: Environmental Issues*, Woodhead Publishing, 1994.
15 M. Aresta and A. Dibenedetto, *Dalton Trans.*, 2007, 2975–2992.
16 U. Jayarathne, P. Chandrasekaran, H. Jacobsen, J. T. Mague and J. P. Donahue, *Dalton Trans.*, 2010, **39**, 9662–9671.
17 Y. F. Zhao, Y. Yang, C. Mims, C. H. F. Peden, J. Li and D. H. Mei, *J. Catal.*, 2011, **281**, 199–211.
18 T. Riedel, M. Claeys, H. Schulz, G. Schaub, S. S. Nam, K. W. Jun, M. J. Choi, G. Kishan and K. W. Lee, *Appl. Catal., A*, 1999, **186**, 201–213.
19 A. A. Gokhale, J. A. Dumesic and M. Mavrikakis, *J. Am. Chem. Soc.*, 2008, **130**, 1402–1414.
20 L. C. Grabow and M. Mavrikakis, *ACS Catal.*, 2011, **1**, 365–384.
21 S. Hunger and L. G. Benning, *Geochem. Trans.*, 2007, **8**, 1–20.
22 L. G. Benning, R. T. Wilkin and H. L. Barnes, *Chem. Geol.*, 2000, **167**, 25–51.
23 R. T. Wilkin and H. L. Barnes, *Geochim. Cosmochim. Acta*, 1996, **60**, 4167–4179.
24 M. J. Dekkers, H. F. Passier and M. A. A. Schoonen, *Geophys. J. Int.*, 2000, **141**, 809–819.
25 M. J. Dekkers and M. A. A. Schoonen, *Geophys. J. Int.*, 1996, **126**, 360–368.
26 A. R. Lennie, S. A. T. Redfern, P. E. Champness, C. P. Stoddart, P. F. Schofield and D. J. Vaughan, *Am. Mineral.*, 1997, **82**, 302–309.
27 U. Frank, N. R. Nowaczyk and J. F. W. Negendank, *Geophys. J. Int.*, 2007, **168**, 904–920.
28 U. Frank, N. R. Nowaczyk and J. F. W. Negendank, *Geophys. J. Int.*, 2007, **168**, 921–934.
29 S. Mann, N. H. C. Sparks, R. B. Frankel, D. A. Bazylinski and H. W. Jannasch, *Nature*, 1990, **343**, 258–261.
30 M. Posfai, P. R. Buseck, D. A. Bazylinski and R. B. Frankel, *Am. Mineral.*, 1998, **83**, 1469–1481.
31 M. Posfai, P. R. Buseck, D. A. Bazylinski and R. B. Frankel, *Science*, 1998, **280**, 880–883.
32 W. Martin, J. Baross, D. Kelley and M. J. Russell, *Nat. Rev. Microbiol.*, 2008, **6**, 805–814.
33 M. J. Russell and A. J. Hall, *Evolution of Early Earth's Atmosphere, Hydrosphere, and Biosphere: Constraints from Ore Deposits*, 2006, vol. 198, pp. 1–32.
34 W. Martin and M. J. Russell, *Philosophical Transactions of the Royal Society of London Series B-Biological Sciences*, 2003, vol. 358, pp. 59–83.
35 C. Huber and G. Wachtershauser, *Science*, 1997, **276**, 245–247.
36 G. Wachtershauser, *Prog. Biophys. Mol. Biol.*, 1992, **58**, 85–201.
37 G. Wachtershauser, *Microbiol. Rev.*, 1988, **52**, 452–484.
38 C. J. Cramer and D. G. Truhlar, *Phys. Chem. Chem. Phys.*, 2009, **11**, 10757–10816.

39 J. K. Norskov, T. Bligaard, J. Rossmeisl and C. H. Christensen, *Nat. Chem.*, 2009, **1**, 37–46.
40 G. Kresse and J. Hafner, *Phys. Rev. B: Condens. Matter Mater. Phys.*, 1993, **47**, 558.
41 G. Kresse and J. Furthmüller, *Comput. Mater. Sci.*, 1996, **6**, 15.
42 G. Kresse and D. Joubert, *Phys. Rev. B: Condens. Matter Mater. Phys.*, 1999, **59**, 1758.
43 J. P. Perdew, J. A. Chevary, S. H. Vosko, K. A. Jackson, M. R. Pederson, D. J. Singh and C. Fiolhais, *Phys. Rev. B: Condens. Matter Mater. Phys.*, 1992, **46**, 6671.
44 S. H. Vosko, L. Wilk and M. Nusair, *Can. J. Phys.*, 1980, **58**, 1200–1211.
45 S. Grimme, *J. Comput. Chem.*, 2006, **27**, 1787–1799.
46 S. Irrera, A. Roldan, G. Portalone and N. H. de Leeuw, *J. Phys. Chem. C*, 2013, **117**, 3949–3957.
47 N. Y. Dzade, A. Roldan and N. H. de Leeuw, *J. Chem. Phys.*, 2013, **139**, 124708.
48 S. S. Tafreshi, A. Roldan, N. Y. Dzade and N. H. de Leeuw, *Surf. Sci.*, 2014, **622**, 1–8.
49 S. Haider, A. Roldan and N. H. de Leeuw, *J. Phys. Chem. C*, 2014, **118**, 1958–1967.
50 N. Dzade, A. Roldan and N. de Leeuw, *Minerals*, 2014, **4**, 89–115.
51 F. Zhang, J. D. Gale, B. P. Uberuaga, C. R. Stanek and N. A. Marks, *Phys. Rev. B: Condens. Matter Mater. Phys.*, 2013, **88**, 054112–054117.
52 N. Mermin, *Phys. Rev.*, 1965, **137**, 1441–1443.
53 A. Roldan, D. Santos-Carballal and N. H. de Leeuw, *J. Chem. Phys.*, 2013, **138**, 204712–204716.
54 D. Santos-Carballal, A. Roldan, R. Grau-Crespo and N. H. de Leeuw, *Phys. Rev. B: Condens. Matter Mater. Phys.*, 2015, **91**, 195106.
55 V. I. Anisimov, M. A. Korotin, J. Zaanen and O. K. Andersen, *Phys. Rev. Lett.*, 1992, **68**, 345–348.
56 S. L. Dudarev, G. A. Botton, S. Y. Savrasov, C. J. Humphreys and A. P. Sutton, *Phys. Rev. B: Condens. Matter Mater. Phys.*, 1998, **57**, 1505–1509.
57 A. J. Devey, R. Grau-Crespo and N. H. de Leeuw, *Phys. Rev. B: Condens. Matter Mater. Phys.*, 2009, **79**, 195126–195133.
58 L. Chang, B. D. Rainford, J. R. Stewart, C. Ritter, A. P. Roberts, Y. Tang and Q. W. Chen, *J. Geophys. Res.: Solid Earth*, 2009, **114**, 1–10.
59 L. Chang, A. P. Roberts, Y. Tang, B. D. Rainford, A. R. Muxworthy and Q. Chen, *J. Geophys. Res.*, 2008, **113**, 1–16.
60 I. D. R. Moreira, F. Illas and R. L. Martin, *Phys. Rev. B: Condens. Matter Mater. Phys.*, 2002, **65**, 155102–155116.
61 I. Ciofini, F. Illas and C. Adamo, *J. Chem. Phys.*, 2004, **120**, 3811–3816.
62 F. Illas and R. L. Martin, *J. Chem. Phys.*, 1998, **108**, 2519–2527.
63 D. Munoz, N. M. Harrison and F. Illas, *Phys. Rev. B: Condens. Matter Mater. Phys.*, 2004, **69**, 085115–085124.
64 H. J. Monkhorst and J. D. Pack, *Phys. Rev. B: Solid State*, 1976, **13**, 5188–5192.
65 J. M. D. Coey, M. R. Spender and A. H. Morrish, *Solid State Commun.*, 1970, **8**, 1605.
66 M. R. Spender, J. M. D. Coey and A. H. Morrish, *Can. J. Phys.*, 1972, **50**, 2313–2326.
67 D. J. Vaughan and J. A. Tossell, *Phys. Chem. Miner.*, 1983, **9**, 253–262.

68 D. J. Vaughan and J. A. Tossell, *Am. Mineral.*, 1981, **66**, 1250–1253.
69 D. J. Vaughan and J. R. Craig, *Am. Mineral.*, 1985, **70**, 1036–1043.
70 K. K. Surerus, M. C. Kennedy, H. Beinert and E. Munck, *Proc. Natl. Acad. Sci. U. S. A.*, 1989, **86**, 9846–9850.
71 D. J. Vaughan and M. S. Ridout, *J. Inorg. Nucl. Chem.*, 1971, **33**, 741–746.
72 M. Braga, S. K. Lie, C. A. Taft and W. A. Lester, *Phys. Rev. B: Condens. Matter Mater. Phys.*, 1988, **38**, 10837–10851.
73 D. J. Vaughan and J. A. Tossell, *Am. Mineral.*, 1981, **66**, 1250–1253.
74 D. Rickard and G. W. Luther, *Chem. Rev.*, 2007, **107**, 514–562.
75 G. W. Watson, E. T. Kelsey, N. H. deLeeuw, D. J. Harris and S. C. Parker, *J. Chem. Soc., Faraday Trans.*, 1996, **92**, 433–438.
76 P. W. Tasker, *J. Phys. C: Solid State Phys.*, 1979, **12**, 4977–4984.
77 G. Henkelman and H. Jonsson, *J. Chem. Phys.*, 1999, **111**, 7010–7022.
78 A. Heyden, A. T. Bell and F. J. Keil, *J. Chem. Phys.*, 2005, **123**, 224101–224115.
79 G. Henkelman and H. Jonsson, *J. Chem. Phys.*, 2000, **113**, 9978–9985.
80 G. Henkelman, B. P. Uberuaga and H. Jonsson, *J. Chem. Phys.*, 2000, **113**, 9901–9904.
81 I. Chorkendorff and J. W. Niemantsverdriet, *Concepts of Modern Catalysis and Kinetics*, Wiley-VCH, Weinheim, 2005.
82 A. Roldan, G. Novell, J. M. Ricart and F. Illas, *J. Phys. Chem. C*, 2010, **114**, 5101–5106.
83 P. Stoltze, *Prog. Surf. Sci.*, 2000, **65**, 65–150.
84 H. Eyring, *J. Chem. Phys.*, 1935, **3**, 107–115.
85 M. G. Evans and M. Polanyi, *Trans. Faraday Soc.*, 1935, **31**, 0875–0893.
86 H. S. Bengaard, J. K. Nørskov, J. Sehested, B. S. Clausen, L. P. Nielsen, A. M. Molenbroek and J. R. Rostrup-Nielsen, *J. Catal.*, 2002, **209**, 365–384.
87 L. C. Grabow, B. Hvolbaek and J. K. Norskov, *Top. Catal.*, 2010, **53**, 298–310.
88 N. Hollingsworth, A. Roffey, H.-U. Islam, M. Mercy, A. Roldan, W. Bras, M. Wolthers, C. R. A. Catlow, G. Sankar, G. Hogarth and N. H. de Leeuw, *Chem. Mater.*, 2014, **26**, 6281–6292.
89 Y. Yang, C. A. Mims, R. S. Disselkamp, J. H. Kwak, C. H. F. Peden and C. T. Campbell, *J. Phys. Chem. C*, 2010, **114**, 17205–17211.
90 A. A. Peterson, F. Abild-Pedersen, F. Studt, J. Rossmeisl and J. K. Norskov, *Energy Environ. Sci.*, 2010, **3**, 1311–1315.
91 L. Z. Gao and C. T. Au, *J. Catal.*, 2000, **189**, 1–15.
92 P. Ferrin and M. Mavrikakis, *J. Am. Chem. Soc.*, 2009, **131**, 14381–14389.
93 E. Vesselli, M. Rizzi, L. de Rogatis, X. L. Ding, A. Baraldi, G. Comelli, L. Savio, L. Vattuone, M. Rocca, P. Fornasiero, A. Baldereschi and M. Peressi, *J. Phys. Chem. Lett.*, 2010, **1**, 402–406.
94 J. Yoshihara and C. T. Campbell, *J. Catal.*, 1996, **161**, 776–782.

Faraday Discussions

PAPER

A unified view on heterogeneous and homogeneous catalysts through a combination of spectroscopy and quantum chemistry†

Dimitrios Maganas,[a] Annette Trunschke,[b] Robert Schlögl*[ab] and Frank Neese*[a]

Received 25th November 2015, Accepted 4th January 2016
DOI: 10.1039/c5fd00193e

Identifying catalytically active structures or intermediates in homogeneous and heterogeneous catalysis is a formidable challenge. However, obtaining experimentally verified insight into the active species in heterogeneous catalysis is a tremendously challenging problem. Many highly advanced spectroscopic and microscopic methods have been developed to probe surfaces. In this discussion we employ a combination of spectroscopic methods to study two closely related systems from the heterogeneous (the silica-supported vanadium oxide VO_x/SBA-15) and homogeneous (the complex K[$VO(O_2)$Hheida]) domains. Spectroscopic measurements were conducted strictly in parallel for both systems and consisted of oxygen K-edge and vanadium L-edge X-ray absorption measurements in conjunction with resonance Raman spectroscopy. It is shown that the full information content of the spectra can be developed through advanced quantum chemical calculations that directly address the sought after structure–spectra relationships. To this end we employ the recently developed restricted open shell configuration interaction theory together with the time-dependent theory of electronic spectroscopy to calculate XAS and rR spectra respectively. The results of the study demonstrate that: (a) a combination of several spectroscopic techniques is of paramount importance in identifying signature structural motifs and (b) quantum chemistry is an extremely powerful guide in cross connecting theory and experiment as well as the homogeneous and heterogeneous catalysis fields. It is emphasized that the calculation of spectroscopic observables provides an excellent way for the critical experimental validation of theoretical results.

[a]*Max Planck Institute for Chemical Energy Conversion, Stiftstr. 34-36, D-45470 Mülheim an der Ruhr, Germany. E-mail: Frank.Neese@cec.mpg.de; rs01@fhi-berlin.mpg.de*
[b]*Fritz-Haber Institut, Faradayweg 4-6, 14195 Berlin, Germany*

† Electronic supplementary information (ESI) available: Experimental and fitting procedure, computational details, electronic structure, Fig. S1–S9, Table S1, and bibliography. See DOI: 10.1039/c5fd00193e

Introduction

The impact of heterogeneous catalysis on industrial scale processes hardly needs to be emphasized. In fact the majority of catalytic processes that are implemented on a large-scale industrial basis, are based on heterogeneous catalysis.[1] Developing new catalysts that are efficient, stable and selective is a major challenge. Rather than finding such systems through trial and error, one would hope to take a rational approach that is based on a detailed understanding of the catalytic mechanism. To achieve the latter requires obtaining insight into the reactive species that exist on surfaces under *in operando* conditions.[2]

Theoretical calculations are in very widespread use in order to obtain insight into heterogeneous catalysts. The theoretical treatment of surfaces and solids is strongly dominated by the method of density functional theory (DFT).[3] While this method has definite and important strengths and advantages, it also suffers from some significant shortcomings, as will be elaborated below. In many cases the connection between theory and experiment is rather indirect. For example, volcano plots are frequently used to correlate structure and reactivity by connecting substrate binding energies at ideal surfaces with kinetic rates.[4,5] Due to several strong assumptions needed for the connection between thermodynamics, kinetics and mechanism, the necessity for a more immediate connection between theory and experiment emerges. It is the aim of this paper to emphasize that such strong ties between theory and experiment can be established by combining a range of spectroscopic methods with the results of carefully calibrated theoretical spectroscopy.

Combined experimental and theoretical multi-method protocols have been used to great advantage in (bio)inorganic chemistry and catalysis. For example, a theoretical spectroscopy approach has proven instrumental in characterizing exotic reactive species in model systems (*e.g.* genuine Fe(V)[6] and the first Fe(VI) complex[7]), enzymes[8–10] (*e.g.* the identification of a unique carbide center in the active site of the dinitrogen activating enzyme nitrogenase[11]), or in clarifying the structure[12] and oxidation states[13] of the oxygen evolving complex in photosynthesis. It should be emphasized that in order to successfully apply a theoretical spectroscopy approach, it is of utmost importance that the theoretical methods are properly calibrated, that is, the error bars of the theoretical predictions must be known from studying a series of known and understood systems. Hence, all predictions of theoretical methods are only valid within the confidence intervals defined by the calibration procedure. All contemporary calculations of homogeneous and heterogeneous systems likewise are afflicted with methodological errors and errors stemming from the inevitable oversimplification of real systems. It is for this reason that one is well advised to connect theory with as many spectroscopic experiments as one can possibly obtain. It should be emphasized that, once broad agreement between theory and experiment has been achieved, one can: (a) cross correlate structure–spectra relationships between vastly different experimental methods and (b) obtain experimentally calibrated insight into not only the geometric but also the all-important electronic structure of the systems under investigation. The latter is instrumental for obtaining deep insight into the reactivity of the system. This might be impressively exemplified with the case of the reactive Fe(IV)–oxo $(FeO)^{2+}$ core, the complex reactivity of which might well be considered as understood.[9,10]

In this work, we will try to establish a strong connection between the domains of homogeneous and heterogeneous catalysis by using parallel spectroscopic protocols for a heterogeneous system, namely silica supported vanadia catalysts,[14,15] and a closely related homogeneous system, the coordination complex $K[VO(O_2)Hheida]$. The chemical importance of these systems is presented in the ESI† while representative structures are shown in Fig. 1.

The experimental methods that were chosen for the current study involve vanadium L-edge and oxygen K-edge X-ray absorption (XAS) spectroscopy, absorption (ABS) and resonance Raman (rR) spectroscopy. The theoretical methodology required to calculate L-edge XAS and rR spectra is not part of the standard arsenal of theoretical solid state physics or quantum chemistry. In fact, both techniques require extensive method development. In particular, the field of L-edge spectroscopy cannot be successfully approached on the basis of DFT at all. Hence, our theoretical approaches deviate strongly from the currently employed DFT methods in that they are largely based on ideas from wavefunction-based *ab initio* quantum chemistry. We emphasize the properties of the N-particle states of the system (N being the number of electrons) in favour of the ubiquitous one-particle (orbital) picture that is implied by interpretation based on DFT. The drawback of focusing on wavefunction approaches is that truly periodic calculations are not available for the approaches that we are using. The next best alternative, that we follow, is to employ embedded clusters. Once the clusters are sufficiently large, the results of the periodic calculations are approached. However, convergence with cluster size is an issue that must be carefully monitored. Fortunately, cluster calculations with a few hundred atoms are feasible with modern hard- and software technologies such that the limitations of using cluster models are not nearly as severe as they were in the past.

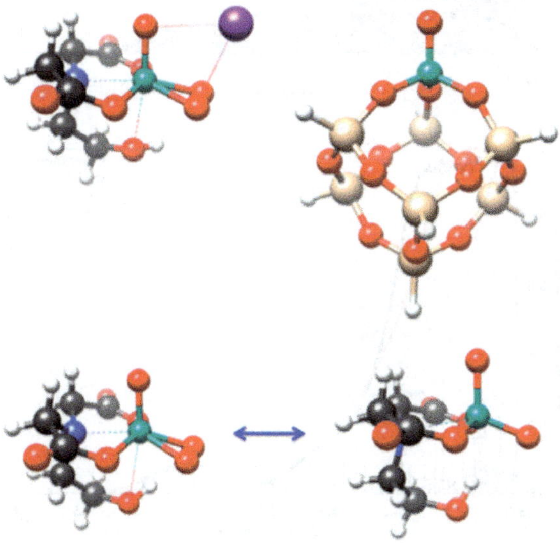

Fig. 1 Top left: The molecular structure of $K[VO(O_2)Hheida]$. Top right: Representative model structure of monomeric silica supported vanadia catalyst, $VSi_7H_7O_{13}$. Bottom: Two possible active forms of the molecular catalyst; $[VO(O_2)Hheida]^-$ (left) and oxygen deficient $[VO(O)Hheida]^-$ (right). Atom colouring: H (white), C (gray), O (red), N (blue), V (cyan), Si (light yellow) and K (purple).

We have recently introduced the restricted open shell configuration interaction singles (ROCIS) method as a method of general applicability to interpret metal L-edge spectra of a large class of molecules, ranging from mononuclear complexes up to polymetallic clusters.[16–19] The method is able to predict metal L-edge spectra for a variety of metal complexes in different oxidation states and coordination environments. In addition it has been calibrated for a large set of vanadium mononuclear complexes and has proven successful in interpreting V L-edge and O K-edge spectra of V_2O_5 (ref. 16) as well as the Ti L-edge spectra of TiO_2 (rutile and anatase) and Ca L-edge spectra of CaF_2 solids.[20] Furthermore, we have also presented a highly efficient method to treat the band-shape of optical absorption spectra and resonance Raman spectra for large molecules.[21,22]

Experimental section

Synthesis

VO_x/SBA-15 catalysts (V loading of 2–14% wt%/0.6–13.7 V-atoms nm^{-2}) were prepared by an ion-exchange approach as described elsewhere.[23] Further details

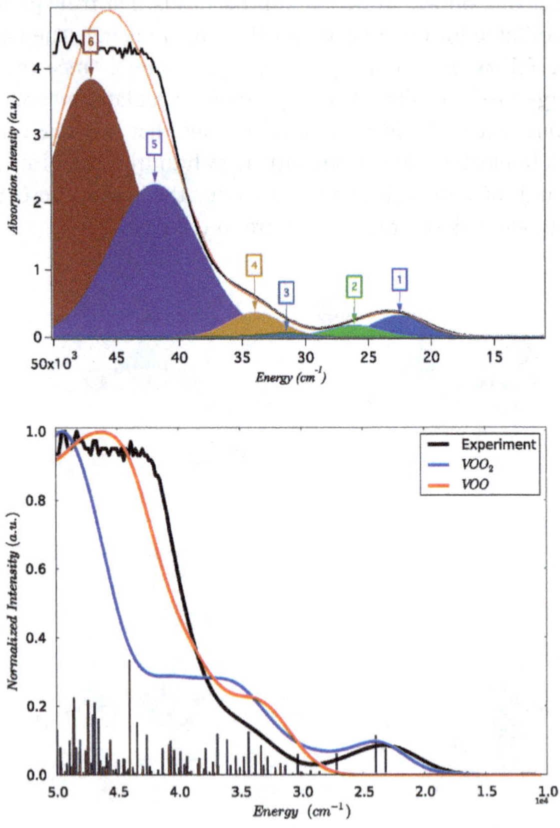

Fig. 2 Black line: experimental absorption spectrum of K[VO(O_2)Hheida]. Top: Gaussian deconvolution of the absorption spectrum. The Gaussian fit is constrained to constant widths FWHM = 4000 and 7500 cm^{-1} for bands 1–4 and 5–6 respectively. Bottom: B3LYP TD-DFT calculated spectra for [VO(O_2)Hheida]$^-$ (blue) and [VO(O)Hheida]$^-$ (red) complexes. Blue and red sticks indicate the corresponding calculated non-relativistic states.

regarding spectroscopic characterization and theoretical calculations are summarized in the ESI.†

Spectroscopic data

The absorption spectrum of K[VO(O_2)Hheida] is presented in Fig. 2. It is characterized by a low energy and a broad signal located at 26 000 cm^{-1}, as well as by a higher energy shoulder showing a maximum at 34 000 cm^{-1}. This later signal grows at the rising edge of an even broader and partially saturated signal centered at 45 000 cm^{-1}. On the other hand the absorption spectrum of VO_x/SBA-15 (vanadium loading 2% and 8%) is presented in Fig. S2.† It is characterized by a low energy shoulder located at 25 000 cm^{-1} followed by a high-energy broad signal centered at about 40 000 cm^{-1}. It should be noted that the absorption spectra show pronounced changes at low vanadium loadings, in fact the spectrum manifold shifts to higher energies as the vanadium loading decreases. This phenomenon merits further investigation.

Furthermore, for both catalytic systems Raman spectra were obtained under near resonance conditions. For the molecular complex, this was achieved using a 532 nm laser.[24] As can be seen in Fig. 3 the spectrum shows characteristic enhanced rR bands located at 970, 922, 569, 556, 509 and 485 cm^{-1}. The corresponding Raman spectrum of dehydrated VO_x/SBA-15 at low vanadium loading (about 2%), was obtained upon excitation[25,26] with a 488 nm laser[23] and is presented in Fig. S2.† For such a low vanadium loading value, the only Raman active signal with substantial intensity is located in the region between 950–1100 cm^{-1}. This signal has been assigned to a variety of V, O and Si functionalities ranging between monomeric to oligomeric vanadia species as are shown in Scheme 1.[14,25,27–33] However, no consensus has emerged. It is clear that based on these data, this signal can only be assigned to a mixture of overlapping features of monomeric and oligomeric vanadyl species coexisting on SBA-15 already at small vanadium loadings (*e.g.* 0.6 V atoms per nm^2).

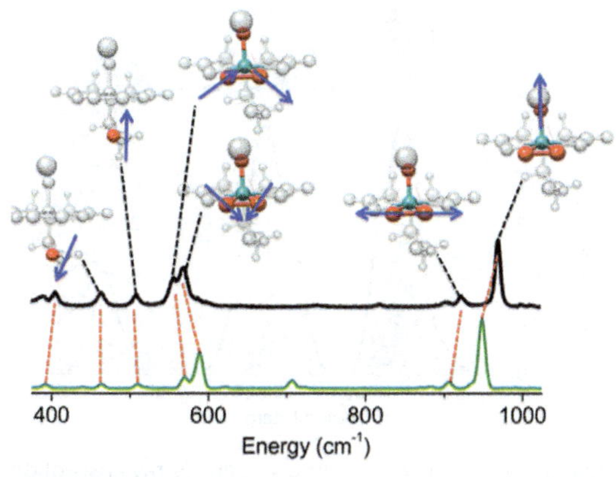

Fig. 3 Experimental (black) rRaman spectrum of K[VO(O_2)Hheida] observed at 18 700 cm^{-1} excitation energy (532 nm). Calculated (green) TD-DFT rRaman spectrum. Coloured groups and blue arrows indicate the active vibrational modes dominating certain experimental (and calculated) signals.

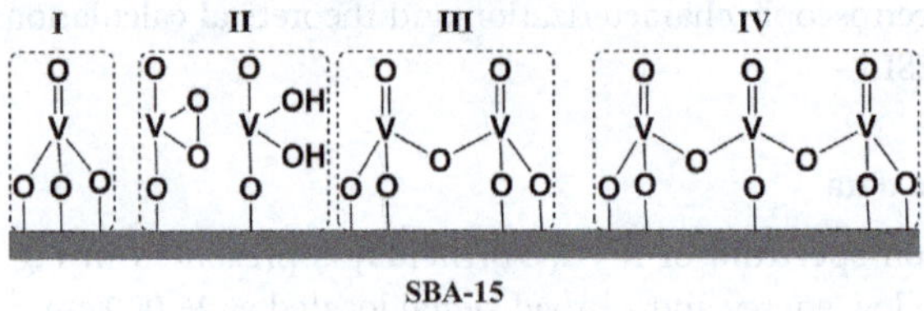

Scheme 1 Four possible scenarios for VO_x growing over a 2D silica surface (SBA-15). (I) Vanadyl type, (II) umbrella type, (III) dimeric and (IV) oligomeric species.

In addition, the experimental V L-edge and O K-edge XAS spectra for K[VO(O_2)-Hheida] and VO_x/SBA-15 at various vanadium loadings are presented in Fig. 4 and 5, respectively. In general VO_x species are characterized by characteristic V L-edge and O K-edge spectra.[34,35] In particular, for the case of the K[VO(O_2)Hheida] complex the V L_3-edge spectrum is characterized by an intense feature located at 518 eV, as well as three shoulders located at 515, 516 and 519 eV. The V L_2-edge spectrum is even broader showing two features centered at 523 and 524 eV. It

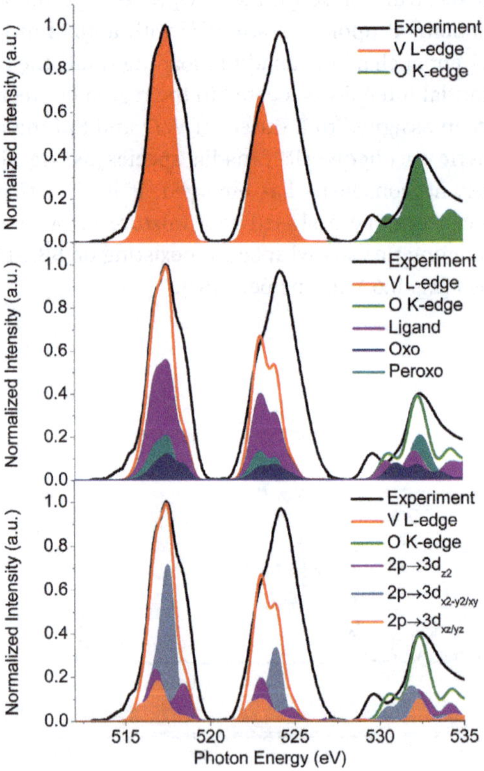

Fig. 4 Black line: experimental V L-edge and O K-edge X-ray absorption spectra. Red and green lines and areas correspond to the B3LYP/ROCIS calculated spectra for 80% [VO(O_2) Hheida] + 20% [VO(O)Hheida] active species composition (top). Deconvolution of the calculated spectra in terms of participating functional groups (middle) and single electron excitations (bottom).

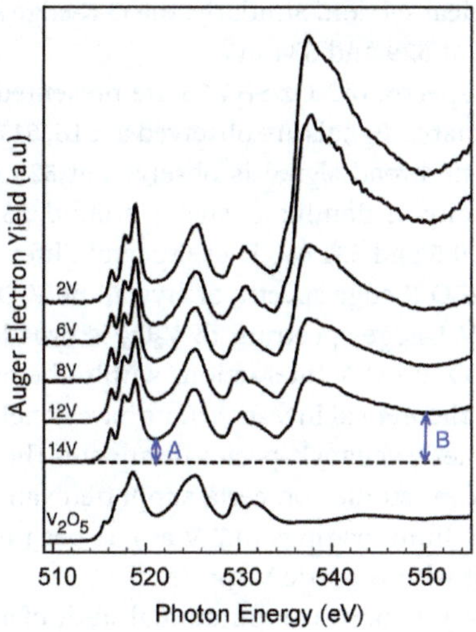

Fig. 5 V $L_{3,2}$ and O K-NEXAFS of dehydrated VO_x/SBA-15 catalysts with various V loadings as indicated. Spectra were recorded in 50 Pa O_2 at 673 K. A reference spectrum of V_2O_5 powder is shown as well. The V edge jump "A" and the total edge jump "B" are marked with arrows.

must be emphasized that in total the V L-edge spectrum reveals a much broader bandwidth with respect to the corresponding spectra of other oxo-peroxo $V^{(V)}$ complexes[17] indicating the coexistence of more than one species, as will be

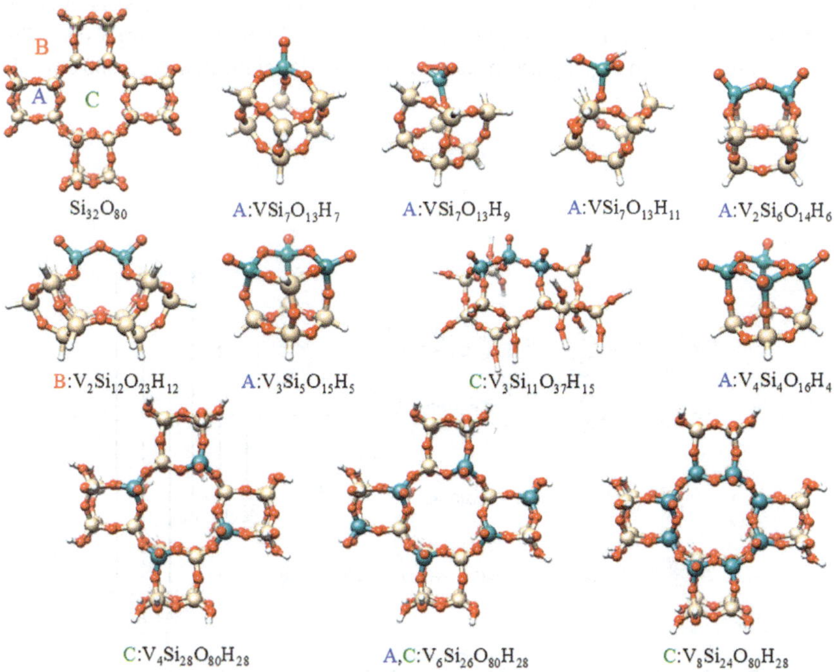

Fig. 6 Graphical representation of the cluster models used to model the different silica supported vanadia centers. More information is provided in the ESI (Fig. S1†).

analyzed in the theoretical section. Similarly, the O K-edge spectrum also has two broad features located at 529 and 534 eV.

The corresponding spectra of VOx/SBA-15 are presented in Fig. 5. In the V L_3 region three distinguishable signals are observed at 516, 517.5 and 519 eV while in the V L_2 region only one broad signal is observed at 525 eV. In addition the O K-edge spectrum contains a dominant signal around 530 eV with two broad shoulders located at 529.5 and 531 eV. These spectra differ substantially from the respective V L-edge and O K-edge spectra of crystalline V_2O_5 shown also in Fig. 5 (bottom). In fact, the V L-edge spectrum of V_2O_5 is broader due to bulk effects originating from weak O–V···O–V interactions within the bilayer crystallographic structure.[16] Previously, theoretical investigations have concluded that a mixture of monomeric and oligomeric vanadyl species dominates the O K-edge spectrum.[36] On the other hand, recent studies on ceria supported vanadia species have suggested that at low vanadium loadings (0.7 V atoms per nm^2) there is a uniform coverage of the catalyst with trimeric V species.[37]

In the next section a systematic computational study of all these properties will be carried out based on model structures described in the ESI and shown in Fig. 6 and S1.†

Theoretical calculations

Absorption spectroscopy

On the basis of the particle/hole excitation patterns shown in Fig. 7 and S3† the absorption spectrum of $[VO(O_2)Hheida]^-$ can be satisfactorily fitted with a minimum of six Gaussian bands. In the second step of the analysis we perform TD-DFT calculations of the absorption spectra assuming the coexistence of $[VO(O_2)Hheida]^-$ and the oxygen deficient $[VO(O)Hheida]^-$ complexes. The results are shown in Fig. 2 (bottom). Evidently, the signal located at 26 000 cm^{-1} is

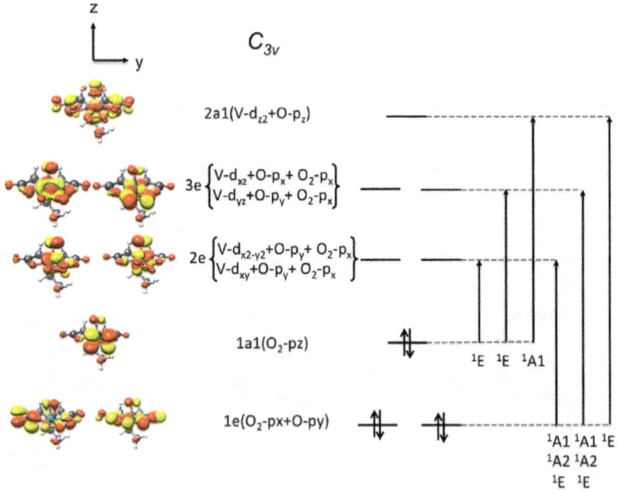

Fig. 7 The MO splitting diagram of the local V–OO_2 (C_{3v}) core. In addition, the adapted 2p–3d transitions in the one electron particle/hole approximation are visualized assuming ideal C_{3v} ligand field splitting.

solely dominated by the V–OO_2 functional group. In fact the Gaussian bands 1 and 2 correspond to the LMCT $^1E(1a_1 \rightarrow 2e)$, $^1E(1a_1 \rightarrow 3e)$ and $^1A_1(1a_1 \rightarrow 2a_1)$ one-electron excitations, respectively. On the other hand, the bands located at 34 000 and 45 000 cm^{-1} (Gaussians 3–6) are dominated by LMCT and Ligand to Ligand CT (LLCT) electron excitations arising from the V–OO_2 and V–OO functional groups. Likewise the changes observed in the VO*x*/SBA-15 spectra (Fig. S2,† top-left) are nicely reproduced in the calculations, as shown in Fig. S2† (top-right). Indeed, by proceeding from monomeric to oligomeric species, there are spectral shifts at lower energies. As long as the vanadium loading remains limited (about 2%), it is necessary to investigate whether a unique structure can primarily dominate the experimental spectra as will be shown below by analyzing the rRaman and XAS spectra.

rRaman spectroscopy

By setting the excitation frequency to pre-resonance with the absorption band at 26 000 cm^{-1}, *e.g.* an excitation frequency of 450 nm, the rRaman bands of K[VO(O_2)Hheida] presented in Fig. 3 are calculated. We notice that certain deficiencies in the calculated energies of the band positions and relative intensities with respect to experiment are observed, most likely due to the fact that the IMDHO model does not account for terms of second or higher order between excited and ground vibrational Hamiltonians.[21,22] Despite the fact that the agreement between theory and experiment is only a semiquantitative agreement for some of the rRaman bands, the IMDHO approach still enables a conclusive assignment of the experimental Raman bands. It should be noted that at this region of excitation energies, the observed absorption band is solely dominated by the $[VO(O_2)Hheida]^-$ species. Following the discussion above, this band corresponds to LMCT states arising solely from the V–OO_2 functional group of the $[VO(O_2)Hheida]^-$ complex. Due to the excellent agreement between theory and experiment the assignment of the rRaman spectral features is straightforward. The signals located at 970 and 922 cm^{-1} are assigned to the stretching vibrations of the V–O (vanadyl–oxo) and the O–O (peroxo) groups, respectively. In addition, the signals centered at 569 and 556 cm^{-1} are dominated by symmetric and antisymmetric V–O (peroxo) stretching vibrations. Finally, the low intensity signals located at 509 and 565 cm^{-1} are dominated by hydroxyl stretching vibrations.

Similarly, the rRaman[23,25,30] spectra were evaluated for low lying excited states of all the model structures representing VO*x*/SBA-15 in the range between 300–500 nm (20 000–35 000 cm^{-1}). The corresponding spectra are presented in Fig. S4 and S5.† This strategy was followed for all model structures and will be illustrated below for the $V_3Si_5H_5O_{15}$ structure. The calculated absorption spectrum is characterized by three low-lying groups of bands located at 22 500 (orange), 25 000 (cyan) and 27 000 cm^{-1} (purple) depicted in Fig. 8 (bottom-left, higher lying states are drawn in gray). These bands originate from either single-electron excitations within the V=O and V–O–V fragments (orange) and the V=O and Si–O–V structural motifs (cyan and purple). Based on those calculated excitation energies the corresponding rRaman spectra can be calculated (Fig. 8, top-right).

The calculated rRaman spectra reproduce the observed signals around 1000 cm^{-1} and they are in very good agreement with previous studies.[25,26,30] Inspection

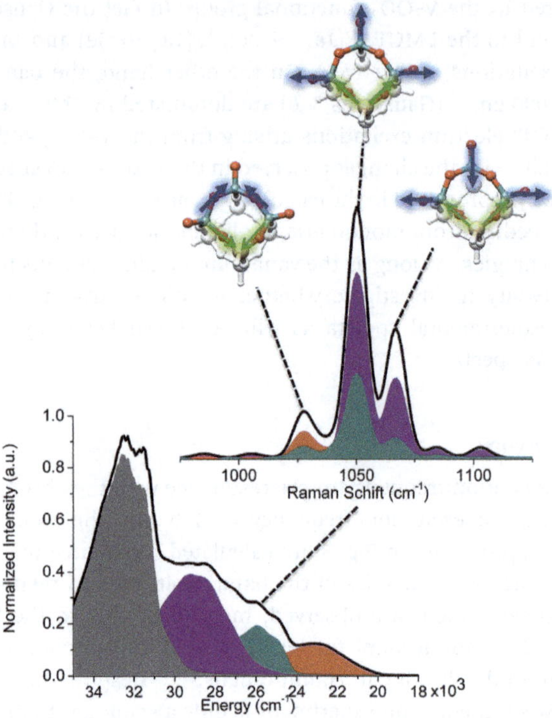

Fig. 8 Bottom-left: Calculated absorption spectra accounting for the band shape of the dominating states for $V_3Si_5H_5O_{15}$. The states are grouped according to the character of the dominating single-electron excitations involving either V=O and V–O–V sequences (orange), or V=O and Si–O–V sequences (cyan and purple). Top-right: Calculated Raman spectra. The spectra are calculated at an absorption energy of 26 000 cm^{-1}. Orange, cyan and purple bands indicate active vibrational modes pointed with arrows.

of Fig. S4 and S5† reveals that, although the presence of monomeric umbrella type structures cannot be excluded, they certainly do not dominate the intensity of the signal around 1000 cm^{-1}, this conclusion follows from the absence of the characteristic O=O and Si–O–V signals expected at <1000 cm^{-1} that are key features for this type of structure. Based on these data it appears that the structures which contain vanadyl V=O bonds and bridging V–O–V units are always in better agreement with experiment for the signals located around 1000 cm^{-1}. At this stage of analysis, the Raman technique alone is not conclusive with respect to the dominant structural feature. Nevertheless, we note that the calculations for the trimeric structure presented in Fig. 8 provide a very good agreement with experiment. Analysis of the predominant vibrational modes indicates that the signal at around 1000 cm^{-1} is dominated by combinations of V–O–V and V–O–Si (1025 cm^{-1}) as well as V=O and V–O–Si (1050, 1075 cm^{-1}) stretching vibrations (Fig. 8, top-right). More information can be obtained by deconvoluting the calculated rRaman signals in terms of the dominating excited state. As can be seen in Fig. 8 (bottom-left) the group of states are heavily overlapping as each absorption band contains tails that overlap with the other two. The signal at 1025 cm^{-1} is mainly dominated by states involving V=O and V–O–V (orange), as well as the V=O and

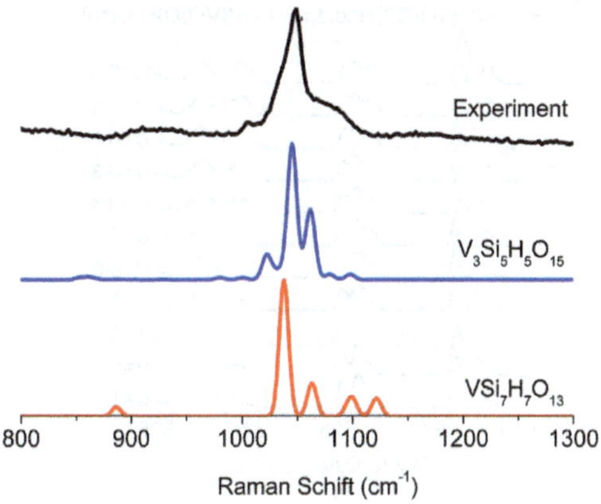

Fig. 9 Comparison between experimental (black) and calculated $VSi_7H_7O_{13}$ (red) and $V_3Si_5H_5O_{15}$ (blue) rRaman spectra.

Si–O–V structural motifs (cyan). On the other hand, the signals at 1050 and 1075 cm^{-1} are primarily dominated by states involving the V═O and Si–O–V fragments (cyan and purple). Furthermore, as is seen in Fig. 9 the experimental features at 1025 cm^{-1} are better reproduced by the trimeric $V_3Si_5H_5O_{15}$, rather than the monomeric $VSi_7H_7O_{13}$ due to the dominating V–O–V vibrations.

X-ray spectroscopy

In an effort to obtain further insight into the predominant species, the V L-edge and O K-edge spectra of all models for both the molecular and the heterogeneous catalytic systems were calculated on the DFT/ROCIS level. It has been discussed recently[24] that the shape of the O K-edge spectrum of K[VO(O_2)Hheida], can be explained by considering the TD-DFT calculation molecular admixtures of $[VO(O_2)Hheida]^-$ and the oxygen deficient $[VO(O)Hheida]^-$ ions according to the relation:[24]

$$\alpha^*[VO(O_2)Hheida] + (1 - \alpha)^*[VO(O)Hheida] \quad (1)$$

Moreover, a quantitative measure of the sample heterogeneity can be obtained when relation (1) is combined with the DFT-ROCIS calculated V L-edge and O K-edge spectra. The corresponding spectra for α-values ranging between 1 and 0 are presented in Fig. 10.

Subsequently, these spectra were used to simultaneously fit the V L_3-edge and O K-edge spectra parts, according to a fitting procedure, described in the ESI.† The best agreement between theory and experiment is observed for $\alpha = 0.8$ (highlighted red in Fig. 10), indicating a 20% admixture of the oxygen deficient species. Additional information can be obtained by analyzing the relativistic many particle states in terms of contributions arising from the oxo, the peroxo and the ligand oxygen functional groups, as well as the most important single electron excitations. This has been proven to be useful in describing bonding features for both molecular complexes[17] and solids.[20] As can be seen in Fig. 4 (middle) all three groups strongly participate in the intensity mechanism of both V L-edge and O K-edge spectra, due to the 20% admixture of the $[VO(O)Hheida]^-$

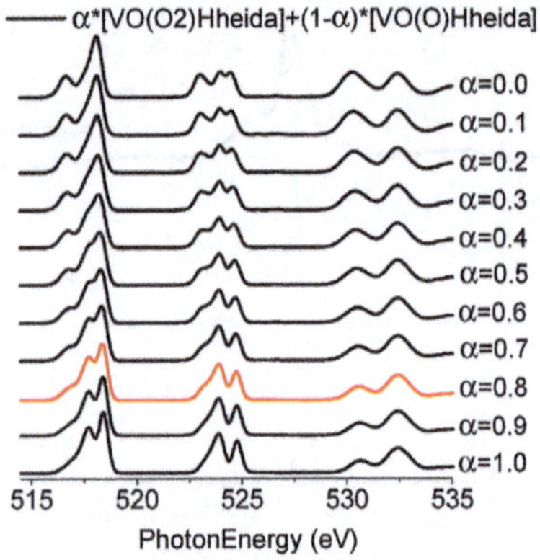

Fig. 10 B3LYP/ROCIS calculated spectra for admixtures of $[VO(O_2)Hheida]^-$ and the oxygen deficient $[VO(O)Hheida]^-$ ions according to the relation (1) for α values ranging between 1 and 0. The red spectrum at $\alpha = 0.8$ indicates the best agreement between theory and experiment.

species, as well as the strongly anisotropic environment around the VO_6 vanadium centers. In particular, oxo, peroxo and ligand oxygen atoms dominate the main experimental features of the O K-edge spectrum located at 529 and 534 eV in accordance with a previous analysis.[24] Moreover, both bands are dominated by combinations of σ- and π-single electron excitations $2p \rightarrow 3d_{z^2}$, $2p \rightarrow 3d_{x^2-y^2/xy}$, and $2p \rightarrow 3d_{xz/yz}$. Furthermore, a similar analysis of the corresponding V L_3-edge reflects the π-character of the signals located at 515 and 518 eV, as well as the σ-character of the signal positioned at 519 eV and the mixed σ- and π-character of the shoulder centered at 516 eV.

In following, the same protocol is applied to the more challenging case of silica supported vanadia catalysts. The resulting V L-edge spectra are shown in Fig. S6 and S7† for 2% and 8% V loadings respectively. By inspection of Fig. S6 and S7,† it becomes evident that the A and B-domain types of structures containing V–O–V bridges ($V_2Si_6H_6O_{14}$, $V_2Si_{12}H_{12}O_{23}$, $V_3Si_5H_5O_{15}$ and $V_4Si_4H_4O_{16}$), provide the best agreement for both vanadium loadings, in terms of relative intensities and intensity distribution, especially for the L_3 resonances located at 516, 517.5 and 519 eV. In addition, the $V_3Si_5H_5O_{15}$ calculated spectrum reproduces equally well both the 2% and the 8% vanadium loading spectra, reflecting the limited oligomerization at these levels of vanadium loading. Evidently by averaging over the above four structures (Fig. S8†) the average spectrum has a closer resemblance with the V_2O_5 spectrum. On the other hand, the 'open' trimeric structure $V_3Si_{11}H_{15}O_{37}$ does not show relevant agreement, indicating the preference of VO_x to form more strain configurations. This is in accordance with the recent observation for silica supported molybdena catalysts that showing that such a strain effect has been directly associated with an increase in reactivity.[38] Collectively, by combining the results from the three individual spectroscopic experiments: rRaman, V L-edge and O K-edge (Fig. 9, 11 and S9†) it is the trimeric structure $V_3Si_5H_5O_{15}$ that provides consistently the best possible agreement. Furthermore, it should be emphasized that the $V_3Si_5H_5O_{15}$ structure provides in addition very good agreement for the relative intensities of the V L-edge and O K-edge spectra

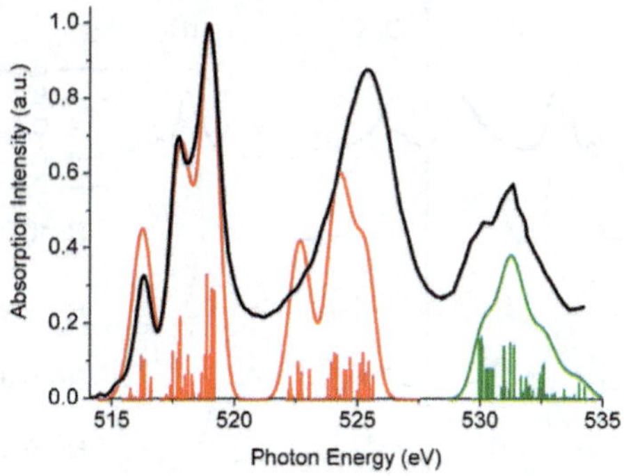

Fig. 11 DFT/ROCIS (B3LYP/def2-TZVP), V L-edge (red solid line) and O K-edge (green solid line) calculated spectra for $V_3Si_5H_5O_{15}$. The black thick line represents the background-subtracted experimental powder spectrum. Red and green stick lines correspond to SOC-corrected states. All calculated spectra were energy shifted. A constant line shape broadening of 0.8 eV was applied.

(Fig. 11). This is important as in contrast to the situation with V_2O_5, these two edge spectra do not provide complementary information about the V–O bonds since there are additional Si–O bonds that lead to recognizable features in the spectrum. Hence, it appears that this trimeric structure provides the best balance between vanadyl bonds V═O and bridging units V–O–V, Si–O–V and Si–O–Si which is of paramount importance for correct spectral predictions.

Degree of oligimerization

As a final step, the effect of the oligomerization on the calculated spectra was investigated. Following the same protocol as for the molecular catalyst, the analysis was accomplished by a simultaneous fit of the weighted superposition of all calculated spectra to the V L_3-edge experimental spectrum (V loading of 2%). It should be highlighted that the V L_3-spectrum serves as a good candidate for this analysis as it is composed of three clearly visible signals. By contrast, the respective Raman and O K-edge spectra were not used in the fitting due to the relatively broad nature of the spectral features. These spectra were used however, in the second step of the analysis in order to validate the fitting results. Initially, the fit was performed over the whole dataset. It was found however that better fits

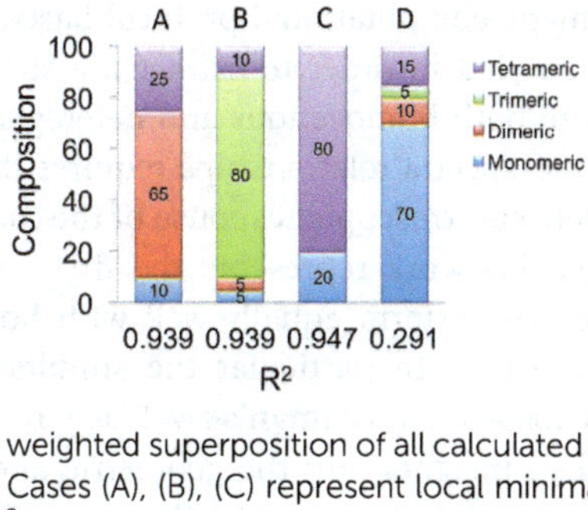

Fig. 12 Fit results of the weighted superposition of all calculated spectra to the V L_3-edge experimental spectrum. Cases (A), (B), (C) represent local minima of the fit function corresponding to the best R^2 values achievable by starting the fit from a pure dimer (A), trimer (B), tetramer (C) or monomer (D) structure.

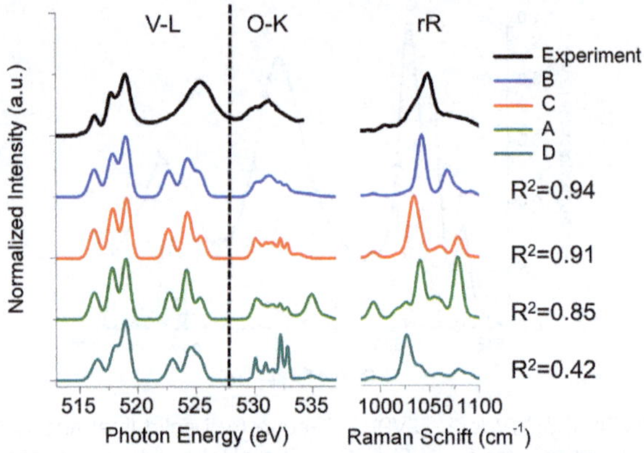

Fig. 13 Experimental (black) *versus* calculated weighted V L-edge, O K-edge and resonance Raman spectra. Weighting schemes, (A): 10% monomeric, 65% dimeric and 25% tetrameric (green), (B): 5% monomeric, 5% dimeric, 80% trimeric and 10% tetrameric (blue), (C): 20% monomeric and 80% tetrameric (red) and (D): 70% monomeric, 10% dimeric, 5% trimeric and 15% tetrameric (cyan). The corresponding R^2 values are provided at the right-hand side.

could be obtained by considering only the monomeric ($VSi_7H_7O_{13}$, $VSi_7H_9O_{13}$ and $VSi_7H_{11}O_{13}$), dimeric ($V_2Si_6H_6O_{14}$ and $V_2Si_{12}H_{12}O_{23}$), trimeric ($V_3Si_5H_5O_{15}$) and tetrameric ($V_4Si_4H_4O_{16}$) model structures. The results of the fitting are visualized in Fig. 12, while the corresponding weighted O K-edge and resonance Raman spectra together with the respective V L-edge spectra are shown in Fig. 13.

The B weighting scheme involving 5% monomeric, 5% dimeric, 80% trimeric and 10% tetrameric structures provides the best agreement with all the three spectra, V L-edge, O K-edge and rRaman ($R^2 = 0.94$). The respective weighting schemes D, A and C provide significantly worse agreement ($R^2 = 0.42$, 0.85 and 0.91, respectively). Thus, the combined analysis of three different spectroscopic methods demonstrates that under limited vanadium loading conditions, trimeric structures are dominant. This is in agreement with recent computational studies that address the reactivity of these systems.[39]

Conclusions

In conclusion, a systematic computational protocol based on advanced theoretical spectroscopy was applied in order to extract the structural information of active catalytic centers in both homogenous and heterogeneous catalysis. Establishing detailed structure–spectra relationships requires the simultaneous interpretation of the complete spectroscopic response of the catalytic material. To the best of our knowledge, this work represents the first study of this kind. The approach was validated to perform equally well with both homogeneous and heterogeneous catalytic cases. In particular the simultaneous analysis of the available experimental absorption, rRaman as well as V L-edge and O K-edge XAS spectra provided detailed insights into the geometric and electronic structural origin of the observed spectral features. In the case of the K[VO(O_2)Hheida] catalyst, the combined analysis revealed that the low-lying absorption LMCT band, as well as the rRaman profile of this excited chromophore, are dominated by the actual $[VO(O_2)Hheida]^-$ complex. On the other hand the respective XAS spectra contain a 20% admixture of the oxygen deficient $[VO(O)Hheida]^-$ species. The same protocol was applied to interpret the spectroscopic response of the heterogeneous silica-supported vanadium oxide VO_x/SBA-15 catalyst. It was shown that under limited vanadium loadings molecular species with predominant trimeric structures provide the best agreement between theory and experiment for all the available spectra. Although a identification of a catalytic center in a heterogeneous catalyst is a much more complicated task, having a tool with predictive power forms an excellent basis for the design of dedicated spectroscopic experiments. Further experimental and theoretical studies based on multiwavelength rRaman and X-ray measurements, as well as high-level theory, are currently underway in order to determine the reaction mechanism of these catalysts.

Acknowledgements

The authors gratefully acknowledge financial support from the Max Planck Society. The Helmholtz-Zentrum Berlin (HZB) staff are acknowledged for continuously supporting the synchrotron based near ambient pressure spectroscopy experiments of the Fritz Haber Institute at BESSY II. We thank Dr Philipp Gruene, Dr Till Wolfram, and Daniel Brennecke for preparation of the supported vanadium catalysts and Dr Olaf Timpe, Dr Johannes Noack, and Anne Kärgel for preparation of the molecular complex. Raman spectroscopy was performed by Dr Philipp Gruene, Dr Oliver Korup, and Dr Johannes Noack. UV-vis spectra were taken by Dr Genka Tzolova-Müller. X-ray absorption spectra were recorded by Dr Michael Hävecker and Dr Till Wolfram. The reviewers of the manuscript are gratefully acknowledged for their constructive comments.

Notes and references

1 R. Schloegl, *Nat. Mater.*, 2008, **7**, 772–774.
2 T. W. Hansen, J. B. Wagner, P. L. Hansen, S. Dahl, H. Topsøe and C. J. H. Jacobsen, *Science*, 2001, **294**, 1508–1510.
3 J. K. Norskov, T. Bligaard, J. Rossmeisl and C. H. Christensen, *Nat. Chem.*, 2009, **1**, 37–46.
4 T. Bligaard, J. K. Nørskov, S. Dahl, J. Matthiesen, C. H. Christensen and J. Sehested, *J. Catal.*, 2004, **224**, 206–217.
5 A. A. Balandin, in *Advances in Catalysis*, ed. H. P. D. D. Eley and B. W. Paul, Academic Press, 1969, vol. 19, pp. 1–210.
6 T. Petrenko, S. DeBeer George, N. Aliaga-Alcalde, E. Bill, B. Mienert, Y. Xiao, Y. Guo, W. Sturhahn, S. P. Cramer, K. Wieghardt and F. Neese, *J. Am. Chem. Soc.*, 2007, **129**, 11053–11060.
7 J. F. Berry, E. Bill, E. Bothe, S. D. George, B. Mienert, F. Neese and K. Wieghardt, *Science*, 2006, **312**, 1937–1941.
8 J. C. Schöneboom, F. Neese and W. Thiel, *J. Am. Chem. Soc.*, 2005, **127**, 5840–5853.
9 S. Ye, C.-Y. Geng, S. Shaik and F. Neese, *Phys. Chem. Chem. Phys.*, 2013, **15**, 8017–8030.

10 S. Ye and F. Neese, *Proc. Natl. Acad. Sci. U. S. A.*, 2011, **108**, 1228–1233.
11 K. M. Lancaster, M. Roemelt, P. Ettenhuber, Y. Hu, M. W. Ribbe, F. Neese, U. Bergmann and S. DeBeer, *Science*, 2011, **334**, 974–977.
12 D. A. Pantazis, W. Ames, N. Cox, W. Lubitz and F. Neese, *Angew. Chem., Int. Ed.*, 2012, **51**, 9935–9940.
13 N. Cox, M. Retegan, F. Neese, D. A. Pantazis, A. Boussac and W. Lubitz, *Science*, 2014, **345**, 804–808.
14 J. E. Molinari and I. E. Wachs, *J. Am. Chem. Soc.*, 2010, **132**, 12559–12561.
15 C. Hess, *ChemPhysChem*, 2009, **10**, 319–326.
16 D. Maganas, M. Roemelt, M. Havecker, A. Trunschke, A. Knop-Gericke, R. Schlogl and F. Neese, *PCCP Phys. Chem. Chem. Phys.*, 2013, **15**, 7260–7276.
17 D. Maganas, M. Roemelt, T. Weyhermuller, R. Blume, M. Havecker, A. Knop-Gericke, S. DeBeer, R. Schlogl and F. Neese, *PCCP Phys. Chem. Chem. Phys.*, 2014, **16**, 264–276.
18 M. Roemelt, D. Maganas, S. DeBeer and F. Neese, *J. Chem. Phys.*, 2013, **138**, 204101.
19 M. Roemelt and F. Neese, *J. Phys. Chem. A*, 2013, **117**, 3069–3083.
20 D. Maganas, S. DeBeer and F. Neese, *Inorg. Chem.*, 2014, **53**, 6374–6385.
21 T. Petrenko and F. Neese, *J. Chem. Phys.*, 2012, **137**, 234107.
22 T. Petrenko and F. Neese, *J. Chem. Phys.*, 2007, **127**, 164112.
23 P. Gruene, T. Wolfram, K. Pelzer, R. Schlogl and A. Trunschke, *Catal. Today*, 2010, **157**, 137–142.
24 L. L. Sun, K. E. Hermann, J. Noack, O. Timpe, D. Teschner, M. Hävecker, A. Trunschke and R. Schlögl, *J. Phys. Chem. C*, 2014, **118**, 24611–24622.
25 A. E. Stiegman, *J. Phys. Chem. C*, 2011, **115**, 10917–10924.
26 Z. Wu, S. Dai and S. H. Overbury, *J. Phys. Chem. C*, 2009, **114**, 412–422.
27 D. E. Keller, T. Visser, F. Soulimani, D. C. Koningsberger and B. M. Weckhuysen, *Vib. Spectrosc.*, 2007, **43**, 140–151.
28 N. Magg, B. Immaraporn, J. B. Giorgi, T. Schroeder, M. Bäumer, J. Döbler, Z. Wu, E. Kondratenko, M. Cherian, M. Baerns, P. C. Stair, J. Sauer and H.-J. Freund, *J. Catal.*, 2004, **226**, 88–100.
29 C. Moisii, M. D. Curran, L. J. van de Burgt and A. E. Stiegman, *J. Mater. Chem.*, 2005, **15**, 3519–3524.
30 C. Moisii, L. J. van de Burgt and A. E. Stiegman, *Chem. Mater.*, 2008, **20**, 3927–3935.
31 J. P. Thielemann, T. Ressler, A. Walter, G. Tzolova-Mueller and C. Hess, *Appl. Catal., A*, 2011, **399**, 28–34.
32 J. N. J. van Lingen, O. L. J. Gijzeman, B. M. Weckhuysen and J. H. van Lenthe, *J. Catal.*, 2006, **239**, 34–41.
33 J. Döbler, M. Pritzsche and J. Sauer, *J. Phys. Chem. C*, 2009, **113**, 12454–12464.
34 D. Maganas, M. Roemelt, M. Havecker, A. Trunschke, A. Knop-Gericke, R. Schlogl and F. Neese, *Phys. Chem. Chem. Phys.*, 2013, **15**, 7260–7276.
35 D. Maganas, M. Roemelt, T. Weyhermuller, R. Blume, M. Havecker, A. Knop-Gericke, S. DeBeer, R. Schlogl and F. Neese, *Phys. Chem. Chem. Phys.*, 2014, **16**, 264–276.
36 M. Cavalleri, K. Hermann, A. Knop-Gericke, M. Haevecker, R. Herbert, C. Hess, A. Oestereich, J. Doebler and R. Schloegl, *J. Catal.*, 2009, **262**, 215–223.

37 M. Baron, H. Abbott, O. Bondarchuk, D. Stacchiola, A. Uhl, S. Shaikhutdinov, H.-J. Freund, C. Popa, M. V. Ganduglia-Pirovano and J. Sauer, *Angew. Chem., Int. Ed.*, 2009, **48**, 8006–8009.

38 K. Amakawa, L. Sun, C. Guo, M. Haevecker, P. Kube, I. E. Wachs, S. Lwin, A. I. Frenkel, A. Patlolla, K. Hermann, R. Schloegl and A. Trunschke, *Angew. Chem., Int. Ed.*, 2013, **52**, 13553–13557.

39 X. Rozanska, R. Fortrie and J. Sauer, *J. Am. Chem. Soc.*, 2014, **136**, 7751–7761.

Faraday Discussions

PAPER

Quantum chemistry of the oxygen evolution reaction on cobalt(II,III) oxide – implications for designing the optimal catalyst†

Craig P. Plaisance,*[a] Karsten Reuter[a] and Rutger A. van Santen*[bc]

Received 7th December 2015, Accepted 4th January 2016
DOI: 10.1039/c5fd00213c

Density functional theory is used to examine the changes in electronic structure that occur during the oxygen evolution reaction (OER) catalyzed by active sites on three different surface terminations of Co_3O_4. These three active sites have reactive oxo species with differing degrees of coordination by Co cations – a μ^3-oxo on the (311) surface, a μ^2-oxo on the (110)-A surface, and an η-oxo on the (110)-B surface. The kinetically relevant step on all surfaces over a wide range of applied potentials is the nucleophilic addition of water to the oxo, which is responsible for formation of the O–O bond. The intrinsic reactivity of a site for this step is found to increase as the coordination of the oxo decreases with the μ^3-oxo on the (311) surface being the least reactive and the η-oxo on the (110)-B surface being the most reactive. A detailed analysis of the electronic changes occurring during water addition on the three sites reveals that this trend is due to both a decrease in the attractive local Madelung potential on the oxo and a decrease in electron withdrawal from the oxo by Co neighbors. Applying a similar electronic structure analysis to the oxidation steps preceding water addition in the catalytic cycle shows that analogous electronic changes occur during this process, explaining a correlation observed between the oxidation potential of a site and its intrinsic reactivity for water addition. This concept is then used to specify criteria for the design of an optimal OER catalyst at a given applied potential.

Introduction

The oxygen evolution reaction (OER), in which two molecules of water are oxidized to give molecular oxygen, generates the electrons needed for the

[a]*Chair of Theoretical Chemistry and Catalysis Research Center, Technical University of Munich, D-85747, Garching, Germany. E-mail: craig.plaisance@tum.de*
[b]*Institute for Complex Molecular Systems, Eindhoven University of Technology, 5612 AZ, Eindhoven, The Netherlands. E-mail: R.A.v.Santen@tue.nl*
[c]*Laboratory of Inorganic Materials Chemistry, Schuit Institute of Catalysis, Eindhoven University of Technology, 5612 AZ, Eindhoven, The Netherlands*

† Electronic supplementary information (ESI) available. See DOI: 10.1039/c5fd00213c

photocatalytic or electrocatalytic production of transportation fuels by the reduction of water or CO_2.[1,2] Despite significant research in developing catalytic materials for carrying out this reaction, the most active materials known at this time are RuO_2 and IrO_2.[3,4] Due to the high cost and scarcity of Ru and Ir, there is much interest in finding active catalysts for the OER based on the earth-abundant 3d transition metal oxides.[3–9]

One of the 3d transition metal oxides, Co_3O_4, has recently received particular attention as a suitable OER catalyst.[10–13] Studies by Frei and coworkers[14] have shown that this material exhibits strong structure sensitivity, with less than 1/1000 of the exposed sites responsible for most of the observed activity. They postulate that this highly active site consists of two Co(IV) cations connected by bridging oxos that carries out O–O bond formation by nucleophilic addition of water to a terminal oxo. We recently used density functional theory (DFT) and a simplified kinetic model to show that this highly active site likely corresponds to a dual-Co site on a step edge of the (311) surface that forms the O–O bond by water addition to a bridging μ^3-oxo, as shown in Scheme 1.[15] The turnover frequency (TOF) calculated for this site at the experimental applied potential was within two orders of magnitude of the one measured by Frei and coworkers, and it was found

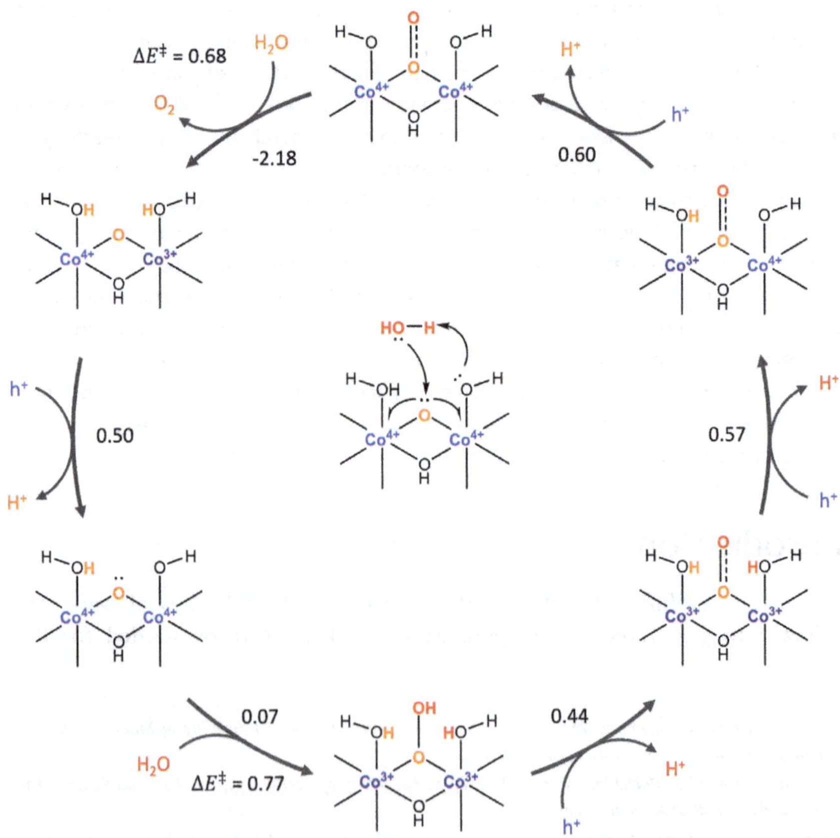

Scheme 1 Catalytic cycle and energetics (eV at $\eta = 0$) of the OER for the dual-Co site on the (311) surface. Reproduced from ref. 15.

that both water addition and O_2 release are kinetically relevant, in agreement with the observation of a superoxo intermediate on the surface in experiments. Sites consisting of a single Co center with a terminal η-oxo and dual-Co sites on the (001) terrace were found to be several orders of magnitude less active by our kinetic model at the experimental applied potential.

One of the key results of our previous study was that the surface containing the most active site changes as the overpotential (applied potential with respect to the equilibrium potential of the OER) is increased. As shown in Fig. 1, at overpotentials below 0.77 V, the dual-Co μ^3-oxo site on either the (311) or (001) surface was found to give the highest TOF, while at overpotentials higher than this, a single-Co η-oxo site on the B termination of the (110) surface was found to take over. We explained this behavior by noting that the single-Co site has a lower intrinsic activation barrier for water addition than the dual-Co site but requires a higher overpotential to oxidize to the active state for this reaction. When the overpotential is <0.77 V, the single-Co site exists in the inactive reduced state so that the dual-Co site has a higher TOF. At higher overpotentials, a significant fraction of the single-Co sites are oxidized to the active state so that this site dominates due to its higher intrinsic reactivity for water addition.

Based on the comparison between the single-Co and dual-Co sites, we postulated that a correlation exists between the intrinsic reactivity of a site for water

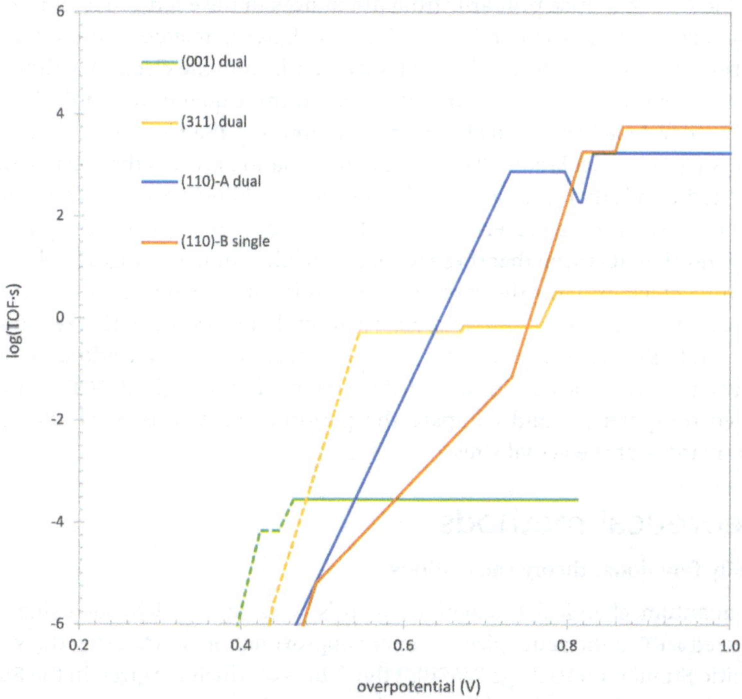

Fig. 1 TOF of the OER on the μ^3-oxo dual-Co sites on the (001) and (311) surfaces, the μ^2-oxo dual-Co site on the (110)-A surface, and the η-oxo single-Co site on the (110)-B surface as a function of overpotential. Solid curves show the regions where H_2O addition is rate limiting. Dashed curves show the regions where O_2 release is rate limiting. Data for (001), (311), and (110)-B surfaces taken from ref. 15.

addition and the applied potential at which it is oxidized to the active state for this reaction. While we noted that water addition and the oxidation occurring prior to it are related by the fact that the Co center(s) oxidized in the prior step are reduced during water addition, we did not provide a rigorous quantum chemical justification for this correlation. Providing such a justification will be a central topic of the current manuscript. Additionally, we now identify a third active site on the A termination of the (110) surface (the polar (110) surface has two terminations, A and B) containing a μ^2-oxo. This site is found to have an intrinsic water addition reactivity and oxidation potential intermediate to those of the μ^3-oxo and η-oxo sites. It gives a higher TOF than the other sites at overpotentials between 0.64 and 0.80 V, as shown in Fig. 1. Thus, there appears to also be a correlation between the coordination of the oxo undergoing water addition and its intrinsic reactivity and oxidation potential.

The aim of this manuscript is to unravel the quantum chemical factors that underlie this relation between the structure of the active site and its water addition reactivity and use these insights to develop a set of design principles for the optimal active site at a given applied overpotential. First, we will examine the changes in electronic structure that occur during water addition to the μ^3-oxo on the (311) surface, the μ^2-oxo on the (110)-A surface, and the η-oxo on the (110)-B surface. It will be seen that analogous electronic changes occur on all three sites but that the energetics of these processes are very different. We then examine the quantum chemical details underlying the differences in these electronic changes and find that they arise primarily from differences in the energy to localize a hole on the oxo prior to O–O bond formation. The latter is related to the local coordination of the oxo by the local Madelung potential and the electron withdrawing properties of the Co neighbors. In order to study these quantum chemical details, we have developed several techniques for examining the electronic structure of plane wave DFT calculations. The O–O bond formation process that occurs during water addition is then compared to the process of adding a hydrogen atom to the same oxo, which is the reverse of one of the oxidation steps occurring prior to water addition. It is seen that these two steps involve similar changes in electronic structure which explains the correlation noted in our previous work between the intrinsic water addition reactivity of a site and the overpotential required to oxidize it to the active state for this reaction. Based on this realization, we finally identify the properties of an ideal active site giving the highest TOF at a given applied overpotential and compare the performance of this ideal site to the performances of the actual sites.

Theoretical methods

Density functional theory calculations

The quantum chemical calculations in this work are carried out using spin-polarized DFT in the generalized gradient approximation (GGA) using the Vienna Ab initio Simulation Package (VASP),[16] the details of which are given in the ESI.† A particular issue in applying this method to the study of 3d transition metal oxides is an over-delocalization of the 3d electrons due to the presence of electron self-interaction in all GGA functionals.[17] An effective and low-cost method to correct for this is by applying an on-site self-interaction correction to the 3d electrons.[18–20] In this method, called GGA + U, the effective coulombic repulsion integral between the 3d electrons (U) must be specified. For this work, we use a value of U determined from a first-principles linear response method as detailed in our previous work.[15]

All calculations were performed on periodic slabs separated by vacuum, not including the electrolyte which is present in the actual system. Although solvation of the transition states by the electrolyte were partially accounted for by the inclusion of one or two extra water molecules in the system (details given in the ESI†), this is a crude approximation and does not include the electrostatic screening effects of the electrolyte. This latter effect is important as it allows for charged surfaces and pathways involving sequential proton and electron transfer processes in addition to the concerted proton and electron transfer processes examined in this work.[21] It is thus possible that a more accurate model of the electrolyte could lead to different results. However, our neglect of charged surface states is supported by experimental observations suggesting that all proton and electron transfers involved in the transition state and pre-equilibria are coupled.[22]

While standard methods have been used to locate minimum energy structures (details given in the ESI†), it was necessary for us to develop a special technique for finding the transition states for the water addition reaction. This is due to the fact that the transition state corresponds to the crossing of potential energy surfaces for two electronic states. As such, the transition state is not a saddle point on a single potential energy surface and cannot be located using the typical methods such as the nudged elastic band and dimer methods that require this. A detailed description of the method we use is given in the ESI† and was also used in our study in ref. 15.

Electronic structure analysis

Quasiatomic orbitals. All of the electronic structure analysis carried out in this work makes use of the quasiatomic orbital method developed by Qian and coworkers[23] and implemented by us in the VASP code. This method allows for the transformation of the Bloch orbitals from a plane wave basis to a minimal atomic orbital basis (the quasiatomic orbitals or QOs) that contains the complete subspace of occupied Bloch orbitals. Because of this latter property, a description of the system in the QO basis exactly reproduces all of the ground state properties of the system in the original plane wave basis. Furthermore, the QOs also have the property of possessing maximum similarity to a set of original atomic orbitals used as input for the procedure while still reproducing the occupied Bloch orbitals. Further details of the QO method are given in the ESI.†

Constrained DFT calculations. In order to determine the energy required to localize a hole on the oxo during the water addition reaction, we employ a constrained DFT method, which we have implemented in the VASP code and which is based on the QO representation. This method allows one to freeze the shape and occupation of any number of Bloch orbitals in the calculation while allowing all of the other Bloch orbitals to relax (while remaining orthogonal to the frozen orbitals). The frozen orbitals are specified as linear combinations of QOs determined from a previous calculation. In the current implementation of this method, the unfrozen Bloch orbitals are restricted to lie in the subspace spanned by the QOs rather than the entire plane wave basis. This limitation is not expected to significantly affect the results since the geometric structure is identical in the constrained calculation and the unconstrained calculation from which the QOs are taken.

Localized hole orbitals. The hole density plots presented in several of the figures throughout this manuscript correspond to localized hole orbitals that were extracted from the QO representation of the DFT calculations. They are constructed from a linear combination of unoccupied Bloch orbitals in a way that maximizes the hole density on a selected atom or set of atoms. The method has similarities to the Natural Bond Orbital method of Weinhold and Landis.[24–26] Further details are given in the ESI.†

Results and discussion

The goal of this work is to examine the quantum chemistry of the water addition reaction on Co_3O_4 to gain insight into the properties that lead to an OER catalyst having maximum activity. We first present a detailed examination of the mechanism of water addition to the (311) μ^3-oxo, the (110)-A μ^2-oxo, and the (110)-B η-oxo. Based on the computed mechanisms, we decompose the water addition reaction into a proton coupled electron transfer (PCET) step and an O–O bond formation step and find that the differences in intrinsic activity between the three sites are primarily due to differences in the O–O bond formation step. We next apply methods we have developed for examining the electronic structure based on constrained DFT calculations and quasiatomic orbitals to further examine the O–O bond formation step occurring during water addition. This shows that the energetic differences in the step between the three sites are primarily a result of differences in the energy to localize a hole on oxos in different coordination environments. A similar electronic structure analysis is then carried out for the hydrogen atom removal occurring in the oxidation step that precedes water addition and it is found that the energy of this process is related to the energy of O–O bond formation by the hole localization energy, which is common to both processes. Based on this correlation between oxidation potential and water addition activity, we finally propose criteria for the design of the optimal OER catalyst at a given applied overpotential.

Mechanism and electronic transitions during water addition

The water addition reaction involves deprotonation of the attacking water by a basic oxygen atom on the active site and formation of an O–O bond between the resulting hydroxide and an oxo. In the process, the oxygen of the water and the oxo are both oxidized from O(−II) to O(−I) with the two electrons being transferred to holes on the Co center(s) of the active site. As will be seen in the following section, this reaction can be decomposed into two steps. The first step is a PCET step in which a proton and electron are transferred to the active site, forming a hydroxyl radical. The second step is the formation of the O–O bond between the hydroxyl and the oxo with a possible electron transfer from the oxo to a hole on one of the Co centers. It will be seen that the relative difficulties of these two steps are very different for the three active sites discussed below and this then leads to the differences in activity as a function of overpotential seen in Fig. 1.

(311) dual-Co μ^3-oxo. The active site on the (311) surface, shown in Fig. 2, consists of two octahedrally coordinated Co centers (Co(1) and Co(2)) connected

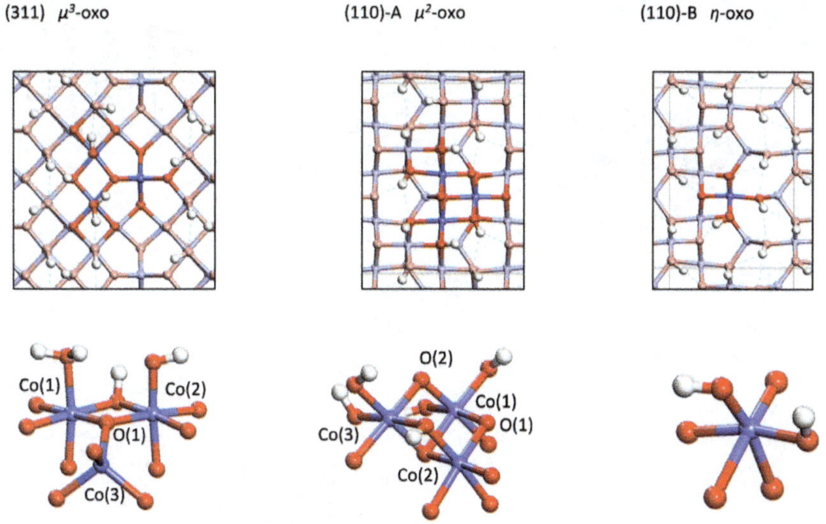

Fig. 2 Top and perspective views of the structures of the active sites on the three surfaces (Co = blue, O = red, H = white spheres).

by bridging oxo and hydroxo ligands. The oxo (O(1)) is also coordinated to a tetrahedrally coordinated Co(II) cation, making it a μ^3-oxo. In the oxidized state that is active for water addition,[15] both octahedrally coordinated Co centers are in the +IV oxidation state with one coordinated by an η-OH and the other by a water. As shown in structure I-a of Fig. 3, a hole resides on each of the Co(IV) centers, localized in a 3d t_{2g} orbital. The holes are strongly localized on the Co(IV) cations, with 83–85% of the hole density located on these cations. The two holes have opposite spins as is required for formation of the O–O bond. This initial electronic state is denoted as (I) in the figure.

The reaction begins by transfer of a proton from the attacking water to an η-OH on the active site. This proton transfer is initially very unfavorable as can be seen by the sharp rise in energy at the start of the potential energy profile in Fig. 3. A second electronic state (II) exists in which the hole originally on the Co(1) center is delocalized into the O 2p band so that this Co is in the +III oxidation state. This state is initially higher in energy than state I, but the proton transfer is considerably more favorable in state II than in state I. This is because, as the proton is transferred, the hole in the O 2p band localizes into an occupied σ* orbital formed between lone pairs on the hydroxide (being formed from the deprotonating water) and the oxo, which can be seen in structures II-b and II-c of Fig. 3. This withdraws negative charge from the hydroxide, stabilizing it and facilitating proton transfer. As a result, the potential energy surface of state II eventually crosses below that of state I as the proton is transferred. This process is therefore a PCET. The crossing of the two potential energy surfaces occurs at a point that is 0.26 eV above the energy of the oxidized site.

Completion of the PCET results in the intermediate shown in structure II-c of Fig. 3, where one hole still resides on a Co(IV) (Co(2)) while the other is localized in the σ* orbital formed between the oxo and the hydroxyl. The hole in the σ* orbital has 89% of its density on the two oxygen atoms. This intermediate is not a local

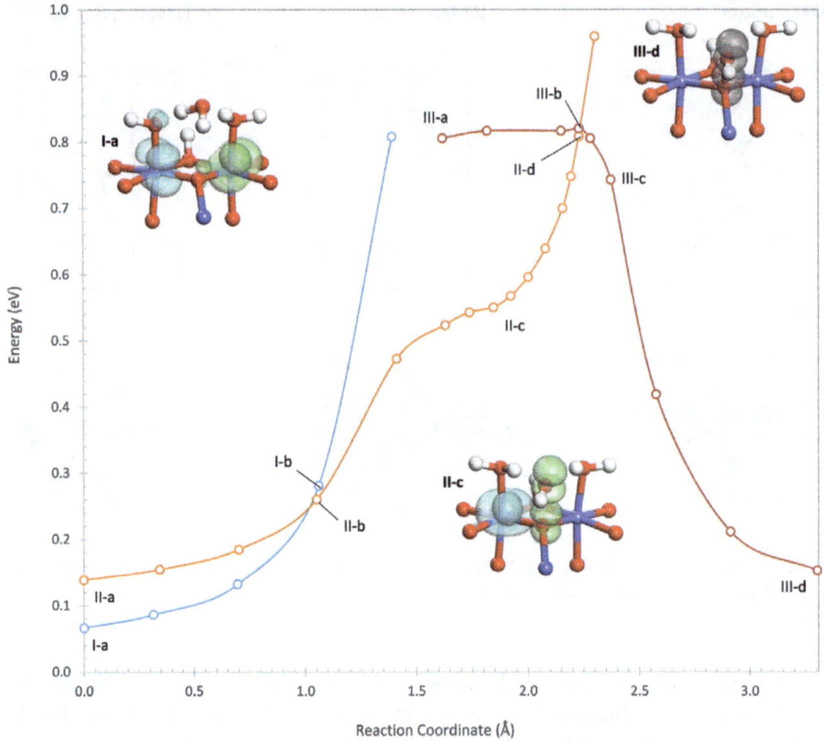

Fig. 3 Potential energy profiles for the three electronic states involved in water addition to the μ^3-oxo on the (311) dual-Co site. The insets show structures along the reaction pathway together with densities of the two holes involved in the reaction (green = spin-up, blue = spin-down, grey = both spins). Energies are relative to the oxidized state of the active site preceding water addition. Additional structures are shown in the ESI.†

minimum on the potential energy surface but exists as an inflection point in the reaction profile in Fig. 3, 0.55 eV above the energy of the oxidized site. In the intermediate, the hydroxyl forms strong hydrogen bonds with the two water ligands coordinating the Co cations in the active site, with O–H distances of 1.52 and 1.77 Å. These hydrogen bonds withdraw 0.18 electrons from the hydroxyl – this quantity (and all further charge quantifications below) is determined using a population analysis discussed in the ESI† based on the quasiatomic orbital representation.

From this intermediate, water addition continues by formation of the O–O bond between the oxo and the hydroxyl. As the O–O bond distance decreases, the energy of state II increases sharply due to Pauli repulsion because of the single electron still occupying the σ* orbital. At an O–O distance of 1.96 Å, the potential energy surface of state II crosses above that of a third electronic state (III) in which the hole residing on Co(2) in state II is delocalized into the O 2p band. The crossing point is 0.27 eV higher in energy than the intermediate and 0.82 eV higher in energy than the oxidized site. In electronic state III, O–O bond formation occurs easily since, as the O–O distance decreases, the hole in the O 2p band localizes into the σ* orbital so that both holes now reside in the σ* orbital. In the product state, both holes are strongly localized in the σ* orbital with 97% of the hole density on the two oxygen atoms.

(110)-A dual-Co μ^2-oxo. The active site on the (110)-A surface consists of three octahedrally coordinated Co cations connected in a ring by three oxo or hydroxo ligands, as shown in Fig. 2. Like the active site on the (311) surface, water addition on this site also occurs to an oxo (O(1)) bridging two octahedrally coordinated Co cations (Co(1) and Co(2)). The oxo is not, however, coordinated to an additional tetrahedrally coordinated Co(II) and is thus a μ^2-oxo rather than a μ^3-oxo. As will be seen, the absence of the Co(II) leads to very different reactivity than the μ^3-oxo on the (311) surface.

In the oxidized state prior to water addition, two of the Co cations (Co(1) and Co(2)) are in the +IV state while the third Co (Co(3)) is in the +III state when the overpotential is <0.81 V and in the +IV state when it is greater (discussed in more detail in the ESI†). The oxidation of Co(3) at 0.81 V is coupled with deprotonation of the bridging hydroxyl. A hole resides in a 3d t_{2g} orbital on each of the two Co cations (Co(1) and Co(2)) coordinated by the bridging oxo (O(1)) which forms the O–O bond. These holes, shown in structure I-a of Fig. 4, have opposite spins as required for the water addition reaction. The hole on Co(2) is well localized in a 3d t_{2g} orbital, with 87% of the hole density residing on the Co cation. The hole on

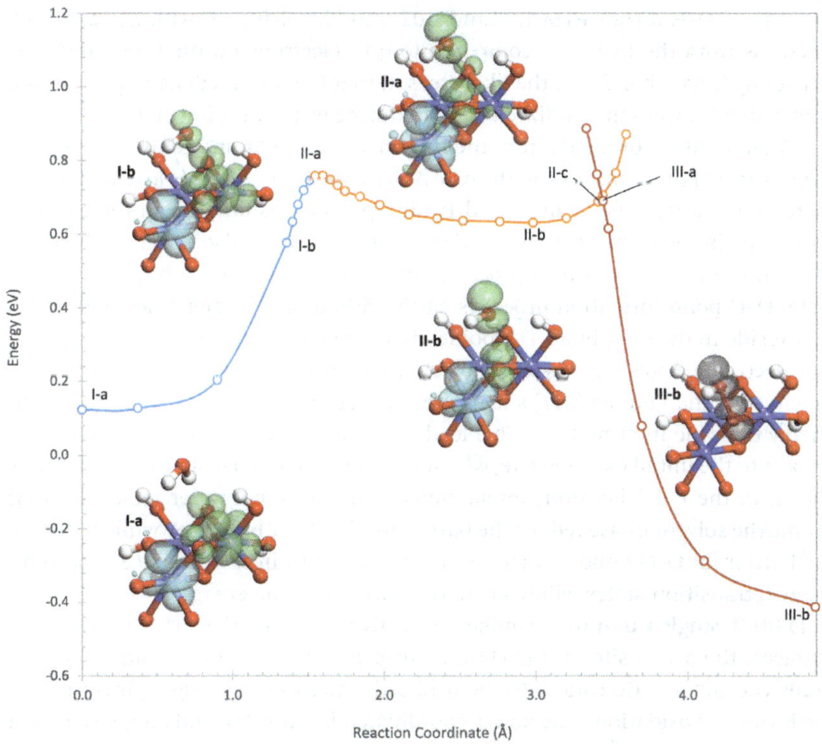

Fig. 4 Potential energy profiles for the three electronic states involved in water addition to the μ^2-oxo on the (110)-A dual-Co site. The insets show structures along the reaction pathway together with densities of the two holes involved in the reaction (green = spin-up, blue = spin-down, grey = both spins). Energies are relative to the oxidized state of the active site preceding water addition. Additional structures are shown in the ESI.†

Co(1) is significantly delocalized onto the two bridging oxo ligands coordinating it, with only 66% of the hole density residing on the Co cation.

Water addition initiates by the transfer of a proton from the attacking water to one of the μ^2-oxos O(2). The potential energy profile is shown in Fig. 4 where it can be seen that, unlike the (311) site, the proton transfer on the (110)-A site does not involve a crossing between two potential energy surfaces. Instead, the hole transfers smoothly from Co(1) to the oxygen on the water and the bridging oxo O(1). A transition state occurs at the point where only 9% of the hole density remains on Co(1), labeled II-a in Fig. 4. The transition state is 0.76 eV higher in energy than the initial oxidized site. In this transition state, the proton is in transit between the hydroxyl and the oxo.

As the proton transfer completes, an intermediate (II-b) is formed in which the hole becomes almost completely localized in a σ* orbital formed between the hydroxyl and the oxo, with 96% of the hole density on these two atoms. Thus, an electron is transferred from the σ* orbital to an empty orbital centered on a Co(IV) so that this first step is a PCET. The energy of the intermediate is 0.63 eV above the energy of the oxidized site prior to water addition. Unlike the intermediate on the (311) site which was identified as an inflection point on the potential energy surface, this intermediate is a true local minimum. As with the intermediate on the (311) site, this intermediate also forms two hydrogen bonds with nearby hydroxo ligands. These hydrogen bonds are weaker than they were on the (311) site, having O–H distances of 1.87 and 2.02 Å and involving the withdrawal of 0.08 electrons from the hydroxyl (compared to 0.18 electrons on the (311) site). The weaker hydrogen bonds are the likely reason that the intermediate requires more energy to form from the oxidized site than during water addition on the (311) site.

Water addition proceeds from the intermediate II-b by shortening of the O–O bond between the hydroxyl and the oxo. As with the (311) site, this leads to a sharp increase in energy due to increased Pauli repulsion resulting from the electron remaining in the σ* orbital. A second electronic state (III) also exists in which the hole on the remaining Co(IV) (Co(2)) is transferred to the σ* orbital. From this state, O–O bond formation proceeds easily without any barrier since both holes now reside in the σ* orbital. The point where the potential energy surfaces of the two electronic states cross is only 0.07 eV higher in energy than the intermediate, much lower than on the (311) site where the crossing point was 0.27 eV higher in energy than the intermediate. This leads to an activation barrier of 0.70 eV with respect to the initial oxidized site, which is lower than the barrier for the PCET. As shown in the ESI,† however, interactions with additional water molecules that mimic the solvent phase reduce the barrier for the PCET by 0.18 eV while reducing the barrier for O–O bond formation by 0.11 eV. Thus, in the presence of solvent, the two transition states will likely have nearly the same energy.

(110)-B single-Co η-oxo. Unlike the active sites on the (311) and (110)-A surfaces, the active site on the (110)-B surface consists of only a single octahedrally coordinated Co cation. In the oxidized state, shown in Fig. 2, this Co is in the formal +V oxidation state and is coordinated by an η-oxo and an η-OH as well as by a μ^3-oxo and a μ^2-OH. In the lowest energy state, there are two holes residing in a pair of π* orbitals between the Co and the η-oxo, shown in structure I of Fig. 5. These holes have parallel spins so that the Co=O is in a triplet state. If we assign the holes in the π* orbitals to the Co cation, then it is in the +V oxidation state and the oxo is in the −II oxidation state.

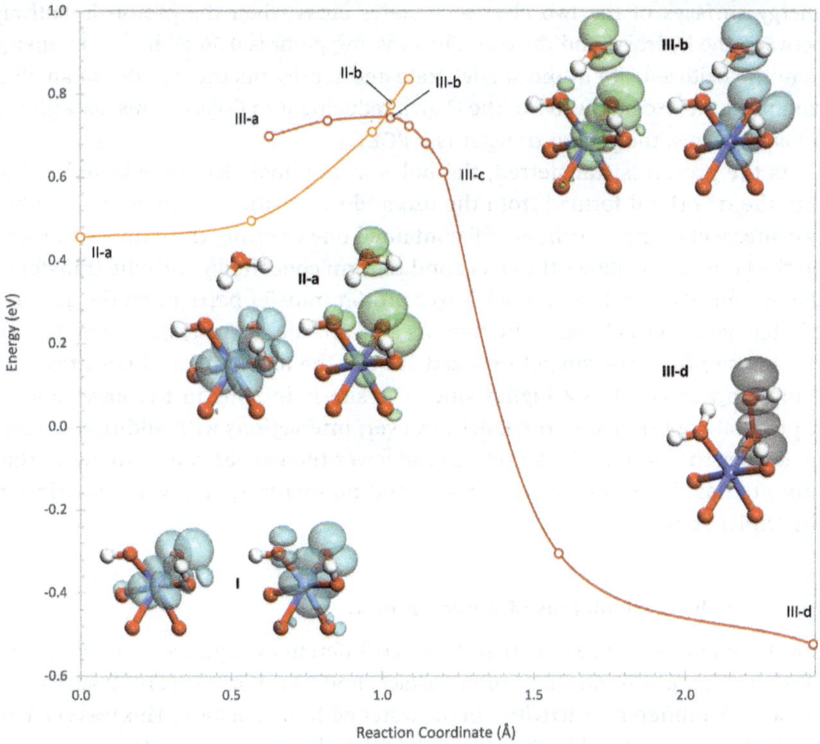

Fig. 5 Potential energy profiles for the three electronic states involved in water addition to the η-oxo on the (110)-B single-Co site. The insets show structures along the reaction pathway together with densities of the two holes involved in the reaction (green = spin-up, blue = spin-down, grey = both spins). Energies are relative to the oxidized state of the active site preceding water addition. Additional structures are shown in the ESI.†

Since formation of the O–O bond requires two holes of opposite spin, the Co=O must first transition to a singlet state that is 0.18 eV higher in energy than the triplet state. In the singlet state, the spin-up hole localizes onto the oxo while the spin-down hole localizes onto the Co, seen in structure II-a of Fig. 5. The spin-up hole has 78% of its density on the oxo, 11% on the water, and only 4% on the Co while the spin-down hole has 79% of its density on the Co and 12% on the oxo. This is best described qualitatively as an O(−I) bound to a Co(IV). Despite the formal transfer of an electron from the oxo to the Co during the spin transition, the actual electron populations on both change by less than 0.02 electrons. This is due to the fact that the holes were initially in π* orbitals that had nearly equal density on the Co and the oxo.

Like the other two sites, water addition begins by transfer of a proton from the water to the η-OH ligand, leading to an initial increase in energy (Fig. 5) since the electrophilic Co(IV) coordinated by the η-OH likely makes this ligand a weak base. A second electronic state (III), initially higher in energy, exists in which both holes reside in orbitals centered on the oxo (52% of the hole density) with significant delocalization onto the Co (28%) and the water (10%) as shown in structures III-a and III-b in Fig. 5. The proton transfer proceeds easily and with almost no activation barrier from this second electronic state (III) so that the potential energy surfaces of the two electronic states cross when the proton is halfway between the hydroxyl and the oxo. The crossing point is 0.56 eV higher in energy than the oxidized site in the singlet state and represents the transfer of an electron from the oxo and water to the Co(IV), reducing it to Co(III) – thus, as with the other two sites, the proton transfer is a PCET.

As the proton is transferred, the holes in the final electronic state localize into the σ* orbital formed from the oxo and the hydroxyl. Unlike on the other two sites where the σ* orbital still contained one electron, the empty σ* orbital on the (110)-B site allows the O–O bond to form concertedly with the transfer of the proton. This leads to a much lower proton transfer barrier and the absence of a barrier for O–O bond formation. As a result, the intrinsic barrier of 0.56 eV for this site from the singlet oxidized state is the lowest of all three sites. The actual barrier is 0.18 eV higher since the site is initially in the lower energy triplet state. As shown in the ESI,† however, interactions with additional water molecules that mimic the solvent phase lower the barrier substantially so that this site has the lowest intrinsic water addition barrier, even when starting in the triplet state.

Quantum chemical analysis of water addition

It was seen in the last section that the three different oxo species — μ^3-oxo on the (311) surface, μ^2-oxo on the (110)-A surface, and η-oxo on the (110)-B surface — possess very different reactivities in the water addition reaction. This was seen, in particular, to be related to the ease with which the oxo forms an O–O bond with the hydroxyl radical formed during the initial PCET process. In this section, we show that the O–O bond formation process is responsible for the differences in water addition energetics on the different sites and that this is due to differences in the cost of localizing a hole on the oxo. The hole localization process is examined using the constrained DFT method we have developed in the quasiatomic orbital formalism and is shown to be related to the coordination of the oxo by Co cations *via* the local Madelung potential and electron withdrawal effects.

In order to separate the PCET and O–O bond formation processes occurring during the water addition reaction, we decompose it into the two separate steps shown in Fig. 6. The first step is the transfer of a proton and electron from a molecule of water in vacuum to the active site, leaving a hydroxyl radical in vacuum. The proton is transferred to a hydroxo ligand on the active site and the electron is transferred to a Co(IV) center, reducing it to Co(III). This step corresponds to the PCET observed in the actual reaction mechanism. In the second step, the hydroxyl radical produced in the first step forms an O–O bond with an oxo on the active site.

The energies of the two steps shown in Fig. 6 confirm that it is the O–O bond formation step rather than the PCET step that is responsible for the differences in reactivity between the sites. This step is most exothermic (−2.36 eV) on the (110)-B η-oxo site which has the highest reactivity. The energy of this step is significantly less exothermic (−1.96 eV) on the less reactive (110)-A μ^2-oxo site and even less (−1.66 eV) on the least reactive (311) μ^3-oxo site. Although the PCET step also accounts for a large part of the difference in reaction energies between the (311) μ^3-oxo site and the (110)-A μ^2-oxo site, this step is more endothermic on the more reactive (110)-B η-oxo site than the less reactive (110)-A μ^2-oxo site.

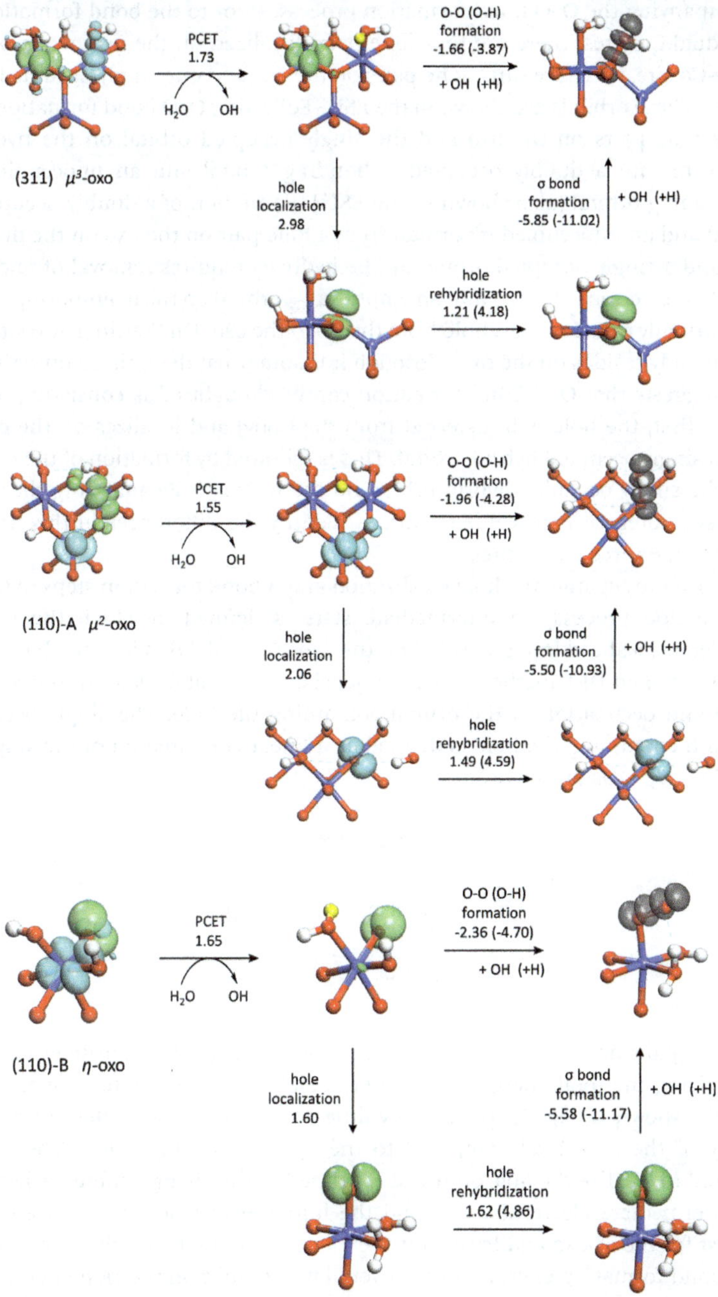

Fig. 6 Energy decomposition of the water addition reaction on the three active sites along with the density of the hole(s) at each step (green = spin-up, blue = spin-down, grey = both spins). The energy decomposition of hydrogen atom addition is also shown (energies in parentheses). All energies are in eV.

From the above analysis, it appears that as the coordination of the oxo decreases, the O–O bond formation step becomes more exothermic. To better understand this relationship, we examine the changes in the electronic structure accompanying the O–O bond formation process. Prior to the bond formation on both dual-Co sites, there are four lone pairs localized on the oxo, while on the single-Co site, there are three lone pairs on the oxo and one singly-occupied O 2p orbital. These orbitals are shown in the ESI.† Following O–O bond formation, one of the lone pairs on the oxo and the singly occupied orbital on the hydroxyl transform into a doubly occupied σ bonding orbital and an unoccupied σ* antibonding orbital (also shown in the ESI†). Formation of a doubly occupied σ orbital and an unoccupied σ* orbital from a lone pair on the oxo on the dual-Co sites and a singly occupied orbital on the hydroxyl requires removal of one electron. This electron is placed into an empty 3d t_{2g} orbital on the neighboring Co(IV), thus formally transferring a hole from the Co to the oxo. On the single-Co site, the hole already resides on the oxo, although it is somewhat delocalized onto the Co. This suggests that O–O bond formation can be thought of as consisting of two steps – first, the hole is transferred from the Co(IV) and localizes on the oxo to form a singly occupied hybrid orbital. This is followed by formation of the σ bond with the singly occupied hybrid orbital on the hydroxyl. By analyzing these two steps separately, it is possible to further identify the differences in this process between the three active sites.

In order to separate the hole localization and σ bond formation steps of the O–O formation process, an intermediate state is defined in which the hole is completely localized into an orbital on the oxo. To model this intermediate state, the constrained DFT method based on quasiatomic orbitals is used to freeze the shape and occupation of the orbital containing the hole. The singly occupied orbital h containing the hole is defined as a linear combination of the single 2s and three 2p QOs on the oxo.

$$\begin{aligned} \mathrm{h} &= \frac{1}{\sqrt{1+a_\mathrm{s}^2}}(a_\mathrm{s}\mathrm{s}+\mathrm{p}) \\ \mathrm{p} &= \frac{1}{\sqrt{N_\mathrm{p}}}(a_x\mathrm{p}_x+a_y\mathrm{p}_y+a_z\mathrm{p}_z) \\ N_\mathrm{p} &= a_x^2+a_y^2+a_z^2 \end{aligned} \quad (1)$$

The expansion coefficients of the orbital containing the hole are determined by minimizing the total energy of the intermediate electronic state. The resulting orbitals (shown in Fig. 6) contain very little 2s character due to the much lower energy of the 2s orbital compared to the 2p orbitals (13–14 eV). The energy required to localize the hole on the oxo, defined as the energy difference between the intermediate electronic state and the initial electronic state, is seen to be highest for the μ^3-oxo and lowest for the η-oxo, consistent with the trends in the O–O bond formation energy and the overall water addition reaction energy. This indicates that the differences in hole localization energy are responsible for most of the activity differences between the three sites. The least reactive μ^3-oxo requires 2.98 eV to localize the hole. In contrast, the most reactive η-oxo requires only 1.60 eV since the hole is already mostly localized on the oxo following the PCET.

Although the hole localization energy seems to explain the trends in the O–O bond formation energy and the water addition reactivity, it overestimates these differences. The differences in O–O bond formation energy between the μ^3-oxo and μ^2-oxo and the μ^2-oxo and η-oxo are 0.30 eV and 0.40 eV while the differences in hole localization energy are 0.91 eV and 0.47 eV, respectively. This is due to the fact that the hole on the oxo has almost entirely 2p character, while the hybrid forming the O–O σ bond has a significant contribution from the 2s orbital (7–8%). This higher 2s character is due to the increased concentration of the hybrid towards the hydroxyl fragment upon mixing with the 2s orbital which leads to increased orbital overlap and a stronger O–O bond. Although the energy to localize the hole in an orbital with significant 2s character is higher, the energy released by formation of the O–O σ bond in the next step is greater.

In order to quantify this rehybridization energy, a second intermediate electronic state is determined in which a_s in eqn (1) is set to 0.3, which is approximately the value of a_s in the hybrid orbitals forming the σ bond (0.28–0.30). As with the first intermediate, the expansion coefficients for the 2p part are determined by minimizing the total energy. Interestingly, the rehybridization energy, shown in Fig. 6, is highest for the η-oxo and lowest for the μ^3-oxo, reducing the differences in hole localization energy so that they are closer to the differences in O–O bond formation energy. This explains the overestimation of these differences when localizing the hole into an orbital with almost entirely 2p character. The reason for these differences in rehybridization energy can be seen in Fig. 6 which shows the change in hole density upon rehybridization. On the (311) surface, this has the effect of shifting the hole density away from the positively charged Co cations, leading to reduced repulsion and greater donation of electron density from the orbitals not containing the hole. A similar effect is seen for the μ^2-oxo on the (110)-A surface due to polarization of the hole away from a nearby proton. The effect is much weaker, however, because the proton is rather far away from the oxo (2.24 Å). No such effect exists for the η-oxo on the (110)-B surface.

Hole transfer energy

By decomposing the energy of water addition to an oxo into components related to PCET, hole localization, and σ bond formation, it was seen that most of the difference in reactivity between the μ^3-, μ^2-, and η-oxo sites is due to the energy to localize a hole on the oxo. This energy was seen to be highest on the least reactive μ^3-oxo and lowest on the most reactive η-oxo. In this section, we further analyze the hole localization process in order to determine why the energy decreases as the coordination of the oxo decreases.

To analyze the hole localization process, we carry out additional constrained orbital calculations for the process shown in Fig. 7. This process consists of the transfer of a hole from a singly occupied 3d orbital on the Co(IV) to a doubly occupied 2p orbital on the oxo. The constrained orbital on the oxo is the 2p component of the hybrid containing the hole in the intermediate state in eqn (1).

$$\mathrm{p} = \frac{1}{\sqrt{N_\mathrm{p}}}\left(a_x\mathrm{p}_x + a_y\mathrm{p}_y + a_z\mathrm{p}_z\right) \tag{2}$$

The constrained orbital on the Co is, likewise, a linear combination of the 3d quasiatomic orbitals on the Co cation

$$\mathrm{d} = \frac{1}{\sqrt{N_\mathrm{d}}}\left(a_{xy}\mathrm{d}_{xy} + a_{xz}\mathrm{d}_{xz} + a_{yz}\mathrm{d}_{yz} + a_{z^2}\mathrm{d}_{z^2} + a_{x^2-y^2}\mathrm{d}_{x^2-y^2}\right) \tag{3}$$

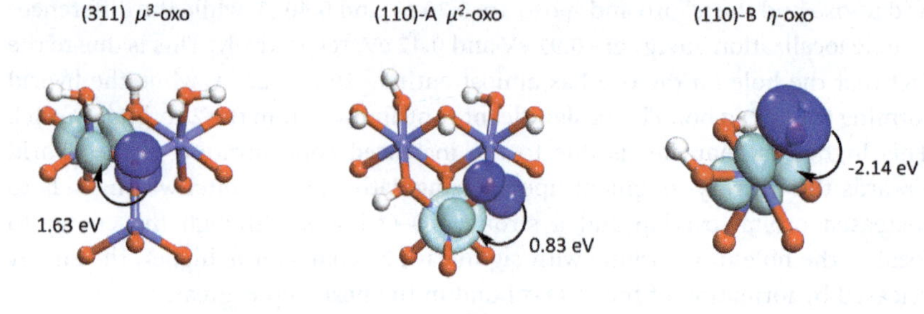

Fig. 7 Orbitals used on the three sites to model the hole transfer process from a 3d orbital on a Co cation (light blue) to a 2p orbital on the oxo (dark blue). Energies of the hole transfer process are given for each active site.

with the expansion coefficients taken from the eigenvector with the lowest eigenvalue of the 5 × 5 density matrix in the subspace of the 3d QOs calculated from the intermediate state following the PCET in Fig. 6. During the calculation, only these two orbitals are constrained and the rest are allowed to relax.

The energy to transfer the hole from the 3d orbital on the Co(IV) to the 2p orbital on the oxo in the constrained system (shown in Fig. 7) follows the same trend as the energy to localize the hole on the oxo in the actual system, being highest for the μ^3-oxo and lowest for the η-oxo. The hole transfer energy in the constrained system is lower than in the actual system, however, due to delocalization of the lone pair on the oxo in the initial state of the actual system. Constraining the lone pair to reside completely on the 2p orbital on the oxo raises the energy of the initial state considerably. The delocalization of the lone pair in the initial state is greater on the less coordinated oxos so that the differences in hole transfer energy in the constrained system overestimate the differences in hole localization energy in the actual system. The hole transfer energy is even negative for the η-oxo since the hole is more stable on the oxo than on the Co in the actual system. Nevertheless, the hole transfer process in the constrained system does provide insight into the differences in hole localization energy in the actual system as will be seen below.

The differential energy to transfer the hole from the Co to the oxo is given by the difference in the energies of the two orbitals containing the holes

$$\frac{\mathrm{d}E}{\mathrm{d}q} = \varepsilon_\mathrm{d} - \varepsilon_\mathrm{p} \tag{4}$$

where q is the hole transfer coordinate with $q = 0$ when the hole is on the Co and $q = 1$ when it is on the oxo.

We make the approximation that the orbital energies depend linearly on the electron population (n_Co and n_O) and magnetic moment (m_Co and m_O) of the respective atoms, with the slopes of the relationships related to the Coulomb integral U and the exchange integral J

$$\begin{aligned} \varepsilon_\mathrm{d} &= \varepsilon_\mathrm{d}^0 + \left(U_\mathrm{d} - \frac{1}{2}J_\mathrm{d}\right)\left(n_\mathrm{Co} - n_\mathrm{Co}^0\right) - \frac{1}{2}J_\mathrm{d}\left(m_\mathrm{Co} - m_\mathrm{Co}^0\right) \\ \varepsilon_\mathrm{p} &= \varepsilon_\mathrm{p}^0 + \left(U_\mathrm{p} - \frac{1}{2}J_\mathrm{p}\right)\left(n_\mathrm{O} - n_\mathrm{O}^0\right) - \frac{1}{2}J_\mathrm{p}\left(m_\mathrm{O} - m_\mathrm{O}^0\right) \end{aligned} \tag{5}$$

We make the further approximation that the electron populations and magnetic moments of the Co and oxo depend linearly on the occupancy of the constrained 3d and 2p orbitals, respectively. For convenience, the reference populations (n^0_{Co} and n^0_O) and magnetic moments (m^0_{Co} and m^0_O) are taken from the state where half of the hole resides in each of the two orbitals ($q = 1/2$). The resulting quantities are given by:

$$\begin{aligned} n_{Co} &= n^0_{Co} + (1 - \chi^n_{Co})\left(q - \frac{1}{2}\right) \\ n_O &= n^0_O - (1 - \chi^n_O)\left(q - \frac{1}{2}\right) \\ m_{Co} &= m^0_{Co} + (1 - \chi^m_{Co})\left(q - \frac{1}{2}\right) \\ m_O &= m^0_O - (1 - \chi^m_O)\left(q - \frac{1}{2}\right) \end{aligned} \tag{6}$$

In these expressions, χ is the screening constant for the electron population and magnetic moment of the respective atom which accounts for changes in the occupancies of the non-constrained atomic orbitals as the occupancy of the constrained orbital varies. Substituting eqn (5) and (6) into eqn (4) and integrating from $q = 0$ to $q = 1$ gives the overall hole transfer energy of

$$\Delta E = \varepsilon^0_d - \varepsilon^0_p \tag{7}$$

So the hole transfer energy is equal (in the linear approximation we have made) to the difference in the two orbital energies when half of the hole is residing on each orbital.

We would now like to address why the difference in orbital energies increases as the coordination of the oxo increases. In general, there are two possible causes for this. The first is the increase in Madelung potential on the oxo as the number of positively charged Co neighbors increases and the second is the decrease in electron population on the oxo due to donation to an increasing number of Co neighbors. In order to separate these two components, we first evaluate the Madelung potential by plotting the energy of the constrained orbital (referenced to the unscreened energy of the Co 3d orbital as explained in the ESI†) with respect to the electron population of the oxo. The electron population on the oxo is varied by changing the occupancy of the constrained orbital. The result is shown in the ESI† where it can be seen that the curve for the η-oxo has the highest energy and the μ^3-oxo has the lowest energy. At an electron population of 6.6 electrons, the orbital energies on the μ^3-, μ^2-, and η-oxo sites are −5.89, −2.72, and 0.96 eV, respectively. This confirms that the Madelung potential becomes significantly more attractive as the number of Co neighbors increases, making it more difficult to transfer an electron from the oxo to the Co.

The electron withdrawing ability of the Co neighbors is more difficult to evaluate. It is not appropriate to quantify it by the electron population on the oxo since this quantity is strongly correlated with the Madelung potential – a more repulsive Madelung potential will cause more electrons to be donated to the Co neighbors even when their electron withdrawing ability is unchanged. Therefore, we quantify the electron withdrawing ability by plotting the number of electrons donated with respect to the unscreened orbital energy of the constrained 2p orbital on the oxo (referenced to the unscreened energy of the Co 3d orbital as explained in the ESI†). The unscreened orbital energy is determined from the self-consistent orbital energy by subtracting off the contribution due to the decrease in electron population from donation to the Co neighbors

$$\varepsilon_{\mathrm{p}} = \varepsilon_{\mathrm{p}} + \left(U_{\mathrm{p}} - \frac{1}{2}J_{\mathrm{p}}\right)(5 + n_{\mathrm{p}} - n_{\mathrm{Co}}) - \frac{1}{2}J_{\mathrm{p}}\left(m_{\mathrm{p}} - m_{\mathrm{Co}}\right) \tag{8}$$

where n_p is the total population of the constrained 2p orbital on the oxo and m_p is the difference in spin-up and spin-down populations. The values of U_p and J_p were determined by the method described in the ESI.†

The plot in Fig. 8 shows that electron withdrawal by the Co neighbors is nearly equal for the μ^3- and μ^2-oxos, but is much weaker for the η-oxo. The weaker electron withdrawal for the η-oxo is expected since it has only one Co neighbor. One would, however, naively expect that the electron withdrawal from the μ^3-oxo should be greater than from the μ^2-oxo since it has one more Co neighbor. It turns out that the extra Co(II) cation coordinated to the μ^3-oxo is not a very strong electron acceptor. This can be seen by comparing the number of electrons donated by the oxo to each Co neighbor at a given value of the unscreened orbital

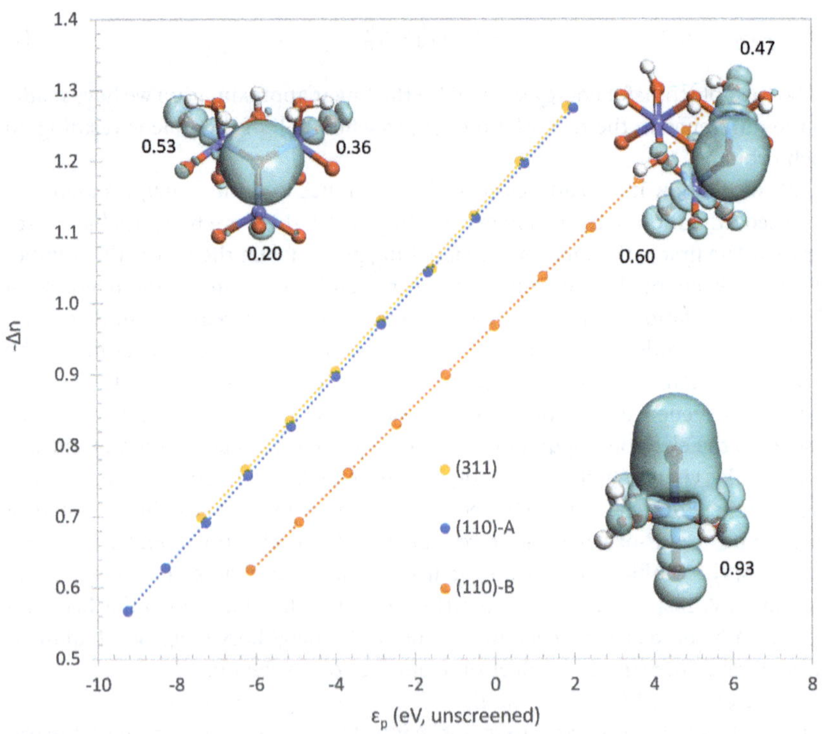

Fig. 8 Number of electrons donated from lone pairs on the oxo to neighboring atoms as a function of the unscreened energy of the 2p orbital containing the hole on the oxo (see text). The total density distribution of the delocalized lone pairs is shown for each site along with the number of electrons donated to each Co neighbor.

energy. To compute this quantity, we first determine the delocalized lone pairs $\tilde{\phi}_i$ by projecting the three non-constrained quasiatomic orbitals ϕ_i of the oxo onto the density matrix P

$$|\tilde{\phi}_i\rangle = \frac{P|\phi_i\rangle}{\sqrt{\langle\phi_i|P|\phi_i\rangle}} \tag{9}$$

We then calculate the population of this orbital on each Co neighbor

$$\Delta q_\mathrm{A} = \sum_{j\in \mathrm{A}} |\langle\phi_j|\tilde{\phi}_i\rangle| + \sum_{\mathrm{B}\in \mathrm{A}} w_\mathrm{B} \sum_{j\in \mathrm{B}} |\langle\phi_j|\tilde{\phi}_i\rangle|^2 \tag{10}$$

where the sums over j consist of all atomic orbitals on the Co cation A and its oxygen neighbors B (excluding the oxo). For oxygen neighbors that are shared between two of the Co cations coordinating the oxo, the population is divided equally between the two Co cations by setting $w_\mathrm{B} = \frac{1}{2}$ in eqn (10). The results of this analysis are given in Fig. 8 where we can see that only 0.20 electrons are donated to the Co(II) coordinated to the μ^3-oxo. This confirms that the Co(II) is a weak electron acceptor compared to the Co(III) and Co(IV) cations which withdraw 0.36 and 0.53 electrons, respectively.

The above analysis makes it clear that the main effect of coordinating the oxo by an additional Co(II) cation in the (311) surface is to increase the attractive Madelung potential on the oxo. This lowers the energy of the orbitals on the oxo and increases the energy needed to localize a hole in one of these orbitals. The presence of the Co(II) cation coordinating the oxo does not lead to a significant increase in electron withdrawal from the oxo. In contrast, the additional Co(III) coordinating the μ^2-oxo on the (110)-A surface leads to significantly greater electron withdrawal compared to the η-oxo on the (110)-B surface coordinated by only a single Co(IV). As with the Co(II) coordinated to the μ^3-oxo on the (311) surface, this Co(III) also increases the attractive Madelung potential on the μ^2-oxo compared to the η-oxo on the (110)-B surface. Thus, localization of the hole on the μ^2-oxo requires significantly more energy than localization on the η-oxo due to both an increase in the attractive Madelung potential on the μ^2-oxo as well as increased electron withdrawal by the Co(III) cation.

Quantum chemical relationship between water addition and oxidation of the active site

Now that we have developed a profound quantum chemical understanding of the O–O bond formation process occurring during water addition, we apply a similar approach to understanding the oxidation step that occurs in the catalytic cycle (Scheme 1) prior to water addition and which determines the overpotential at which the site becomes active. It will be seen that the energetics of the two processes are both largely controlled by the energy to localize a hole on the oxo. This then, is responsible for the correlation observed between the intrinsic reactivity of a site for water addition and the overpotential at which it becomes oxidized to the active state that we observed in our previous work.[15]

The relationship between oxidation and O–O bond formation can be seen by examining the oxidation process in the same context as O–O bond formation. The latter process was seen to proceed by localization of a hole on the oxo followed by formation of a σ bond with a hydroxyl radical. Likewise, as shown in Fig. 9, the oxidation process can be thought of as involving cleavage of an O–H σ bond to give a hydrogen atom and a hole on the resulting oxo which then transfers to a neighboring Co cation. Thus, the reverse of the oxidation process can be thought of as O–H bond formation between an oxo and a hydrogen atom, analogous to the O–O bond formation between the oxo and a hydroxyl. If we decompose the O–H bond formation process in the same way as the O–O bond formation process in Fig. 6, we see that the hole localization energy is exactly the same for the two processes. The rehybridization energy is much higher for O–H bond formation compared to O–O bond formation since the resulting hybrid has significantly higher 2s character ($a_s \approx 0.6$ *vs.* $a_s \approx 0.3$) although the trend is still similar, with the rehybridization energy increasing as coordination of the oxo decreases. The σ bond formation energy is much larger than the O–O bond formation energy but is similar for the three different sites and thus is not the main reason for the differences in O–H bond formation energy.

From the comparison between the O–O and O–H bond formation processes, it can be concluded that the correlation between O–H bond formation (reverse of oxidation) energy and O–O bond formation energy is based on the similar requirements for localization of a hole on the oxo in both processes. The hole localization requires the most energy on the (311) μ^3-oxo so that O–O bond formation is least favorable on this site but oxidation (O–H bond cleavage) is easier than on the other sites. As a result, the (311) μ^3-oxo is the most active site at low overpotentials, as shown in Fig. 1. The hole localization requires less energy on the (110)-A μ^2-oxo and (110)-B η-oxo but oxidation is harder so that these sites lead to higher TOFs than the (311) site at high overpotentials. A similar result was observed by Busch *et al.* who found a roughly linear correlation between the endothermicity of the oxidation step and the exothermicity of the subsequent O–O bond formation step on a binuclear model catalyst containing different 3d

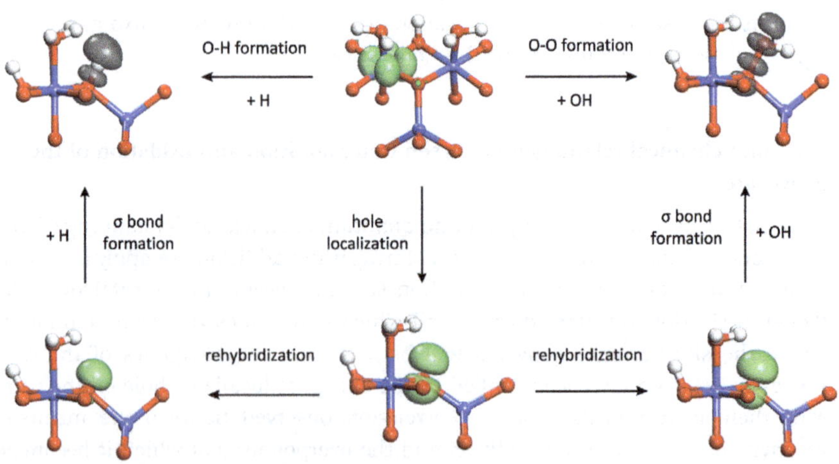

Fig. 9 Comparison of the changes in electronic structure occurring during O–O bond formation during water addition and O–H bond formation during the reverse of the oxidation step preceding water addition.

transition metal centers.[27] Such a correlation was also found by Man *et al.* between the binding energies of hydroxyl and hydroperoxyl species to oxide surfaces, leading to similar behavior.[28]

Design principles for the ideal OER active site at a given overpotential

We would now like to apply the quantum chemical understanding we have developed concerning the water addition reaction catalyzed by different active sites on Co_3O_4 to gain insights into how to design the optimal OER catalyst. From the results showing the TOF as a function of applied overpotential in Fig. 1, it is obvious that it is difficult to say what is the optimal active site since this changes with overpotential. By going to higher overpotentials, one can in general achieve higher TOFs but at the cost of electrochemical efficiency. Thus, one will always have to consider the tradeoff between activity and efficiency. Here, we will focus on the design of the optimal catalyst (the one giving the highest TOF) at a given overpotential.

To calculate the TOF as a function of overpotential, we assume that water addition is the rate limiting step (which is true at overpotentials above 0.53 V (ref. 15)) and that the active site exists predominantly in one particular state, the dominant resting state i, which has a lower free energy than all of the other possible states. We also make the assumption that water addition occurs through a single transition state that proceeds from a state j and that state j (the active state) is in quasi-equilibrium with the dominant resting state i. The TOF is then given by the following expression[15]

$$\mathrm{TOF} = \frac{k_\mathrm{B}T}{h}\exp\left(-\frac{\Delta G_\mathrm{j}^{\ddagger} + \Delta G_\mathrm{ij} - e\,\eta\,\Delta n_\mathrm{ij}}{k_\mathrm{B}T}\right) = \frac{k_\mathrm{B}T}{h}\exp\left(-\frac{\Delta G_\mathrm{app}^{\ddagger}}{k_\mathrm{B}T}\right) \tag{11}$$

where $\Delta G_\mathrm{j}^{\ddagger}$ is the intrinsic activation free energy barrier for water addition with respect to the active state j preceding the transition state, ΔG_ij is the free energy difference between the active state j and the dominant resting state i at zero overpotential, and Δn_ij is the number of coupled electron and proton removal steps occurring between states i and j which determines how the TOF depends on the overpotential η. The numerator in the argument of the exponential is the apparent activation free energy, $\Delta G_\mathrm{app}^{\ddagger}$.

For simplicity, we consider a model consisting of the four possible states of the active site depicted in Scheme 2. The first of these states (0) corresponds to the active state j. In the next two states, one of the Co(IV) cations in the dual-Co active sites is reduced to Co(III) or the Co(V) in the single-Co site is reduced to Co(IV) and a proton is added to either the basic oxygen that deprotonates the water during the PCET part of water addition (1) or the oxo undergoing water addition (2). In the remaining state (3), both Co(IV) cations in the dual-Co site or the Co(V) in the single-Co site are reduced to Co(III) and protons are added to both the oxo and the basic oxygen.

The free energy differences ΔG_ij and values of Δn_ij in eqn (11) for the four states are given in Table 1. In defining the free energy difference ΔG_3 between state (3) and the active state (0) we have made the further assumption that the oxidations occurring between states (1) or (2) and the active state are additive so that $\Delta G_3 = \Delta G_1 + \Delta G_2$. With these definitions, the apparent activation free energy in eqn (11) can be written as

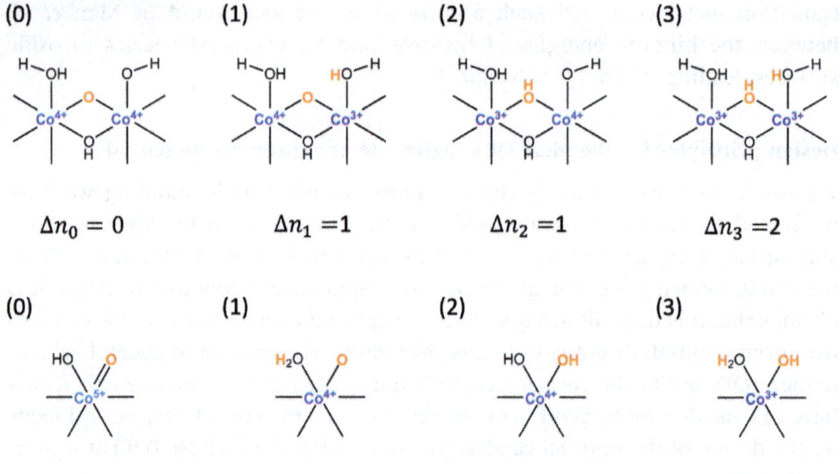

$\Delta G_1 = -\Delta G_{PCET} + \beta_1$ $\quad \Delta G_2 = -\alpha_2 \Delta G_{O-O} + \beta_2$ $\quad \Delta G_3 = \Delta G_1 + \Delta G_2$

Scheme 2 States involved in the two oxidation steps preceding water addition along with definitions of parameters appearing in eqn (17).

$$\Delta G^{\ddagger}_{\text{app}} = \Delta G^{\ddagger}_{\text{J}} + \max\{(\Delta G_1 - e\eta), 0\} + \max\{(\Delta G_2 - e\eta), 0\} \quad (12)$$

From this expression, we can see that if one of the two oxidations converting states (1) and (2) to the active state has a positive free energy, then the active site will exist predominantly in the reduced state and the free energy to oxidize this state to the active state must be added to the intrinsic activation barrier. If both oxidation free energies are positive, then the active site will exist predominantly in state (3) and both oxidation free energies must be added to the intrinsic activation barrier.

We now make an approximation that the intrinsic free energy activation barrier is linearly related to the free energy of the water addition reaction ΔG_{rxn} by a Brønsted–Evans–Polanyi relationship

$$\Delta G^{\ddagger} = \alpha_0 \Delta G_{\text{rxn}} + \beta_0 \quad (13)$$

A plot showing this relationship is included in the ESI† and gives values of $\alpha_0 = 0.42$ and $\beta_0 = 0.68$ eV. As shown previously, the reaction free energy of water

Table 1 Oxidation overpotentials (*V*) for the different sites

	(001)	(311)	(110)-A	(110)-B	(110)-B[f]
$\eta_{\text{ox},1}$[a]	0.58	0.50	0.68	0.58	0.64
$\eta_{\text{ox},2}$[b]	0.25	0.21	0.63	1.04	1.01
$\eta_{\text{ox,id}}$[c]	0.41	0.36	0.66	0.81	0.83
$\eta_{\text{ox,corr}}$[d]	0.66	0.50	0.73	1.00	1.05
η_{ox}[e]	0.42	0.50	0.73	0.82	0.87

[a] $\eta_{\text{ox},1} = \Delta G_1/e$. [b] $\eta_{\text{ox},2} = \Delta G_2/e$. [c] $\eta_{\text{ox,id}} = (\eta_{\text{ox},1} + \eta_{\text{ox},2})/2$. [d] $\eta_{\text{ox,corr}} = \eta_{\text{ox}} + \Delta G_0/e$. [e] Actual calculated oxidation potential. [f] More oxidized (110)-B surface.

addition is the sum of the PCET free energy ΔG_{PCET} and the O–O bond formation energy $\Delta G_{\text{O–O}}$

$$\Delta G_{\text{rxn}} = \Delta G_{\text{PCET}} + \Delta G_{\text{O–O}} \tag{14}$$

The PCET process is simply the reverse of the oxidation process occurring between state (1) and the active state so that

$$\Delta G_{\text{PCET}} = -\Delta G_1 + \beta_1 \tag{15}$$

where $\beta_1 = 2.23$ eV accounts for the different reference states used in the two processes (hydrogen atom from water *versus* an electron and proton at zero overpotential). We then make the further assumption that the free energy of the O–O bond formation process is linearly related to the oxidation process occurring between state (2) and the active state (0).

$$\Delta G_{\text{O–O}} = -\alpha_2 \Delta G_2 + \beta_2 \tag{16}$$

This is based on the relationship between the two quantities discussed earlier and a plot showing the relationship is included in the ESI.† The correlation is not excellent, but there is definitely a trend which leads to values of the parameters in eqn (16) of $\alpha_2 = 1.00$ and $\beta_2 = -1.34$ eV. Substituting eqn (13)–(16) into eqn (12) finally gives

$$\Delta G^{\ddagger}_{\text{app}}/e = -\alpha'_1 \eta_{\text{ox},1} - \alpha'_2 \eta_{\text{ox},2} + \beta + \max\{(\eta_{\text{ox},1} - \eta), 0\} + \max\{(\eta_{\text{ox},2} - \eta), 0\} \tag{17}$$

with

$$\eta_{\text{ox},1} = \Delta G_1/e$$

$$\eta_{\text{ox},2} = \Delta G_2/e$$

$$\alpha'_1 = \alpha_0 = 0.42$$

$$\alpha'_2 = \alpha_0 \alpha_2 = 0.42$$

$$\beta = \beta_0 + \alpha_0 \beta_1 + \alpha_0 \beta_2 = 1.05 \text{ eV}$$

This expression allows us to define the optimal active site at a given overpotential, which is illustrated in Fig. 10. When both of the oxidation potentials of the active site ($\eta_{\text{ox},1}$ and $\eta_{\text{ox},2}$) are less than the overpotential, the site exists predominantly in the active state so that the last two terms on the RHS of eqn (17) are zero. In this case, increasing either of the oxidation potentials is correlated with a decrease in the intrinsic activation barrier by the first two terms on the RHS of eqn (17) so that the TOF increases. If one or both of the oxidation potentials is greater than the overpotential, then the active site will exist predominantly in a reduced state so that the oxidation energy must be added to the intrinsic activation barrier. Since α'_1 and α'_2 are both less than unity, the increase in the last two terms on the RHS of eqn (17) corresponding to an increase in the oxidation

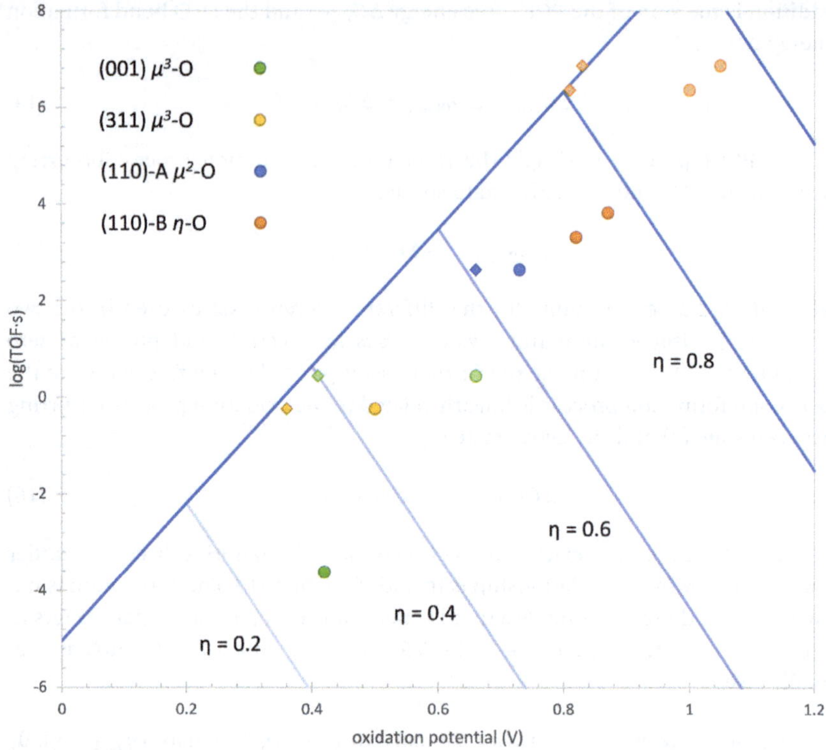

Fig. 10 TOF of water addition as a function of the oxidation potential of the active site at different overpotentials η (V) for the ideal site based on eqn (17). The TOF and oxidation potentials of the actual sites are also shown (dark colored circles). Light colored circles for the (001) and (110)-B sites show the TOF and oxidation potentials when the lower energy state preceding the active state is not considered. Diamonds show the performance of the sites if both oxidation potentials depicted in Scheme 2 are equal. Two active states of the site on the (110)-B surface are shown, one operating at overpotentials above 0.86 V and the other operating at overpotentials lower than this.

potentials will dominate the decrease in intrinsic activation barrier in the first two terms, leading to a decrease in the TOF. This means that the active site giving the maximum TOF at a given overpotential, the Sabatier maximum, will have both oxidation potentials exactly equal to the overpotential.

Fig. 10 also shows the performance of the active sites on the three surfaces examined in this work plus the dual-Co site on the (001) surface examined in our previous work[15] in comparison to the ideal performance. The values of the oxidation potentials used to determine these results are given in Table 1. The oxidation potential is taken as the potential at which the active site is first oxidized to the active state for water addition (the first point where the TOF in Fig. 1 becomes flat with respect to overpotential) and the TOF is calculated at this overpotential. One can see that all of the real active sites are less active than the ideal active site. There are two causes of this, the first being that the two oxidation potentials are not equal and the second being the existence of a state that has the same degree of oxidation as the active state but has a lower free energy. The first of these affects all four active sites while the second affects only the (001) and (110)-B sites. We will now discuss these causes of lower activity in more detail.

For the sites on the (001) and (110)-B surfaces, the lower activity is partially due to the existence of a state that has the same degree of oxidation as the active state but has a lower free energy. As a result, the intrinsic activation barrier increases by an amount equal to the free energy difference between these two states. Since the two states have the same degree of oxidation, increasing the overpotential has no effect on this free energy difference. On the (001) site, the lower energy state has the proton from the μ^3-OH of the active state attached to the η-OH instead (shown in the ESI†). The free energy of this state is 0.24 eV below that of the active state. This was discussed in detail in our previous work[15] and is due to the high acidity of the μ^3-OH. The lower energy state on the (110)-B site has the Co=O fragment in the triplet state rather than the active singlet state, which is lower in energy by 0.18 eV. In Fig. 10, points for these two sites are also plotted in which the lower energy state has been removed. This is done by increasing the oxidation potentials and decreasing the activation barriers by the energy difference ΔG_0 between the two states (0.24 eV for the (001) site and 0.18 eV for the (110)-B site). This moves the points for these two sites closer to the ideal activity.

Even after eliminating the performance decrease due to the presence of lower energy states for the (001) and (110)-B sites, the activities for these sites and the other two sites are still about 2–3 orders of magnitude lower than the ideal activity. This is caused by the two oxidation potentials depicted in Scheme 2 and listed in Table 1 not being equal. The overpotential at which the site becomes oxidized to the active state is determined by the higher of these two potentials. If one of them is lower than the other, then the intrinsic activation barrier is not as low as it would be if the lower oxidation potential was equal to the higher oxidation potential. This is similar to the concept put forward in the purely thermodynamic model of the OER by Rossmeisl, Nørskov, and others[29] where it is proposed that the optimal catalyst will have the same oxidation potential for all four oxidations occurring during the reaction. It is interesting that this aspect of the simpler model still appears in our more complete kinetic model.

Fig. 10 shows the activities of the four sites at an oxidation potential equal to the average of the two values $\eta_{ox,1}$ and $\eta_{ox,2}$ in Table 1 (called the ideal oxidation potential, $\eta_{ox,id}$). Except for the (110)-A site, all of the other sites have activities very close to the ideal activity, demonstrating that this is indeed the primary cause of the lower activity. The (110)-A site has an activity that is still about one order of magnitude lower than the ideal activity, possibly due to steric effects since the bridging oxo and hydroxo ligands on this site are more rigid than the terminal hydroxo and water ligands on the other sites.

Conclusions

We have examined the mechanism and quantum chemistry of the nucleophilic addition of water to oxo species on three different surface terminations of Co_3O_4. This particular elementary step was examined because it was found to be the rate controlling step on these sites over the largest range of applied overpotentials in our previous study.[15] We find that the intrinsic activation barrier and reaction energy of water addition decreases as the coordination of the oxo decreases, being highest for a μ^3-oxo on the (311) dual-Co site and lowest for an η-oxo on the (110)-

B single-Co site, with a μ^2-oxo on the (110)-A dual-Co site lying in between. Water addition occurs on all three sites by the coupled transfer of a proton and electron (PCET) from the attacking water to a basic oxygen and a Co(IV) on the active site followed by formation of an O–O σ bond between the resulting hydroxyl and the oxo. Differences in the energetics of the O–O bond formation step are responsible for the differences in intrinsic reactivity between the three sites, with this step being more difficult on the two dual-Co sites due to the required transfer of a hole from a Co(IV) center to the oxo prior to O–O bond formation. On the single-Co site, the hole already resides on the oxo so that O–O bond formation occurs concertedly with the PCET step. Furthermore, O–O bond formation following the PCET has a considerable activation barrier on the μ^3-oxo of the (311) site while it has only a small activation barrier on the μ^2-oxo of the (110)-A site.

To determine the origin of the more favorable O–O bond formation step on sites with a less coordinated oxo, we examined the energy required to localize a hole in a 2p orbital on the oxo on each of the sites. This energy was found to be highest on the least active (311) μ^3-oxo and lowest on the most active (110)-B η-oxo, with the (110)-A μ^2-oxo lying in between. By examining the orbital energies in a model process involving the transfer of a hole from a Co 3d orbital to an O 2p orbital, we conclude that the higher hole localization energy on the μ^3-oxo compared to the μ^2-oxo is due to a more attractive Madelung potential on the μ^3-oxo from the additional Co(II) cation coordinating it. The lower hole localization energy on the η-oxo compared to the μ^2-oxo is also due to a more attractive Madelung potential on the more highly coordinated μ^2-oxo as well as greater electron withdrawal from the oxo by the two Co neighbors of the μ^2-oxo compared to the single Co neighbor of the η-oxo.

We used the same technique to examine the hydrogen atom removal occurring during oxidation of the active site prior to water addition in order to identify the quantum chemical origin of the correlation seen between oxidation potential and intrinsic activity for water addition. It is seen that the hydrogen atom addition occurring during the reverse of the oxidation process involves similar changes in electronic structure as the hydroxyl radical addition occurring during O–O bond formation. Both processes involve localization of a hole on the oxo followed by σ bond formation. The correlation between the energies of the two processes is related to the hole localization energy, which is common for both.

Based on this correlation between intrinsic water addition reactivity and oxidation potential, we investigate the properties leading to the most active catalyst at a given applied potential. We find that for the ideal active site at a given applied potential, the oxidation potentials of the two steps preceding water addition must be equal to each other and to the applied potential. All of the real active sites examined in this work have sub-optimal activity due to the two oxidation potentials being unequal and, on some sites, the existence of an inactive state that has the same degree of oxidation but a lower free energy than the active state. This suggests a possible strategy for engineering an active site with improved activity.

Acknowledgements

C. P. acknowledges partial support from NRSC-Catalysis and the Institute for Complex Molecular Systems at the Technical University of Eindhoven and the

Solar Technologies Go Hybrid initiative of the State of Bavaria. We also acknowledge NWO for access to supercomputing facilities.

References

1 M. G. Walter, E. L. Warren, J. R. McKone, S. W. Boettcher, Q. Mi, E. A. Santori and N. S. Lewis, *Chem. Rev.*, 2010, **110**(11), 6446–6473.

2 A. Marshall, B. Børresen, G. Hagen, M. Tsypkin and R. Tunold, *Energy*, 2007, **32**(4), 431–436.

3 E. Fabbri, A. Habereder, K. Waltar, R. Kötz and T. J. Schmidt, *Catal. Sci. Technol.*, 2014, **4**(11), 3800–3821.

4 W. T. Hong, M. Risch, K. a. Stoerzinger, A. Grimaud, J. Suntivich and Y. Shao-horn, *Energy Environ. Sci.*, 2015, **8**, 1404–1427.

5 M. E. G. Lyons and M. P. Brandon, *J. Electroanal. Chem.*, 2010, **641**(1–2), 119–130.

6 M. Dinca, Y. Surendranath and D. G. Nocera, *Proc. Natl. Acad. Sci. U. S. A.*, 2010, **107**(23), 10337–10341.

7 J. Suntivich, K. J. May, H. a. Gasteiger, J. B. Goodenough and Y. Shao-Horn, *Science*, 2011, **334**(6061), 1383–1385.

8 D. G. Nocera, *Acc. Chem. Res.*, 2012, **45**(5), 767–776.

9 M. Zhang and H. Frei, *Catal. Lett.*, 2015, **145**(1), 420–435.

10 F. Jiao and H. Frei, *Angew. Chem., Int. Ed.*, 2009, **48**(10), 1841–1844.

11 D. A. Lutterman, Y. Surendranath and D. G. Nocera, *J. Am. Chem. Soc.*, 2009, **131**(11), 3838–3839.

12 M. Bajdich, M. García-mota, A. Vojvodic, J. K. Nørskov and A. T. Bell, *J. Am. Chem. Soc.*, 2013, **135**(36), 13521–13530.

13 M. García mota, M. Bajdich, V. Viswanathan, A. Vojvodic, A. T. Bell and J. K. Nørskov, *J. Phys. Chem. C*, 2012, **116**(39), 21077–21082.

14 M. Zhang, M. de Respinis and H. Frei, *Nat. Chem.*, 2014, **6**(4), 362–367.

15 C. P. Plaisance and R. A. van Santen, *J. Am. Chem. Soc.*, 2015, **137**(46), 14660–14672.

16 G. Kresse and J. Furthmüller, *Phys. Rev. B: Condens. Matter Mater. Phys.*, 1996, **54**(16), 11169–11186.

17 A. J. Cohen, P. Mori-Sanchez and W. Yang, *Science*, 2008, **321**(5890), 792–794.

18 V. I. Anisimov and O. Gunnarsson, *Phys. Rev. B: Condens. Matter Mater. Phys.*, 1991, **43**(10), 7570–7574.

19 A. I. Liechtenstein, V. I. Anisimov and J. Zaanen, *Phys. Rev. B: Condens. Matter Mater. Phys.*, 1995, **52**(8), R5467–R5470.

20 V. I. Anisimov, F. Aryasetiawan and A. I. Lichtenstein, *J. Phys.: Condens. Matter*, 1997, **9**(4), 767–808.

21 M. T. M. Koper, *Phys. Chem. Chem. Phys.*, 2013, **15**(5), 1399–1407.

22 Y. Surendranath, M. W. Kanan and D. G. Nocera, *J. Am. Chem. Soc.*, 2010, **132**(14), 16501–16509.

23 X. Qian, J. Li, L. Qi, C.-Z. Wang, T.-L. Chan, Y.-X. Yao, K.-M. Ho and S. Yip, *Phys. Rev. B: Condens. Matter Mater. Phys.*, 2008, **78**(24), 245112.

24 A. E. Reed and F. Weinhold, *J. Chem. Phys.*, 1985, **83**(4), 1736.

25 F. Weinhold and C. R. Landis, *Valency and Bonding: A Natural Bond Orbital Donor–Acceptor Perspective*, Cambridge University Press, 2005.

26 J. P. Foster and F. Weinhold, *J. Am. Chem. Soc.*, 1980, **102**(24), 7211–7218.

27 M. Busch, E. Ahlberg and I. Panas, *Catal. Today*, 2013, **202**, 114–119.
28 I. C. Man, H.-Y. Su, F. Calle-Vallejo, H. A. Hansen, J. I. Martínez, N. G. Inoglu, J. Kitchin, T. F. Jaramillo, J. K. Nørskov and J. Rossmeisl, *ChemCatChem*, 2011, 3(7), 1159–1165.
29 J. Rossmeisl, Z.-W. Qu, H. Zhu, G.-J. Kroes and J. K. Nørskov, *J. Electroanal. Chem.*, 2007, **607**(1–2), 83–89.

Faraday Discussions

PAPER

Brønsted activity of two-dimensional zeolites compared to bulk materials

Joachim Sauer*

Received 2nd December 2015, Accepted 15th January 2016
DOI: 10.1039/c5fd00207a

Different reactivity parameters yield different results for the relative acidity of zeolitic Brønsted sites in thin films and in bulk materials. Whereas the adsorption energies of ammonia and pyridine are about the same, the energy of deprotonation is much lower for two-dimensional systems than for three-dimensional systems. It is shown that this is due to the smaller effective dielectric constant of two-dimensional systems, which leads to much lower deprotonation energies, but also to much lower interaction energies between the protonated molecule and the negatively charged surface site. In the total adsorption energies, both effects nearly compensate each other.

1. Introduction

The catalytic activity[1,2] of the proton forms of zeolites originates from bridging Al–O(H)–Si groups which represent Brønsted acid sites.[3,4] They are located at the internal surface of the channels and cavities of the crystalline zeolite structures. In recent years two-dimensional zeolites, *i.e.* nanosheets as thin as single unit cells of, *e.g.* MFI,[5,6] have found much interest.[7] To understand if "being two-dimensional" affects the catalytic activity, we compare the protonation ability of bridging Al–O(H)–Si groups of three-dimensional (3d, bulk) and two-dimensional (2d) zeolites. As a model for the latter, we consider an aluminosilicate bilayer with a two-dimensional arrangement of hexagonal prisms (H-2dH) and compare it with the protonated form of H-chabazite (H-CHA)[8,9] and other zeolites. Such a protonated aluminosilicate bilayer has indeed been grown as an ultra-thin film under UHV conditions by deposition of O_2, Si and Al on the Ru(0001) surface and subsequent adsorption and desorption of water.[8,10] It has also been speculated that low-density large-pore zeolites can be assembled from such double-layer aluminosilicate sheets.[11]

A flat surface may be considered as the limiting case of a pore with an infinitely large curvature. The surface curvature effect of Derouane[12] (Fig. 1) has been invoked to explain the smaller adsorption energies for CO, C_2H_4, and pyridine on the two-dimensional system with a flat surface compared to the microporous bulk systems.[8] If this effect (largely due to dispersion) is approximately eliminated

Humboldt-Universität, Institut für Chemie, Berlin, Germany. E-mail: js@chemie.hu-berlin.de

Fig. 1 Worsening of the fit of a molecule (red circle) to the surface from a small pore to a large pore 3d system and to a flat surface.

from the adsorption system, the remaining "specific" interaction is larger for Brønsted sites on the 2d system as compared to the 3d system.[8]

IR spectroscopy has shown that the OH frequency-shifts on adsorption of carbon monoxide and ethene are indeed somewhat larger for the proton forms of aluminosilicate bilayers than for acidic zeolites (H-CHA and H-ZSM-5).[8] Both the 2d and 3d systems are able to protonate ammonia and pyridine.[8] In contrast, much lower deprotonation energies have been obtained for the two-dimensional systems (H-2dH) compared to the bulk system (H-CHA).[9] Hence, the deprotonation energy points to a special effect of being two-dimensional, different from the surface curvature effect, and the question arises: Why is this not seen in the energies of adsorption of ammonia or pyridine that involve proton transfer? Answering this question is the aim of this study.

2. Methods

There are different measures of acidity strength,[13–15] *i.e.* parameters that are related to the (free) energy for the protonation of a substrate molecule S by a zeolitic hydroxyl group ZO–H,

$$\text{Z–OH} + \text{S} \rightarrow \text{Z–O}^- \cdots \text{HS}^+. \quad (1)$$

They may be categorized according to different stages of reaction (1), ranging from little mutual perturbation of the reactants ("early" transition state) to completed proton transfer ("late" transition state). Parameters that describe the intrinsic acidity without any interaction with a substrate molecule are the OH frequency[13,16] (lower frequency – weaker OH bond – stronger acid) and the ^{1}H-NMR shift[17,18] (larger shift – more positive charge on proton site – stronger acid) for the initial state, as well as the deprotonation energy (or the proton affinity of the deprotonated Brønsted site) for the final state.[4]

Although the deprotonation energy neglects interactions between the substrate and the catalyst surface, often it predicts reactivity trends correctly. An example is the H-SSZ-13/H-SAPO-34 comparison: two catalysts with the same active site and framework structure (CHA), but with different framework compositions, SiO_2/$AlPO_4$. The calculations yielded deprotonation energies that were, depending on the specific method, 30–40 kJ mol^{-1} higher for H-SAPO-34 than for H-SSZ-13, indicating a lower acidity of the former. This fits the statement "Another key feature of the SAPO-34 molecular sieve is its optimized acidity relative to aluminosilicate based zeolitic materials" in a publication on the UOP/hydro methanol-to-olefin process.[19]

The OH frequency shift on adsorption of weak bases such as CO and C_2H_2 is among the parameters that take the interaction with a probe molecule (model substrate) into account and represent an "early" transition state. In contrast, adsorption of strong bases like ammonia or pyridine implies a late transition state with the proton transferred to the reaction partner.

The advantage of a computational approach is that all these parameters can be calculated and their relation can be analysed, see ref. 18, 20 and 21 for comparisons between different parameters. The calculations use density functional theory (DFT) and apply periodic boundary conditions.

The hybrid QM/MM calculations are performed with the B3LYP hybrid functional (QM) and shell-model ion-pair potential for the periodic structure (MM).[22] Harmonic frequencies are calculated for the QM cluster models at the QM:MM structures. They are scaled with a factor of 0.9716, obtained by comparison of the B3LYP frequencies for water and methanol with experimental values.[22] When calculating deprotonation energies, a constant is added to eliminate the effect of different systematic errors in different computational models. Values of −46 and 0 kJ mol^{-1} for Hartree–Fock and DFT (B3LYP) results, respectively, are obtained from calibration calculations for the related molecules methanol and silanol for which very accurate deprotonation energies are known.[9,23] For further details on QM/MM calculations see ref. 9.

Details of the full DFT (PBE functional + dispersion) slab calculations of the interaction of CO, C_2H_4, NH_3, and NC_5H_5 with acidic sites are given in ref. 8. The OH vibrational wavenumbers for the 2dH system have been calculated from the DFT bond distances making use of the ω/r correlation with anharmonicity corrections proposed by Nachtigall.[24]

3. Results and discussion

Table 1 shows the OH stretching frequency of the H-2dH system and compares it with those of other zeolite frameworks. In the CHA and FAU frameworks there is

Table 1 OH stretching frequencies, ν_{OH} (cm^{-1}), both observed and calculated, and calculated deprotonation energies, E_{DP} (kJ mol^{-1})

Zeolite	Site[a]	ν_{OH}, obsd[b]	ν_{OH}, calcd	E_{DP}[f]	E_{DP}[g]
FAU	O1H	3623	3626[c]	1206	1198[h]
MFI	O7–H	3610	3608[c]	1235	
CHA	O1H	3603	3606[c]	1225	1233[i]
FER	O7–H	3609	3598[d]		1218[j]
2dH	O2–H	3594	3600[e]		1069[i]
CHA	O3–H	3579	3588[c]		
FAU	O3–H	3550	3563[c]	1203	

[a] Crystallographic position. [b] See computational articles for original references to experimental results. [c] Ref. 22. [d] Ref. 25. [e] Ref. 9 – the Ru(0001) surface was included in the model. [f] Ref. 26 and 27 – corrected by −46 kJ mol^{-1}, *cf.* ref. 28. [g] The correction term is zero. [h] Ref. 23 – Al/Si = 1/11 for CHA. [i] Ref. 9 – Al/Si = 1/7 for 2dH, the Ru(0001) surface was not included in the 2dH model. [j] Ref. 25.

only one crystallographic position for substitution of Si by Al, but there are two different oxygen positions at which the protons can sit. This gives rise to a high-frequency and a low-frequency band for H-CHA and H-FAU. Table 1 shows that the result for H-2dH is bracketed by the high- and low-frequencies of H-CHA, which themselves are embraced by the corresponding frequencies for H-FAU. The calculations reproduce the sequence of the experimental frequencies and, on both scales, the H-2dH result is very close to the result for H-ferrierite (H-FER).

Whereas the unperturbed OH group of H-2dH does not show any peculiarity among the zeolitic OH groups, this changes when probe molecules are used. The OH frequency shift measured on adsorption of CO on H-2dH is 379 cm^{-1}, 63 cm^{-1} larger than the shifts for H-CHA.[8] The calculated shift is also larger, but only 23 cm^{-1}.[8] If one uses the relation between frequency shift and deprotonation enthalpy suggested by Paukshtis and Yurchenko,[29,30] these frequency shifts translate into deprotonation enthalpies for H-2dH that are 35 and 12 kJ mol^{-1}, respectively, lower than the values for H-CHA, indicating a higher acidity of the two-dimensional system.[9]

The surprise comes when the deprotonation energy is explicitly calculated. The value for H-2dH is as much as 164 kJ mol^{-1} lower than that for H-CHA (Table 2). Obviously, the correlation between an initial state parameter (OH frequency shift on complex formation with the weak base CO) and a final state parameter (deprotonation energy) breaks down. The exceptionally low value is a true consequence of the low dimensionality of the thin film that results in a much lower effective dielectric constant for the thin film (1.8) than for the bulk material of the same chemical composition (2.9). As a result, in the thin film, the charge created on deprotonation is less screened and better stabilized by electrostatic interactions with the surrounding ions, as illustrated in Fig. 2. The acidity increase will be less strong if the 2d-system is not a free-standing film in vacuum, but a film on a metal substrate,[9] or if the 2d zeolite sheet is not used for gas phase catalysis, but is surrounded by solvent molecules.

Such a high acidity seems to be in contradiction with the protonation ability of Brønsted sites in the H-2dH films. As other zeolites, they are found to protonate NH_3 and NC_5H_5, whereas on adsorption of CO and C_2H_4 the proton is not transferred.[8] Moreover, when looking at the adsorption energy differences for the

Table 2 Deprotonation energies, ΔE_{DP}, binding energy of a protonated base molecule on the deprotonated zeolite surface, ΔE_{IP}, and energies of adsorption, ΔE_{Ads}

Zeolite	H-CHA		H-2dH		Difference	
$\Delta E_{DP}(ZOH)^a$	1234		1069		−165	
S	NH_3	NC_5H_5	NH_3	NC_5H_5	NH_3	NC_5H_5
$-\Delta E_{PA}(S)^b$	854	929	854	929	854	929
$+\Delta E_{IP}$	−539	−487	−375	−285	163	202
$\Delta E_{Ads}(S)^c$	−159	−182	−160	−145	−2	37
Dispersion^c	−27	−66	−29	−47	+2	19
"Active site"^{d}	−131	−116	−131	−98	0	18

^{a} Ref. 9. ^{b} Ref. 31. ^{c} Ref. 8. ^{d} $E_{Ads}(S)$ – dispersion.

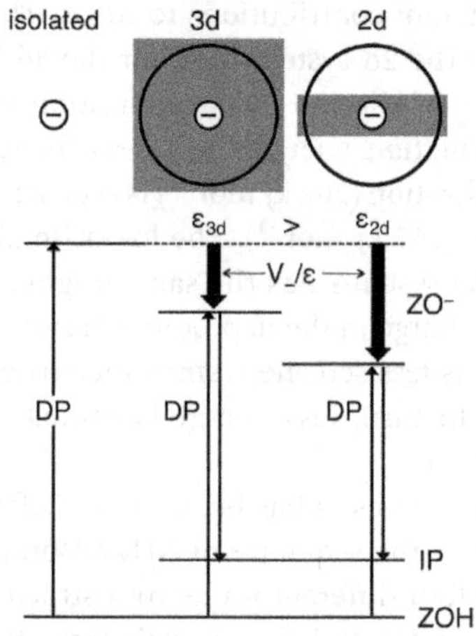

Fig. 2 Comparison of deprotonation and ion-pair formation of an isolated site in the gas phase and a site embedded in a three-dimensional (3d) or two-dimensional (2d) dielectric medium. V_{-}/ε is the interaction of the negative charge with surrounding charges, ε the effective dielectric constant, DP the deprotonation energy, and IP the ion-pair energy.

molecules that undergo protonation, we find almost no change for NH_3, whereas for NC_5H_5 we find a 37 kJ mol^{-1} less favourable reaction energy on the H-2dH film. We can use the dispersion contribution to the binding energy to estimate the surface curvature effect[12] which yields larger energies for adsorption in a micropore as compared to a flat surface (Table 2). If we remove the dispersion contribution, we will get an interaction energy that relates to the "active site". For NH_3 this parameter does not indicate any difference between the Brønsted site reactivity in a 3d bulk material and a 2d material, while for NC_5H_5 it indicates an even higher activity for the 3d bulk material.

How can this be understood in view of the much lower deprotonation energy of the H-2dH system? The energy of adsorption, ΔE_{Ads}, for the formation of an interacting ion pair (eqn (1)) can be decomposed[27,32] into the energy of the interaction, ΔE_{IP}, between the protonated species and the deprotonated surface site,

$$\text{Z–O}^- + \text{HS}^+ \rightarrow \text{Z–O}^-\cdots\text{HS}^+ \tag{2}$$

the deprotonation energy, ΔE_{DP}, of the Brønsted site,

$$\text{Z–OH} \rightarrow \text{Z–O}^- + \text{H}^+ \tag{3}$$

and the proton affinity, ΔE_{PA}, of the base molecule

$$\text{S} + \text{H}^+ \rightarrow \text{HS}^+. \tag{4}$$

Hence,

$$\Delta E_{Ads} = \Delta E_{IP} + \Delta E_{DP} + \Delta E_{PA}. \tag{5}$$

Table 2 shows the different contributions to ΔE_{Ads}. The ion-pair stabilization ΔE_{IP} is much smaller for the 2d system than for the 3d bulk system. For NH_4^+ formation, the difference of the ion-pair stabilization energies outweighs the difference of the deprotonation energies, whereas for HNC_5H_5 formation, the decrease in ion-pair stabilization (202 kJ mol^{-1}) is even larger than the decrease of the deprotonation energy (165 kJ mol^{-1}). The lower (in absolute terms) ion-pair stabilization energy for 2d systems has the same origin as the lower deprotonation energy: the negative charge in the deprotonated system is more stabilized by surrounding charges if it is less screened (small dielectric constant), and it costs less energy to create it. In turn, less energy is gained when annihilating this charge by ion-pair formation.

For different zeolite structures, it has been shown before that the sequence of deprotonation energies and the sequence of NH_3 adsorption energies is not the same (Table 3).[27] For the four different framework structures, the deprotonation energies vary over 29 kJ mol^{-1} and the ion-pair formation energies over 23 kJ mol^{-1}, whereas the energies of NH_3 adsorption vary over 10 kJ mol^{-1} only. The differences of the ion-pair formation energies partially outweigh differences of the deprotonation energies. This general compensation effect is connected with the properties of conjugated acid–base pairs. The anion Z–O$^-$ formed on deprotonation of a stronger acid Z–OH is a weaker base and, hence, binds the "acid" NH_4^+ less strongly. In addition, there are effects connected with the specific local surface structure around the active site.

For the isomorphous substitution of metal ions into microporous aluminium phosphates (AlPO-34) and chabazite, the trend has been confirmed that NH_3 adsorbs more strongly if the deprotonation energy becomes smaller, but the correlation is not quantitative.[21] Although a good correlation between the heat of NH_3 adsorption and the enthalpy barrier for the methylation of propene could be established,[21] this may not always be the case, for example different reactivity sequences of H-MFI and H-MOR have been predicted if substituted amines and pyridines are used as probe molecules.[33]

4. Conclusions

Calculated deprotonation energies indicate a much higher intrinsic acidity for Brønsted sites in two-dimensional aluminosilicates than in bulk zeolites, whereas the energy of ammonia adsorption shows hardly any difference. The difference between the two reactivity parameters is due to the interaction of the NH_4^+ cation formed on protonation of NH_3 with the negatively charged surface site. The differences of this interaction between the two- and three-dimensional case outweigh the differences in the deprotonation energies.

Table 3 Deprotonation energies, ΔE_{DP}, ion-pair formation energies, ΔE_{IP}, and energies of adsorption of NH_3, ΔE_{Ads}[27]

Zeolite	H-FAU	H-CHA	H-MOR	H-MFI
ΔE_{DP}	1206	1225	1230	1235
ΔE_{IP}	−457	−476	−484	−480
ΔE_{Ads}	−109	−109	−116	−106

Whereas the deprotonation energy is a perfect measure of the intrinsic acidity for gas phase reactions, it is not a suitable reactivity parameter for the design of solid acids.

Acknowledgements

This work has been supported by the German Research Foundation (DFG) within the Collaborative Research Center (CRC 1109) and by the "Fonds der Chemischen Industrie". I thank Dr Marcin Rybicki for comments on the manuscript and discussions.

References

1 A. Corma, *Chem. Rev.*, 1995, **95**, 559.
2 A. Corma, *J. Catal.*, 2003, **216**, 298.
3 W. O. Haag, R. M. Lago and P. B. Weisz, *Nature*, 1984, **309**, 589.
4 W. J. Mortier, J. Sauer, J. A. Lercher and H. Noller, *J. Phys. Chem.*, 1984, **88**, 905.
5 M. Choi, K. Na, J. Kim, Y. Sakamoto, O. Terasaki and R. Ryoo, *Nature*, 2009, **461**, 246.
6 Y. Seo, K. Cho, Y. Jung and R. Ryoo, *ACS Catal.*, 2013, **3**, 713.
7 W. J. Roth, P. Nachtigall, R. E. Morris and J. Čejka, *Chem. Rev.*, 2014, **114**, 4807.
8 J. A. Boscoboinik, X. Yu, E. Emmez, B. Yang, S. Shaikhutdinov, F. D. Fischer, J. Sauer and H.-J. Freund, *J. Phys. Chem. C*, 2013, **117**, 13547.
9 M. Rybicki and J. Sauer, *Phys. Chem. Chem. Phys.*, 2015, **17**, 27873.
10 J. A. Boscoboinik, X. Yu, B. Yang, F. D. Fischer, R. Włodarczyk, M. Sierka, S. Shaikhutdinov, J. Sauer and H.-J. Freund, *Angew. Chem., Int. Ed.*, 2012, **51**, 6005.
11 C. J. Dawson, M. A. B. Pope, M. O'Keeffe and M. M. J. Treacy, *Chem. Mater.*, 2013, **25**, 3816.
12 E. G. Derouane, J.-M. Andre and A. A. Lucas, *J. Catal.*, 1988, **110**, 58.
13 J. Sauer, *J. Mol. Catal.*, 1989, **54**, 312.
14 W. E. Farneth and R. J. Gorte, *Chem. Rev.*, 1995, **95**, 615.
15 E. G. Derouane, J. C. Védrine, R. R. Pinto, P. M. Borges, L. Costa, M. A. N. D. A. Lemos, F. Lemos and F. R. Ribeiro, *Catal. Rev. Sci. Eng.*, 2013, **55**, 454.
16 H. Kriegsmann, *Z. Phys. Chem. (Leipzig)*, 1988, **269**, 178.
17 H. Pfeifer, D. Freude and M. Hunger, *Zeolites*, 1985, **5**, 274.
18 U. Fleischer, W. Kutzelnigg, A. Bleiber and J. Sauer, *J. Am. Chem. Soc.*, 1993, **115**, 7833.
19 J. Q. Chen, A. Bozzano, B. Glover, T. Fuglerud and S. Kvisle, *Catal. Today*, 2005, **106**, 103.
20 M. Sierka, U. Eichler, J. Datka and J. Sauer, *J. Phys. Chem. B*, 1998, **102**, 6397.
21 C.-M. Wang, R. Y. Brogaard, B. M. Weckhuysen, J. K. Nørskov and F. Studt, *J. Phys. Chem. Lett.*, 2014, **5**, 1516.
22 M. Sierka and J. Sauer, *J. Phys. Chem. B*, 2001, **105**, 1603.
23 M. Sierka and J. Sauer, *Faraday Discuss.*, 1997, **106**, 41.
24 P. Nachtigall, O. Bludsky, L. Grajciar, D. Nachtigallova, M. R. Delgado and C. O. Arean, *Phys. Chem. Chem. Phys.*, 2009, **11**, 791.
25 V. Nieminen, M. Sierka, D. Y. Murzin and J. Sauer, *J. Catal.*, 2005, **231**, 393.

26 U. Eichler, M. Brändle and J. Sauer, *J. Phys. Chem. B*, 1997, **101**, 10035.
27 M. Brändle and J. Sauer, *J. Am. Chem. Soc.*, 1998, **120**, 1556.
28 J. Sauer, in *Modelling of Structure and Reactivity in Zeolites*, ed. C. R. A. Catlow, Academic Press, London, 1992, p. 183.
29 E. A. Paukshtis, R. I. Soltanov and E. N. Yurchenko, *React. Kinet. Catal. Lett.*, 1981, **16**, 93.
30 E. A. Paukshtis and E. N. Yurchenko, *Russ. Chem. Rev.*, 1983, **52**, 242.
31 E. P. Hunter and S. G. Lias, in *NIST Chemistry WebBook, NIST Standard Reference Database Number 69*, ed. P. J. Linstrom and W. G. Mallard, National Institute of Standards and Technology, Gaithersburg MD, 2005, vol. 20899.
32 M. Brändle and J. Sauer, *J. Mol. Catal. A: Chem.*, 1997, **119**, 19.
33 C. Lee, D. J. Parrillo, R. J. Gorte and W. E. Farneth, *J. Am. Chem. Soc.*, 1996, **118**, 3262.

Faraday Discussions

PAPER

Modelling metal centres, acid sites and reaction mechanisms in microporous catalysts

Alexander J. O'Malley,*[ac] A. J. Logsdail,[ab] A. A. Sokol[a] and C. R. A. Catlow*[abc]

Received 1st February 2016, Accepted 3rd February 2016
DOI: 10.1039/c6fd00010j

We discuss the role of QM/MM (embedded cluster) computational techniques in catalytic science, in particular their application to microporous catalysis. We describe the methodologies employed and illustrate their utility by briefly summarising work on metal centres in zeolites. We then report a detailed investigation into the behaviour of methanol at acidic sites in zeolites H-ZSM-5 and H-Y in the context of the methanol-to-hydrocarbons/olefins process. Studying key initial steps of the reaction (the adsorption and subsequent methoxylation), we probe the effect of framework topology and Brønsted acid site location on the energetics of these initial processes. We find that although methoxylation is endothermic with respect to the adsorbed system (by 17–56 kJ mol^{-1} depending on the location), there are intriguing correlations between the adsorption/reaction energies and the geometries of the adsorbed species, of particular significance being the coordination of methyl hydrogens. These observations emphasise the importance of adsorbate coordination with the framework in zeolite catalysed conversions, and how this may vary with framework topology and site location, particularly suited to investigation by QM/MM techniques.

I. Introduction

Computational methods are now very widely used in catalytic science and are increasingly powerful in obtaining an understanding of catalysis at the molecular level, where they yield models for both the structure and mechanism that assist and complement the interpretation of experimental data. The range and scope of the field is illustrated by the recent review of their applications in zeolite catalytic science given in ref. 1.

[a]*Department of Chemistry, University College London, 20 Gordon St., London WC1 HOAJ, UK*
[b]*Cardiff Catalysis Institute, School of Chemistry, Cardiff University, CF10 3AT, UK*
[c]*UK Catalysis Hub, Research Complex at Harwell, Science and Technology Facilities Council Rutherford Appleton Laboratory, Harwell Science and Innovation Campus, Oxon OX11 0QX, UK*

The majority of contemporary computational studies in catalysis use electronic structure methods, especially Density Functional Theory (DFT) which is very widely applied, employing methods based on periodic boundary conditions (PBCs). Such methods have enjoyed considerable success and have indeed become increasingly predictive. An alternative approach is to use "QM/MM" or embedded cluster methods, which have been extensively used in computational chemistry and biomolecular sciences and in modelling localised states in solids. Such methods describe a finite region (which in the case of catalysis includes the active site, the reacting species and surrounding atoms) by a quantum mechanical method, while describing more distant regions of the solid by a more approximate method, often based on interatomic potentials. They have a number of advantages which will be discussed in greater detail in the following section; but it can be argued that they are inherently more appropriate for modelling a localised state, such as an active site on a surface or in an enzyme, as they focus the computational effort on the region containing the active site rather than the whole molecule or solid.

This paper will present recent applications of QM/MM techniques to the important and topical problem of catalytic methanol conversion in ZSM-5 and where we obtain new information on mechanistic aspects inspired by recent observations from neutron scattering studies.

In the next section we describe in greater detail the basis of the methodologies employed and illustrate briefly their application by reference to earlier studies of metal centres in zeolites. The subsequent section presents our study of methanol/ZSM-5 which demonstrates well the power of the techniques and yields new insight into the interaction of methanol with acidic sites in zeolites.

II. QM/MM techniques

1. Methodology

In our work, we have employed a hybrid QM/MM model of zeolites developed and implemented by Paul Sherwood and colleagues[2–6] in the computational quantum-chemical environment software ChemShell[7–10] as illustrated in Fig. 1.

The model draws on the strengths of simple molecular cluster approaches that terminate a cluster carved out of the material with hydrogen, but also accounts for steric constraints and the elastic response of the framework. The cluster is treated with an appropriate level QM method, which is chosen as a compromise between the required accuracy and available compute resources. The ChemShell zeolite model thus has much in common with alternative subtractive mechanical embedding approaches by Sauer[11–13] (implemented in QM-POT) and Morokuma[14] (ONIOM). The ChemShell methodology, however, goes beyond such schemes and provides an accurate description of the electrostatic interactions between the QM region and the framework, including polarisation and embedding effects. More recently, similar approaches (although more focused on a careful termination of the QM region with specially calibrated link atoms) have been employed, for example, by Nasluzov *et al.*[15,16] (cov-EPE) and Sushko *et al.*[17,18] (GUESS). The native ChemShell zeolite model briefly described in this section is complemented in ChemShell with support for both the QM-POT, cov-EPE and GUESS methods. The flexible, modular approach adopted in ChemShell extends these methodologies to other classes of material, from metal oxides to enzymes. As argued above, these

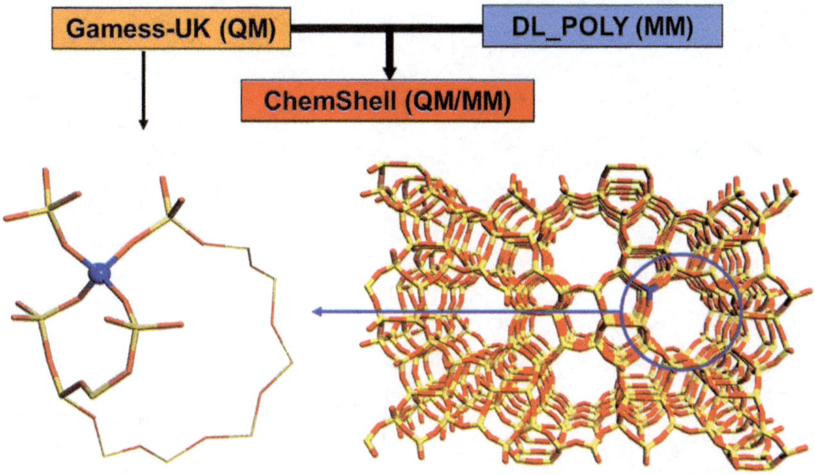

Fig. 1 ChemShell hybrid QM/MM embedded cluster diagram. A QM region incorporates a metal centre serving as an active catalytic site.

schemes serve as a computationally cheaper and at the same time chemically more accurate alternative to popular DFT PBC approaches.

To describe the long-range polarisation effects in these materials, the ChemShell model employs the Hill–Sauer molecular mechanical force-field[19,20] with rigid ions described by modified constant point charges $q_{\mathrm{Si}} = 1.2e$ and $q_{\mathrm{O}} = -0.6e$, and atoms assumed to bind to each other *via* polar covalent bonds. These charges have been fitted to reproduce the electrostatic field in the pores of the zeolitic framework materials described using PBCs. When we define a QM cluster embedded in a siliceous system, we have to cut through Si–O bonds that connect atoms, one in the QM and the other in the MM regions. As under-coordinated QM atoms on the QM region boundary are terminated with hydrogens, an artificial strain on the system is introduced due to spurious interactions between the hydrogen and the nearest neighbours in the MM region. To remedy this artefact, a charge on the corresponding MM atoms is shifted outwards, further onto the next nearest neighbours, whereas the dipole created by such a shift is compensated locally by a pair of point charges placed along the corresponding bonds. Further information on the method can be found in the original publications, and relevant model construction details are given below.

Next, we provide a short outline of our earlier work on metal centres in zeolites, which illustrate the power of our approach and its potential in the elucidation of the structure and properties of catalytically active sites in zeolites.

2. Applications to metal centre in zeolites

Incorporation of metal sites in zeolitic frameworks has been the subject of numerous investigations. In particular, Ti doped silicalite-1 (TS-1),[21] a siliceous counterpart of ZSM-5, has been of great interest, since it is used with aqueous H_2O_2 as the primary oxidant in a large number of important reactions such as the epoxidation of linear olefins, oxidation of linear alkenes to alcohols and ketones,

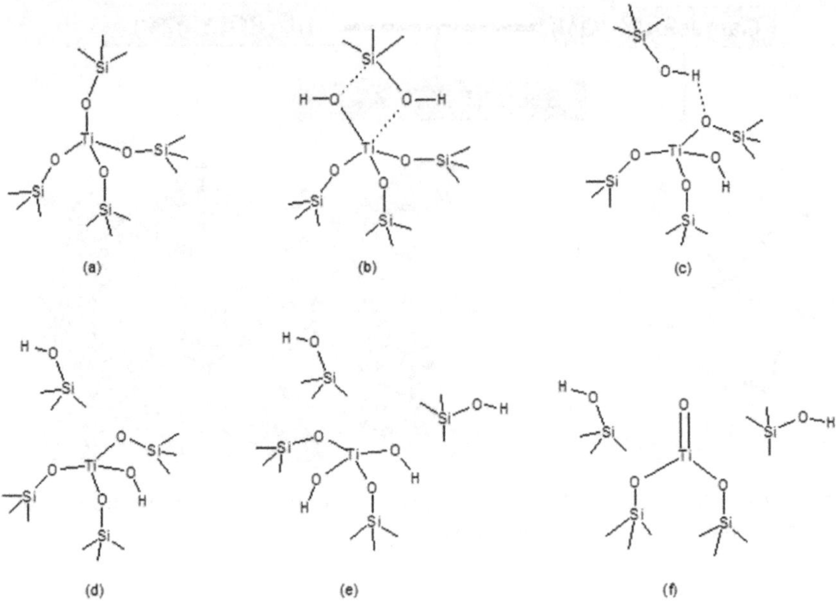

Fig. 2 The six models used to represent the active Ti sites in TS-1: (a) tetrapodal, (b) 2MR, (c) tripodal (1I), (d) tripodal (2I), (e) bipodal, and (f) titanyl.

oxidation of alcohols, hydroxylation of aromatics, oxidation of amines, and oxidation of sulphur compounds and ethers.[22]

In our work,[23–25] we have proposed a number of structural models for a Ti species that would result from the framework hydrolysis and site inversion that can take place during the synthesis and post-synthetic treatment and/or, depending on the catalytic process, occur as a part of the catalytic transformation. The models of interest are shown in Fig. 2.

The geometrical parameters of the tripodal sites obtained with ChemShell calculations proved to be in excellent agreement with experimental EXAFS data,[26] and therefore could be usefully employed to rationalise the difference in the observed spectroscopic signature of metal centres in differently treated samples, as summarised in Table 1.

Our analysis of site transformation in the presence of water, illustrated in Fig. 3,[27] allowed us to draw the crucial conclusion that the catalytically active sites in TS-1 are hydrated tripodal rather than tetrapodal, which is strikingly similar to Ti doped mesoporous materials that had been previously explored using simple cluster models.[28]

Further, we consider the site reactivity towards H_2O_2 in anhydrous and hydrous conditions according to the reaction pathways drawn in Fig. 4 and 5.[29]

Our calculations confirmed the close competition between the η^1 and η^2 complexes as candidates for the oxygen donating species in TS-1. Curiously, a much higher energy (by *ca.* 400–450 kJ mol^{-1}) superoxide radical species has also been reported (from EPR studies) to emerge as a result of interaction of aqueous H_2O_2 and non-aqueous urea H_2O_2.[30] Our calculations unambiguously identified the superoxide supported on tripodal Ti as the magnetically active

Table 1 Comparison of selected structural parameters of (a) the tripodal (1I) and (b) tripodal (2I) Ti models

(a)

	Ti–O distance [Å]		Ti–Si distance [Å]		Ti–O–Si angle [°]	
	Tripodal (1I)	Expt.[a] untreated	Tripodal (1I)	Expt.[a] untreated	Tripodal (1I)	Expt.[a] untreated
Ti	1.76	1.81 ± 0.02	3.16	3.18 ± 0.02	133	140 ± 5
	1.79		3.27	3.30 ± 0.02	153	152 ± 5
	1.83		3.37	3.39 ± 0.02	164	167 ± 5
	1.75[b]					

(b)

	Ti–O distance [Å]		Ti–Si distance [Å]		Ti–O–Si angle [°]	
	Tripodal (2I)	Expt.[a] NH_4Ac treated	Tripodal (2I)	Expt.[a] NH_4Ac treated	Tripodal (2I)	Expt.[a] NH_4Ac treated
Ti	1.78	1.81 ± 0.02	3.16	3.23 ± 0.02	137	142 ± 5
	1.79		3.16	3.23 ± 0.02	137	142 ± 5
	1.80		3.34	3.37 ± 0.02	157	161 ± 5
	1.75[b]					

[a] Ref. 26. [b] Refers to the Ti–OH bond.

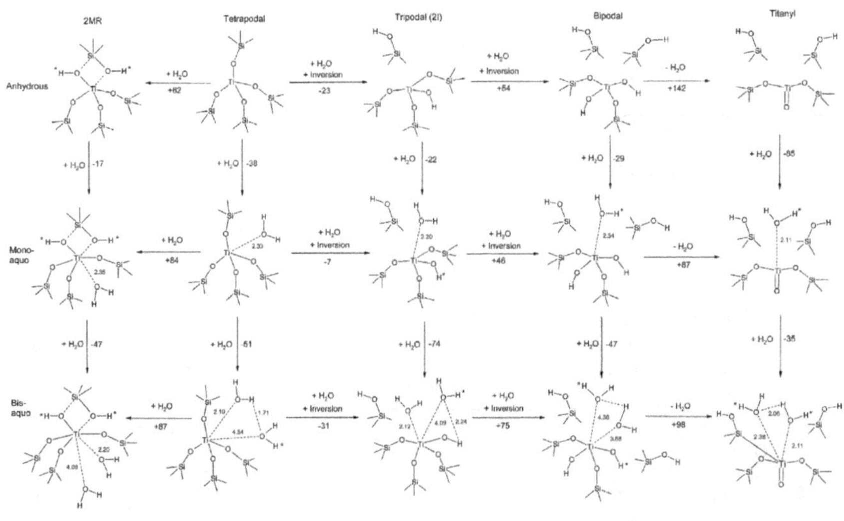

Fig. 3 Calculated energy differences in kJ mol^{-1} of five models chosen as representatives of Ti sites in TS-1, with and without one and two molecules of water. The dashed lines highlight the bonding between Ti and the oxygen atoms in water and the hydrogen bonding within the framework. *H indicate atoms that are hydrogen bonding with framework oxygen atoms.

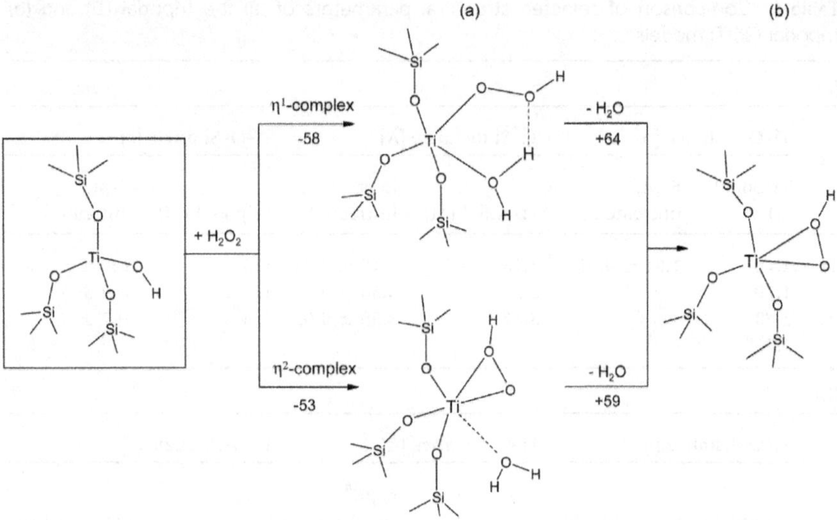

Fig. 4 Reaction pathways for the Ti–η^1(OOH) and Ti–η^2(OOH) formation (a) with and (b) without one water. Energies in kJ mol^{-1}.

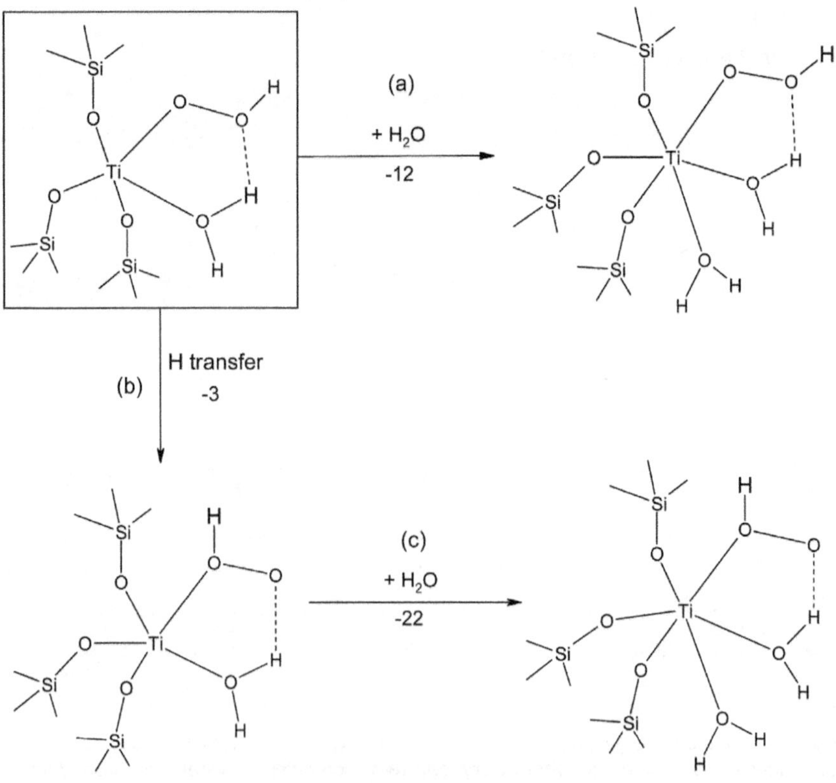

Fig. 5 Reaction pathways for the formation of (a) six-coordinated Ti–η^1(OOH), (b) five-coordinated Ti–η^1(OOH) and (c) six-coordinated Ti–η^1(O(H)OH) species. Energies in kJ mol^{-1}.

Table 2 EPR spin Hamiltonian parameters of the Ti(OO•) radical species

g tensor	Calculated (B3LYP) at BB1K geometry	Experiment[a]	
		Aqueous H_2O_2	Urea–H_2O_2
g_x	2.003	2.002	2.004
g_y	2.010	2.009	2.010
g_z	2.025	2.026	2.028

[a] Ref. 30.

centre with the spectroscopic signature in close agreement with the experiment as shown in Table 2.

From this short outline, we conclude that the hybrid QM/MM methodology implemented in ChemShell provides us with a reliable tool to investigate both the structure and physical and chemical properties of active sites in zeolites. In the next section we employ ChemShell in a study of methoxylation of two zeolites with very different behaviour.

III. Methanol-active site interactions in zeolites

The zeolite catalysed conversion of methanol to hydrocarbons (MTH) has been researched significantly since its initial discovery by Chang and Silvestri,[31] with the first commercialization of the MTH process in New Zealand in 1985.[32] Demand for light olefins has also lead to extensive research into the use of zeolitic catalysts for converting methanol to olefins (MTO).[33]

An early step of interest in the reaction mechanism is the formation of framework methoxy species after initial physisorption of methanol through H-bonding to the zeolite Brønsted acid site.[34–36] However, studying such species experimentally is difficult, as secondary reactions dominate rapidly at higher temperatures.[37] Recent studies using neutron scattering[38] have observed the total conversion of methanol to framework methoxy at room temperature in commercial samples of H-ZSM-5 (in contrast to previous experiments discussed in ref. 38, which suggest that elevated temperatures are necessary), which is, however, not observed in commercial samples of H-Y with the same Si/Al ratio. This difference was attributed to the necessary dealumination procedure for high silica H-Y synthesis (creating defects which decrease the activity). However, detailed theoretical studies into the effect of framework topology on methanol behaviour in zeolite catalysts are necessary for improved understanding and further catalyst design. A short discussion of theoretical studies of methanol behaviour in zeolite catalysts will now follow.

Though the MTG/MTO process has received a significant amount of attention from a theoretical perspective,[39,40] the aforementioned earlier steps are not understood in detail. *Ab initio* studies of methanol adsorption began with small aluminosilicalite clusters.[41,42] Numerous studies using three tetrahedral site clusters have concluded that the hydrogen bonding of the methanol oxygen with the Brønsted acid hydroxyl is exothermic, generally calculating adsorption energies of between 50 and 100 kJ mol^{-1}.[35,36,43–45] It has also been found that there is an energetic difference between 'side-on' and 'end-on' adsorption, with the end-on geometry (where the methyl group protrudes almost perpendicularly from the aluminosilicate cluster) calculated to be more favourable, and methoxylation taking place *via* the rapid equilibration between the end-on and side-on geometry.[43,46] All studies into the methoxylation process using clusters conclude a barrier to methoxylation with values of ~180–225 kJ mol^{-1}.[35,36,43,45,46] Methoxylation and adsorption has also been studied in periodic zeolite structures, with static calculations showing endothermic methoxylation in chabazite (with a barrier of 56 kJ mol^{-1}),[47] and DFT-MD giving an adsorption energy of −94 kJ mol^{-1}.[48] Andzelm and Govind[49] studied the methoxylation process in a periodic ferrierite structure, finding a methoxylation barrier of 226 kJ mol^{-1}. However, it was found that when a second methanol molecule was added to the system, the barrier to methoxylation was then reduced by 42 kJ mol^{-1} due to the formation of hydrogen bonding networks spanning the width of the channel, which significantly stabilise the methoxylation transition states.

We note that the use of a discrete cluster model is less accurate than the use of a periodic zeolite structure, due to the lack of any long-range electrostatic potential influencing the system energetics. However, as noted earlier, in addition to being computationally expensive in calculations involving large unit cells, the use of periodic boundary conditions can bring problems when calculating reaction barriers and adsorption energetics, such as incomplete relaxation of the system or the influence of the interaction between sorbate mirror images. This problem can be especially significant when comparing the effect of the framework structure on sorbate behaviour. A large zeolite unit cell would minimise the effect of these sorbate–sorbate interactions on the methoxylation process compared to, say, chabazite where they may become significant. These complications can be bypassed using the quantum mechanical/molecular mechanical (QM/MM) embedded cluster technique[9,50] outlined in the previous section, where large framework structures may be modelled relatively cheaply, providing an accurate electrostatic potential that will affect the behaviour and reactivity at an active site.

We present a comparison of the deprotonation, methanol adsorption, and methoxylation energies in H-ZSM-5 and H-Y zeolite frameworks, in an attempt to investigate the effect of framework topology and active site location on reactive sorbate behaviour. The deprotonation energy is first investigated as it is a major factor controlling the Brønsted acidity of the differing frameworks; our calculations include a comparison of three different sites in the ZSM-5 structure. The adsorption energy of methanol at these sites is then compared to quantify the differences in interaction strength between methanol and the differing acidic hydroxyl sites, after which the methoxylation energy is then calculated to quantify the effect of these differing environments on the reaction.

1. Methodology

We first discuss the construction of the embedded cluster models for the H-Y and H-ZSM-5. We begin with the purely siliceous spherical structure generated from the periodic experimental unit cell structure of silicalite[51] and siliceous faujasite.[52] The use of the experimental unit cell is assumed to have minimal effects on the geometric and electronic structure. Each spherical cluster centres on a Si T-site.

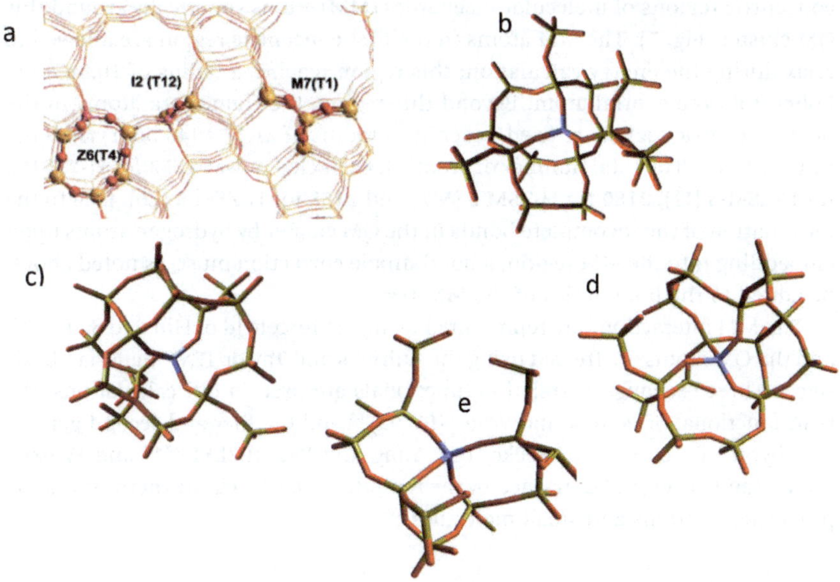

Fig. 6 (a) The locations of the I2, Z6 and M7 clusters in the MFI framework (reproduced from ref. 54 with permission from the PCCP Owner Societies), and the optimised QM clusters for the (b) I2, (c) M7, (d) Z6 sections of the MFI framework, along with the optimised QM cluster in the zeolite Y calculations (e). The central T-site is coloured purple.

Only one cluster is generated for the H-Y structure as the tetrahedral sites are equivalent, which, however, is not the case for H-ZSM-5. Therefore, three embedded clusters are generated based on three sites in the ZSM-5 structure:[53] a T-site at the straight channel (M7-T1), the sinusoidal channel (Z6-T4) and the channel intersections (I2-T12). The clusters, including their locations in the MFI structure are shown in Fig. 6.[54]

The QM region is expanded from the central atom to include the fifth neighbour, *i.e.* the third oxygen atom away from the T-site as shown in Fig. 6. The cluster is then terminated with hydrogen atoms. We also include two

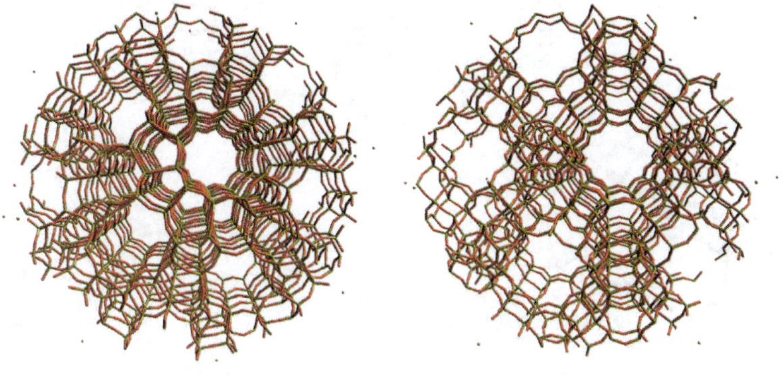

Fig. 7 Spherical embedded cluster models of the I2-T12 centred H-ZSM-5 system (left) and the zeolite Y system (right).

concentric regions of molecular mechanics (MM) atoms spherically around this QM cluster (Fig. 7). The MM atoms in the first concentric region are allowed to relax during the energy calculation; this region reaches a radius of 10.58 Å (20 bohr) from the central atom. Beyond this region, the framework atoms in the next concentric region are fixed, reaching a radius of 21.17 Å (40 bohr) from the central T-site. The total number of atoms in each cluster was 1653 for H-Y, 2165 for H-ZSM-5 (I2), 2180 for H-ZSM-5 (M7) and 2155 for H-ZSM-5 (Z6). Due to the termination of the incomplete bonds in the QM cluster by hydrogen atoms upon embedding into the MM region, a bond-dipole correction must, as noted above, be added to the boundaries of the MM region.[5]

The MM interactions are represented using the forcefield of Hill and Sauer,[19,20] and the QM atoms are treated using the Ahlrichs and Taylor TZVP Gaussian basis sets.[55] Three exchange–correlation functionals are used in our calculations: the GGA functional of Perdew and Wang (PW91),[56] and the first and second generation hybrid functionals of Becke, Lee, Yang and Parr (B3LYP)[57–60] and Wilson, Bradley and Tozer (B97-2) trained by the reproduction of accurate thermodynamic properties of atoms and small molecules.[61]

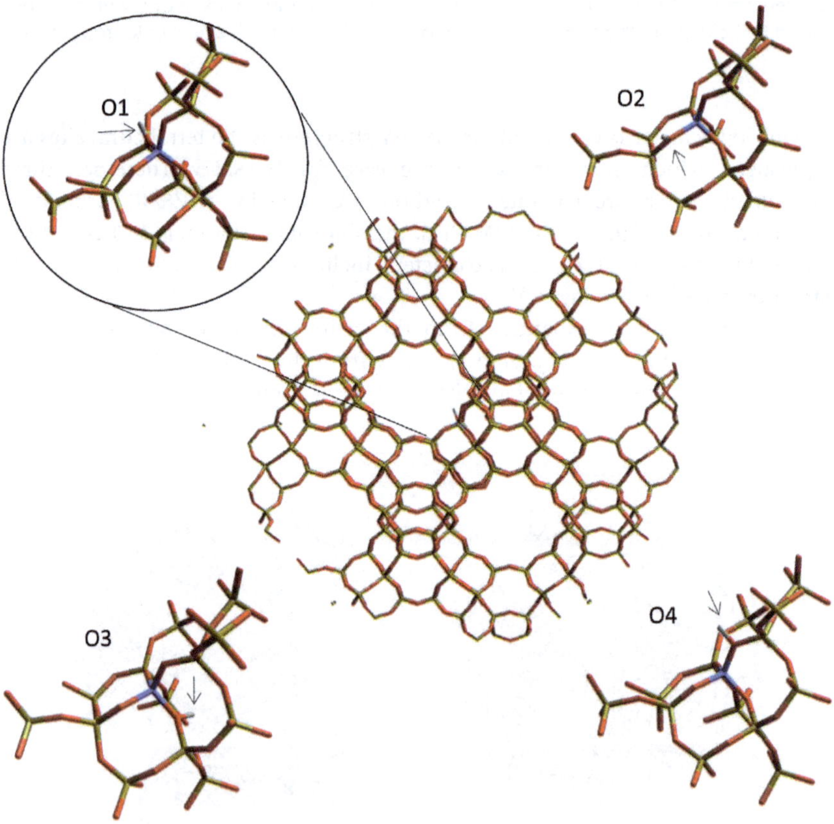

Fig. 8 Oxygen sites from which the deprotonation energy is calculated around the aluminium substitution in the H-Y cluster.

The calculation of the energy and gradients of the QM region used the Gamess-UK package,[62] and the calculation of the energy and gradients of the MM region employed the DL_POLY package,[63] with the QM and MM calculations coupled in the ChemShell environment.[53] Electronic minimisation was deemed complete for the QM system when the energy variation was below a threshold of 2.72×10^{-6} eV was achieved (1×10^{-7} Ha). The geometries were optimised using the Limited-Memory Broyden–Fletcher–Goldfarb–Shanno (L-BFGS)[64–67] algorithm with a convergence threshold of 0.015 eV Å^{-1}.

2. Results and discussion

2.1. Calculation of deprotonation energies. In our first comparison of H-Y and H-ZSM-5, we investigate the deprotonation energy differences between the structures. The deprotonation energy influences acidity[68–70] as a lower deprotonation energy will favour protonation of any sorbate. We begin by determining the oxygen site around the central aluminium T-site, that has the lowest deprotonation energy. This site will then be used for further studies into the methanol adsorption, and the later methoxylation studies. We define the deprotonation energy as:

$$E_{\mathrm{dp}} = E_{\mathrm{zeoH}} - E_{\mathrm{zeo}^-}. \quad (1)$$

Here, E_{zeoH} is the energy of the protonated zeolite and E_{zeo^-} is the energy of the deprotonated zeolite. Polarisation effects outside of our active region, where all

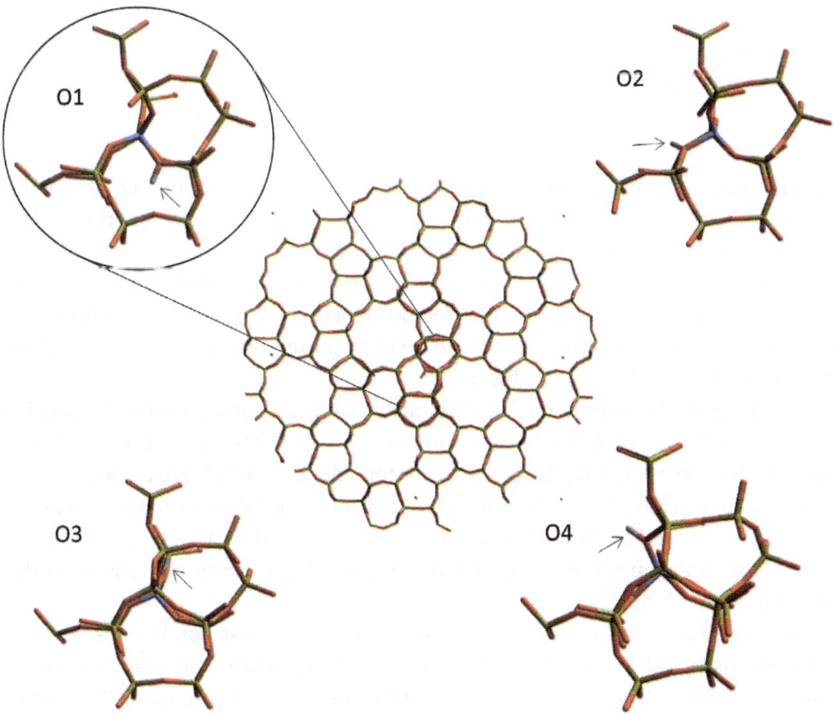

Fig. 9 Oxygen sites around the aluminium substitution in the H-ZSM-5 (I2-T12) cluster for which the deprotonation energy is calculated.

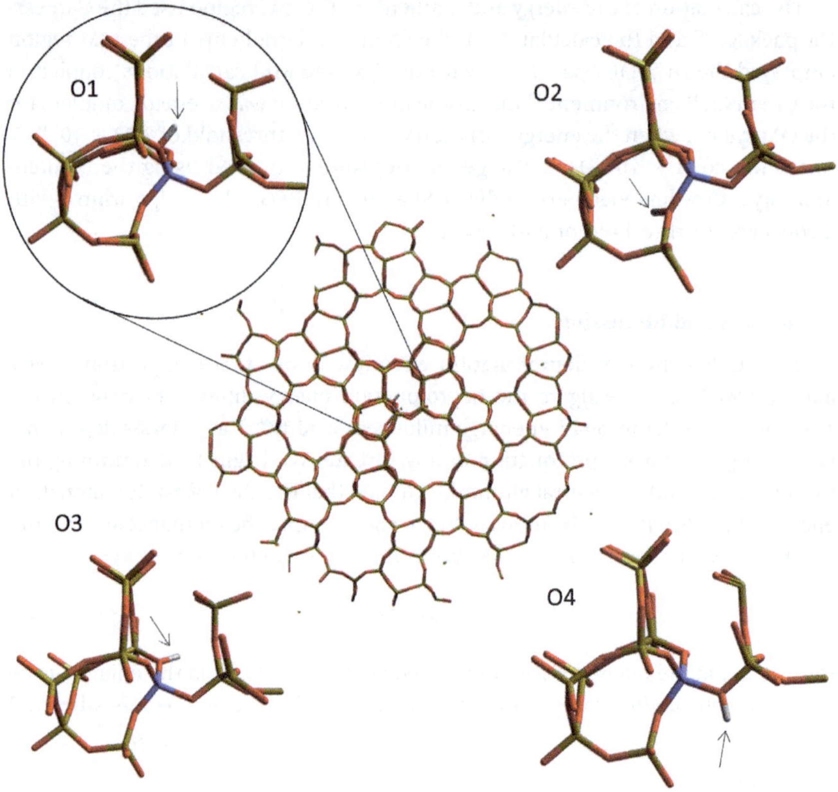

Fig. 10 Oxygen sites around the aluminium substitution in the H-ZSM-5 (M7-T1) cluster for which the deprotonation energy is calculated.

atoms relax during geometry optimisation, are neglected in the following discussion as they represent a relatively constant shift in the deprotonation energies for all Brønsted sites considered (< 50 kJ mol^{-1} using Born and Jost's formulae[9,71]). Fig. 8–11 show geometries of the four protonated systems around the central Al T-site in each zeolite system, labelled O1–O4. The deprotonation energies of each oxygen site were calculated using the PW91 exchange correlation functional and are listed in Table 3.

Using the PW91 functional, the most acidic oxygen sites are O1 for H-Y, O3 for H-ZSM-5 (I2), O3 for H-ZSM-5 (M7) and O2 for H-ZSM-5 (Z6). The difference between the lowest and highest deprotonation energies, which are respectively H-Y and H-ZSM-5 (M6), is 90.2 kJ mol^{-1}. H-Y has the lowest deprotonation energy of the series, followed by ZSM-5 (Z6) located in the sinusoidal channel, ZSM-5 (I2) in the intersections, and the T-site with the highest deprotonation energy is in the straight channel (M7).

We now study the deprotonation energies of the most acidic sites in each structure using the B3LYP and B97-2 hybrid exchange–correlation functionals, as listed in Table 4. The deprotonation energies calculated using the B3LYP functional were higher than when using the PW91 functional, and higher still using the B97-2 functional, which can be attributed to the superior representation of the

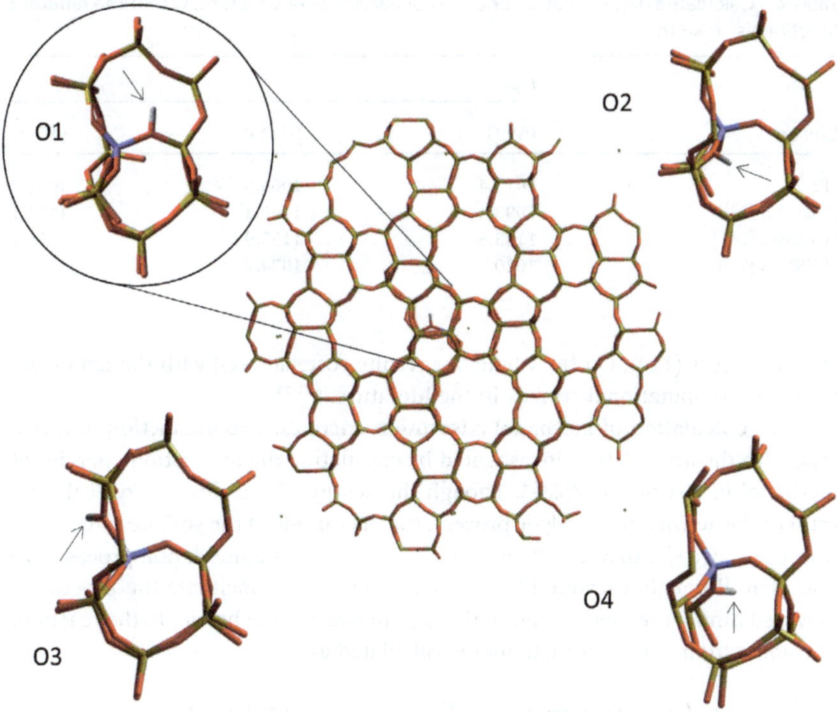

Fig. 11 Oxygen sites around the aluminium substitution in the H-ZSM-5 (Z6-T4) cluster for which the deprotonation energy is calculated.

Table 3 Calculated deprotonation energies of each zeolite structure, with the most acidic oxygen atoms highlighted in bold

	E_{dp} kJ mol^{-1}			
Cluster	O1	O2	O3	O4
HY	**1013.4**	1041.0	1031.8	1023.0
H-ZSM-5 (I2)	1099.2	1106.3	**1093.4**	1122.0
H-ZSM-5 (M7)	1123.6	1114.7	**1103.6**	1109.5
H-ZSM-5 (Z6)	1059.5	**1020.9**	1042.9	1061.0

localised electrons in the zeolite due to the exact exchange interaction from Hartree–Fock methods incorporated into the hybrid functionals. However, the ranking of the deprotonation energy is maintained as H-Y < H-ZSM-5 (Z6) < H-ZSM-5 (I2) < H-ZSM-5 (M7) for all functionals. The deprotonation energy of 1093.4 kJ mol^{-1} calculated for ZSM-5 (I2) is 122 kJ mol^{-1} lower than obtained by *ab initio* molecular dynamics for the MFI structure,[70] using the same PW91 functional. Possible reasons for this difference could be the aforementioned sorbate–sorbate interactions, the more complete representation of polarisation afforded using the QM/MM method, or the charge compensation parameters needed for periodic DFT calculations. We also note that ref. 70 used a different intersectional T-site

Table 4 Calculated deprotonation energies of each zeolite structure, with three different functionals, in kJ mol^{-1}

	E_{dp}		
Cluster	PW91	B3LYP	B97-2
HY	1013.4	1065.9	1081.3
H-ZSM-5 (I2)	1093.4	1100.4	1114.1
H-ZSM-5 (M7)	1103.6	1155.9	1166.1
H-ZSM-5 (Z6)	1020.9	1084.7	1101.9

(T7) from ours (T12). On the whole our results correlate well with the data from the other computational studies in the literature.[16,72,73]

2.2. Calculation of methanol adsorption energies. The interaction of methanol with the active site is investigated by calculating the adsorption energies of methanol in H-Y and H-ZSM-5. Though the acidity of a zeolite is a typical indicator of the activity in a catalytic process, the magnitude of the sorbate interaction with the active site may also be a deciding factor in a methoxylation process. We note that although a favourable adsorption energy can facilitate the process, it may also hinder the methoxylation through increasing the barrier to the reaction. The adsorption energy of methanol is calculated as:

$$E_{ads} = (E_{MeOH+ZeoH} - E_{BSSE}) - (E_{MeOH} + E_{ZeoH}), \quad (2)$$

where E_{MeOH} is the calculated energy of a methanol molecule *in vacuo*, E_{ZeoH} is the energy of the protonated embedded zeolite cluster, $E_{MeOH+ZeoH}$ is the energy of the methanol adsorbed onto the protonated embedded zeolite cluster, and E_{BSSE} is the basis set superposition error calculated using the counterpoise correction method.[74] The geometry of the adsorbed methanol for each cluster, after optimisation with the B3LYP functional, is shown in Fig. 12.

We have studied the 'side-on' adsorption in all zeolite systems. Though we note in previous studies that 'end-on' adsorption is preferred,[43,46] the side-on adsorption is more conducive to methoxylation, as is the emphasis of this study, and the barriers to this switching between the geometries are very small and may be overcome under experimental conditions. From Fig. 12 it is observed that the H-bond length between the hydroxyl site and the methanol hydroxyl is 1.476 Å for the H-Y cluster and 1.452–1.478 Å for the H-ZSM-5 clusters, with the shortest H-bond being in the M7-T1 cluster despite having the highest deprotonation energy. However, we note that the range of bond distances of 0.026 Å is quite small and so a correlation with deprotonation energies is not necessarily meaningful.

We should note that the BSSE values were calculated with the B3LYP functional only. As an approximation, we have then taken the percentage error (5.5%) that these values contribute to each system and applied it to the adsorption energy calculations using the PW91 and B97-2 functionals. The calculated adsorption energies are listed in Table 5 after subtraction of this percentage error.

For all functionals, the order of methanol adsorption energy is H-ZSM-5 (M7) < H-Y < H-ZSM-5 (I2) < H-ZSM-5 (Z6). We note that the difference in adsorption between the most favourable site, H-ZSM-5 (M7), and the least favourable site, H-ZSM-5 (Z6), is between 60 and 70 kJ mol^{-1}, depending on the functional used. The

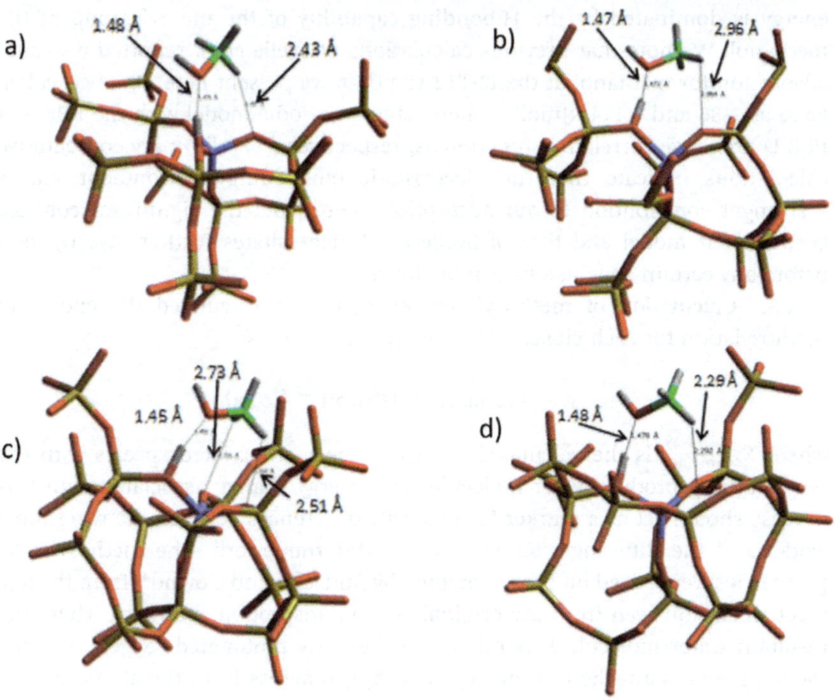

Fig. 12 Optimised adsorbed geometry of methanol in (a) H-Y, (b) H-ZSM-5 (I2), (c) H-ZSM-5 (M7), and (d) H-ZSM-5 (Z6).

methanol in the H-ZSM-5 (M7) system is H-bonded to two framework oxygens by two C–H hydrogens (depicted in Fig. 12c), suggesting that this geometry is more favourable than a short, single C–H···O bond, as is demonstrated by the less favourable adsorption energy in the H-ZSM-5 (Z6) system. It may be significant that the most favourable adsorption process also occurs at the site with the highest deprotonation energy, but we also see that the site with the lowest deprotonation energy (H-Y) has the second lowest adsorption energy. The calculated adsorption energy for methanol in ZSM-5 (M7) appears anomalous when considering the overall correlation between deprotonation and methanol adsorption energies for FAU and MFI, however it is perhaps more reasonable to consider that there should not be a correlation at all, and in fact the adsorption

Table 5 Corrected adsorption energies of methanol (E_{ads}), in kJ mol^{-1}, using three different functionals

Cluster	PW91	B3LYP	B97-2
HY	−180.2	−179.6	−180.4
H-ZSM-5 (I2)	−186.8	−177.5	−168.9
H-ZSM-5 (M7)	−215.1	−194.8	−206.9
H-ZSM-5 (Z6)	−147.24	−136.6	−136.7

energy is dominated by the H-bonding capability of the methyl group of the methanol. We note that previous calculations of Svelle *et al.* reported a weaker adsorption for methanol at the I2-T12 site than we present here: E_{ads} was calculated as −86 and −114 kJmol^{-1} when using a periodic model with the PBE and PBE-D exchange-correlation functionals, respectively.[75] Preliminary comparative calculations indicate that the electrostatic embedding environment makes a stronger contribution to our adsorption energy, but the significant contrast between our model and that of Svelle *et al.* necessitates further investigation before any certain conclusions can be drawn.

2.3. Calculation of methoxylation energies. We calculated the energy of methoxylation for each cluster. The energy is defined as:

$$E_{\text{methox}} = E_{\text{ZeoMe(aq)}} - (E_{\text{MeOH}} + E_{\text{ZeoH}}), \tag{3}$$

where $E_{\text{ZeoMe(aq)}}$ is the calculated energy of the methoxylated species with the neighbouring product water molecule. The energy change associated with this process should act as a marker for reactivity differences between the two frameworks and the differing sites within the MFI framework. The methoxylation geometries were based on those obtained by Andzelm and Govind[49] from the S_N2 mechanism followed from the original side-on adsorption geometry, where the resultant water molecule H-bonds to the formerly protonated oxygen, and the methyl group is attached to the adjacent oxygen across from the aluminium T-site. The geometries were optimised and energies calculated using solely the B3LYP functional. The values of E_{methox} are listed in Table 6, along with the difference between E_{ads} and E_{methox}. The optimised methoxylation geometries are shown in Fig. 13.

Firstly, we note that all the methoxylation energies are exothermic overall but are endothermic with regard to the adsorbed methanol molecule, as observed in previous cluster simulations[35] and in the periodic chabazite structure.[42] We observe that the order of the methoxylation energy is H-Y < H-ZSM-5 (I2) < H-ZSM-5 (M7) < H-ZSM5-I2 (Z6), in agreement with deprotonation energies but in contrast to the order of methanol adsorption energies shown in the previous section. This suggests that the tendency of methoxylation is not related directly to the adsorption energy, at least when only one methanol molecule is present in the immediate environment.

Finally, we turn our attention to the geometries of the methoxylated system. The most exothermic methoxylation takes place in H-Y, which is also the more coordinated system: the water molecule is coordinated to three framework oxygen

Table 6 Methoxylation energies (in kJ mol^{-1}) of the four zeolite systems and the energy difference between the adsorbed state and the methoxylated state

System	E_{methox}	$E_{\text{ads}} - E_{\text{methox}}$
Me-Y + H_2O	−162.0	+17.6
Me-ZSM-5 (I2) + H_2O	−150.4	+27.1
Me-ZSM-5 (M7) + H_2O	−138.6	+56.2
Me-ZSM-5 (Z6) + H_2O	−118.0	+18.6

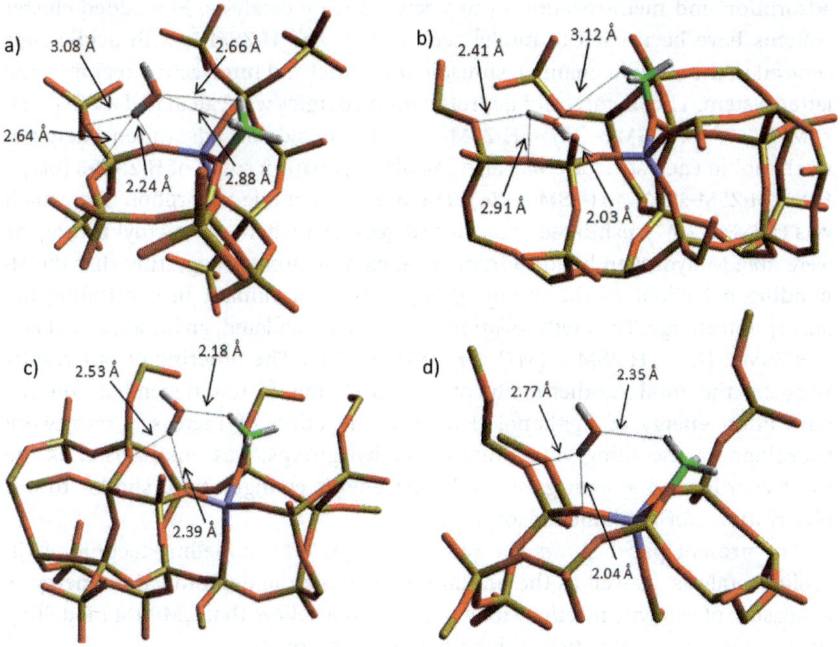

Fig. 13 Optimised geometries of the methoxylated structures of (a) HY (b) H-ZSM-5 (I2), (c) H-ZSM-5 (M7), and (d) H-ZSM-5 (Z6).

sites. Threefold coordination is also the case for the ZSM-5 (I2) system, which has the second most exothermic methoxylation energy, but for H-Y the water molecule is also coordinated to two of the methyl hydrogens. The H-bonds associated with the H-Y system are not the shortest of all the configurations: the shortest is 2.24 Å, compared to 2.04 Å in the ZSM-5 (Z6) system. However, the longer H-bonds in H-Y are compensated by their increased quantity, thus highlighting the importance of highly coordinated structures in stabilising the system when considering the interaction of sorbates such as methanol with the active sites of acidic zeolite catalysts.

Using the QM/MM methodology we have, therefore, obtained new insights into the interaction, and reaction, of methanol with zeolite acidic sites and discussed how this may change with framework structure and aluminium location. Future work will use the embedded cluster models to investigate the barriers of methoxylation in zeolites using transition state searches. The effect of including multiple methanol molecules will also be investigated as the formation of H-bonded networks reduces the total methoxylation barrier for the FER framework,[49] thus further emphasising the importance of highly coordinated systems in controlling the barrier of such reactive processes.

IV. Summary and conclusions

QM/MM methods clearly have a major role to play in the study of catalytic systems. Earlier work demonstrated their efficacy in modelling active sites in zeolites. Here, we investigated the energetics of deprotonation, methanol adsorption and methoxylation in two acidic zeolite catalysts. Embedded cluster systems have been used to model zeolites H-Y and H-ZSM-5, with acidic sites centred in the straight channel, sinusoidal channel and intersection region of the latter system. The hierarchy of deprotonation energies was calculated as H-Y < H-ZSM-5 (Z6) < H-ZSM-5 (I2) < H-ZSM-5 (M7). The side-on adsorption energy of methanol in each structure was also calculated, giving a trend of H-ZSM-5 (M7) < H-Y < H-ZSM-5 (I2) < H-ZSM-5 (Z6). The most favourable adsorption site, which was H-ZSM-5 (M7), exhibited an adsorbed geometry where two methyl hydrogens were able to hydrogen bond to framework oxygen atoms, suggesting that the H-bonding behaviour of the methyl group may be dominant in controlling the adsorption energy. The methoxylation energy was calculated, giving a trend of H-Y < H-ZSM-5 (I2) < H-ZSM 5 (M7) < H-ZSM5-I2 (Z6). The ordering of our results suggests the total exothermicity of methoxylation is not dependent on the adsorption energy of methanol. However, the effect of increased framework coordination, including H-bonding of methyl groups, was emphasised as the most coordinated system gave the lowest energy configuration, similar to our observations for methanol adsorption.

The present paper shows the role of the QM/MM modelling techniques in zeolite catalysis, as well as the limitations in using the deprotonation energy as a measure of catalytic reactivity for a material. We believe that QM/MM modelling in catalytic science can, however, be made much broader.

Acknowledgements

We are indebted to our early collaborator Judy To for her work on metal doped zeolites, reviewed in this paper, and ChemShell developers Paul Sherwood and Tom Keal for support with ChemShell calculations. *Via* our membership of the UK's HEC Materials Chemistry Consortium, which is funded by the EPSRC (EP/L000202), this work used the ARCHER UK National Supercomputing Service (http://www.archer.ac.uk). The authors acknowledge further the use of the UCL Legion High Performance Computing Facility (Legion@UCL), and associated support services, in the completion of this work. The authors would like to acknowledge the Engineering and Physical Sciences Research Council, Grant No. EP/K014668/1 and Grant No. EP/G036675/1 for financial support under their Centres for Doctoral Training scheme, the STFC and the ISIS Pulsed Neutron and Muon Source for funding and sponsorship. AJL is grateful to the Ramsay Memorial Fund and UCL Chemistry for research fellowship funding. UK Catalysis Hub is kindly thanked for resources and support provided *via* our membership of the UK Catalysis Hub Consortium and funded by the EPSRC (grants EP/K014706/1, EP/K014668/1, EP/K014854/1 EP/K014714/1 and EP/M013219/1).

References

1 V. Van Speybroeck, K. Hemelsoet, L. Joos, M. Waroquier, R. G. Bell and C. R. A. Catlow, *Chem. Soc. Rev.*, 2015, **44**, 7044–7111.

2 S. P. Greatbanks, P. Sherwood and I. H. Hillier, *J. Phys. Chem.*, 1994, **98**, 8134–8139.

3 S. P. Greatbanks, P. Sherwood, I. H. Hillier, R. J. Hall, N. A. Burton and I. R. Gould, *Chem. Phys. Lett.*, 1995, **234**, 367–372.

4 S. P. Greatbanks, I. H. Hillier, N. A. Burton and P. Sherwood, *J. Chem. Phys.*, 1996, **105**, 3770–3776.
5 P. Sherwood, A. H. de Vries, S. J. Collins, S. P. Greatbanks, N. A. Burton, M. A. Vincent and I. H. Hillier, *Faraday Discuss.*, 1997, **106**, 79–92.
6 S. P. Greatbanks, I. H. Hillier and P. Sherwood, *J. Comput. Chem.*, 1997, **18**, 562–568.
7 P. Sherwood, A. H. de Vries, M. F. Guest, G. Schreckenbach, C. R. A. Catlow, S. A. French, A. A. Sokol, S. T. Bromley, W. Thiel, A. J. Turner, S. Billeter, F. Terstegen, S. Thiel, J. Kendrick, S. C. Rogers, J. Casci, M. Watson, F. King, E. Karlsen, M. Sjøvoll, A. Fahmi, A. Schäfer and C. Lennartz, *J. Mol. Struct.: THEOCHEM*, 2003, **632**, 1–28.
8 S. Metz, J. Kaestner, A. A. Sokol, T. W. Keal and P. Sherwood, *Wiley Interdiscip. Rev.: Comput. Mol. Sci.*, 2014, **4**, 101–110.
9 A. A. Sokol, S. T. Bromley, S. A. French, C. R. A. Catlow and P. Sherwood, *Int. J. Quantum Chem.*, 2004, **99**, 695–712.
10 D. Berger, A. J. Logsdail, H. Oberhofer, M. R. Farrow, C. R. A. Catlow, P. Sherwood, A. A. Sokol, V. Blum and K. Reuter, *J. Chem. Phys.*, 2014, **141**, 024105.
11 M. Sierka and a. Joachim Sauer, *Faraday Discuss.*, 1997, **106**, 41–62.
12 U. Eichler, M. Brandle and J. Sauer, *J. Phys. Chem. B*, 1997, **101**, 10035–10050.
13 J. Sauer, U. Eichler, U. Meier, A. Schafer, M. von Arnim and R. Ahlrichs, *Chem. Phys. Lett.*, 1999, **308**, 147–154.
14 L. W. Chung, W. M. C. Sameera, R. Ramozzi, A. J. Page, M. Hatanaka, G. P. Petrova, T. V. Harris, X. Li, Z. F. Ke, F. Y. Liu, H. B. Li, L. N. Ding and K. Morokuma, *Chem. Rev.*, 2015, **115**, 5678–5796.
15 V. A. Nasluzov, E. A. Ivanova, A. M. Shor, G. N. Vayssilov, U. Birkenheuer and N. Rösch, *J. Phys. Chem. B*, 2003, **107**, 2228.
16 E. A. Ivanova Shor, A. M. Shor, V. A. Nasluzov, G. N. Vayssilov and N. Rösch, *J. Chem. Theory Comput.*, 2005, **1**, 459–471.
17 A. S. Mysovsky, P. V. Sushko, S. Mukhopadhyay, A. H. Edwards and A. L. Shluger, *Phys. Rev. B: Condens. Matter Mater. Phys.*, 2004, **69**, 10.
18 A. H. Edwards, P. V. Sushko, A. L. Shluger and V. B. Sulimov, *IEEE Trans. Nucl. Sci.*, 2002, **49**, 1383–1388.
19 J. R. Hill and J. Sauer, *J. Phys. Chem.*, 1994, **98**, 1238–1244.
20 J. R. Hill and J. Sauer, *J. Phys. Chem.*, 1995, **99**, 9536–9550.
21 M. Taramasso, G. Perego and B. Notari, *US Pat.*, 4410501, 1983.
22 S. Bordiga, A. Damin, G. Berlier, F. Bonino, G. Ricchiardi, A. Zecchina and C. Lamberti, *Int. J. Mol. Sci.*, 2001, **2**, 167.
23 S. A. French, A. A. Sokol, J. To, C. R. A. Catlow, N. S. Phala, G. Klatt and E. van Steen, *Catal. Today*, 2004, **93–5**, 535–540.
24 J. To, P. Sherwood, A. A. Sokol, I. J. Bush, C. R. A. Catlow, H. J. J. van Dam, S. A. French and M. F. Guest, *J. Mater. Chem.*, 2006, **16**, 1919–1926.
25 J. To, A. A. Sokol, S. A. French, C. R. A. Catlow, P. Sherwood and H. J. J. van Dam, *Angew. Chem., Int. Ed.*, 2006, **45**, 1633–1638.
26 D. Gleeson, G. Sankar, C. R. A. Catlow, J. M. Thomas, G. Spano, S. Bordiga, A. Zecchina and C. Lamberti, *Phys. Chem. Chem. Phys.*, 2000, **2**, 4812–4817.
27 J. To, A. A. Sokol, S. A. French and C. R. A. Catlow, *J. Phys. Chem. C*, 2007, **111**, 14720–14731.

28 C. M. Barker, D. Gleeson, N. Kaltsoyannis, C. R. A. Catlow, G. Sankar and J. M. Thomas, *Phys. Chem. Chem. Phys.*, 2002, **4**, 1228–1240.
29 J. To, A. A. Sokol, S. A. French and C. R. A. Catlow, *J. Phys. Chem. C*, 2008, **112**, 7173–7185.
30 V. N. Shetti, P. Manikandan, D. Srinivas and P. Ratnasamy, *J. Catal.*, 2003, **216**, 461–467.
31 C. D. Chang and A. J. Silvestri, *J. Catal.*, 1977, **47**, 249–259.
32 C. Maiden, *Stud. Surf. Sci. Catal.*, 1988, **36**, 1–16.
33 U. Olsbye, S. Svelle, M. Bjørgen, P. Beato, T. V. Janssens, F. Joensen, S. Bordiga and K. P. Lillerud, *Angew. Chem., Int. Ed.*, 2012, **51**, 5810–5831.
34 R. Shah, J. D. Gale and M. C. Payne, *J. Phys. Chem.*, 1996, **100**, 11688–11697.
35 S. R. Blaszkowski and R. A. van Santen, *J. Phys. Chem. B*, 1997, **101**, 2292–2305.
36 P. E. Sinclair and C. R. A. Catlow, *J. Chem. Soc., Faraday Trans.*, 1997, **93**, 333–345.
37 Z.-M. Cui, Q. Liu, S.-W. Bain, Z. Ma and W.-G. Song, *J. Phys. Chem. C*, 2008, **112**, 2685–2688.
38 A. J. O'Malley, S. F. Parker, A. Chutia, M. R. Farrow, I. P. Silverwood, V. García-Sakai and C. R. A. Catlow, *Chem. Commun.*, 2016, **52**, 2897–2900.
39 D. Lesthaeghe, V. Van Speybroeck, G. B. Marin and M. Waroquier, *Ind. Eng. Chem. Res.*, 2007, **46**, 8832–8838.
40 K. Hemelsoet, J. Van der Mynsbrugge, K. De Wispelaere, M. Waroquier and V. Van Speybroeck, *ChemPhysChem*, 2013, **14**, 1526–1545.
41 R. Vetrivel, C. Catlow and E. Colbourn, *J. Phys. Chem.*, 1989, **93**, 4594–4598.
42 J. Gale, C. Catlow and J. Carruthers, *Chem. Phys. Lett.*, 1993, **216**, 155–161.
43 C. Zicovich-Wilson, P. Viruela and A. Corma, *J. Phys. Chem.*, 1995, **99**, 13224–13231.
44 P. Sinclair and C. Catlow, *J. Chem. Soc., Faraday Trans.*, 1996, **92**, 2099–2105.
45 S. R. Blaszkowski and R. A. van Santen, *J. Am. Chem. Soc.*, 1996, **118**, 5152–5153.
46 S. Blaszkowski and R. Van Santen, *J. Phys. Chem.*, 1995, **99**, 11728–11738.
47 J. Gale, R. Shah, M. Payne, I. Stich and K. Terakura, *Catal. Today*, 1999, **50**, 525–532.
48 F. Haase, J. Sauer and J. Hutter, *Chem. Phys. Lett.*, 1997, **266**, 397–402.
49 J. Andzelm, N. Govind, G. Fitzgerald and A. Maiti, *Int. J. Quantum Chem.*, 2003, **91**, 467–473.
50 D. Berger, A. J. Logsdail, H. Oberhofer, M. R. Farrow, C. R. A. Catlow, P. Sherwood, A. A. Sokol, V. Blum and K. Reuter, *J. Chem. Phys.*, 2014, **141**, 024105.
51 G. Artioli, C. Lamberti and G. Marra, *Acta Crystallogr., Sect. B: Struct. Sci.*, 2000, **56**, 2–10.
52 J. Hriljac, M. Eddy, A. Cheetham, J. Donohue and G. Ray, *J. Solid State Chem.*, 1993, **106**, 66–72.
53 P. Sherwood, A. H. de Vries, M. F. Guest, G. Schreckenbach, C. R. A. Catlow, S. A. French, A. A. Sokol, S. T. Bromley, W. Thiel and A. J. Turner, *J. Mol. Struct.: THEOCHEM*, 2003, **632**, 1–28.
54 D. Nachtigallova, P. Nachtigall and O. Bludský, *Phys. Chem. Chem. Phys.*, 2004, **6**, 5580–5587.
55 R. Ahlrichs and P. Taylor, *J. Chim. Phys. Phys.-Chim. Biol.*, 1981, **78**, 315–324.

56 J. P. Perdew and Y. Wang, *Phys. Rev. B: Condens. Matter Mater. Phys.*, 1992, **45**, 13244.
57 A. D. Becke, *J. Chem. Phys.*, 1993, **98**, 5648–5652.
58 C. Lee, W. Yang and R. G. Parr, *Phys. Rev. B: Condens. Matter Mater. Phys.*, 1988, **37**, 785.
59 S. Vosko, L. Wilk and M. Nusair, *Can. J. Phys.*, 1980, **58**, 1200–1211.
60 P. Stephens, F. Devlin, C. Chabalowski and M. J. Frisch, *J. Phys. Chem.*, 1994, **98**, 11623–11627.
61 P. J. Wilson, T. J. Bradley and D. J. Tozer, *J. Chem. Phys.*, 2001, **115**, 9233–9242.
62 M. F. Guest, I. J. Bush, H. J. Van Dam, P. Sherwood, J. M. Thomas, J. H. Van Lenthe, R. W. Havenith and J. Kendrick, *Mol. Phys.*, 2005, **103**, 719–747.
63 W. Smith, C. Yong and P. Rodger, *Mol. Simul.*, 2002, **28**, 385–471.
64 C. Broyden, *Math. Comput.*, 1970, **24**, 365–382.
65 R. Fletcher, *Comput. J.*, 1970, **13**, 317–322.
66 D. Goldfarb, *Math. Comput.*, 1970, **24**, 23–26.
67 D. F. Shanno, *Math. Comput.*, 1970, **24**, 647–656.
68 W. Mortier, J. Sauer, J. Lercher and H. Noller, *J. Phys. Chem.*, 1984, **88**, 905–912.
69 J. Sauer, *J. Mol. Catal.*, 1989, **54**, 312–323.
70 F. Haase and J. Sauer, *Microporous Mesoporous Mater.*, 2000, **35**, 379–385.
71 W. Jost, *J. Chem. Phys.*, 1933, **1**, 466–475.
72 M. Sierka, U. Eichler, J. Datka and J. Sauer, *J. Phys. Chem. B*, 1998, **102**, 6397–6404.
73 A. J. Jones and E. Iglesia, *ACS Catal.*, 2015, **5**, 5741–5755.
74 S. F. Boys and F. d. Bernardi, *Mol. Phys.*, 1970, **19**, 553–566.
75 S. Svelle, C. Tuma, X. Rozanska, T. Kerber and J. Sauer, *J. Am. Chem. Soc.*, 2008, **131**, 816.

Faraday Discussions

PAPER

The bifurcation point of the oxygen reduction reaction on Au–Pd nanoalloys†

Jakub Staszak-Jirkovský,‡*[ab] Elisabet Ahlberg,[b] Itai Panas*[c] and David J. Schiffrin[a]

Received 21st December 2015, Accepted 7th January 2016

DOI: 10.1039/c5fd00233h

The oxygen reduction reaction is of major importance in energy conversion and storage. Controlling electrocatalytic activity and its selectivity remains a challenge of modern electrochemistry. Here, first principles calculations and analysis of experimental data unravel the mechanism of this reaction on Au–Pd nanoalloys in acid media. A mechanistic model is proposed from comparison of the electrocatalysis of oxygen and hydrogen peroxide reduction on different Au–Pd ensembles. A H_2O production channel on contiguous Pd sites proceeding through intermediates different from H_2O_2 and OOH^{σ} adsorbate is identified as the bifurcation point for the two reaction pathway alternatives to yield either H_2O or H_2O_2. H_2O_2 is a leaving group, albeit reduction of H_2O_2 to H_2O can occur by electrocatalytic HO–OH dissociation that is affected by the presence of adsorbed OOH^{σ}. Similarities and differences between electrochemical and direct synthesis from $H_2 + O_2$ reaction on Au–Pd nanoalloys are discussed.

Introduction

Hydrogen peroxide is an important chemical with both a very wide range of industrial applications and uses in domestic products.[1] H_2O_2 can be produced by the electrochemical reduction of O_2 or by what is termed the direct synthetic route involving the chemical reaction of O_2 with H_2.[2–4] These two syntheses might follow very different mechanisms. For the former, the rate determining step is electron transfer from an electrocatalytic surface to an oxygen molecule whereas for the latter, chemical reactions, for instance between adsorbed hydrogen atoms and molecular oxygen, determine reduction rates. There are, however, similarities between these two reaction mechanisms. For the Au–Pd alloys usually employed,

[a]*Chemistry Department, University of Liverpool, L69 7ZD, UK. E-mail: mrjikra@gmail.com*

[b]*Department of Chemistry and Molecular Biology, University of Gothenburg, SE-412 96, Gothenburg, Sweden*

[c]*Chalmers University of Technology, Department of Chemistry and Chemical Engineering, SE-412 96, Gothenburg, Sweden. E-mail: itai.panas@chalmers.se*

† Electronic supplementary information (ESI) available: Extended methods, electrochemical results analysis and additional discussion of computational results. Fig. S1–S5. See DOI: 10.1039/c5fd00233h

‡ Current address: CEO, Machavert Pharma, 12635 E Montview Blvd., Aurora, CO 80045, USA.

when in contact with a H_2–O_2 mixture under pressure in an acid solution, the fast reaction equilibrium:

$$\tfrac{1}{2}H_2 \rightleftharpoons H^+ + e^- \tag{I}$$

establishes a potential difference between the metallic catalyst particles and the solution. This will be a mixed potential determined by the rates of the oxidation reaction (I) (standard potential = −0.241 V *vs.* SCE) and the reduction of oxygen:

$$O_2 + 2H^+ + 2e^- \rightleftharpoons H_2O_2 \tag{II}$$

(standard potential = 0.441 V *vs.* SCE; potentials are referred throughout this work to the Saturated Calomel Electrode, SCE). Thus, in the case of the direct peroxide synthesis, the actual interfacial conditions do not differ from those in the electrochemical case. The main difference is that this potential difference can only be changed by altering the hydrogen pressure whereas in an electrochemical experiment the applied electrical potential determines the reaction driving force (but see below).

Whether the above distinction in the reaction mechanism is adequate is a matter of debate. The present work concentrates on the electrochemical mechanism and investigates electrocatalytic reactions for peroxide synthesis with a long term view of its implementation using fuel cell technology.[5–7] By contrast with the direct route, carrying out the reaction electrochemically might offer the interesting possibility of utilising part of the $H_2 + O_2$ Gibbs energy of the reaction to generate electricity, which will appear as a production cost credit, whilst producing simultaneously a valuable chemical. Recent developments of macrocylic compound catalysts[8] have demonstrated the possibility of achieving the production of an average concentration of 11–12% peroxide solution in a fuel cell.[7] In this cogeneration approach the applied potential for oxygen reduction is determined by reaction (I) carried out on a separate appropriate high efficiency electrode. In this case, the electrolyte is an ion exchange membrane, for instance Nafion®, and the production unit component is the membrane electrode assembly of a polymer electrolyte fuel cell (PEMFC) that is currently commercially available for transport vehicles.

The difference between the electrochemical and the direct route is that for the former, the resulting electrical power can be employed externally from the physical separation of the two reactions, hydrogen oxidation and oxygen reduction. For either production strategy, understanding of the mechanism of the electrochemical oxygen reduction reaction (ORR) is essential. This question has attracted a great deal of attention not only for its significant practical importance, for instance, in fuel cells, but also as an archetypal example of a reaction involving multiple electron and proton transfers.[9] The oxygen reduction reaction kinetics has been the subject of much debate due to its high activation energy and the consequent large overpotentials needed for this reaction to proceed at a reasonable rate, thus decreasing the high theoretical efficiency attainable in fuel cells.

The efficiency of catalysts for H_2O_2 production is determined by the rate of further reduction of the peroxide produced and of the direct reduction of oxygen to water. A high degree of conversion has been demonstrated by the use of AuPd alloys as a catalyst.[10] Unexpectedly, the rate of subsequent hydrogenation of the peroxide produced was found to be substantially reduced by acid pre-treatment of the carbon support.[3] Establishing the point in the reduction mechanism where O–O bond breaking takes place, namely the bifurcation point in the reduction mechanism, and hence predicting behaviour, are essential for determining if the reduction product is H_2O_2 or H_2O. If the catalytic properties of the catalyst surface are such that the rate of bond breaking of the reaction intermediate(s) is enhanced, the reduction product will be H_2O and a $4e^-$ mechanism will prevail, a desirable outcome for traction applications. If the reduction intermediate does not compromise the stability of the O–O bond, the main reduction product will be H_2O_2, of interest in the possibility of cogeneration of peroxide and energy production. The ability of directing product distribution to H_2O or H_2O_2 has, therefore, large technological impact, *e.g.* in the electrochemical[11] or the direct heterogeneous[1–3,10] H_2O_2 industrial syntheses, in energy storage and for applications in high energy density batteries.[12]

Metal alloys and composites are advanced electrocatalytic materials for many applications of intense current interest, not only for oxygen reduction[13] but also for many other (electro)catalytic reactions such as alcohol oxidation[14] or hydrogen production.[15] The surface structure and composition of a metal alloy critically influences its electrocatalytic properties by ligand and ensemble effects.[16–18] Ligand effects have been described by changes in the energy of the metal d-band centre upon alloying, *e.g.* on alloys terminated by a single component skin.[18] In the case of ensemble effects, reaction selectivity can be assessed by comparing adsorption energies of reactants and products on individual alloy components at the surface. For the ORR, placing individual transition metal centres into a Au host can lead to significant changes in the energy required for O–O bond breaking as shown for gold-transition metal nanoalloys.[11,19]

The surface of AuPd alloys was investigated by Behm *et al.*[16,20] who demonstrated the presence of equilibrium ensembles of Pd atoms within a surface Au matrix by STM measurements. At low Pd concentrations the alloy surface contains a high proportion of monoatomic Pd sites, a result that was theoretically analysed by Jacob *et al.*[21] Their presence leads to exceptional catalytic properties such as, for instance, a pronounced increase in selectivity for the direct H_2O_2 synthesis and other reactions of industrial interest.[1,4,20,22,23] The presence on the alloy surface of single Pd atoms surrounded by Au has been demonstrated by low-energy ion scattering spectroscopy and Auger electron spectroscopy,[24] and the importance of these monoatomic sites in electrocatalysis has also been recently confirmed for Hg–Pt nanoalloys.[25]

The mechanism of the direct H_2O_2 synthesis has been investigated using Density Functional Theory (DFT) modelling with particular reference to the relationship between ensemble and electronic effects in enhancing the H_2O_2 yield for Pd monomers present on the alloy surface.[26] The direct synthesis of H_2O_2 on Pd(111) and Au(211) surfaces with a single Pd atom incorporated on the latter was investigated and a dissociation to water of $O{-}OH^{\sigma}$, a key O_2 reduction intermediate, was identified.[27] Different possible O_2 reduction channels by reaction with adsorbed H (H_{ads}) resulting from the dissociative adsorption of H_2 on Pd sites have been modelled and activation energy barriers calculated.[28,29] These calculations provide a useful guide for the interpretation of a reduction mechanism involving H_{ads} as the reducing agent.

Previous reaction modelling[11] assumed that the activation barrier for O–O bond splitting maps the calculated reaction energies and as presented here, activation energy calculations are required. The present work investigates, both experimentally and by DFT modelling, O_2 reduction mechanisms on Au–Pd nanoalloys taking place at ensembles constituting individual Au sites, monoatomic Pd sites and contiguous Pd surface atoms present in a Au matrix. In addition to these structural effects, water molecules are not simple spectator species in the ORR mechanism as shown recently when comparing experimental results in superdry solvents and those with controlled addition of water.[30] For this reason, the influence of water molecules in the rate determining step of reduction has been incorporated in these calculations employing a microsolvation model to determine the main features of their participation in the reduction mechanism. Another important assumption that is usually made is that the near surface atom arrangement does not change with reaction conditions. Although not addressed in the present work, the surface composition of nanoalloys does depend on applied potential[31] and there is strong evidence of changes in segregation energy of alloy components caused by adsorption of reactants and/or reaction intermediates.[32,33]

The paper is divided into four sections. The methods used for calculating electrochemical rate constants and corresponding experimental results and how these relate to the peroxide production current efficiency are first presented. Density functional theory calculations are then employed for energy and activation energy calculations for different surface ensembles to identify the reaction bifurcation point, *i.e.*, the step in the reduction sequence where oxygen reduction can result in the production of either water or hydrogen peroxide. Microsolvation calculations are included to assess the importance of water molecules in the reaction activation energies, the intermediate present at the bifurcation point is identified and finally, a comparison of this work with the direct peroxide synthesis is discussed.

Electrochemical methods

The kinetics of the ORR is usually analysed considering the overall mechanism given in Scheme 1;[34,35] this implies the intermediate formation of H_2O_2 that can

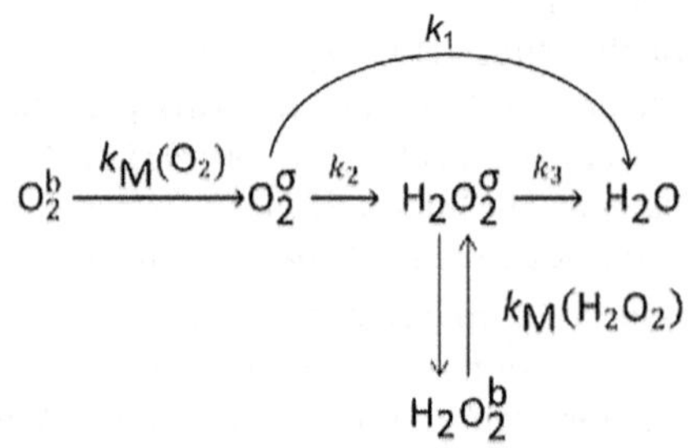

Scheme 1 Conventional ORR scheme.[19] b = solution bulk, σ = species either adsorbed or at the electrode surface, $k_M(O_2)$ and $k_M(H_2O_2)$ are the mass transfer coefficients of oxygen and hydrogen peroxide, respectively, and k_i (i = 1, 2, 3) are the electrochemical rate constants.

be either desorbed into solution or further reduced to water with a rate constant k_3. In addition, a parallel direct§ reduction of O_2 to water (k_1) is usually included. The rate constants of the individual irreversible electron transfer steps are given by:[36]

$$k_i = k_i^0 \mathrm{e}^{\frac{-\alpha_i F\left(E-E_i^0\right)}{RT}} \tag{1}$$

k_i^0 = standard electrochemical heterogeneous rate constant (cm s^{-1}) at the standard potential, E_i^0; α = transfer coefficient and F = Faraday constant. Since these reactions are irreversible, the reverse reactions are not taken into account in this reaction scheme.

The rotating ring disk electrode (RRDE) technique was employed in the present work. For a rotating disk electrode, k_M, the mass transfer coefficient for species i derived from the Levich equation is given by:[36]

$$k_M = 0.2006 n D_i^{2/3} \nu^{-1/6} \omega^{1/2} \tag{2}$$

D_i is the diffusion coefficient of i ($cm^2\ s^{-1}$), ω is the rotation rate (rpm) and ν is the kinematic viscosity of the solution ($cm^2\ s^{-1}$). The measured current at the ring electrode poised at a potential where all the captured peroxide is oxidised provides a simple route for quantifying the amount of peroxide that is produced at the disk electrode and that is not further reduced. The collection efficiency (N) of peroxide produced at the disk and measured at the ring is given by:

$$N = -\frac{I_D}{I_R} \tag{3}$$

which is independent of rotation rate. I_D and I_R are the disk and ring currents, respectively, with their corresponding signs (negative for a reduction process). The values of the rate constants as a function of potential can be obtained from the RRDE results following well established procedures:[32,35,37]

$$k_1 + k_2 = k_M(O_2)\frac{[(I_R/AN) - j_D]}{\left[j_D + 4Fc_{O_2}^b k_M(O_2) - (I_R/AN)\right]} \tag{4}$$

where A is the area of the disk electrode and j_D is the current density at the disk electrode. For the calculation of k_3, the diagnostic criteria method of Damjanovic *et al.*[38] and Wroblowa *et al.*[35] based on the analysis of the rotation rate dependence of ($-NI_D/I_R$) was followed. Details of these calculations and equations employed are given in the ESI and in ref. 8.†

Two parameters are important for characterising the electrocatalytic properties of the nanoalloys investigated, n, the number of electrons exchanged per O_2 molecule reduced and $S_{H_2O_2}$, and the H_2O_2 production selectivity, *i.e.*, the number of moles of peroxide produced per mole of oxygen reduced. Both quantities relate to the same property but the selectivity is a more useful practical quantity for visualising catalyst efficiency.

§ The nomenclature used in the literature can be confusing since the term "direct" is used in both the electrochemical reduction of O_2 and the peroxide synthesis of H_2O_2. For the former, it means the direct $4e^-$ conversion of O_2 to water whereas for the latter it means the synthesis of peroxide from the reaction of H_2 and O_2. We make this distinction clear in the text.

Computational methods

The plane-wave code CASTEP[39] as implemented in the Material Studio suite[40] was used in all calculations, which were performed at the GGA PBE[41] level of theory using ultrasoft pseudopotentials[42] in conjunction with a 300 eV cutoff energy and a $3 \times 3 \times 1$ Monkhorst and Pack[43] k-point set for sampling the Brillouin zone. In order to extract the ensemble effect, first a 5 layer slab comprising $3a \times 3b \times 5c$ supercells was employed in conjunction with a 20 Å vacuum space separating the slabs. (111) surfaces were considered throughout. Comparative calculations, addressing four different Pd distributions, were performed in order to quantify the drive for Pd–O(ads) co-aggregation.

Electrochemical results

The kinetics of the ORR is usually analysed considering the overall mechanism given in Scheme 1A.[34,35] The results are first analysed with this scheme and then the rate constants derived are critically discussed. Typical RRDE results are illustrated in Fig. 1 showing the detection in the ring of the H_2O_2 produced in the disk. A full set of results is given in the ESI.†

The effect of Pd concentration on the value of n and the resulting yield of peroxide at 800 rpm obtained from the RRDE data is shown in Fig. 2 and 3. A full set of results can be found in the ESI.† As previously observed,[11] peroxide selectivity is strongly dependent on Pd composition and increases to high values up to an 8% Pd content. Selectivity is at a maximum when Pd atoms are present as monomers on the surface but decreases dramatically as the concentration of dimers and trimers increases.[22] The corresponding decrease of the value of n can be clearly seen in Fig. 3.

Before calculating individual rate constants, it is important to ascertain the suitability of the overall model described in Scheme 1. The physical meaning of the derived electrochemical rate constants is discussed further on. Classical diagnostic plots[35] were employed for these calculations and an example of the results obtained for the 8% Pd nanoalloy is shown in Fig. 4. Other data is shown in the ESI.†

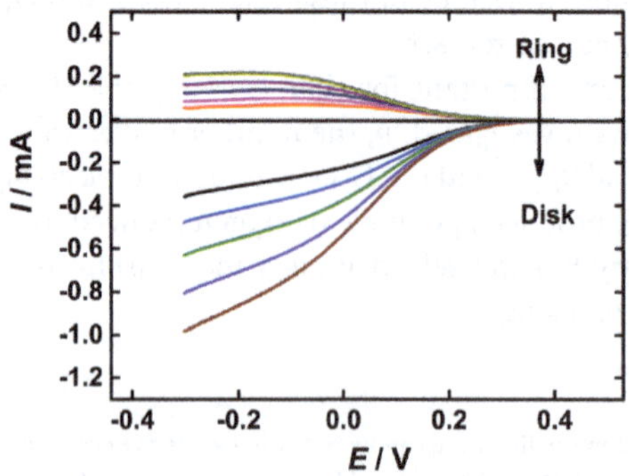

Fig. 1 Potential dependence of the disk and ring currents for O_2 reduction at different rotation rates for $Au_{0.92}Pd_{0.08}$ nanoalloys.

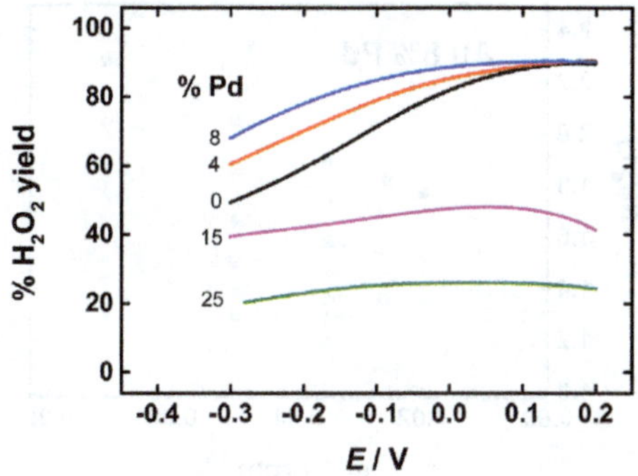

Fig. 2 Potential and Pd concentration dependence of peroxide yield at $\omega = 800$ rpm.

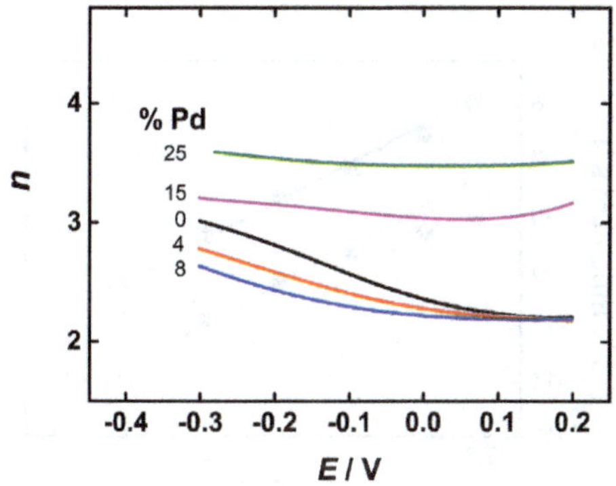

Fig. 3 Potential and Pd concentration dependence of n at $\omega = 800$ rpm.

These diagnostic plots have been extensively employed in the O_2 reduction literature.[37] The relationship between the RRDE measurements and the constants in Scheme 1 is given by:[35,37]

$$-\left(\frac{NI_{\mathrm{D}}}{I_{\mathrm{R}}}\right) = \left(1 + \frac{2k_1}{k_2}\right) + \frac{2k_3}{B_{\mathrm{H_2O_2}}}\left(1 + \frac{k_1}{k_2}\right)\omega^{-\frac{1}{2}} \tag{5}$$

with

$$B_{\mathrm{H_2O_2}} = 0.2006nFD_{\mathrm{H_2O_2}}^{2/3}\nu^{-1/6}. \tag{6}$$

A limiting value greater than 1 is observed for the limiting value of $NI_{\mathrm{D}}/I_{\mathrm{R}}$ at $\omega^{-0.5} = 0$ indicating a contribution to the O_2 reduction current from a parallel channel directly to water, *i.e.*, $k_1 \neq 0$ (see Scheme 1)[35] and therefore the full eqn (5) must be employed to calculate the rate constants. The above method could not be used for materials with a high Pd content for which the slope of $-(NI_{\mathrm{D}}/I_{\mathrm{R}})$ *vs.* $\omega^{-0.5}$ was negative, indicating that the simple model shown in Scheme 1 ceases to be applicable at high Pd concentrations.

A comparison of the potential dependence of the rate constant for peroxide reduction to water (k_3) during O_2 reduction on Au and $Au_{0.92}Pd_{0.08}$ nanoparticles is shown in Fig. 5. The values of k_3 were calculated from eqn (SI-4) in the ESI.†

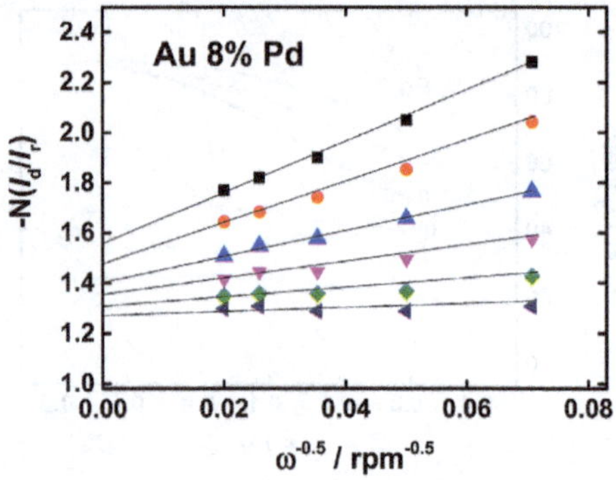

Fig. 4 Diagnostic plots for O_2 reduction at different potentials.[35] Applied potentials (from top to bottom): −0.293, −0.258, −0.207, −0.158, −0.105 and −0.051 V.

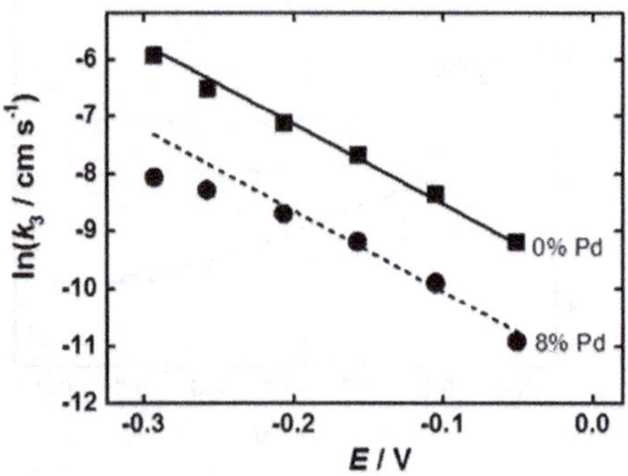

Fig. 5 Comparison of the potential dependence of H_2O_2 reduction on Au and on a $Au_{0.92}Pd_{0.08}$ nanoparticle electrode.

A linear logarithmic dependence of k_3 on potential is observed for Au corresponding to eqn (2), as expected for a simple rate determining electron transfer step. Alloying has two important consequences: (1) a clear decrease in this rate constant and (2) a significant onset of a non-linear response as the potential is made more negative, *i.e.*, as the rate of O_2 reduction increases, demonstrating a potential dependent blocking of nanoparticle active sites.

Thus, these results show that the increase in the peroxide yield observed on alloying (Fig. 2) for the electrochemical peroxide synthesis can be a consequence of the inhibition of H_2O_2 reduction. This observation is in agreement with extensive studies on different promoters tested for the direct synthesis of peroxide involving anions that are known to adsorb on gold.[10] The potential dependence of the electrochemical constants k_1 and k_2 were calculated from eqn (4) and (5) employing the experimentally observed limiting value of NI_D/I_R for $\omega^{-0.5} = 0$ (see, *e.g.*, Fig. 4). A comparison of these rate constants for 0% Pd and 8% Pd is shown in Fig. 6 and 7.

Eqn (4) and (5), and those presented in the ESI†, are applicable regardless of the intermediate stages of the reaction since they represent global

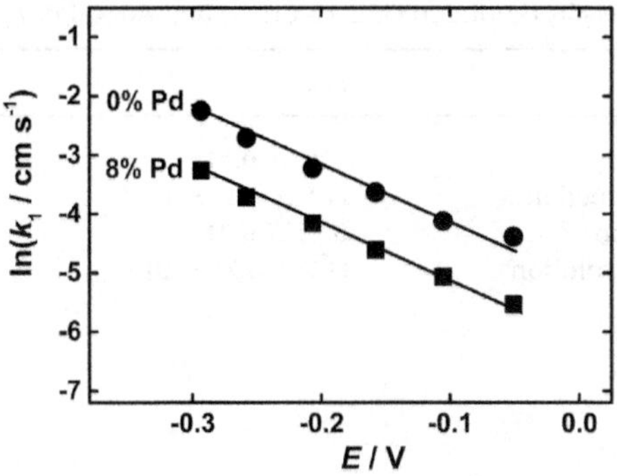

Fig. 6 Comparison of the potential dependence of ln k_1 for Au and $Au_{0.92}Pd_{0.08}$ nanoparticle electrodes.

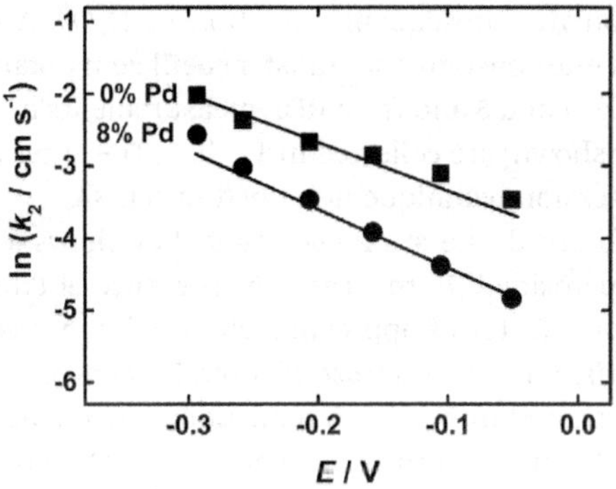

Fig. 7 Comparison of the potential dependence of ln k_2 for Au and $Au_{0.92}Pd_{0.08}$ nanoparticle electrodes.

relationships derived from disk and ring current measurements. These electrochemical measurements and the diagnostic criteria employed provide experimental quantitative measurement of the amount of peroxide that is formed per mole of O_2 reduced and indicate, as shown in the present work, the presence of parallel reduction pathways. In the simple model usually employed in these studies, the chemical nature of the "direct" electrochemical reduction process with the corresponding rate constant k_1 is, however, unclear. The inclusion of this channel attempts to capture reactions in the adsorbed state for which the breaking of the O–O bond occurs without the prior formation of H_2O_2.

The formation of water as the final reduction product in this model does not need to follow a strict sequential channel involving peroxide. Electrochemical measurements alone, however, cannot distinguish between these different possibilities and for this reason the present work aims at disentangling the reduction mechanism on nanoalloys combining classical electrochemical measurements with quantum chemical modelling.

Table 1 Results for H_2O_2 reduction kinetics on Au and $Au_{0.92}Pd_{0.08}$ nanoparticles

% Pd	0%	8%
α_3 from O_2 reduction	0.33 ± 0.01	0.32 ± 0.03
k_3^0 (cm/s) from O_2 reduction	$(1.8 \pm 0.1) \times 10^{-7}$	$(4.5 \pm 1) \times 10^{-8}$
α_3 from H_2O_2 solution	0.24 ± 0.01	0.27 ± 0.02
k_3^0 (cm/s) from H_2O_2 solution	$(1.7 \pm 0.7) \times 10^{-5}$	$(3.6 \pm 0.1) \times 10^{-5}$

Discussion

The kinetic parameters for the oxygen reduction reaction are analysed first in relation to the selectivity towards H_2O_2. The yield of peroxide shows Pd content dependent maxima between 0 and 0.3 V. The reduction of H_2O_2 in oxygen free 0.1 M $HClO_4$ was measured for Au and $Au_{0.92}Pd_{0.08}$ nanoparticle electrodes to establish the influence of the intermediates formed in the initial electron transfer step of the ORR on the subsequent reduction of H_2O_2. A comparison of the peroxide reduction rate constant and transfer coefficient obtained from the RRDE results shown in Fig. 4 and 5 and from RDE measurements in 0.6 mM H_2O_2 in 0.1 M $HClO_4$ (data not shown) are collected in Table 1. The latter were obtained from the non-linear regression technique described in ref. 34.

As previously observed, the 8% Pd–Au nanoalloy shows a high selectivity for the production of peroxide.[11] In this case, the presence of Pd monomers leads to an increase of k_3^0 by a factor of approximately two for the reduction from H_2O_2 solutions but surprisingly, to a *decrease* of several orders of magnitude when this rate constant is measured from oxygen reduction experiments (see Table 1). This is a surprising result since according to Scheme 1, H_2O_2 should be an obligatory intermediate in the ORR and therefore, the rate constant k_3 should be the same regardless of whether it is measured in the presence of oxygen or by direct reduction of H_2O_2 added to the solution. It is proposed that the difference in rates is due to the adsorption of ORR intermediates at the nanoalloy surface with dispersed Pd single-atoms that block H_2O_2 reduction sites, as compared with Au sites in the presence of O_2. This would imply that surface adsorption of reaction intermediates, although facilitating the conversion of O_2 to H_2O_2, hinders, at the same time, the rate of O–O bond splitting to yield H_2O from the peroxide formed.

ORR intermediates

The adsorption of molecular oxygen and of other possible adsorbed oxygenated species formed during the ORR, *e.g.*, $O\text{–}OH^{\sigma}$ and $H_2O_2^{\sigma}$, are considered. The absence or extremely weak adsorption of molecular oxygen on gold from the gas phase has been known for a long time and extensively verified from surface science measurements[44–47] and DFT calculations.[47] Although the adsorption of oxygen atoms on Au is thermodynamically favourable, adsorption of O_2 from the gas phase only occurs at a significant rate above 500 °C due to the high activation energy for O–O bond breaking.[44] DFT calculations confirm that molecular oxygen is not adsorbed on Au(111).[48] For Pd alloys, when dimers are present, adsorption at a bridge position can take place with a binding energy of −0.17 eV.[49] In these experiments, temperature programmed desorption demonstrated the lability of these structures since no adsorbed O_2 could be detected for temperatures above ~220 K. For strained Au surfaces and other surface orientations, significant adsorption of O_2 is predicted.[48] Adsorption of O_2 on Au in acid media during electrochemical experiments is suggested from *in situ* FTIR by the observation of a vibration band at 1540 cm^{-1} that decreases in intensity at negative potential where $O–O^{\sigma}$ would be reduced.[50]

The formation and stabilisation of the intermediate $O–OH^{\sigma}$ on Au at potentials where the ORR proceeds has also been previously proposed as the main adsorbed intermediate formed directly at the first O_2 reduction step.[51,52] Surface Enhanced Infrared Absorption Spectroscopy (SEIRA) has provided direct evidence for the presence of $O–OH^{\sigma}$ by the observation of a potential dependent band at 1220 cm^{-1}.[53] SEIRA measurements in deuterated solutions identified this vibration as being due to $O–OH^{\sigma}$ and not to an adsorbed superoxo ($O–O^{-}$) species. This band was also detected on Ge at 1210 cm^{-1} for O_2 reduction in 0.1 M $HClO_4$ and the assignment confirmed from the measured isotopic shift with $^{18}O_2$.[53] These results show that after the first electron and proton transfers, the resulting OOH^{σ} intermediate is likely to be strongly adsorbed on gold. Additional experimental evidence that supports adsorption of ORR intermediates on Au from acid media has been recently obtained by Uosaki *et al.* from electrochemical quartz crystal microbalance measurements.[54] In agreement with these observations, surface enhanced Raman spectroscopy measurements on a roughened Au electrode displayed a vibration at 1165 cm^{-1} that was assigned to the O–O stretch mode of protonated superoxide, OOH.[47] Importantly, neither signals corresponding to the O–O stretching of O_2 nor vibrations corresponding to adsorbed peroxide, $H_2O_2^{\sigma}$, were detected.[47] This suggests zero or very low coverage by O_2^{σ} or $H_2O_2^{\sigma}$, although for the latter the sensitivity of these measurements suffers from high background due to strong absorbance by the base electrolyte anion. For the above reasons, the formation of stable $H_2O_2^{\sigma}$ has been disregarded and the properties of adsorbed $O–OH^{\sigma}$ have been primarily considered in the DFT modelling presented below. Calculations for $O–O^{\sigma}$ have also been carried out to verify changes in bond breaking energy when Pd atoms form part of the surface ensembles.

DFT modelling of the O_2 reduction reaction

We attempt to rationalise these findings by relating experimental data to DFT modelling of the likely transition states of the elementary steps by analysing the calculated activation energies of O–O bond splitting for different oxygenated species involved in the ORR. The species/intermediates considered are: (1) $O–O^{\sigma}$, the adsorbed dioxygen molecule, (2) $O–OH^{\sigma}$ and (3) $H_2O_2^{\sigma}$; σ denotes an adsorbed species bound to a surface metal reaction centre(s).

Possible reaction channels for O–O bond breaking were analysed for the surface ensembles investigated by calculating the corresponding optimised transition state (TS) geometries and the resulting activation energies:[11,31] (1) a Au atom present at a Au surface (TS1, Au@Au), (2) Pd "hotspots" consisting of a Pd monomer within a gold matrix (TS2, Pd@Au) and (3) two contiguous surface Pd atoms (TS3, Pd_2@Au). Calculations were carried out both in the absence and presence of water molecules for reactions involving proton transfer by taking microsolvation effects into account (see later). The DFT calculation results are

Table 2 Calculated energies for the O–O bond splitting reactions considered[a]

Entry no.	Reaction	Reaction energy (eV)	Activation energy (eV)
1	HOOH–Au@Au → O(ads) + H_2O	−0.717	1.564
2	HOOH–Pd@Au → O(ads) + H_2O	−0.751	0.820
3	HOOH(aq)–Au@Au → O(ads) + H_2O	−0.913	1.477
4	HOOH(aq)–Pd@Au → O(ads) + H_2O	−1.11	0.973
5	HOO–Au@Pd → O(ads) + OH(ads)	−0.152	1.102
6	HOO–Au@Au → O(ads) + OH(ads)	−0.020	1.211
7	HOO–Pd_2@Au → O(ads) + OH(ads)	−0.392	0.670
8	HOO(aq)–Au@Au → O(ads) + OH(ads)	−0.311	1.564
9	HOO(aq)–Pd@Au → O(ads) + OH(ads)	−0.199	1.695
10	HOO(aq)–Pd_2@Au → O(ads) + OH(ads)	−0.159	1.457
11*	HOO(aq)–Pd_2@Au → O(ads) + OH(ads)	−0.159	0.957
12	OO–Au@Au → 2O(ads)	0.577	1.690
13	OO–Pd@Au → 2O(ads)	0.588	1.853
14	OO–Pd_2@Au → 2O(ads)	0.362	1.792

[a] The reactions correspond to slab calculations and for simplicity only the relevant dimer is given, the rest of the slab is composed of Au; the crystallographic orientation was (111); (aq) means the results of microsolvation calculations; (ads) denotes the adsorbed state. @ denotes that the intermediate is attached to given moiety on the (111) metal surface. *This reaction step does not include proton transfer as in entries 8–10.

summarised in Table 2 where total reaction energies and activation energies for O–O bond breaking of the adsorbed species followed by adsorption of reaction products are presented. An example of the calculated geometry of the transition state for breaking the O–O bond in O–OH^{σ} is illustrated in Fig. 8 for the surface ensembles investigated.

Table 2 shows that on Au@Au a direct O–O bond splitting of O–O^{σ} requires a very high activation energy, of ~1.7 eV, and this increases to ~1.8 eV for the Pd@Au sites (entries 12–14 in Table 2). In addition, the calculated reaction energy is positive and therefore this dissociative channel for adsorbed O_2 is very unlikely compromising, therefore, the feasibility of the direct electrochemical reduction channel shown in Scheme 1. For Au@Au and monoatomic Pd sites within a Au matrix, Pd@Au, the activation energies for HO–O^{σ} dissociation of the monodentate adsorbed O–OH^{σ} moiety are generally lower by ~0.6 eV (Table 2, entries 5 and 6) than those for O–O^{σ}, with negative values calculated for the reaction energies. Thus, a dissociative channel for HOO–Pd@Au with high activation energy emerges, implying the preferential formation of $H_2O_2{}^{\sigma}$–Pd@Au as the final reduction product rather than the splitting of the O–O bond that would result in a $4e^-$ reduction pathway. In turn, the formation of this likely product/intermediate is to be followed by either dissociation or desorption of H_2O_2 in the solution bulk. For the $O_2 + H_2$ reduction on a Pd colloid, the original O–O group is retained in the obligatory reaction pathway from O–OH^{σ} to H_2O_2. Elegant experiments for the reduction of a $^{16}O_2/^{18}O_2$ mixture demonstrated that reduction does not involve bond reorganisation since the peroxide produced did not contain any $H_2{}^{16}O^{18}O$.[55]

Therefore, at low Pd concentrations, the bifurcation point in the mechanism leading to H_2O or H_2O_2 should involve the adsorbed O–OH^{σ} species.[50,53] In the

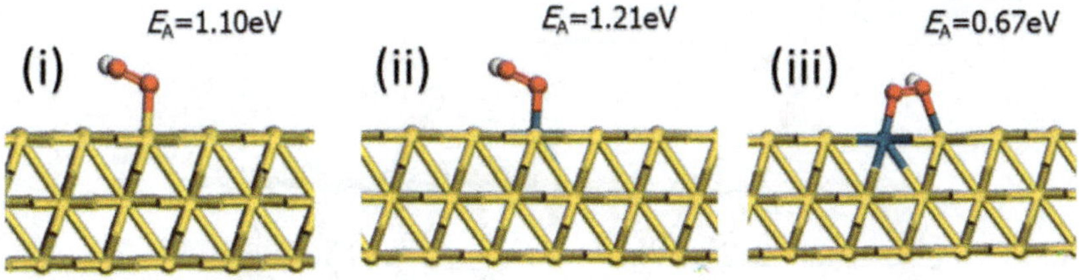

Fig. 8 Geometry of favoured transition states for the O–O bond splitting step of $HO–O^σ$ at (i) Au (yellow) (TS1), (ii) Pd@Au (blue) (TS2) and (iii) Pd_2@Au (TS3). The activation energies (E_A) are given in the figure.

case of nearest neighbour Pd surface atoms, the Pd_2@Au sites, a bi-dentate bonding to the surface is achieved, (see Fig. 8) and the corresponding activation barriers for O(ads) + HO(ads) formation is ~0.5 eV lower for the bi-dentate moiety (entry 7 in Table 2). In this case, the O–O bond breaking activation energy is significantly lower than that of other ensembles and therefore, at high Pd surface concentrations where Pd dimers and trimers predominate,[16] the $4e^-$ channel should be dominant, as is experimentally observed (Fig. 2 and 3). This analysis agrees with the observed higher yield of H_2O_2 on dilute Au–Pd alloys although, as discussed below, inhibition of the reduction rate of H_2O_2 after its formation has also been identified as an important contribution to the high yield of peroxide observed.

Solvation effects

The results in Fig. 5 and the above discussion demonstrate that the observed high yield of peroxide can be due either to changes in the reduction rate of H_2O_2 (ref. 34) or the high activation barrier for O–O bond breaking. However, as recently demonstrated,[30] the influence of solvent on the elementary steps for the ORR cannot be ignored and for this reason the effect of water on reaction kinetics was analysed considering microsolvation effects. This method has been previously successfully employed for the analysis of homogenous reactions such as those involving the hydronium radical,[56] or heterogeneous processes for the analysis of hydrocarbon epoxidation reactions by hydrogen peroxide.[57,58] Microsolvation effects have been included here to estimate the impact of non-covalent interactions on the individual reaction steps with particular reference to the dependence of the activation energy of reactions at the ensembles investigated and the role of hydrogen bonding on the transition state.[58] A full modelling of the solvent environment would be very challenging and beyond the scope of the present work. Two water molecules were included in the analysis of the O–O bond breaking reaction and the results of these calculations are shown in entries 3, 4, 8–10 and 11 of Table 2. The objective of these calculations was to establish general trends of the influence of water molecules involved in the reaction that would lead to a H_2O formation channel for different surface ensembles. The difference between the results in entries 3, 4 and 8, 9 is in the final products considered, but it is important to note that the trend in the activation barriers is the same.

Reactions of $HO–OH^σ$ at surface ensembles. Fig. 9 shows an example of calculated geometries of the transition and final states for the interaction of

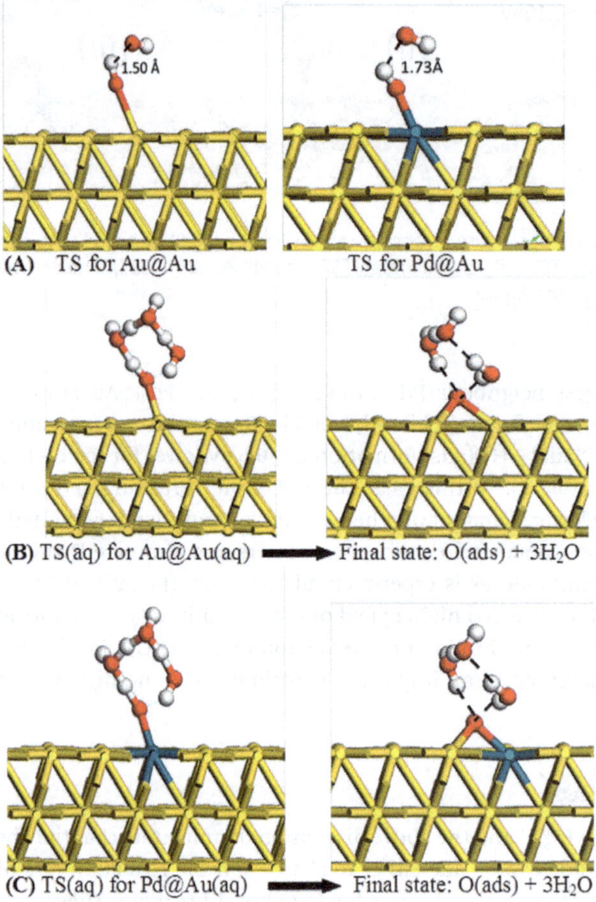

Fig. 9 Geometry of the transition state for hydrogen peroxide reduction. DFT modelling for different ensembles (A) TS = transition state without solvent interactions; (B) and (C) TS(aq) = microsolvated transition state at Au@Au and Pd@Au, respectively, leading to the final product indicated in the figure (see entries 1 to 4 in Table 2). Pd is dark blue, Au yellow, oxygen red and hydrogen white.

peroxide with the Au@Au and Pd@Au sites with and without microsolvation. The geometries of these transition states are very similar and lead to the same products, a surface oxide and water. At the potentials where these reactions proceed, the former will be rapidly reduced to regenerate the reaction site and produce another water molecule leading to a CE (chemical-electrochemical) mechanism[39] for H_2O_2 reduction. The corresponding activation energies are presented in entries 1–2 and 3–4 of Table 2.

When microsolvation is taken into account, the energy of O–O bond breaking for H_2O_2 at a Pd@Au site is 0.2 eV more negative than that for a Au@Au site; most of this difference originates from the formation of a more stable Pd–O bond in the reaction product. These calculations predict a difference in activation energies of −0.5 eV between the Au@Au and Pd@Au ensembles (compare entries 3 and 4 in Table 2). The trend in the calculated activation barriers for O–O bond splitting for

$HO–OH^\sigma$ from peroxide solutions in the absence of oxygen are in agreement with previous work[11] and H_2O_2 experimental reduction data in Table 1, indicating that the activation barriers for O–O bond splitting for $HO–OH^\sigma$ from peroxide solutions and in the absence of oxygen are lower at the Pd monomer sites than for pure Au.

Thus, the reduction of peroxide should be *faster* on Pd@Au than on pure Au and this effect should be more pronounced for alloys containing a higher proportion of monomeric Pd sites. This is a surprising result since the yield of H_2O_2 from the ORR *is larger on dilute* Au–Pd nanoalloys than on pure Au although the calculated activation energy predicts that the opposite should be observed.

The observed difference in the potential dependence of peroxide selectivity between the Au@Au and Pd@Au ensembles could arise from an electron transfer rate determining step to the H_2O_2 LUMO or from the potential dependence of coverage by oxygenated species blocking surface sites. Although it is not possible to distinguish unambiguously between these mechanisms from RRDE measurements, spectroscopic and other evidence[47,50,53,54] indicates a strong potential dependence of surface coverage by $O–OH^\sigma$ on Au surfaces. It is proposed, therefore, that the difference between H_2O_2 reduction rates from O_2 saturated and O_2 free solutions must be related to the presence of these adsorbed species, especially on Pd monoatomic sites, blocking the peroxide reduction channel. Indeed, O_2 adsorption is enhanced by ~0.6 eV on Pd@Au sites with respect to Au@Au and in addition, pDOS diagrams indicate a substantial Pd–OO interaction, with both metallic and local Pd orbital contributions.[11]

Reactions of $O–OH^\sigma$ at surface ensembles. The microsolvation calculations described above were also applied to the initial steps of the ORR, for O–O bond splitting of the $O–OH^\sigma$ and $O–O^\sigma$ species. The symmetries of the activation barriers for O–O bond breaking at Au@Au, Pd@Au and Pd_2@Au sites can be compared to elucidate the experimentally observed trends. In addition, the possibility of two alternative reaction pathways for $O–OH^\sigma$ subject to microsolvation can be considered: (1) the O–O breaking proceeds directly to surface OH and oxide or (2) a proton reorganisation occurs by a Grotthuss mechanism in the microsolvation shell during the splitting of $O–O^\sigma$.

It should be stressed that the purpose of these calculations was to assess the importance of solvent interaction (*i.e.*, non-covalent interaction effects) for a given transition state and provide an estimate of the extent of molecular reorganization necessary for the reaction to proceed. These results are illustrated in Fig. 10, S4 and S5.† First, the effect of solvation is introduced by including two water molecules in the O–O bond breaking reaction. It is proposed that the weaker the interaction of the adsorbed reaction intermediates with the surface, the greater the influence of solvation is and thus the transfer coefficient (shape of the activation barrier) will be larger. The validity of this argument was tested by considering two different mechanisms for O–O bond breaking including either a water dimer bridge offering a Grotthuss proton transfer path or the reaction proceeding from an already protonated OOH^σ intermediate.

The Grotthuss proton transfer path comprises an alternative dissociation channel which relates to that of the bi-nuclear ensemble Pd_2@Au, *i.e.*, a peroxo species bound to the Pd centres by both oxygen atoms of the intermediate. When employing this reaction pathway for all the considered surface atom ensembles, the activation energies calculated are only marginally lower than the case of the

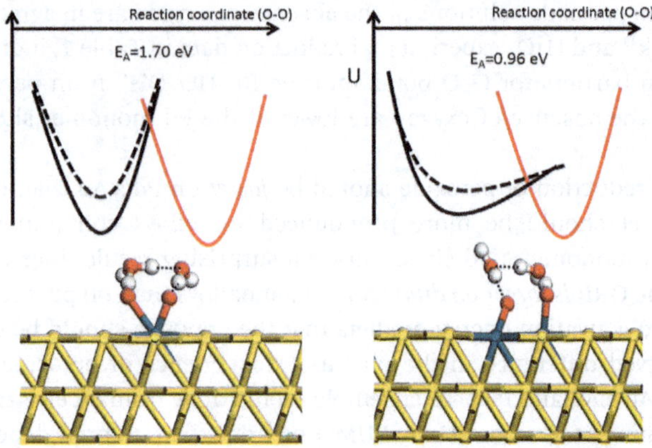

Fig. 10 Micro-solvation effect on transition states during the reduction of OOH, (i) at Pd@Au (left panel) and (ii) at Pd_2@Au site (right panel). Above are corresponding schematic diagrams of potential dependent activation barriers where the dashed lines depict the effect of changing electrode potential. These diagrams show that a higher impact of increased potential (dashed lines) on rate of dissociation should be expected in the case of a higher activation energy and different geometry of the transition state. This applies for the experimentally observed dependence of H_2O_2 production on potential where for Au and Pd@Au sites the H_2O_2 yield decreases strongly with potential, whereas in the case of Pd_2@Au there is a weak dependence (see Fig. 2).

bi-dentate structure as compared to pure gold or Pd monomer within gold. Besides the similarity of energies calculated for the TS structures (see Fig. S4†) it is observed that for both Au@Au and a Pd@Au sites, O–O bond dissociation occurs after proton transfer has taken place. In the case of two contiguous Pd sites, this occurs simultaneously with proton transfer implying a different chemistry at the latter site (compare entries 8–10 in Table 2).

In the case of the Pd bi-nuclear micro-solvated surface moiety, O–OH dissociation can take place directly without involving a Grotthuss proton transfer, *i.e.*, in the absence of micro-solvation. This reaction path for Pd dimers that does not involve proton transfer was found to be much more favourable than that with proton transfer. For Au and Pd@Au, however, such a reaction path is not available and O–O splitting is energetically more favourable upon solvent assisted proton transfer. As experimentally observed, the trend in activation energies favouring H_2O_2 production on Pd@Au ensembles is preserved. Micro-solvation increases the barrier for dissociation of O–OH$^\sigma$ by ~0.5 eV for Pd@Au and Au but only by ~0.25 eV for Pd_2@Au, which shows (see Table 2, entries 8–11) that it has a much smaller impact on the corresponding transition state (TS3; Fig. SI5†) for Pd_2@Au sites. This suggests that a smaller molecular reorganization is required for this reaction step at the Pd dimers, *i.e.*, *the transition state geometry is closer to the product than to the reactant* and rationalises the slower decrease of H_2O_2 production with potential for materials with higher Pd content (see Fig. 2). This highlights a different mechanism for the O–O splitting reaction for Au@Au and Pd@Au compared to Pd_2@Au sites.

It is interesting to note, however, that the dissociative channel for HOO–Pd_2@Au is complicated by a Grotthuss process (see Table 2, entries 10–11), while the dissociative channel for H_2O_2 at a Pd_2@Au site is straight-forward. Thus, the Pd_2@Au ensemble is not as bad as it appears in that it also benefits, to some extent at least, from the blocking of H_2O_2 reduction resulting from the adsorption of O_2 reduction intermediates, this time to form HOO–Pd_2@Au.

The ORR mechanism and the reaction bifurcation point

Microsolvation appears to have a significant effect on transition states on different surface ensembles mostly related to the reduction of OOH^σ, whereas it does not have such an influence on the direct reduction of H_2O_2 where no stable surface adsorbed intermediates were found. The opposing impact of microsolvation on the activation energies of O–O bond breaking for H_2O_2 and O–OH^σ reflects the different involvement of the surface active sites in the reaction. These calculations suggest, therefore, that the stronger interaction of O–OH^σ with surface sites than that of H_2O_2 can shift the product distribution of the ORR toward H_2O_2. Thus, *the adsorbates formed from the initial stages of O_2 reduction appear to have a dual function, both as ORR intermediates and in blocking surface sites; the balance between these two effects determines product distribution.* The trends in the DFT results in Table 2 provide additional support for the proposal of the blocking of the Pd@Au sites by O–OH^σ. Comparison of entries 9 and 4 indicates that the activation energy for the reduction channel of O–OH^σ to water is 0.7 eV *higher* than that of HO–OH, *i.e.*, O–OH^σ should be much more stable than HO–OH on the Pd sites, thus blocking the $4e^-$ channel for further reduction of H_2O_2. From the all the above analyses, it is concluded that the O–OH^σ species represents the bifurcation point of the ORR. As argued previously, "direct" electrochemical reduction of adsorbed O_2 in Scheme 1 is unlikely due to the large energy that would be required for O–O bond breaking (see Table 2) and the adsorbed O–OH^σ is the route to either water or peroxide. Therefore, the following ORR reaction scheme is proposed:

Noting that H_2O_2 can be regarded as a leaving group and does not form a stable adsorbate at the surface,[47] it is proposed that H_2O_2 is reduced directly with a lower activation barrier on Pd@Au sites through the $H_2O_2^\sigma$ transition state. The peroxide reaction rate through this channel is, however, hindered by blocking of active sites by the adsorption of O–OH^σ.

The proposed mechanism differs from the classical Scheme 1 by identifying the bifurcation step as O–O bond splitting of O–OH^σ. Three different reduction channels can be distinguished, directly to H_2O (I), to H_2O_2 and desorption to the solution bulk (II) and reduction to water of the previously formed H_2O_2 (III). For the Au–Pd alloys, there is an alternative by-pass to the H_2O production channel if surface Pd dimers or trimers are present. This by-pass and reaction (I) would be equivalent to the "direct" reduction pathway in Scheme 1 with just one rate constant k_1, *i.e.*, there is no "direct" $4e^-$ reduction involving the breaking of the O–O bond prior to electron transfer.[35,37,38] A justification for this can be seen by comparing the results shown in Fig. 6 and 7 for the 8% nanoalloy. The transfer coefficients for the process described by the rate constants k_1 and k_2 are indistinguishable and there is no justification that two *different* channels should contribute to a parallel reaction implied in the results shown in Fig. 4, indicating that only a single adsorbed intermediate needs to be considered for the putative "direct" electrochemical O_2 reduction.

Comparison of the direct and electrochemical peroxide synthesis mechanism

It is important to ask again the question formulated in the Introduction; if the success of AuPd alloys for the direct chemical synthesis of H_2O_2 is determined by an electron transfer step in a reaction akin to a corrosion process occurring at a mixed potential, or whether the reaction can be entirely described by chemical steps involving only bond reorganisations without separate electron transfers. The direct chemical synthesis has been extensively analysed[26–29] following the traditional Horiuti–Polanyi (H–P) hydrogenation mechanism,[59] involving the dissociative adsorption of molecular hydrogen to give adsorbed H atoms, followed by the sequential addition of individual hydrogen atoms to the substrate. A schematic diagram of these two strategies is shown in Fig. 11. An essential requirement for the H–P mechanism is the dissociative adsorption of H_2 on the catalyst (Fig. 11(a)) and this condition is fulfilled by Pd. This was originally observed for the hydrogenation of ethane and subsequently for many other reactions, although the universal applicability of this reaction scheme has recently been questioned.[60] The electron is the reducing agent in the electrochemical mechanism (Fig. 11(b)).

For the direct synthesis, use of AuPd alloys leads to high efficiency in the H_2O_2 production but an essential requirement would be the presence of dimers or higher Pd aggregates at the surface otherwise dissociative adsorption of H_2 could not take place. Hayden *et al.*[61] demonstrated that hydride formation is only observed for Pd concentrations greater than 30 at% and voltammetric measurements showed no evidence of hydrogen adsorption for an 8% Pd alloy for which the surface is mainly populated by monomers. This is a curious observation showing a significant difference between the direct and electrochemical routes. High conversion rates have been observed for the former for AuPd catalyst containing 50 at% Pd (ref. 1, 2, 10 and 23) or higher, whereas for the latter, the yield of H_2O_2 is highest at ~8 at% Pd (present work and ref. 11) decreasing sharply for higher concentrations of Pd in the alloy.

Here there is conflicting experimental evidence: (1) there is no doubt that the electrochemical route does produce peroxide in high yields when the H–P mechanism cannot be followed due to the absence of the required surface hydride; (2) the high peroxide yield, as argued in the present work, results from a combination of two effects, the rate of formation of $O{-}OH^{\sigma}$ and the blocking of surface sites by this intermediate for the electrochemical reduction of H_2O_2; (3) although the Nernstian potential that must be established in the course of the direct synthesis with high Pd concentration catalyst should be close to the

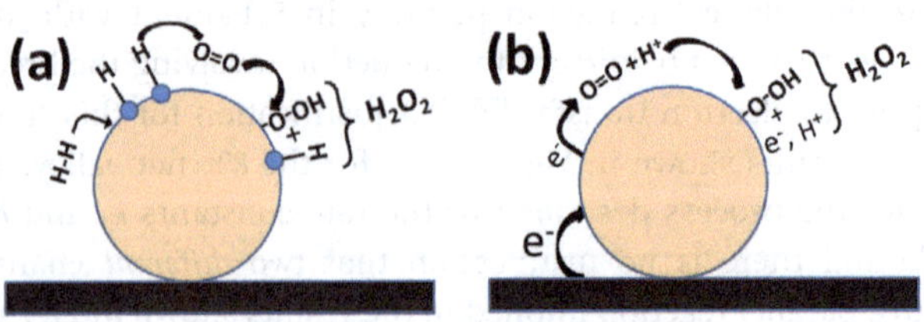

Fig. 11 Schematic diagram of the mechanisms for the reduction of oxygen to produce hydrogen peroxide: (a) direct and (b) electrochemical synthesis.

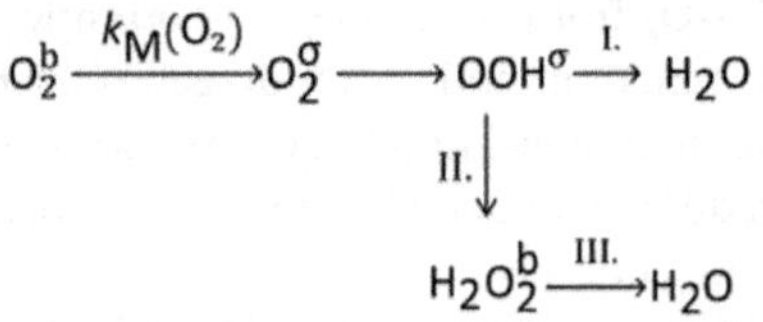

Scheme 2 The proposed ORR scheme: $O{-}OH^{\sigma}$ is the bifurcation point to yield H_2O (I) or H_2O_2 (II) as primary reaction products although the latter can also lead to H_2O (III).

reversible potential of the H_2/H^+ couple under the conditions of the synthesis and hence, from the electrochemical evidence, very low yields of peroxide should result, the opposite is observed. A possible explanation for this apparent contradiction is the efficient blocking of surface sites for the hydrogenation of the peroxide produced, *i.e.*, blocking of channels I and III after the reaction bifurcation point in Scheme 2. The switching from a $4e^-$ to a $2e^-$ O_2 reduction mechanism in acid media by a monolayer of adsorbed atomic hydrogen has been observed for Pt[51,62] and recently for a carbon supported Pd catalyst.[63] This switching of the reduction mechanism is a consequence of the adsorption of hydrogen atoms occupying the μ-peroxo adsorption O_2 sites that lead to the full reduction of oxygen to water, thus forcing the reduction to follow exclusively the peroxide channel. It is proposed that this competition for the surface ensembles appropriate for the $4e^-$ O_2 reduction is a large contributory factor to the success of the direct peroxide synthesis, although the large influence of the support on peroxide yield still needs explanation.

A convergence of the two synthesis routes discussed above emerges from the recent results by Yook *et al.*[64] In a direct synthesis configuration, these authors physically separated the locations where hydrogen equilibrium (I) is located from the sites at which reaction (II) takes place by coating Pt particles within stable nanoscale amorphous carbon layers. The reduction of O_2 on carbon is well known to lead to H_2O_2 and this dual catalyst led to significant yields of peroxide. Thus, in this case, the organisation of the reactants and products precluded the applicability of the H–P hydrogenation mechanism and the function of dissolved H_2 is to provide *in situ* the driving force to carry out the reduction of O_2.

Conclusions

The unravelling of reaction mechanisms for complex multistep reactions such as the ORR only from simple electrochemical measurements presents difficulties since these essentially provide a simple route for measuring the amount of peroxide that is produced for a given transformation rate of oxygen. An example of these general problems is the question of the presence or otherwise of the direct electrochemical reaction pathway with rate constant k_1 (see Scheme 1). As predicted by the diagnostic criteria discussed in relation to Fig. 4, a reduction pathway that does not follow a sequential route involving the necessary formation of H_2O_2 as an intermediate is present and the H_2O_2/H_2O reaction bifurcation point for oxygen reduction has been identified as the adsorbed $O{-}OH^{\sigma}$ radical. The identification of this key intermediate is of practical importance for guiding the synthesis of new suitable materials either for the electrochemical or the chemical synthesis of H_2O_2 from H_2 and O_2. Importantly, this adsorbed species appears to have a dual function, both as an oxygen reduction intermediate that can facilitate further reduction events whilst at the same time, blocking surface sites required for the reduction of peroxide; the balance between these two effects determines product distribution.

The analysis of the kinetics of the ORR on Au–Pd nanoalloys presented here shows a good consistency between theoretical modelling and experimental data although many uncertainties regarding the individual electron and proton transfer steps remain. A clear trend in the activation energies and the experimental rate constants for the ORR on different Au–Pd ensembles was observed, suggesting that the highly efficient H_2O_2 production on dilute Au–Pd nanoalloys involves the presence of adsorbed reaction intermediates such as O–OH$^\sigma$ on Pd@Au ensembles inhibiting further reduction of H_2O_2 to water, in agreement with the known promoting effect of adsorbed promoters for increasing the peroxide yield such as Br^- or Cl^- in the direct synthesis of H_2O_2.[10] The present conclusions can be likely extended to other alloy materials and similar reactions, highlighting the importance of ensemble effects in electrocatalysis.

DFT calculations point to the existence of a particularly fragile balance between forces influenced predominantly by the electrode material (*i.e.* adsorption strength and whether mono- or bi-dentate adsorption sites are present), interactions with solvent/electrolyte and competition between reactants and products for catalytic sites. The design of an ideal catalyst for H_2O_2 production must depend critically on finding this balance and recent unprecedented advances in techniques and methods in computational science makes possible the fruitful blending of theory and experiment for the rational design of new catalytic materials.

Acknowledgements

Support from Solvay R&D, Belgium and from the European Union through the ELCAT Marie-Curie Initial Training Network, Proposal No. 214936-2, 2008–2012, is gratefully acknowledged.

References

1 J. K. Edwards and G. J. Hutchings, *Angew. Chem., Int. Ed.*, 2008, **47**, 9192–9198.
2 P. Landon, P. J. Collier, A. J. Papworth, C. J. Kiely and G. J. Hutchings, *Chem. Commun.*, 2002, 2058–2059.
3 J. K. Edwards, B. Solsona, E. Ntainjua, A. F. Carley, A. A. Herzing, C. J. Kiely and G. J. Hutchings, *Science*, 2009, **323**, 1037–1041.
4 L. Ouyang, G.-J. Da, P.-F. Tian, T.-Y. Chen, G.-A. Liang, J. Xu and Y.-F. Han, *J. Catal.*, 2014, **311**, 129–136.
5 E. Lobyntseva, T. Kallio, N. Alexeyeva, K. Tammeveski and K. Kontturi, *Electrochim. Acta*, 2007, **52**, 7262–7269.
6 I. Yamanaka, S. Tazawa, T. Murayama, R. Ichihashi and N. Hanaizumi, *ChemSusChem*, 2008, **1**, 988–990.
7 M. I. Q. R. Campos, S. G. Lopez and R. J. Potter, Patent WO2013045895–A1, GB2509466-A, Johnson Matthey PLC, 2013.

8 M. Campos, W. Siriwatcharapiboon, R. J. Potter and S. L. Horswell, *Catal. Today*, 2013, **202**, 135–143.
9 C. Costentin, V. Hajj, M. Robert, J.-M. Savéant and C. Tard, *Proc. Natl. Acad. Sci. U. S. A.*, 2011, **108**, 8559.
10 J. K. Edwards, S. J. Freakley, R. J. Lewis, J. C. Pritchard and G. J. Hutchings, *Catal. Today*, 2015, **248**, 3–9.
11 J. S. Jirkovský, I. Panas, E. Ahlberg, M. Halasa, S. Romani and D. J. Schiffrin, *J. Am. Chem. Soc.*, 2011, **133**, 19432–19441.
12 Z. Peng, S. A. Freuenberger, Y. Chen and P. Bruce, *Science*, 2012, **337**, 563–566.
13 C. Wang, *et al.*, *Nano Lett.*, 2011, **11**, 919–926.
14 J. S. Spendelow and A. Wieckowski, *Phys. Chem. Chem. Phys.*, 2007, **9**, 2654–2675.
15 R. Subbaraman, D. Tripkovic, D. Strmcnik, K.-C. Chang, M. Uchimura, A. P Paulikas, V. R. Stamenkovic and N. M. Markovic, *Science*, 2011, **334**, 1256–1260.
16 F. Maroun, F. Ozanam, O. M. Magnussen and R. J. Behm, *Science*, 2001, **293**, 1811–1814.
17 P. Liu and J. K. Nørskov, *Phys. Chem. Chem. Phys.*, 2001, **3**, 3814–3818.
18 V. R. Stamenkovic, B. S. Mun, M. Arenz, K. J. J. Mayrhofer, C. A. Lucas, G. Wang, P. N. Ross and N. M. Markovic, *Nat. Mater.*, 2007, **6**, 241–247.
19 R. B. Rankin and J. Greeley, *ACS Catal.*, 2012, **2**, 2664–2672.
20 M. Ruff, N. Takehiro, P. Liu, J. K. Nørskov and R. J. Behm, *ChemPhysChem*, 2007, **8**, 2068–2071.
21 S. Venkatachalam and T. Jacob, *Phys. Chem. Chem. Phys.*, 2009, **11**, 3263–3270.
22 M. S. Chen, D. Kumar, C. W. Yi and D. W. Goodman, *Science*, 2005, **310**, 291–293.
23 Y. Pluntke, L. A. Kibler and D. M. Kolb, *Phys. Chem. Chem. Phys.*, 2008, **10**, 3684–3688.
24 C.-W. Yi, K. Luo, T. Wei and D. W. Goodman, *J. Phys. Chem. B*, 2005, **109**, 18535–18540.
25 S. Siahrostami, *et al.*, *Nat. Mater.*, 2013, **12**, 1137–1143.
26 H. C. Ham, G. S. Hwang, J. Han, S. W. Nam and T. H. Lim, *J. Phys. Chem. C*, 2009, **113**, 12943–12945.
27 R. Todorovic and R. J. Meyer, *Catal. Today*, 2011, **160**, 242–248.
28 J. Li, T. Ishihara and K. Yoshizawa, *J. Phys. Chem. C*, 2011, **115**, 25359–25367.
29 J. Li and K. Yoshizawa, *Catal. Today*, 2015, **248**, 142–148.
30 J. Staszak-Jirkovský, R. Subbaraman, D. Strmcnik, K. L. Harrison, C. E. Diesendruck, R. Assary, O. Frank, L. Kobr, G. K. H. Wiberg, B. Genorio, J. G. Connell, P. P. Lopes, V. R. Stamenkovic, L. Curtiss, J. S. Moore, K. R. Zavadi and N. M. Markovic, *ACS Catal.*, 2015, **5**, 6600–6607.
31 J. S. Jirkovsky, I. Panas, S. Romani, E. Ahlberg and D. J. Schiffrin, *J. Phys. Chem. Lett.*, 2012, **3**, 315–321.
32 E. Völker, F. J. Williams, E. J. Calvo, T. Jacob and D. J. Schiffrin, *Phys. Chem. Chem. Phys.*, 2012, **14**, 7448–7455.
33 H. Liao, A. Fisher and Z. J. Xu, *Small*, 2015, **11**, 3221–3246.
34 J. S. Jirkovský, M. Halasa and D. J. Schiffrin, *Phys. Chem. Chem. Phys.*, 2010, **12**, 8042–8052.
35 H. S. Wroblowa, Y.-C. Pan and G. Razumney, *J. Electroanal. Chem.*, 1976, **69**, 195–201.

36 A. J. Bard and L. R. Faulkner, *Electrochemical Methods—Fundamentals and Application*, John Wiley & Sons, New York, 2001.
37 Z. Jia, G. Yin and J. Zhang in *Rotating Electrode Methods and Oxygen Reduction Electrocatalysts*, ed. W. Xing, G. Yin and J. Zhang, Elsevier, New York, NY 10010 USA, 2014, ch 6.
38 A. Damjanovic, M. A. Genshaw and J. O'M. Bockris, *J. Chem. Phys.*, 1966, **45**, 4057–4059.
39 S. J. Clark, M. D. Segall, C. J. Pickard, P. J. Hasnip, M. J. Probert, K. Refson and M. C. Payne, *Z. Kristallogr.*, 2005, **220**, 567–570.
40 Accelrys Inc., *Materials Studio 5.0, Simulation software. 15.*
41 J. P. Perdew, K. Burke and M. Ernzerhof, *Phys. Rev. Lett.*, 1996, **77**, 3865–3867.
42 D. Vanderbil, *Phys. Rev. B: Condens. Matter Mater. Phys.*, 1990, **41**, 7892.
43 H. J. Monkhorst and J. D. Pack, *Phys. Rev. B: Condens. Matter Mater. Phys.*, 1976, **13**, 5188.
44 M. A. Chesters and G. A. Somorjai, *Surf. Sci.*, 1975, **52**, 21–28.
45 P. Légaré, L. Hilaire, M. Sotto and G. Maire, *Surf. Sci.*, 1980, **91**, 175–186.
46 N. D. S. Canning, D. Outka and R. J. Madix, *Surf. Sci.*, 1984, **141**, 240–254.
47 X. Li and A. A. Gewirth, *J. Am. Chem. Soc.*, 2005, **127**, 5252–5260.
48 Y. Xu and M. Mavrikakis, *J. Phys. Chem. B*, 2003, **107**, 9298–9307.
49 W.-Y. Yu, L. Zhang, G. M. Mullen, G. Henkelman and C. B. Mullins, *J. Phys. Chem. C*, 2015, **119**, 11754–11762.
50 N. Ohta, K. Nomura and I. Yagi, *J. Phys. Chem. C*, 2012, **116**, 14390–14400.
51 A. M. Gómez-Marín and J. Feliu, *ChemSusChem*, 2013, **6**, 1091–1100.
52 P. Vassilev and M. T. M. Koper, *J. Phys. Chem. C*, 2007, **111**, 2607–2613.
53 S. Nayak, P. U. Biedermann, M. Stratmann and A. Erbe, *Phys. Chem. Chem. Phys.*, 2013, **15**, 5771–5781.
54 S. F. Tong, H. Noguchi, T. Masuda and K. Uosaki, *Electrochem. Commun.*, 2013, **34**, 33–36.
55 D. P. Dissanayake and J. H. Lunsford, *J. Catal.*, 2003, **214**, 113–120.
56 A. L. Sobolewski and W. Domcke, *Phys. Chem. Chem. Phys.*, 2002, **4**, 4–10.
57 A. Lundin, I. Panas and E. Ahlberg, *J. Phys. Chem. A*, 2007, **111**, 9080–9086.
58 A. Lundin, I. Panas and E. Ahlberg, *Phys. Chem. Chem. Phys.*, 2007, **9**, 5997–6003.
59 I. Horiuti and M. Polanyi, *Trans. Faraday Soc.*, 1934, **30**, 1164–1172.
60 B. Yang, X.-Q. Gong, H.-F. Wang, X.-M. Cao, J. J. Rooney and P. Hu, *J. Am. Chem. Soc.*, 2013, **135**, 15244–15250.
61 F. A. Al-Odail, A. Anastasopoulos and B. E. Hayden, *Phys. Chem. Chem. Phys.*, 2010, **12**, 11398–11406.
62 N. M. Markovic, H. A. Gasteiger and P. N. Ross Jr, *J. Phys. Chem.*, 1995, **99**, 3411–3415.
63 R. Rahul, R. K. Singh, B. Bera, R. Devivaraprasad and M. Neergat, *Phys. Chem. Chem. Phys.*, 2015, **17**, 15146–15155.
64 S. Yook, H. Shin, H. Kim and M. Choi, *ChemCatChem*, 2014, **6**, 2836–2842.

Faraday Discussions

DISCUSSIONS

Catalyst design from theory to practice: general discussion

Charles Campbell, Rutger van Santen, Michail Stamatakis, Nicola Collis, Hans-Joachim Freund, Craig Plaisance, Joachim Sauer, Bruce Garrett, Elad Gross, Andrzej Kotarba, Bert Weckhuysen, Wataru Ueda, Christian Reece, Richard Catlow, Annette Trunschke, Ludovic Briquet, Viet Thang Ho, Itai Panas, David Willock, Cynthia Friend, Michael Bowker, Neil Robinson, Albert Bruix, Shelaka Gupta, Bruce Gates, Avelino Corma, Robert Madix, Dimitrios Manganas, Alberto Roldan, Alexander O'Malley and Jakub Staszak-Jirkovský

DOI: 10.1039/C6FD90016J

Hans-Joachim Freund opened the discussion of the paper by Alberto Roldan: How is the atomic hydrogen produced on the greigite surface? In the paper (DOI: 10.1039/C5FD00186B) there is no comment whether you studied dissociate hydrogen adsorption.

Alberto Roldan answered: We agree with Prof. Freund that we have not studied the hydrogen dissociation on the surface. However, the experiments are carried out on small nanoparticles exposing different surfaces and edges where H_2 molecules dissociate, as DRIFTS tests (not published) have suggested by the formation of reduced carbonated species.

Rutger van Santen remarked: According to Wachterhauser, the reactive phase is Mackinawite. This is a layered FeS phase. The S may also have a higher reactivity than that of Greigite. How does the activation of H_2 on Greigite compare with that of Mackinawite, and would the reactivity of S be such that SO groups form in the surface that helps to generate CO? The SO group can have a subsequent reaction with H_2S to give S_2, which would help to transform the Mackinawite to pyrite.

Alberto Roldan replied: Although the reactivity of Mackinawite is not within the scope of this paper (DOI: 10.1039/C5FD00186B), we know that CO_2 adsorbs stronger on its surfaces than on greigite[1] and therefore it is worth considering FeS as a catalyst. We are currently also investigating the reactivity of Mackinawite but a detailed comparative study is still missing.

1 N. Y. Dzade, A. Roldan and N. H. de Leeuw, *J. Chem. Phys.*, 2015, **143**, 094703.

Michail Stamatakis said: Including the zero point energy (ZPE) corrections for the initial, transition and final states of the reactions will shift these energies to higher values. Are these shifts uniform among these different states, or do you see significant differences in the ZPE of the reactants *versus* the products, for instance? Since hydrogenation steps involve (presumably mobile) H adatoms being incorporated into the C1 fragments, it is conceivable that the vibrational frequencies change significantly, thereby resulting in non-uniform shifts.

Alberto Roldan responded: The energy contribution of the ZPE to the total energy slightly shifts the profile and only the activation energies would change (but by less than 0.1 eV). Thus, the general trend and viability of the process is not compromised.

David Willock continued: You show that there are two iron species on the surface, Fe(A) and Fe(B), and it appears that Fe(B) responds much more strongly to adsorbates. Is this due to the co-ordination of these cations on the surface? Did you consider competitive adsorption of other molecules (especially water) at these sites that may change the effective calculated adsorption energies?

Alberto Roldan replied: That is right, both FeA and FeB at the surface are under coordinated, especially FeB which occupies an octahedral site but it is only 3-coordinated. Both sites FeA and FeB respond to the presence of molecules leading to competitive adsorptions, *e.g.* carbonate and H_2O.[1]

1 A. Roldan, N. Hollingsworth, A. Roffey, H.-U. Islam, J. B. M. Goodall, C. R. A. Catlow, J. A. Darr, W. Bras, G. Sankar, K. B. Holt, G. Hogartha and N. H. de Leeuw, *Chem. Commun.*, 2015, **51**, 7501.

Albert Bruix asked: The *U*-correction provides a computationally feasible improvement to the poor description of systems with strongly correlated electrons by standard LDA and GGA exchange–correlation functionals. However, some calculated properties, including adsorption energies, have been found to depend on the choice of the *U* parameter. The effect on calculated adsorption and reaction energies is particularly significant when these processes involve changes in the occupation of the d and f orbitals in transition and rare-earth oxides, respectively. Could you comment on how the choice of *U* may affect the calculated mechanisms? Are there any steps, such as hydrogen adsorption, which involve charge transfer to or from the Fe 3d orbitals?

Alberto Roldan answered: We followed the Hubbard approach to accurately compute the d-Fe electron correlation. Although small (U_{eff}= 1 eV), this approach provides a better description of the electronic structure (DOS) of the material and therefore its redox properties. However, the Fe surface is not the species responsible for the transfer of electrons to CO_2 of the H-adatoms, the nearby S is the electron transfer site, as explained in the manuscript.

Wataru Ueda remarked: Heterogeneous catalysis researchers often try to mimic bio-enzymatic systems with solid-state materials with the same elemental constituents. However, catalytically active sites in the enzymatic systems always assume discrete structures. On the other hand, simple solid-state materials contain repeated structures of discrete sites. Therefore, solid-state materials cannot mimic enzymatic systems, unless solids are constructed with structurally isolated discrete units. Thus, we need new design concepts and a new methodology of materials synthesis. What do you think about this?

Alberto Roldan replied: I do agree that enzymes and extended materials cannot be directly compared, not only because enzymatic catalytic centres are discrete structures but also because they have the backbone around these structures to stabilise the sites and the reaction intermediates. Nevertheless, enzymes can provide useful hints to explain the behaviour of extended materials. Particularly in our case, we followed the suggestions in Origin of Life theories, which hypothesized that extended [Fe-S] structures could have been predecessors of contemporary enzymes.

Elad Gross commented: Can you identify any similarities between the heterogeneous and enzymatic catalysts as you decrease the size of Fe_3S_4, making it comparable to the core size of the enzymatic catalyst?

Alberto Roldan replied: The enzymatic cluster is surrounded by a backbone that may stabilise intermediates in the reduction process. This is lacking on the pristine surface and therefore the mechanism is not necessarily the same. It has been shown, for instance, that small clusters of Fe_4S_4 are not stable once they are exposed to reactants. Another example are the chalcogels,[1] whose poor electronic conductivity reduces their activity towards reductive processes. Thus, although the active centre is very similar, the backbone surrounding them makes enzymes able to catalyse the reduction in very specific conditions.

1 B. D. Yuhas, C. Prasittichai, J. T. Hupp and M. G. Kanatzidis, *J. Am. Chem. Soc.*, 2011, **133**, 15854.

Joachim Sauer opened the discussion of the paper by Dimitrios Manganas: I understand that from including the spectra of all models and fitting to the observed spectra you reach the conclusion that different sites are present and that dimeric structures are dominating. The fit will of course get better with an increasing number of structures included in the fit.

What did you do to make sure that the structures you include are significant in the statistical sense?

Dimitrios Manganas replied: All calculated spectra were normalized and constant energy shifts and broadening Gaussian parameters were applied as described in the text. Comparison with the experimental spectra was performed *via* a least squares fitting procedure as described in the supporting information, using R^2 as a measure of the goodness of fit. In fact, it was found that this process is satisfactorily sensitive with respect to the degree of oligomerization and the number of structures included in the statistical analysis. For example, as it is discussed in the section 'X-ray spectroscopy', the individual V L-edge spectra of the A and B-domain type structures containing V–*O*–V bridges ($V_2Si_6H_6O_{14}$, $V_2Si_{12}H_{12}O_{23}$, $V_3Si_5H_5O_{15}$ and $V_4Si_4H_4O_{16}$), provide the best agreement between theory and experiment, in terms of relative intensities and intensity distribution, especially for the L_3 resonances located at 516, 517.5 and 519 eV. Moreover, by averaging over the above four structures (Fig. S8 DOI: 10.1039/C5FD00193E), the average spectrum has a closer resemblance with the V_2O_5 spectrum. There are of course limitations. Although comparison of individual spectra can be conclusive for specific coordinating groups (*e.g.* umbrella type monomers), it cannot clearly distinguish between monomers and oligomers. The situation is improved when a simultaneous comparison between the three experiments is performed.

Once again, it should be highlighted that this procedure describes a molecular protocol that is developed to provide spectra-structure correlations. Hence, in this present form it cannot provide safe interpretations for a particular catalytic center unless spectra under *in operando* conditions are provided and reaction intermediates through accurate energetic studies are identified.

David Willock continued: How well are the calculations and spectroscopic measurements able to distinguish between the trimer structures and larger oligomers of VO_x species?

Dimitrios Manganas responded: A quantitative estimation regarding the degree of oligomerization is a rather challenging task that would require much more experimental information for conclusive interpretations. Nevertheless, it was observed that trimeric structures in 'closed form' are not only able to reproduce three individual spectroscopic experiments (Raman, V L-edge and O K-edge), but they also provide very good agreement for the relative intensities of the V L-edge and O K-edge spectra. Moreover, by averaging over the monomers, dimers, trimers and tetramers, the resulting V L-edge spectrum has a close resemblance with the V_2O_5 spectrum, rather than with the V loading spectra shown in Fig. S8 (DOI: 10.1039/C5FD00193E). Based on these observations, it seems that the presented combined experimental and theoretical protocol is sensitive with respect to the degree of oligomerization at least to some extent.

Bert Weckhuysen remarked: I appreciate the combination of different experimental methods, as well as their theoretical validation in order to shed new insights in the field of supported vanadium oxide catalysis. The novel aspect is certainly related to the interpretation of the L-edge vanadium X-ray absorption spectroscopy, as well as in the combination of the different methods, culminating in Fig. 13 of DOI: 10.1039/C5FD00193E. I have, however, a few questions.

1) Did you try to use another support oxide to test your hypothesis? In other words, what would be the effect of (additional) hydroxyl groups in the way the species are attached to the surface, and what would it mean for the relative ratio of isolated *vs.* oligomeric vanadium oxide species, differing in their number of vanadium oxide species.

2) Did you try to use oxygen-18 and how this would influence the Raman spectra, as well as the other spectra? It would be interesting to notice then what the relative changes are in the different spectra. One would also be able to calculate such spectral changes. It would be another way to validate the conclusion you have made.

3) Fig. 12 (DOI: 10.1039/C5FD00193E) is very absolute in terms of what the relative amount is of specific oligomeric vanadium oxide species. How sure are we

that this is indeed the situation? What would happen when *e.g.* water is added to the system?

Dimitrios Manganas replied: As is stated above, multiwavelength Raman studies have been used to identify species on silica supported vanadia oxide catalysts including analysis of overtones, ^{18}O labeling and discussing the resonance enhancement mechanism of particular signals (ref. 25 DOI: 10.1039/C5FD00193E). However, even such integrated Raman analysis cannot be conclusive about the catalytic active sites, leaving this concept an open discussion field (ref. 26 DOI: 10.1039/C5FD00193E). This is exactly why multimethod experimental and theoretical spectroscopic protocols with even higher predictive power need to be developed.

The current work (DOI: 10.1039/C5FD00193E) provides a first step towards this direction. We are in the process of extending the available experimental data towards spectroscopic investigations that are more related to the actual structural conditions of the catalytic centers. Moreover, we are planning to transfer this problem in the field of reactivity and study the energetics of the catalytic systems on particular reactions. It will be only then, when structural combinations as those presented in Fig. 12 (DOI: 10.1039/C5FD00193E) may be thought to be relevant to the active catalytic center. For the time being they just provide confident structure-spectra correlations.

Bruce Gates asked: You referred to 'spectroscopic response to structural correlation'. Could you please explain what this means?

Dimitrios Manganas answered: In this work (DOI: 10.1039/C5FD00193E), by choosing a vanadium molecular catalyst of well-defined geometry and oxidation state and by calculating its main spectroscopic properties in terms of rRaman and X-ray absorption, we could define an analysis protocol that relates structural information to electronic structure information to spectroscopic properties. The same protocol could be applied to identify the structure and composition of the molecular species that dominate the spectroscopic properties of a silica-supported vanadia catalyst. Of course, such observation cannot be translated to direct information about the active catalytic species, as this requires further studies by transferring the problem in the field of reactivity. This is, however, one more example that demonstrates that molecular protocols, which are widely used in field of homogenous catalysis, are also applicable to a large extent in the more challenging field of heterogeneous catalysis.

David Willock enquired: The Raman modes responsible for spectral lines are often thought of as quite localised effects, bond stretching, bending, *etc.* How much information on extended structure are you likely to be able to obtain?

Dimitrios Manganas answered: It is perhaps safe to say that actually the majority of the spectroscopic techniques can be considered to a large degree local in nature, probing the local structure of a studied system. Interestingly, in the case of V_2O_5, the shape of the V L-edge XAS spectrum has been proven sensitive to the bulk structure of the solid, but not the corresponding O K-edge one (ref. 33 and 35 DOI: 10.1039/C5FD00193E). Hence, it seems that the investigation of the effects due to the extended structure would require further information by addressing such questions, *e.g.* in the field of reactivity.

Joachim Sauer continued: I do not think that you have made sufficiently clear what is new in your approach compared to well-established concepts in surface science and heterogeneous catalysis. The comparison of surface structures with inorganic model compounds (the similarities and differences) has a long history. I mention only the early work of Muetterties on metal carbonyls[1] and a 1989 Chemical Reviews article with the title 'Molecular Models in ab Initio Studies of Solids and Surfaces'.[2] As far as a local probe is concerned, the spectroscopic techniques may differ in the locality for an individual site, but there may be site-site interactions, and after all you observe a signal that is an average for the sample you are looking at.

1 E. L. Muetterties, *Science*, 1977, **196**, 839.
2 J. Sauer, *Chem. Rev.*, 1989, **89**, 199.

Dimitrios Manganas answered: Indeed combining theoretical and experimental multimethod protocols is a practice with long history that has been successfully applied in the fields of homogeneous and heterogeneous catalysis. In general, the more spectroscopic information using various spectroscopic techniques is available, the more reliable the derived structure/activity relationships will be. In this work (DOI: 10.1039/C5FD00193E) apart from rRaman and O K-edge spectroscopies that have been successfully used in the past for structure to spectra correlations, V L-edge spectroscopy is introduced in addition.

It must be highlighted that interpretation of metal L-edge spectra is a challenging task as the open core p-shell is subject to strong spin orbit coupling (SOC), which splits the spectrum into L_3 and L_2 edges and strongly mixes the 'non-relativistic' final states. In that respect, it is only possible to achieve first principles quantitative description of the experimental spectra by a) constructing all possible electron configurations involved in the L-edge excitation process, b) accounting for electron dynamic correlation and c) properly describing the spin–orbit coupling and covalent interactions. In addition, since metal L-edge spectroscopy is widely applied in the field of solid-state chemistry and heterogeneous catalysis, a successful methodology should also be able to operate on sufficiently large cluster models and in a predictive fashion. This has all been taken into account in the DFT/ROCIS method development.[1–3]

In this concept, this work (DOI: 10.1039/C5FD00193E) presents a computational protocol based on advanced theoretical spectroscopy combining rRaman, O K-edge and V L-edge X-ray spectroscopies, that are used to establish structure–spectra relationships in the field of vanadia catalysts. In a more general spirit, the need of a simultaneous interpretation of the available spectroscopic data of the studied systems is emphasized.

1 M. Roemelt and F. Neese, *J. Phys. Chem. A*, 2013, **117**, 3069.
2 D. Maganas, S. DeBeer and F. Neese, *Inorg. Chem.*, 2014, **53**, 6374.
3 M. Roemelt, D. Maganas, S. DeBeer and F. Neese, *J. Chem. Phys.*, 2013, **138**, 204101.

Bert Weckhuysen said: What would happen if you use another Raman laser excitation frequency? I notice that you have a 450 nm band as the sample seems to have a light orange colour. This absorption band is considered to originate from oligomeric species. What would happen if you excite with another Raman laser source? What would be the effect of the (dominant) molecular structure of the supported vanadium oxide species you have proposed.

Dimitrios Manganas answered: This is a very valid question that needs to be explored. We are currently in the process analyzing multiwavelength rRaman spectra of the system.

Annette Trunschke remarked: An additional comment from the experimentalists regarding ongoing experimental work: currently we are studying vanadium oxide species with variable degrees of oligomerisation at the surface of different supports by application of a set of complementary spectroscopic techniques, including UV-vis, X-ray absorption, and Raman spectroscopy. The Raman spectra are measured by applying different excitation energies in the range between 325 and 633 nm.

Bert Weckhuysen replied: This is indeed a very good strategy to give further understanding to this important catalyst system, *i.e.* multiple characterization techniques, including Raman spectroscopy with different excitation energies, as applied on supported vanadium oxide catalysts varying in their V oxide loading (and therefore most probably a different degree of oligomerization/polymerization).

Bert Weckhuysen continued: What would happen if you used molybdate instead of tungstate? How unique is this catalytic behavior in relation to the tungstate in your material?

Annette Trunschke answered: Tungstates and molybdates have been studied before in alkane oxidation, but usually reaction temperatures above 500 °C have been applied to achieve appreciable conversion. Hydrothermally prepared $MnMoO_4$ composed of large primary particles (not nano-structured) showed selectivity to propene between 60 and 80% (depending on the preparation technique) in the oxidative dehydrogenation of propane at conversions between 3 and 10%. However, a reaction temperature of 500 °C was necessary to achieve these conversions due to the low specific surface area of the materials (1–7 $m^2\ g^{-1}$). Consequently, high selectivity can also be achieved over manganese molybdate at conversions comparable to those adjusted for manganese tungstate. The unique feature of the presented $MnWO_4$ catalyst is its high activity and selectivity at temperatures as low as usually applied for vanadium oxide catalysts. Efforts in the synthesis may lead to optimized properties of $MnMoO_4$ catalysts as well.

Joachim Sauer remarked: The use of a different excitation wavelength in Raman spectra proved crucial in an early study of VO_x/SiO_2 and VO_x/Al_2O_3 systems.[1]

1 N. Magg, B. Immaraporn, J. B. Giorgi, T. Schroeder, M. Bäumer, J. Döbler, Z. Wu, E. Kondratenko, M. Cherian, M. Baerns, P. C. Stair, J. Sauer, H.-J. Freund, *J. Catal.*, 2004, **226**, 88.

Dimitrios Manganas responded: I fully agree that studies like the one by Magg *et al.* have been proven instrumental in establishing structure-spectra correlations in the field of the vanadia oxide catalysts.

Andrzej Kotarba asked: Different excitation energies will lead to laser beam penetration of an investigated material to a different extent, so the Raman spectrum would change accordingly. Do you include this effect in your modelling when constructing the structure-spectra relationship?

Dimitrios Manganas answered: Indeed multiwavelength Raman studies have been used to identify active species on silica supported vanadia oxide catalysts (see for example ref. 26 DOI: 10.1039/C5FD00193E. We are in the process of revisiting such experiments and extending the analysis towards this direction.

Bruce Gates opened the discussion of the paper by Craig Plaisance: Could you make a precise statement of what your catalyst design principles are?

Craig Plaisance replied: We want to design an active site in which the two oxidation steps ($H^+ + e^-$ removal) preceding water addition have associated oxidation potentials equal to the operating potential of the anode. These two steps effectively 'prime' the catalyst to carry out water addition by storing the energy supplied by the external potential in the form of holes localized on the active site. The more endothermic these two steps are, the more energy is stored in the holes, leading to more exothermic water addition. If these two steps have oxidation potentials equal to the operating potential of the anode, then we are effectively storing in the active site 100% of the energy supplied by the external potential in order to drive the water addition step.

David Willock said: The quasiatomic orbitals look like a very useful tool to extract chemically intuitive information from periodic DFT. But, could you explain how the quasiatomic orbitals exactly reproduce all of the ground state properties? Surely the plane wave basis can be extended indefinitely and so must be more complete than the minimal basis used for the quasiatomic orbitals.

Craig Plaisance answered: The key point is that they exactly reproduce the ground state properties, *i.e.* the occupied orbitals, not the entire band structure. As long as there are more quasiatomic orbitals (QOs) than there are electrons, one can describe all of the occupied states within the subspace spanned by the QOs.

Ludovic Briquet asked: How accurate is your method to compute energy barriers? And how does it compare to other methods such as Nudged Elastic Bands?

Could your model be used to screen the activity of several different materials in a reasonable time frame? The kinetic treatment requires evaluation of the energy barriers, which could be computationally intensive.

Craig Plaisance answered: The method we use is accurate as long as the reorganization energy of the modes normal to the reaction coordinate between the two potential energy surfaces is not significant. We have not tested this assumption, but it is something we are planning to look at in the future. It would actually not be difficult to include this reorganization energy using Marcus theory at a fixed point along the reaction coordinate and we will likely do this in future studies. NEB calculations are not applicable for calculating these barriers since the transition state is not a saddle point on a single potential energy surface.

The calculation of barriers is much more expensive than calculating only the reaction energies. It is possible, as we show in the paper (DOI: 10.1039/C5FD00213C), to determine a linear free energy relationship between the barrier and the reaction energy that could be used in the screening studies. The design principle we present in the paper (DOI: 10.1039/C5FD00213C) requires only reaction free energies, although it is derived from kinetic treatment.

Andrzej Kotarba said: The cobalt spinel is a very good catalyst for N_2O decomposition. We have investigated the system and optimized it by doping and engineering the nanocrystallite morphology for several years. In the deN_2O process, after reactive N_2O adsorption resulting in N_2 release, the recombination of surface oxygen is the rate limiting step. In real conditions, the catalyst works in the presence of water vapour. So, similarly to your process, the affinity to oxygen and water is a key factor for the cobalt spinel's catalytic performance. Obviously, they are different at various facets. For deN_2O we have found Co_3O_4 (100) as the optimal one. The DFT calculated surface energies of the facets increased in the following order: 1.39 J m^{-2} (100) < 1.48 J m^{-2} (111) < 1.65 J m^{-2} (110). At the same time these facets are most often observed experimentally for Co_3O_4 nanocrystals.[1] Was there a particular reason for the selection of (110) and (311) surface? One of the conclusions in the paper (DOI: 10.1039/C5FD00213C) is that the system is sub-optimal. What do you have in mind? Does the molecular modelling give us any hints or guidance for the optimization of the catalyst's performance *e.g.* by bulk or surface doping?

1 F. Zasada, W. Piskorz, P. Stelmachowski, A. Kotarba, J.-F. Paul, T. Płociński, K. J. Kurzydłowski and Z. Sojka, *J. Phys. Chem. C*, 2011, **115**, 6423.

Craig Plaisance responded: The (110)-B surface was chosen because it does not contain pairs of adjacent Co cations in the surface like the other terminations, but only isolated Co cations. Such a site corresponds to the 'slow' site proposed by Frei and coworkers,[1] while the 'fast' site they propose consists of the pair of coupled Co cations. One goal of our work (DOI: 10.1039/C5FD00213C) was to examine whether such a dual-Co site was in fact more active, so it was important to also examine a single-Co site.

The (110)-A surface was chosen because it is the only termination on which a μ^2-oxo species can be formed.

The reason for choosing the (311) surface is rather complex and it took us some time to come up with it. The problem with the (001) surface is that one of the protons in the active state prefers to sit on the terminal hyroxo, which is needed to deprotonate the incoming water, rather than on a μ^3-oxo. By introducing a step edge, the μ^3-oxo becomes a much more basic μ^2-oxo so that the proton stays on this site, leaving the terminal hydroxo free to deprotonate the water. The (311) surface is the simplest surface that contains such a step edge. This is described in detail in our JACS paper.[2]

The site on the (311) surface is sub-optimal because the oxygen atom that deprotonates the water and the oxygen atom that forms the hydroperoxo product have different proton affinities. The former is part of an η-hydroxo, the latter is a μ^3-oxo. To improve the activity, one needs to increase the proton affinity of the μ^3-oxo so that it is closer to that of the η-hydroxo. One way to do this may be to replace the Co(II) cation coordinating the μ^3-oxo with a less acidic cation, although I am not sure what a good candidate would be.

1 M. Zhang, M. de Respinis and H. Frei, *Nat. Chem.*, 2014, **6**, 362.
2 C. P. Plaisance and R. A. van Santen, *J. Am. Chem. Soc.*, 2015, **137**, 14660–14672.

David Willock continued: Your calculated barriers always seem to have a maximum point at the crossover point between different electronic states. Does this mean that the electron jumps before the proton moves in your PCET process or is more interpretation required?

Craig Plaisance answered: The electron jumps when the proton is approximately in-between the two oxygen atoms. The reaction coordinate in this part of the plot is mostly associated with the motion of the proton between the two oxygen atoms.

Bert Weckhuysen remarked: You provide us with new insights from theoretical chemistry. I am not so familiar with this field but what is the experimental backing/information to make the statement that one out of thousands are behaving in this way? What experiments are done to support this finding/conclusion?

Craig Plaisance answered: The experiments of Frei and coworkers[1] measured the turnover frequency of the active site by measuring the lifetime of an isotopically labeled superoxo intermediate detected by FTIR. The turnover frequency was found to be about 150 times greater than the rate per exposed Co on a nm-sized catalyst and 2500 times greater than the rate per exposed Co on a μm-sized catalyst.

1 M. Zhang, M. de Respinis and H. Frei, *Nat. Chem.*, 2014, **6**, 362.

Charles Campbell enquired: I read that paper by Frei[1] to which you referred, where he simply divides the steady state rate per unit area by the first-order rate constant for a single elementary step to estimate the site density to be $1/1000^{th}$ of the full surface area. I do not accept the logic of his analysis. At steady state, when one solves a microkinetic model for such reactions, one often finds such small ratios as being simply due to the presence of other adsorbates which block most of the active sites at steady state. When they are temporarily occupied in such a way, they cannot make a product, but that does not mean they are not active sites.

1 M. Zhang, M. de Respinis and H. Frei, *Nat. Chem.*, 2014, **6**, 362.

Craig Plaisance replied: What you are saying is correct in principle, but I do not think this is how Frei and coworkers[1] determined the active site density. They divided the steady state rate per unit area by the turnover frequency of the active site, not by the first-order rate constant for a single elementary step. The low active site density is supported by the 16-fold increase in rate per unit surface area they observe on a nm-sized particle *vs.* a μm-sized particle, indicating a high degree of structure sensitivity. While on the μm-sized particle, about 1/2500 surface sites are active, on the nm-sized particle, about 1/150 surface sites are active.

1 M. Zhang, M. de Respinis and H. Frei, *Nat. Chem.*, 2014, **6**, 362.

Neil Robinson said: An interesting parallel between the mechanisms at the three sites you have studied is the role of surface hydroxyl groups in 'directing' the approach of water molecules through the formation of hydrogen bonding interactions. Such surface–adsorbate interactions are of course ubiquitous in many real-world catalytic systems due to the wide ranging coverage of surface hydroxyls on many catalytic materials and supports.

You have noted within the Introduction of your work (DOI: 10.1039/C5FD00213C) that only a very small number of active sites are responsible for the overall activity of such a catalyst. One might therefore expect some degree of competition to arise between the formation of hydrogen bonds with these active hydroxyls and with passive hydroxyls decorating the rest of the oxide surface. Do you believe any such competitive process will impact the kinetics of your reaction mechanism? Have you, for example, investigated the energetics of water adsorption to passive hydroxyls on the cobalt oxide surface in order to ascertain whether this adsorption competition has relevance from an adsorption energy point of view?

Craig Plaisance responded: I do not think that adsorption of water on the other hydroxyls will have a large effect on the TOF. A water molecule hydrogen bonding to these other hydroxyls will not block another water molecule from binding to the hydroxyls at the active site. There may be small perturbations due to changes in the complex network of hydrogen bonding in the solvent, but I do not expect these to be large.

Christian Reece enquired: In your model you use the microsolvation model to account for the effect of the solvent in your reaction, have you tried increasing the complexity of your model? Perhaps using dynamics to simulate the effect of a solvent? Do you feel that 'frozen' DFT simulations are still applicable when we know that a catalyst surface is a highly dynamic system?

Craig Plaisance answered: No, we have not tried a more complex solvation model as of now. We plan in the future to model the electrolyte using an implicit solvation model based on the Poisson–Boltzmann equation. Molecular dynamics is very expensive using DFT and hybrid QM/MM methods are rather involved to parameterize correctly, so these implicit solvation models are a good compromise.

As to the second question, I want to make a distinction between the dynamics related to solvent motion and the dynamics related to changes in the morphology of the catalyst particle. I believe that for surface catalytic reactions, transition state theory gives a good approximation of the rates of elementary processes.

Therefore, we can calculate these rates using statistical sampling rather than dynamics. In relation to solvent interactions, this would entail calculating the transition state free energy by sampling over many solvent configurations, while keeping the surface and adsorbates in the same local saddle point. This can be done using techniques such as thermodynamic integration and metadynamics. Even these techniques are very expensive, however, limiting their large-scale application such as in screening studies.

Concerning dynamics related to changes in the morphology of the catalyst particle, these processes happen on much longer timescales than the surface reactions that govern activity and selectivity. An active site will turn over many times before any changes in its structure can occur. Therefore, I believe that it is a very good approximation to treat the active site as a static entity. There is still, however, a very large distribution of different active sites on the catalyst particle which would require statistical sampling for a rigorous treatment.

In my work (DOI: 10.1039/C5FD00213C), I am more interested in using theory to understand how the structure of an active site influences its activity and selectivity, not so much about calculating a TOF for the entire catalyst particle. This allows us to determine the optimal active site for a given reaction. Once this is known, the question then becomes how can we maximize the number of these optimal active sites on a given catalyst particle.

Michail Stamatakis opened a general discussion: Catalysis is a multiscale problem and at each scale there exists a variety of modelling methodologies which we need to choose from. For instance, at the molecular scale one has several options of both static and dynamic approaches (DFT, configuration interaction, coupled cluster, *ab initio* MD, classical MD, *etc.*) At the catalyst scale one can formulate mean-field micro-kinetic models or carry out kinetic Monte Carlo (KMC) simulations. Dr Roldan made a choice to perform micro-kinetic modelling instead of KMC, and Dr Manganas chose the restricted open shell configuration interaction singles instead of other *ab initio* approaches. Are there any general guidelines on how we choose the most suitable approach for a system, given the complexity, and what information we need to obtain?

Dimitrios Manganas responded: As is stated in the second discussed paper (DOI: 10.1039/C5FD00193E), in order to successfully apply the theoretical spectroscopy approach it is of utmost importance that the theoretical methods are properly calibrated, meaning that the error bars of the theoretical predictions must be known from studying a series of known and understood systems. Hence, all predictions of the theoretical methods are only valid within the confidence intervals defined by the calibration procedure. All contemporary calculations of homogeneous and heterogeneous systems likewise are inflicted with methodological errors and those errors stem from the inevitable oversimplification of the real systems. It is for this reason that one is well advised to connect the theory with as many spectroscopic experiments as one can possibly obtain. It should be emphasized that, once a broad agreement between the theory and experiment has been achieved, one can: a) cross correlate the structure/spectra relationship between vastly different experimental methods and b) obtain experimentally calibrated insight into not only the geometric- but also the all-important electronic structure of the systems under investigation. The latter is instrumental for obtaining deep insight into the reactivity of the system. The theoretical treatment of surfaces and solids is strongly dominated by the method of density functional theory (DFT). While this method has definite and important strengths and advantages, it also suffers from some significant shortcomings, as will be elaborated below. In particular, the field of L-edge spectroscopy cannot be successfully approached on the basis of DFT at all. Hence, our theoretical approaches deviate strongly from the currently employed DFT methods in that they are largely based on ideas from wavefunction-based *ab initio* quantum chemistry. We emphasize the properties of the N-particle states of the system (N being the number of electrons) in favour of the ubiquitous one-particle (orbital) picture that is implied by interpretation based on DFT. The drawback of focusing on wavefunction approaches is that truly periodic calculations are not available for the approaches that we are using. The next best alternative, that we follow, is to employ embedded clusters. Once the clusters are sufficiently large, the results of the periodic calculations are approached. However, convergence with cluster size is an issue that must be carefully monitored. Fortunately, cluster calculations with a few hundred atoms are feasible with modern hard- and software technologies, such that the limitations of using cluster models are not nearly as severe as they were in the past.

With all this in mind we have recently introduced the restricted open shell configuration interaction singles (ROCIS) method as a method of general applicability to interpret metal L-edge spectra of a large class of molecules, ranging from mononuclear complexes up to polymetallic clusters.[1-5]

1 M. Roemelt and F. Neese, *J. Phys. Chem. A*, 2013, **117**, 3069.
2 M. Roemelt, D. Maganas, S. DeBeer and F. Neese, *J. Chem. Phys.*, 2013, **138**, 204101.
3 D. Maganas, M. Roemelt, M. Havecker, A. Trunschke, A. Knop-Gericke, R. Schlögl and F. Neese, *Phys. Chem. Chem. Phys.*, 2013, **15**, 7260.
4 D. Maganas, M. Roemelt, T. Weyhermuller, R. Blume, M. Havecker, A. Knop-Gericke, S. DeBeer, R. Schlögl and F. Neese, *Phys. Chem. Chem. Phys.*, 2014, **16**, 264.
5 D. Maganas, S. DeBeer and F. Neese, *Inorg. Chem.*, 2014, **53**, 6374.

Alberto Roldan also replied: Atomistic simulation is helpful to determine certain parameters of materials and catalysts. However, similar to characterisation techniques, there are different levels of calculations to determine, for instance, the electronic structure of a catalyst or the dynamic formation of a solid-liquid interface. Therefore, the selection of a particular atomistic technique depends on what you wish to analyse, similar to the choice between, for example, XPS and TEM to characterise a sample.

Bruce Garrett addressed Craig Plaisance, Alberto Roldan and Dimitrios Manganas: Calculations start with the definition of the structural model of the system. The structural models are always approximations of the system probed experimentally and should represent configurations contributing to the observable of interest. These configurations can include different sites on a catalyst that may have different compositions (*e.g.*, the presence or absence of lattice vacancies) as well as changes for a single site (*e.g.* those induced by thermal fluctuations of the geometry). It is possible that the most stable structures are not the ones that contribute the most to the experimental signal. For example, for reactive processes it is the transition state with the lowest free energy of activation and not the equilibrium geometry with the lowest free energy that will contribute the most to the rate. What evidence do you have that your structural models are really representative of the ensembles contributing to the property of interest?

Dimitrios Manganas answered: As it is stated above, with no doubt, safe interpretations for a particular catalytic center can only be provided when spectra under operando conditions are collected and reaction intermediates through accurate energetic studies are identified. However, the scope of the present study (DOI: 10.1039/C5FD00193E) is a bit more fundamental, targeting spectra-structure correlations on catalytically relevant systems. The goal was to develop a combined experimental and theoretical protocol that involves a) state of the art experimental techniques such as multiwavelength Raman and multimetal soft X-ray spectroscopies and b) theoretical methods with strong predictive performance, that are applied in combination. In fact, combined experimental and theoretical multi-method protocols have been used to great advantage in (bio) inorganic chemistry and catalysis to unravel otherwise unknown structural topologies in enzymes (*e.g.* the identification of a unique carbide center in the active site of the dinitrogen activating enzyme nitrogenase), or in clarifying the structure and oxidation states of the oxygen evolving complex in photosynthesis (see references in the text of the manuscript). Hence in this study (DOI: 10.1039/C5FD00193E), a systematic computational protocol based on advanced theoretical spectroscopy was applied in order to extract the structural information of catalytically relevant centers, in both homogenous and heterogeneous catalysis. Certainly, as it is also stated in the conclusions, in the field of heterogeneous catalysis a universal structural identification of the catalytic center is a much more complicated task as there are additional factors that strongly influence the predominant structure of the active catalytic center, such as the support environment and the reaction conditions. In fact, in the case of silica supported vanadia catalysts, it has been shown that with a different support to SBA-15 and at certain conditions, a monomeric structure can dominate. Nevertheless, in our view, having a tool with such a high predictive capability forms an excellent basis for the design of dedicated spectroscopic experiments.

Craig Plaisance responded: In our study (DOI: 10.1039/C5FD00213C), we examined several active sites with very different structures and calculated the turnover frequency for each of them. Only one of these sites is present on the thermodynamically most stable (001) surface while another is present on a step edge defect. This latter site gives a calculated turnover frequency within 1–2 orders of magnitude of the experimentally measured value while the other sites have much lower turnover frequencies at the experimental overpotential. Furthermore, on this particular site, both water addition and O_2 release are calculated to be kinetically relevant, in agreement with the transient kinetics of a superoxo species observed in experiments.

More important than identifying the active site operating during the experiment, however, is to understand how the structure of the active site influences its activity. These kinds of questions are what theory is best at answering. Once we know what is the optimal active site, we can then try to synthesize a catalyst with a large number of these active sites.

Alberto Roldan also responded: We agree that the definition of the model is crucial for a reliable description of the catalyst. For this reason, a computational study requires as much experimental input as possible or needs to explore different options, *e.g.* when a reaction mechanism occurs on an idealised extended surface, it is not representative of the kinetics, as defects such as steps or vacancies should be considered as well.

Richard Catlow asked: Yours (DOI: 10.1039/C5FD00186B) and the other modelling papers show clearly the power of modelling in catalytic science. One crucial issue is the calculation of activation energies. Can you comment on the level of accuracy that we can now achieve, as accurate values are crucial if they are to be implemented in kinetic schemes?

Alberto Roldan answered: State-of-the-art computational methods allow the highly accurate calculation of energetics. However, the high accuracy often requires reducing the model size to a few atoms. Less accurate calculations, *e.g.* DFT, allow us to simulate more realistic systems and even include implicit/explicit solvents. These methodologies still provide reliable trends at an affordable cost and relatively short timescales.

Andrzej Kotarba addressed Craig Plaisance, Dimitrios Manganas, Alberto Roldan and David Willock: One of the important issues that I would like to point out is the adequate selection of the surface for molecular modelling that practical people deal with. More and more often we use DFT results to get molecular insights, to understand and interpret our experimental results. In this session we discussed the termination of Co_3O_4 nanocrystals. The most probable facets and most often obtained experimentally are (100) and (111). (The DFT calculated surface energies of the facets increased in the following order 1.39 J m^{-2} (100) < 1.48 J m^{-2} (111) < 1.65 J m^{-2} (110).[1]) The calculations described in the paper (DOI: 10.1039/C5FD00213C) were done for (311) and (110). One of the experimental parameters that can be easily compared with the calculated one is the activation energy. Obviously, we measure the apparent activation energy, but still it can be compared with the highest activation barrier present in the calculated energetic profile. Shall we be satisfied when the difference between the theoretical and experimental values is below 10%? What are the other experimental parameters that can be directly calculated and compared to validate the theoretical model? Is it necessary to go into microkinetic modelling? How about the experimental average value of the electron work function, which we found as a good descriptor of the electrodonor properties of the catalytic surface (*e.g.* strong correlation with catalytic activity in a soot oxidation reaction)?

1 F. Zasada, W. Piskorz, P. Stelmachowski, A. Kotarba, J.-F. Paul, T. Płociński, K. Kurzydłowski and Z. Sojka, *J. Phys. Chem. C*, 2011, **115**, 6423.

David Willock responded: It is interesting to see that the DFT calculated surface energies give a nice rationale for the experimentally observed facets of real crystals. You are correct that, in calculations, we should consider the major facets but also the step edges that will be present as extended defects as these may have atomic geometries required for some reaction steps. In paper DOI: 10.1039/

C5FD00213C the (311) surface can be thought of as a periodic model of a step edge on the (111) facet of a real crystal. The ability of DFT and its modifications to recover exact values of the reaction barriers in surface reactions is still limited and an absolute accuracy of 10% should still be regarded as a good agreement. The power of modelling is to allow the effect of alterations to the surface composition and/or structure on the calculated barriers to be assessed, *i.e.* will a particular modification make the catalyst more or less efficient? Microkinetic modelling allows the overall kinetics of a chain of elementary steps to be evaluated and brings in the effects of site densities and non-Langmuir behaviour into play.

Dimitrios Manganas replied: As was also mentioned in this discussion, knowledge of reliable and quantitative experimental values for energetics, on as many steps as possible, along a catalytic cycle together with the key spectroscopic properties of the key reaction intermediates, offers significant information for a valid and useful theoretical study. This requires of course that the theoretical methods are properly calibrated against well-known systems.

Craig Plaisance also responded: I would like to point out that for the Co_3O_4 system, the active site is very likely not on the most probable termination, as suggested by the results of Frei and coworkers,[1] which show that only 0.1–1% of the exposed sites are active. Thus, our active site on the (311) step edge is in good agreement with this high degree of structure sensitivity. In their experimental work, they do not determine an activation barrier but do determine a turnover frequency, which is within 1–2 orders of magnitude of the one we calculate on the (311) surface. I would consider this to be a very good agreement, especially considering that the TOF we calculate for the site on the more stable (001) surface is 3–4 orders of magnitude lower than on the (311) site at the experimental overpotential.

Other experimental parameters that could be compared are the kinetic orders with respect to reactant pressures/concentrations. Another one is the presence of reaction intermediates that are detected in the experiment. For example, Frei and coworkers[1] observe a superoxo intermediate using FTIR on Co_3O_4. The superoxo intermediate will also have a large concentration on the (311) active site since its formation is thermodynamically favorable and the reaction which consumes it (O_2 release) is kinetically relevant at the experimental overpotential. This is not the case for any of the other active sites we studied.

Concerning microkinetic modelling, I think this becomes necessary whenever the reaction network is too complex to analyze 'on paper'. For the OER, this is not the case since it essentially consists of two kinetically relevant steps occurring in series. Cycles with many branch points (Fischer Tropsch, for example) are probably too complicated to analyze on paper and require microkinetic modelling to identify the dominant pathways and kinetically relevant steps.

1 M. Zhang, M. de Respinis and H. Frei, *Nat. Chem.*, 2014, **6**, 362.

Alberto Roldan replied: Different computational approaches may provide very good agreement with the experimental values, although they are intensive and demanding. We should therefore compromise between accuracy and practicability without losing reliability. DFT calculations usually provide acceptable energy values and can define adequate trends on the behaviour of the catalyst. There are many experimental parameters to compare with, from atomic distances and geometrical analysis to spectroscopy techniques such as infrared, XPS, *etc.* and certainly the electron work function is a clear, standard comparison between experiments and electronic structure calculations.

Nicola Collis addressed Craig Plaisance and Alberto Roldan: To what extent do you consider defects in a surface? We know that surface defects are important factors for both metal particle–support interactions in supported catalysts and surface reactivity in general so modelling perfect surfaces doesn't really capture the whole story of catalyst behaviour in real systems. All of the defects are important, oxygen vacancies and terminations (imperfect coordination at edges and corners) play a big role in the nature of the final catalyst.

Craig Plaisance answered: One of the active sites that we examine is located on a step edge and we find that this site gives the highest turnover frequency at intermediate overpotentials. I would imagine that oxygen vacancies are not present to a large extent in these systems because they are operating at a very high positive potential. Cation defects are likely to be more common as Co_3O_4 is known to be a p-type semiconductor. These defects certainly play a role in the electrical conductivity which is vital to the overall performance of the catalyst, although it is beyond the scope of the current study. It may be interesting to examine the effect of cation defects near the active site in future studies.

Alberto Roldan replied: We agree on the important effect of defects in heterogeneous catalysis and it has been shown that low-coordinated atoms present a different activity than extended surfaces.[1] Nevertheless, defects are localised situations and extended surfaces are the starting point of an investigation if experimental data are not available to identify the catalytic site.

1 S. S. Tafreshi, *J. Phys. Chem. C*, 2014, **118**, 26103.

Annette Trunschke addressed Alberto Roldan and Craig Plaisance: Generally, it is assumed that only a small number of sites at the surface of a heterogeneous catalyst are active and responsible for catalysis (concept originally proposed by Sir H. S. Taylor in 1925). These sites exist in a non-balanced state (*e.g.*, low-coordinated species, defects, strained configurations). Theory tends to model the catalyst surface by minimizing the total energy of the considered system. In this way the thermodynamic equilibrium is taken into account, but metastable configurations are neglected. It has been shown, however, that high-energy sites account for catalytic activity. How does theory deal with metastable states at catalyst surfaces? Is it possible to use deviations from the thermodynamically stable state as another descriptor for catalytic activity?

Craig Plaisance answered: Generally, I think it is better to use theory to try and understand the relationship between the structure of the active site and its activity and/or selectivity, rather than use it to calculate a turnover frequency for a specific catalyst. This type of study brings a conceptual understanding that can lead to principles for designing the optimal active site for a given reaction. Once we identify the structure of the active site giving the optimal theoretical performance, whether it be a low energy structure or a metastable defect site, we can try to synthesize a catalyst having a maximum concentration of these sites.

Alberto Roldan also responded: Static modelling provides insights into the thermochemistry of a system, usually considering periodic methodologies and well-defined systems. However, many reactive sites are located in defective non-ideal surfaces making their computation demanding. Recent advances in computational methods have developed reactive interatomic potentials that allow us to explore larger systems, including defects and dynamic aspects of the catalytic site. It shows the trend in simulation methodologies to aim at closer agreement with experimental conditions. Hopefully one day we will able to model, within affordable timescales, stable and metastable states along reaction pathways.

Shelaka Gupta addressed Craig Plaisance: Do the solvation models predict the activation barrier correctly in DFT compared to the gas phase? Compared to calculations in the gas phase in DFT, are the solvent phase calculations using implicit and explicit models correct? How accurate are these models?

Craig Plaisance replied: I am not sure what the phrase 'compared to the gas phase' means since an activation barrier with a solvent present will obviously not be comparable to the activation barrier in the gas phase. Thank you for the clarification, though I still do not understand why one would want to use gas phase calculations to evaluate the accuracy of solvation methods. If you are asking whether these calculations with the solvation methods are more accurate than those in the gas phase, then yes, I would expect them to be more accurate. I do not have a great deal of experience using these solvation methods, however, so this is just my impression.

As far as the more general question "Do the solvation models predict the activation barrier correctly in DFT?", I would say that good solvation models can accurately capture the electrostatic and dispersion interactions between the solute and the solvent. What they cannot capture accurately is hydrogen bonding. This is because these solvation models are based on a dielectric screening model, while hydrogen bonding is a chemical phenomenon involving the transfer of electron density between the solute and solvent. One way to better capture the effects of hydrogen bonding is to include the first shell of solvent molecules in the simulation and use the implicit solvent model to approximate the remainder of the solvent. This way, we model these hydrogen bonds explicitly while the implicit solvation model takes care of the interactions between the explicit solvent molecules and the bulk solvent. If the implicit solvent model accurately predicts the free energy of solvation of the solvent molecule (*e.g.* a water molecule solvated in water), then one can probably assume that the interactions between the explicit solvent molecules and the implicit solvent are accurate.

Joachim Sauer commented: Knowing the atomic structure of a solid catalyst in its as prepared form is an important first step in understanding heterogeneous catalysis. For enzymes, every mechanistic understanding started from the availability of an X-ray structure. With solid catalysts it is more difficult because morphological defects and point defects may play a very important role. Care must be taken, however, when postulating defects in computational models. It is obvious that the reactivity of defects is higher. Therefore, the computations must include the full catalytic cycle with the regeneration of defects as the active site.

Alberto Roldan replied: We do agree with this comment. I would like to add that the information exchange between the experiment and the model is very important as it reduces the number of possible defects or structures to simulate, hence accelerating the outcomes.

Rutger van Santen remarked: To predict catalyst performance computationally remains a large challenge. One of the issues is best paraphrased as: 'the snake bites its tail'. During reaction conditions the surface becomes in contact with adsorbate. The surface is typically highly covered with the adsorbate. This causes substantial lateral interactions, that are difficult to model and to include in microkinetics modelling. Also, these lateral interactions may induce surface reconstruction, which implies that catalytic reaction centers are added or destructed. Such longer time events require the inclusion of instantaneous interactions of many particles at long length and time scales that can now only be done using molecular dynamics force field approaches, that are not yet very precise. The temperature of a reaction is also strongly dependent on the adsorption energies of the co-reactants, since the surface phase may alter during the reaction and depends on the reaction conditions. Proper simulation requires an iteration process of reactivity prediction, computation of altered local conditions due to product formation and recalculation of the surface phase.

Advances in the molecular understanding of heterogeneous catalyst reactivity are momentous and are due to advances in theory, spectroscopy and catalyst preparation. Currently, we are at the stage where the complexity of the interplay of catalytic reactivity events can become addressed.

Michail Stamatakis answered: Combining cluster expansions with Brønsted–Evans–Polanyi (BEP) relations is emerging as a successful approach for addressing the challenge of lateral interactions and their effects on kinetics (still on static surfaces though, no reconstructions).[1] Cluster expansion Hamiltonians are used to model in detail the energetics of the adsorbate overlayer, providing a way to calculate the energy for the initial and final state of a reaction in the presence of spectators. Then, BEP relations correlate the activation energy with the reaction energy (the latter computed by the cluster expansion), thereby providing a way to make the connection with kinetics. My lab has implemented this approach in an on-lattice kinetic Monte Carlo framework,[2] which has recently been applied to understand coverage effects on the CO–O titration experiment on Pd(111).[3] The latter study provides an explanation of the observed half-order kinetics with respect to O at high temperatures on the basis of such coverage effects, rather than O-island formation as previously thought.

1 C. Wu, D. J. Schmidt, C. Wolverton and W. F. Schneider, *J. Catal.*, 2012, **286**, 88.
2 J. Nielsen, M. d'Avezac, J. Hetherington and M. Stamatakis, *J. Chem. Phys.*, 2013, **139**, 224706.
3 M. Stamatakis and S. Piccinin, *ACS Catal.*, 2016, **6**, 2105.

Robert Madix responded: Professor van Santen raises a very important point. Let me extend the conversation somewhat. It is now well known that metal atoms on surfaces form small cluster complexes with adsorbed atoms. For example, hydrogen atoms mobilize Ni atoms on Ni(110) *via* the formation of NiH_x entities and oxygen on Ag(111) or (110) surfaces forms local surface 'oxides', even at low coverage. In these cases the species may be involved in hydrogenation or oxidation reactions, respectively. The extent to which this occurs is a matter of the relative time scale for the reaction encounter between the second reactant and the adsorbed species *vs.* the time scale for the formation of these complexes. Obviously, if one first adsorbs either hydrogen or oxygen and then adsorbs the second reactant, the complex phases will form prior to the reaction. However, if the reactants are present simultaneously in the steady state, reaction pathways involving the 'naked' adsorbed species and the complexed species will compete. To the best of my knowledge this question has received very little, if any, attention. It is of obvious conceptual importance.

Alberto Roldan also replied: Indeed, there is work ongoing to address some of these difficulties. I would like to point out also that the reactivity is directly linked to the electronic structure of the catalyst, *e.g.* the d-band centre. In combination with BEP, this allows us to extrapolate static energy profiles to similar systems (*e.g.* dopants or different metals), where the mechanism takes place along the same intermediates. These and similar approaches help us to shorten timescales from basic research to manufacturing as recognised by the European Commission and the National Research Council of US.[1,2]

1 *Nat. Mater.*, 2016, **15**, 365.
2 Integrated Computational Materials Engineering: A Transformational Discipline for Improved Competitiveness and National Security, Washington, DC: The National Academies Press; 2008.

David Willock commented: All researchers in catalysis have a working 'model' of how they think the system is working at the molecular level, this is how we decide which materials to test for a given reaction. Computer simulation is a way to test out these models through calculations alongside experimental work. It links the structural characterisation of the materials to the elementary steps that occur in the surface catalysed reactions, giving a potential energy surface with which to test the likely validity of our conceptual models.

Avelino Corma opened the discussion of the paper by Joachim Sauer: Regarding the Brønsted sites, in those systems, there should be a lot of silanol groups, have you considered that? In reality, when you have these systems you should have plenty of silanol groups. What do you expect from the calculations, considering your existing knowledge?

Joachim Sauer answered: The synthesized nano-sheets of ZSM-5,[1] of course, are terminated with silanol groups. We are running calculations on models for such sheets. The presence of the silanol groups turns out to be a challenge for optimizing the structures. The surfaces of the silica-bilayers I was discussing in my contribution feature only (hydrophobic) Si–O–Si bonds, similar to the internal surfaces of the pores of all-silica zeolites.

1 M. Choi, K. Na, J. Kim, Y. Sakamoto, O. Terasaki and R. Ryoo, *Nature*, 2009, **461**, 246.

David Willock remarked: It should be possible to calculate the acidity of the silanol defect site using the same methodology that you have presented here.

Joachim Sauer replied: Yes, it is.

Michael Bowker asked: Is the influence of the metal surface included in these calculations? Most importantly the metal conduction electrons will screen out charges on the oxide, and image charges will also be present in the metal.

Joachim Sauer replied: For the explicit calculations of the OH frequency shift on CO adsorption on the Brønsted sites we have compared the free standing film in vacuum with the film supported by the metal surface. There are only minor changes. Note, that there are only van der Waals' interactions between the film and the metal layer which means that the screening effects will be damped.

Let me also add that we use the dielectric considerations only for rationalizing the results that have been obtained by explicit DFT calculations. The analytical model can be set up by taking a different dielectric constant for the film into account, for details see our paper.[1]

1 M. Rybicky and J. Sauer, *Phys. Chem. Chem. Phys.*, 2015, **17**, 27873.

David Willock commented: You mention the increased dispersion interaction between a molecule and the surface of a pore compared to a flat surface. The dispersion is increased because the molecule is effectively closer to a larger area of the surface. The pore will also restrict the configurations available to the molecule, particularly as the dimensions of the pore approach the kinetic diameter of the molecule. So although the enthalpy of adsorption becomes more negative, the entropy of the molecule will also become less positive and so the effect of the pore structure on the free energy of adsorption is more difficult to predict.

Joachim Sauer replied: It is the loss of translational (and rotational) entropy of the gas that counteracts the energy gain on adsorption, and the balance of the two terms for specific pressure and temperature determines the adsorption equilibrium. Indeed, on a flat surface the loss of translational entropy will be less.

I have discussed energy effects only. There are two that go in opposite directions: On the flat film, stabilization of the adsorbate by dispersion is less because of the missing pore fit. On the contrary, the intrinsic binding strength may be higher because of a higher acidity which is due to the smaller dielectric constant of the thin film.

Viet Thang Ho asked: You investigated Brønsted acidity of 2D and 3D systems based on the effective dielectric constant and this parameter dependence is lower in 2D than in 3D systems, what are the effects of the thickness of the 2D material on the dielectric constant? Have you investigated the Brønsted acidity of 2D zeolites on different materials, for example MFI or MCM-22 zeolite?

Joachim Sauer replied: The thinner the layer the smaller the effective dielectric constant.[1] So far we have done calculations for 2D systems only on the hexagonal double layer described in our paper (DOI: 10.1039/C5FD00207A). There are calculations in progress for ZSM-5 thin sheets as experimentally studied.[2]

1 M. Rybicki and J. Sauer, *Phys. Chem. Chem. Phys.*, 2015, **17**, 27873.
2 Y. Seo, K. Cho, Y. Jung and R. Ryoo, *ACS Catal.*, 2013, **3**, 713.

Charles Campbell remarked: Regarding the last comment: Since the zeolite is an insulator, I assume that the charge screening length is much, much longer than the film thickness you are modelling.

Joachim Sauer replied: Indeed, the difference between the bulk materials and thin films will only show up if the film thickness is below the screening length.

David Willock continued: We have had quite a discussion on the representation of the dielectric constant of materials in the type of computer model that you use. Do the materials discussed in your paper (DOI: 10.1039/C5FD00207A) have particularly high dielectric constants?

Joachim Sauer answered: The dielectric constant for the bulk zeolite (2.9) is close to or somewhat above the values reported for dense silica polymorphs, *e.g.* 2.4 for α-quartz.[1] Our average value for the thin film (1.8) is significantly lower, and this is the very point in my discussion.

1 L. Levien, C. T. Previtt, D. J. Weidner, *Am. Mineral.*, 1980, **65**, 920.

Richard Catlow enquired: You point out that in general the absolute deprotonation energy is not a good measure of acidity, but in the case of the results on the SAPO system which you presented, there is clearly a correlation. Can you comment on the circumstances or systems for which the deprotonation will be a useful guide to acidity

Joachim Sauer replied: The deprotonation energy probes the final state of the reaction. It will be a useful measure, only if the interaction of the probe molecule in the initial state and the protonated molecule in the final state are negligible or cancel. This is expected to be the case for a similar local surface environment (SAPO-34 and SSZ-13 both have the CHA structure), but not if this environment is as different as between a planar surface (2-dimensional film) and a cavity. For gas phase molecules, there is no interaction and the deprotonation energy is generally used as a measure of acidity.

Itai Panas commented: We learned about the change in acidity when going from a zeolite bulk to a supported two-layer structure, and that crucial to the resulting acidity is the polarizability of the solid acting as a 'solvent' to the remaining charge on deprotonation. Upon performing periodic boundary condition calculations, inter-unit cell repulsion (owing to the translational symmetry) between intra-unit cell induced dipoles (monopole contribution removed) should act to artificially suppress the deprotonation. Was any dependence of acidity on supercell size (coverage) observed for the freestanding two-layer system? If so, would it be expected that a metallic support would act to reduce this effect?

Joachim Sauer responded: For the free-standing silica bilayer we have performed calculations for different cell sizes,[1] and the effective dielectric constant has been derived from a plot of the calculated deprotonation energy as a function of the inverse distance between the charges created on deprotonation. We did not do such calculations in the presence of the metal substrate, but we expect changes compared to the free standing film. The changes may turn out to be modest, because the distance between the silica bilayer and the metal support is in the range of van der Waals' bonds.

1 M. Rybicki and J. Sauer, *Phys. Chem. Chem. Phys.*, 2015, **17**, 27873.

Viet Thang Ho opened the discussion of the paper by Alexander O'Malley: There are 12 different T sites in MFI zeolite, how can you choose 3 representative T sites in your study? This means that you cannot conclude the acidity of MFI is lower than that of FAU zeolite. Also, I disagree completely with your results of the deprotonation energies without considering the stability of the Brønsted site at each Al position because the lowest deprotonation energies are not always the most stable Brønsted sites.

Alexander O'Malley replied: The number of symmetry inequivalent Brønsted sites in MFI is, in fact, significantly higher as the interaction between Al substituted T sites as well as differing proton positions result in numerous configurations and disorder (the full unit cell is also twice as large and low *T* NMR clearly shows a larger number of T sites).[1,2] The most stable, *i.e.* the lowest energy, Brønsted sites will necessarily have the highest deprotonation energy. Energy distribution of Al over a number of zeolitic frameworks has been the subject of numerous previous investigations reported in the literature.[3,4,5] Importantly, in the real material, Al siting would most probably be controlled by kinetic factors that are determined by the history of any particular sample, including the synthesis method and post-synthetic treatments, but not their thermodynamic stability. In particular, proton forms of zeolites often obtained by ion exchange in metal zeolites (*e.g.* Na or K), where siting of the extraframework cations has a crucial effect on the preferential siting of framework Al, has also been previously studied computationally.[6] Despite the plethora of possibilities, the sites of interest would most significantly differ in their local environment and accessibility by guest molecules, with the basic mechanism of acidic catalysis shared by 'similar' sites. From this point of view it is entirely reasonable to pick and investigate representative active sites, which has been done in our study (DOI: 10.1039/C6FD00010J), and is common practice in this field.

We emphasise that deprotonation energies are not a direct measure of acidity of respective Brønsted sites and are indeed only one factor amongst several that contribute to the acidity of a site.

1 C. A. Fyfe, J. H. O'Brien and H. Strobl, *Nature*, 1987, **326**, 281.
2 S. Sklenak, J. Dedecek, C. Li, B. Wichterlova, V. Gabova, M. Sierka and J. Sauer, *Phys. Chem. Chem. Phys.*, 2009, **11**, 1237.
3 K. P. Schröder, J. Sauer, M. Leslie and C. R. A. Catlow, *Zeolites*, 1992, **12**, 20.

4 R. Grau-Crespo, A. G. Peralta, A. R. Ruiz-Salvador, A. Gómez and R. López-Cordero, *Phys. Chem. Chem. Phys.*, 2000, **2**, 5716.
5 G. Sastre, V. Fornes and A. Corma, *J. Phys. Chem. B*, 2002, **106**, 701.
6 S. A. French, R. Coates, D. W. Lewis and C. R. A. Catlow, *J. Solid State Chem.*, 2011, **184**, 1484.

David Willock remarked: Your models of the adsorption of methanol to the zeolite surface suggest a significant methyl C-H interaction with the oxygen basic sites. Did you try to calculate the barrier to methyl rotation in these structures to estimate the degree to which this is hindered by these interactions?

Alexander O'Malley answered: We did not explore the rotational barrier of methyl groups interacting with basic oxygen sites. However, this development would be a sensible extension of our work (DOI: 10.1039/C6FD00010J) in assessing the strength of C–H interactions with the framework. The rotation of the methyl group would be hindered by the Brønsted acid site and the nearby framework as noted by Prof. Sauer, and torsional interactions with the methanol O–H would also affect the rotational barrier. The study of methyl rotation barriers is possible experimentally through NMR and techniques based on quasielastic neutron scattering, which can then be compared to computed values and so would be suitable complementary methods for future investigations.

Michael Bowker asked: As in the vein of Prof Sauer's paper (DOI: 10.1039/C5FD00207A), this is an atomic scale simulation, with a small spatial extent, so is the screening length factor embedded in the calculation or does it not matter? As the dielectric constant is a bulk measure is it not an appropriate factor in such calculations which operate on the nanoscale? Surely at this level a 'local dielectric constant' is both different and directional?

Alexander O'Malley answered: The approach both we and Prof. Sauer use describes the dielectric response including any anisotropic components, fully using an atomistic model for all regions near the charged species that are created. As a large region is modelled atomistically, the response of the remaining distant regions will be small and can be handled by a correction based on a continuum dielectric model

Richard Catlow commented: Regarding the calculation of the deprotonation energy, this is best considered in terms of the creation of a charged defect with the energy being determined largely by the electrostatic energy of the interaction of the charge on the defect (in this case the negatively charged deprotonated species) and the surrounding lattice with the (stabilising) polarisation energy, *i.e.* the dielectric response of the lattice.

Joachim Sauer commented: In our calculations, we explicitly included everything at the atomic level, *i.e.*, we performed supercell calculations with an increasing distance between the negatively charged defects. We then fitted an analytical expression that includes the dielectric constant[1] and obtained our effective values for the dielectric constant. That means the dielectric constant does not enter our results, it is rather used to rationalize the numeric results.

1 M. Leslie and M. J. Gillan, *J. Phys. C: Solid State Phys.*, 1985, **18**, 973.

Cynthia Friend opened the discussion of the paper by Jakub Staszak Jirkovský: Did you consider and/or measure compositional changes as the pH or the potential changed? It is possible, even likely, that composition at the surface changes and the state of the material is not necessarily at equilibrium.

Jakub Staszak-Jirkovský answered: These are important questions for the analysis of the electrocatalytic properties of alloys. There are several effects that can be foreseen. These are related to changes in the surface segregation of Pd due to the adsorption of OH and/or oxygenated species, to the stability of surface oxides/adsorbed oxygenated species and to the reduction mechanisms itself. The segregation energies of Pd in a Au host for bulk phases[1] and nanoparticles[2] calculated under vacuum are 0.28 and 0.89 eV per atom, respectively, *i.e.*, the replacement of a surface Au atom by Pd should be unfavourable. It is well established, however, that this behaviour is radically altered by the binding of adsorbates to alloy components. For Au1–*x*Pd*x* nanoalloys, electrochemical measurements showed that Pd atom surface segregation leading to oxide formation is only observed for nanoalloys with $x > 0.15$, whereas no significant oxide phase formation was observed at lower Pd concentrations at potentials more negative than 0.2 V *vs.* SCE.[3] Regarding the possibility of non-equilibrium conditions in the nanoalloys investigated, previous work showed that indeed, there is an asymmetry in the rates of Pd oxidation from the nanoalloy due to the formation of surface oxides with the corresponding abstraction of Pd atoms from the bulk of the nanoparticles and the reduction. The rate of abstraction of Pd to the surface is very fast but its re-incorporation rate to the bulk is slow.[4] The origin of this asymmetry is not clear but to avoid uncertainties in the composition, the electrodes used in this work (DOI: 10.1039/C5FD00233H) were never poised at potentials where Pd oxide formation could take place (E > 0.6 V *vs.* SCE) that could result in hysteresis effects due to the irreversibility of oxide reduction. As an additional measure, the scans from which the rate constants have been calculated always started at the negative end, just before hydrogen evolution, to ensure that the measurements corresponded to a reduced surface.

For comparison purposes, all the present work (DOI: 10.1039/C5FD00233H) was carried out at the same acid concentration, 0.1 M $HClO_4$. In experiments with nanoalloys, care has to be taken to ensure that oxide adsorption does not take place in the potential range of interest to avoid large Pd concentration changes and this introduces a severe limitation on the pH at which the measurements can be carried out. Thus, increasing the solution pH reduces the available potential range where oxide-free surfaces can be investigated.

1 A. V. Ruban, H. L. Skriver and J. K Nørskov, *Phys. Rev. B*, 1999, **59**, 15990.
2 L.-L. Wang and D. D. Johnson, *J. Am. Chem. Soc.*, 2009, **131**, 14023.
3 J. S. Jirkovský, I. Panas, E. Ahlberg, M. Halasa, S. Romani and D. J. Schiffrin, *J. Am. Chem. Soc.*, 2011, **133**, 19432.
4 J. S. Jirkovský, I. Panas, S. Romani, E. Ahlberg and D. J. Schiffrin, *J. Phys. Chem. Lett.*, 2012, **3**, 315.

Cynthia Friend continued: Under the reaction conditions, you may have metastable phases present, not only the thermodynamic phases. Have you considered this possibility and considered how you might model them?

David Willock replied: If we have structures for the metastable phases then we should be able to simulate them in the same way as any other structure. The difficulty is estimating the relative importance of the thermodynamically preferred and the large number of possible alternative structures. Experiments can help here if we have some structural information on the active form of the catalyst.

Craig Plaisance asked: In your talk, you mentioned some of the similarities and differences between electrochemical and non-electrochemical H_2O_2 production. Do you think there is fundamentally any difference between these processes at the microscopic level?

More specifically, consider we have a non-electrochemical system with an H_2 pressure exactly equivalent to the operating potential of the electrochemical system with respect to the SHE and occurring at the same pH. In both of these systems, the following equilibrium is present between the protons in solution, hydrogen atoms on the metal surface (H*), and electrons in the metal bulk:

(1) $H^+ + e^- \rightleftharpoons H^*$

In the non-electrochemical system, we also have the equilibrium between H_2 in the gas phase, protons in solution, and electrons in the bulk metal:

(2) $H_2 \rightleftharpoons 2\,H^+ + 2\,e^-$

In the electrochemical system, this reaction can be considered to occur in the reference electrode. Fundamentally, it does not seem like it would matter whether reaction (2) occurs on the catalyst surface or in the reference electrode. This would give identical chemical potentials of H^+ and e^- in the two systems, and therefore, identical H* concentrations on the catalyst surface.

Jakub Staszak-Jirkovský responded: There are two considerations here. The thermodynamics for the overall potential determining reaction is as you describe in equation (2) and the role of H* in the reduction of O_2 to peroxide subsequently in reaction (1). The rate of the chemical reaction following the Horiuti–Polanyi (H–P) hydrogenation mechanism[1] will depend both on the surface coverage by adsorbed hydrogen (M–H) and hence, on the Gibbs energy of reaction (1), and on the Gibbs energy of the reaction of H* with O_2 (or O–OH*). Electrochemical hydrogen adsorption (termed Hupd in electrochemical literature), however, does not take place on Au and no surface hydride adsorbed film can be detected on gold surfaces.[2,3] In spite of this, the electrochemical reduction of O_2 on both bulk[4] and nanoparticulate gold[5] does lead to the production of peroxide with significant yields without any evidence of the presence of Hupd on the electrode surface. Thus, in these cases, the prevalent electrochemical mechanism involves electron transfer from the nanoparticle's Fermi level to the oxygen species, either adsorbed or in solution, and not hydride transfer. The chemical and electrochemical mechanisms are dissimilar. Although for the Au–Pd alloys the surface coverage by H* at the active sites will be the same regardless of the way the potential is applied, the rate determining steps for hydride and electron transfer are very different. The former requires H_2 dissociative adsorption followed by M–H bond breaking, whereas the latter implies electron transfer. In this respect, mechanistic considerations are important to decide how to investigate possible improvements to alternatives, if any, to the current anthraquinone production method. At potentials close or below 0 V against the reversible hydrogen electrode (that correspond to the potentials where reaction (2) occurs reversibly) there could be an overlap between the electrochemical and heterogeneous mechanism. At these potentials, H* can exist at the surface and the reaction could proceed *via* a similar mechanism as in the heterogeneous case. The difference, however, is that in electrochemical systems electrons can travel through the external electrical circuit, whereas in heterogeneous systems this cannot happen. In order to preserve charge balance, both oxidation and reduction reactions must take place at the same rate, akin to a corrosion process. Please also note that in the electrochemical case only oxygen reduction reactions prevail at the Au–Pd electrode.

1 I. Horiuti and M. Polanyi, *Trans. Faraday Soc.*, 1934, **30**, 1164.
2 M. A. Rizatti and K. Jüttner, *J. Electroanal. Chem.*, 1983, **144**, 351.
3 A. Hamelin and A. M. Martins, *J. Electroanal. Chem.*, 1996, **407**, 13.
4 T. Inasaki and S. Kobayashi, *Electrochim. Acta*, 2009, **54**, 4893.
5 J. S. Jirkovský, M. Halasa and D. J. Schiffrin *Phys. Chem. Chem. Phys.*, 2010, **12**, 8042.

Michael Bowker enquired: Do you know your surface composition, and do you know what kind of surface you want? What is the role of Pd? Presumably it is in the surface in order to dissociate the oxygen?

Jakub Staszak-Jirkovský responded: We comment on the surface composition in our reply to the question from Dr Cynthia M. Friend. The overall aim was to obtain a catalytic material that could be incorporated into a fuel cell so as to allow the partial use of the Gibbs energy of the oxygen reduction reaction to generate electricity, thus providing a route for a co-generation process where the electrical energy produced would appear as a hydrogen peroxide production credit. Our original idea was to obtain metallic electrocatalytic alloy surfaces containing metal centres (the so-called 'hot spots'[1]) that would provide sites for the end-on adsorption of oxygen, since this geometry would lead naturally to the electrochemical reduction of oxygen to peroxide. Pd was an obvious choice, although DFT modelling indicated that Pt–Au alloys would be equally satisfactory. However, due to the risk of catalytic decomposition of the peroxide produced by Pt, Pd was used throughout as the gold alloying element. The role of Pd was to provide the single atom catalytic centre for O_2 binding that would result in an appropriate ensemble for the central reaction sequence described in this paper (DOI: 10.1039/C5FD00233H), thus avoiding the dissociation of the O–O bond on further reduction to produce water.

1 J. S. Jirkovský, I. Panas, E. Ahlberg, M. Halasa, S. Romani and D. J. Schiffrin, *J. Am. Chem. Soc.*, 2011, **133**, 19432.

Cynthia Friend said: Although you assert that there are isolated, single Pd atoms, that has not been demonstrated. A single Pd atom may not be sufficient. I understand that you base you assertion on that Pd is isolated, because you do not observe behavior indicating that PdO forms. On the other hand, small clusters of Pd can still have oxygen bound to them, but not form an oxide. Did you consider other possible configurations of Pd in the catalyst and specifically calculate the free energy? It is likely that there is a distribution of Pd configurations and arrangements that may contribute to various extents.

Jakub Staszak-Jirkovský answered: We did not attempt to measure the surface concentration of Pd monomers but relied on abundant literature information of the well-known properties of Au–Pd alloys. It has been known for a long time that these alloys have unusual surface properties and in particular, that ensembles involving Pd monomers, in contrast with those containing neighbouring Pd atoms, are the most stable. This property is important for understating the strong dependence of catalytic activity on alloy composition. The evidence for this has been obtained from a variety of techniques. From STM measurements, Behm *et al.*[1] demonstrated that monoatomic ensembles predominate in the low Pd concentration range (7 and 15%). Goodman *et al.*[2] confirmed these results employing low-energy ion scattering spectroscopy (LEISS), infrared absorption spectroscopy (IRAS), temperature-programmed desorption (TPD) and X-ray photoelectron spectroscopy (XPS) and concluded that these are responsible for the observed catalytic properties of these alloys.[3,4] More recent STM studies by Baber *et al.*[5] have also demonstrated that Pd monomers constitute the most stable configuration for dilute Au–Pd alloys.

Many modelling studies of this important catalytic system confirm these experimental observations. The energetically favourable presence of Pd monomers in Au–Pd alloys compared with those of dimer or trimer ensembles has been demonstrated from first principles calculations by Jacob *et al.*[6] and by Yuan *et al.*[7] The role of Pd monomers in Pd@Au ensembles for the direct $H_2 + O_2$ peroxide synthesis has been highlighted by Ham *et al.*[8] employing DFT calculations and showing the effect of a surface Pd atom selectivity in preserving the integrity of the O–O bond during hydrogenation.

Previous work on similar electrocatalysis showed that for Au and Pd, both surface oxides and adsorbed oxygenated species are electrochemically reduced at the potentials investigated for O_2 reduction (see Fig. 4 in ref. 9 DOI: 10.1039/C5FD00233H). Although we agree that other ensembles could be present in our experiments, at a low concentration of Pd nanoalloys, their surface concentration is very likely to be small not only for the reasons presented above but also due to the peroxide yields observed for the low Pd concentration nanoalloys. The presence of vicinal surface Pd atoms leads to water formation and hence decreases the peroxide yield, in agreement with the experimental observations in the present work (DOI: 10.1039/C5FD00233H) for increasing concentrations of Pd in the nanoalloys investigated.

1 F. Maroun, F. Ozanam, O. M. Magnussen and R. J. Behm, *Science*, 2001, **293**, 1811.
2 C.-W. Yi, K. Luo, T. Wei and D. W. Goodman, *J. Phys. Chem. B*, 2005, **109**, 18535.
3 M.S. Chen, K. Luo, T. Wei, Z. Yan, D. Kumar, C.-W. Yi and D.W. Goodman, *Catal. Today*, 2006, **117**, 37.
4 M. Chen, D. Kumar, C.-W. Yi and D. W Goodman, *Science*, 2005, **310**, 291.
5 A. E. Baber, H. L. Tierney and E. C. H. Sykes, *ACS Nano*, 2010, **4**, 1637.
6 S. Venkatachalam and T. Jacob, *Phys. Chem. Chem. Phys.*, 2009, **11**, 3263.
7 D. Yuan, X. Gong and R. Wu, *Phys. Rev. B*, 2007, **75**, 085428.
8 H. C. Ham, G. S. Hwang, J. Han, S. W. Nam and T. H. Lim, *J. Phys. Chem. C*, 2009, **113**, 12944.
9 J. S. Jirkovský, I. Panas, E. Ahlberg, M. Halasa, S. Romani and D. J. Schiffrin, *J. Am. Chem. Soc.*, 2011, **133**, 19432.

Michael Bowker opened a general discussion: We must believe in scientific approach and in our ability to accurately measure properties experimentally, and to predict them from theory. Using theory, we can, in the future, tackle the design of new catalysts. If we identify the intermediates and their binding energies (TPD spectroscopy), and the barriers to conversion at the rate determining step (*e.g.* using TPD or molecular beam methods), and plug these numbers into micro-kinetic models, then we can prescribe the improvement in rate for that particular step. This could be by, for example, the decreasing of the binding of a critical intermediate. In turn, theory could establish whether this would happen for a different metal, or in the presence of 'poisons' or promoters, for the experimentalist to then verify. I believe these kinds of developments are just around the corner.

Joachim Sauer remarked: I do not believe in *in silico* design of new catalysts, I believe in the powerful computational methods and tools we have today when used in close collaboration with experimental techniques. 'Theory' is behind both experiment and computation. An important issue is the relevance of the computational models for the experiments performed. The question has been raised why can't computational chemistry be more predictive for catalytic reaction steps.

Within the limitations of the Born–Oppenheimer approxiation, stable structures are the local minima on the potential energy surface and we have methods to find the global minimum structures, *e.g.* genetic algorithms. Because of the errors connected with the existing methods for calculating the potential energy surface, these structures need to be verified by experiments. We also have methods of statistical thermodynamics which allow us to predict the most stable phases under given external conditions (partial pressure) and for a given temperature. The situation is much less favorable for transition structures between minima which define the elementary reaction steps.

We are missing general techniques for identifying all possible elementary reaction steps from a given intermediate, which means that we do not have the methods for a complete exploration of the whole reaction mechanism.

Again, feedback loops between computational predictions and experiments are necessary. There are developments in the right direction, *e.g.* the work of Maeda, Taketsugu, and Morokuma.[1]

1 S. Maeda, T. Taketsugu and K. Morokuma, *J. Comput. Chem.*, 2014, **35**, 166.

Richard Catlow continued: I agree with you that *in silico* design is not a useful concept. We use modelling together with experiment to develop a molecular understanding of catalysis. It is worth noting, however, that global optimisation methods are now very effective at predicting both bulk and surface structures.

Faraday Discussions

PAPER

Supports and modified nano-particles for designing model catalysts

C. P. O'Brien,[a] K.-H. Dostert,[b] M. Hollerer,[c] C. Stiehler,[b] F. Calaza,[b] S. Schauermann,[bd] S. Shaikhutdinov,[b] M. Sterrer[c] and H.-J. Freund*[b]

Received 9th October 2015, Accepted 21st October 2015
DOI: 10.1039/c5fd00143a

In order to design catalytic materials, we need to understand the essential causes for material properties resulting from its composite nature. In this paper we discuss two, at first sight, diverse aspects: (a) the effect of the oxide–metal interface on metal nanoparticle properties and (b) the consequences of metal particle modification after activation on the selectivity of hydrogenation reactions. However, these two aspects are intimately linked. The metal nanoparticle's electronic structure changes at the interface as a catalyst is brought to different reaction temperatures due to morphological modifications in the metal and, as we will discuss, these changes in the chemistry lead to changes in the reaction path. As the morphology of the particle varies, facets of different orientations and sizes are exposed, which may lead to a change in the surface chemistry as well. We use two specific reactions to address these issues in some detail. To the best of our knowledge, the present paper reports the first observations of this kind for well-defined model systems. The changes in the electronic structure of Au nanoparticles due to their size and interaction with a supporting oxide are revealed as a function of temperature using CO_2 activation as a probe. The presence of spectator species (oxopropyl), formed during an activation step of acrolein hydrogenation, strongly controls the selectivity of the reaction towards hydrogenation of the unsaturated C=O bond *vs.* the C=C bond on Pd(111) when compared with oxide-supported Pd nanoparticles.

Introduction

The properties of dispersed metal catalysts are dominated by two factors: the size, morphology, and chemical modification of the metal nanoparticles after having been activated, and the interaction of these nanoparticles, or small metal clusters, with the support, *i.e.* the metal–support interface. These two factors are strongly

[a]*US Army Research Laboratory, USA*
[b]*Fritz Haber Institute of the Max Planck Society, Department of Chemical Physics, Faradayweg 4-6, 14195 Berlin, Germany. E-mail: freund@fhi-berlin.mpg.de*
[c]*Institute of Physics, University of Graz, Austria*
[d]*Institut für Physikalische Chemie, Christian-Albrechts-Universität zu Kiel, Germany*

interlinked and need to be addressed in model studies if we want to understand their influence at the atomic level and find design principles for new catalytic materials. This is described in this paper and exemplified by investigating two, seemingly different, model systems.

There is no doubt that the metal–support interface is one of the determining factors in catalysis, as documented by many examples in the literature,[1] yet, it is not considered in most model studies. The electronic structure of the metal–support interface changes as the morphology of the metal particle may vary as a function of temperature, for example, due to sintering. This may lead to the formation of particles with a variety of facets, which, in turn, changes the surface chemistry.[2–5]

Plausible explanations have been presented over the years for the "structure sensitivity", defined early on by Michel Boudart in connection with the function of the catalyst. It mainly refers to the size of a metal nanoparticle and relates to the various possible terminations with different facets.[4–6] This concept was taken as a basis for the surface science approach based on metal single crystal surfaces with respect to experiments,[7,8] and theoretical slab calculations,[9] with varying terminations, the result of which could be used by superposition to explain the observed catalytic behavior. While in this approach the support might be taken as the cause of the particular appearance of the nanoparticle, the support's chemical nature and specific interaction with the nanoparticle are not taken into account in such approaches.[10] The appearance of particles of different sizes and shapes may cause a modification of the catalyst during the activation process. A series of attempts has been made in the literature to consider material modification by the overgrowth of oxide films[11] or the incorporation of carbon in the surface[2] as well as subsurface areas, which only then is turned into an effective hydrogenation catalyst.[12] Robert Schlögl[13] has recently presented a comprehensive review on the dynamic catalyst.

In the present paper we would like to address both the influence of the metal–oxide interface, as well as the influence of spectator species on selective hydrogenation, in an attempt to exemplify design principles for new catalytic materials. The present work is based on and is a substantial extension of previous work on CO_2 activation on supported Au nanoparticles and on the competitive, selective hydrogenation of C=C *vs.* C=O double bonds on oxide-supported Pd nanoparticles as a function of particle size. The paper is organized as follows: after a short description of the experiments, we present and discuss the new results in comparison to the results obtained so far. We conclude with a summary and an outlook to further experiments.

Experimental conditions

The experiments on supported Au nanoparticles on ultrathin MgO(100) films were carried out using different ultrahigh-vacuum (UHV) set-ups, all of them equipped with basic sample cleaning and preparation (sputtering, heating, evaporation, gas dosing) as well as characterization (low-energy electron diffraction, LEED) facilities. One of them had, in addition, a polarization modulation infrared reflection absorption spectroscopy (IRAS) and X-ray photoelectron spectroscopy (XPS) set-up attached. Infrared spectra were acquired with a Bruker IFS66v FTIR spectrometer using a liquid nitrogen cooled MCT detector. The XPS spectrometer was a SPECS type hemispherical analyzer (Phoibos 150) with a dual anode (Mg/Al) X-ray source. Scanning tunneling microscopy (STM) experiments were carried out both at the Fritz Haber Institute using a home-built low-temperature STM operated at liquid helium temperature as well as in the Physics Department of the University of Graz using a CreaTec low-temperature STM operated at liquid nitrogen temperature. The surface of the Ag(100) crystal used as a support for the MgO(100) films was cleaned by repeated sputter (Ar^+, 800 V, 5 μA)–anneal (700 K) cycles, and the cleanliness and surface quality was checked using LEED, as well as Auger spectroscopy and XPS. Preparation of the MgO films was carried out according to published methods by reactive deposition of Mg in an oxygen atmosphere (1×10^{-6} mbar).[14] CO_2 was dosed from the background *via* a leak valve.

The experiments on acrolein hydrogenation on iron oxide supported Pd were performed with an ultra-high vacuum molecular beam machine that has been described in detail previously.[15] Molecular beams of acrolein and H_2 were directed at the sample simultaneously while the sample was held at a constant temperature. The effusive molecular beams were produced by doubly differentially pumped multi-channel array sources. Acrolein (Sigma-Aldrich, 95% purity) was purified prior to each experiment by repeated freeze–pump–thaw cycles. During all reactivity experiments the flux of H_2 on the sample surface was 4.8×10^{15} molecules per cm^2 per s. The sample was exposed to H_2 for five minutes prior to acrolein exposure. The flux of acrolein on the sample surface was 1.5×10^{13} molecules per cm^2 per s. Gas-phase fragments m/z = 56, 57, and 58 were detected with a quadrupole mass spectrometer (QMS) (ABB Extrel). Acrolein and the hydrogenation products propenol and propanal were identified by their fragmentation patterns in the QMS. Acrolein generates signals at m/z = 56, 57, and 58 with an intensity distribution of 1 : 0.04 : 0.01. Propenol causes signals at m/z = 57 and 58 with an intensity ratio of 1 : 0.1, and propanal was identified by signals at the same masses with an intensity distribution of 0.3 : 1. Surface species were detected simultaneously with the gas-phase products using an infrared spectrometer (Bruker IFS 66v) with an MCT detector and a spectral resolution of 2 cm^{-1}.

A well-ordered ~10 nm thick Fe_3O_4 film was grown on a Pt(111) substrate (see ref. 16 and 17 for details), followed by Pd deposition onto the Fe_3O_4 film at 120 K by physical vapor deposition of Pd (Goodfellow, >99.9%) using a commercial evaporator (Focus EFM 3). After depositing Pd, the sample was annealed at 600 K and the Pd nanoparticles were stabilized by repeated cycles of oxidation and reduction at 500 K.[18] The size of the Pd nanoparticles was controlled by the nominal thickness of the Pd film deposited onto the Fe_3O_4 substrate at 120 K, in comparison with the STM images reported previously (see ref. 19 for details). The Pd(111) crystal was cleaned by repeated cycles of Ar^+ sputtering at room temperature, annealing at 1000 K, and oxidation in 1×10^{-6} mbar O_2 at 750 K. The cleanliness of the Pd/Fe_3O_4 and Pd(111) samples was verified prior to every experiment by infrared reflection absorption spectroscopy (IRAS) of adsorbed CO.

Results and discussion

Activation of carbon dioxide[20,21]

In a previous study[21] we observed a scenario schematically represented in Fig. 1. A two-layer MgO film was grown on an Ag(100) substrate, and Au was deposited

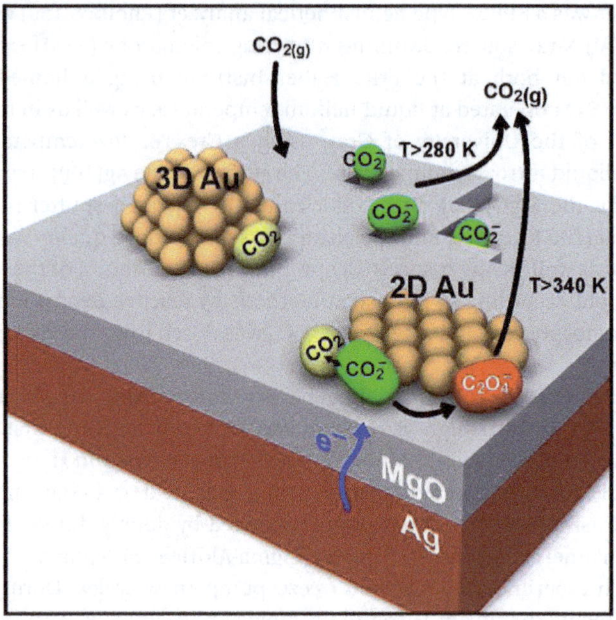

Fig. 1 Schematic showing the individual steps of oxalate formation upon chemisorption of CO_2 on the rim of 2D Au islands on thin MgO(001)/Ag(001) films.[21] Carboxylate species (green) are formed both on defect sites of the MgO film and on the rim of the 2D Au islands by electron transfer. Only on the latter, is additional CO_2 (yellow) able to solvate the carboxylate species, yielding a $(CO_2)_2^-$ dimer ion, which, after an additional electron transfer, results in adsorbed oxalate $C_2O_4^{2-}$ (red). This reaction does not occur on 3D Au particles.

from an Au evaporator at 77 K, and annealed consecutively at 300 K. It was noticed previously that deposition at a low temperature led to the two-dimensional growth of Au nanoparticles and islands,[22] as suggested by theoretical calculations.[23] The driving force for this unusual wetting of the oxide surface is an electron transfer from the Ag substrate through the film and collected by the Au islands due to the large electron affinity of Au.[23] STM investigations, together with model calculations, suggested that the charge in this metal–insulator–metal (MIM) system localizes at the rim of the nanoparticles, *i.e.* at the nanoparticle–oxide interface.[24] Having noticed this phenomenon, a study of the molecular adsorption was started. The first attempt was made using CO.[25] However, it turned out that direct imaging of the molecules was not possible, but inelastic spatially resolved electron tunneling spectra of the CO hindered translation at 45 meV provided evidence that these molecules do interact exclusively with the rim of the nanoparticle. It is interesting to note that the CO molecules do not exhibit an appreciable IR intensity.[25] This is indicative of a geometry leading to a small dipole moment perpendicular to the surface. Following up on this, we investigated the adsorption of a bigger molecule, isophorone, promising easier constant current imaging. It was possible to image the molecule at the rim, and, in addition, to study the consequences of adsorption on the quantum well states of the nanoparticle, which develop due to its finite size.[26] In fact, the energetic position of the quantum well states of the same nanoparticle was compared with those of the particle interacting with isophorone. Isophorone, as revealed by infrared spectroscopy, physisorbs on Au nanoparticles. This leads to a characteristic enhancement of the effective electron mass of the Au electrons, in line with physisorption. While this is important to understand the details of the molecule–nanoparticle interaction, these observations do not provide evidence for a reaction. This came from studies of CO_2 adsorption on these systems.[21] When CO_2 adsorbs on the surface, it is located at the rim of the clusters and it transforms to CO_2^- (see Fig. 1) by electron transfer from the Au nanoparticle. While this process is energetically uphill by 0.6 eV, the presence of additional CO_2, able to "solvate" the carboxylate to a $(CO_2)_2^-$ dimer ion, leads to a 1.6 eV downhill process. Spectroscopic studies, further elaborated and extended on below, suggest, together with model calculations, that this finally results in an oxalate ion by a second electron transfer, all happening at the rim of the nanoparticle.

Fig. 2 shows a set of infrared spectra of the surface as a function of annealing temperature. First, we concentrate on the trace recorded at the lowest temperature. The peak at 1295 cm^{-1} is due to CO_2 adsorption on the oxide support forming a carboxylate, *i.e.* not a carbonate(!), at specific sites on the MgO film. The two bands at 1220 cm^{-1} and 1440 cm^{-1} have been identified as being due to the oxalate species. This spectrum is entirely consistent with the spectra recorded before.[21] The other spectra in Fig. 2 have been recorded after annealing the Au

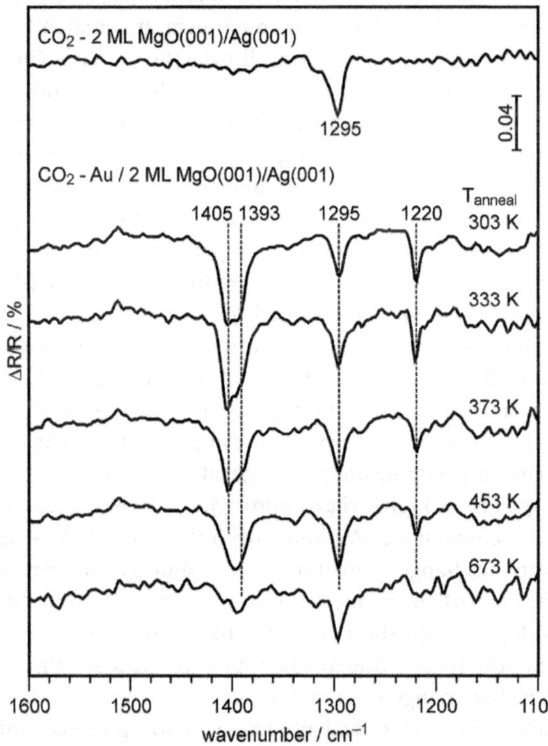

Fig. 2 IRA spectra of bare (top) and Au-loaded 2 monolayer (ML) thick MgO(001)/Ag(001) samples recorded after a saturation dose of CO_2 at 223 K. Au was deposited at 100 K and the samples were subsequently annealed at the indicated temperature prior to CO_2 adsorption.

deposits at higher temperatures. Obviously, annealing at a higher temperature does not influence the intensities in the infrared spectra massively, unless the system is annealed at or higher than 500 K. This is also consistent with our previous findings using temperature programmed desorption, where oxalate dissociation into two CO_2 molecules started to occur well below this temperature. If one annealed not quite at this temperature, then the process of oxalate formation was entirely reversible. This hinted towards a kinetically controlled process that changes the system.

Ricci *et al.*,[23] in a landmark paper, demonstrated that for nanoparticles on ultrathin oxide films, the electron transfer through the film depends strongly on the dimensionality of the Au particle on the oxide. Clearly, a two-dimensional particle was prone to accepting more electronic charge than a three-dimensional particle, and it also depended considerably on the contact area of the particle. Given this scenario, one might envision that by increasing the mobility of the Au atoms on the oxide film through annealing at a higher temperature, three-dimensional particles start to grow on the film. These particles would exhibit a small enough charge transfer that the chemical potential for forming the CO_2 anion and the oxalate is impossible. This would establish the first proof of a morphology driven change in a reaction at the oxide metal interface. In order to see whether this would be a valid interpretation, we performed and present here a STM study that is summarized in Fig. 3. This figure shows a set of STM images taken at different annealing temperatures. Fig. 3a has been recorded at 77 K and reveals Au adsorbed as single atoms or small clusters. Annealing at room temperature or slightly above (Fig. 3b) causes two-dimensional islands to form all over the surface. Their average size varies between 3 and 5 nm with a variation in shape ranging from roundish (hexagonal) islands to more rectangular (raft-like) ones. This is consistent with our previous findings.[22] Increasing the temperature to 400 K increases the average island diameter to 4–8 nm but does not change the growth mode (Fig. 3c). A drastic change occurs if the annealing temperature is raised to 500 K (Fig. 3d). We find a cross-over to three-dimensional growth with particle diameters between 5 and 10 nm, and heights of 1–2 nm. There are still a few two-dimensional islands observed, but most of the Au is contained in 3D particles.

Using this information we may now return to a discussion of Fig. 2, and try to understand the changes in the oxalate signals as a function of temperature. As stated above, the lowest temperature trace represents the oxalate species at the rim of the 2D nanoparticles, as proven in previous reports.[21] The infrared spectra start to attenuate and significantly change after the system has been either prepared at temperatures higher than approx. 500 K, or annealed for a sufficiently long time at that temperature. We know from the above STM study that this is concomitant with the change from 2D to 3D morphology. As referred to above, the formation of the 3D particles reduces the electron transfer from the Ag substrate, and, we conclude, must be the reason for the reduced ability to form oxalate species. Note that the species due to adsorption on the MgO film at 1295 cm^{-1} is not influenced by the change in growth mode.

At the present time, theoretical modeling of the process, which would be highly desirable, is not possible. Hannu Häkkinen and his group have greatly contributed to our understanding of the oxalate formation, as documented in ref. 21. However, a full, detailed description involving nanoparticles of the size observed experimentally seems to be out of range at present. Schematically, as

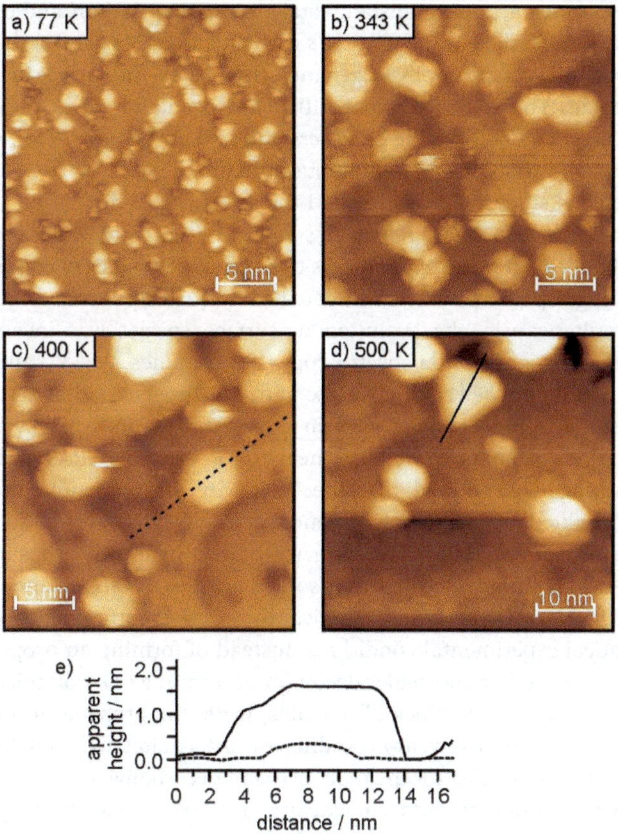

Fig. 3 STM images of (a) Au deposited on 2 ML MgO(001)/Ag(001) at 77 K, and after subsequent annealing of the system at 343 K (b), 400 K (c) and 500 K (d). All images were taken at 77 K. Scan area: (a)–(c) 25 nm × 25 nm; (d) 50 nm × 50 nm. $U_{bias} = +(0.5–0.75)$ V. $I_t = 30$ pA. (e) Height profiles of representative islands/particles from the 400 K (dashed line) and 500 K (solid line) annealed samples, highlighting the transition from 2D to 3D geometries.

depicted in Fig. 1, this means a change from reacting CO_2 at nanoparticles (2D) towards unreactive particles with a 3D morphology. To the best of our knowledge, this is the first clear experimental indication for such behavior.[27]

In order to relate the findings reported in this paper to the design of real powder-based catalysts, we refer to our study showing doping of bulk oxides with transition metals,[28] which provide an electron source to trigger a similar control of morphology as the Ag support of the ultrathin film. It will be a future task to experimentally establish this in model systems and in powder samples. However, based on the findings reported here, one may envision the control and design of CO_2 activation catalysts by doping induced charge transfer, and thus this establishes a design principle for supported Au catalysts.

Hydrogenation of acrolein[29]

Hydrogenation of C═C double bonds has been shown to depend on the presence of hydrogen atoms absorbed inside the metal nanoparticles.[2] The absorbed hydrogen provides a pool to sustain catalytic activity. Carbon and/or carbonaceous deposits adsorbed at the particle's corners and edges control the diffusion of the hydrogen necessary for hydrogenation from below the surface to the surface of the particle. The selective, competitive hydrogenation of acrolein, *i.e.* the hydrogenation of a C=C double bond *versus* the hydrogenation of a C=O double bond, represents an even more ambitious situation.[29]

In a previous study,[29] we investigated the selectivity of the partial hydrogenation of acrolein on a Pd(111) single crystal and on Pd/Fe_3O_4 model catalysts by isothermal molecular beam experiments under well-defined UHV conditions. The formation of gas-phase products was detected by quadrupole mass spectrometry (QMS); simultaneously, the evolution of surface species was investigated by infrared reflection absorption spectroscopy (IRAS) studies. Pd(111) and Pd/Fe_3O_4 showed very different selectivities in the partial hydrogenation of acrolein. Over Pd/Fe_3O_4, selective conversion of acrolein to propanal occurs, while over a Pd(111) single crystal, propenol is formed with near 100% selectivity. IRAS studies on the surface turning over showed that the selectivity for propenol formation on Pd(111) critically depends on the presence of a dense overlayer of an oxopropyl species formed at an initial stage of acrolein and hydrogen exposure. On the modified Pd(111) surface, acrolein is adsorbed *via* the C=O bond and reacts with propenol. On Pd/Fe_3O_4, however, significantly different surface chemistry occurs under identical experimental conditions. Instead of forming an oxopropyl layer, a fraction of the acrolein molecules decomposes, forming CO molecules and C_xH_y fragments that eventually block all Pd sites, while the propanal formation rate decreases to zero. Most likely, low-coordinated surface sites and (100) facets of the Pd clusters are responsible for the rapid acrolein decarbonylation. The reason for the significantly different surface chemistry on Pd/Fe_3O_4 and Pd(111), however, needs further investigation. In particular, a detailed kinetic analysis of all possible reaction pathways on the different Pd surfaces would be required. Nevertheless, our investigations unambiguously show that the modification of the Pd(111) surface with a dense oxopropyl overlayer correlates with a change in selectivity from propanal to propenol formation. Moreover, the absence of the oxopropyl layer on Pd/Fe_3O_4 seems to be related to the decarbonylation of acrolein.

Based on our previous results, we have addressed the question of whether it is possible to also produce propenol on Pd/Fe_3O_4 model catalysts. We could think about a large number of different approaches, such as modifying the Pd clusters prior to acrolein conversion, *e.g.* by pre-adsorbed CO or hydrocarbons or by C modification of low-coordinated Pd sites. In the present study, however, we focus on the effect of Pd particle size. We have investigated the hydrogenation of acrolein on 7 nm and 12 nm Pd particles using molecular beam techniques under isothermal conditions. The formation of the gas-phase products has been detected by QMS and the surface composition has been simultaneously investigated by IRAS. In all experiments, the surface was pre-exposed to 4.8×10^{15} H_2 per cm^2 per s for 300 s before the acrolein beam with 1.5×10^{13} molecules per cm^2 per s was switched on.

Fig. 4 shows the formation rates of the partial hydrogenation products on Fe_3O_4-supported 7 nm and 12 nm Pd particles detected in the gas phase. With both particles sizes, similar propanal formation rates are observed, which pass through a maximum after about 30 s and then decrease to zero. On the 12 nm particles, however, a small amount of propenol is also detected. The propenol production rate rapidly increases after about 40 s, when the propanal formation has almost stopped, passes through a maximum after about 60 s and finally decreases to zero. In general, both partial hydrogenation products have been identified by their characteristic fragmentation patterns in QMS. The clearly different time dependence of the two product formation rates in this study, however, provides an additional possibility to unambiguously distinguish between propanal and propenol. Nevertheless, the total amount of propenol that is formed on the 12 nm particles is much smaller than on a Pd(111) single crystal. It should be noted that the formation of gas-phase propenol has only been detected in a narrow temperature range near 250 K. It was not observed in our previous studies on Pd particles under slightly different conditions, such as a lower or higher surface temperature or when using a pulsed acrolein beam instead of continuous exposure.

Fig. 5 illustrates the correlation between the evolution of propenol in the gas phase detected by QMS and the formation of surface species on the 12 nm Pd particles turning over, as studied by IRAS. A time-dependent series of IR spectra obtained with a time resolution of 45 s is shown in Fig. 5a. IR absorption features appear near 1855 cm^{-1}, 1755 cm^{-1}, and 1670 cm^{-1}. The vibration at 1670 cm^{-1} is most likely associated with the C=O stretching in molecularly adsorbed acrolein. It has been assigned to acrolein adsorbed on a Pd(111) single crystal in a previous study.[29] The adsorbate giving rise to the IR absorption near 1755 cm^{-1} accumulates at the beginning of the acrolein exposure and saturates during the first 45 s. Interestingly, this band has previously been related to the oxopropyl species modifying the Pd(111) surface for propenol production. The IR vibration at 1855 cm^{-1} most likely shows CO adsorbed on the Pd(111) facets. The slowly increasing IR absorption intensity indicates a rather slow accumulation of CO, most likely from acrolein decarbonylation. Hence, the decarbonylation of acrolein is significantly less efficient at 250 K compared to our previous studies at 270 K. Fig. 5b illustrates the simultaneously recorded propenol evolution in the gas phase. A

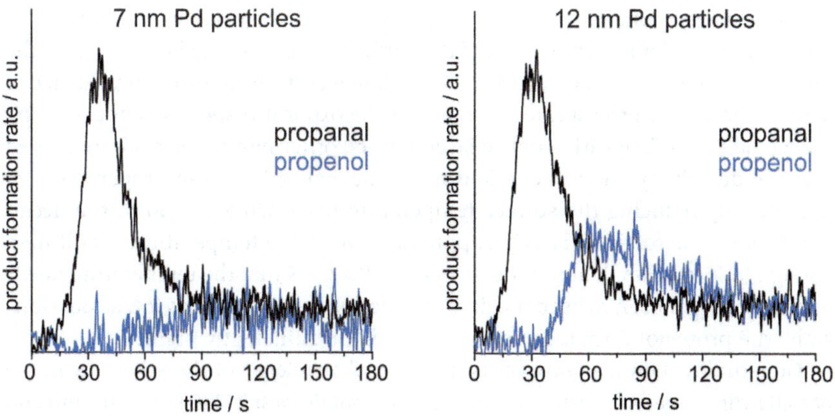

Fig. 4 Formation rates of propanal (black lines) and propenol (blue lines) over Fe_3O_4-supported Pd particles with diameters of 7 nm (left) and 12 nm (right) at 250 K, detected by QMS.

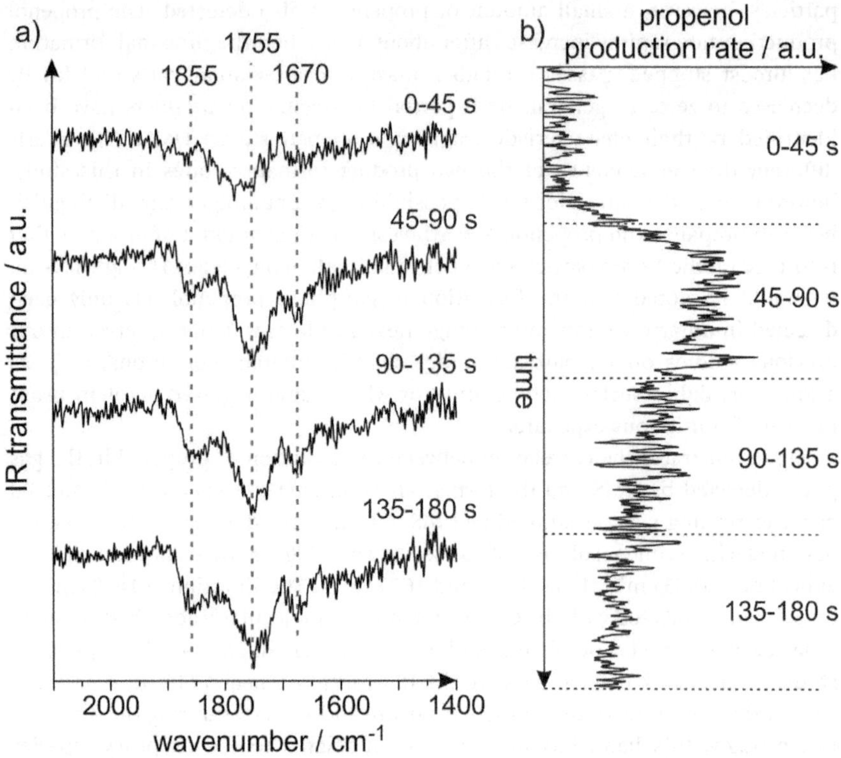

Fig. 5 (a) Time-resolved IR spectra monitoring the evolution of surface species on the 12 nm Pd particles turning over and (b) QMS measurements showing the formation rate of propenol in the gas phase.

clear onset is observed after about 40 s, which is approximately when the oxopropyl species saturates.

The simultaneously performed IRAS and QMS studies on the 12 nm Pd particles turning over at 250 K give detailed insights into the mechanism of the acrolein conversion. On the one hand, our results strongly reveal that on the 12 nm particles a modification of the surface by an oxopropyl layer triggers the propenol production. In contrast to Pd(111), however, the additional presence of acrolein and CO on the surface reveals that the oxopropyl species is not covering the whole Pd surface with a dense layer. On the other hand, it is again indicated that the decarbonylation of acrolein has to be avoided to form propenol on Pd particles. By reducing the surface temperature from 270 K to 250 K, the decarbonylation of acrolein is largely suppressed, while the temperature is still high enough to form a small amount of propenol. It seems that the temperature needs to be carefully chosen to keep the decarbonylation reaction rate as low as possible, while the propenol formation rate should still be sufficiently high.

Our studies show a significant particle size dependence of the selectivity in the partial hydrogenation of acrolein over Pd model catalysts at 250 K. On both 7 nm and 12 nm Pd particles, a similar amount of propanal is formed. Hydrogenation of acrolein to propenol, in contrast, appears possible only on 12 nm particles. The clearly different time dependence of propanal and propenol evolution agrees well with our previous findings, showing propanal formation as long as pristine Pd is available and propenol production, in contrast, after a dense oxopropyl overlayer is formed. It seems that the selectivity towards propenol formation increases with increasing Pd cluster size from 0% on 7 nm Pd particles and a small amount over 12 nm Pd particles to approximately 100% on a Pd(111) single crystal. Possibly, the formation of an oxopropyl layer on sufficiently large domains is necessary to produce propenol, which is more likely to happen on larger Pd clusters having larger (111) facets and a lower concentration of low-coordinated sites. Low-coordinated sites, such as edges, corners, and defects, may catalyze acrolein decarbonylation.

Synopsis

The present study clearly points to two important factors influencing the catalytic activity of supported metal nanoparticles, which influence the design of new catalytic materials:

• The interaction with the support controls the size and shape of the nanoparticle, and, in particular, the charge exchange and transfer at the oxide–metal interface. Specifically designed model systems allow us to isolate details of the problem and study them at the atomic level. Here, the surface science approach turns out to be the method of choice to image and characterize molecules interacting with the metal particle at the metal–oxide interface using a combination of scanning probe and spectroscopic techniques. This has allowed, for the first time, the direct identification of the consequence of molecular adsorption and reaction on the electronic structure of a metal nanoparticle. Temperature is decisive in determining the mobility of metal atoms and particles on supports. We show in this paper that a change in particle morphology controls the reactivity of the system at the oxide metal interface, as the transition from two-dimensional Au particles at a lower temperature to three-dimensional growth at a higher temperature reduces the charge transfer from the support to the metal nanoparticle.

• Adsorbed species formed during a reaction, which are not turned over, have been identified as being able to modify a supported metal nanoparticle, and *via* this modification, control the selectivity of a hydrogenation reaction. The selective hydrogenation of acrolein has been studied as an example. In particular, an oxopropyl species formed from adsorbed acrolein modifies the surface in such a way that acrolein hydrogenation leads to propenol and the hydrogenation of the C=C double bond is avoided. We have demonstrated before that Pd(111) single crystals selectively hydrogenate the C=O bond in acrolein, while supported Pd nanoparticles exclusively hydrogenate the C=C bond. We show in this paper that there is a particle size effect. The particle needs to have a certain size in order to expose (111) facets of a sufficient size to favor the effect. If particles are smaller, presumably the presence of edges, corners and irregularities in the particles favor the formation of CO, which poisons the oxopropyl effect.

The knowledge derived from both of these observations can be used to design catalytic materials and control catalytic reactions.

Acknowledgements

We are grateful to the German Science Foundation through the cluster of Excellence Unicast (administered by the TU Berlin) as well as the Fonds der

Chemischen Industrie for financial support. F. Calaza thanks the Alexander von Humboldt foundation for a Georg Forster Research Fellowship. C. Stiehler thanks the Studienstiftung des Deutschen Volkes for a fellowship. S. Schauermann is grateful to the European Research Council for a Starting Grant "Enantioselective Reactions on Model Chirally Modified Surfaces". M. Sterrer acknowledges support from the European Research Council through Grant Agreement No. 280070 (STRUBOLI).

References

1 H. K. G. Ertl, F. Schüth and J. Weitkamp, *Handbook of Heterogeneous Catalysis*, VCH, Wiley-VCH, Weinheim, compl. rev. and enlarged ed. edn., 2008.
2 M. Wilde, K. Fukutani, W. Ludwig, B. Brandt, J.-H. Fischer, S. Schauermann and H. J. Freund, *Angew. Chem., Int. Ed.*, 2008, **47**, 9289–9293.
3 S. J. Thomson and G. Webb, *J. Chem. Soc., Chem. Commun.*, 1976, 526–527.
4 G. C. Bond, *Metal-Catalysed Reactions of Hydrocarbons*, Springer, US, New York, 2005.
5 J. Sauer and H.-J. Freund, *Catal. Lett.*, 2015, **145**, 109–125.
6 M. Boudart and G. Djéga-Mariadassou, *La cinétique des réactions en catalyse hétérogène*, Masson, Paris, New York, 1982.
7 G. Ertl, *Angew. Chem., Int. Ed.*, 2008, **47**, 3524–3535.
8 G. Ertl and H. J. Freund, *Phys. Today*, 1999, **52**, 32–38.
9 J. Sauer, *Chem. Rev.*, 1989, **89**, 199–255.
10 H. J. Freund, N. Nilius, T. Risse and S. Schauermann, *Phys. Chem. Chem. Phys.*, 2014, **16**, 8148–8167.
11 S. Shaikhutdinov and H.-J. Freund, *Annu. Rev. Phys. Chem.*, 2012, **63**, 619–633.
12 D. Teschner, J. Borsodi, A. Wootsch, Z. Révay, M. Hävecker, A. Knop-Gericke, S. D. Jackson and R. Schlögl, *Science*, 2008, **320**, 86–89.
13 R. Schlögl, *Angew. Chem., Int. Ed.*, 2015, **54**, 3465–3520.
14 J. Pal, M. Smerieri, E. Celasco, L. Savio, L. Vattuone and M. Rocca, *Phys. Rev. Lett.*, 2014, **112**, 126102.
15 J. Libuda, I. Meusel, J. Hartmann and H.-J. Freund, *Rev. Sci. Instrum.*, 2000, **71**, 4395.
16 W. Weiss and W. Ranke, *Prog. Surf. Sci.*, 2002, **70**, 1–151.
17 C. Lemire, R. Meyer, V. E. Henrich, S. Shaikhutdinov and H. J. Freund, *Surf. Sci.*, 2004, **572**, 103.
18 T. Schalow, B. Brandt, D. Starr, M. Laurin, S. Schauermann, S. Shaikhutdinov, J. Libuda and H. J. Freund, *Catal. Lett.*, 2006, **107**, 189–196.
19 T. Schalow, B. Brandt, D. E. Starr, M. Laurin, S. K. Shaikhutdinov, S. Schauermann, J. Libuda and H. J. Freund, *Phys. Chem. Chem. Phys.*, 2007, **9**, 1347–1361.
20 H.-J. Freund and M. W. Roberts, *Surf. Sci. Rep.*, 1996, **25**, 225–273.
21 F. Calaza, C. Stiehler, Y. Fujimori, M. Sterrer, S. Beeg, M. Ruiz-Oses, N. Nilius, M. Heyde, T. Parviainen, K. Honkala, H. Häkkinen and H.-J. Freund, *Angew. Chem., Int. Ed.*, 2015, **54**, 12484.
22 M. Sterrer, T. Risse, U. Martinez Pozzoni, L. Giordano, M. Heyde, H.-P. Rust, G. Pacchioni and H.-J. Freund, *Phys. Rev. Lett.*, 2007, **98**, 096107.
23 D. Ricci, A. Bongiorno, G. Pacchioni and U. Landman, *Phys. Rev. Lett.*, 2006, **97**, 036106.

24 X. Lin, N. Nilius, M. Sterrer, P. Koskinen, H. Haekkinen and H.-J. Freund, *Phys. Rev. B: Condens. Matter Mater. Phys.*, 2010, **81**, 153406.

25 X. Lin, B. Yang, H. M. Benia, P. Myrach, M. Yulikov, A. Aumer, M. Brown, M. Sterrer, O. Bondarchuk, E. Kieseritzky, J. Rocker, T. Risse, H. Gao, N. Nilius and H. J. Freund, *J. Am. Chem. Soc.*, 2010, **132**, 7745–7749.

26 C. Stiehler, F. Calaza, W.-D. Schneider, N. Nilius and H.-J. Freund, *Phys. Rev. Lett.*, 2015, **115**, 036804.

27 M. Chen and D. W. Goodman, *Chem. Soc. Rev.*, 2008, **37**, 1860–1870.

28 X. Shao, S. Prada, L. Giordano, G. Pacchioni, N. Nilius and H.-J. Freund, *Angew. Chem., Int. Ed.*, 2011, **50**, 11525–11527.

29 K.-H. Dostert, C. P. O'Brien, F. Ivars-Barcelo, S. Schauermann and H.-J. Freund, *J. Am. Chem. Soc.*, 2015, **137**, 13496–13502.

Faraday Discussions

PAPER

Effects of particle size and edge structure on the electronic structure, spectroscopic features, and chemical properties of Au(111)-supported MoS_2 nanoparticles

Albert Bruix,* Jeppe V. Lauritsen and Bjørk Hammer

Received 30th November 2015, Accepted 13th January 2016
DOI: 10.1039/c5fd00203f

Materials based on MoS_2 are widely used as catalysts and their structure usually consists of single-layered MoS_2 nanoparticles whose edges are known to constitute the catalytically active sites. Methods based on density functional theory are used in this work to calculate the electronic structure of representative computational models of MoS_2 nanoparticles supported on Au(111). By considering nanoparticles with different edge-terminations, compositions, and sizes, we describe how the electronic structure, Mo3d core-level shifts, and chemical properties (*i.e.* H adsorption and S vacancy formation) depend on the MoS_2 nanoparticle size and structure. In addition, site-specific properties, largely inaccessible when using only slab models of MoS_2 edges, are reported, which reveal that the edge sites are not uniform along the nanoparticle and largely depend on the proximity to the corners of the triangular NPs, especially when interacting with a metallic support. Furthermore, a structural motif where H atoms adsorb favourably in a bridging position between two Mo atoms is proposed as an active site for the hydrogen evolution reaction.

Introduction

During the last five decades, materials based on nanostructured MoS_2 have become standard as hydrotreating catalysts for applications in fossil fuel refinement for removing sulphur and nitrogen heteroatoms from the hydrocarbons in crude oil.[1,2] More recently, different forms of nanostructured MoS_2 (and other transition metal dichalcogenides[3]) have been proven to be active for the hydrogen evolution reaction (HER).[4] In the most stable bulk polymorph (2H stacking), MoS_2 forms layered structures consisting of stacked S–Mo–S sheets bound weakly by van der Waals interactions, which facilitate the preparation of single-layer (SL) structures. The versatility of materials based on MoS_2 is related to interesting alterations of the electronic structure of MoS_2 upon changes in its dimensionality.

Interdisciplinary Nanoscience Center (iNANO), Department of Physics and Astronomy, Aarhus University, DK-8000 Aarhus C, Denmark. E-mail: abruix@phys.au.dk

In particular, MoS_2 becomes a direct gap photoluminescent semiconductor upon the separation to single layers,[5,6] in contrast to the indirect gap featured by the bulk crystal. In turn, the under-coordinated edges of MoS_2 nanoparticles (NPs), which are significantly reconstructed with respect to the bulk truncations of the single layers, have metallic character[7,8] and are generally considered as the active sites of catalysts based on MoS_2,[9,10] whereas the basal plane of these NPs is considered to be more chemically inert.[11]

The higher activity of the edge sites of MoS_2 has spurred efforts directed at designing nanostructured MoS_2-based materials exposing solely active edge-sites,[12,13] and model systems consisting of supported MoS_2 nanoparticles with high edge concentrations have typically been used in fundamental experiments for characterizing the (catalytic) properties of the edges.[8,14,15] These studies have revealed that SL MoS_2 is predominantly terminated by the so-called Mo-edges $(10\bar{1}0)$ and S-edges $(\bar{1}010)$ due to their lower edge-formation energies,[16] which leads to a strong preference for triangular and hexagonal nanoparticle shapes, depending on the synthesis conditions.[14,15,17–19] In addition, the stability of each termination also depends on their composition, *i.e.* the amount of sulphur covering the edges of the nanoparticle.[16,20–22] With respect to edge-stability, different theoretical studies based on density functional theory (DFT) agree that, at least in a gas-phase environment, the most stable termination corresponds to Mo-edges,[16,21] and that their edge composition depends on the working conditions.[16,20–22] Therefore, under a high chemical potential of S (*i.e.* in S-abundant conditions), the most stable Mo-edge composition corresponds to fully sulphided edges where an edge S dimer is coordinated to each edge Mo (Fig. 1 – Mo-edge 100% S). However, at a reduced chemical potential of S (S-lean conditions),

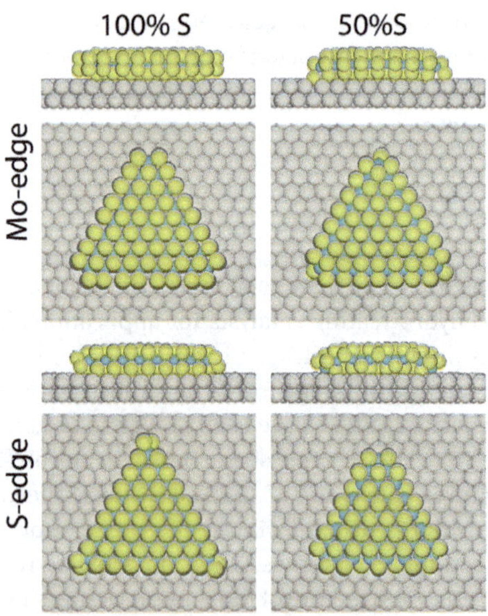

Fig. 1 MoS_2 nanoparticle models ($n = 8$) supported on the Au(111) surface. Both S- and Mo-edge type nanoparticles are shown with either 100% S or 50% S-edge compositions. Periodic images of the nanoparticles have been removed for clarity.

a situation where only half of the edge S atoms remain is preferred (Fig. 1 – Mo-edge 50% S). In fact, the Mo-edge with 50% S was recently experimentally observed in model systems of MoS_2 on Au(111) when modifying the preparation conditions towards lower S abundance.[23] In turn, for the S-edge, the transition from 100% S to 50% S (Fig. 1 – S-edge 100% S and 50% S) is also predicted to take place, but at much lower S chemical potentials. Interestingly, MoS_2 nanoparticles in industrial-style catalysts were recently found, using atom-resolved transmission electron microscopy (TEM), to predominantly expose the Mo-edge with 50% S,[24,25] which highlights the relevance of this particular termination.

In a recent study combining *in situ* X-ray photoemission spectroscopy (XPS) and DFT calculations, we have shown that the metallic character of the 100% S Mo-edges of Au(111)-supported MoS_2 nanoparticles affects the spectral shape of the Mo3d core-level.[26] In particular, the presence of a metallic state along the 100% S Mo-edges stabilises the formation of core-holes through screening effects, which reduces the binding energies of the 3d core electrons of the edge Mo atoms. Furthermore, reduction to the industrially relevant 50% S covered Mo-edge involves the loss of the metallic character of the edge. As a result, such reduction results in shifts towards a higher core-level binding energy (BE) of the edge Mo3d levels, which contrasts with conventional interpretations of redox processes and XPS. In the present work, we extend this analysis to the core-level shifts (CLS) of MoS_2 NPs with S-edge terminations, which may also become exposed under industrially relevant conditions,[15] and compare the evolution upon reduction of such shifts to the Mo-edge NP case. Moreover, by considering nanoparticles of different lateral sizes, we explore the role of size-effects in the electronic structure properties of Au-supported MoS_2 NPs.

During the hydrodesulphurisation (HDS) process on MoS_2 catalysts, H_2 molecules adsorb and dissociate on S atoms located at the edges of MoS_2 nanoparticles. The adsorbed H atoms then participate in the hydrogenation of the heteroatom of an S-containing hydrocarbon molecule, which is bound through the heteroatom in favourable anchoring sites, such as S vacancies or the so-called brim sites. The catalytic cycle is completed with the S–C bond scission and the subsequent desorption of an S-free hydrocarbon molecule and H_2S.[27–29] The bonding strengths of S and H on the edge sites therefore play an important role in such reaction mechanisms and, interestingly, have been shown to be good descriptors for the desulphurisation and HER activity of MoS_2-based catalysts. Following Sabatier's principle, Mo–S bonds in active catalysts should be strong enough to efficiently bind S heteroatoms, but not so strong as to prevent S removal as H_2S.[30,31] Similar trends govern the HER activity, for which volcano plots have been derived, indicating that the more active sites are those where H adsorption is moderate, *i.e.* with close to null free energy of hydrogen adsorption.[32–36] However, some studies have suggested that instead of equilibrium thermodynamics, reaction barriers should be used as the descriptor for the HER activity. Anyhow, determining the ability of the different sites in MoS_2 nanoparticles to bind hydrogen and S atoms is crucial for assessing their catalytic potential. In order to relate the electronic structure of the edges to their chemical properties, in this work we have also studied how changes in nanoparticle type (S-edge or Mo-edge), edge composition, and size affect their ability to bind H and S atoms. It should also be noted that previous studies have modelled MoS_2 NP edges by means of infinite periodic stripe models,[20,22,27,28,32,37] and only in some cases, supported nanoparticles have been modelled,[38,39] although limited to small $n = 3$ sizes. Therefore, the models used here (and in ref. 26) are the first to simultaneously address size-representative nanoparticles and the presence of a support. This allows consideration of site-specific electronic and chemical properties within supported MoS_2 nanoparticles, which are inaccessible when using freestanding stripe-models.

Computational details

MoS_2 nanoparticle models supported on Au(111)

Periodic models have been used to study the structure and properties of MoS_2 NPs supported on the Au(111) surface. In particular, we have considered triangular single-layer MoS_2 NPs exposing either Mo-edge $(10\bar{1}0)$ or S-edge $(\bar{1}010)$ terminations (Fig. 1). In model catalyst studies, the mean size of the triangular MoS_2 NPs supported on Au(111) corresponded to $n = 8$ for the selected coverage,[14] where n is the number of Mo atoms along each edge. For NPs with completely sulphided edges, such size corresponds to a stoichiometry of $Mo_{36}S_{104}$ and $Mo_{36}S_{90}$ for the Mo-edge and S-edge particles, respectively. Models representing these systems and their reduced 50% S counterparts (Mo-edge: $Mo_{36}S_{80}$, S-edge: $Mo_{36}S_{63}$ – also shown in Fig. 1) have thus been chosen to study the electronic structure of the MoS_2 nanoparticles on Au(111). In order to investigate the differences in chemical properties for the four different types of supported MoS_2 NPs (Mo- and S-edge, 100% and 50% S), we have calculated S vacancy formation and H adsorption energies at different positions for MoS_2 NPs represented by models of size $n = 6$ instead. These triangular NPs have equivalent structures to the larger models shown in Fig. 1, but are smaller ($Mo_{21}S_x$) and therefore more computationally feasible for a full systematic study of S and H binding energies. Independent of their size, the four different NP types are labelled according to their edge and S coverage as Mo-100, Mo-50, S-100, or S-50. The H-adsorption $E_{ads}(H)$ and S vacancy formation $\Delta E(S_{vac})$ energies are calculated by

$$E_{ads}(H) = E(H/MoS_2/Au) - \frac{1}{2}E(H_2) - E(MoS_2/Au), \quad (2.1)$$

$$\Delta E(S_{vac}) = E(MoS_{2-x}/Au) + E(H_2S) - E(H_2) - E(MoS_2/Au), \quad (2.2)$$

where $E(MoS_2/Au)$, $E(H/MoS_2/Au)$, $E(MoS_{2-x}/Au)$, $E(H_2)$, and $E(H_2S)$ are the energies of the supported MoS_2 NP, the NP with an adsorbed H atom, the NP with a sulphur vacancy, a single H_2 molecule, and a single H_2S molecule, respectively. The free energy of H adsorption ΔG^0_H, which can be used as a description for the HER activity,[32–36] is calculated by

$$\Delta G^0_H = E_{ads}(H) + \Delta E_{ZPE} - T\Delta S_H, \quad (2.3)$$

where ΔE_{ZPE} and ΔS_H are the differences in the zero point energy potential and entropy, respectively, between the adsorbed H and the gas-phase H_2 states. To a good approximation and as calculated in previous studies involving similar systems,[32–34,40] these contributions to the free energy under standard conditions (atmospheric pressure, 300 K) can be included by the addition of a constant offset value ($\Delta G^0_H = E_{ads}(H) + 0.29$ eV).

In addition, in order to investigate size effects in supported MoS_2 NPs, $\Delta E(S_{vac})$ and CLS have been compared for Mo-edge models with $n = 6$, 8, and 10 (where n is the number of Mo atoms at each edge of the triangular MoS_2 NPs). The underlying Au slab used for supporting the $n = 6$, 8, and 10 models consists of (8 × 8), (10 × 8), and (12 × 10) supercells of two-layer Au(111), respectively. The Au(111) slabs were constructed with the optimised lattice parameter (4.17 Å) at the level of theory used (*vide infra*).

Density functional theory calculations

DFT. The DFT calculations were performed using the PBE exchange–correlation functional, the Projector Augmented Wave (PAW) method,[41] and a real space grid for the expansion of the wave functions, as implemented in the GPAW code[42,43] and supported by the Atomic Simulation Environment (ASE).[44] The geometries for the different models used were optimised until the forces on each atom were lower than 0.025 eV $Å^{-1}$ and the electronic structure at each geometry step was self-consistently converged with energy, density, and eigenvalue thresholds of 5×10^{-4} eV, 1×10^{-4}, and 5×10^{-8} eV, respectively. A grid spacing of 0.175 Å was used and the reciprocal space was sampled using only the gamma-point. The upper layer of the two-layer slab used to represent the Au(111) substrate was also allowed to relax.

STM simulations. The STM simulations are based on the Tersoff–Hamann approximation, which relates the intensities of the STM images with the iso-surfaces of the local density of states at the Fermi level.[45,46] The potential used for the simulations is $V = -0.80$ V and the current intensity is $I = 0.0017$ nA, which, to a good approximation, corresponds to a charge density contour of 8.2×10^{-6} $Å^{-3}$,[47] the same as previously used in STM simulations of MoS_2.[15,26]

Core level shifts calculations. In XPS, a beam of photons is used to photo-excite a core-level electron. The binding energy (BE) of the core electron is the energy difference between the excited state with a core-hole and the fundamental ground state, which can be determined experimentally by measuring the kinetic energy of the photo-excited electron. Each element has characteristic core-level BEs that also depend on the local environment of each atom. Therefore, for a particular element, the measurement of such chemical shifts provides information about the oxidation state, coordination environment and bonding (hybridisation). In order to calculate the core-level shifts (CLS) for the different Mo species present in Au(111)-supported MoS_2 NPs, we have used the full core-hole approach as implemented in GPAW,[48] which consists of a final state approximation explicitly introducing a core-hole in the selected Mo atoms and neglecting the excited electron (*i.e.* removing it to infinity). This is achieved by using a different PAW-type setup with an electron hole in the selected core-level for describing the atom the electron is to be removed from. Since the removed electron is not explicitly reintroduced in the computational cell, a compensating homogeneous background density is implicitly included as the calculations assume charge neutrality. In this paper, we only focused on the shifts in BE, which can be calculated more accurately than the absolute BE. Mo3d CLS between Mo atoms at different positions of the same model can be simply calculated as the differences in the total energy between creating the core-hole in different Mo atoms. Furthermore, we have calculated all Mo3d chemical shifts relative to an

Au(111)-supported Mo_{21} nanoparticle, which represents metallic Mo. In order to compare the CLS of Mo atoms in different models (those in MoS_2 nanoparticles relative to the metallic Mo_{21} nanoparticle), we used a spectator Mo atom located at the opposite side of the slab of each system (Fig. 2). Since this spectator atom is identical in all the different systems considered, this approach partially eliminates the errors arising from calculating the relative BE using different models. Furthermore, instead of calculating the shifts with respect to just one Mo atom, we considered the average shift with respect to all atoms of the Mo_{21} NP. This method for calculating the CLS was successfully used in combination with *in situ* XPS on model MoS_2/Au to study the reduction of Au-supported MoS_2.[26]

Results and discussion

Electronic structure of the different models

The electronic structure of the Mo-edge MoS_2 nanoparticles was found to change significantly when the edge composition was reduced from the 100% S state (corresponding to complete sulphidation) to 50% S,[22,26] which results from removing half of the edge S atoms from the NP. Such reduction involves the substitution of $S_2^{\delta-}$ dimers by more oxidizing S^{2-} species and was found to lead to the loss of the one-dimensional metallic state along the edges (see side views in Fig. 1, which also reveal that S^{2-} monomers are shifted towards bridging sites on the Mo-edge). In addition, it is found that the electronic structure of the MoS_2 NPs also depends on the type of edge they are terminated with (*i.e.* S-edge or Mo-edge). STM studies on hexagonally shaped MoS_2 nanoparticles produced under sulforeductive conditions revealed very different features for S-edge and Mo-edge terminatons.[14] In order to investigate the differences between NPs with different terminations, we have compared the evolution of the electronic structure of the S-edge and Mo-edge particles upon reduction from the 100% to the 50% S state. In particular, we compared the simulated STM images (Fig. 3) as well as the density of states projected on the valence 4d orbitals of the Mo atoms (pDOS-Mo4d, Fig. 4), the density of states projected on the valence 3p states of the S atoms (pDOS-S3p, Fig. 4), and the calculated CLS (Fig. 5). As shown in previous work,[26] the changes in the electronic structure of the Mo-edges upon reduction from 100% S to 50% S appear as a lower pDOS-Mo4d near the Fermi level (Fig. 4) and also in the simulated STM images as a significant reduction of the bright brim near the edges of the MoS_2 NPs (Fig. 3 – Mo-edge). The pDOS for the Mo-100 NP is very similar to those calculated for a freestanding MoS_2 NP,[49] although the

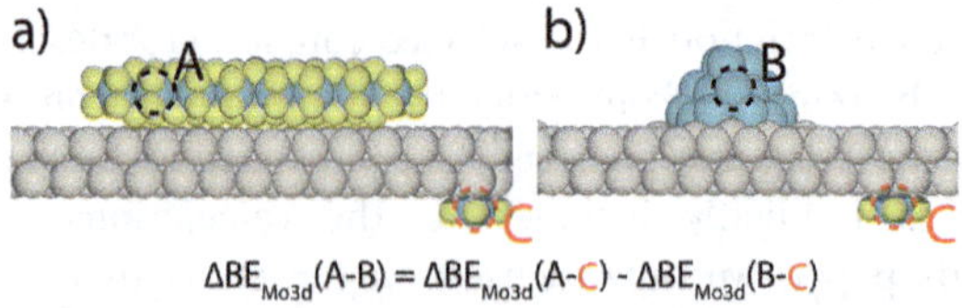

Fig. 2 Models for $Mo_{36}S_{104}$ ($n = 8$) and Mo_{21} nanoparticles supported on Au(111) with the spectator Mo atom present at the opposite side of the slab. The equation used to calculate the core-level shifts with respect to a metallic NP is shown, where A–C correspond to Mo atoms indicated by dashed circles.

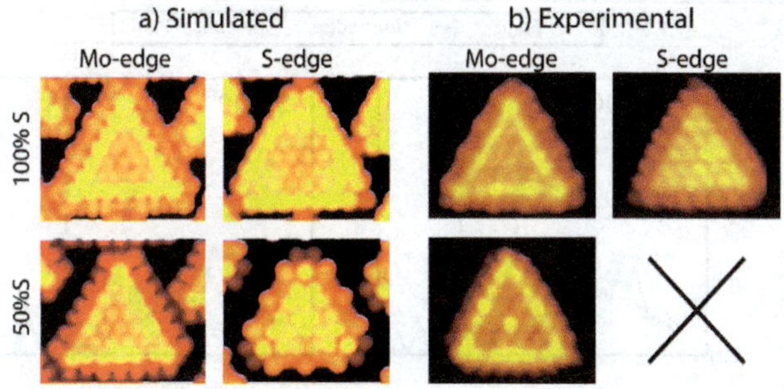

Fig. 3 (a) Simulated STM images for S-edge and Mo-edge 100% S and 50% S generated with the Tersoff–Hamann approximation (with $I = 0.017$ nA, $V = -0.80$ V). The bright edge-brim of the 100% S Mo-edge partially disappears upon reduction of the edge, whereas the removal of S from the 100% S S-edge lowers (increases) the intensity of the edge (basal plane) S atoms. (b) Experimental STM images of the different $n = 8$ NPs. The Mo-100 image is taken from ref. 26. The Mo-50 and S-100 images result from the data set reported in ref. 23 for MoS_2 synthesized under different sulfiding conditions (which promote the synthesis of such structures). An experimental image of the S-edge 50% S particle is not included, as it has not been experimentally observed.

interaction with the Au(111) surface seems to lead to the formation of gap states for the basal plane atoms, revealed as a small, albeit non-zero, pDOS near the Fermi level. Remarkably, the brim in the 100% S Mo-edge NP is related to two metallic states, which are localised at edge Mo and S atoms.[22] In accordance, the pDOS-S3p around the Fermi level for the near-edge S atoms forming the bright brim is larger than for the rest of the S atoms of the NP. In turn, the STM simulation of the S-edge 100% S NP reveals a higher intensity of the S atoms right at the edges.

A metallic state is present along the S-edges as well,[22] and the edge S atoms that appear brighter in the STM simulations also feature a larger pDOS-3p around the Fermi level (Fig. 3). It should also be noted that the higher simulated STM intensity of such edge S atoms is an electronic effect which STM is sensitive to and it does not imply that their height is elevated from the MoS_2 plane. In fact, the edge S atoms (and also the corner S atoms, which also appear brighter) of the S-edge 100% S NP are geometrically ~0.2 Å lower in height than the basal plane S atoms. Furthermore, in contrast to the situation at the Mo-edge, where reduction eliminates the metallic character, the reduction of the S-edge from 100% S to 50% S leads to a higher pDOS-Mo4d near the Fermi level (Fig. 4 – S-edge 50% S). In addition, for this NP, the basal plane S atoms appear brighter in the STM simulations and their pDOS-S3p is also slightly larger. The different electronic structure changes of the S-edge can also be traced back to the reducing nature of the removed S species. S atoms at the S-edges are similar, in terms of charge (oxidation state), to the S^{2-} monomers at the edges of the 50% S Mo-edges. Therefore, removing such species reduces the Mo atoms at the edges, making them more metallic. In addition, the Mo pDOS of the S-100 NP is significantly different from that reported for a freestanding S-100 NP,[50] which suggests that NPs with S-edges interact more strongly with the Au substrate compared to the Mo-edge NPs.

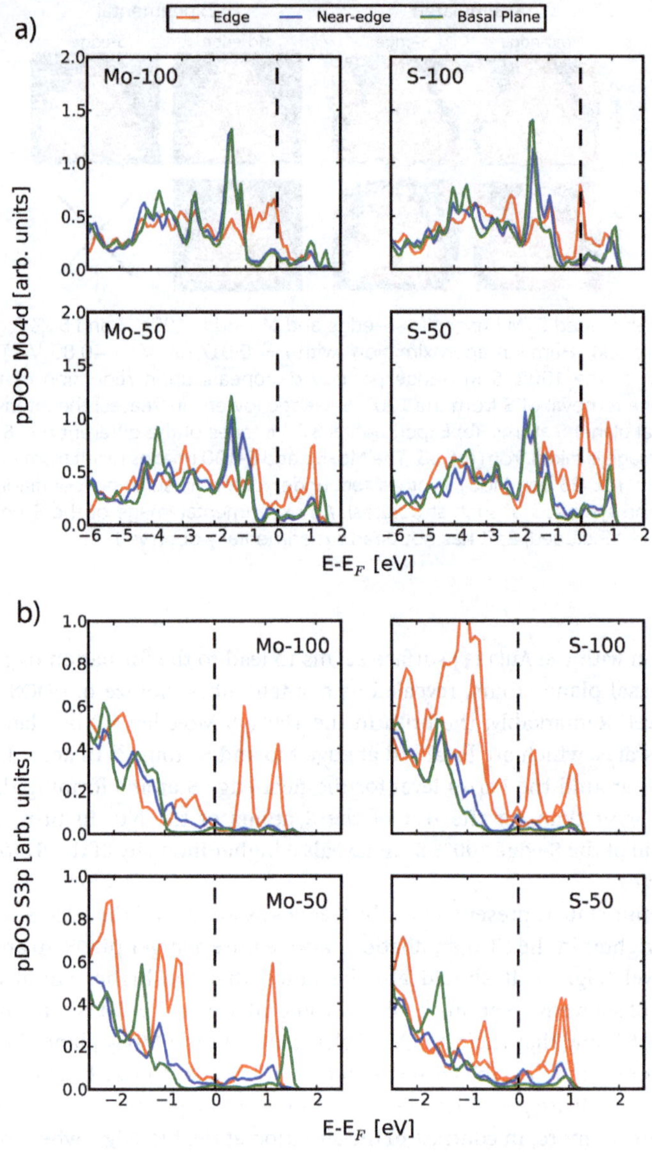

Fig. 4 Density of states projected on the valence (a) 4d states of Mo atoms and (b) 3p states of S atoms of Mo- and S-edge NPs with 100% S and 50% S compositions at the edges. The metallic character of the 100% S-edges for both NP types manifests as a higher pDOS for Mo atoms located at the edge with respect to those at the near-edge or basal plane positions. Upon reduction to the 50% S-edge state, the metallic character of the Mo-edge (S-edge) disappears (strengthens).

In order to evaluate the quality of the simulated STM results, a set of images of the same MoS_2 NPs from STM experiments under UHV conditions[26] are shown in Fig. 3b. The agreement between the simulated and experimental results is good, especially for the Mo-edge terminated NPs. In particular, the bright brim and the off-registry protrusions at the edge of the Mo-100 NP are well reproduced, as well

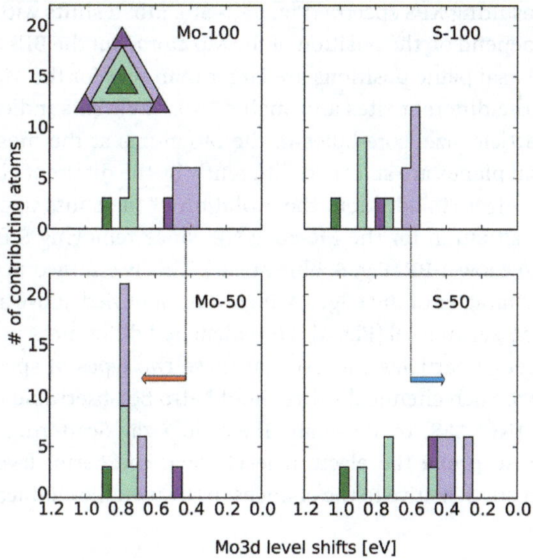

Fig. 5 Mo3d CLS for the Mo- and S-edge MoS_2 nanoparticles supported on Au(111) with either 100% or 50% edge composition. The different colors within the histograms denote the location of the Mo and S atoms (corner, edge, near-edge or basal plane) in the MoS_2 triangle contributing to the different Mo3d shifts, according to the color-coding indicated within the triangle drawing. Reducing the Mo-edge MoS_2 NPs merges the contributions from Mo atoms at different positions in the same BE region (red arrow), whereas reduction of the S-edge NPs leads to opposite shifts (blue arrow), *i.e.* towards lower BE.

as their decrease in brightness upon reduction to the Mo-50 state. For the S-100 NP, the in-registry position of the protrusions at the edges is well reproduced by the simulations, but the greater brightness of the S atoms at the basal plane with respect to those at the edge is not.

Mo3d core-level shifts

The differences in the electronic structure described above are reflected in the calculated Mo3d CLS for the four MoS_2 NP models. The metallic state along the 100% S Mo-edges gives rise to site-specific Mo3d BE shifts within the NP. In particular, the Mo3d BE for the atoms located at the edges is ~0.4 eV lower than for atoms located at the basal plane (Fig. 5 – Mo-100 and S-100). In addition, the disappearance of the edge metallic state upon reduction leads to shifts in the 3d states of the Mo atoms towards higher BEs, which merges contributions from the edge and basal plane Mo atoms into the same BE region (Fig. 5, red arrow). These calculated shifts agree well with the XPS experiments of the MoS_2/Au(111) model catalyst, where two peaks separated by 0.4 eV are observed on highly sulphided samples, corresponding to the edge and basal plane Mo species.[26] Furthermore, the calculated shifts of the edge Mo atoms towards a higher BE when removing S atoms from the Mo-edge can be detected experimentally as an increase in the relative intensity of the high BE peak with respect to the lower BE one when reducing the model MoS_2/Au(111) catalyst samples.[26] In turn, the calculated shifts for the S-edge NPs show that the electronic structure differences described above will affect the resulting XPS spectra (Fig. 5 – S-50). Mo3d shifts within the 100% S S-edge NP also depend on the position of the Mo atom, but the BEs of the atoms in both edge and basal plane positions are larger than that for the Mo-edge and the shifts between the different sites are smaller (~0.3 eV). This indicates that, even for this nanoparticle size, core-holes in the Mo atoms at the most central positions of the basal planes are screened differently by the different charge densities of the S- or Mo-edges. In addition, the evolution of the shifts upon reduction is opposite to the situation for the Mo-edge NP, since removing S from the edges leads to shifts to a lower BE (Fig. 5, blue arrow). This is clearly due to the increase in the metallic character of the edges, which was identified above as an increased pDOS around the Fermi level (Fig. 4). The calculated differences in the CLS of the Mo-edge and S-edge particles suggest that these two types of species are distinguishable by XPS. Such chemical shifts should also be observable in near-edge X-ray absorption (NEXAFS) or Resonant Inelastic X-ray Scattering (RIXS) experiments, which also probe the electron levels near the Fermi level, and should therefore be directly sensitive to the changes in the metallic character of the edges upon reduction.

Chemical properties of Au(111)-supported MoS_2 nanoparticles

The binding energies of S and H atoms in MoS_2-based catalysts are directly relevant in the steps of different catalytic mechanisms such as HDS and the HER. In order to investigate the ability of different sites of Au(111)-supported MoS_2 NPs to bind S and H, we have calculated the adsorption energy of single H atoms (E_{ads}(H)) and the S vacancy formation energy ($\Delta E(S_{vac})$) at every inequivalent S site of four different $n = 6$ MoS_2 nanoparticle models. In particular, we have considered the same NP types as in the previous section (Mo-edge and S-edge, 100% and 50% S) but with a more computationally feasible particle size. Therefore, these smaller models present the same edge configuration as those shown in Fig. 1, but with shorter edges and fewer basal plane atoms.

S vacancy formation. Basal plane S atoms have been removed from both the upper and lower S layers, thereby assessing how the proximity to the Au surface affects vacancy formation. In turn, vacancies formed at the edges would lead to the same structure, independent of whether the S atom is removed from the upper or lower layer and, therefore, only one $\Delta E(S_{vac})$ value is given for every edge position. Nevertheless, we must point out that local relaxations from starting geometries created after removing either the upper or lower layer S atoms from the edge do not always converge to the same structure, in which case the reported values correspond to the most stable relaxed geometry.

All $\Delta E(S_{vac})$ values were found to be positive, which indicates that the formation of S vacancies is unstable with respect to consuming H_2 and forming H_2S for all the different S positions and NP types considered (Fig. 6). However, in strong contrast to the uniform $\Delta E(S_{vac})$ energies obtained from semi-infinite slab calculations,[20,22,27,37,51] some interesting trends emerge showing that the S vacancy formation energy strongly depends on the edge termination, its composition, and the position of the removed S atom within the MoS_2 NP. This important result indicates that the sites along the edges of MoS_2 NPs are not uniform and that, in principle, the reactivity is modulated along the edge positions. Thus, for the completely sulphided nanoparticles, it is clearly more favourable to remove under

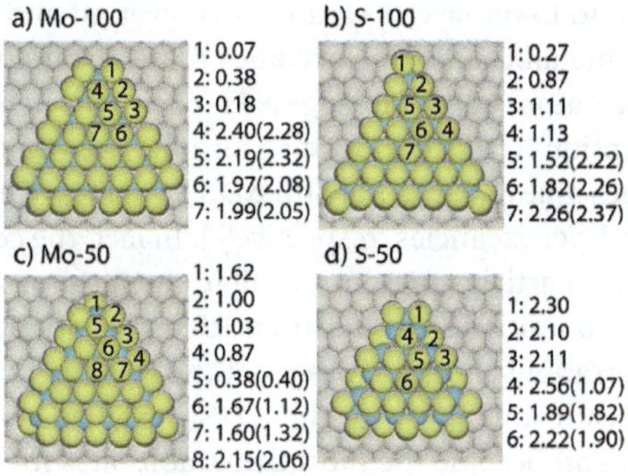

Fig. 6 $\Delta E(S_{vac})$ values in eV for the sulphur vacancies at different positions of $n = 6$ Au(111)-supported MoS_2 nanoparticles. The values in parenthesis correspond to the $\Delta E(S_{vac})$ values for the S atoms removed from the lower layer of S closer to the Au(111) surface.

coordinated S atoms from the edges than more coordinated S atoms from near-edge or basal plane positions. Given this trend, it is not surprising that the most stable S vacancy results from removing the least coordinated S atoms, *i.e.* those found in the corners of the triangular nanoparticle, with $\Delta E(S_{vac})$ values of 0.07 eV and 0.27 eV for the Mo-edge 100% S and S-edge 100% S NPs, respectively (Fig. 6a and c). This is in very good agreement with the high resolution STM studies of MoS_2 NPs on Au(111), which revealed that S vacancies are easily formed at the corners of the MoS_2 NPs, facilitating the adsorption of refractory hydrocarbons.[52] We note that S vacancies at the basal planes of extended MoS_2 single layers have been recently shown to be active for the HER,[36] but on the basis of their high energy, such defects would hardly be present at the less extended basal planes of MoS_2 NPs. Another interesting result is the lower $\Delta E(S_{vac})$ value associated with removing an S atom from position 3 with respect to position 2 in the Mo-100 NP. Such a difference in energy is attributed to the stabilising interaction between neighbouring S_2 dimers at the Mo-edge; when an S vacancy is formed at position 2, the corner S_2 dimer does not have any neighbouring analogue to interact with. In contrast to the completely sulphided NPs, for the reduced 50% S NPs the formation of S vacancies at the edge is not preferred and the most stable situation corresponds to removing an S atom from the near-edge position closest to the NP corner (positions 5 and 4 for Mo-edge 50% S and S-edge 50% S, respectively – Fig. 6b and d). Remarkably, the absolute $\Delta E(S_{vac})$ value and the relative stability between removing the upper and lower S atoms from near-edge positions depend on the NP type and composition. This also affects, although to a smaller degree, the $\Delta E(S_{vac})$ value of the basal plane positions, which indicates that this particle size ($n = 6$) is not large enough to include sites that are fully representative of more extended basal plane regions present in larger particles. Thus, for the Mo-edge 100% S, the upper and lower layer vacancies in the near-edge and basal plane positions are similar in energy, but the formation of S vacancies in the upper layer is slightly preferred (except in the corner basal plane position – 4). On the other hand, for the S-edge 100% S NP, it is significantly more favourable to remove S atoms from the upper S layer than removing those in contact with the Au(111) surface. In contrast, for the 50% S NPs with the S-edge, the relative stability between the upper and lower layer S vacancies is reversed and removing S atoms in contact with the metallic substrate becomes more favourable (Fig. 6b and d). However, the corner S atom of the Mo-edge NPs (position 5 in the 50% S structure – Fig. 6b) is an exception to this trend, since forming the vacancy at either of the layers has almost the same energy cost (with $\Delta E(S_{vac})$ values of 0.38 and 0.40 eV for the upper and lower layer vacancies, respectively). In fact, the corner basal plane S atoms of the reduced particles present peculiar chemical properties, which are related to the lower coordination of the atoms that they are bound to and to the larger structural relaxations that the corners of the NPs can undergo. These sites thus present significant differences in the $\Delta E(S_{vac})$ values with respect to other near-edge positions and become the most favourable sites for the formation of S vacancies for the reduced nanoparticles.

The comparison of the calculated $\Delta E(S_{vac})$ values for the different NPs also reveals that, not surprisingly, the more sulphided NPs (*i.e.* with more S atoms) are easier to reduce that NPs with fewer S atoms. Therefore, the different systems studied here are ordered from lower to higher $\Delta E(S_{vac})$ values as follows: Mo-100 ($Mo_{21}S_{66}$ – 0.07 eV) < S-100 ($Mo_{21}S_{56}$ – 0.27 eV) < Mo-50 ($Mo_{21}S_{48}$ – 0.38 eV) < S-50 ($Mo_{21}S_{35}$ – 1.07 eV).

The chemical properties of the sites near and at the corners of the MoS_2 NPs are relevant, but in both model and industrial-style catalysts the MoS_2 NPs are generally larger than the $n = 6$ model used here. This means that the MoS_2 nanoparticles studied experimentally expose longer edges and therefore a larger number of edge-sites. It therefore makes sense to compare the chemical properties of just the most edge-like sites (*i.e.* those farthest from the corners) for each of the different NPs studied. Furthermore, these are comparable to the sites exposed by stripe models of extended MoS_2 edges used in previous studies.[20,27,28,37,53] The most edge-like S atoms correspond to atoms 3, 4, 4, and 3 (in Fig. 6) for the Mo-100, Mo-50, S-100, and S-50 NPs, respectively. The trends for the $\Delta E(S_{vac})$ values in such sites indicate that the formation of S vacancies is easier at the Mo-edges than at the S-edges, independent of the edge composition. Therefore, the different edges studied here are ordered from lower to higher $\Delta E(S_{vac})$ values as follows: Mo-100 (0.18 eV) < Mo-50 (0.87 eV) < S-100 (1.13 eV) < S-50 (2.11 eV). The edge $\Delta E(S_{vac})$ value obtained for the Mo-100 NP (0.18 eV) is significantly lower than the one reported for the Mo-100 edge of an unsupported stripe model ($\sim$0.45 eV).[51] In turn, the edge $\Delta E(S_{vac})$ value calculated for the S-100 NP (1.13 eV) is twice as large as that for the S-100 edge of an unsupported stripe model (0.54 eV).[37] In the case of the S-100 edge, the difference in energy required for removing an S atom can be attributed to the different edge terminations exposed by the Au(111)-supported NP and the freestanding stripe model. The latter undergoes a significant reconstruction,[37,54] which may facilitate the formation of S vacancies. The presence of the Au support and the finite size of the NP limit the mobility of the edge, thereby preventing such reconstructions and the exposure of sites more prone to S removal. In addition, it should be noted that removing a single S atom from a stripe model results in a vacancy concentration and stoichiometry that depend on the unit cell size. Thus, defect concentration differences probably also contribute to the reported $\Delta E(S_{vac})$ differences. Since the Mo-100 NP is expected to only weakly interact with the Au(111) substrate, such coverage effects and the proximity to the corner of the Mo-100 NP are likely to lead to the differences in $\Delta E(S_{vac})$ with respect to an unsupported Mo-100 slab.

However, for the reduced NPs, the calculated edge $\Delta E(S_{vac})$ values are very similar to those reported for stripe models with the same edge composition;[37] 0.87 (0.79) eV and 2.11 (2.09) eV for the Mo-50 and S-50 NPs (stripe model), respectively. Furthermore, and less surprisingly, the $\Delta E(S_{vac})$ value calculated here for the Mo-50 edge (0.87 eV) is also in good agreement with those calculated for MoS_2 NPs with the same edge composition (~1 eV).[21] This good agreement indicates that, for these edge terminations, some properties of the most edge-like sites of the supported MoS_2 NP are quite well represented by the widely used stripe models even when the presence of the support is not taken into account. It also implies that the edge sites of the MoS_2 nanoparticles exposing the Mo-50 edges (as those reported in ref. 23) are representative of those in free-standing (weakly interacting) MoS_2 nanoparticles. Nevertheless, and as clearly demonstrated in the present discussion, considering the presence of the Au substrate as well as the explicit zero-dimensionality of the MoS_2 NPs provides site-specific chemical properties largely inaccessible when using solely one-dimensional stripe models. In addition, the importance of including the support in MoS_2 NP models is further highlighted by comparing the $\Delta E(S_{vac})$ value for the corner S atoms of the Au-supported Mo-50 particle (position 1) calculated in this work (1.62 eV) with that calculated for the same position of the corresponding freestanding nanoparticle (0.5 eV).[21] It is thus clear that these corner S atoms bind strongly to the Au(111) surface and that omitting this interaction leads to an inverted $\Delta E(S_{vac})$ order with respect to the edge S positions.

H adsorption. Inequivalent S atoms of the $n = 6$ Mo-100, Mo-50, S-100, and S-50 models have been considered as adsorption sites for single H atoms. Only the upper layer S atoms have been examined for adsorption at the near-edge, basal plane, or edge positions. Similar to the just described trends in the formation energy of S vacancies (*vide supra*) and in contrast to the slab calculations, the adsorption of H atoms results in $E_{ads}(H)$ values that strongly depend on the edge termination, its composition, and the position of the adsorption site within the MoS_2 NP (Fig. 7). Remarkably, the $E_{ads}(H)$ values range from −0.48 to 1.20 eV,

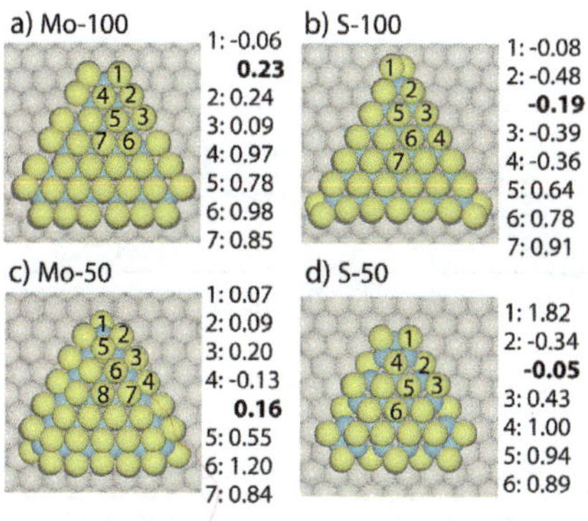

Fig. 7 $E_{ads}(H)$ values in eV for H atoms adsorbed at different positions of $n = 6$ Au(111)-supported MoS_2 nanoparticles. The corresponding ΔG^0_H values (also in eV) are included in bold for the most stable adsorption sites of each NP. For the other sites, ΔG^0_H can also be calculated straightforwardly by adding a 0.29 eV off-set to the $E_{ads}(H)$ values, as an approximation for the zero-point energy and entropic contributions.[33,34]

which indicates that, depending on the NP and hydrogenated site, the adsorption of a single H atom can be more (E_{ads}(H) < 0) or less (E_{ads}(H) > 0) stable with respect to a gas-phase H_2 molecule. In particular, for all the NPs studied, H adsorption on S atoms at the near-edge or basal plane positions is clearly unfavourable, with E_{ads}(H) ranging from 0.55 to 1.20 eV. The basal plane ΔG^0_H values (1.23 to 1.30 eV) are significantly lower than those recently reported for extended layers of MoS_2 without S vacancies,[36] which suggests that these sites have different chemical properties. Furthermore, there is at least one adsorption site in each NP with E_{ads}(H) < 0, and only the S-100 NP has several adsorption sites capable of adsorbing H. This is in good agreement with previous theoretical modelling studies reporting stable hydrogenation of S-edges.[22,27,32,37] In addition, the E_{ads}(H) differences between the edge sites within each NP show that the under-coordination of corner sites as well as the proximity to such sites significantly affect H adsorption. Notably, the most stable adsorption site within the Mo-100 NP corresponds to the corner S atom and, in contrast, for the Mo-50 NP, the most stable situation corresponds to hydrogenating the S atom at the edge position farthest from the corner (Fig. 8a).

As was done for the S vacancy formation in the previous subsection, H adsorption on the most edge-like sites for each NP can be compared to the results obtained with Au(111)-supported MoS_2 stripe models.[32] Such edge-like S sites correspond to atoms 3, 4, 4, and 3 (in Fig. 7) for the Mo-100, Mo-50, S-100, and S-50 NPs, respectively, and the resulting ΔG^0_H values at these edge sites are 0.38, 0.16, −0.07 and 0.72 eV, respectively. The ΔG^0_H value for the Mo-50 NP (0.16 eV) agrees quite well with that reported by Tsai *et al.* for the same edge type at low coverage (0.20 eV). However, the Au support-induced partial occupation of the antibonding orbitals reported by these authors leading to weaker S–H bonds is not reproduced for the supported Mo-50 NP (Fig. 8a), which may be what leads to the slightly more favourable H adsorption on the latter. The same authors do not calculate H adsorption on S-100 or S-50 edges, but their reported ΔG^0_H values for the S-75 edge (0.35 eV) lies right in between those calculated in this work for the edges of S-100 (0.07 eV) and S-50 (0.72 eV) NPs. According to previously established trends,[32–34,40] the exothermic ΔG^0_H value (−0.05 eV) calculated for site number 2 of the S-50 NP indicates that it should be active for the HER.

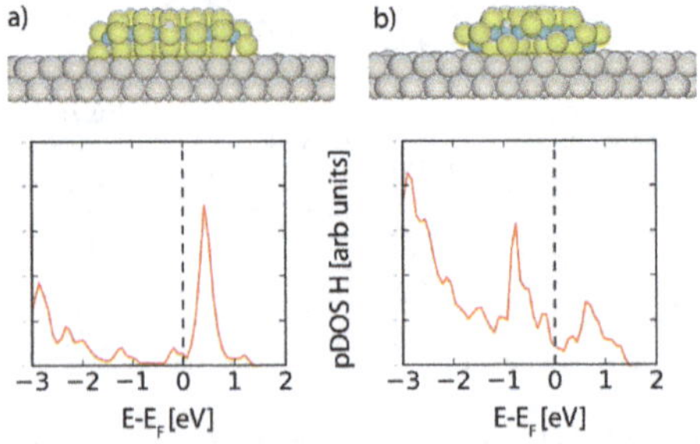

Fig. 8 The most stable structures after the adsorption of a single H atom on the Mo-50 (a) and S-50 (b) MoS_2 nanoparticles, leading to the formation of an S–H bond or to H in a bridging position between two under-coordinated Mo atoms, respectively. The DOS projected on the H atoms is shown under each structure.

Furthermore, it should be noted that when H is adsorbed at this site, it does not form a S–H bond with the sulphur. Instead, the structural relaxation leads to H in a bridging position between the two exposed Mo atoms above the sulphur atom (Fig. 8b). This suggests that MoS_2-based catalysts featuring edges with under-coordinated Mo atoms that can accommodate H in such bridging positions would be particularly active for evolving hydrogen, although such catalysts may be difficult to synthesise according to the trends in $\Delta E(S_{vac})$ described above. This result constitutes an addition to the recent proposals of sites specifically designed for the HER, such as those found in dimeric Mo_2S_x (ref. 35) and trimeric Mo_3S_x (ref. 13) clusters or on S vacancies on strained MoS_2 basal planes,[36] whose activity has been rationalised on the basis of their ΔG^0_H values. However, it should be pointed out that, although active sites for the HER require moderate ΔG^0_H values, this does not imply that all sites with such ΔG^0_H values are active.

Size-effects

The particle size of nanostructured materials is known to play an important role in their catalytic activity. Such effects are often related just to the larger number of surface (or other under-coordinated) sites that smaller particles expose to reactants. However, in some notorious cases, such as that of Au-based catalysts, size-effects operate in ways that go beyond what can be attributed to an increased surface area. Activity towards some reactions is therefore only achieved for very reduced nanoparticle sizes.[55] In the case of model catalysts of MoS_2 NPs supported on the Au(111) surface, the structure of the MoS_2 nanoparticles was shown to depend on their size.[14] In particular, smaller particles are preferentially terminated with S-edges, whereas larger particles are terminated with Mo-edges instead. Consequently, these changes also lead to interesting size-related reactivity, as smaller particles terminated with S-edges bind HDS reactants more efficiently.[52] Here, we investigate how some spectroscopic features and chemical properties of Mo-100 and Mo-50 MoS_2 NPs are affected by their size. This is relevant, since the experiments in ref. 26 reflect a certain size distribution and thus the spectral shape in XPS is a convolution of these sizes. In particular, we have calculated the Mo3d CLS for MoS_2 NPs with $n = 6$, 8, and 10 (where n is the number of Mo atoms at each edge). In addition, the effect of particle size on the reactivity of the edges of these NPs has also been analysed by calculating $\Delta E(S_{vac})$ resulting from removing an S atom from the centre of the edge (*i.e.* the farthest position from the corner of the NP).

Effect of size on the Mo3d CLS. The Mo3d CLS of the Mo atoms within the $n = 8$ Mo-100 NPs have been shown to depend on the position of the Mo atom within the MoS_2 NP and for the Mo atoms located at the edges, the shifts also depend strongly on the edge composition (*vide supra*). The calculated Mo3d CLS for the smaller ($n = 6$) and larger ($n = 10$) particle sizes are presented in Fig. 9 together with the results for the $n = 8$ nanoparticle. For the three 100-Mo NPs, the shifts for the corner and edge Mo atoms (corresponding to dark and light purple bins, respectively) appear at similar BEs (0.4–0.5 eV) and the same ~0.4 eV difference between the edge and basal plane shifts is also reproduced, indicating that the metallic character of the edges and the resulting CLS are not affected by the size of the NP. In fact, the CLS within the Mo-100 NPs clearly increase with the distance of the Mo atoms to the particle edge. This indicates that for each discrete increase

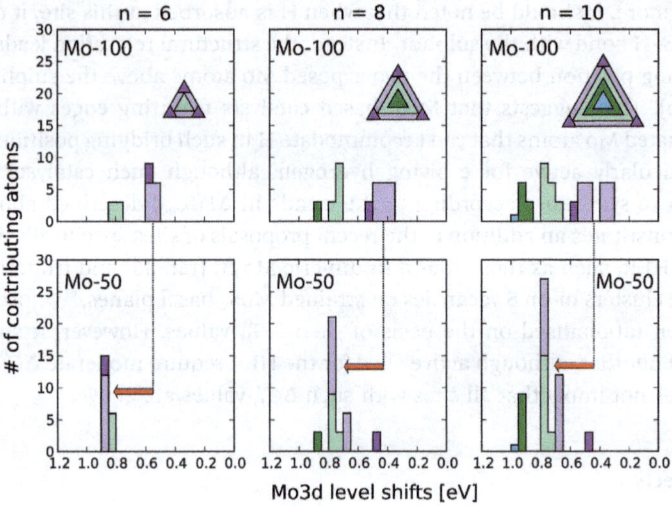

Fig. 9 Mo3d CLS for Mo-100 and Mo-50 NPs of increasing size ($n = 6$, 8, and 10). The different colors within the histograms denote the location of the Mo atoms (corner, edge, near-edge or basal plane) within the MoS_2 triangle contributing to the different Mo3d shifts, according to the color-coding indicated within the triangle drawing for each different size. For all sizes, reducing the Mo-edge MoS_2 NPs leads to shifts towards higher BEs for the edge Mo atoms, which merges contributions to the Mo3d spectra from atoms in different positions in the same high BE region. Increasing the size of the NPs does not affect the behavior of the edge shifts and just increases the number of atoms contributing to the high BE regions.

in particle size, new basal plane sites with subsequently larger CLS are added to the NP. This is more clearly illustrated in the triangle drawings in Fig. 9, where different (same) colors denote different (equivalent) sites within the triangular nanoparticles. In addition, the relative abundance of Mo atoms at the basal plane, which are characterised by a higher Mo3d BE, increases with size. As a consequence, differences in the particle size between MoS_2 samples should be directly discernible by means of XPS spectra, since larger particles should give rise to a more intense contribution at the high Mo3d BE regions. Remarkably, the desulphurisation of the three Mo-100 NPs resulting in Mo-50 type structures leads to similar shifts in CLS, independent of the particle size. For all the cases considered, the disappearance of the edge metallic state upon reduction leads to shifts in the 3d states of the Mo atoms towards higher BEs, merging contributions from the edge and basal plane Mo atoms into the same BE region (Fig. 9, red arrows). Therefore, the increase in the relative intensity of the XPS peak at a higher BE upon reduction of the Mo-100 edges, which was detected experimentally by means of *in situ* XPS measurements at elevated H_2 pressures,[26] is predicted to be independent of the nanoparticle size. However, it should be noted that, according to the calculated shifts presented here, both the increase in particle size and the edge reduction would lead to similar changes in the resulting XPS spectra, *i.e.* as shifts in the Mo3d spectra towards higher BEs. Nevertheless, sintering processes giving rise to larger MoS_2 NPs have not been observed in MoS_2/Au(111) model systems below 723 K. Therefore, we suggest that XPS measurements can be used after the preparation of the samples to evaluate the particle size of Mo-100 NPs, and then used *in situ* in order to follow the transformation into the active Mo-50 edge state upon reduction.

Effect of size on the formation energy of S vacancies. The similar electronic structure of the edges of MoS_2 nanoparticles has been shown to lead to similar edge Mo3d fingerprints for particles of different sizes. One would expect that such similarities extend to the chemical properties of the edges as well. To investigate this, we have calculated $\Delta E(S_{vac})$ in MoS_2 NPs of different sizes for the removal of an S atom from the edge position most distant to the NP corner. As expected, the resulting $\Delta E(S_{vac})$ values do not differ much with size, either for the particles with Mo-100 edges or for the ones with Mo-50 edges (Fig. 10). Near the corners of the triangular MoS_2 NPs of any size, the chemical properties should be similar to those described in previous sections for the $n = 6$ NPs. Therefore, the lack of significant size effects within the investigated NPs suggests that once a particle is larger than a certain threshold size and there are edge sites that do not feel the presence of the corner, larger particles will only differ on the number of equivalent edge sites exposed. Given the little differences in the calculated $\Delta E(S_{vac})$ values seen in Fig. 10, we suggest such size threshold to be at $n = 6$. However, it is worth mentioning again that for this particle size, all other corner and edge sites display markedly different energies that may be much more active than those at the middle of the edge, which implies that larger particles with a lower concentration of such sites will have a very different overall activity, *i.e.* a size effect. It should also be noted that the differences in chemical properties for the different sites of the $n = 6$ particles indicate that even smaller particles are not representative of or have the same chemical properties as periodic models of extended edges, in disagreement with the equivalence suggested by Huang *et al.*[56]

Conclusions

We have investigated four different types of MoS_2 nanoparticles supported on Au(111) and how differences in edge termination, composition, and size affect the electronic structure, the chemical properties, and the core-level shifts. Namely, the properties of the NPs exposing Mo-100, Mo-50, S-100, and S-50 edges of

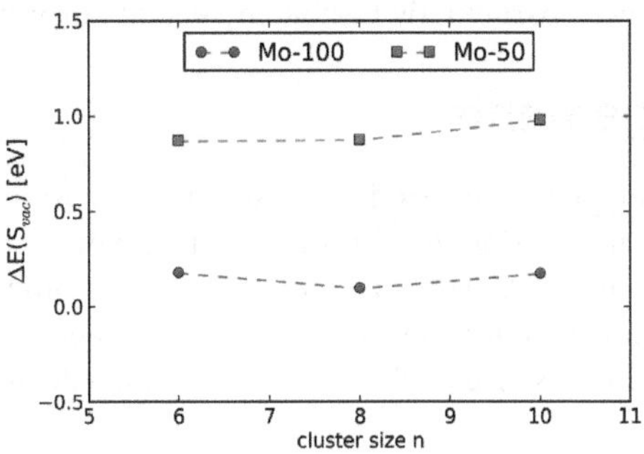

Fig. 10 $\Delta E(S_{vac})$ values calculated for S atoms removed from the most edge-like sites (*i.e.* farthest from the corner) in Mo-100 and Mo-50 NPs of increasing size, namely for $n = 6$, 8, and 10. The similarity between NPs of different sizes indicates that edge-sites are more or less equivalent, independent of the NP size.

various sizes (with 6, 8, or 10 Mo atoms at each edge) have been scrutinised. We have shown that the edges of completely sulphided MoS_2 nanoparticles with either S-edge or Mo-edge terminations are metallic, leading to the presence of brighter features in simulated STM images and to shifts towards lower BEs in the core 3d levels of the edge Mo atoms. These results are in very good agreement with STM and XPS studies of model MoS_2/Au(111) systems, which also reveal bright brims at the edges of Mo-100 NPs and the presence of low Mo3d BE peaks in XPS corresponding to the edge Mo atoms. Remarkably, upon reduction to 50% S compositions, the metallic character of the Mo-100 and S-100 edges evolves differently; it disappears when the Mo-100 edges are reduced to Mo-50 but not when S-100 edges are reduced to the S-50 state. These differences correlate with the calculated Mo3d CLS, which indicates that the different NP types can be distinguished on the basis of their XPS spectra. In particular, it should be possible to straightforwardly detect the Mo-50 edges (considered the active sites in HDS and the HER) as they lead to the presence of single XPS peaks at high BEs, in contrast to the two peaks featured by other NP types corresponding to the metallic edges and to the semi-conducting basal plane sites.

Furthermore, we have demonstrated strong site dependence in the H and S adsorption energies within the different MoS_2 NPs. Thus, the chemical properties are not only found to depend on the NP structure and edge composition, but also on whether S and Mo atoms are located at the corner, edge, near-edge or basal plane positions. In addition, the properties of the edge sites also depend on the proximity to the NP corners. This study thus highlights the importance of studying the properties of MoS_2 catalysts with nanoparticle models, which provide site-specific information largely inaccessible when using only slab models of MoS_2 edges. Nevertheless, the edges of MoS_2 nanoparticles of different sizes are quite similar and $n = 6$ NPs already expose some edge-sites with local properties that are representative of more extended edges. Interestingly, for the S-50 NPs of such lateral size, a structural motif where H atoms adsorb favourably in a bridging position between two Mo atoms is proposed as an active site for the hydrogen evolution reaction.

Finally, we highlight that the NP models presented here will enable future work aimed at clarifying how molecular adsorption varies with cluster size and structure. Exploring such effects should further contribute to the rational design of HDS (among other) catalysts, whose efficiency may be optimised by reproducing active sites with particularly flexible molecular adsorption.

Acknowledgements

We acknowledge support from Haldor Topsøe A/S, Innovation Fund Denmark (CAT-C), and the Danish Council for Independent Research|Natural Sciences (Sapere Aude Grant no. 0602-02566B). AB acknowledges support from the European Research Council under the European Union's Seventh Framework Programme (FP/2007–2013)/Marie Curie Actions/Grant no. 626764 (Nano-DeSign).

References

1 H. Toulhoat and P. Raybaud, *Catalysis by Transition Metal Sulphides*, Editions Technip, Paris, 2013.

2 H. Topsøe, B. S. Clausen and F. E. Massoth, *Hydrotreating Catalysis*, New York, 1996.

3 X. Chia, A. Y. S. Eng, A. Ambrosi, S. M. Tan and M. Pumera, *Chem. Rev.*, 2015, **115**, 11941–11966.

4 J. D. Benck, T. R. Hellstern, J. Kibsgaard, P. Chakthranont and T. F. Jaramillo, *ACS Catal.*, 2014, **4**, 3957–3971.

5 A. Splendiani, L. Sun, Y. Zhang, T. Li, J. Kim, C.-Y. Chim, G. Galli and F. Wang, *Nano Lett.*, 2010, **10**, 1271–1275.

6 K. F. Mak, C. Lee, J. Hone, J. Shan and T. F. Heinz, *Phys. Rev. Lett.*, 2010, **105**, 136805.

7 M. Bollinger, J. Lauritsen, K. Jacobsen, J. Nørskov, S. Helveg and F. Besenbacher, *Phys. Rev. Lett.*, 2001, **87**, 196803.

8 S. Helveg, J. Lauritsen, E. Laegsgaard, I. Stensgaard, J. Norskov, B. Clausen, H. Topsoe and F. Besenbacher, *Phys. Rev. Lett.*, 2000, **84**, 951–954.

9 J. V Lauritsen, M. Nyberg, R. T. Vang, M. V Bollinger, B. S. Clausen, H. Topsøe, K. W. Jacobsen, E. Laegsgaard, J. K. Nørskov and F. Besenbacher, *Nanotechnology*, 2003, **14**, 385–389.

10 T. F. Jaramillo, K. P. Jørgensen, J. Bonde, J. H. Nielsen, S. Horch and I. Chorkendorff, *Science*, 2007, **317**, 100–102.

11 H. Wang, Q. Zhang, H. Yao, Z. Liang, H.-W. Lee, P.-C. Hsu, G. Zheng and Y. Cui, *Nano Lett.*, 2014, **14**, 7138–7144.

12 J. Kibsgaard, Z. Chen, B. N. Reinecke and T. F. Jaramillo, *Nat. Mater.*, 2012, **11**, 963–969.

13 J. Kibsgaard, T. F. Jaramillo and F. Besenbacher, *Nat. Chem.*, 2014, **6**, 248–253.

14 J. V Lauritsen, J. Kibsgaard, S. Helveg, H. Topsøe, B. S. Clausen, E. Laegsgaard and F. Besenbacher, *Nat. Nanotechnol.*, 2007, **2**, 53–58.

15 J. V. V. Lauritsen, M. V. V. Bollinger, E. Lægsgaard, K. W. W. Jacobsen, J. K. K. Nørskov, B. S. S. Clausen, H. Topsøe and F. Besenbacher, *J. Catal.*, 2004, **221**, 510–522.

16 D. Cao, T. Shen, P. Liang, X. Chen and H. Shu, *J. Phys. Chem. C*, 2015, **119**, 4294–4301.

17 J. Kibsgaard, J. V Lauritsen, E. Laegsgaard, B. S. Clausen, H. Topsøe and F. Besenbacher, *J. Am. Chem. Soc.*, 2006, **128**, 13950–13958.

18 J. V. Lauritsen and F. Besenbacher, *Advances in Catalysis Volume 50*, Elsevier, 2006, vol. 50.

19 J. Kibsgaard, B. S. Clausen, H. Topsøe, E. Lægsgaard, J. V. Lauritsen and F. Besenbacher, *J. Catal.*, 2009, **263**, 98–103.

20 P. Raybaud, J. Hafner, G. Kresse, S. Kasztelan and H. Toulhoat, *J. Catal.*, 2000, **189**, 129–146.

21 H. Schweiger, P. Raybaud, G. Kresse and H. Toulhoat, *J. Catal.*, 2002, **207**, 76–87.

22 M. Bollinger, K. Jacobsen and J. Nørskov, *Phys. Rev. B: Condens. Matter Mater. Phys.*, 2003, **67**, 085410.

23 H. G. Füchtbauer, A. K. Tuxen, Z. Li, H. Topsøe, J. V. Lauritsen and F. Besenbacher, *Top. Catal.*, 2014, **57**, 207–214.

24 C. Kisielowski, Q. M. Ramasse, L. P. Hansen, M. Brorson, A. Carlsson, A. M. Molenbroek, H. Topsøe and S. Helveg, *Angew. Chem., Int. Ed.*, 2010, **49**, 2708–2710.

25 L. P. Hansen, Q. M. Ramasse, C. Kisielowski, M. Brorson, E. Johnson, H. Topsøe and S. Helveg, *Angew. Chem., Int. Ed.*, 2011, **50**, 10153–10156.
26 A. Bruix, H. G. Füchtbauer, A. K. Tuxen, A. S. Walton, M. Andersen, S. Porsgaard, F. Besenbacher, B. Hammer and J. V Lauritsen, *ACS Nano*, 2015, **9**, 9322–9330.
27 P. Moses, B. Hinnemann, H. Topsoe and J. Norskov, *J. Catal.*, 2007, **248**, 188–203.
28 B. Hinnemann, P. G. Moses and J. K. Nørskov, *J. Phys.: Condens. Matter*, 2008, **20**, 064236.
29 S. Cristol, J. F. Paul, E. Payen, D. Bougeard, F. Hutschka and S. Clémendot, *J. Catal.*, 2004, **224**, 138–147.
30 H. Toulhoat, P. Raybaud, S. Kasztelan, G. Kresse and J. Hafner, *Catal. Today*, 1999, **50**, 629–636.
31 H. Toulhoat, *J. Catal.*, 2003, **216**, 63–72.
32 C. Tsai, F. Abild-Pedersen and J. K. Nørskov, *Nano Lett.*, 2014, **14**, 1381–1387.
33 B. Hinnemann, P. G. Moses, J. Bonde, K. P. Jørgensen, J. H. Nielsen, S. Horch, I. Chorkendorff and J. K. Nørskov, *J. Am. Chem. Soc.*, 2005, **127**, 5308–5309.
34 J. K. Nørskov, T. Bligaard, A. Logadottir, J. R. Kitchin, J. G. Chen, S. Pandelov and U. Stimming, *J. Electrochem. Soc.*, 2005, **152**, J23.
35 Z. Huang, W. Luo, L. Ma, M. Yu, X. Ren, M. He, S. Polen, K. Click, B. Garrett, J. Lu, K. Amine, C. Hadad, W. Chen, A. Asthagiri and Y. Wu, *Angew. Chem.*, 2015, **127**, 15396–15400.
36 H. Li, C. Tsai, A. L. Koh, L. Cai, A. W. Contryman, A. H. Fragapane, J. Zhao, H. S. Han, H. C. Manoharan, F. Abild-Pedersen, J. K. Nørskov and X. Zheng, *Nat. Mater.*, 2015, **15**, 48–53.
37 P. Y. Prodhomme, P. Raybaud and H. Toulhoat, *J. Catal.*, 2011, **280**, 178–195.
38 C. Arrouvel, M. Breysse, H. Toulhoat and P. Raybaud, *J. Catal.*, 2005, **232**, 161–178.
39 D. Costa, C. Arrouvel, M. Breysse, H. Toulhoat and P. Raybaud, *J. Catal.*, 2007, **246**, 325–343.
40 J. Bonde, P. G. Moses, T. F. Jaramillo, J. K. Nørskov and I. Chorkendorff, *Faraday Discuss.*, 2009, **140**, 219–231.
41 P. E. Blöchl, *Phys. Rev. B: Condens. Matter Mater. Phys.*, 1994, **50**, 17953–17979.
42 J. Enkovaara, C. Rostgaard, J. J. Mortensen, J. Chen, M. Dułak, L. Ferrighi, J. Gavnholt, C. Glinsvad, V. Haikola, H. A. Hansen, H. H. Kristoffersen, M. Kuisma, A. H. Larsen, L. Lehtovaara, M. Ljungberg, O. Lopez-Acevedo, P. G. Moses, J. Ojanen, T. Olsen, V. Petzold, N. A. Romero, J. Stausholm-Møller, M. Strange, G. A. Tritsaris, M. Vanin, M. Walter, B. Hammer, H. Häkkinen, G. K. H. Madsen, R. M. Nieminen, J. K. Nørskov, M. Puska, T. T. Rantala, J. Schiøtz, K. S. Thygesen and K. W. Jacobsen, *J. Phys.: Condens. Matter*, 2010, **22**, 253202.
43 J. J. Mortensen, L. B. Hansen and K. W. Jacobsen, *Phys. Rev. B: Condens. Matter Mater. Phys.*, 2005, **71**, 035109.
44 S. R. Bahn and K. W. Jacobsen, *Comput. Sci. Eng.*, 2002, **4**, 56–66.
45 J. Tersoff and D. R. Hamann, *Phys. Rev. Lett.*, 1983, **50**, 1998–2001.
46 J. Tersoff and D. R. Hamann, *Phys. Rev. B: Condens. Matter Mater. Phys.*, 1985, **31**, 805–813.
47 W. A. Hofer, A. S. Foster and A. L. Shluger, *Rev. Mod. Phys.*, 2003, **75**, 1287–1331.

48 M. P. Ljungberg, J. J. Mortensen and L. G. M. Pettersson, *J. Electron Spectrosc. Relat. Phenom.*, 2011, **184**, 427–439.

49 T. Li and G. Galli, *J. Phys. Chem. C*, 2007, **111**, 16192–16196.

50 J.-O. Joswig, T. Lorenz, T. B. Wendumu, S. Gemming and G. Seifert, *Acc. Chem. Res.*, 2015, **48**, 48–55.

51 N. Dinter, M. Rusanen, P. Raybaud, S. Kasztelan, P. da Silva and H. Toulhoat, *J. Catal.*, 2009, **267**, 67–77.

52 A. Tuxen, J. Kibsgaard, H. Gøbel, E. Lægsgaard, H. Topsøe, J. V. Lauritsen, F. Besenbacher, E. Laegsgaard, H. Topsøe, J. V. Lauritsen and F. Besenbacher, *ACS Nano*, 2010, **4**, 4677–4682.

53 P. G. Moses, J. J. Mortensen, B. I. Lundqvist and J. K. Norskov, *J. Chem. Phys.*, 2009, **130**, 104709.

54 B. Hinnemann, J. K. Nørskov and H. Topsøe, *J. Phys. Chem. B*, 2005, **109**, 2245–2253.

55 M. Haruta, T. Kobayashi, H. Sano and N. Yamada, *Chem. Lett.*, 1987, **16**, 405–408.

56 Y. Huang, R. J. Nielsen, W. A. Goddard III and M. P. Soriaga, *J. Am. Chem. Soc.*, 2015, **137**, 6692–6698.

Faraday Discussions

PAPER

High spatial resolution mapping of chemically-active self-assembled N-heterocyclic carbenes on Pt nanoparticles

Y. Levratovsky and E. Gross*

Received 25th November 2015, Accepted 27th November 2015

DOI: 10.1039/c5fd00194c

The properties of many functional materials critically depend on the spatial distribution of surface active sites. In the case of solid catalysts, the geometric and electronic properties of different surface sites will directly impact their catalytic properties. However, the detection of catalytic sites at the single nanoparticle level cannot be easily achieved and most spectroscopic measurements are performed with ensemble-based measurements in which the reactivity is averaged over millions of nanoparticles. It is hereby demonstrated that chemically-functionalized N-heterocyclic carbene molecules can be attached to the surfaces of Pt nanoparticles and utilized as a model system for studying catalytic reactions on single metallic nanoparticles. The formation of a carbene self-assembled layer on the surface of a Pt nanoparticle and its stability under oxidizing conditions were investigated. IR nanospectroscopy measurements detected the chemical properties of surface-anchored molecules on single nanoparticles. A direct correlation was identified between IR nanospectroscopy measurements and macroscopic ATR-IR measurements. These results demonstrate that high spatial resolution mapping of the catalytic reactivity on single nanoparticles can be achieved with this approach.

Introduction

Vibrational spectroscopy is one of the most powerful techniques for analysis of catalytic processes since it provides direct chemical information, from which the chemical structure of reactants, intermediates and products can be determined.[1–3] A variety of vibrational spectroscopy methods have been developed in the last few decades and they have been utilized to directly analyse the mechanism of catalytic processes. Among these methods are surface sensitive techniques, such as SFG (Sum Frequency Generation) vibrational spectroscopy[4] and PM-IRAS (Polarization Modulation Infrared Reflection Absorption Spectroscopy),[5] that can identify surface-adsorbed intermediates in catalytic reactions. High temporal resolution IR

Institute of Chemistry and the Centre for Nanoscience and Nanotechnology, The Hebrew University of Jerusalem, Jerusalem 91904, Israel. E-mail: elad.gross@mail.huji.ac.il

spectroscopy is another tool to detect transient species by tracking the formation and consumption of intermediates under reaction conditions.[6–8] It has been widely demonstrated that these *in situ* methods provide essential details about the mechanism and kinetics for a wide array of catalytic processes.

However, all the above-mentioned techniques are limited in their spatial resolution and therefore cannot detect processes on single nanoparticles.[9] Uncovering the complex factors that direct catalytic processes at the single particle level is a key element for optimizing chemical reactions and designing superior catalysts.[10–12] The spatial resolution of spectroscopic measurements is controlled by the diffraction limit of the light and therefore the optimum spatial resolution that can be gained in IR measurements is in the range of ~10 μm.[2] Due to their limited spatial resolution, IR measurements can only detect the averaged vibrational properties of billions of atoms and cannot spatially resolve the properties of reactants and products on a single catalytically-active nanoparticle.

High spatial resolution infrared microscopy has recently become possible with the introduction of scattering-type scanning near-field optical microscopy (s-SNOM).[13–16] This method is based on an atomic force microscopy (AFM) tip that locally enhances the scattered IR signal. An illuminating infrared beam is focused onto a metallic AFM tip that acts as an antenna and concentrates the incident light at its apex. Recording the scattered light, while scanning the sample surface, will yield a near-field IR image of the sample. The resolution of the IR mapping does not depend on the illumination wavelength and is directly controlled by the size of the tip apex, which is typically in the size range of 20 nm.[17,18] In many studies that involve IR nanospectroscopy, synchrotron radiation was used as an IR source since it is spectrally broad, bright, and spatially coherent, with a spectral irradiance 1000 times higher than that of a thermal source. It was recently demonstrated that high spatial resolution IR spectroscopy of solids,[19–24] polymers[25–27] and proteins[28,29] can be achieved with IR nanospectroscopy. In this study we demonstrate that IR nanospectroscopy can be utilized for high spatial resolution detection of OH-functionalized N-heterocyclic carbene molecules that are anchored to the surfaces of catalytically-active Pt nanoparticles.

In the field of molecular ligands, carbon-based molecules known as N-heterocyclic carbenes (NHC) have been widely used as ligands in organometallic complexes.[30] NHC molecules have two heteroatoms adjacent to the carbene carbon, which increase their stability so that they can be prepared, distilled, functionalized and stored for long periods of time.[31–33] These properties make the NHC ligands highly suitable for functionalization of metallic surfaces.[34,35]

It is hereby demonstrated that chemically-functionalized NHC molecules can be attached to the surfaces of supported Pt nanoparticles. The carbene molecules show high stability under oxidizing conditions. IR nanospectroscopy measurements reveal that the vibrational spectra of surface-anchored NHC molecules can be detected on single Pt nanoparticles. These measurements validate the fact that surface-anchored NHC molecules are a suitable model system for the detection of catalytic reactions at the nanoscale.

Results and discussion

Pt nanoparticles were prepared by the deposition of a Pt film on a Si (110) surface, which was coated with a native layer of SiO_2. An equivalent of 3 nm of Pt was evaporated on the surface of a Si wafer at a deposition rate of 0.3 nm s^{-1}. HR-SEM imaging revealed that the deposition of Pt atoms on the Si surface at room temperature led to the formation of an almost continuous flat film that covered 90% of the native SiO_2 surface (Fig. 1a). The fraction of the Pt covered area was reduced to 65% following annealing of the sample at 673 K for 5 hours. However, no separated Pt nanoparticles were detected after this treatment (Fig. 1b). When the sample was annealed at 823 K for 5 hours, the Pt covered area was further reduced to 30%. At this stage, disconnected Pt nanoparticles in the size range of 100 ± 30 nm were detected (Fig. 1c).

The synthesis of the free carbene molecules, functionalized with an OH group, was performed in a glove box and was based on a recently published protocol.[36] The freshly-prepared carbene solution was transferred into a vial in which the Pt/SiO_2 sample was kept. After 48 hours, the silica sample was removed from the glove box and washed with THF (5 mL) and distilled water (5 mL), intermittently. No changes in the morphology of the Pt nanoparticles were detected following the anchoring of carbene molecules on their surfaces (Fig. 1d).

The IR spectra of the inactivated carbene and the surface-anchored carbene were measured with FTIR and ATR-FTIR, respectively (Fig. 2a and b). Absorption peaks of the free carbene were detected at 1050 and 1600 cm^{-1} and at 2950 cm^{-1}, correlating to C–OH and C–H vibrations (Fig. 2a). For the surface-anchored carbene (Fig. 2b), a dominant peak was detected at 1050–1300 cm^{-1}, which correlates to C–OH. Interestingly, a minor peak was detected at 1850 cm^{-1}; this peak was not detected in the free carbene IR spectrum. The presence of this peak, which correlates to the formation of a carbonyl bond (C=O) indicates that some of the

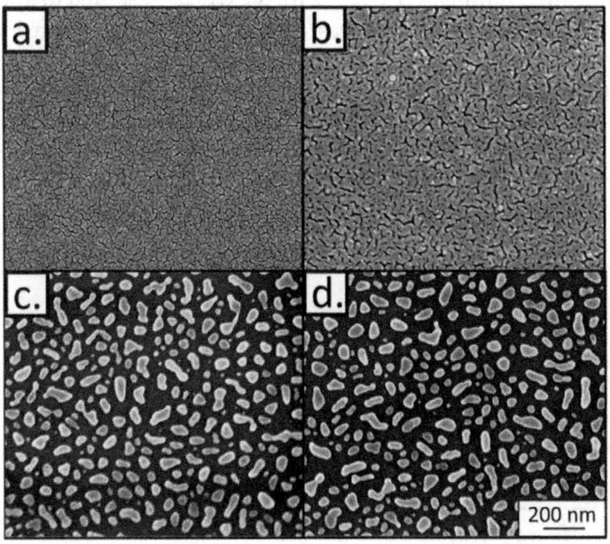

Fig. 1 HR-SEM image of the Pt film deposited on the Si (110) surface that is coated with a layer of native silica (a). The sample was annealed at 673 K and 823 K for five hours and then imaged by HR-SEM ((b) and (c), respectively). The structure of the Pt nanoparticles was not modified following the anchoring of the NHC molecules on the nanoparticles' surfaces (d).

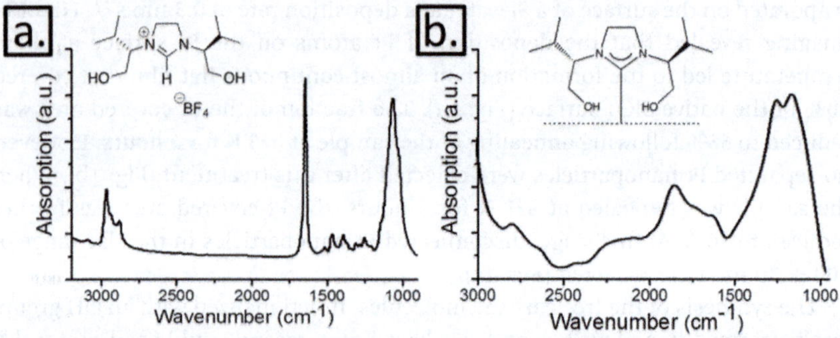

Fig. 2 FTIR (a) and ATR-FTIR (b) spectra of the carbene-OH molecules before and after its attachment to the Pt surface, respectively.

OH groups were oxidized following their attachment to the surfaces of the Pt nanoparticles.

The attachment of the carbene molecules to the Pt surface of the Pt nanoparticles was monitored by XPS (X-ray Photoelectron Spectroscopy) (Fig. 3). Changes in the C1s and N1s signals were detected following the attachment of the carbene molecules to the surfaces (Fig. 3a and b, respectively).[34] The XPS signal of carbon was shifted to higher energies following the attachment of the carbene molecules (Fig. 3a, curves 1 and 2). These changes indicated that molecules with C–O functionality were attached to the surfaces. After exposure of the sample to oxidizing conditions (1 atm O_2, 5 h, 70 °C), the XPS signal was further shifted to higher energies, which indicated that the surface-anchored OH group was further oxidized under these conditions (Fig. 3a, curve 3). After exposure of the sample to harsher oxidizing conditions (H_2O_2, rt, 7 h), the C1s peak shifted to lower energies and resembled the C1s peak that was detected from the uncoated Pt surfaces (Fig. 3a, curve 4). These changes reveal that some of the carbene molecules were detached from the surfaces following their exposure to harsher oxidizing conditions.[34]

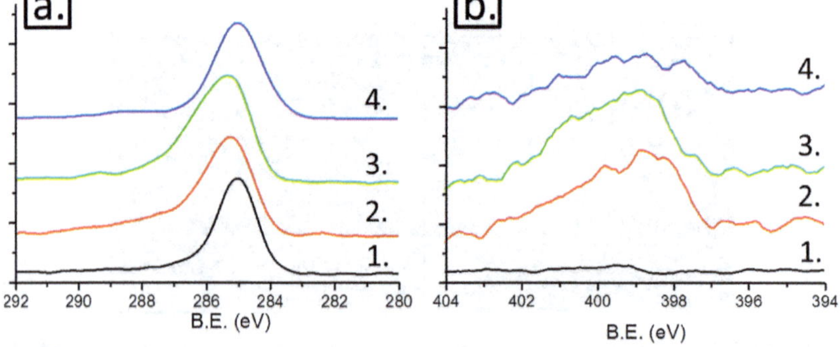

Fig. 3 C1s (a) and N1s (b) XPS spectra of the Si supported Pt nanoparticles without a carbene layer (1) and with a carbene layer (2). The XPS spectra were measured again after the carbene-coated sample was exposed to oxidizing conditions of O_2, 70 °C, 7 h (3) and H_2O_2, rt, 7 h (4).

The detection of the N1s XPS signal following the immersion of Pt in the carbene solution verified that carbene molecules were anchored on the Pt surfaces (Fig. 3b, curves 1 and 2). No change in the N1s signal was obtained after exposure of the sample to mild oxidizing conditions (Fig. 3b, curve 3). However, under harsher oxidizing conditions (H_2O_2, rt, 7 h), a decrease of about 50% in the N1s peak area was detected (Fig. 3b, curve 4). These changes reveal that under these conditions some of the carbene molecules were detached from the surface of Pt nanoparticles.

The stability of the surface-anchored carbene molecules on the Pt nanoparticles was further analysed by calculating the N1s/Pt4f XPS signals ratio (Fig. 4). An increase in the N1s/Pt4f signals ratio following exposure of the sample to carbene molecules indicated that the Pt surfaces were coated with carbene molecules (Fig. 4a and b). Almost no change in the N1s/Pt4f XPS signals ratio was detected following exposure of the coated sample to 1 atm of O_2 at 70 °C for 7 h (Fig. 4c), indicating that no detachment of carbene molecules from the surfaces occurred under these conditions. However, after exposure of the sample to H_2O_2 (rt, 7 h), the N1s/Pt4f XPS signals ratio decreased by more than 25%, showing that some of the surface-anchored molecules were detached from the Pt surfaces (Fig. 4d). These results show that the surface-anchored carbene molecules are stable on the Pt surface under mild oxidizing conditions.

AFM-IR nanospectroscopy measurements were conducted in order to characterize the properties of the surface-anchored molecules on the single nanoparticles. Synchrotron IR light (provided by the Advanced Light Source, Lawrence Berkeley National Laboratory) was focused onto the apex of an oscillating Pt–Si AFM tip (Nanosensors) at frequency ω in a modified commercial AFM (Innova, Bruker).[13,20,29] A modified commercial FTIR spectrometer (Nicolet 6700, Thermo Scientific) collected the IR nanospectroscopy spectra. The near-field signal can be differentiated from the far-field background by detecting the high harmonics of 2ω, which are induced by tip oscillations, with a lock-in amplifier (Fig. 5).[37]

AFM topography imaging of the surface was conducted (Fig. 6a) and followed by broadband IR spectroscopy measurements of the same area (Fig. 6b). The broadband measurements identify areas in which the IR light is being absorbed by the sample. Comparison between the topography AFM image and the

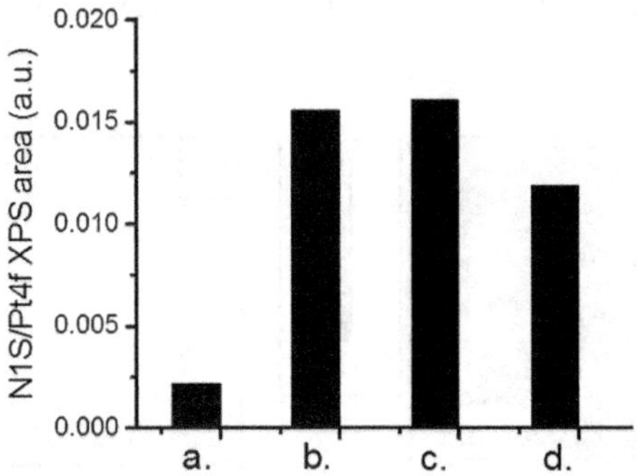

Fig. 4 The N1s/Pt4f XPS signals ratio analysis of the bare Pt nanoparticles (a) and carbene-coated nanoparticles (b). The N1s/Pt4f XPS signals ratio was analysed after the carbene-coated sample was exposed to oxidizing conditions of O_2, 70 °C, 7 h (c) and H_2O_2, rt, 7 h (d).

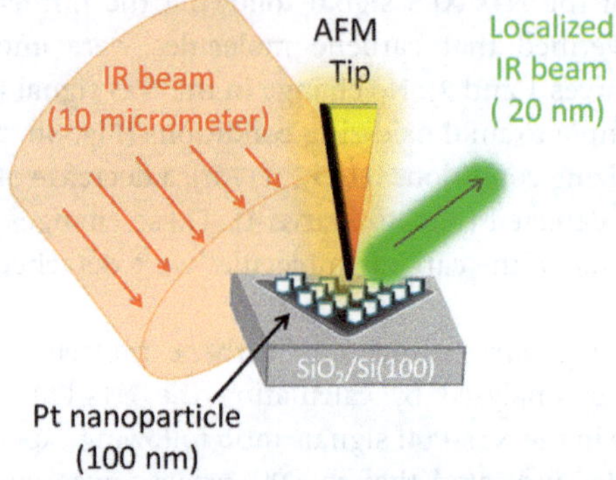

Fig. 5 OH-functionalized carbene molecules (marked with blue pentagons) were anchored to the surfaces of Pt nanoparticles (black triangle). The nanoparticles were supported on a Si wafer (grey cube). The Pt–Si AFM tip acts as an optical antenna by localizing the diffraction limited IR beam (~10 μm) and inducing an IR scattering signal with a tip-limited spatial resolution of ~20 nm.

broadband AFM-IR image (Fig. 6a and b, respectively) reveal that IR absorption was solely induced by the Pt nanoparticles. These images demonstrate that absorption occurs only on areas that are coated with Pt nanoparticles and show that the carbene molecules were specifically deposited on the Pt surfaces and were not adsorbed on the bare Si surface.

IR nanospectroscopy point measurements were conducted at the surface of a single Pt nanoparticle (Fig. 7). The near-field spectrum is reported in the form of a normalized scattering amplitude using the bare SiO_2/Si surface as a reference point: $A_{(\nu)} = A_{(\nu)\text{sample}}/A_{(\nu)\text{reference}}$. Topography imaging of the surface was performed by AFM mapping (Fig. 7a). While the AFM tip was positioned above the bare SiO_2 surface, a featureless IR spectrum was detected. However, once the AFM tip was positioned above the surface of the Pt nanoparticle (black circle in Fig. 7a), an IR signal was detected with a dominant peak at 1250 cm^{-1} and a minor peak at 1800 cm^{-1} (black curve in Fig. 6b). The IR signal from 1900 to 2500 cm^{-1} is not shown since in this region the IR beam is mostly absorbed by a diamond window that separates the UHV part of the beam from the low vacuum area.

The dominant 1250 cm^{-1} peak is associated with C–OH stretching and shows that most of the surface-adsorbed molecules at this site were functionalized with an

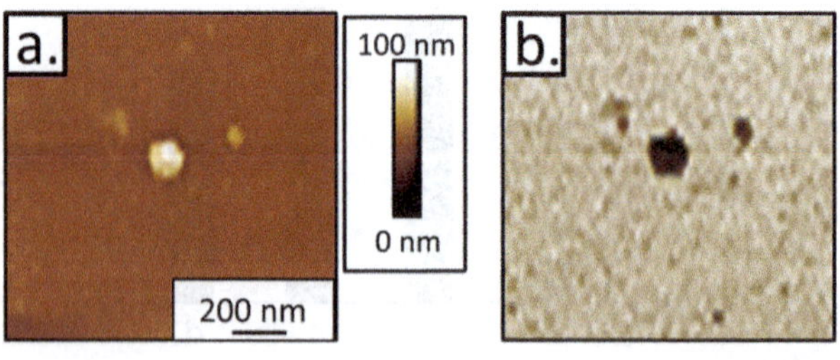

Fig. 6 (a) AFM topography image of the carbene-coated Pt nanoparticles (bright spots) that were deposited on a Si surface. (b) Broadband AFM-IR nanospectroscopy image. The darker areas in this image reflect structures that absorb the IR light.

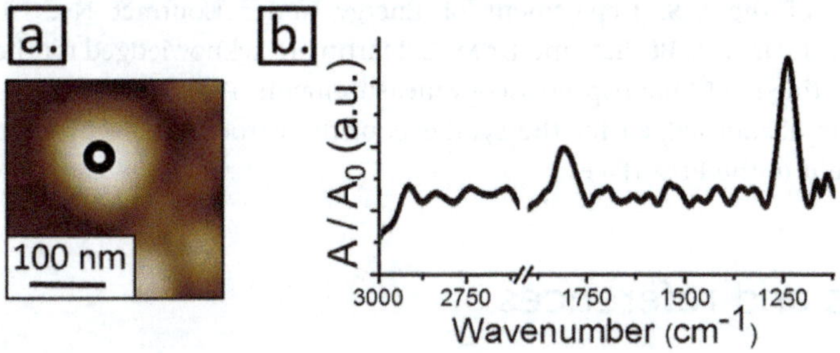

Fig. 7 (a) AFM topography image of the carbene-coated Pt nanoparticle on the Si surface after exposure of the coated sample to an environment of 1 atm O_2 at 40 °C for 7 h. (b) IR nanospectroscopy point measurements were conducted above the surface of a single Pt nanoparticle (black circle in Fig. 7a and the corresponding IR spectrum in Fig. 7b).

OH group. The minor peak at 1800 cm^{-1} correlates to a carbonyl stretch and shows that some of the surface-anchored ligands were oxidized following their attachment to the catalytic surface. The peak positions in the IR nanospectroscopy measurements correlate well to the peak positions in the far-field IR-ATR measurements. These similarities in the IR peaks position demonstrate that IR nanospectroscopy can detect the chemical properties of chemically-active self-assembled molecules that are anchored on the surfaces of catalytic nanoparticles. Therefore, this approach can be utilized to detect structure–reactivity correlations at the nanoscale.

Conclusions

It is demonstrated that chemically-functionalized NHC molecules can be attached to the surfaces of Pt nanoparticles. The surface-anchored molecules are highly stable under mild oxidizing conditions. However, detachment of some of the ligands was observed when the sample was exposed to harsh oxidizing conditions. Broadband IR nanospectroscopy measurements proved that the ligands were solely anchored to the surfaces of the Pt nanoparticles. IR nanospectroscopy measurements indicated that the functional OH groups of some of the surface-anchored ligands were oxidized following their deposition on the surfaces of the Pt nanoparticles and exposure to an environment of 1 atm O_2 at 40 °C for 7 h. These initial results demonstrate that IR-nanospectroscopy measurements can be utilized to detect the properties of surface-anchored molecules on single nanoparticles. Based on these measurements, future studies are planned in which the impact of oxidizing and reducing conditions on the properties of surface-anchored molecules will be examined. It is expected that the analysis of the catalytic reactivity on different nanoparticles with IR nanospectroscopy will uncover novel structure–reactivity correlations at the nanoscale, which cannot be detected by ensemble-based measurements.[38]

Acknowledgements

The synchrotron measurements were performed at the Advanced Light Source (ALS) at Lawrence Berkeley National Laboratory (LBNL). The Advanced Light Source is supported by the Director, Office of Science, Office of Basic Energy

Sciences, of the U.S. Department of Energy under Contract No. DE-AC02-05CH11231. Dr H. A. Bechtel and Dr M. C. Martin are acknowledged for their help in conducting the IR nanospectroscopy measurements. Prof. F. D. Toste and Dr C. Y. Wu are acknowledged for the synthesis of the carbene molecules and their attachment to the Pt surface.

Notes and references

1 J. M. Andanson and A. Baiker, *Chem. Soc. Rev.*, 2010, **39**, 4571–4584.
2 E. Stavitski and B. M. Weckhuysen, *Chem. Soc. Rev.*, 2010, **39**, 4615–4625.
3 I. E. Wachs and C. A. Roberts, *Chem. Soc. Rev.*, 2011, **40**, 5922.
4 L. R. Baker, G. Kennedy, M. van Spronsen, A. Hervier, X. J. Cai, S. Y. Chen, L. W. Wang and G. A. Somorjai, *J. Am. Chem. Soc.*, 2012, **134**, 14208–14216.
5 F. Gao, Y. L. Wang and D. W. Goodman, *J. Phys. Chem. C*, 2010, **114**, 4036–4043.
6 N. Sivasankar and H. Frei, *J. Phys. Chem. C*, 2011, **115**, 7545–7553.
7 N. Sivasankar, W. W. Weare and H. Frei, *J. Am. Chem. Soc.*, 2011, **133**, 12976–12979.
8 M. Zhang, M. de Respinis and H. Frei, *Nat. Chem.*, 2014, **6**, 362–367.
9 I. L. C. Buurmans and B. M. Weckhuysen, *Nat. Chem.*, 2012, **4**, 873–886.
10 E. Gross, X. Z. Shu, S. Alayoglu, H. A. Bechtel, M. C. Martin, F. D. Toste and G. A. Somorjai, *J. Am. Chem. Soc.*, 2014, **136**, 3624–3629.
11 G. A. Somorjai and Y. Li, *Introduction to surface chemistry and catalysis*, Wiley, Hoboken, N.J., 2nd edn, 2010.
12 E. Gross and G. A. Somorjai, *J. Catal.*, 2015, **328**, 91–101.
13 H. A. Bechtel, E. A. Muller, R. L. Olmon, M. C. Martin and M. B. Raschke, *Proc. Natl. Acad. Sci. U. S. A.*, 2014, **111**, 7191–7196.
14 N. Kumar, B. Stephanidis, R. Zenobi, A. J. Wain and D. Roy, *Nanoscale*, 2015, **7**, 7133–7137.
15 E. M. van Schrojenstein Lantman, T. Deckert-Gaudig, A. J. G. Mank, V. Deckert and B. M. Weckhuysen, *Nat. Nanotechnol.*, 2012, **7**, 583–586.
16 R. Zhang, Y. Zhang, Z. C. Dong, S. Jiang, C. Zhang, L. G. Chen, L. Zhang, Y. Liao, J. Aizpurua, Y. Luo, J. L. Yang and J. G. Hou, *Nature*, 2013, **498**, 82–86.
17 F. Huth, A. Govyadinov, S. Amarie, W. Nuansing, F. Keilmann and R. Hilenbrand, *Nano Lett.*, 2012, **12**, 3973–3978.
18 E. A. Muller, B. Pollard and M. B. Raschke, *J. Phys. Chem. Lett.*, 2015, **6**, 1275–1284.
19 M. Ayache, S. F. Lux and R. Kostecki, *J. Phys. Chem. Lett.*, 2015, **6**, 1126–1129.
20 G. Dominguez, A. S. Mcleod, Z. Gainsforth, P. Kelly, H. A. Bechtel, F. Keilmann, A. Westphal, M. Thiemens and D. N. Basov, *Nat. Commun.*, 2014, 5.
21 P. Hermann, A. Hoehl, G. Ulrich, C. Fleischmann, A. Hermelink, B. Kastner, P. Patoka, A. Hornemann, B. Beckhoff, E. Ruhl and G. Ulm, *Opt. Express*, 2014, **22**, 17948–17958.
22 F. Huth, M. Schnell, J. Wittborn, N. Ocelic and R. Hillenbrand, *Nat. Mater.*, 2011, **10**, 352–356.
23 O. Khatib, J. D. Wood, A. S. McLeod, M. D. Goldflam, M. Wagner, G. L. Damhorst, J. C. Koepke, G. P. Doidge, A. Rangarajan, R. Bashir, E. Pop, J. W. Lyding, M. H. Thiemens, F. Keilmann and D. N. Basov, *ACS Nano*, 2015, **9**, 7968–7975.

24 I. T. Lucas, A. S. McLeod, J. S. Syzdek, D. S. Middlemiss, C. P. Grey, D. N. Basov and R. Kostecki, *Nano Lett.*, 2015, **15**, 1–7.

25 A. M. Katzenmeyer, G. Holland, K. Kjoller and A. Centrone, *Anal. Chem.*, 2015, **87**, 3154–3159.

26 S. Mastel, A. A. Govyadinov, T. V. A. G. de Oliveira, I. Amenabar and R. Hillenbrand, *Appl. Phys. Lett.*, 2015, 106.

27 B. Pollard, E. A. Muller, K. Hinrichs and M. B. Raschke, *Nat. Commun.*, 2014, 5.

28 I. Amenabar, S. Poly, W. Nuansing, E. H. Hubrich, A. A. Govyadinov, F. Huth, R. Krutokhvostov, L. B. Zhang, M. Knez, J. Heberle, A. M. Bittner and R. Hillenbrand, *Nat. Commun.*, 2013, 4.

29 S. Berweger, D. M. Nguyen, E. A. Muller, H. A. Bechtel, T. T. Perkins and M. B. Raschke, *J. Am. Chem. Soc.*, 2013, **135**, 18292–18295.

30 M. N. Hopkinson, C. Richter, M. Schedler and F. Glorius, *Nature*, 2014, **510**, 485–496.

31 A. J. Arduengo, R. L. Harlow and M. Kline, *J. Am. Chem. Soc.*, 1991, **113**, 361–363.

32 A. Igau, H. Grutzmacher, A. Baceiredo and G. Bertrand, *J. Am. Chem. Soc.*, 1988, **110**, 6463–6466.

33 M. Niehues, G. Erker, G. Kehr, P. Schwab, R. Frohlich, O. Blacque and H. Berke, *Organometallics*, 2002, **21**, 2905–2911.

34 C. M. Crudden, J. H. Horton, I. I. Ebralidze, O. V. Zenkina, A. B. McLean, B. Drevniok, Z. She, H. B. Kraatz, N. J. Mosey, T. Seki, E. C. Keske, J. D. Leake, A. Rousina-Webb and G. Wu, *Nat. Chem.*, 2014, **6**, 409–414.

35 A. V. Zhukhovitskiy, M. G. Mavros, T. van Voorhis and J. A. Johnson, *J. Am. Chem. Soc.*, 2013, **135**, 7418–7421.

36 V. Jurcik, M. Gilani and R. Wilhelm, *Eur. J. Org. Chem.*, 2006, **84**, 5103–5109.

37 A. Dazzi, C. B. Prater, Q. C. Hu, D. B. Chase, J. F. Rabolt and C. Marcott, *Appl. Spectrosc.*, 2012, **66**, 1365–1384.

38 E. Gross, J. M. Krier, L. Heinke and G. A. Somorjai, *Top. Catal.*, 2012, **55**, 13–23.

Faraday Discussions

PAPER

Designing for selectivity: weak interactions and the competition for reactive sites on gold catalysts†

C. G. F. Siler,[a] R. J. Madix[a] and C. M. Friend*[ab]

Received 24th November 2015, Accepted 25th November 2015
DOI: 10.1039/c5fd00192g

A major challenge in heterogeneous catalysis is controlling reaction selectivity, especially in complex environments. When more than one species is present in the gas mixture, the competition for binding sites on the surface of a catalyst is an important factor in determining reaction selectivity and activity. We establish an experimental hierarchy for the binding of a series of reaction intermediates on Au(111) and demonstrate that this hierarchy accounts for reaction selectivity on both the single crystal surface and under operating catalytic conditions at atmospheric pressure using a nanoporous Au catalyst. A partial set of measurements of relative binding has been measured by others on other catalyst materials, including Ag, Pd and metal oxide surfaces; a comparison demonstrates the generality of this concept and identifies differences in the trends. Theoretical calculations for a subset of reactants on Au(111) show that weak van der Waals interactions are key to predicting the hierarchy of binding strengths for alkoxides bound to Au(111). This hierarchy is key to the control of the selectivity for partial oxidation of alcohols to esters on both Au surfaces and under working catalytic conditions using nanoporous gold. The selectivity depends on the competition for active sites among key intermediates. New results probing the effect of fluorine substitution are also presented to extend the relation of reaction selectivity to the hierarchy of binding. Motivated by an interest in synthetic manipulation of fluorinated organics, we specifically investigated the influence of the $-CF_3$ group on alcohol reactivity and selectivity. 2,2,2-Trifluoroethanol couples on O-covered Au(111) to yield $CF_3CH_2O-C(=O)(CF_3)$, but in the presence of methanol or ethanol it preferentially forms the respective 2,2,2-trifluoroethoxy-esters. The ester is not the dominant product in any of these cases, though, indicating that the rate of β-H elimination from adsorbed trifluoroethoxy is slower than that for either adsorbed methoxy or ethoxy, consistent with their relative estimated β-C–H bond strengths. The measured equilibrium constants for the competition for binding to the surface are 2.9 and 0.38 for ethanol and methanol, respectively, *vs.* 2,2,2-trifluoroethanol, indicating that the binding

[a]*School of Engineering and Applied Science, Harvard University, 12 Oxford St, Cambridge, MA 02138, USA. E-mail: friend@fas.harvard.edu*
[b]*Dept. of Chemistry and Chemical Biology, Harvard University, 12 Oxford St, Cambridge, MA 02138, USA*

† Electronic supplementary information (ESI) available: See DOI: 10.1039/c5fd00192g

strength of 2,2,2-trifluoroethoxy is weaker than ethoxy, but stronger than methoxy. These results are consistent with weakening of the interactions between the surface and the alkyl group due to Pauli repulsion of the electron-rich CF_3 group from the surface, which offsets the van der Waals attraction. These experiments provide guiding principles for understanding the effect of fluorination on heterogeneous synthesis and further demonstrate the key role of molecular structure in determining reaction selectivity.

Introduction

An important facet of the design of efficient catalytic processes is understanding the factors that determine reaction selectivity, especially in complex environments. One factor in determining reaction selectivity in heterogeneous catalysis is competition for active sites among the various reactants and products of reactions. The competition for active sites will depend on the relative binding strengths of these disparate species. The relative binding strengths, in turn, are known to depend on the composition and structure of the surface and the functionality of the primary moiety bound to the surface. Generally, periodic trends in relative binding strengths have been described in terms of the relative metal–atom binding strengths calculated using Density Functional Theory (DFT), leading to so-called linear scaling relationships.[1] While there is no question that many aspects of reactivity trends are predicted using these scaling relationships, this approach does not capture variability in reactivity within a homologous series of molecules, *e.g.* alcohols with different chain lengths.

We have shown that variation in molecular structure, even for a homologous series, has a significant impact on the competition for active sites. Experimentally, we established a hierarchy for the binding of a broad range of molecules on Au single crystals by studying displacement reactions (Fig. 1 and Scheme 1).[2] Clearly, the functional group through which the molecule is primarily bound to the surface is important in determining the binding strength, *e.g.* carboxylates are more strongly bound that alkoxides. There are also important differences in the binding strength depending on the alkyl group within a homologous series, *e.g.* the alkoxides. Generally, an increase in chain length strengthens binding. On the other hand, fluorination of the terminal group in ethoxy weakens the binding.

In selected cases, the equilibrium constants for this competition (Fig. 1) were measured in order to quantify the relative binding affinity. Specifically, the

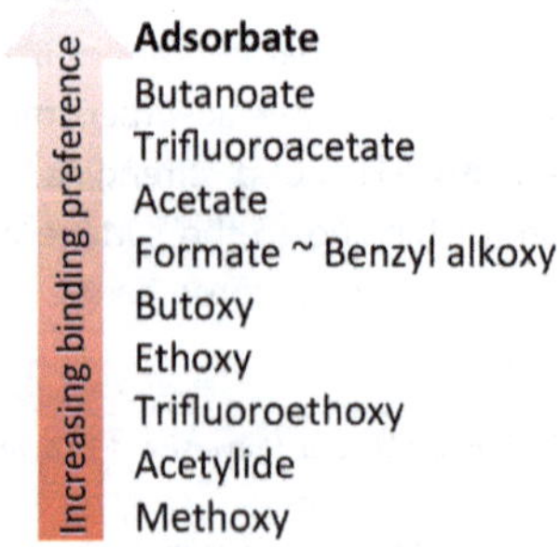

Fig. 1 Hierarchy of the relative binding affinities of various molecules on Au(111).

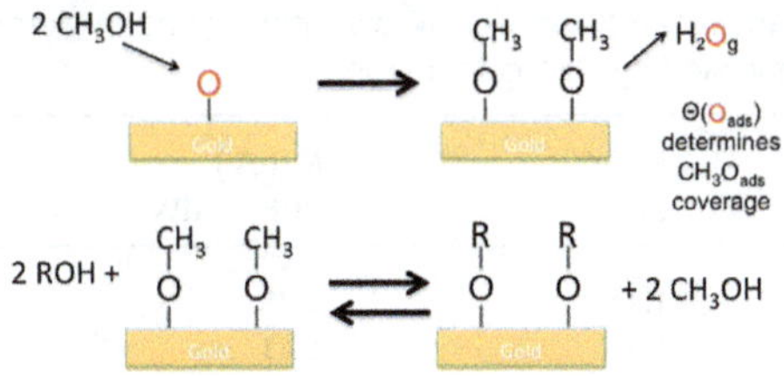

Scheme 1 The displacement reactions on Au(111). Adsorbed O is required to activate an alcohol on Au (top). Competition for binding sites occurs through displacement, which establishes an equilibrium concentration of two dissimilar alcohols because of their different adsorption energies.

equilibrium constants for the displacement of methoxy (CH_3O_{ads}) by ethanol or 1-butanol on Au(111) were measured to be 5 and 10, respectively.[3] These studies raised the important question of the underlying physical basis of these differences in the relative binding strengths and also suggested that additional quantitative measurements of equilibrium constants would be valuable in predicting reaction selectivity when one or more of these intermediates competes for reactive sites.

The generality of the use of displacement reactions to establish a hierarchy of the relative binding affinities of various molecules on surfaces has been demonstrated by other studies. On coinage metals, Ag(110) and Cu(110), the experimentally determined hierarchy of binding is similar to that on Au(111) and is related to the gas phase acidities of the molecules. On Ag(110), the order of binding of adsorbed species was determined to be $HCOO \approx CH_3COO > C_2H_5O > C_2H > CH_3O$.[4] A somewhat different set of molecules was studied on Cu(110) but with a similar hierarchy: HCOO > 1-propoxy ~ 2-propoxy > $C_2H > C_2H_5O > CH_3O$.[5] Analogous studies on ZnO and MgO established a different order of binding affinities.[6] Although the carboxylates (HCOO and CH_3COO) still had higher binding affinities to these oxide surfaces than any of the alcohols investigated, similar to the coinage metals, the relative behaviour of methanol and ethanol was different. On ZnO, methoxy and ethoxy were similarly bound, and on MgO, methanol displaced ethoxy – the opposite of the coinage metals. The behaviour on the oxides was explained in terms of charge stabilization of the anion of the gas-phase acid by cation sites on these oxides; however, advanced theoretical methods were not yet available to test this postulate. Nevertheless, these investigations illustrate the generality of using displacement reactions to establish a qualitative hierarchy for binding affinity.

Theoretical studies of a selected set of alkoxides on Au(111) demonstrated the key role of weak van der Waals interactions in determining the relative binding for the specific case of alkoxide binding (Table 1).[2] The van der Waals interactions of the alkyl group with the surface increase with increasing alkyl chain length by ~0.13 eV per CH_2 group. The presence of an electron-rich group, *e.g.* CF_3, also affects the relative binding strength *via* Pauli repulsion with the surface, leading to a decrease in the binding energy of 0.27 eV for trifluorethoxy *vs.* ethoxy. The combination of model studies on gold single crystals and theoretical investigations establish the important parameters that dictate the relative binding energy – functional group binding and weak van der Waals forces. The similarities in

Table 1 The significant role of van der Waals interactions in determining the relative binding strengths is revealed by DFT calculations[2]

Adsorbate	E_b (eV) PBE	E_b (eV) PBE + vdW	Difference due to vdW (eV)
CH_3O	1.15	1.29	0.14
CF_3CH_2O	1.11	1.41	0.30
CH_3CH_2O	1.38	1.64	0.28
1-$CH_3(CH_2)_3O$	1.33	1.80	0.47

binding trends on Ag and Cu suggest that weak interactions are also important on these metals. On the other hand, the experimental hierarchies on the oxides, ZnO and MgO, suggest that the van der Waals interactions play less of a role on the oxides due to lower polarizability. This important point needs further study using theoretical tools.

The relationship between reaction selectivity and competitive binding is specifically illustrated by the oxidative coupling of dissimilar alcohols on Au. Detailed mechanistic studies on Au(111) using surface science techniques established the mechanism for selective oxidative coupling of methanol and subsequently of other longer-chain alcohols, *e.g.* ethanol (Scheme 2).[7,8] Adsorbed O is required to activate alcohols on gold,[9,10] the first step being hydrogen transfer from the O–H to the adsorbed O to yield an adsorbed alkoxy (RO_{ads}) and hydroxyl. An aldehyde is formed in an ensuing β-C–H scission step and three reaction pathways are available thereafter: (1) coupling to a neighbouring unreacted alkoxy to form an ester; (2) reaction with excess adsorbed O to form a carboxylate; or (3) desorption from the surface. The relative contributions of these pathways depend on the ratio of adsorbed O to alkoxy on the surface and on the molecular structure of the alkyl group in the alkoxy.[2,7,8]

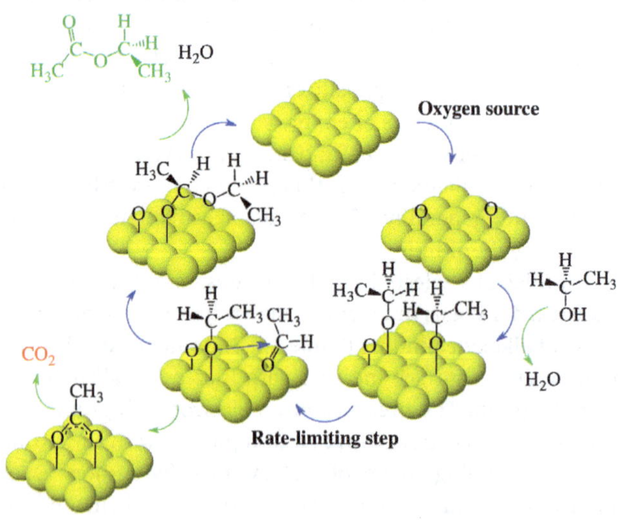

Scheme 2 The mechanism for oxidative coupling of ethanol on Au activated by adsorbed O.

The same general mechanism leads to the coupling of dissimilar alcohols on gold. The *selectivity* for the production of specific esters from a mixture of different alcohols depends strongly on the composition of the reaction mixture both in ultrahigh vacuum and under catalytic flow conditions. For example, the optimum selectivity for the production of methyl acetate from coupling of methanol with ethanol on gold has essentially the same dependence on reactant mixture both under the controlled conditions of ultrahigh vacuum and catalytic flow conditions (Fig. 2).[3,11,12] A similar correspondence was also measured for methanol coupling with 1-butanol on Au(111) and over nanoporous Au.[12] The two main factors that control reaction selectivity are: (1) the relative rate of β-H elimination from the adsorbed alkoxy species to form the aldehyde; and (2) the relative concentrations of the reactant alkoxys, which is, in turn, determined by the equilibrium between the adsorbed intermediates and the ambient reactant gas phase.

In this work, we extend the study of reaction selectivity and its relationship to competitive binding to the study of fluorinated ethanol. Fluorinated organic compounds are widely used in the synthesis of specialized polymers and pharmaceuticals. The fascinating chemistry of fluorinated organics is due to the stability of the C–F bond and the high electronegativity and low polarizability of

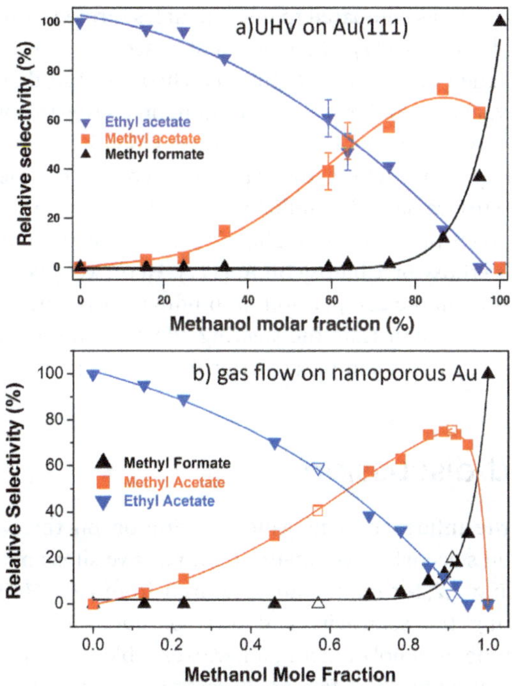

Fig. 2 The selectivity as a function of the mole fraction of methanol in the oxygen-assisted reaction of ethanol and methanol is similar for Au(111) under ultrahigh vacuum conditions (a) and over a nanoporous Au catalyst under steady state reaction conditions at atmospheric pressure (b). (a) Reprinted with permission from ref. 3. Copyright 2010 American Chemical Society. (b) Reprinted (color altered) from ref. 12 with permission from Elsevier.

fluorine, which can change the electron density distribution, molecular conformations, and intramolecular forces of organic compounds, among other effects, altering their physical and chemical properties.[13] These unique interactions of fluorine have been applied to catalysis, enhanced medicinal effectiveness of drugs, pesticide potency, and more favourable environmental compatibility, to name a few examples.[13–15] Hydrofluoroalcohols, in particular, are used as solvents for enantioselective catalysis, in lubricants, in electronic components, and as refrigerants.[16–18] They are environmentally-friendly alternatives to chlorofluorocarbons or hydrofluorocarbons because they cause less greenhouse warming and no stratospheric ozone depletion.[14,16,18–20]

Considerable effort has been made to develop homogeneous catalysts for stereoselective transformations of fluorinated compounds; fluorine has also been used as a steering group to build novel homogeneous catalysts.[14,19,20] In contrast, there has been little effort to develop heterogeneous catalysts, which have potential advantages from a process standpoint.[21] Therefore, we set out to develop a fundamental understanding of the surface chemistry of fluorinated organics on gold in order to provide the understanding necessary for controlling heterogeneous catalysis involving these fascinating systems. We specifically investigated the effect of fluorination in 2,2,2-trifluoroethanol on O-covered Au(111) by generalizing the mechanism for alcohol transformations on gold.

This work was also motivated by our recent study of the relative stability of surface intermediates, in which the stability of 2,2,2-trifluoroethoxy formed from 2,2,2-trifluoroethanol was compared to intermediates formed from other primary alcohols and acids on Au(111).[2] Displacement reactions showed that fluorine substitution weakened the binding on Au(111). Theoretical studies demonstrated that the van der Waals forces that are important in predicting the order of stability for alkoxys with varying chain lengths were essentially the same for ethoxy and 2,2,2-trifluoroethoxy (Table 1) but the repulsion between the electron-rich CF_3 group and the Au surface has a destabilizing effect. The net result is that adsorbed 2,2,2-trifluoroethoxy is predicted to be slightly more stable than methoxy but less stable than either ethoxy or 1-butoxy on Au(111). Here, we quantify the effect of fluorine substitution on the competition for binding to the surface by measuring the equilibrium and demonstrate the steering effect of fluorine substitution on the reaction selectivity in the complex reaction environment of coadsorbed alkoxy intermediates.

Results and discussion

We established the influence of fluorine substitution on the two factors that determine reaction selectivity – competition for reactive sites and the rate of β-H elimination – using 2,2,2-trifluoroethoxy, derived from CF_3CH_2OH, as a prototype. By measuring the selectivity and total product yields for mixtures of CF_3CH_2OH and either ethanol or methanol, we were able to determine the relative rates of β-H elimination for the different alkoxy species, equilibrium constants for competitive binding (Scheme 3) and the overall trends in selectivity. First, for reference, we separately address the oxidation of the fluorinated ethanol and ethanol on gold. Then we explore the mixtures of alcohols, first addressing the difference in β-H elimination, then the binding equilibrium constants, and finally the overall selectivity trends.

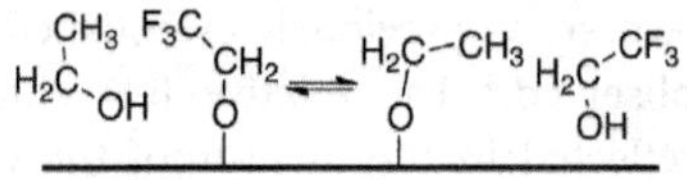

Scheme 3 Competitive equilibrium of alkoxys on a gold surface, shown for 2,2,2-trifluoroethoxy and ethoxy.

Reaction of CF_3CH_2OH on 0.1 monolayers (ML) O/Au(111) produces 2,2,2-trifluoroacetaldehyde and 2,2,2-trifluoroethyl-2,2,2-trifluoroacetate at 240–300 K and 277 K, respectively, and CO_2 and CF_3 were produced *via* trifluoroacetate decomposition at 595 K (Fig. 3a). These results demonstrate that the same overall mechanism for selective alcohol oxidation on Au applies to the fluorinated species.[8,22] However, there are interesting selectivity differences, which can be clearly observed by placing the oxidation of both 2,2,2-trifluoroethanol and ethanol side-by-side under the same oxidizing conditions (Fig. 3a and b).

In contrast to ethanol oxidation, for which our results are consistent with previous results,[8] oxidation of trifluoroethanol results in a much larger molar ratio of aldehyde *vs.* ester production: ~16 : 1 *vs.* ~1.6 : 1 with and without F substitution, respectively. Furthermore, there is also a considerable amount of CO_2 production at high temperature for $CF_3CH_2O_{ads}$ but no detectable CO_2 for the reaction of $CH_3CH_2O_{ads}$ under the same conditions. This pattern of CO_2 production indicates the presence of trifluoroacetate from over-oxidation on the surface.[22]

There are four possible ester products that can be derived from the oxidative coupling of mixtures of two different alcohols: two cross-coupling esters and two

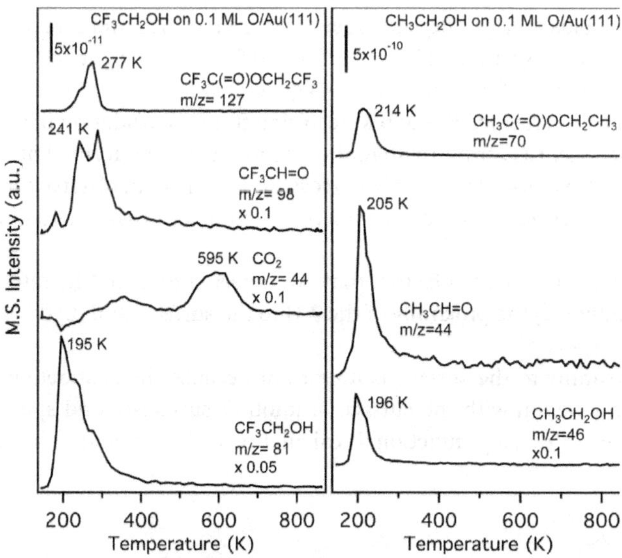

Fig. 3 Oxidation of 2,2,2-trifluoroethanol (left) and ethanol (right) on 0.1 ML O/Au(111). Each alcohol was dosed in excess at 120 K, and the heating rate was 5 K s^{-1}. The *m*/*z* traces have been corrected for fragmentation and ionization cross-section, and then the multiplication factor was applied so that all products could be seen. Also, note the different scales. For more experimental and calculation details see the ESI.†

self-coupling esters. However, as previously established, only one of the cross-coupling products is observed.[3] The relative β-H elimination rates between different alcohols are reflected in the identity of the coupling products – the alcohol with the faster β-H elimination rate makes up the acyl portion of the alkyl-ester product (Scheme 4).

Oxidation of CF_3CH_2OH in the presence of ethanol or methanol on O/Au(111) yields coupling products, as exemplified for specific mixtures (Fig. 4). In the example shown for the CH_3CH_2OH and CF_3CH_2OH mixture, the only esters formed are trifluoroethylacetate and ethylacetate; no trifluoroethyltrifluoroacetate or ethyltrifluoroacetate is detected. Both of the esters formed are derived from acetaldehyde formed from ethanol oxidation (Scheme 4) – clear evidence that β-H elimination from the ethoxy is significantly faster than from 2,2,2-trifluoroethoxy. If β-C–H bond breaking occurred at a similar rate for the two alkoxys, all coupling products would be observed. This faster rate of β-H elimination from the ethoxy is observed for all compositions tested (see below). It appears that acetaldehyde also preferentially adds to $CF_3CH_2O_{ads}$ over $CH_3CH_2O_{ads}$ based on the lower temperature for the production of trifluoroethylacetate relative to ethylacetate (Fig. 4a), with a peak onset approximately 20 K lower. There is a qualitatively similar shift in the peak onset for neat desorption of the products, but only by 10 K. Notably, there is no CO_2 produced.

Selective oxidation of a mixture of methanol and 2,2,2-trifluoroethanol yields 2,2,2-trifluoroethylformate at 238 K and 2,2,2-trifluoroethyltrifluoroacetate at 270 K on O/Au(111) (0.1 ML initial O coverage) for a mixture dilute in methanol (Fig. 4b). Therefore, the rate of β-H elimination is faster for methoxy than for 2,2,2-trifluoroethoxy, as there was no methyltrifluoroacetate detected in any experiment.

In contrast, coupling of methoxy with other longer-chain alkoxys on O/Au(111) exclusively yields methyl esters because of the generally slower rate of β-C–H bond scission for methoxy. Formaldehyde is formed from methoxy at ~210 K, below the temperature for the formation of $CF_3CH(=O)$ from CF_3CH_2O (~240 K). A large amount of CO_2 is formed at ~580 K, resulting from the oxidation of $CF_3CH=O$ to form an acetate. CO_2 from methanol oxidation would be formed at ~295 K (see ESI,† Fig. S1). Notably, there is also excess oxygen present, observed by the $m/z = 32$ peak at ~520 K. This feature was present for all of the methanol mixtures tested.

There are two principal factors that are expected to affect the rate of the C–H bond breaking: (1) the proximity of the β-H to the surface; and (2) the β-C–H bond strength.

The proximity to the surface is important because the abstraction of the β-H requires interaction with the surface or another surface-bound species.[23] Based on the recent density functional calculations that include van der Waals

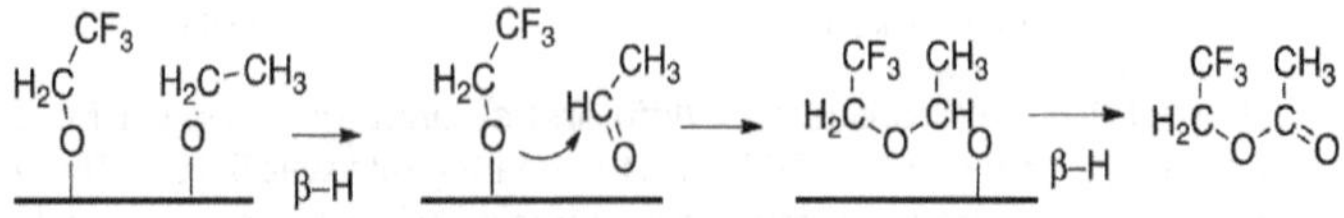

Scheme 4 Coupling of dissimilar alkoxys depends on the relative rates of β-H elimination, shown for 2,2,2-trifluoroethanol and ethanol.

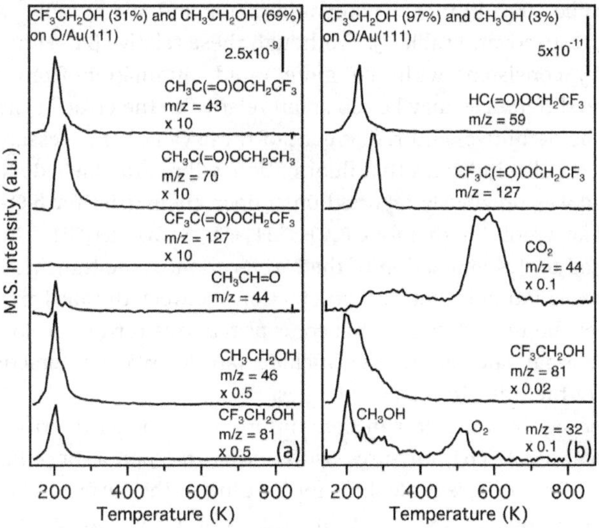

Fig. 4 Oxidation of a mixture of 2,2,2-trifluoroethanol and either ethanol (a) or methanol (b) on 0.1 ML O/Au(111). The mixtures used were 0.31 : 0.69 ratio of CF_3CH_2OH : CH_3CH_2OH and 0.97 : 0.03 ratio of CF_3CH_2OH : CH_3OH. The mixtures of alcohols were dosed in excess at 120 K to ensure that equilibrium was reached. The heating rate was 5 K s^{-1}. The relative amounts of the alcohols were determined from condensation of the alcohol mixture onto clean Au(111) followed by desorption. The *m*/*z* traces have been corrected for fragmentation and ionization cross-section, and then the multiplication factor was applied to emphasize the ester products as indicated. No trifluoroacetaldehyde was detected in the experiment with ethanol, while a small amount was detected in the experiment with methanol, but not shown here. Experimental and calculation details available in the ESI.†

interactions, the distance between the β-H and the Au surface is smaller in the non-fluorinated alkoxy species (as measured by the distance from the C1 carbon of the alkoxy to the Au surface (Table 2)) because of repulsion between the CF_3 group and the surface.[2] Hence, the rate of β-C–H dissociation should be faster for ethoxy and methoxy in comparison to trifluoroethoxy, all other factors being equal.

The second factor, the β-C–H bond strength, is not experimentally known for the CF_3CH–HOH bond; however, from recent calculations for ΔH_f for CF_3CHOH and CF_3CH_2OH,[24] the bond dissociation energy for the CF_3CH–HOH bond is

Table 2 Homolytic bond dissociation energies for the relevant β-C–H bonds. Bond dissociation energies are taken from ref. 25 except for the value for F_3CH_2COH, which was estimated using the calculated enthalpies of formation from ref. 24. Distance data taken from ref. 2.

Bond broken (bold)	Bond dissociation energy (kJ mol^{-1})	Distance between C1 of the alkoxy and the Au surface (Å)
H_2C–**HOH**	402	2.83
H_3CHC–**HOH**	401	2.81
F_3CHC–**HOH**	409	2.86

estimated to be 409.3 kJ mol^{-1}, compared to 402 and 401 kJ mol^{-1} for methoxy and ethoxy, respectively (Table 2).[25] Although these relative β-C–H bond energies are generally consistent with the more rapid elimination from ethoxy and methoxy, the differences may be too small relative to the error in the calculated energies to be definitive, and refinement of these values is necessary. Nevertheless, this factor also indicates that fluorination will retard aldehyde formation. A similar trend was previously reported on copper: the barrier for β-C–H dissociation was 8.4 kcal mol^{-1} higher for $CF_3CH_2CH_2$–Cu *vs.* $CH_3CH_2CH_2$–Cu. This effect was attributed to destabilization of the carbo-cation in the transition state.[26] An alternative explanation is that weak interactions between the alkyl groups and the surface affect the relative rates. This concept requires further exploration using density functional calculations that include van der Waals interactions for the calculation of the transition state energies.

The other significant factor in determining the product distribution for the oxidation of CF_3CH_2OH mixtures with either methanol or ethanol is the competition of the different alcohols for binding to the surface. This factor was quantitatively evaluated by measuring the equilibrium constants for binary mixtures of CF_3CH_2OH with either CH_3CH_2OH or CH_3OH exposed to O/Au(111) over a range of compositions. A series of mixtures of the two alcohols was dosed in excess on the gold surface with 0.1 monolayers (ML) of pre-adsorbed atomic O. The adsorbed O is required for initial activation of the O–H bonds in alcohols. The alcohols then form an equilibrium mixture consisting of adsorbed alcohols and alkoxys (Scheme 3), and the equilibrium constant was determined from quantitative analysis of the respective product formed (Fig. 5a and b) using a method described previously.[3] The equilibrium constant, K, was measured for the displacement reaction

$$RO_{ads} + R'OH \leftrightarrow ROH + R'O_{ads}, \qquad (1)$$

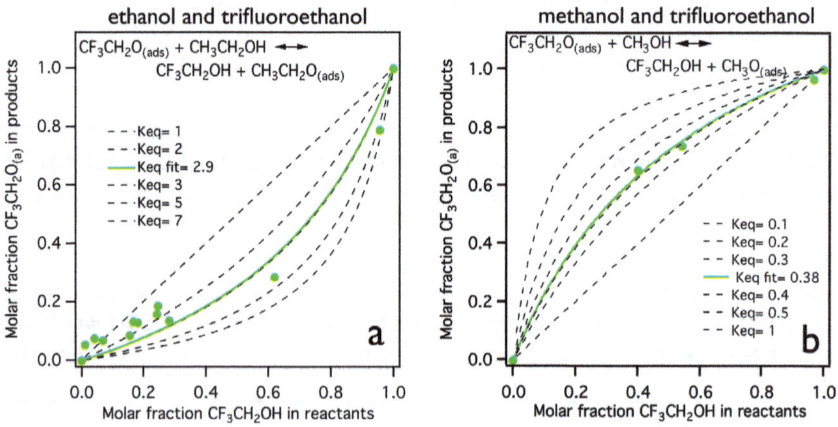

Fig. 5 Variation in the reactant composition affects the product distributions, measurements of which were used to calculate the equilibrium constants for the competitive binding of the two different pairs of alcohols. The dashed curves are the predicted behaviour for various values of K_{eq} and the solid curve represents the best fit for the experimental data.

assuming equilibrium of the alcohols with the condensed phase of co-dosed (in excess) alcohols. Therefore,

$$K = \frac{\theta_{R'O}\theta_{ROH}}{\theta_{RO}\theta_{R'OH}}, \tag{2}$$

where θ_{RO} and $\theta_{R'O}$ are the relative coverages of reacted surface alkoxy species and are determined by quantitative analysis of the product distribution of the coupling reactions, and θ_{ROH} and $\theta_{R'OH}$ are the relative amounts of alcohols in the mixture dosed on the surface. The relative amounts of reacted alkoxys, in turn, reflect the outcome of the competitive equilibrium and are used to calculate the equilibrium constant. In this work, ROH will always refer to 2,2,2-trifluoroethanol, while R′OH will be either ethanol or methanol.

The equilibrium constants for 2,2,2-trifluoroethanol and ethanol (Fig. 5a) and 2,2,2-trifluoroethanol and methanol (Fig. 5b) were determined to be 2.9 and 0.38, respectively, in qualitative agreement with the trend indicated by DFT calculations that included van der Waals interactions, which indicated the relative binding strengths of 1-butoxy > ethoxy > trifluoroethoxy > methoxy (Table 3). Using our values for the equilibrium constants between methanol or ethanol and 2,2,2-trifluoroethanol, the $K_{eq.}$ value for ethanol with methanol can then be estimated by taking the ratio of these numbers (2.9/0.38), which gives $K_{eq.} \sim 7.5$. The previously measured value of 5 indicates that, while not perfect, our results are internally consistent to within reasonable error. In these quantitative calculations, error can be introduced several ways, including calculations of the fragmentation correction, estimation of the ionization cross section, and deconvolution of various fragments. At most, we estimate error bars of 2.9 ± 1 and 0.38 ± 0.1. For discussion of values used and calculation details, refer to the ESI.†

Although the *relative* $K_{eq.}$ values are qualitatively consistent with the order of the calculated binding energies for the adsorbed alkoxy intermediates, additional calculations are required to theoretically estimate the equilibrium constants. Based on recent work evaluating the entropy of bound species, we expect the relative differences in entropy between the three alkoxys to have negligible effect.[27] In order to theoretically determine the equilibrium constants, the different stabilities and entropies of the gas-phase alcohols must also be included.

Table 3 The relative $K_{eq.}$ values are consistent with the differences in the calculated binding energies (BE) between alkoxys

Alcohol pair $RO_{ads} + R'OH \leftrightarrow ROH + R'O_{ads}$	Computed difference in BE of the alkoxys (eV)	Measured $K_{eq.}$
$CH_3O_{(a)} + CH_3(CH_2)_3OH \leftrightarrow CH_3OH + CH_3(CH_2)_3O_{(a)}$	0.51	10 (ref. 3)
$CH_3O_{(a)} + CH_3CH_2OH \leftrightarrow CH_3OH + CH_3CH_2O_{(a)}$	0.35	5 (ref. 3)
$CF_3CH_2O_{(a)} + CH_3CH_2OH \leftrightarrow CF_3CH_2OH + CH_3CH_2O_{(a)}$	0.23	2.9
$CF_3CH_2O_{(a)} + CH_3OH \leftrightarrow CF_3CH_2OH + CH_3O_{(a)}$	−0.12	0.38

The preference for ethoxy adsorption over trifluoroethoxy on the Au surface is manifested in the product distributions as well. There are few fluorinated products formed below 0.2 mole fraction of CF_3CH_2OH in the mixture (Fig. 6a). On the other hand, fluorinated products were observed from all of the CF_3CH_2OH/CH_3OH mixtures studied, starting at a CF_3CH_2OH mole fraction of ~0.4 (Fig. 6b).

Unfortunately, the selectivity for production of fluorinated esters is not especially high in any case studied (Fig. 6). The best selectivity for $CF_3CH_2O{-}C({=}O)CH_3$ cross-coupling product is ~40% and is achieved for a mole fraction of the CF_3CH_2OH of ~0.65, following the trend expected given the equilibrium constant between trifluoroethanol and ethanol. The best selectivity for the $CF_3CH_2O{-}C({=}O)H$ is between 11–13% for the mixtures studied. However, given the equilibrium constant between trifluoroethanol and methanol, a higher selectivity may be expected with mole fractions of trifluoroethanol lower than 0.4, which were not studied here. It is worth noting that the oxygen coverage was not adjusted to optimize for selectivity toward esters in this study. In previous studies, sometimes an oxygen coverage of lower than 0.1 ML O/Au was required to obtain optimum selectivity.[7]

The difference in binding energy and hence efficiency of alkoxy formation may explain part of the lower selectivity for the methanol case compared to the ethanol case. The higher rates of CO_2 production may be explained, in part, by the presence of excess oxygen. In our experiments, there was excess O for all of the methanol/trifluoroethanol experiments. However, there was not excess O for the ethanol/trifluoroethanol experiments. Interestingly, there is also a small amount of excess oxygen in the oxidation of pure methanol (see Fig. S1†), but although the trifluoroethanol oxidation is much less selective than ethanol oxidation, we did not observe excess oxygen. This supports the hypothesis that as the binding energy of the alkoxy increases, it competes more effectively and uses the surface oxygen more completely. When using less strongly bound alkoxys, therefore, one

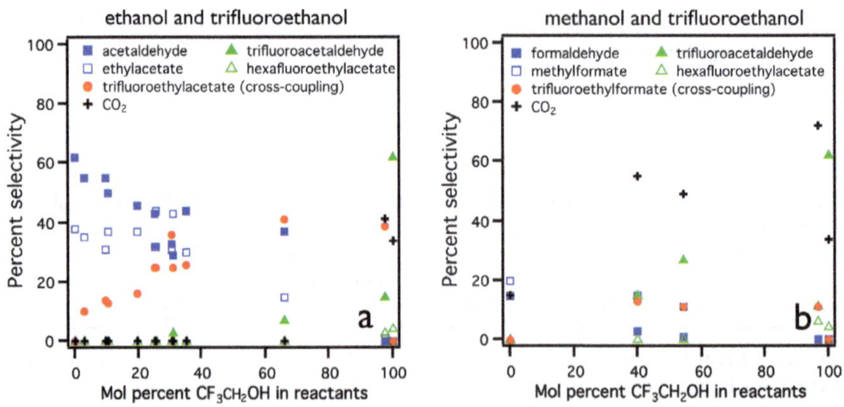

Fig. 6 Variation in the reactant composition affects the product distributions for coupling of CF_3CH_2OH with ethanol (a) or with methanol (b). All measurements were performed after exposing an excess of the alcohols mixtures onto Au(111) containing 0.1 ML of O_{ads} initially. In the case of methanol, only a few data points were measured in the region of composition exhibiting the biggest differences in curvature for different values of K_{eq}. Experimental and calculation details available in the ESI.†

might expect to improve the selectivity by using a lower initial oxygen coverage with excess alcohol.

In previous studies of aliphatic alcohols, it has been found that the longer chain alcohols have both a stronger binding energy and a faster rate of β-H elimination.[3] The case of trifluoroethanol having a binding energy between ethanol and methanol, yet having a slower β-H elimination than either of them, stands out as a unique contrast. As discussed above, this is due to the fluorination of the methyl carbon, part of the effect of which is to lengthen the distance between the C1 carbon and the surface due to Pauli repulsion. The observation that such a subtle change in distance between a carbon and the surface can have such a marked effect on the reaction rate, and thereby control the reaction pathway, may lead to new ways of understanding and designing reactions in the future, particularly when using fluorinated compounds.

Conclusions

By establishing a hierarchy of binding for a range of adsorbates on gold, we have established that small differences in the binding of reactive species have a significant impact on reaction selectivity by affecting the competition for binding sites. The coupling of dissimilar alcohols is used to illustrate that the dependence of the selectivity for the production of various esters on the composition of the reactant mixture is the same on O-covered single crystals and under working catalytic conditions using nanoporous Au catalysts. Density functional theory calculations show that weak van der Waals interactions play an important role in determining this competition for binding of a homologous series of reactants, thus providing an explanation for the observed trends in selectivity and binding.

We further extended the concept of competitive binding to the investigation of fluorinated alcohols. Fluorinated organics are becoming prominent in the chemical industry and in synthesis because of their unique properties and myriad of applications. In this study we established principles for reactive processes involving fluorinated alcohols on a surface, with a specific aim towards ester synthesis. Subtle differences between β-H elimination rates and equilibrium constants between alkoxys on a surface have a marked effect on the products of the reaction between two alcohols, and we have here evaluated these factors for trifluoroethanol, compared to ethanol and methanol. 2,2,2-Trifluoroethoxy has a slower rate of β-H elimination than both methoxy and ethoxy. These results are consistent with the estimated β-H bond dissociation energies of the alcohols and with the geometric structures of the adsorbed alkoxys predicted by DFT including van der Waals forces. The binding energy of adsorbed 2,2,2-trifluoroethoxy lies between ethanol and methanol, manifesting differences between the equilibrium constants for the formation of 2,2,2-trifluoroethoxy in the presence of ethanol or methanol.

Acknowledgements

We gratefully acknowledge the support of this work by the U.S. Department of Energy, Basic Energy Sciences, Grant no. DE-FG02-84ER13289.

Notes and references

1 J. K. Norskov, T. Bligaard, J. Rossmeisl and C. H. Christensen, *Nat. Chem.*, 2009, **1**, 37–46.
2 J. C. F. Rodriguez-Reyes, C. G. F. Siler, W. Liu, A. Tkatchenko, C. M. Friend and R. J. Madix, *J. Am. Chem. Soc.*, 2014, **136**, 13333–13340.
3 B. Xu, R. J. Madix and C. M. Friend, *J. Am. Chem. Soc.*, 2010, **132**, 16571–16580.
4 M. A. Barteau and R. J. Madix, *Surf. Sci.*, 1982, **120**, 262–272.
5 S. W. Jorgensen and R. J. Madix, *Surf. Sci.*, 1983, **130**, L291–L293.
6 R. N. Spitz, J. E. Barton, M. A. Barteau, R. H. Staley and A. W. Sleight, *J. Phys. Chem.*, 1986, **90**, 4067–4075.
7 B. Xu, X. Liu, J. Haubrich, R. J. Madix and C. M. Friend, *Angew. Chem., Int. Ed.*, 2009, **48**, 4206–4209.
8 X. Liu, B. Xu, J. Haubrich, R. J. Madix and C. M. Friend, *J. Am. Chem. Soc.*, 2009, **131**, 5757–5759.
9 D. A. Outka and R. J. Madix, *J. Am. Chem. Soc.*, 1987, **109**, 1708–1714.
10 B. Xu, R. J. Madix and C. M. Friend, *Acc. Chem. Res.*, 2014, **47**, 761–772.
11 K. J. Stowers, R. J. Madix and C. M. Friend, *J. Catal.*, 2013, **308**, 131–141.
12 L.-C. Wang, K. J. Stowers, B. Zugic, M. L. Personick, M. M. Biener, J. Biener, C. M. Friend and R. J. Madix, *J. Catal.*, 2015, **329**, 78–86.
13 R. Berger, G. Resnati, P. Metrangolo, E. Weber and J. Hulliger, *Chem. Soc. Rev.*, 2011, **40**, 3496–3508.
14 D. Cahard and V. Bizet, *Chem. Soc. Rev.*, 2014, **43**, 135–147.
15 T. J. Wallington, W. F. Schneider, D. R. Worsnop, O. J. Nielsen, J. Sehested, W. J. Debruyn and J. A. Shorter, *Environ. Sci. Technol.*, 1994, **28**, 320A–326A.
16 T. J. Wallington and W. F. Schneider, *Environ. Sci. Technol.*, 1994, **28**, 1198–1200.
17 J. P. Bégué, D. Bonnet-Delpon and B. Crousse, *Synlett*, 2004, 18–29.
18 M. Lily, D. Sutradhar and A. K. Chandra, *Comput. Theor. Chem.*, 2013, **1022**, 50–58.
19 C. Hollingworth and V. Gouverneur, *Chem. Commun.*, 2012, **48**, 2929–2942.
20 S. Y. Lee, S. Neufeind and G. C. Fu, *J. Am. Chem. Soc.*, 2014, **136**, 8899–8902.
21 L. E. Manzer, *Catal. Today*, 1993, **18**, 199–207.
22 C. G. F. Siler, T. Cremer, J. C. F. Rodriguez-Reyes, C. M. Friend and R. J. Madix, *ACS Catal.*, 2014, **4**, 3281–3288.
23 B. Xu, T. A. Baker, E. Kaxiras and C. M. Friend, *J. Phys. Chem. C*, 2011, **115**, 3703–3708.
24 Y. Wang, J.-Y. Liu, Z.-S. Li, L. Wang and C.-C. Sun, *J. Comput. Chem.*, 2007, **28**, 802–810.
25 Y.-R. Luo, *Comprehensive Handbook of Chemical Bond Energies*, CRC Press, Boca Raton, 2007.
26 J. G. Forbes and A. J. Gellman, *J. Am. Chem. Soc.*, 1993, **115**, 6277–6283.
27 C. T. Campbell and J. R. V. Sellers, *J. Am. Chem. Soc.*, 2012, **134**, 18109–18115.

Faraday Discussions

PAPER

The curious case of zeolite–clay/binder interactions and their consequences for catalyst preparation†

Gareth T. Whiting,‡* Abhishek Dutta Chowdhury,‡ Ramon Oord, Pasi Paalanen and Bert M. Weckhuysen*

Received 30th November 2015, Accepted 2nd December 2015
DOI: 10.1039/c5fd00200a

Zeolite-based catalyst bodies are commonly employed in a range of important industrial processes. Depending on the binder and shaping method chosen, vast differences in the reactivity, selectivity and stability are obtained. Here, three highly complementary micro-spectroscopic techniques were employed to study zeolite ZSM-5–binder interactions in SiO_2-, Al_2O_3-, $SiO_2 : Al_2O_3$- (2 : 1 mix) and kaolinite-bound catalyst pellets. We establish how their preparation influences the zeolite–clay/binder interactions. Using thiophene as an acid-catalyzed staining reaction, light absorbing oligomers produced in each sample were followed. To our surprise, kaolinite decreased the overall reactivity of the sample due to the phase change of the binder, creating a hard impenetrable outer layer. Aluminum migration to the zeolite was observed when Al_2O_3 was selected as a binder, creating additional Brønsted acid sites, which favored the formation of ring-opened thiophene oligomers compared to the larger oligomer species produced when SiO_2 was used as a binder. In the latter case, the interaction of the Si–OH groups in the binder with thiophene was revealed to have a large impact in creating such large oligomer species. Furthermore, the combination of a $SiO_2 : Al_2O_3$ mix as a binder enhanced the reactivity, possibly due to the creation of additional Brønsted acid sites between the two binder components during pellet preparation. It is evident that, independent of the shaping method, the intimate contact between the zeolite and binder heavily impacts the reactivity and product selectivity, with the type of binder playing a vital role.

1. Introduction

The use of catalyst materials in the petrochemical and bulk chemical synthesis industries is vast, and plays a vital role in supplying many of our everyday

Inorganic Chemistry and Catalysis Group, Debye Institute for Nanomaterials Science, Utrecht University, Universiteitsweg 99, 3584 CG Utrecht, The Netherlands. E-mail: g.t.whiting@uu.nl; b.m.weckhuysen@uu.nl

† Electronic supplementary information (ESI) available. See DOI: 10.1039/c5fd00200a

‡ Authors contributed equally to this work.

needs.[1–3] In industrial processes such as hydrocarbon and/or aromatic isomerization- and alkylation-type reactions, zeolites provide the main active phase of the catalyst materials.[4,5] However, in order for the zeolite to function at its optimum level, it is often dispersed non-homogeneously in a binder or matrix, *i.e.* SiO_2, Al_2O_3 and/or clay, and shaped to the desired needs. Such catalysts are commonly referred to as 'catalyst bodies' or 'technical bodies', and take the form of extrudates, pellets or granules.[6] Binder/matrix components offer numerous advantages to the zeolite (as highlighted in a perspective article by Hargreaves *et al.*[7]), but their overriding role is to increase the mechanical strength in order to resist attrition loss within industrial reactor conditions. However, our preconceived ideas that the zeolite's physicochemical properties and catalytic performance will be fully maintained during/after the preparation of catalyst bodies (*i.e.* shaping with binders, heat treatment, *etc.*) would be naive, with the binder known to interact with the zeolite and also known to play a vital role in the catalyst's performance.[4,7–9]

Until recently, limited academic research has been reported on such industrial catalysts, but a spate in the development of advanced characterization techniques able to probe the multi-dimensional parameters of these materials with high spatiotemporal resolution (and also non-invasively) has provided significant insights into the structure–performance relationships of these closely guarded materials.[10–20] Particularly in the field of Fluid Catalytic Cracking (FCC) particles and zeolite-based binder-bound extrudates and pellets for *e.g.* methanol-to-hydrocarbon (MTH) catalysis, combinations of chemical imaging and structural imaging techniques, together with standard physicochemical characterization of the materials, have enhanced our knowledge into how these complex materials function.[17,18,21–23] For instance, staining probe molecule reactions (*i.e.* acid-catalyzed fluorostyrene oligomerization), coupled with visible micro-spectroscopy, confocal microscopy and X-ray absorption micro-spectroscopy, have shown that the distribution of the zeolite, the presence of metal poisons, and the type/amount of binder present, all significantly influence the catalytic performance and, in particular, the catalyst's ability to resist deactivation.[12,16,20,24,25] Overall, the information gained on binder effects from such advanced characterization studies of industrially relevant catalyst materials is key to improving the catalyst body design, which could play a vital role in the future of these catalysts. This is particularly prominent, given the ever stricter and challenging societal, environmental and economical demands enforced on catalytic industrial reaction processes.

The use of inexpensive inorganic binders, such as SiO_2, Al_2O_3 and clay, for zeolite-based catalyst bodies is common in industrial processes, such as xylene isomerization and benzene alkylation (among others), but several reports note that the zeolite–binder interactions can indeed influence the reactivity, selectivity and stability of the overall catalyst material. When employing clay (*i.e.* bentonite, kaolin and attapulgite) as a binder with zeolites, high-temperatures are used in the pelletization process in order to eliminate the activity of the clay and destroy the surface area, whilst creating meso- and/or macro-porosity.[26] However, it has been reported that a substantial decrease in the inter- and intra-crystalline acidity of the H-zeolite occurs during pelletization due to the neutralization of acid sites with mobile alkaline or alkaline-earth cations from the clay binder.[27] Clay has also been shown to decrease weak acidity and increase high strength acid site density,[28] as well as influence product selectivity[29] and improve coke stability (by trapping coke precursors in the kaolinite binder).[30] Al_2O_3 as a binder for zeolite-based catalyst bodies has also been known to influence product selectivity and overall activity due to the migration of aluminum from the binder to the zeolite framework, creating additional Brønsted acid sites.[17,31,32] It is also used as a heat sink and a trap for metal poisons in petroleum feedstocks, hence retaining zeolite acidity.[33] A similar trend occurs for SiO_2-bound zeolite materials, whereby a more unexpected solid-state ion exchange takes place, with Michels *et al.* and Lee *et al.* reporting a loss of Brønsted acidity due to dealumination from the framework of the zeolite.[34,35] Gélin *et al.* also witnessed the opposite trend, with Si species from the SiO_2–Al_2O_3 matrix migrating to the zeolite framework, healing defects and preserving crystallinity.[36] It is clear that, although some binders are often considered as inert, the intimate contact between the zeolite and binder in catalyst bodies can produce remarkable consequences or advantages and their interactions therefore require a greater degree of attention, if improvements in the catalyst design are to be made.

In our previous work, we have focused on SiO_2- and Al_2O_3-bound zeolite-based catalysts in the form of extrudates.[17] However, in this work, we set out to synthesize four ZSM-5–binder-bound pellets, using either SiO_2, Al_2O_3, $SiO_2 : Al_2O_3$ (mix 2 : 1) or kaolinite as a binder, and study whether any zeolite–binder interactions have taken place during their physical mixing, and how these parameters influence the overall catalytic properties of the catalysts prepared. To do this, a staining probe reaction, namely thiophene oligomerization, was employed along with several highly complementary non-invasive *in situ* micro-spectroscopic techniques: UV-Visible, confocal fluorescence and IR micro-spectroscopy. Using these three techniques, we are able to follow variations in product selectivity as well as in the location and distribution of the light absorbing oligomers formed in each sample and correlate these to any zeolite–clay/binder effects.

2. Experimental

2.1. Materials

Four batches of SiO_2- (purchased from Aldrich, Davisil™), Al_2O_3- (Catapal®), SiO_2/Al_2O_3- (2 : 1 mix of Davisil™ : Catapal®) and kaolinite- (purchased from Sigma Aldrich) bound ZSM-5 samples were prepared by a pelletization procedure: NH_4–ZSM-5 powder (purchased from Zeolyst CBV 8014) with a $SiO_2 : Al_2O_3$ ratio of 80 was thoroughly mixed with the relevant binder at a ZSM-5 : binder wt% ratio of 30 : 70 (to ensure noticeable binder effects) before pressing into pellets. Each pellet underwent heat treatment under a N_2 atmosphere (to form H–ZSM-5) from 30 °C to 600 °C (at 15 °C min^{-1}) and it was then held at this temperature for 3 h, before cooling to 30 °C. Sample notations are as follows: ZSM-5–*a*: where '*a*' corresponds to either: Si (SiO_2); Al (Al_2O_3); SiAl (SiO_2–Al_2O_3 mix, 2 : 1); Kao (kaolinite).

2.2. Ar physisorption

Ar physisorption on all samples was performed with an automated gas sorption system Micromeritics TriStar 3000. Prior to the measurements at 77 K, all samples were dried at 300 °C for 20 h. The surface areas were calculated using the BET model, whilst the external surface area, micropore surface area and micropore volume were determined using the *t*-plot method (Table 1 and Fig. 1).

2.3. X-ray powder diffraction

All samples were measured with a Bruker D8 Discover diffractometer equipped with a 1-dimensional LynxEye XE energy dispersive detector, in transmission capillary geometry with a Mo ($K_{\alpha1}$ 0.709 Å) source. The sample powders were loaded into 980 μm inner diameter quartz capillaries (1000 μm outer diameter) so that the sample bed length was about ≈20 mm. The measurement parameters used were $2\theta = 2.5–24°$ with a step size of 0.008 and an exposure time of 3.5 s per step. The Mo source was run with a power of 2500 W (50 kV × 50 mA) (Fig. 2).

2.3.1. Rietveld analysis. Rietveld analysis on the measured samples was performed using the TOPAS 5 software package. Fitting was done with structures from the ICCD PDF-4+ 2015 database. Individual peaks were used to fit the amorphous peaks ($2\theta \approx 10–11$) and the primary beam ($2\theta = 2.0–2.5$), and the rest of the background was corrected with a software provided polynomial. Sample displacement and absorption path length error in the capillary geometry were corrected with software provided functions (Table 2).

2.4. NH_3-TPD

The acidity of the samples was measured by temperature-programmed desorption (TPD) with ammonia as the probe molecule using Micrometrics AutoChem 2920 apparatus. Prior to the adsorption of ammonia, 100 mg of the sample placed in a quartz tube reactor was preheated at 550 °C under 50 mL min^{-1} of He flow for 15 min (rate 10 °C min^{-1}). Adsorption of ammonia (10 vol% in He) was performed in a pulse wise manner at 100 °C. In order to remove the physisorbed ammonia, the sample was flushed with a He stream (25 mL min^{-1}) at 100 °C for 1 h. The ammonia desorption was carried out under a He flow in the temperature range 100–700 °C at a heating rate of 5 °C min^{-1}. The amount of adsorbed ammonia was monitored with a thermal conductivity detector (TCD) (Fig. 3 and Table 2).

Table 1 Textural properties of the ZSM-5–clay/binder-bound pellets determined by Ar physisorption

Sample	BET surface area (m^2 g^{-1})	*t*-Plot micropore area (m^2 g^{-1})	Pore volume (cm^3 g^{-1})	*t*-Plot micropore volume (m^2 g^{-1})
ZSM-5	371	306	0.18	0.12
ZSM-5–Si	252 (296)[a]	90 (110)[a]	0.46 (0.80)[a]	0.03 (0.04)[a]
ZSM-5–Al	269 (316)	93 (97)	0.30 (0.33)	0.03 (0.04)
ZSM-5–SiAl	272	106	0.35	0.04 (0.04)
ZSM-5–Kao	176 (113)	150 (97)	0.09 (0.06)	0.06 (0.04)
SiO_2	264	26	1.06	0
Al_2O_3	293	33	0.39	0
Kaolinite	3	7	0.01	0

[a] Predicted values calculated from the pellet composition and the experimental values of the pure binder and pure zeolite components.

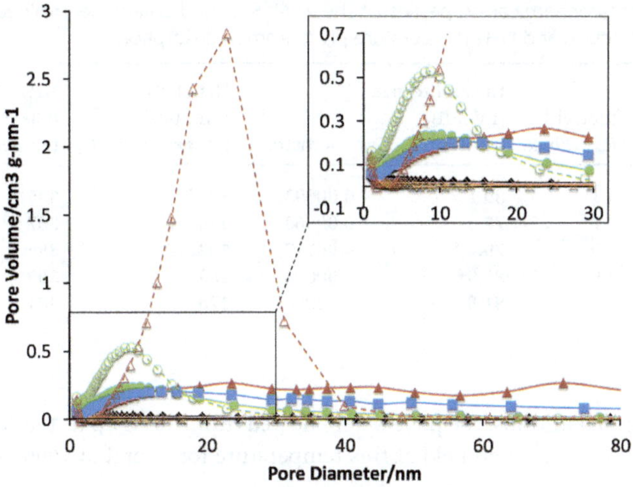

Fig. 1 Pore volume distributions, determined by the Barrett–Joyner–Halenda adsorption method with Ar, for each selected binder (dashed lines) and its corresponding zeolite–clay/binder pellets (solid lines): ZSM-5 (black diamonds); ZSM-5–Si (red triangles); ZSM-5–Al (green circles); ZSM-5–SiAl (blue squares); ZSM-5–Kao (orange crosses).

2.5. UV-Vis absorption micro-spectroscopy

The UV-Vis micro-spectroscopy measurements were performed with a CRAIC 20/30 PV™ UV-Vis-NIR micro-spectrophotometer using a 15× objective. A 75 W xenon lamp was used for illumination. The pellets were placed on the heating stage of an *in situ* Linkam cell (THMS600) equipped with a temperature controller (Linkam TMS94), where they were impregnated with 8 μL of thiophene. During each experiment, after complete adsorption of 8 μL of thiophene under

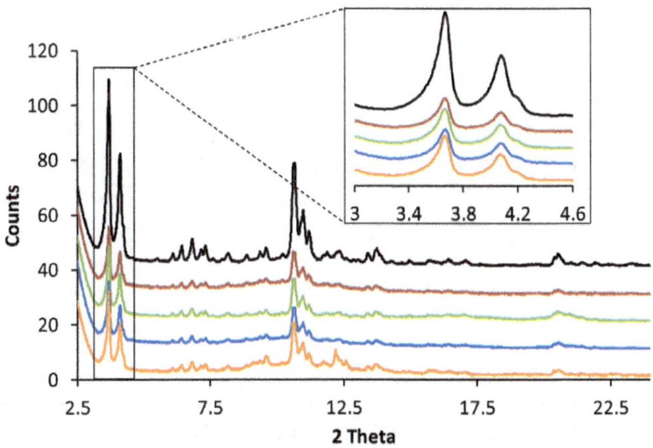

Fig. 2 Powder X-ray diffraction (XRD) patterns of the following ZSM-5-bound pellets (from top to bottom): ZSM-5 (black); ZSM-5–Si (red); ZSM-5–Al (green); ZSM-5–SiAl (blue); and ZSM-5–Kao (orange). The inset corresponds to the main zeolite diffraction peaks, noting no noticeable shifts in 2θ between the samples under study.

Table 2 Physicochemical properties of the ZSM-5-bound pellets determined by X-ray powder diffraction and NH_3 temperature programmed desorption

Sample	Rietveld R_{wp} values	Crystallite size (LVol-IB) (nm)	Microstrain	NH_3-TPD min. peak temperature (°C)	NH_3-TPD max. peak temperature (°C)
ZSM-5	3.1	89.1	0.00093	191	385
ZSM-5–Si	2.7	77.6	0.00063	175	367
ZSM-5–Al	3.0	79.7	0.00117	192, 290	386
ZSM-5–SiAl	2.6	85.8	0.00094	183	365
ZSM-5–Kao	2.5	81.8	0.00096	175	384

atmospheric conditions, the pellets were heated (from 30 °C up to 120 °C at a rate of 15 °C min^{-1}, and then held at this temperature for a total of 1000 s).

2.6. Confocal fluorescence microscopy

A Nikon Eclipse 90*i* confocal microscope with a 100 × 0.73 NA dry objective was used for the fluorescence microscopy investigations. Excitation light was provided by focusing two specific laser lines, 488 and 561 nm, simultaneously on the desired sample, located in an open *in situ* cell (Linkam Instruments, FTIR 600). The microscope was equipped with a Nikon A1 scan head, accommodating the optics, which couple fiber optics for excitation and emission light with the

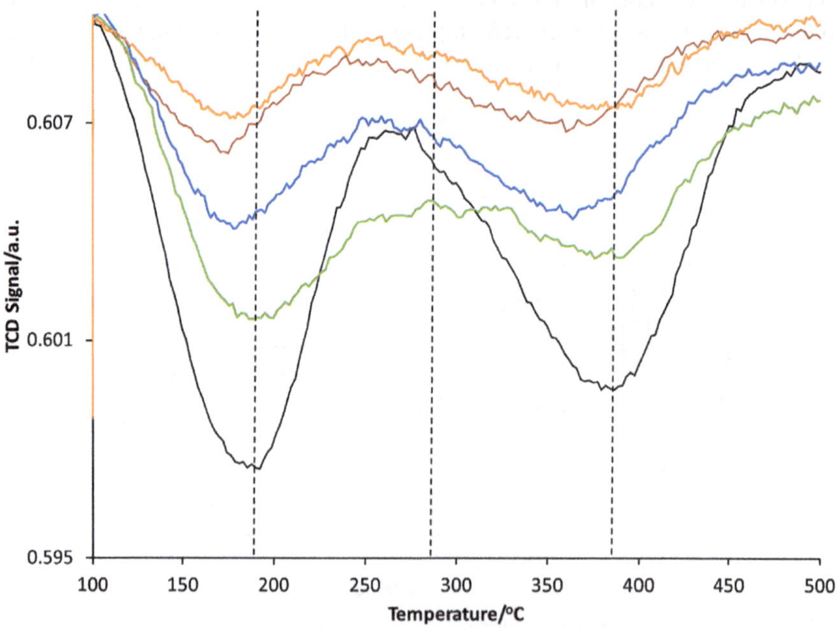

Fig. 3 NH_3-temperature programmed desorption (TPD) profiles of the following ZSM-5-bound pellets: ZSM-5 (black); ZSM-5–Si (red); ZSM-5–Al (green); ZSM-5–SiAl (blue); and ZSM-5–Kao (orange).

microscope. A spectral analyzer in the Nikon A1 system was equipped with 32 photomultiplier tubes (PMTs) set to collect emission light in the region of 480–730 nm, with a resolution of 6 nm. 3D confocal fluorescence microscopy images were recorded at a similar focal depth in each sample using an identical laser power. Prior to microscopic analysis, to each sample, 8 μL of the desired probe molecule was impregnated at 30 °C, before ramping the temperature (at 15 °C min^{-1}) to 120 °C, holding at this temperature for 15 min, and cooling to 30 °C.

2.7. Infrared micro-spectroscopy

In situ infrared micro-spectroscopy experiments were performed using a Perkin Elmer Spotlight 400 FT-IR microscope with an aperture size of 100 × 100 μm. Each zeolite–clay/binder-bound pellet was placed under the microscope and spectra were obtained in reflectance mode before, during and after thiophene oligomerization (from 30 °C to 120 °C at 15 °C min^{-1}).

3. Results and discussion

3.1. UV-Vis absorption micro-spectroscopy

In order to probe possible zeolite–clay/binder interactions created during the catalyst preparation procedure (*i.e.* pelletization and/or heat treatment), a suitable reaction was chosen. The oligomerization of thiophene has previously been utilized by our research group for such purposes on zeolite–binder-bound catalyst extrudates, and offers an excellent means to gain insight into characteristics such as how the nature and/or number of Brønsted acid sites within the samples influence the catalytic reactivity and intermediate/product selectivity.[17,18] A particular highlight of the chosen reaction is that the majority of oligomer species produced on the acid sites within the samples are light absorbing (Scheme 1). Therefore, UV-Vis absorption spectroscopy allows us to obtain, *in situ*, the formation of absorption bands and hence the nature and relative amount of oligomer species formed, as a function of reaction time. Scheme 1 provides a general overview of the possible light absorbing species produced (measured using Time-Dependent Density Functional Theory (TD-DFT)), either during an oligomerization reaction pathway, or during a ring-opening pathway (leading to the formation of thiol-like species), with the pathway chosen determined by the strength and number of Brønsted acid sites of the zeolite.

Given the complex nature of the overall UV-Vis absorption spectra of all the samples (Fig. S1 and S2†), absorption spectra of each sample were compared at various reaction times during the oligomerization reaction (Fig. 4). Fig. 4(a) displays the absorption spectra for pellets of ZSM-5, ZSM-5–Si, ZSM-5–Al, ZSM-5–SiAl and ZSM-5–Kao after 10 s of reaction time. Even at this early stage of the reaction, visible differences between each catalyst's absorption spectrum are apparent. The absorption band formed at 275 nm can be assigned to α-protonated thiophene monomers (Scheme 1B) on zeolite Brønsted acid sites.[37] The band at 310 nm is attributed to a protonated dimeric carbocation, formed by the interaction of two thiophene molecules linked together in the zeolite pores (C).[37] The more intense absorption bands at 410, 500, and 550 nm are assigned to molecules formed *via* the oligomerization route, namely: a protonated trimeric carbocation species, *i.e.* $C_{12}H_{13}S_3^+$; a protonated trimeric carbocation species, *i.e.*

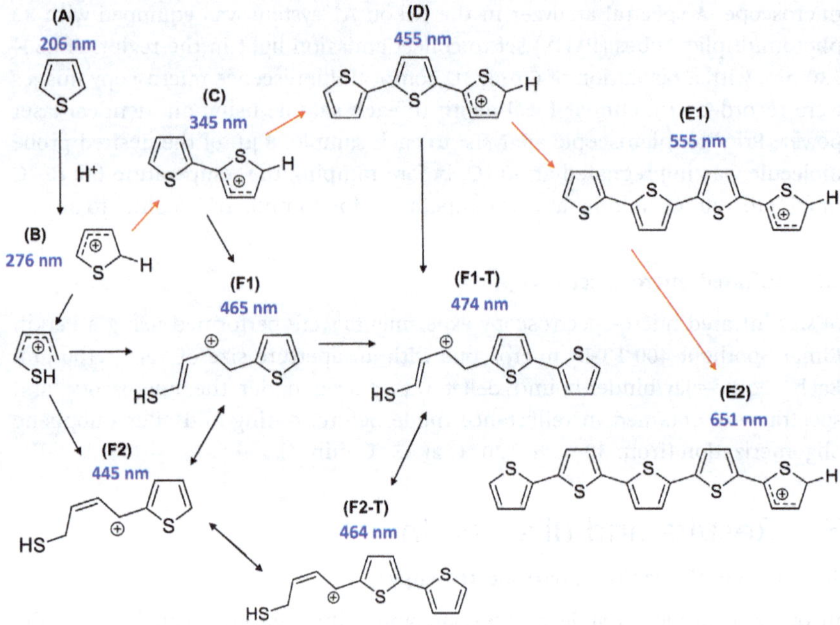

Scheme 1 Proposed reaction pathway of thiophene oligomerization on zeolite Brønsted acid sites. Thiophene monomer (A) undergoes protonation on the Brønsted acid sites of H–ZSM-5 to form the protonated monomer (B). Two different reaction pathways are possible at this stage: opening of the thiophene ring and subsequent reaction with a second monomer to form a thiol-like carbocation (F1/F1-T/F2/F2-T), or, alternatively, dimerization (to C) can occur followed by further oligomerization (to D) or opening of the thiophene ring, forming a thiol-like carbocation (F1/F1-T/F2/F2-T). The trimeric carbocationic species (D) can undergo further oligomerization to form more extended/conjugated species (E1/E2). [Reproduced from ref. 17].

$C_{12}H_9S_3^+$; and a tetrameric carbocation species (E1), respectively.[37] Finally, the band at 445 nm can be assigned to a thiol-like species (F2) formed *via* a ring-opening reaction pathway.

Taking each sample into consideration, a feature that stands out in the spectra at 10 s of reaction time (Fig. 4(a)) is the surprising lack of intense absorption bands for the kaolinite-bound sample, ZSM-5–Kao. This suggests a clear inactivity when ZSM-5 crystals are bound to clay, compared with ZSM-5 bound to an amorphous binder. As NH_3-TPD showed (Fig. 3), the sample bound to kaolinite displayed the lowest amount of weak and strong Brønsted acid sites, which could explain the low thiophene oligomerization reactivity, but this is not likely considering the similar profile for ZSM-5–Si (which shows a far higher reactivity, and even has slightly weaker acid sites due to a slight shift to lower temperatures in the NH_3 desorption peaks). X-ray diffraction of the ZSM-5–Kao material (Fig. 2) showed a clear phase change from kaolinite (Fig. S3†) to a combination of muscovite and orthoclase upon preparation of the ZSM-5–Kao pellet. This feature is due to the loss of hydroxyl groups from kaolinite upon heat treatment of the pellet, which propagates the phase change of the clay,[38] and could play a vital role in the inactivity of the sample. Further evidence to support this claim is observed in Fig. S2(a),† where the heat-treated ZSM-5 powder pelletized with

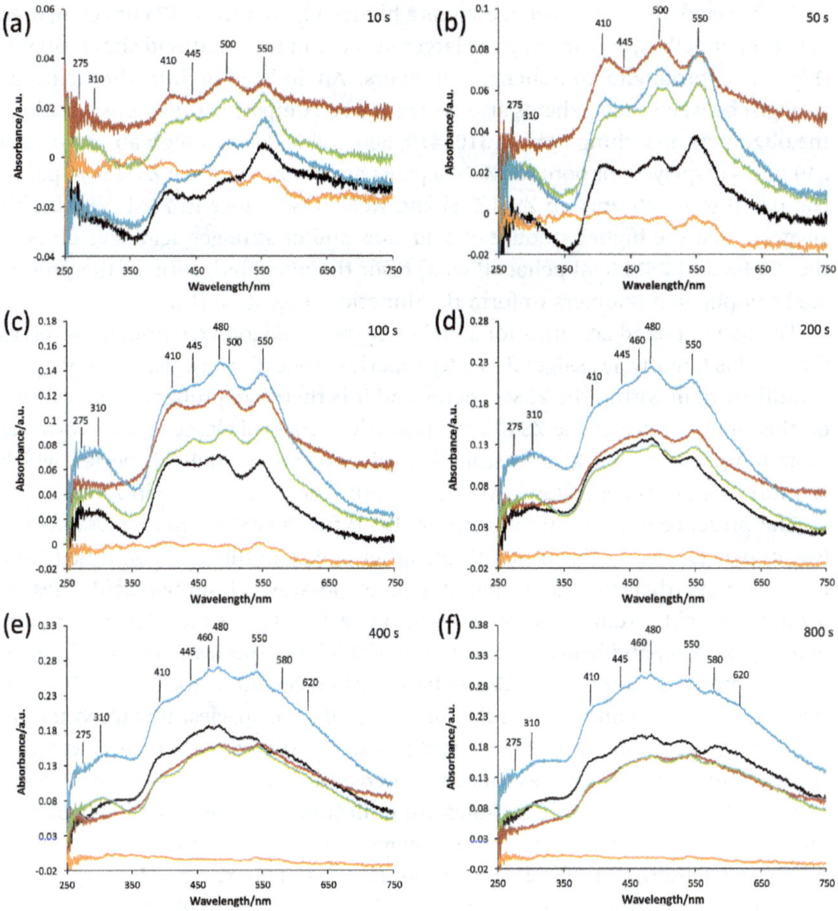

Fig. 4 UV-Vis absorption spectra of ZSM-5 (black line), ZSM-5–Si (red), ZSM-5–Al (green), ZSM-5–SiAl (blue), and ZSM-5–Kao (orange) pellets reacted with thiophene after: (a) 10 s; (b) 50 s; (c) 100 s; (d) 200 s; (e) 400 s; and (f) 800 s of reaction time.

non-heat-treated kaolinite reveals a far higher reactivity towards thiophene oligomerization (due to the unchanged phase of kaolinite). Looking towards the Mohs hardness properties of the newly developed binder in ZSM-5–Kao, an increase from 2.5 (kaolinite) to 6 (orthoclase) gives an indication that thiophene impregnation into the pellet could be hindered. Ar physisorption results provide further support that the phase change of the material is detrimental to the textural properties of the catalyst, with limited porosity available in ZSM-5–Kao, and even a striking elevation in the expected micropore area and micropore volume values (Table 1). Given the temperature of the heat treatment of the pellets was 600 °C, it is highly likely that upon the loss of hydroxyl groups and H_2O, partial steaming of the zeolite takes place, exposing a greater micropore area. Also of importance is to take into account solid-state interfacial reactions between the clay material and the zeolite, leading to a decrease in the number of Brønsted acid sites.

As expected, the ZSM-5 pellet obtains a higher absorbance at 275 nm compared to the other pellets, considering the larger number of Brønsted acid sites available (Fig. 3) to protonate thiophene monomers. An indication into the acid site strength between each pellet is also evident when comparing the intensity ratio of the oligomers absorbing light at 310, 410, 500 and 550 nm. A high absorbance at 310 nm is displayed for both the ZSM-5 pellet and more so for the ZSM-5–Al pellet, compared to its absence in ZSM-5–Si and weak absorbance in ZSM-5–SiAl. This suggests that the higher amount of acid sites and/or stronger acid sites on both the ZSM-5 and ZSM-5–Al pellets (Fig. 3) favor the oligomerization of the protonated thiophene monomers to form the dimeric carbocation (C).

The 550 nm band accounts for a higher degree of oligomerization, however, as the band is tentatively assigned to a tetrameric carbocation species (Scheme 1), it is unlikely to fit within the ZSM-5 pores and it is therefore probable that it forms on the outer surface of the ZSM-5 crystals. Given that this large species forms at room temperature, it is also feasible that this is a cationic radical species, which can form on Lewis acid sites (extra framework aluminum, EFAL) in zeolites.[39]

The presence of a 445 nm absorption band for the ZSM-5–Al and ZSM-5–SiAl (weak) pellets, clearly absent for all the other pellets at this reaction time (10 s), also suggests that the Al_2O_3-bound pellets possess additional acid sites of a different acid strength (most likely Lewis acid sites), which favor the ring-opening pathway and hence the formation of thiol-like species. This was observed in our previous studies on ZSM-5-based Al_2O_3-bound extrudates, where Al migration occurred and favored the formation of such species. It is important to note that Al_2O_3 alone is inactive for the reaction (Fig. S2†) and therefore this suggests that Al migration has indeed created additional acid sites in the ZSM-5–Al and ZSM-5–SiAl pelletized samples. An indication that this has taken place for ZSM-5–Al is presented using Rietveld analysis of the ZSM-5 and ZSM-5–Al X-ray diffraction patterns (Table 2) as well as the NH_3-TPD results (Fig. 3). When comparing the microstrain values between the two samples (0.00093 and 0.00117, respectively), there is a clear increase for the Al_2O_3-bound sample, which could suggest the incorporation of Al in the zeolite framework. This indication is enhanced further by the NH_3-TPD profile of ZSM-5–Al, which not only presents two peaks corresponding to weak (192 °C) and strong (386 °C) acid sites, but also a new peak (compared to ZSM-5) centered at 290 °C, suggesting Al migration creates additional acid sites of moderate acid strength, also reported by Zhang *et al.*[40]

The decrease in absorbance of the 275 nm band and the rise in the 310 nm band for all the pellets (except ZSM-5–Kao) after 50 s (Fig. 4(b)) and 100 s (Fig. 4(c)) show that the protonated thiophene monomers are interacting within the zeolite pores, leading to the formation of the dimeric carbocationic species. This is accompanied by an overall increase in absorbance across the whole wavelength range as expected, due to the increase in the formation of carbocationic species as the reaction time increases. However, it is noticeable that the ratio of the 550 nm band (tetramer oligomer species, E1) to the 410 nm band (trimeric oligomer species, D) is decreasing, accompanied by a rise in the 445 (except for ZSM-5–Si) and 480 nm bands. This suggests that alternative reaction pathways are now taking place, with an increase in the formation of thiol-like species (F2 and F1-T) and trimeric species (D), at the expense of the formation of larger oligomers (E1).

However, the lower ratio of the 550 nm absorbance band can be explained in Fig. 4(d) at a reaction time of 200 s, whereby an increase in a broad band >550 nm is now visible, suggesting that these species are oligomerizing to even larger species over time. Again, the increase in the absorbance of the 445, 460 and 480 nm bands shows that ring-opening is now one of the main reaction pathways for all the catalysts (except ZSM-5–Kao), possibly due to the lack of strong acid sites at this point of the reaction, which are involved in forming larger oligomer species. It is also important to note the much higher overall absorbance of the ZSM-5–SiAl pellet compared with ZSM-5, ZSM-5–Si and ZSM-5–Al (which are all relatively similar), suggesting a binder effect in this sample. A possible explanation could be the interaction between the two components of the binder (*i.e.* Al migration to SiO_2, or Si migration to Al_2O_3), creating additional acid sites to perform the thiophene oligomerization reaction. This is again backed by the NH_3-TPD profile of ZSM-5–SiAl (Fig. 3), which shows a far higher ratio of stronger acid sites to weaker sites compared to ZSM-5 and a greater relative amount of both weak and strong acid sites compared to ZSM-5–Si. However, further studies are needed to develop a better understanding of how this mixed binder promotes such a high product formation.

Similar reactivity and selectivity trends continue towards the termination of the reaction at 400 s (Fig. 4(e)) and 800 s (Fig. 4(f)), with a decrease in the absorbance of both the 275 and 310 nm bands and a continued rise in bands >400 nm, particularly towards the formation of thiol-like species and large oligomers. The appearance of both 580 and 620 nm bands and the broadness of the UV-Vis spectrum background (particularly in the case of ZSM-5–Si) suggests the formation of complex larger molecules and/or coke in each pellet. Regarding ZSM-5–Si, it is important to note at this stage, that a SiO_2 pellet produces a very broad band in the >580 nm region (Fig. S2(c)†), which explains the high absorbance in the ZSM-5–Si spectra. The reason for SiO_2 reacting with thiophene is, at this stage, not clear, as SiO_2 is not expected to provide acid sites with enough strength to convert thiophene monomers.

Confocal fluorescence microscopy corresponds well with UV-Vis micro-spectroscopy and is an excellent method to provide visible insight into the oligomer species formed and how they are distributed over each ZSM-5–clay/binder-bound pellet. Fig. 5 presents the 3D confocal fluorescence microscopy volumes of each individual sample excited with both 488 and 561 nm lasers simultaneously, after the completion of the thiophene oligomerization reaction. These two laser line wavelengths were chosen to gain an understanding into the relative amount and nature of 'smaller' to 'larger' oligomer species formed. As expected for the ZSM-5 pellet (Fig. 5(a)), a homogeneous distribution of fluorescent oligomers are obtained, with an orange fluorescence indicating the formation both <3-thiophene ring oligomers/thiol-like species and >3-thiophene ring oligomers, with a higher relative amount of the latter (Fig. S4†).

ZSM-5–Al (Fig. 5(c)) and ZSM-5–SiAl (Fig. 5(d)) show similar results to that of ZSM-5, with a more heterogeneous distribution of ZSM-5 crystals/agglomerates due to the binder distribution. However, ZSM-5–Si reveals a higher fluorescence intensity than that for all the other pellets, showing its higher ratio of larger oligomers/coke species (Fig. S4†), as evidenced by UV-Vis absorption (Fig. 4(f)), and this is again explained by the unexpected reactivity of SiO_2 (Fig. S2(c)†). The lack of reactivity for the ZSM-5–Kao pellet leads to an almost blank image, with

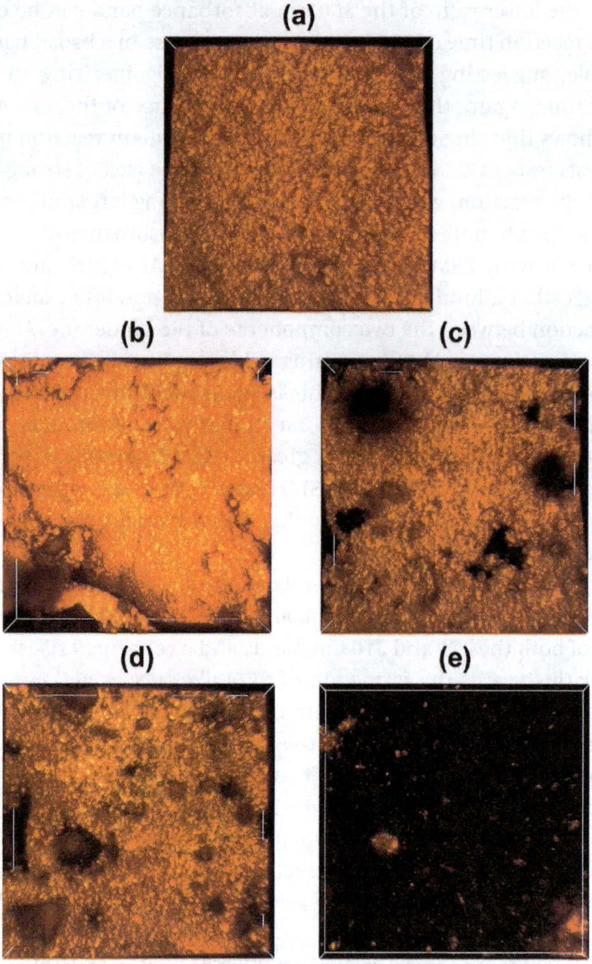

Fig. 5 3D confocal fluorescence microscopy images of the following ZSM-5–binder-bound pellets after reaction of thiophene at 120 °C: (a) ZSM-5; (b) ZSM-5–Si; (c) ZSM-5–Al; (d) ZSM-5–SiAl; and (e) ZSM-5–Kao. The *XYZ* axis values correspond to: 127 × 127 × 27 µm.

only a few visible fluorescent ZSM-5 crystals, confirming that the binder has a detrimental effect to the active zeolite phase by blocking access to the active zeolite beneath the hard outer layer.

In order to develop a greater understanding of the chemical species formed in each ZSM-5–binder-bound pellet upon thiophene oligomerization, *in situ* FT-IR micro-spectroscopy was performed. Given the high concentration of moisture present in the ZSM-5 pellet (Fig. S5†), it was difficult to draw detailed conclusions from the whole range of FT-IR spectra before, during and after thiophene oligomerization. Therefore, we will focus on the results of the other pellets (which have only 30 wt% ZSM-5 present).

Fig. 6(a) and (b) present the FT-IR spectra of ZSM-5–Si before, during and after thiophene oligomerization, allowing a correlation to those timescales in the UV-

Vis absorption spectra (Fig. 4). Although it is not possible to observe a band at 3610 cm^{-1} corresponding to SiOHAl groups (Brønsted acid sites) in the fresh pellet and after thiophene impregnation, it is already known (and expected) that this band will decrease in intensity or become perturbed upon the adsorption of thiophene monomers.[41] Two other prominent features expected in the fresh ZSM-5–Si pellet are visible at 3746 cm^{-1} and between 1250 and 950 cm^{-1}, corresponding to the Si–OH groups of SiO_2 and zeolite overtones, respectively.

Upon thiophene addition and reaction, numerous bands appear over the whole FT-IR range (4000–750 cm^{-1}). Focusing first on the 4000–2250 cm^{-1} region (Fig. 6(a)), the most noticeable change, which helps explain the reactivity of SiO_2, is the disappearance of the 3746 cm^{-1} band, indicating that thiophene monomers not only interact with ZSM-5 acid sites, but also with the binder surface. The presence of a broad band around 3620 cm^{-1} has previously been attributed to the hydrogen bonding of Si–OH groups with π-electron donor molecules, enforcing this theory.[41,42] The bands present at 3110, 3076 (unsaturated =C–H stretching vibrations) and 1408 cm^{-1} (symmetric ring stretching vibrations of C=C)[43] are assigned to thiophene rings hydrogen bonded to the Brønsted acid sites and Si–OH groups, corresponding well with the 275 nm UV absorption band observed previously (Fig. 4(a)). The bands visible at 1588 and 760 cm^{-1} also refer to the thiophene ring, as C–H in-plane bending vibrations and neighboring aromatic C–H, respectively.

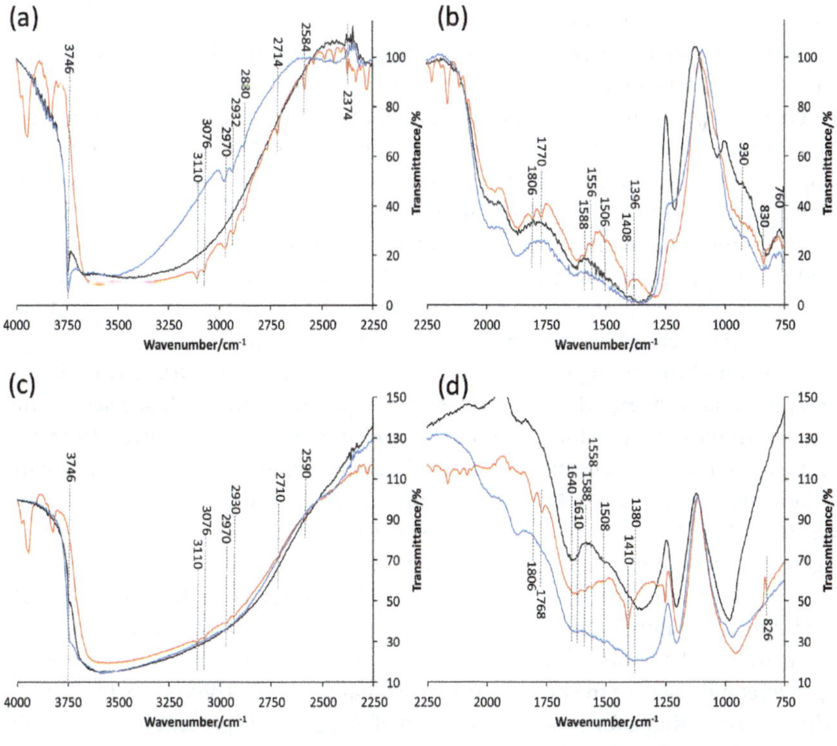

Fig. 6 *In situ* FT-IR spectra of (a and b) ZSM-5–Si and (c and d) ZSM-5–Al before (black), during (red) and after (blue) reaction with thiophene.

However, there are not only bands due to intact/adsorbed thiophene present, but also additional bands even at a low reaction temperature, suggesting several species are formed. The bands observed at 2970, 2932 and 2880 cm^{-1} are attributed to CH_2 groups in proximity to a double bond, *i.e.* asymmetric *CH_2–($CH{=}CH_2$) stretching vibrations, or a neighboring S atom (CH_2–S–).[41] Therefore, these are attributed to thiol-like species with further bands at 1806–1770 and 930 cm^{-1} also indicating their formation (C=C–H stretch and bending, respectively). However, Geobaldo *et al.* propose an alternative theory, suggesting that the bands present at 1800–1600 cm^{-1} and 2980–2800 cm^{-1} correspond with carbocationic oligomer species, also reported in our UV-Vis absorption spectra (Fig. 4).[37] Both proposals correspond well with our results in this work, however, we can prove the presence of thiol-like species given that perturbed S–H bands are present at 2714, 2584 and 2374 cm^{-1}.[41]

Towards the termination of the reaction, it is clear that no bands corresponding to adsorbed thiophene monomers are present (*i.e.* 3110 and 3076 cm^{-1}). This is accompanied by the re-appearance of the Si–OH groups (3746 cm^{-1}), suggesting full conversion of thiophene to its respective intermediate/products, recognized by bands still present at 2970, 2932, 930 cm^{-1}. Another noticeable feature in the region of 1250–900 cm^{-1} is the shift towards lower wavenumbers for the zeolite overtone bands upon thiophene reaction. Given the higher ratio of larger oligomers/coke species formed in ZSM-5–Si compared to in the other samples (also displayed by the higher orange fluorescence intensity), it could be that such large species formed (inside the pores) are distorting the zeolite framework, but further studies with *in situ* XRD (Rietveld analysis) are needed to confirm this finding.

The *in situ* FT-IR spectra of ZSM-5–Al (Fig. 6(c) and (d)) show similar features but also distinct differences compared to the ZSM-5–Si spectra. The decrease in intensity of the 3746 cm^{-1} band of fresh ZSM-5–Al is expected considering the significant decrease in Si–OH groups compared to ZSM-5–Si, but it is also apparent that bands present upon thiophene impregnation are much lower in intensity in the 4000–2250 cm^{-1} region. This could be due to the presence of H_2O in the sample (as with ZSM-5), but yet it is still possible to distinguish vital additional bands. The presence of bands at 3110, 3076, 2970, 2930, 2710 and 2950 cm^{-1} all agree with the proposed formation of both oligomer and thiol-like species. However, additional bands at 1610 and 1380 cm^{-1}, as well as the large increase and broadening of the 1410 cm^{-1} band, confirm perturbed vibration of thiophene molecules, which are strongly interacting with EFAL species.[41] This again enforces our previous proposal that Al migration has occurred from the Al_2O_3 binder to ZSM-5, creating additional acid sites that are now interacting with thiophene. As with ZSM-5–Si, the re-appearance of the 3746 cm^{-1} band and absence of the 3110 and 3076 cm^{-1} bands towards the termination of the reaction again suggest the full conversion of thiophene monomers.

As in the case of ZSM-5–Si, very distinct and clear bands are formed with ZSM-5–SiAl upon its impregnation with thiophene (Fig. 7(a) and (b)). The increase in intensity of all relevant bands related to thiophene and intermediate/product species could relate to the high absorbance obtained in the UV-Vis spectra (Fig. 4). This indicates that a far greater absorption of thiophene is present in this sample compared with the other pellets. Other than the main bands relating to oligomer and thiol-like species, similar bands to ZSM-5–Al are also present (1620 cm^{-1}),

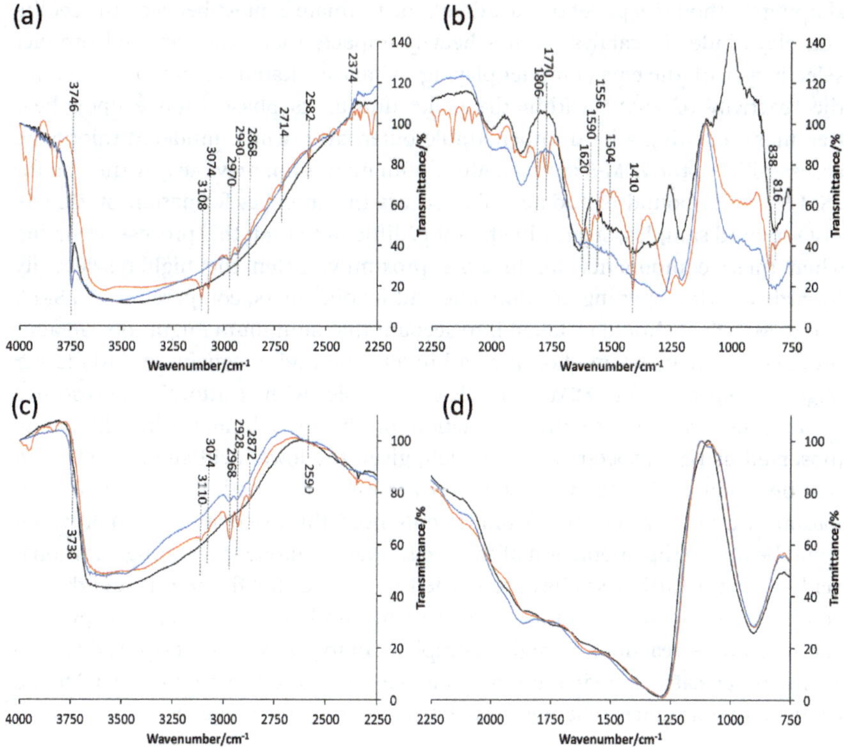

Fig. 7 *In situ* FT-IR spectra of (a and b) ZSM-5–SiAl and (c and d) ZSM-5–Kao, before (black), during (red) and after (blue) reaction with thiophene.

again suggesting the interaction of thiophene with EFAL species created from Al migration in ZSM-5–SiAl.

Finally, the *in situ* FT-IR spectra of ZSM-5–Kao confirm the limited reactivity of the catalyst pellet. Taking the bands present in the region of 4000–3800 cm^{-1} as an 'internal standard' between samples (relating to the adsorption of thiophene), it is apparent that a far lower concentration of thiophene is adsorbed on ZSM-5–Kao, confirming the earlier proposal that the phase change upon pelletization/heat treatment creates a hard layer hindering the access of thiophene to the zeolite active phase. Nevertheless, the vibrational bands present around 2968, 2928 and 2872 cm^{-1} suggest the formation of oligomers and thiol-like species, possibly due to some accessible zeolite on the surface, but the absence of observed bands in the 2800–750 cm^{-1} region do not allow further interpretation.

4. Conclusions

In relation to our previous studies on ZSM-5–binder extrudates, four ZSM-5–binder-bound pellets were prepared using SiO_2, Al_2O_3, $SiO_2 : Al_2O_3$ (2 : 1 mix) and kaolinite as binder materials, in order to determine how zeolite–clay/binder interactions upon physical mixing influence the catalytic performance. Using complementary micro-spectroscopic techniques, coupled with thiophene oligomerization as a staining probe reaction, it is evident that, independent of the shaping method (*i.e.* pellet or extrudate), the intimate contact between the zeolite and clay/binder in catalyst bodies heavily impacts their reactivity and product selectivity, with the type of binder playing a vital role. Kaolinite was found to limit the reactivity of ZSM-5 within the pellet due to its phase change upon heat treatment, creating a hard impenetrable outer layer, which hindered thiophene accessibility to the ZSM-5 crystals. Aluminum migration, previously witnessed for ZSM-5–binder-bound extrudates, also occurs in the pellet formation of ZSM-5–Al_2O_3-bound samples, suggesting that it is difficult to limit this process occurring when these components are in close proximity. Given the higher selectivity towards the ring opening of thiophene and its oligomers, compared with ZSM-5 and ZSM-5–Si, it should be taken into account that aluminum migration can alter product selectivity when chosen as a binder. The high reactivity towards larger oligomer species for ZSM-5–Si, clearly visible with confocal fluorescence microscopy, was due to the interaction of the SiO_2 binder with thiophene (observed using IR spectroscopy), which, given the lower acid site strength and number compared to ZSM-5 and ZSM-5–Al, was unexpected. Finally, the combination of a $SiO_2 : Al_2O_3$ mixed binder enhanced the reactivity of ZSM-5, which could be due to the interaction of the two binder components creating additional acid sites, but further studies are needed to resolve this finding. Evidently, the combination of UV-Vis, confocal fluorescence and IR micro-spectroscopy with staining probe reactions is highly complementary, providing new insights into zeolite-based catalyst bodies and proving that the choice of binder is crucial to the catalytic performance, whether it be detrimental or advantageous.

Acknowledgements

The authors would like to thank Lennart Webber (Utrecht University, UU) for the Ar physisorption analysis, and Joris Goetze (UU) for fruitful discussions. This work was financially supported by a NWO personal 'Veni' grant awarded to G. T. W. and a Top Grant of NWO to B. M. W.

References

1 A. Corma, *Chem. Rev.*, 1995, **95**, 559–614.
2 M. J. Climent, A. Corma and S. Iborra, *Chem. Rev.*, 2011, **111**, 1072–1133.
3 A. Corma, M. J. Díaz-Cabañas, J. Martínez-Triguero, F. Rey and J. Rius, *Nature*, 2002, **418**, 514–517.
4 A. de Lucas, J. L. Valverde, P. Sánchez, F. Dorado and M. J. Ramos, *Ind. Eng. Chem. Res.*, 2004, **43**, 8217–8225.
5 C. Perego and P. Ingallina, *Catal. Today*, 2002, **73**, 3–22.
6 S. Mitchell, N.-L. Michels and J. Pérez-Ramírez, *Chem. Soc. Rev.*, 2013, **42**, 6094–6112.
7 J. S. J. Hargreaves and A. L. Munnoch, *Catal. Sci. Technol.*, 2013, **3**, 1165.
8 P. Sánchez, F. Dorado, A. Fúnez, V. Jiménez, M. J. Ramos and J. L. Valverde, *J. Mol. Catal. A: Chem.*, 2007, **273**, 109–113.
9 E. T. C. Vogt, G. T. Whiting, A. Dutta Chowdhury and B. M. Weckhuysen, *Adv. Catal.*, 2015, **58**, 143–314.
10 F. Meirer, D. T. Morris, S. Kalirai, Y. Liu, J. C. Andrews and B. M. Weckhuysen, *J. Am. Chem. Soc.*, 2015, **137**, 102–105.

11 Z. Ristanović, J. P. Hofmann, G. de Cremer, A. V. Kubarev, M. Rohnke, F. Meirer, J. Hofkens, M. B. J. Roeffaers and B. M. Weckhuysen, *J. Am. Chem. Soc.*, 2015, **137**, 6559–6568.
12 F. Meirer, S. Kalirai, D. Morris, S. Soparawalla, Y. Liu, G. Mesu, J. C. Andrews and B. M. Weckhuysen, *Sci. Adv.*, 2015, **1**, 1400199.
13 F. Meirer, S. Kalirai, J. Nelson Weker, Y. Liu, J. C. Andrews and B. M. Weckhuysen, *Chem. Commun.*, 2015, **51**, 8097–8100.
14 S. Kalirai, U. Boesenberg, G. Falkenberg, F. Meirer and B. M. Weckhuysen, *ChemCatChem*, 2015, **7**, 3674–3682.
15 J. Ruiz-Martínez, A. M. Beale, U. Deka, M. G. O'Brien, P. D. Quinn, J. F. W. Mosselmans and B. M. Weckhuysen, *Angew. Chem., Int. Ed.*, 2013, **52**, 5983–5987.
16 Z. Ristanović, M. M. Kerssens, A. V. Kubarev, F. C. Hendriks, P. Dedecker, J. Hofkens, M. B. J. Roeffaers and B. M. Weckhuysen, *Angew. Chem., Int. Ed.*, 2015, **54**, 1836–1840.
17 G. T. Whiting, F. Meirer, M. M. Mertens, A.-J. Bons, B. M. Weiss, P. A. Stevens, E. de Smit and B. M. Weckhuysen, *ChemCatChem*, 2015, **7**, 1312–1321.
18 G. T. Whiting, F. Meirer, D. Valencia, M. M. Mertens, A.-J. Bons, B. M. Weiss, P. A. Stevens, E. de Smit and B. Weckhuysen, *Phys. Chem. Chem. Phys.*, 2014, **16**, 21531–21542.
19 S. Mitchell, N.-L. Michels, K. Kunze and J. Pérez-Ramírez, *Nat. Chem.*, 2012, **4**, 825–831.
20 I. L. C. Buurmans, J. Ruiz-Martínez, W. V. Knowles, D. van der Beek, J. A. Bergwerff, E. T. C. Vogt and B. M. Weckhuysen, *Nat. Chem.*, 2011, **3**, 862–867.
21 E. T. C. Vogt and B. M. Weckhuysen, *Chem. Soc. Rev.*, 2015, **44**, 7342–7370.
22 I. L. C. Buurmans, F. Soulimani, J. Ruiz-Martínez, H. E. van der Bij and B. M. Weckhuysen, *Microporous Mesoporous Mater.*, 2013, **166**, 86–92.
23 I. L. C. Buurmans and B. M. Weckhuysen, *Nat. Chem.*, 2012, **4**, 873–886.
24 M. Ibáñez, M. Gamero, J. Ruiz-Martínez, B. M. Weckhuysen, A. T. Aguayo, J. Bilbao and P. Castaño, *Catal. Sci. Technol.*, 2016, **6**, 296–306.
25 P. Castaño, J. Ruiz-Martínez, E. Epelde, A. G. Gayubo and B. M. Weckhuysen, *ChemCatChem*, 2013, **5**, 2827–2831.
26 R. V. Jasra, B. Tyagi, Y. M. Badheka, V. N. Choudary and T. S. G. Bhat, *Ind. Eng. Chem. Res.*, 2003, **42**, 3263–3272.
27 V. R. Choudhary, P. Devadas, A. K. Kinage and M. Guisnet, *Appl. Catal., A*, 1997, **162**, 223–233.
28 P. Cañizares, A. Durán, F. Dorado and M. Carmona, *Appl. Clay Sci.*, 2000, **16**, 273–287.
29 F. Dorado, R. Romero and P. Cañizares, *Appl. Catal., A*, 2002, **236**, 235–243.
30 J. M. Fougerit, N. S. Gnep and M. Guisnet, *Microporous Mesoporous Mater.*, 1999, **29**, 79–89.
31 M. W. Kasture, P. S. Niphadkar, V. V. Bokade and P. N. Joshi, *Catal. Commun.*, 2007, **8**, 1003–1008.
32 G. de la Puente, E. Falabella Sousa-Aguiar, A. Figueiredo Costa and U. Sedran, *Appl. Catal., A*, 2003, **242**, 381–391.
33 F. V. Pinto, A. S. Escobar, B. G. de Oliveira, Y. L. Lam, H. S. Cerqueira, B. Louis, J. P. Tessonnier, D. S. Su and M. M. Pereira, *Appl. Catal., A*, 2010, **388**, 15–21.
34 N.-L. Michels, S. Mitchell and J. Pérez-Ramírez, *ACS Catal.*, 2014, **4**, 2409–2417.

35 K.-Y. Lee, H.-K. Lee and S.-K. Ihm, *Top. Catal.*, 2010, **53**, 247–253.
36 P. Gélin and T. Des Courières, *Appl. Catal.*, 1991, **72**, 179–192.
37 F. Geobaldo, T. Palomino, S. Bordiga, C. Fisica and V. Pietro Giuria, *Phys. Chem. Chem. Phys.*, 1999, **1**, 561–569.
38 M. Alkan, Ç. Hopa, Z. Yilmaz and H. Güler, *Microporous Mesoporous Mater.*, 2005, **86**, 176–184.
39 X. Liu, K.-K. Iu, J. K. Thomas, H. He and J. Klinowski, *J. Am. Chem. Soc.*, 1994, **116**, 11811–11818.
40 Y. Zhang, Y. Zhou, A. Qiu, Y. Wang, Y. Xu and P. Wu, *Ind. Eng. Chem. Res.*, 2006, **45**, 2213–2219.
41 C. L. Garcia and J. A. Lercher, *J. Phys. Chem.*, 1992, **96**, 2669–2675.
42 H. Knozinger, *The Hydrogen Bond*, ed. P. Schuster, G. Zundel and C. Sandorfy, North Holland, Amsterdam, 1976, vol. 3, p. 1275.
43 Z. Wu, C. Li, Z. Wei, P. Ying and Q. Xin, *J. Phys. Chem. B*, 2002, **106**, 979–987.

Faraday Discussions

PAPER

Methanol oxidation on Fe_2O_3 catalysts and the effects of surface Mo

M. Bowker,[ab] E. K. Gibson,*[ac] I. P. Silverwood[ad] and C. Brookes[ab]

Received 15th December 2015, Accepted 22nd December 2015
DOI: 10.1039/c5fd00225g

The adsorption of methanol on haematite has been investigated using temperature programmed methods, combined with *in situ* DRIFTS. Model catalysts based on this material have then been made with a shell–core configuration of molybdenum oxide monolayers on top of the haematite core. These are used as models of industrial iron molybdate catalysts, used to selectively oxidise methanol to formaldehyde, one of the major chemical outlets for methanol. Haematite itself is completely ineffective in this respect since it oxidises it to CO_2 and the DRIFTS shows that this occurs by oxidation of methoxy to formate at around 200 °C. The decomposition behaviour is affected by the absence or presence of oxygen in the gas phase; oxygen destabilises the methoxy and enhances formate production. In contrast, when a monolayer of molybdena is placed onto the surface by incipient wetness, and it remains there after calcination, the pathway to formate production is blocked and formaldehyde is the main gas phase product in TPD after methanol dosing.

Introduction

This paper concerns the adsorption and reaction of methanol with iron oxide (haematite) and model catalysts of Mo monolayers deposited on that oxide. This is of importance for a variety of reasons. Firstly, iron oxide is one on the most abundant materials readily available on earth and so represents an interesting base material for catalysis from both an economic and environmental point of view. Due to its abundance it is relatively cheap and environmentally cleaner as it is usually mined from near the surface of the earth, therefore little energy is wasted in separation and refining, in contrast to the situation for precious metals. It is also used in a variety of catalytic applications, such as ammonia synthesis, the

[a]*UK Catalysis Hub, Research Complex at Harwell (RCaH), Rutherford Appleton Laboratory, Harwell, Oxon, OX11 0FA, UK. E-mail: emma.gibson@rc-harwell.ac.uk*

[b]*Cardiff Catalysis Institute, School of Chemistry, Cardiff University, Main Building, Park Place, Cardiff, CF10 3AT, UK*

[c]*Department of Chemistry, University College London, 20 Gordon St., London, WC1H 0AJ, UK*

[d]*ISIS Neutron and Muon Facility, Science and Technology Facilities Council, Rutherford Appleton Laboratory, Harwell Science and Innovation Campus, Oxon OX11 0QX, UK*

high temperature water-gas shift reaction (HTWGS), Fischer–Tropsch synthesis of hydrocarbons and styrene synthesis by ethyl benzene dehydrogenation.[1–4] It must be noted, however, that in the active phase of ammonia synthesis it is in metallic form, in FTS, it is probably carbidic, while in HTWGS, although prepared as haematite, it is in the magnetite form during reaction. However, it is also used as part of the process for the selective oxidation of methanol, which consists of iron in compound form with molybdenum[5] and in its ferric state. In connection with the latter, we are interested in the way in which methanol interacts not only with iron oxide, but as we report below, we have investigated also model methanol oxidation catalysts by placing monolayers of Mo at the surface. We report the changes in behaviour of reactivity and surface species formed when such model catalysts are fabricated using temperature programmed methods and DRIFTS.

Experimental

The Fe_2O_3 catalyst used in this study was a commercial Fe_2O_3 (Sigma-Aldrich, <50 nm particle size). A catalyst of the type MoO_x/Fe_2O_3 was synthesised by doping 6 monolayer equivalents of Mo oxide onto the Fe_2O_3. The desired amount of aqueous ammonium heptamolybdate was dosed onto the surface of the Fe_2O_3 by incipient wetness impregnation. The sample was dried at 120 °C for 24 h, before being calcined in a muffle furnace in air at 500 °C, for 24 h.

Reactor data was obtained by use of a CATLAB reactor (Hiden Ltd, Warrington, U.K.). For the catalytic reaction (TPO/TPR), 1 μL of liquid methanol was injected every 2 min into a flow of 10% O_2/He or He, at a flow rate of 30 mL min^{-1}. The products were determined by the online mass spectrometer in the CATLAB system. For the TPD, approximately 6 injections of 1 μL of methanol were dosed onto the catalyst, at ambient temperature in a flow of 30 mL min^{-1} of 10% O_2/He or He. This was followed by ramping the temperature to 400 °C at a rate of 8 °C min^{-1}, while monitoring the products formed by mass spectrometry.

In situ diffuse reflectance infrared Fourier transform spectroscopy (DRIFTS) was performed in a Harrick high temperature DRIFTS cell fitted with ZnSe windows. The cell was attached to the Praying Mantis Optics and spectra collected with an Agilent Carey 680 FTIR spectrometer, taking 64 scans with a resolution of 4 cm^{-1} using the DTGS detector. The TPD and TPO experiments were performed by injecting approximately 5 injections of 1 μL of methanol at ambient pressure in a flow of 30 mL min^{-1} of 10% O_2/He or He. The temperature was then ramped at 10 °C min^{-1} while collecting spectra every 15 °C.

Results and discussion

Fig. 1 shows TPD data from the adsorption of methanol on the iron oxide catalyst. A simple pattern is observed, and is similar to that reported by us previously.[6] That is, weakly held methanol is desorbed at low temperatures, with a maximum at 100 °C. This is then followed by a major desorption at around 300 °C, when both CO_2 and hydrogen desorb, though the line-shapes for the two are a little different. The hydrogen shows a shoulder (at ~250 °C) on the main peak at 300 °C, while the CO_2 appears to be a single peak at ~305 °C. As reported previously[6–8] this desorption is likely to be due to the presence of a formate species on the surface. This is confirmed by the *in situ* TPDRIFT (temperature-programmed diffuse reflectance

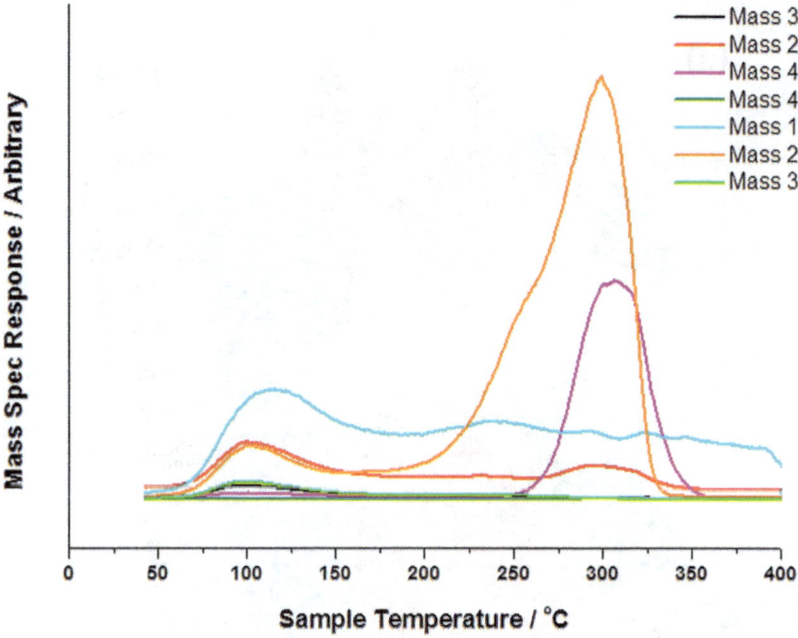

Fig. 1 TPD profile of Fe_2O_3 under a 8 °C min^{-1} ramp in a flow of He after methanol adsorption.

infra-red Fourier transform) spectra in Fig. 2, but additionally here we can identify the temperature range in which the formate is formed. Upon adsorption we see a variety of bands due to a mixture of methanol and methoxy, but after heating to ~100 °C, we are left with just methoxy bands at 1070, 2827 and 2937 cm^{-1} (Table 1). Upon heating above about 175 °C, formate bands (1569, 1377, 1358 and 2866 cm^{-1}, Table 1 and references therein) start to appear and methoxy is slowly lost, leaving mainly formate at 325 °C. Thus the formate is present on the surface before the desorption of CO_2, but is not present upon adsorption, so it is formed during the heating process. Most of the hydrogen desorption is clearly from the methoxy decomposition, the methoxy loss rate from the IR data being maximised at ~250 °C (Fig. 3), beginning at ~200 °C and being 80% diminished by 300 °C. Thus it appears that the low temperature shoulder on the hydrogen curve in Fig. 1 is associated with the methoxy decomposition, but the maximum in hydrogen evolution occurs well after most of the methoxy has been lost.

One thing which is immediately obvious from the TPD data is that very little water is evolved from the catalyst during the experiment, and this is unusual for oxidic catalysts: for instance, similar experiments with MoO_3 yield water and no hydrogen ((ref. 6 and 7) and see below). This further implies that hydrogen has difficulty extracting oxygen from the iron oxide lattice in these circumstances, a feature which is an obvious advantage when considering a reaction such as the water-gas shift reaction in the forward direction, eqn (1). Iron oxide is the catalyst for the high temperature shift reaction,[1] though under operating conditions it is thought to be present in the form of magnetite.

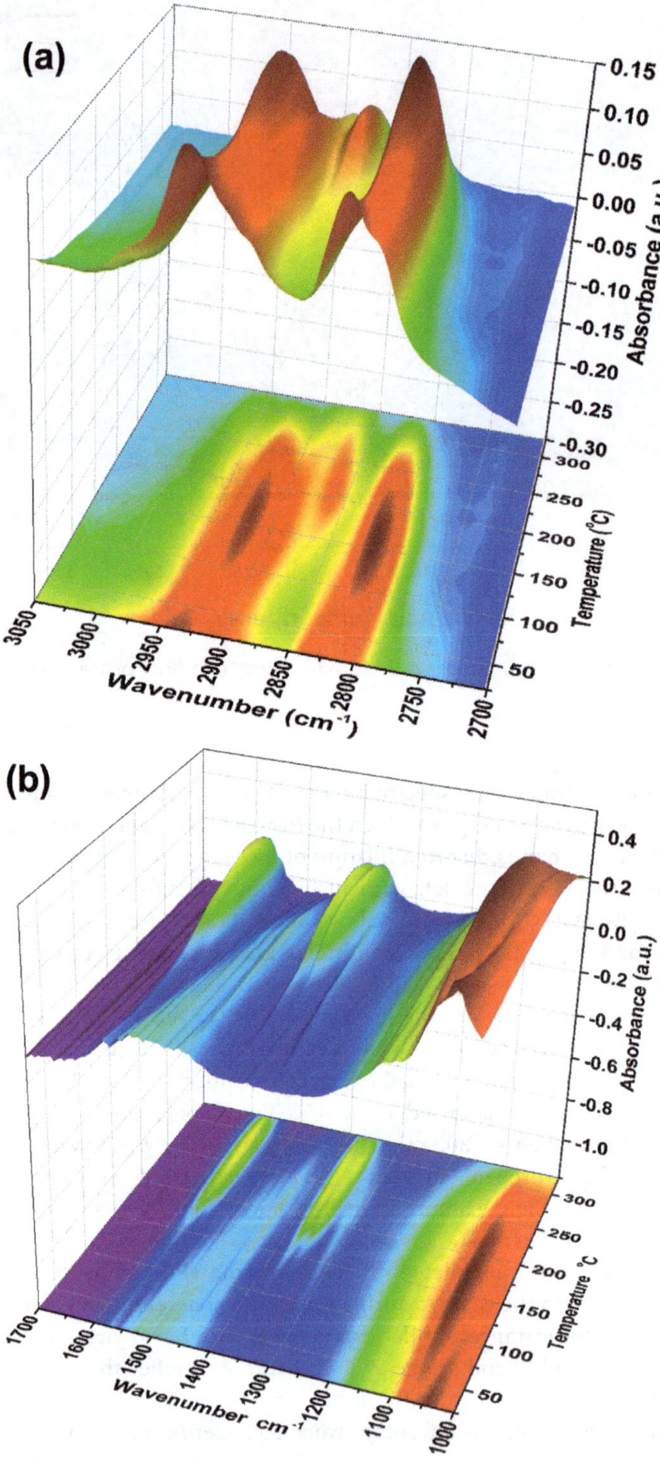

Fig. 2 Stacked TPDRIFTS spectra under He, in the (a) 2700–3050 cm^{-1} and (b) 1000–1700 cm^{-1} range.

Table 1 Assignments for the bands observed in the DRIFTS experiments, [a] bands observed above ~150 °C, [b] bands observed above 175 °C

Assignment	Methoxy[9] (cm^{-1})	Methanol[9] (cm^{-1})	Formate[10] (cm^{-1})	This work, above 100 °C, (cm^{-1}) TPD	TPO
$\nu_{as}(COO) + \delta(CH)$			2960	2960[b]	2964[a]
$\nu_s(CH)$	2898	2925	2880	2922, 2895, 2866[b]	2923, 2895, 2864[a]
$2\delta(CH_3)$	2805	2824		2816	2818, 2808
$\nu_s(COO) + \delta(CH)$			2730	2734[b]	2734[a]
$\nu_{as}(COO)$			1565	1569[b]	1571[a]
$\delta(CH)$	$\delta_{as}(CH)$ 1460, $\delta_s(CH)$ 1439		1379	1377[b]	1376[a]
$\nu_s(COO)$			1358	1358[b]	1355[a]
$\nu(CO)$[11]	1070	1036		1073	1070

$$CO + H_2O \rightarrow CO_2 + H_2 \quad (1)$$

Fig. 4 shows a similar TPD experiment, but this time it is run in a flow of O_2/He, effectively TPO, temperature-programmed oxidation. The results are rather revealing and different from those in Fig. 1. Here we still have the major CO_2 evolution at ~300 °C, but there is little hydrogen coincident with it. Instead the hydrogen evolution is now shifted down to ~200 °C and appears to be

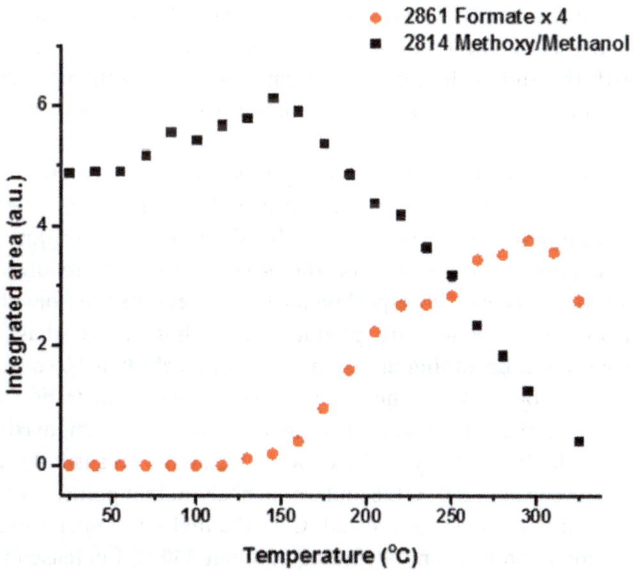

Fig. 3 Integrated intensities of the main bands due to methoxy and formate species on the iron oxide surface as a function of temperature during TPD of methanol under a flow of He.

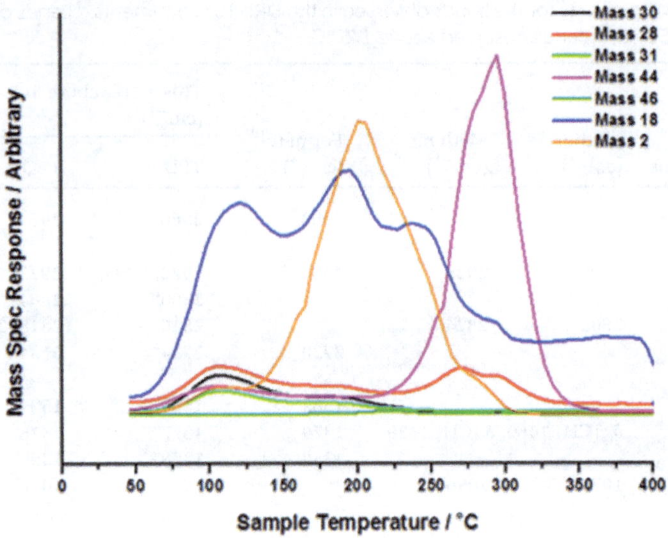

Fig. 4 TPD profile of Fe_2O_3 under a 8 °C min^{-1} ramp in a flow of 10% O_2/He after methanol adsorption.

independent of CO_2 evolution. There is also water desorbing in a rather broad band of peaks from ~100–300 °C, and a small CO evolution at about 270 °C (mass 28 for CO is a cracking fragment of mass 44 for CO_2, but the lineshape of the two here is different). The IR data are shown in Fig. 5 and 6. In Fig. 6 the same main bands are seen as in Fig. 2, that is, the formate and the methoxy, but now the ratio of these two IR bands is different, with much more formate formed. Further, the main hydrogen evolution appears to relate to the methoxy decomposition: the highest rate of loss of methoxy in IR (Fig. 6) is at about 190 °C, corresponding with the peak in the TPD. The methoxy peak is lost in IR by about 250 °C, corresponding with the end of the main hydrogen desorption. Similarly the formate band in IR is maximised at about the same temperature (~300 °C) as the CO_2 desorption.

Experiments were carried out with the pulsed flow reactor described in the experimental section, and the data are displayed in Fig. 7, as a plot of the integrals of products as a function of temperature. In this situation, where only methanol is present during pulses into the He flow, the product profile is broadly similar to the results of the transient TPD experiments above. Here there is some hydrogen production, but the products are predominantly those of combustion, with methanol conversion beginning at around 160 °C, reaching 50% conversion by ~220 °C and 100% by ~260 °C. The first product is that of dehydration, dimethyl ether (46 amu), together with water. This occurs at low conversion, maximising at 200 °C, but with high selectivity at these low conversions. This is followed by CO and H_2 seen between 220–300 °C (28 and 2 amu peaks at that point), and with CO_2 and H_2O becoming dominant above 250 °C. At the highest temperatures there is also some methane production, beginning at about 350 °C (increase in 16 amu, while carbon dioxide decreases). Clearly, since there is significant water production here, we must be reducing the iron oxide catalyst and labilising oxygen from subsurface regions. In the TPD experiment of Fig. 1 this does not

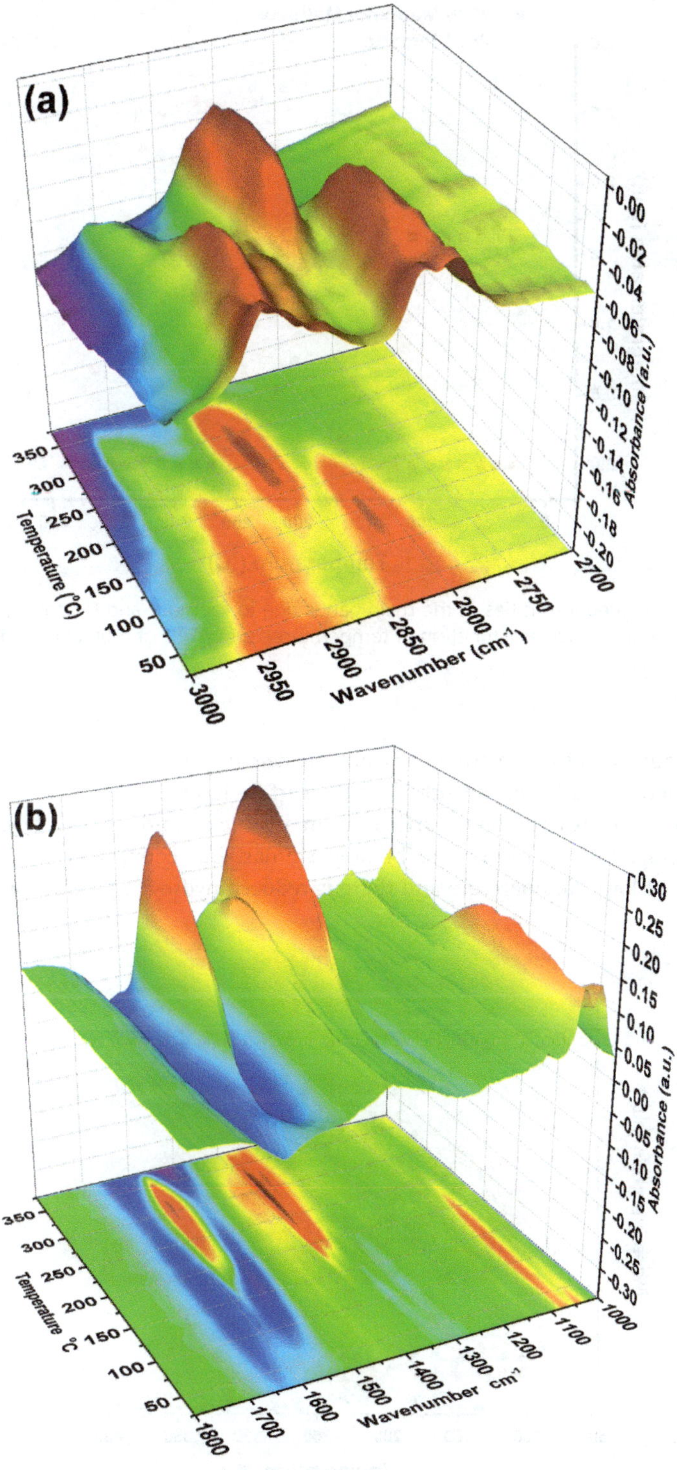

Fig. 5 Stacked TPDRIFTS under 10% O_2/He spectra in the (a) 2700–3000 cm^{-1} and (b) 1000–1800 cm^{-1} range.

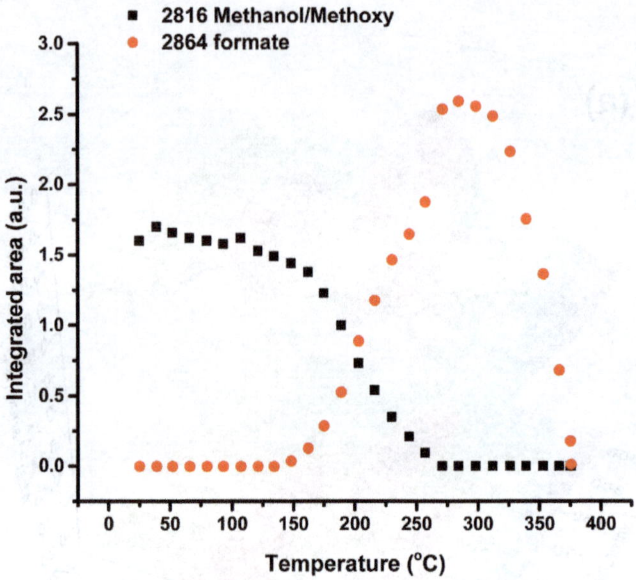

Fig. 6 Integrated intensities of the main bands due to methoxy and formate species on the iron oxide surface as a function of temperature during TPD of methanol under a flow of 10% O_2/He.

occur, but, of course, methanol is not present in the gas phase during those experiments, and is only adsorbed on the surface layer.

If the same experiment is carried out in aerobic conditions, in a flow of 10% O_2/He (Fig. 8), the results are broadly similar, but conversion is shifted to somewhat lower temperature and there is reduced hydrogen production.

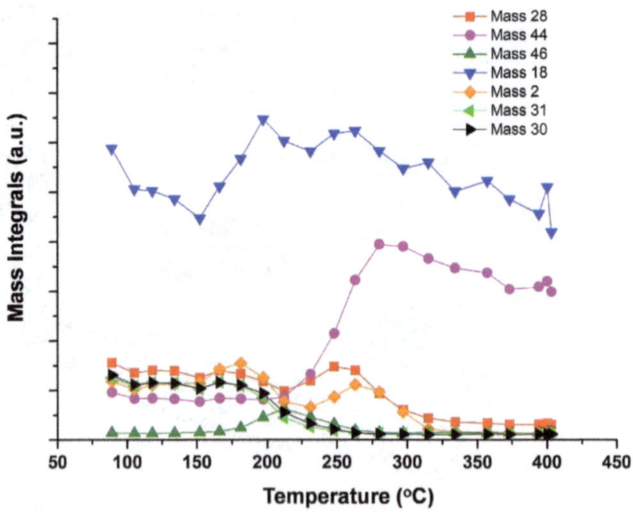

Fig. 7 TPPR (temperature-programmed pulsed reaction) profiles for methanol pulses into a flow of He.

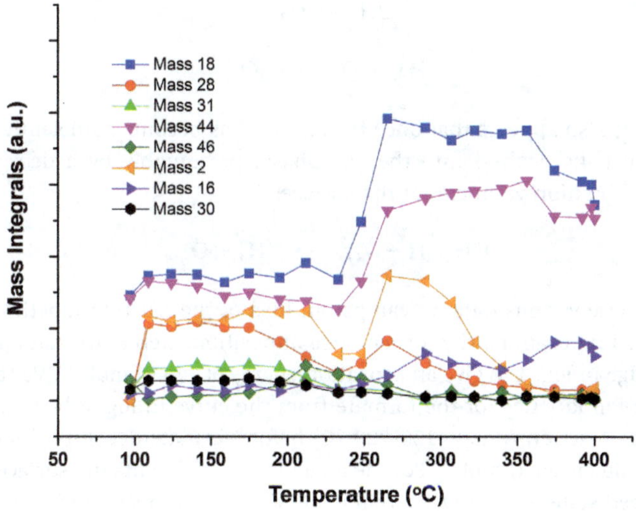

Fig. 8 TPPR (temperature-programmed pulsed reaction) profiles for methanol pulses into a flow of 10% O_2/He.

Clearly then, the following series of reactions takes place on the haemetite surface.

$$CH_3OH + S \rightarrow CH_3OH_a \tag{2}$$

where the subscript a refers to adsorbed species and S is an adsorption site.

$$CH_3OH_a + S \rightarrow CH_3O_a + H_a \tag{3}$$

$$CH_3O_a + CH_3OH \rightarrow CH_3OCH_3 + OH_a \tag{4}$$

The reason for writing DME production in this form, that is, the reaction of surface methoxy with gas phase methanol, is that (i) DME production ceases in Fig. 7 when methanol conversion is very high (*i.e.* little methanol in the gas phase) and (ii) we see no DME in the TPD experiments, where there is no gas phase methanol dosed during the temperature ramp. The other reactions are:

$$CH_3O_a + O_{sl} \rightarrow HCOO_a + 2H_a \tag{5}$$

Here O_{sl} refers to lattice oxygen in the iron oxide, but which is at the surface and so available for reaction (and intrinsically different from lattice oxygen in the bulk). It is formed from the IR experiments during the temperature ramp above 150 °C. The formate can then decompose as follows:

$$HCOO_a \rightarrow CO_2 + H_a \tag{6}$$

and the liberated hydrogen then has two possible routes to loss from the surface, recombination to gas phase hydrogen or reaction with surface hydroxyls to give water.

$$2H_a \rightarrow H_2 \tag{7}$$

$$2H_a + O_{sl} \rightarrow H_2O \tag{8}$$

Finally it also appears that once the surface has become sufficiently reduced, methane can be evolved into the gas phase, presumably by a deoxygenation reaction with anion vacancies at the surface:

$$CH_3OH + V_O^{2-} \rightarrow CH_4 + O_{sl} \tag{9}$$

From these various data we can get an idea of the reaction kinetics involved here. If we first examine the formate decomposition, step 6, we can analyse the leading edge in Fig. 1 to give an activation energy of 128 kJ mol^{-1}. We can obtain the formation kinetics for the formate from the curve in Fig. 6 between 10 and 20 minutes reaction time (~120–220 °C) before formate decomposition occurs, and that value is 60 kJ mol^{-1}. For the anaerobic experiments the surface ends up in a reduced state, since CO_2 is a product, so, at least in the surface region, the overall reaction in the absence of gas phase oxygen is

$$CH_3OH + O_s \rightarrow CO_2 + 2H_2 + O_v \tag{10}$$

and clearly the dominant process in the TPD of Fig. 1 is hydrogen evolution, also seen in Fig. 4. However, in these cases methanol was not present in the gas phase during the TPD. Fig. 7 and 8 also show that when methanol is present in the gas phase, then water production dominates above 300 °C, though some hydrogen is produced at intermediate temperatures. So H_2 can apparently reduce the sample under these conditions:

$$H_2 + O_s \rightarrow H_2O \tag{11}$$

We have gone on to make model catalysts related to industrial formaldehyde synthesis catalysts, which are iron molybdates.[6] These convert methanol by selective oxidation, with high selectivity. We have deposited monolayers of Mo on the surface of iron oxide, because then otherwise bulk spectroscopic techniques (such as XAS[12,13]) can be used to give surface information, if they are sensitive to Mo. When Mo is deposited onto the surface of the iron oxide there is a complete change of behaviour from that described above. TPD results reported in ref. 13 show that the main product is now formaldehyde, with no CO_2 produced. Similarly, reactor results (not shown, but see ref. 6 and 8) show high selectivity to formaldehyde in the temperature range 200 to 350 °C. This in itself shows that once Mo is deposited onto the surface it remains there after calcination, and repeated TPD experiments are reproducible. Spectroscopy and electron microscopy confirm such surface stability.[12,13] The TPD alone implies that the methoxy species dominates the surface and that dehydrogenation of this species yields the formaldehyde. This is confirmed from TPDRIFTS studies, Fig. 9, which show only the presence of methoxy on the surface after the adsorption of methanol. This methoxy species is lost at approximately the same temperature as the formaldehyde peak appears in the TPD.[13] No conversion

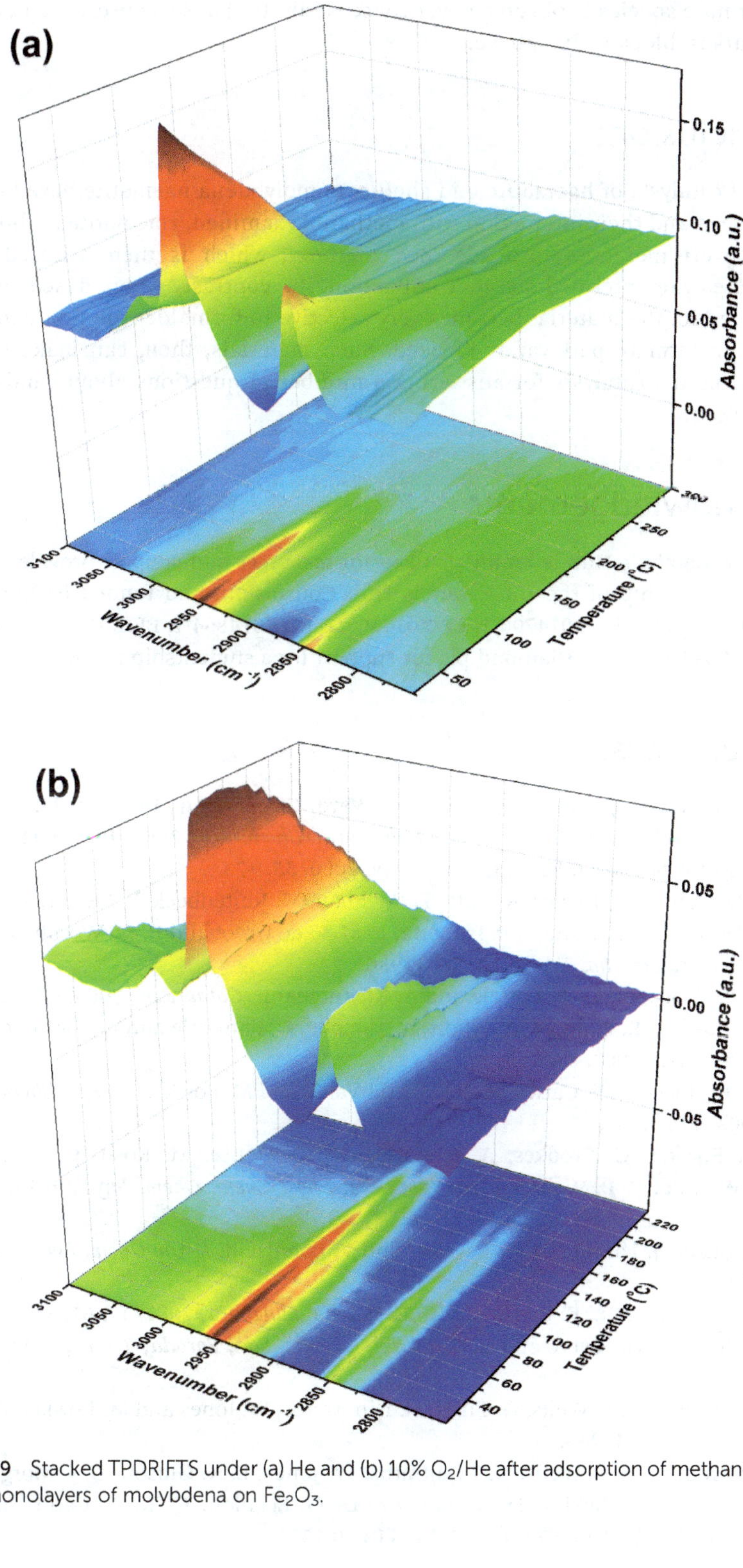

Fig. 9 Stacked TPDRIFTS under (a) He and (b) 10% O_2/He after adsorption of methanol on 6 monolayers of molybdena on Fe_2O_3.

to formate species is observed over these catalysts. Thus the presence of Mo appears to block step 5 above.

Conclusions

Model catalysts of haematite and shell/core molybdena/haematite have been produced and their interaction with methanol examined. Haematite is shown to convert methanol into adsorbed methoxy, which is then oxidised to formate, the intermediate to combustion. In contrast, when dosed with surface Mo, the material becomes very selective to formaldehyde production and the formate pathway is blocked. Such materials, then, can make very useful model catalysts for answering a number of questions about catalytic behaviour.

Acknowledgements

The UK Catalysis Hub is kindly thanked for resources and support provided *via* our membership of the UK Catalysis Hub Consortium and funded by EPSRC (portfolio grants EP/K014706/1, EP/K014668/1, EP/K014854/1, EP/K014714/1, and EP/I019693/1) and to Diamond plc for support for a studentship to CB.

References

1 C. Rhodes, G. J. Hutchings and A. M. Ward, *Catal. Today*, 1995, **23**, 43.
2 A. Y. Khodakov, W. Chu and P. Fongarland, *Chem. Rev.*, 2007, **107**, 1692.
3 N. Mimura and M. Saito, *Catal. Today*, 2000, **55**, 173.
4 I. D. Gonzalez-Jimenez, K. Cats, T. Davidian, M. Ruitenbeek, F. Meirer, Y. Liu, J. Nelson, J. C. Andrews, P. Pianetta, F. M. F. de Groot and B. M. Weckhuysen, *Angew. Chem., Int. Ed.*, 2012, **51**, 11986.
5 A. P. V. Soares, M. F. Portela and A. Kiennemann, *Catal. Rev.*, 2005, **47**, 125.
6 M. Bowker, R. Holroyd, A. Elliott, A. Alouche, C. Entwistle and A. Toerncrona, *Catal. Lett.*, 2002, **83**, 165.
7 M. P. House, A. F. Carley, R. Echeverria-Valda and M. Bowker, *J. Phys. Chem. C*, 2008, **112**, 4333.
8 M. Bowker, C. Brookes, A. F. Carley, M. P. House, M. Kosif, G. Sankar, I. Wawata, P. P. Wells and P. Yaseneva, *Phys. Chem. Chem. Phys.*, 2013, **15**, 12056.
9 G. Busca, J. Lamotte, J.-C. Lavalley and V. Lorenzelli, *J. Am. Chem. Soc.*, 1987, **109**, 5197.
10 L. J. Burcham, L. E. Briand and I. E. Wachs, *Langmuir*, 2001, **17**, 6175.
11 A. Glisenti, G. Favero and G. Granozzi, *J. Chem. Soc., Faraday Trans.*, 1998, **94**, 173.
12 C. Brookes, P. P. Wells, G. Cibin, N. Dimitratos, W. Jones and M. Bowker, *ACS Catal.*, 2014, **4**, 243.
13 C. Brookes, P. P. Wells, N. Dimitratos, W. Jones, E. K. Gibson, D. J. Morgan, G. Cibin, C. Nicklin, D. Mora-Fonz, D. O. Scanlon, C. R. A. Catlow and M. Bowker, *J. Phys. Chem. C*, 2014, **118**, 26155.

Faraday Discussions

DISCUSSIONS

Application of novel catalysts: general discussion

Amy L. Miller, Michael Bowker, Andrés García-Trenco, Joseph Socci, Nia Richards, Graham Hutchings, Nicola Collis, James Earley, Simon Freakley, Robbie Burch, Mark Howard, Elad Gross, Andrzej Kotarba, Miron V. Landau, James Anderson, Bert Weckhuysen, Simon Kondrat, Evgeny Naranov, John Mark Douthwaite, Ram Tiruvalam, Sarwat Iqbal, Luke Parker, Parag Shah, Ewa Nowicka, Wataru Ueda, Alessandro Piovano, Philip Landon, Nico Fischer, Christian Reece, Bruce Gates, Michael Lamb, Eoin Jackman, Tomasz Jakubek, Avelino Corma, Michael Claeys, Henrik Svengren, Cynthia Friend, David Lennon, Joshua Makepeace, Hazel Hunter and Haresh Manyar

DOI: 10.1039/C6FD90018F

Elad Gross opened the discussion of the paper by Simon Kondrat: What are the surface properties of the perovskite materials that you used in your study? Are the surface and catalytic properties of Cr or Mn based perovskites similar to those of CrO_x or MnO_x?

Simon Kondrat answered: As we have heard throughout this discussion, the surfaces of metal oxide systems cannot be assumed to be the same as the bulk. Often we see amorphous or phase segregated oxides on the surface. In our manuscript (DOI: 10.1039/C5FD00187K), we used XPS (Fig. 8) to show that the surface appeared to be a mixture of a perovskite and a phase separated lanthanum oxide. The difference in the B site environment was difficult to determine by XPS. Taking into account the presence of phase separated oxides, we decided to test the AuPt/single B metal oxides (Table 3, DOI: 10.1039/C5FD00187K). We found that when the single oxides were used as a support there was little control over the reaction selectivity. This shows that the catalysis is not simply an effect of the B site oxide and that the perovskite structure is influential. Simplistically, this could be the effect of having discrete B sites with controlled valance.

Parag Shah commented: The different synthesis methods for the perovskite produced catalysts vary the selectivity and activity for the glycerol oxidation reaction. What effect would changing the synthesis method of the Au–Pt nanoparticle, such as impregnation, deposition precipitation or co-precipitation, have on the selectivity and activity of the glycerol oxidation? Using alternative methods, as mentioned above, would lead to an increased variation in the particle sizes compared to the sol immobilisation method used in the paper (DOI: 10.1039/C5FD00187K). Do you think the particle size will affect the selectivity or activity of the glycerol oxidation reaction?

Simon Kondrat responded: Previous studies compare AuPd nanoparticles synthesised by sol immobilisation and wet impregnation.[1] The AuPd composition and size were shown to affect the catalyst activity, with small particles being more active. Under our reaction conditions, where the dehydration pathway is viable, a reduced activity towards oxidation products could be expected to result in a higher lactic acid selectivity. This would suggest that impregnation catalysts would be more selective towards lactic acid. Given the number of potential variables in catalyst design, such as metal nanoparticle composition, size, morphology, reaction conditions and support type, we could not investigate all in a single study. Your suggestion is an excellent idea for further work.

1 N. Dimitratos, J. A. Lopez-Sanchez, J. M. Anthonykutty, G. Brett, A. F. Carley, R. C. Tiruvalam, A. A. Herzing, C. J. Kiely, D. W. Knight and G. J. Hutchings, *PCCP*, 2009, **11**, 4952.

Andrzej Kotarba asked: As mentioned in the paper (DOI: 10.1039/C5FD00187K), the perovskite structure is versatile, where by substitution at the A and B sites we can tune the surface properties and thus change the catalytic performance. Additionally, you introduced the AuPt nanoparticles on the surface. Thus, the surface seems to be quite complex. Have you got any mechanistic model of how the system operates? Is there any synergy between the perovskite support and the metal nanoparticle? What are the active sites or should we consider an active ensemble instead?

Simon Kondrat answered: Several studies have investigated the mechanism of glycerol oxidation of metal nanoparticle surfaces. A good example being Davis and co-worker's isotope and computational study with glycerol and ethanol oxidation.[1] First, the deprotonation of the alcohol to form an alkoxy intermediate requires an OH species on the Au surface to have a suitable activation energy. The subsequent C–H activation, also facilitated by surface OH species, results in the formation of bound aldehyde. Importantly, Davis notes that oxygen from O_2 is not incorporated into the product but instead regenerates OH species. Interestingly, they then considered the sequential oxidation to form glyceric acid both in the aqueous phase with base and on the surface, with the activation energy being much lower on the latter.

This model only considers Au or Pt metals and does not consider metal–support interactions. For many reactions, a metal–support interaction is highly important. We show that the perovskite supports on their own are inactive and the metals are required. However, the nature of the perovskite changes the reaction products. This either points to a metal active site where the nature of the metal is influenced by the perovskite support or an active ensemble where the perovskite facilitates oxygen reduction reactions to maintain surface coverage of the hydroxide species.

1 B. N. Zope, D. D. Hibbitts, M. Neurock and R. J. Davis, *Science*, 2010, **330**, 74.

Nia Richards remarked: Would it be possible to produce gold and platinum nanoparticles on the surface of the perovskite support during the supercritical anti solvent (SAS) preparation process?

Simon Kondrat replied: It would be possible to precipitate La, B metal, Au and Pt. However, as with conventional co-precipitation, the complexity of the SAS system increases dramatically when the number of components is increased. The precipitation yield of each component is dependent on SAS parameters, such as the carbon dioxide pressure, temperature and flow rate. Previously, we have shown that SAS precipitation results in exceptional mixing of components throughout the bulk of the material.[1,2] This would result in Au and Pt being precipitated in the bulk of the material and not on the surface where it is required.

1 S. A. Kondrat, P. J. Smith, P. P. Wells, P. A. Chater, J. H. Carter, D. J. Morgan, E. M. Fiordaliso, J. B. Wagner, T. E. Davies, L. Lu, J. K. Bartley, S. H. Taylor, M. S. Spencer, C. J. Kiely, G. J. Kelly, C. W. Park, M. J. Rosseinsky and G. J. Hutchings, *Nature*, 2016, **531**, 83.
2 R. P. Marin, S. A. Kondrat, T. E. Davies, D. J. Morgan, D. I. Enache, G. B. Combes, S. H. Taylor, J. K. Bartley and G. J. Hutchings, *Catal. Sci. Technol.*, 2014, **4**, 1970.

Philip Landon commented: You try to correlate product selectivity with oxygen adsorption (Fig 10, DOI: 10.1039/C5FD00187K). This plot uses values for selectivity measured after 6 h of reaction time and indeed there seems to be a link. However, the selectivity does change with conversion and therefore reaction time (Fig 7, DOI: 10.1039/C5FD00187K). How well does the correlation hold if the selectivity is determined for other reaction times and plotted against oxygen adsorption?

Simon Kondrat responded: You are correct that there are changes in the product selectivity with conversion for all of the catalysts tested. Interestingly, there is an observable increase in the lactic acid selectivity at the expense of C_3 products at high conversions for most samples. This could be due to the blocking of oxidation sites as the reaction progresses. However, the changes are subtle and the correlation between oxygen adsorption still applies. It should be noted that none of the selectivity trend lines for catalysts cross at any point.

Sarwat Iqbal remarked: My question is related to the regeneration study of your hopcalite materials. The catalyst showed an increase in activity upon the 2nd and 3rd use, but the 4th one showed a decline in activity and a change in the selectivity pattern as well. You have linked this with the removal of PVA on the first uses and an accumulation of organic species on the 4th use. Do you have any evidence for these two effects? Do you expect a change in the hopcalite structure (surface or bulk) upon reuse?

Simon Kondrat replied: We have previously shown, along with other studies, that prolonged glycerol oxidation testing results in the formation of strongly bound organic surface species that block active sites.[1] We have not tested this on our perovskite supported catalysts, due to the difficulty in obtaining IR spectra with the dark coloured supports. We did reflux our AuPt/perovskite catalysts prior to a test reaction and found a modest improvement in the activity, supporting our theory on PVA removal marginally improving the activity.

1 S. A. Kondrat, P. J. Miedziak, M. Douthwaite, G. L. Brett, T. E. Davies, D. J. Morgan, J. K. Edwards, D. W. Knight, C. J. Kiely, S. H. Taylor and G. J. Hutchings, *ChemSusChem*, 2014, 7, 1326.

John Mark Douthwaite asked: Could the increase in the selectivity to lactic acid observed with time indicate that lactic acid is simply stable under the reaction conditions?

Simon Kondrat responded: From some mechanistic work we have carried out, we have found that once the reaction has proceeded down the oxidation pathway the products cannot convert to lactic acid. Studies of lactic acid stability have shown that after 6 h reaction time, a very small amount of lactic acid did convert to the oxidation products, however the amount was almost negligible. The two pathways can be considered separate, so I do not think the changes in the selectivity are explained by the stability of lactic acid.

Michael Lamb said: The perovskite-based nanoparticle catalyst you describe in this paper (DOI: 10.1039/C5FD00187K) was recycled three times. Coke deposition led to the deactivation of this system, and consequently the catalyst could not be re-used further. Are there calcination conditions under which coke is removed from the surface without causing the agglomeration of the nanoparticles?

Simon Kondrat answered: Thermal treatment would remove surface organic species. However, the AuPt nanoparticles would not be stable under such thermal treatments and particle sintering would then result in catalyst deactivation. Another possibility, not with Au but maybe with Pt or Pd, would be that at high temperatures the metal would be incorporated into the perovskite B site.

An interesting strategy would be to deliberately perform a high temperature calcination to incorporate the metal into the perovskite and remove the organic species. We could then perform a reduction step to re-form the metal nanoparticles.

James Earley commented: You mention that the catalyst deactivation observed during the forth cycle is likely due to adsorption of organic compounds blocking the active sites. One way to regenerate your catalyst would be to treat it with scCO_2. You could then re-test your catalyst after the scCO_2 treatment and see if the activity is recovered. This may prove your hypothesis. It may also allow for analysis of the removed organic compounds. This treatment would tie in nicely with your original catalyst preparation method using scCO_2 as an anti-solvent. Have you thought about doing this?

Simon Kondrat responded: This is an excellent suggestion. The use of super-critical fluids in extraction technologies is well established, as in the case of caffeine extraction from coffee. The process could well be a non-invasive method to remove organic compounds that block the catalytic sites.

Wataru Ueda said: There are at least two designing methods to make perovskite materials with a large surface area. One is the introduction of porous structures and the other is the introduction of nano-sized particles. There are many reports for the latter but the methods are not perfect because the methods often cause formation of impurity phases. Therefore, the former seems more desirable. Since the introduction of a mesopore is considered to be simply a similar methodology for making nano-sized particles, designing a method to introduce pore structures during the crystal formation of perovskites is highly necessary. However, methods are still challenging. We need to have a new design concept. What is your opinion on this? Another question is about the use of NaOH for the reaction. NaOH can assist the oxidation of the O–H group over Au catalysts. Once 1,3-dihydroxyacetone is formed by the oxidation, the transformation of 1,3-dihydroxyacetone to lactic acid can take place easily, particularly over a solid-acid surface. Therefore, metal catalysts supported on acidic oxide supports might be better than the currently studied system for the reaction.

Simon Kondrat replied: You are correct that creating porous frameworks would be an alternative method for increasing the perovskite surface area. Both hard and soft templating methods have been attempted.[1,2] It has been found that the complete removal of the template results in collapse of the pore structure. If template impurities can be tolerated, these methods may be a viable way of making porous perovskites. I agree that new design concepts are required. For example, a 3D printing/bottom up lithography could provide highly controllable methods of designing catalysts. However, industrial processes require affordable and highly scalable techniques, which present additional challenges.

With regards to your second question, with Au catalysts, NaOH is required to deprotonate glycerol to form the bound alkoxide, which can then be oxidised. Without the AuPt/perovskite catalyst, we observed no conversion of glycerol to lactic acid. You are correct that the initial oxidation product (glyceraldehyde or dihydroxyacetone) could be catalysed by an acid or a base. Villa and co-workers looked at AuPt nanoparticles on a range of solid acid supports and did not report any lactic acid formation. This could be due to the acid sites not being strong enough to promote the dehydration pathway relative to the oxidation pathway.[3] Fan and co-workers have successfully converted glycerol to lactic acid, base free, using a Pt/Sn–MFI catalyst, showing that your suggestion is feasible.[4]

1 R. Zhang, P. Li, N. Liu, W. Yue and B. Chen, *J. Mater. Chem. A*, 2014, **2**, 17329.
2 H. Lu, P. Zhang, Z.-A. Qiao, J. Zhang, H. Zhu, J. Chen, Y. Chen and S. Dai, *Chem. Commun.*, 2015, **51**, 5910.
3 A. Villa, S. Campisi, K. M. H. Mohammed, N. Dimitratos, F. Vindigni, M. Manzoli, W. Jones, M. Bowker, G. J. Hutchings and L. Prati, *Catal. Sci. Technol.*, 2015, **5**, 1126.
4 H. J. Cho, C.-C. Chang and W. Fan, *Green Chem.*, 2014, **16**, 3428.

Tomasz Jakubek opened the discussion of the paper by Haresh Manyar: The material used in this paper (DOI: 10.1039/C5FD00227C), OMS-2, also known as cryptomelane, contains potassium cations located in the tunnelled structure. In a poster I presented during the Faraday Discussions, I showed results concerning soot combustion where cryptomelane containing different intra-tunnel cations (H^+, Li^+, Na^+, K^+) was used. Two comments come to mind. First, when hydrogen is the intra-tunnel cation, the hydrogen could perhaps not only form on the surface of the material, but also originate from inside the structure, thereby increasing the catalytic activity. Second, I have observed a correlation between the work function of cryptomelane and its catalytic activity in soot combustion.

The smaller the ionization energy of the intra-tunnel cation (a greater cation radius), the lower the work function of the material and the higher the catalytic activity in soot combustion. Since, in your reaction, halogenated arenes are used, the differences in the work function could perhaps influence the adsorption of halogens on the active sites and poisoning of the catalyst surface. Were other intra-tunnel cations considered during the synthetic method?

Haresh Manyar replied: In this study, we have only used K–OMS-2 and we have not investigated OMS-2 doped with other intra-tunnel cations. However, it seems to be an interesting set of materials to investigate for the adsorption of different substrates as well as the catalytic activity.

Eoin Jackman asked: Was the change in stirring speed between stages in the procedure important to the overall reaction process?

Haresh Manyar responded: The change in stirrer speed is not important, however it is preferential to keep the stirrer speed >1000 rpm during the reaction to increase the mass transfer rates and operate the reaction under intrinsic kinetic conditions.

James Anderson remarked: You propose that using the Pd based catalyst the reaction takes place exclusively on the metal surface while on the Pt based catalyst you suggest that the reaction takes place on the metal and on the support surface following hydrogen spillover. Can you elaborate as to why you believe you have two different pathways over the two different materials and do you have any evidence for spillover?

Haresh Manyar replied: From the reaction composition profiles over Pd and Pt supported on OMS-2 catalysts, we observe the reaction proceeding through different reaction pathways, as evident from the reaction intermediates. In the hydrogenation of nitrobenzene there are two main pathways, a direct pathway *via* the formation of nitrosobenzene and an alternate pathway *via* the formation of coupling products, *i.e.* azoxybenzene. We also observe similar product profiles on OMS-2 and Pt supported on OMS-2 catalysts but enhanced reaction rates in the presence of Pt on OMS-2. Furthermore, the reactions show different reaction composition profiles in the presence of Pt supported on alumina, which indicates that in the case of Pt/OMS-2 catalysts, the reaction mainly occurs on the surface of OMS-2 and the enhanced rates are due to spillover of hydrogen from Pt metal.

Graham Hutchings asked: The data in Fig. 6 (DOI: 10.1039/C5FD00227C) appear to be very complicated. Have you investigated competitive adsorption, and have you added the reaction intermediates and products at the start of the reaction? Are some of the effects you observe related to competitive adsorption?

Haresh Manyar responded: The data in Fig. 6 (DOI: 10.1039/C5FD00227C) represents the hydrogenation of intermediate products occurring in both reaction pathways, *i.e.* the direct pathway *via* the formation of nitrosobenzene and the second reaction pathway *via* the formation of coupling products, which are azobenzene and azoxybenzene. In order to reflect on both reaction pathways using the different catalysts used in this study, *i.e.* OMS-2, Pt/OMS-2 and Pd/OMS-2, Fig. 6 should also be viewed along with Fig. S4 and Fig. S5 in the ESI. Competitive adsorption of the reactants and intermediates plays a key role in dictating the product selectivity. Although we did not conduct the hydrogenation of nitrobenzene by the addition of intermediates and products at the start of reaction, we have conducted this experiment using similar catalysts for other hydrogenations of α,β-unsaturated aldehydes and ketones, and we have seen a significant influence of competitive adsorption.

Michael Lamb said: The reaction profile shown in Fig. 5 (DOI: 10.1039/C5FD00227C) shows the consumption of the starting material and shows the formation of a number of intermediates, highlighted in the reaction mechanism before the desired product formation. However, there does not appear to be quantitative conversion to the desired product. There is a significant difference in the concentration of the starting material and the product. Without mass balance, how can you be sure that the catalyst you describe in this paper is 99% selective?

Haresh Manyar answered: Fig. 5 indicates the reaction composition profiles as observed from the analysis of the liquid phase, however this does not reflect the strongly adsorbed species bound on the catalyst surface. Moreover, the mass balances observed were always more than 90–95%, as observed from the calibration curves.

Simon Freakley enquired: How important is the molecular sieve structure to the activity? What makes this structure preferable compared to a mixed valance state oxide of manganese?

Haresh Manyar replied: The molecular sieve structure is important and essential, as the molecular sieve structure offers lattice oxygen vacancies, a crystalline structure with defects, a larger surface area, as well as potential for inducing shape selectivity in the catalytic processes.

Avelino Corma asked: How general is your catalyst for the reaction when you have other ones? How do they compare on selectivity, generality and rate with the state-of-the-art catalysts?

Haresh Manyar responded: Manganese oxide based octahedral molecular sieves are very versatile catalysts both in hydrogenation as well as oxidation reactions. The versatility is evident from the range of functionalities and substrates studied so far. We have explored the catalytic activity of these materials in the selective reduction of functional groups, such as carbonyl, carbon–carbon double bond and nitro groups, in a variety of substrates, such as nitrobenzene, halogenated nitrobenzenes, nitrobenzaldehyde, benzaldehyde, α,β-unsaturated aldehydes and ketones such as benzalacetone, cinnamaldehyde, citral, ketoisophorone, isophorone, cyclohexenone and furfural. We have also explored their catalytic activity for the selective oxidation of alcohols and aldehydes such as benzyl alcohol, cinnamyl alcohol and 5-hydroxymethyl furfural to 2,5-furan dicarboxylic acid. Manganese oxides have shown promising results compared to state-of-the-art catalysts, especially since we can do reactions in the absence of noble metals using water or water/alcohol mixtures as solvents.

We can compare the selectivity and rate with state-of-the-art catalysts. Manganese oxides have often shown high selectivities >90%, but the reaction rates are slower compared to noble metal catalysts, however, by using metal supported on OMS-2, it is possible to enhance the reaction rates by enhancing the hydrogen dissociation followed by spillover to substrates.

Andrzej Kotarba said: The K–Mn–O structure you chose as a base for the catalyst is very specific (*n.b.* of cryptomelane mineral). The octahedral manganese blocks form conductive tunnels where K^+ ions can diffuse relatively freely. In such a structure, the ionic conductivity starts at ~300 °C. So, at the applied calcination temperature (450 °C), I would expect the potassium ions to diffuse from the bulk to the external surface and promote the Pt nanoparticles (such phenomena is commonly applied in electrocatalysis). Therefore, the calcination conditions, *i.e.* the temperature, time, and atmosphere, seem to be of critical importance for the catalyst surface state (potassium coverage) and thus its catalytic performance. Do you observe such effects in your Pt-supported OMS-2 catalyst?

Haresh Manyar answered: So far, we have not studied the influence of the calcination conditions on the catalytic activity, but it is worthwhile studying this criteria in order to correlate the surface coverage of K^+ ions with the catalytic activity. However, we need to investigate the changes in the K^+ ion coverage under the reaction conditions, as in the case of liquid phase reactions, in the presence of solvents, as this may present a different scenario than gas phase reactions.

Graham Hutchings remarked: You mentioned in the discussion of your paper (DOI: 10.1039/C5FD00227C) that when the catalysts are operated under anhydrous conditions you obtain a different selectivity and conversion compared with when water is present. How does the catalyst change when water is present? Does water change the surface structure? You put a lot of stress on the structure so is it different when there is water in the reaction medium?

Haresh Manyar responded: In our previous studies on the liquid phase hydrogenation of α,β-unsaturated aldehydes and ketones using OMS-2 catalysts, we have shown that dissociation of H_2 on the surface of OMS-2 is a water assisted process. From the DFT calculations, weak adsorption of H_2 on OMS-2 (001) was observed. Under anhydrous conditions, the dissociation barriers were found to be 1.90, 1.79 and 2.15 eV over the OMS-2 (001) surface. In contrast, the barrier decreased to 0.10 eV in the presence of water, indicating that the dissociation is water assisted. In this case, H_2 is dissociated by the surface Mn of OMS-2 (001) that has been modified by the O atom from the adsorbed H_2O molecule.

Bruce Gates said: You have investigated catalysis involving a complex reaction network, which can take a lot of effort to resolve. Did you resolve your network into the individual reactions, and did you determine the kinetics of any individual reactions? When faced with the challenge of a complex reaction network, how do you simplify matters and how do you represent kinetic information for a simplified representation of a reaction network?

Haresh Manyar replied: In liquid phase heterogeneous catalysis, we often come across complex reaction networks, which include the possibility of the formation of multiple products and different reaction pathways, including reaction intermediates. In this study (DOI: 10.1039/C5FD00227C), we have multiple functionalities present in halogenated arenes and in previous studies we had α,β-unsaturated aldehydes and ketones. We aim to resolve the reaction networks into individual reactions, which could be parallel or consecutive. Wherever possible, we use as many intermediates as possible to study individual reactions from the network, as well as competitive adsorption under similar reaction conditions. Often, comparison of the reaction rates and reaction composition profiles indicate a lot about possible molecular interactions occurring at the active site. We apply *in situ* spectroscopy studies, such as ATR-FTIR and XANES, as well as computational studies along with reaction kinetic analysis to achieve further insights into the complex reaction networks.

Andrzej Kotarba commented: Mentioning optimization, you have pointed to an essential issue. In my opinion, the optimization of the catalytic systems *via* understanding how the system works at a molecular level (usually experiments in tandem with theoretical modelling), establishing the key reactivity descriptors and finding correlations between the surface structure (active sites) and catalytic activity lead to rational catalytic design. Once we get insights into the reaction mechanism, we propose dedicated experiments where the proof-of-concept can be verified. Of course, since the catalytic process is complex, we are unable to take into account all the variables. Therefore, we have to pass through several iterations. Such methodology, presented in several papers at this conference, is certainly laborious but, in my opinion, this is the current state-of-the-art of catalyst design.

Haresh Manyar responded: Understanding the molecular interactions at the active site is the key step to establishing design criteria for the optimization of the catalytic activity, however there is a complex environment around the active site, especially in the case of liquid phase reactions due to the presence of solvent molecules, which also compete with reactants, intermediates and products. In the near future, the ability to tune the molecular interactions using the liquid structure of solvent + reactants around the active sites will lead to the design of superior catalysts.

Nicola Collis opened the discussion of the paper by David Lennon: I am curious as to why η-alumina was chosen for this catalyst, given that the literature suggests it has stronger acid sites than gamma? Have you done any direct comparisons with other alumina phases which may have fewer or even none of the sites you are removing by capping on the eta?

David Lennon answered: This question directly addresses the main "design" concept for this development of a new methyl chloride synthesis catalyst, with the innovation originating from the industrial centre. γ-Aluminas are typically used as catalysts in the production of chloromethane. η-Alumina possesses a greater Lewis acidity than γ-alumina. Preliminary laboratory measurements indicated the site-selective nature of the product and by-product production, as described in the Introduction section of the paper (DOI: 10.1039/C5FD00202H). Against this background, it was proposed that the use of a more acidic catalyst would enable certain sites to be "capped" (or neutralised); this would then selectively attenuate the by-product formation whilst reducing the rate of product formation to an acceptable degree.

Mark Howard commented: The shape of the DME curves in Fig. 6 and 7 (DOI: 10.1039/C5FD00202H) suggests that DME rises towards equilibrium for un-doped and 0.1 mmol doped materials, but more slowly for the latter where the lower peak level is presumably a consequence of the effect of higher temperature on the methanol/DME equilibrium. At equivalent temperatures and conversions, these two materials might therefore converge on similar DME levels. In contrast, for the 1.0 mmol doped materials, the DME levels remain very low – seemingly well below equilibrium levels. This would suggest a suppression of its rate of formation by some orders of magnitude compared to the other materials – a very large and remarkable effect. Have I understood this correctly?

David Lennon responded: Absolutely. These statements are the origins of the favourable selectivity outcomes reported in Table 1 (DOI: 10.1039/C5FD00202H). Although doping of the η-alumina with 1.0 mmol $g_{(cat)}^{-1}$ reduces methyl chloride production at 560 K by 57% (Fig. 7), it virtually suppresses the rate of dimethyl ether formation (Fig. 7), resulting in a selectivity to methyl chloride of 99.6% under the stated laboratory conditions.

Michael Bowker asked: Do you know anything about the surface composition of the catalyst? Is the surface chlorided during this process? Have you looked at the effect of adding water?

David Lennon answered: The surface composition of the un-doped η-alumina catalyst is described in an earlier publication.[1] A future publication will correlate how the addition of a Group 1 metal salt over the concentration range examined here perturbs the acid site distribution of the η-alumina. Hydrogen chloride dissociatively adsorbs on the activated η-alumina catalyst and exhibits a range of adsorption enthalpies. These specific adsorbate/substrate interactions are described in ref. 2.

No, we have not looked at the effect of adding water to this reaction system. In fact, a lot of effort was expended to ensure that the HCl was anhydrous; this procedure minimised corrosion issues with the experimental apparatus.

1 D. T. Lundie, A. R. McInroy, R. Marshall, J. M. Winfield, P. Jones, C. C. Dudman, S. F. Parker, C. Mitchell and D. Lennon, J. Phys. Chem. B, 2005, 109, 11592.
2 A. R. McInroy, D. T. Lundie, J. M. Winfield, C. C. Dudman, P. Jones, S. F. Parker and D. Lennon, Catal. Today, 2006, 114, 403.

Graham Hutchings asked: Question 1 – I wonder if you can comment on the industrial use of the catalyst? I would expect the CsCl and KCl to have a small but significant vapour pressure under the reaction conditions, which would lead to the mobility of Cs and K down the catalyst bed. Is this observed on long term use? If so, you would need to add more Cs and K than the optimum concentrations to ensure the promotional effect is maintained.

Question 2 – Do the data in Fig. 3 (DOI: 10.1039/C5FD00202H) show a slight curvature of the Arrhenius plot? Or is this within experimental error? If there is curvature, then it might be indicative of a diffusion limited process?

David Lennon replied: Question 1 – The catalyst has operated within a large-scale industrial process for an acceptable period of time. Fig. 8 of the paper (DOI: 10.1039/C5FD00202H) shows that a Group 1 metal salt concentration of *ca.* 0.3 mmol $g_{(cat)}^{-1}$ is sufficient to significantly attenuate the rate of dimethyl ether production in the laboratory. It seems reasonable to assume that a Group 1 metal salt concentration above the threshold value of *ca.* 0.3 mmol $g_{(cat)}^{-1}$ mitigates against the loss of CsCl from areas of the catalyst bed, maintaining local concentrations above the threshold value and thereby sustaining a high methyl chloride selectivity. No laboratory based data are available to support this assumption.

Question 2 – The authors acknowledge that some degree of diffusional limitation towards methyl chloride production may become evident at higher temperatures. However, the linearity of the fit over the temperature range 487–541 K, evident in Fig. 3 (DOI: 10.1039/C5FD00202H), is good and the errors in the apparent activation energies are modest (σ methyl chloride = 1.02%, σ dimethyl ether = 1.16%), indicating good reproducibility of the data. Hence, for temperatures below 540 K, the system is deemed to be under kinetic control.

Evgeny Naranov enquired: Have you considered the reaction between the side product (DME) and hydrochloric acid? Is it possible to convert DME to the final product under your conditions? What about the activation energy of this process?

David Lennon responded: This work has not explored the reaction between dimethyl ether and hydrogen chloride.

Alessandro Piovano asked: After doping alumina with a Group 1 metal halide, did you test the effect of different thermal pre-treatments of the catalyst directly inside the reactor, in order to reduce the OH population on the alumina surface and to increase the amount of exposed Al^{3+} Lewis acid sites?

David Lennon answered: We have attempted a variant of this approach. We wondered whether it would be possible to use various thermal treatments to control the acid site distribution of the catalyst and thereby influence the product distribution in the methyl chloride synthesis. Although we were able to modify the distribution of Lewis acid sites and thereby perturb the product distribution, these treatments were unable to specifically deactivate the sites responsible for by-product formation. These investigations lead to the conclusion that physical methods alone were ineffective in enhancing methyl chloride selectivity; it was only by chemical modification strategies, as outlined in the paper (DOI: 10.1039/C5FD00202H), that industrially desirable methyl chloride selectivity could be obtained.

Andrzej Kotarba said: In Fig. 8 and 9 (DOI: 10.1039/C5FD00202H) similar profiles for DME and MC production, respectively, as a function of K and Cs loading, are presented. Since, as mentioned in the paper, the ionic radius for K^+ and Cs^+ substantially differ, I would expect a shift in the observed profiles. The site blocking effectiveness should be higher for Cs^+ than for K^+ and thus the resulting critical concentrations for each alkali ion should also be different. What is the surface dispersion and coverage for these site blocking agents? How are the observed similarities in the activity profiles explained?

David Lennon answered: The profiles are similar in Fig. 8 and 9 as the addition of a Group 1 metal salt is preferentially "capping" strong and medium-strong Lewis acid sites. A subsequent publication will address this matter specifically. Chemisorbed methoxy species are the intermediates for both product and by-product production, which, for the un-doped catalyst, reside at strong, medium-strong and medium-weak sites. The blocking of the strong and medium-strong Lewis acid sites by doping therefore reduces the concentration of chemisorbed methoxy species, thereby attenuating both product and by-product formation. This is attributed to a first order effect.

Fig. 8 (DOI: 10.1039/C5FD00202H) illustrates a second order effect, with the CsCl doped catalyst displaying reduced dimethyl ether production compared with that observed when using KCl as a dopant. It is this less dramatic second order effect that is tentatively attributed to differences in the ionic radii of the cations. The surface coverage and dispersion of these site blocking agents have not been precisely determined. Moreover, the form of the Group 1 metal salt adsorption complex responsible for the improved product selectivity is also unknown.

Joseph Socci asked: What is the long term stability of the Group 1 metal salt dopants on the surface of the η-alumina with regards to its "capping" ability, and do you have any data on this?

David Lennon responded: We have no laboratory based measurements available to answer this question.

Simon Kondrat opened the discussion of the paper by Henrik Svengren: Under operating conditions you see the formation of an amorphous overlayer that you believe is the active surface in your reactions. Do you know what the composition or local order is of the overlayer? Furthermore, have you considered preparing entirely amorphous materials so that you can probe local structure effects with bulk techniques, such as XAFS?

Henrik Svengren responded: The surface layer was investigated *ex situ* with SEM-EDS to establish the composition of the elements. To avoid a strong signal from the bulk, a small crystallite hanging out over a crack in the electrode was scanned from the empty space, further touching the very tip where the typical morphology, believed to be an amorphous layer, was seen and then further into the crystallite to give a strong signal from the bulk. The relative comparison of the elemental composition showed a quite comparable composition to the crystal bulk, which was $Co_3Sb_4O_6F_6$, as shown in Fig. 10 (DOI: 10.1039/C5FD00169B). The interaction volume of the electron beam is not ideal during such a scan so one has to be tolerant to a larger error between the light, medium and heavy elements than during ideal conditions. An attempt to also see changes in the local order of the thin, presumably amorphous, layer was made with *in operando* EXAFS measurements, however the bulk signal was clearly dominating. The only conclusion from these measurements was that the crystal structure persists throughout electrolysis. Preparing an entirely amorphous material would be a true challenge because the active surface layer is formed from its initial crystalline compound in the presence of the chosen electrolyte under an applied potential. A similar question was put forward by Dr Hutchings (see below), please refer to the latter part of our answer to his question. In our answer, we further develop EXAFS measurements, addressing the question that you have also asked. In addition to these answers, we would like to mention that another possibility would be to perform catalysis with powder of even smaller particle size, for a much longer time than 65 h, and see if the surface layer grows thicker. If so, the surface layer volume to bulk volume will be optimized to a large difference and consequently the EXAFS signal would reflect the matrix with the largest total volume.

Graham Hutchings asked: Following on from the point raised by Dr Kondrat concerning the surface layer, can you say anything about the timescale of its formation? Is there an induction period during which the surface is formed? If the layer can be characterised, it might provide the basis for identifying materials that could be synthesised as improved electrodes.

Henrik Svengren replied: The time lapse for the surface layer to change from smooth to irregular and fluffy that supposedly contains a highly catalytic amorphous phase is most likely rather quick. Relying on the linear scan voltammetry curves in Fig. 2 (DOI: 10.1039/C5FD00169B), the first scan shows a significant oxidation wave deduced for Co(II) to Co(III) that has an insignificant appearance in the following scans. It is likely that the available catalytic sites in the surface have been activated to a higher oxidation state and are no longer in the initial crystal structure but rather in the supposed amorphous state. The results from the chronopotentiometry seen in Fig. 4 (DOI: 10.1039/C5FD00169B) show that the potential stabilizes after 1 min at a given current; this could well be a measure of the fact that the catalytic surface layer has reached a certain equilibrium. The thickness of the activated layer would supposedly be very thin at first and after a long time of constant faradaic current the layer may grow slightly thicker. Characterizing such catalytically active surface layers is a true challenge because of the lack of crystallinity and because the underlying crystalline bulk gives a strong signal. In our specific case, the layer exists in its active form in the presence of both the electrolyte and a potential, consequently characterization should be carried out *in operando*. The best alternative would probably be EXAFS. If the surface layer volume to bulk volume can be optimized to a large difference, the signal would then reflect the matrix with the largest total volume. Another alternative would be to mount a single crystallite so that a thin beam with a low incident angle touches the surface and this will give data with a low signal from the bulk. This might be difficult to practically set up *in operando*. A combination of spectroscopic methods, such as IR, Raman or UV spectroscopy, which have small penetration depths and hence would not give large signals from the bulk, could also be applied, but then the difficulty would be handling the small size of the crystallites and mounting these practically to obtain *in operando* data. For the latter proposal, the presentation by Dimitrios Maganas *et al.* (DOI: 10.1039/C5FD00193E) is highly relevant. The synthesis of such active compounds is another discipline that is yet to be explored. In our specific case, the active surface layer has been formed from its initial crystalline compound in the presence of the chosen electrolyte under an applied potential. To obtain the same active compound using another method would be a question of mimicking these conditions; speculatively, perhaps without the initial crystalline compound and instead using the ingoing elements in solution directly in the electrolyte.

Philip Landon said: Your electrodes were formulated in a way that made it possible to prepare them reproducibly for evaluation within this study. However, you also mention that this means that they are not optimised for bench marking against electrodes manufactured in other ways. If this is the case, how do you compare your results against those reported in the literature?

Henrik Svengren answered: Indeed, the electrochemical performance of a catalytic compound shows significant deviation between electrodes with different designs and compositions. Therefore, effort was put into formulating electrodes with a high reproducibility, having a homogenous composition suitable for our powders. These electrodes have the benefit of accurate relative ranking of the catalyst performance. A good way of comparing the catalysts in our study with other catalysts reported in the literature is to apply the active compounds with our electrode formulation, keeping a similar molar ratio and particle size. Such a comparison was made in our study using Co(II)/Co(III)-oxide powder, which is a well-studied catalyst frequently referred to in the literature with an excellent catalytic performance for water oxidation. In our study of the oxofluorides, the most active compound contained 100% Co in the metal position of the solid solution, $Co_3Sb_4O_6F_6$, which makes the comparison to Co(II)/Co(III)-oxide highly relevant.

Robbie Burch opened a general discussion: Much of the very early work on catalyst discovery in the 19th Century involved catalyst design by trial and error. Then, we had the work of Haber, who tested many thousands of ammonia synthesis catalyst combinations essentially using a high throughput experimentation approach. More recently, we have had input from theoretical models to "predict" new catalysts. In reality, there have been few, if any, examples where modelling has taken us all the way from a blank page to a final catalyst. Where modelling has been useful is in finding novel catalyst combinations, for example bimetallic catalysts, that might not have been expected based on prior knowledge and this may lead, after extensive experimentation, essentially through trial and error, to a useful final catalyst. In your opinion, how useful is theoretical modelling when applied to catalyst design, and at what point is it still best to resort to traditional trial and error experimentation?

Simon Kondrat responded: From my perspective, catalyst discovery requires a good foundation in theory, with respect to the reaction mechanism and the desired catalyst properties. This foundation is greatly aided by theoretical modelling and surface science, which can help rationalise experimental studies to provide concepts for catalyst development. Trial and error using a strong concept, in part based on modelling studies, still has its place. High-throughput techniques still require basic knowledge of the catalyst system so that "hits" can be clearly defined and identified. Once an effect or trend has been identified, we try to fully rationalise it through further theoretical modelling, which can then be used to develop more refined concepts. I think that catalyst design is an iterative process.

John Mark Douthwaite addressed Simon Kondrat and Haresh Manyar: How influential do you think the binding energy and surface lifetime of the substrate with the catalyst are for controlling the selectivity in oxidation and reduction reactions? Should more focus be placed on tuning these parameters during catalyst design?

Simon Kondrat answered: The heat of adsorption, surface coverages and surface lifetimes of species are, of course, fundamental to understanding catalysis. We showed that the selectivity could be correlated with simple oxygen adsorption on a clean perovskite surface (DOI: 10.1039/C5FD00187K). However, things are complicated under liquid phase reactions as surface coverage will have to take into account the solvent, hydroxide species and the substrate. It is likely that the perovskites have different coverages of hydroxide species and they will also facilitate the oxygen reduction reaction depending on the occupancy of the B metal e_g orbital. This would correlate with the basic concept of oxygen adsorption.

Haresh Manyar responded: In liquid phase oxidation and reduction reactions, the adsorption strength and adsorption geometry of the reactants, intermediates and products play key roles in controlling the reaction pathways, reaction rates and product selectivity. Moreover, in liquid phase reactions, surface species are not limited to reaction intermediates. Adsorbed solvent molecules also play key roles in affecting the adsorption geometry, so *in situ* spectroscopic investigation of this phenomenon at the molecular level should be key design criteria for catalyst design.

Michael Bowker said: Liquid phase catalysis often seems to me to be "catalysis in soup", since there can be a wide range of ingredients in the mix. You have a huge amount of base and a large metal to glycerol ratio, so how much design goes into such parameters, or is it simply a recipe passed on from someone else for quite different catalysts? There is clearly an important role for liquid phase bases or acids in such processes, which themselves effectively act as co-catalysts. This leads me to wonder whether we could make more use of these methods in gas–solid catalysis. Perhaps we should add gaseous acids or bases, which adsorb weakly to give an additional tool to manipulate reactions. Of course, they have to be non-reactive in themselves, and probably only weakly acidic and basic molecules could be used. There are some reactions that already use such gas phase additives (*e.g.* ethene epoxidation), which, to some extent, play that role by equilibrium deposition of one component of the molecule.

Simon Kondrat responded: In our paper (DOI: 10.1039/C5FD00187K) we did not change parameters, such as the base : glycerol and glycerol : metal ratio, as we were investigating the effect of the perovskite support on the selectivity of the reaction. However, there have been numerous publications on the effect of changing the base concentration, temperature and glycerol : metal ratio. Ref. 1–4 are studies that investigated the effect of these parameters on the activity and selectivity.

Briefly, with gold nanoparticles, the importance of the base has been highlighted many times, as gold cannot deprotonate the glycerol. Also, increasing the temperature and base concentration opens up the dehydration pathway of the glyceraldehyde to ultimately produce lactic acid. Therefore, they are a good starting place for investigating the current catalysts. Optimisation through adjusting these parameters for the specific catalysts in question is indeed possible.

The concept of using weak bases or acids in gas phase reactions is an interesting one that could well be explored.

1 S. Carrettin, P. McMorn, P. Johnston, K. Griffin and G. J. Hutchings, *Chem. Commun.*, 2002, 696.
2 F. Porta and L. Prati, *J. Catal.*, 2004, **224**, 397.
3 P. Lakshmanan, P. P. Upare, N.-T. Le, Y. K. Hwang, D. W. Hwang, U. H. Lee, H. R. Kim and J.-S. Chang, *Appl. Catal., A*, 2013, **468**, 260.
4 Y. Shen, S. Zhang, H. Li, Y. Ren and H. Liu, *Chem.–Eur. J.*, 2010, **16**, 7368.

Graham Hutchings addressed Avelino Corma and Haresh Manyar: Following on from the points raised by Michael Bowker concerning the use of base in liquid phase reactions, solid bases that are stable under reaction conditions are not many. Is this a research area that should be given attention?

Avelino Corma replied: There are fewer processes catalyzed by solid bases than by solid acids. Solid bases can be deactivated by CO_2 and water. This already limits the uses. If the above is not the case, they have to compete with NaOH, which is a difficult competition. Nevertheless, there are solid base catalysts working in the production of chemicals and fine chemicals, and research is still going on for particular applications.

Haresh Manyar answered: In my view, yes, this research area should be given a high priority. The availability of robust, efficient and reusable solid base catalysts will enable several advantages over homogeneous bases, such as NaOH. The key advantages include: (i) there would be no need for down stream neutralization of the effluent stream, (ii) there would be a lower risk of corrosion and hence, no need for base-tolerant coatings for reactors, (iii) continuous flow processes can be developed, (iv) solventless, high temperature and vapour phase reactions are possible with solid base catalysts but not with NaOH, and (v) tailored multi-functional catalysts for one-pot tandem/cascade reactions are possible with solid base catalysts.

We must apply our collective knowledge developed within the catalysis community, such as *in situ* operando techniques, computational modelling, surface science and reaction engineering, to develop stable and reusable solid base catalysts.

Mark Howard said: In the current commercial catalyst manufacture and implementation, we have a large but known set of "levers" that we can pull – for example, the calcination temperature, start-up and operating conditions, *etc.* "Catalysis by Design" should, of course, be about the more directed, intelligent and efficient application of these "levers". However, are there also new levers emerging that we should think about? Does this also mean wide scale adoption of entirely new catalyst manufacturing methods?

Michael Bowker answered: I suspect that in 50 year's time, catalysts manufactured on the large scale may be produced in very different ways from the present methods. They will take advantage of renewable energy on site for processing energy, and they will produce very little effluent. They may well be gas or liquid spray deposition reel-to-reel processes, which can reduce gas and liquid pollutant emission, technologies already in use today for other types of large volume production.

Cynthia Friend responded: Understanding how to make a material that will form active sites under reaction conditions could lead to paradigms about how to design materials that will, for example, stabilize a site using defects or strain in the material so that its function is optimized and retained.

Simon Kondrat replied: A tool that we are starting to use more frequently is the *in situ* characterisation of catalysts under operating conditions. This helps us understand how the catalyst changes with conditions and ultimately what the active site is. Also, an increasing range of elegant materials chemistry techniques allows us to make powdered catalysts, with well-controlled properties, that can bridge model catalysts and technical catalysts. These methodologies can be used to drive catalyst discovery when combined with relevant characterisation, theory and understanding of the reaction mechanism.

Nicola Collis commented: I would like to respond to a general point about how much attention researchers pay to any impurities in the precursors used for catalyst preparation. In industrial gas phase catalysts, we certainly do pay attention to impurities and the nature of the catalyst precursors. There are several known catalyst poisons that are commonly found in precursors, such as Na, K, and Cl. We have to be extremely careful in selecting ingredients for the synthesis to avoid high levels of such impurities. We often find that literature methods (even those specifically aimed at gas phase oxidation reactions) are unsuitable for large scale catalyst production, so we spend a considerable amount of effort translating literature methods into something that is industrially sustainable.

Avelino Corma replied: I agree with the comment and I strongly believe that closer collaboration between industry and academia can help us to achieve faster catalyst optimization and scale up.

Bert Weckhuysen said: Related to the theme of this Faraday Discussion entitled "Designing New Heterogeneous Catalysts", I would like to refer to an editorial, which just came online in Angewandte Chemie, International Edition. The very nice and instructive editorial was written by Avelino Corma, who is also attending this conference, and has the title "Heterogeneous Catalysis: Understanding for Designing, and Designing for Applications".[1] This is obviously recommended reading material for the participants of this symposium.

1 A. Corma, *Angew. Chem., Int. Ed.*, 2016, **55**, 6112.

Mark Howard opened the discussion of the paper by James Anderson: The hydrogenation experiments are carried out with a very low partial pressure of acetylene and hydrogen with the balance being nitrogen and this is necessary in order to carry out the selectivity measurements. However, the industrial application has an overwhelming excess of ethylene present. What are your insights on how to translate these results to the industrial conditions?

James Anderson replied: We have used relatively simple test conditions (1 Bar total pressure, and an absence of ethylene from the feed) as they provide a straightforward method for comparing samples and providing insight into selectivity trends for the selective hydrogenation of acetylene. In addition to these simple tests, we have conducted a large number of experiments over the years that are more akin to industrial conditions and, in general, we find good agreement between samples that perform well under these conditions and under the more simple test conditions of low pressure/absence of ethylene in the feed.

James Earley enquired: You say that all of your catalytic testing was performed at 100% acetylene conversion, is this also 100% H_2 conversion? If not, this may explain the difference in selectivities of the different catalysts as the relative surface coverage of H_2 would be changing. It would be interesting to change the H_2 ratio of the different catalysts to see how this affects the selectivity. It would also be interesting to look at the conversion *vs.* selectivity plots for these catalysts. The reason being, at less than 100% conversion, the selectivities of the different catalysts may be similar. It may be that the difference in the selectivity that you observe under the conditions tested is due to a difference in the activity, which may directly correlate to the relative Pd weight loadings of the catalysts, which are decreasing with the more Au that is added to the catalyst. If Au is acting as a selectivity modifier and it is not a Pd weight loading effect, do you think that this is geometric or electronic?

James Anderson replied: The catalyst testing was performed at a H_2 : acetylene ratio of 1.8 : 1 and also at 3.6 : 1. We assume, but did not confirm, that we still have unconverted hydrogen in the outlet stream. A comparison of the selectivities as a function of these two hydrogen ratios is shown in Fig. 9 (DOI: 10.1039/C5FD00188A). The higher hydrogen ratio gives a slight increase to ethane selectivity, mainly as a consequence of the reduced oligomer selectivity. Yes, we agree that plots of selectivity *vs.* conversion are a useful way of comparing catalyst performance. At less than 100% acetylene conversion, it is possible that there is a greater degree of similarity in the selectivity patterns, although clearly the real challenge is to obtain a high selectivity as close as possible to 100% acetylene conversion and hence the selection of these criteria for comparison in this study. We think that *e.g.* increasing the space velocity but avoiding a drop in conversion below 100% is unlikely to improve the selectivity for the higher loaded samples, while perhaps reducing the space velocity for the lower loaded samples may have a negative effect on the selectivity. As always, it is difficult to distinguish between electronic and geometric effects, but we feel that the role of Au (at least in the low loaded sample) is structural, it creates a core, and prevents the formation of 3D Pd structures, which will otherwise lead to hydride availability. It also creates a surface on which the size of the Pd ensembles is limited, which we also believe from our work on diluted Pd materials[1] to be an advantage in producing selective catalysts.

1 A. J. McCue, C. J. McRitchie, A. M. Shepherd and J. A. Anderson, *J. Catal.*, 2014, **319**, 127.

Ewa Nowicka commented: As discussed earlier, the role of the stabilizing agent is very important to ensure that the nanoparticles do not agglomerate. However, in this case, the PVP, which was used as a stabilizing agent, was not fully removed from the catalyst surface, and therefore it may be that during the CO-DRIFT analysis, CO adsorption did not take place on certain metal sites due to the presence of PVP. Is that valid? In terms of the active sites, does the metal–support interface play any role in this reaction? Would it be possible to replace PVP with PVA in order to ensure full removal of the capping agent (in hot water, with the temperature below 120 °C) and thus perform CO-DRIFT analysis on a clean surface as we reported in a previous publication?[1]

1 G. M. Lari, E. Nowicka, D. J. Morgan, S. A. Kondrat and G. J. Hutchings, *Phys. Chem. Chem. Phys.*, 2015, **17**, 23236.

James Anderson answered: FTIR analysis of the as-prepared materials does suggest (bands due to CH stretching modes) retention of some PVP. When we compare CO adsorbed on the as-prepared samples (Fig. 10, DOI: 10.1039/C5FD00188A) with samples where we had treated (hydrolysis, photolysis) to remove the PVP (Fig. 11, DOI: 10.1039/C5FD00188A), additional types of sites are revealed, which we assign to edge sites on Pd (1970 cm^{-1}) and edge sites on Au (2110 cm^{-1}). However, it is likely that some PVP is still present after these treatments. Catalytic data (Table 2, DOI: 10.1039/C5FD00188A), however, suggests that removing PVP from these edge sites did not have a major influence on the selectivity, suggesting that they are either inactive or have a low TOF or that only a very small number of additional sites have been made available. The use of a stabilising agent that is more susceptible to hydrolysis, such as PVA, may increase the number of sites liberated following hot water treatment, but assuming that they occupy similar sites, then it might be assumed that similarly, a limited influence on the reaction might ensue. We have not attempted a synthesis involving PVA in the place of PVP. In the work presented, we followed the synthetic methodology described by Wilson *et al.*[1]

We are not of the opinion that the metal–support interface is influential and there is limited support for such a site in the literature in terms of the acetylene hydrogenation reaction. Please note that this preparation method, which prepares the metal particles followed by deposition on the support, is unlikely to create strong interactions between the metal and support, and this is consistent with the fact that we see no alignment of the crystal lattices of the metal and support in the TEM image.

1 A. R. Wilson, K. Sun, M. Chi, R. M. White, J. M. Lebeau, H. H. Lamb and B. J. Wiley, *J. Phys. Chem. C*, 2013, **117**, 17557.

Graham Hutchings asked: Question 1 – Au is selective but not active. Pd is active but less selective. When combined, a synergistic effect is observed. Can the preparation method be used to generate a particle with a Pd core and a Au shell as this may have an improved selectivity?

Question 2 – Often methods by which bimetallic nanoparticles are made lead to nanoparticle variations with the particle size. Is this something you have considered? Also, perhaps you could make a series of catalysts with nanoparticles with different mean particle size diameters so as to explore any effect of the particle size on the selectivity.

James Anderson replied: Yes, it is possible to prepare samples where Pd nanoparticle cores are used to seed other metals and this has been employed to yield *e.g.* core Pd/shell Ag catalysts,[1] which are selective in the semihydrogenation of alkynes (liquid phase). We have not attempted this here as our objective was to prepare thin layers of Pd, which, at least for single layers of Pd, might avoid hydride formation and subsequently provide a high selectivity to ethylene (it is widely believed that the formation of Pd hydride is responsible for over hydrogenation). We assume that the Pd cores with Au or Ag shells would not prevent hydride formation.

For the reasons in my response to your first question, we have tried to ensure that we have prepared particles that are reasonably uniform in terms of particle size distribution (Fig. 8, DOI: 10.1039/C5FD00188A) and that the only variable we are introducing is the thickness of the Pd shell. This means that, in principle, we avoid any size effects that might influence the selectivity. However, your suggestion regarding making samples of varying sizes but a fixed shell thickness is worth following up. We have not attempted this to date.

1 T. Mitsudome, T. Urayama, K. Yamazaki, Y. Maehara, J. Yamasaki, K. Gohara, Z. Maeno, T. Mizugaki, K. Jitsukawa and K. Kaneda, *ACS Catal.*, 2016, **6**, 666.

Miron V. Landau said: If I understood correctly, the main target of the preparation strategy selected in your work for the synthesis of gold nanoparticles modified with palladium atoms was the control of the relative exposure of gold and palladium atoms on the final gold nanoparticle surface covered with an incomplete palladium monolayer. In this strategy the gold nanoparticles were prepared first, with the organic capping agent stabilizer strongly adsorbed at the surface, followed by deposition of palladium. It looks like, in this case, there is a limitation of a very low palladium loading above which undesired separate palladium nanoparticles are formed. A reason for this could be that a large part of the surface of gold nanoparticles is covered by organics, leaving a very low capacity for palladium atoms. Did you try to use another preparation strategy – the co-formation of binary gold–palladium nanoparticles in one step? This would increase the opportunity to control the relative exposure of Au and Pd atoms on the nanoparticle surface over a wide range.

James Anderson answered: The objective was to use a Au core of uniform size and then deposit Pd to produce shells of Pd of increasing layer thickness over the Au. The gold particles are prepared first followed by the attempt to coat these in palladium. The capping agent/stabiliser (PVP) is not added until later and so a lack of available Au atoms due to blocking by PVP cannot be the reason why we have formed separate Pd particles. At this stage, we have not attempted other strategies to prepare similar types of structures, including the single step procedure you suggest. We would imagine though that such a procedure would produce particles where both types of atoms are included in the bulk and surface, which would then not be suitable for testing our hypothesis.

Mark Howard commented: You attribute the higher selectivity to ethylene at a higher temperature for the 0.2 Pd(Au) system to an increased availability of hydrogen, and therefore reduced oligomerisation. You also comment that adsorption of ethylene onto Au has a significant energy barrier, which presumably means that its desorption from the surface will also have a significant energy barrier. Could increased rates of desorption of ethylene from the catalyst surface at higher temperatures be an explanation or contributor to lower oligomers?

James Anderson responded: There are probably a number of factors that may contribute to the interchange between ethylene and oligomer selectivity as a function of reaction temperature (Fig. 9a, DOI: 10.1039/C5FD00188A). However, we believe that a lower acetylene coverage and increased rates of hydrogenation of the acetylene over the gold due to enhanced hydrogen spillover rates are the dominant factors.

Michael Bowker said: The selectivity in the table is not really a fair comparison, is Pd at 100% conversion more selective than Au at 100% conversion? Do you know what the surface looks like *after* the reaction? In that respect, have you done CO adsorption on a used catalyst and does it look different from pre-reaction?

James Anderson replied: Yes, I fully agree. All of the Pd containing samples were pushed to 100% conversion to determine the selectivity under conditions where all the acetylene was converted. Au is not very active and so it is not so easy to attain such conditions even if the space velocity is reduced considerably. I do agree, however, that such a comparison is not appropriate. We have not, as yet, performed any post reaction analysis on the samples. Please note, however, that the highest reaction temperature seen by the samples is 363 K and so we do not expect to see any significant migration of the Pd or Au to/from the surface under these conditions.

Elad Gross asked: Did you observe any reconstruction of the Au–Pd nanoparticles into a more thermodynamically stable structure following their exposure to the reaction conditions?

James Anderson answered: As in my response to Michael Bowker, we have not performed any post-reaction analysis of the samples as yet. However, from our work on bimetallics over the years, we would not envisage any substantive migration of Au or Pd during the reaction given that the maximum temperature experienced by the samples was 363 K.

Ram Tiruvalam said: Sol immobilization techniques, while good at making core–shell particles, have other problems such as secondary nucleation of the shell material and compositional variations with particle size and the presence of a ligand *etc.* There are some other ways to make nanoparticles, such as the use of cluster sources and electroless deposition, which their proponants claim is superior. What do you think is the best way to make and study core–shell nanoparticles? What synthesis procedure will enable us to precisely put a monolayer and bilayer of one metal on another and study them in a test reactor (in other words, is sol immobilization still our best bet)?

James Anderson replied: From the work we present here (DOI: 10.1039/C5FD00188A), we think that making uniform sized particles and the issue of the (retained) ligands/stabilising agents are less of an issue, whereas the problem of secondary nucleation of the shell material really is significant. These are now, of course, fairly routine synthesis methods but the challenges of making particles of uniform composition and of being able to ensure that this is what we have made are still problematic. The advantage of the sol immobilisation method is that it is fairly accessible to most laboratories and we can then transfer them to test (hopefully) these "model" catalysts under realistic conditions.

Christian Reece opened the discussion of the paper by Joshua Makepeace: Relating to eqn (7) in the paper (DOI: 10.1039/C5FD00179J), you state that you can quantitatively calculate the activation energy using a two parameter curve that uses A and E_A. In Table 1 (DOI: 10.1039/C5FD00179J) you state the calculated activation energies, which appear to change with changing species. These data are deceiving as the two parameters you use to fit your curve, A and E_A, are correlated, meaning any change in E_A can be compensated by a change in A. This means you should either combine the two variables into one constant or you should publish the relationship between A and E_A, as, at the moment, it would seem that when you go from $Li_2Ca(NH)_2$ to $Li_2Mg(NH)_2$ the activation energy E_{A2} changes. In fact, when you look at the relationship between E_A and A, you find that the relationship is constant (*i.e.* as A_2 doubles so does E_{A2}), which would indicate that the same process is occurring, not two different processes.

Joshua Makepeace answered: The least-squares fitting of the ammonia decomposition data presented in Fig. 1 of the paper (DOI: 10.1039/C5FD00179J), and the refined values of A and E_A and their estimated uncertainties listed in Table 1 (DOI: 10.1039/C5FD00179J) correspond to the functions described by eqn (7) and (8). Least-squares minimisation was performed using a simplex algorithm. The A and E_A parameters in eqn (7) are quite different from one another with A and E_A having dimensions of entropy (kJ mol^{-1} K^{-1}) and energy (kJ mol^{-1}), respectively:

$$C_i = 1 - \exp(-\exp(A/R - E_A/RT))$$

where $C_i = (1 - \text{ammonia fraction})$. Consider the temperature, $T_e = E_A/A$. At this temperature:

$$C_i = 1 - \exp(-\exp(A/R - E_A/RT)) = 1 - \exp(-\exp(0)) = 1 - \exp(-1) = 0.6231.$$

Thus, the ratio E_A/A is determined graphically when $C_i = 0.6231$. Differentiating C_i with respect to temperature gives:

$$\partial C_i/\partial T = (E_A/RT^2)\exp(A/R - E_A/RT)\exp(-\exp(A/R - E_A/RT))$$

which at $T = T_e$ gives $\partial C_i/\partial T|_{(T = T_e)} = E_A/(\exp RT_e^2)$. Thus, given that T_e is fixed for a given data set, the rate of change of the conversion at $T = T_e$ is directly proportional to E_A, the activation energy of the process. The higher the activation energy, the steeper the slope at $T = T_e$. From this analysis, it is clear that A and E_A have very different impacts on the functional fitting of the ammonia conversion.

Luke Parker said: Fig. 1 (DOI: 10.1039/C5FD00179J) shows that the blank Ni-plated reactor can achieve a conversion close to 100% at 600 °C and that the lithium imide catalysts tested achieve similar conversions but at lower temperatures. You also mention in the discussion that a recent study suggested that the transition metal may influence the decomposition pathway of lithium amide. Is the presence of a transition metal essential for lithium imide catalysts to be active? Or do you believe that, if it were possible to do the reaction in the absence of other metals, these catalysts would still exhibit a similar activity?

Joshua Makepeace responded: As was outlined in the paper (DOI: 10.1039/C5FD00179J), the choice of the steel/nickel-plated steel reactors for this work was made on the basis of concerns regarding the reactivity of the metal amide and imide materials with quartz surfaces. The isotope exchange reactions performed in this and other studies[1] indicate that the bulk of the metal imide/amide interacts with the ammonia present in the reactor. Our previous work on sodium amide[2] reported a significant kinetic isotope effect for the ammonia decomposition reaction when the amide was present, but no such effect for the blank nickel-plated reactor. These results indicate that the decomposition of ammonia is predominantly by the metal imide/amide catalyst, rather than at the interface of the reactor wall and the catalyst powder.

1 J. W. Makepeace, T. J. Wood, H. M. A. Hunter, M. O. Jones and W. I. F. David, *Chem. Sci.*, 2015, **6**, 3805.
2 T. J. Wood, J. W. Makepeace, H. M. A. Hunter, M. O. Jones and W. I. F. David, *Phys. Chem. Chem. Phys.*, 2015, **17**, 22999.

Philip Landon enquired: If you are intending to make a practical and useful device, would you use stainless steel to make the container?

Joshua Makepeace replied: In designing a practical ammonia decomposition reactor, care would need to be taken in the choice of material for the construction in order to minimise the effects of stress corrosion cracking or other degradation phenomena (*e.g.* embrittlement from hydrogen exposure). We are currently investigating a number of alloys for their suitability for a practical device.

Michael Bowker commented: The "blank" experiment shows that the nickel-plated tube is surprisingly active, so it would be useful to put a large area nickel catalyst in there to compare with your own catalysts, since the tube wall will have a very small surface area. This Ni catalyst should have a reasonably large metal area for a fair comparison to be made.

Joshua Makepeace answered: Surface nitriding of the nickel-plated reactor will result in a larger surface area, and therefore a higher ammonia decomposition activity, than the equivalent flat metal surface. In our previously published work,[1,2] we compared the performance of sodium amide and lithium imide–amide with supported nickel (66 ± 5% nickel, Alfa Aesar) and ruthenium (5% ruthenium, Alfa Aesar) catalysts. Nitrogen adsorption isotherm measurements on the Ni- and Ru-containing catalysts indicate that the surface areas are of the order of 130 m^2 g^{-1} and 190 m^2 g^{-1}, respectively. The results of these experiments showed that the metal amide and imide catalysts gave superior ammonia decomposition performances compared to the nickel catalyst across the range of test conditions, and superior high-temperature performances compared to the ruthenium catalyst.

1 W. I. F. David, J. W. Makepeace, S. K. Callear, H. M. A. Hunter, J. D. Taylor, T. J. Wood and M. O. Jones, *J. Am. Chem. Soc.*, 2014, **136**, 13082.
2 J. W. Makepeace, T. J. Wood, H. M. A. Hunter, M. O. Jones and W. I. F. David, *Chem. Sci.*, 2015, **6**, 3805.

Simon Kondrat opened the discussion of the paper by Miron V. Landau: From the TEM image of the final state catalyst, the particles were dispersed in a graphitic matrix. What is the nature of the carbon and how important is it in dictating the activity or selectivity? The carbon layer that you create – can you tune that under activation? Also, do you need an interface between the carbide and oxide nanoparticles?

Miron V. Landau replied: The carbon between the Fe oxide and carbide nanoparticles in the catalytic material is graphitic with highly disordered layers (wide XRD peaks) and nonporous – its excess decreases the surface area. The formation of the carbon layer between the Fe-based nanoparticles is a sequence of proceeding the Fe-carbide oxidation (CO_2, H_2O) and Fe-oxide carburization (H_2/CO) with the production of carbon as a by-product.[1] The content of carbon may be controlled under activation, *i.e.* Fe-oxide may be completely carburized with H_2/CO in the activation step. Under CO_2 hydrogenation conditions, it became oxidized by CO and H_2O, producing a large amount of carbon. In this case, Fe-based nanoparticles in the steady state catalyst became embedded in a bulk carbon matrix. Insertion of Al in the structure of magnetite does not allow deep carburization of iron. The Fe–Al–O spinel is partially carburized by CO/H_2, reaching a steady state with the Fe-oxide/Fe-carbide ratio dictated by thermodynamics. This excludes over-carburization and subsequent formation of a large amount of carbon, increasing the surface area and activity of the catalyst.[2] From the beginning of the insertion of the concept of catalyst polyfunctionality in heterogeneous catalysis,[3] it was assumed that for different active sites at a catalyst surface, the rule "the closer the better" should work. This tells us that the Fe-carbide–Fe-oxide contact interface is sufficient for CO_2 hydrogenation. However, it was recently demonstrated that even classical "metal–acid" catalysts in hydrocarbon hydrocracking catalytic materials where platinum nanoparticles are deposited at the surface of an alumina binder and not in/on zeolite crystals work better.[4] The carbon layer between Fe-carbide and Fe-oxide nanoparticles may play a role in the spillover of intermediates, which should be investigated in future studies.

1 E. de Smit and B. M. Weckhuysen, *Chem. Soc. Rev.*, 2008, **37**, 2758.
2 M. V. Landau, M. Herskowitz and R. Vidruk, *ChemSusChem*, 2014, **7**, 785.
3 P. B.Weisz, *Adv. Catal.*, 1962, **13**, 137.
4 J. Zecevic, G.Vanbutsele, K. P. de Jong and J. A. Martens, *Nature*, 2015, **528**, 245.

Andrés García-Trenco commented: In the set of reactions presented in the paper (DOI: 10.1039/C5FD00174A), the Reverse Water Gas Shift Reaction (RWGSR) is first, leading to the formation of CO and H_2O, and then the Fischer Tropsch synthesis (FTS), producing hydrocarbons and H_2O. Under high conversions of CO_2, a high production of H_2O is expected, being one of the reaction products of the RWGSR. Typically, it is observed that the RWGSR is fast, and thus it is controlled by its thermodynamic equilibrium. Could this H_2O formed under high CO_2 conversion inhibit the RWGSR, reducing the efficiency of the process?

Miron V. Landau answered: It is well-established in most relevant kinetic studies[1,2] that water is a strong inhibitor for both the RWGS and FTS reactions – its partial pressure exists in a denominator of expressions for differential reaction rates with significant coefficients. The RWGSR is strongly thermodynamically limited at temperatures where Fe-carbide phases may efficiently catalyze the FTS. However, consumption (elimination) of CO in the FTS reaction "opens" the thermodynamic system, reducing the thermodynamic limitations for the tandem RWGS–FTS reaction. Therefore, only one factor limits the reactions rate – inhibition by water produced in this process, where the partial pressure rises from the top to the bottom of the plug flow reactor. This is clearly shown in the presented work (Table 4, DOI: 10.1039/C5FD00174A), where carrying out CO_2 hydrogenation in three reactors in series with intermediate separation of water permitted about a two-fold increase in CO_2 conversion compared with that obtained in one reactor loaded with the same catalyst at the same temperature, pressure, H_2/CO_2 ratio and total residence time.

1 T. Riedel, G. Schaub, K. W. Jun and K. W. Lee, *Ind. Eng. Chem. Res.*, 2001, **40**, 1355.
2 N. Meiri, Y. Dinburg, M. Amoyal, V. Koukouliev, R. Vidruk-Nehemya, M. V. Landau and M. Herskowitz, *Faraday Discuss.*, 2015, **183**, 197.

Nico Fischer asked: Could you please comment on the catalyst phase composition after the activation step? You also seem to activate your catalyst under different conditions. Can you please comment on this?

Regarding Table 3 (DOI: 10.1039/C5FD00174A), the composition of the spent catalyst originating from the K/Fe–Cu–delafossite seems peculiar. It has a very high degree of oxidation (40% Fe_3O_4) even though Cu is present and could act as a reduction promoter. Also, the total composition adds up to 135%. Can you comment on this?

Miron V. Landau replied: The activation of the selected Fe-based oxide catalyst precursors was done in such a way depending on their ability for iron reduction with hydrogen. Materials that may be reduced at temperatures ≤300 °C were directly carburized in H_2/CO flow at 300 °C without intensive formation of carbon. Materials that required temperatures >350 °C for iron reduction, like Fe-Ba-hexaaluminate, were reduced in pure hydrogen at 900 °C, converting some of the iron to the metallic state followed by carburization of only this reduced iron in H_2/CO flow at 300 °C. The second case gives better control for the phase composition of steady state catalysts since under CO_2 hydrogenation conditions, the phase composition of activated hardly reducible materials does not change. In Table 3 (DOI: 10.1039/C5FD00174A), the phase composition of the catalyst obtained from the K/Fe–Cu–delafossite precursor (wt% of phases identified by XRD) was given as particle sizes by mistake. The right composition is: Fe_3O_4 – 31%, Cu – 32%, Fe_5C_2 – 28% and CuO – 9%, *i.e.* the values given in the particle size column.

Amy L. Miller said: In the text of the paper, Fig. 8 (DOI: 10.1039/C5FD00174A) was described as showing a linear dependence for the precursors containing 4 wt % potassium. The point that has the highest surface ratio of $Fe_{carbide}$ to Fe_{oxide} was described in your talk as being an outlier, where the deviation from the linear trend was rationalised as arising from a higher potassium concentration. However, if this outlier is included, the shape of all the points would fit a parabola very well. This apparent parabola exhibits a maximum at a surface ratio of about 0.35. Would it be possible to add one or more points with a surface ratio of 0.45 or greater and a potassium concentration of 4 wt%? This would allow the relationship of the reaction rates with a high surface ratio to be found, keeping the potassium concentration as a control variable. This could verify the hypothesis that the K–ferrite catalyst only deviates from the linear trend due to the higher potassium concentration.

Miron V. Landau answered: The potassium ferrite was selected for the present investigation since we hoped to reach a high $Fe_{carbide}/Fe_{oxide}$ surface ratio with this precursor. It is known that a potassium promoter facilitates carburization of Fe-oxide phases and stabilizes Fe-carbides. Potassium ferrite containing potassium as a component of the crystalline phase atomically distributed in the crystal bulk in contrast to a potassium promoter deposited at the surface of the Fe-oxide precursors is expected to yield the highest $Fe_{carbide}/Fe_{oxide}$ surface ratio. However, the activity of the synthesized steady state catalyst in FTS was strongly depressed relative to the RWGSR, yielding a high misbalance of the efficiencies of the catalyst functions. It looks like this happened due to the existence of an extra amount of potassium in this steady state catalyst increasing the strong surface basicity, which favors strong adsorption of CO_2 and catalyst deactivation.[1] This makes potassium ferrite an outlier. I think that there is no reason to expect an extremal trend of the extent of the catalyst functions balancing (RFTS + RM)/RWGSR with increasing the $Fe_{carbide}/Fe_{oxide}$ surface ratio. With increasing the relative contribution of "carbide" active sites, this extent should reach unity, so that all CO produced in the first step will be converted to hydrocarbons.

1 N. Meiri, Y. Dinburg, M. Amoyal, V. Koukouliev, R. Vidruk-Nehemya, M. V. Landau and M. Herskowitz, *Faraday Discuss.*, 2015, **183**, 197.

Mark Howard commented: If the reaction mechanism is constrained to proceed through CO, as shown in Scheme 1 (DOI: 10.1039/C5FD00174A), and the ratio plotted in Fig. 8 (DOI: 10.1039/C5FD00174A) is an absolute rate ratio, rather than a ratio of rate coefficients, then presumably this ratio cannot exceed 1. Could the lower rate ratio for the sample at $Fe_{carbide}/Fe_{oxide} \sim 0.53$ be due to insufficient RWGS activity, causing the FTS and methanation reactions to be starved of CO? In other words, is this a type of volcano plot?

Miron V. Landau responded: With large amounts of potassium, like in K-ferrite, the activity of the Fe-oxide phases in the RWGSR may be inhibited due to stabilization of the surface oxygen vacancies against oxidative filling with oxygen atoms belonging to CO_2 molecules. However, it should result in an increased (RFTS + RM)/RWGS rate ratio, which was not the case in our experiments. It looks like this is not a type of volcano plot, potassium ferrite ($Fe_{carbide}/Fe_{oxide} \sim 0.53$) is an outlier here due to strong inhibition of the Fe-carbide phase with an excess of potassium.

Michael Claeys communicated: You describe this reaction as a consecutive reaction with the formation of CO *via* RWGS followed by FTS conversion to products. Does this mean that more than one catalytic site is required for the reactions, or can these reactions occur on the same site? What do you think are the sites that play a role here? If these are different sites, should one perhaps make an effort to introduce a (R)WGS functionality in order to speed things up?

Miron V. Landau communicated in reply: The results presented in the manuscript (DOI: 10.1039/C5FD00174A) show that the two states of iron existing at the steady state catalyst surface – cationic iron (Fe 2+, 3+) belonging to the Fe-oxide phases and near metallic iron (Fe δ+) at the surface of the Fe-carbide phases – form the active sites involved in the catalytic cycles of the RWGS and FTS reactions, respectively. In the first case, a well established RWGS mechanism, including the formation of oxygen vacancies and surface carbonate intermediates, assuming redox cycles of Fe-sites (surface ions), may be proposed. In the second case, CO is adsorbed at the Fe-carbide surface by the formation of coordinative carbonyl bonds, followed by conversion (hydrogenation) of surface carbonyls to carbine intermediates. The whole process requires both functionalities. The RWGS reaction is strongly limited thermodynamically. In order to get reasonable conversions, the catalyst should have two functions – RWGS and FTS. The consecutive FTS reaction using the RWGS product (CO) "opens" the thermodynamic system that allows shifting of the RWGS to the right, getting high conversions in the tandem reaction. Therefore, to develop an efficient catalyst for CO_2 hydrogenation to hydrocarbon fuels in order to speed things up, one should make efforts to combine both types of functionalities in one catalytic material with maximal efficiency and balance the rates of the two consecutive reactions.

Graham Hutchings opened the discussion of the paper by Bruce Gates: In your concluding remarks, you mentioned that little had been said during the discussion about amorphous materials as catalysts. Perhaps this is because they are very difficult to study and the really elegant techniques at the disposal of scientists today are really designed for crystalline materials. I agree that we should focus more on amorphous materials. For example, with vanadium phosphates most people have studied crystalline vanadium phosphates, such as vanadyl pyrophosphate. Yet it has been known for about 40 years that the surface is very different and has an amorphous nature. Indeed, when a bulk amorphous vanadium phosphate was synthesised and used as a catalyst for butane oxidation, it was found to have a higher intrinsic activity.[1] How can we encourage the field to move on and to address the nature of these amorphous surface layers?

1 G. J. Hutchings, J. A. Lopez-Sanchez, J. K. Bartley, J. M. Webster, A. Burrows, C. J. Kiely, A. F. Carley, C. Rhodes, M. Hävecker, A. Knop-Gericke, R. W. Mayer, R, Schlögl, J. C. Volta and M. Poliakoff, *J. Catal.*, 2002, **208**, 197.

Robbie Burch said: Professor Gates has pointed out some missing topics in our discussions about new catalyst design and it is also clear that virtually all the presentations and discussion have been academic in nature. Would you like to say a little bit about new catalyst design from an industrial perspective?

Mark Howard replied: I would say that catalyst "design" in industry is still largely based on principles of "function", as Prof. Gates described, usually applied in a somewhat iterative fashion. Although this is now routinely supported by good scientific and analytical input, there is usually a good measure of trial and error as well. We often have a good idea of at least some of the desired design parameters, optimum metal particle size, for example, but delivering these through the often complex sequence of practical manufacturing steps, pre-treatments and start-up remains rather iterative, and there are usually plenty of empirical parameters in the mix too.

This is partly the origin of my question earlier about "levers". Practical catalysis is not just about knowing the design parameters, one also has to control these parameters through catalyst manufacturing, start-up and operation – and have practical "levers" available for doing this. In many cases, translating the sort of work we have heard about at this meeting into practical "levers that we can pull" is still a huge challenge with a very large gap – it is likely to require a lot of engagement between industry and academia. There have been some excellent contributions and participants from industry at this meeting, but too few in number.

Bert Weckhuysen commented: An obvious, although not always well remembered, example in which the catalyst literally breaks up during catalysis and in which the active sites are *in situ* generated during this process of catalyst degradation is the olefin polymerization catalyst. A lot of changes are taking place on the surface, and new sites are created as the polymerization continues. May be you can design such a polymerization catalyst as a single site system, but while it is working more and other sites are definitely generated. This system provides another angle to the concept of single site (polymerization) catalysis.

Faraday Discussions

PAPER

The preparation of large surface area lanthanum based perovskite supports for AuPt nanoparticles: tuning the glycerol oxidation reaction pathway by switching the perovskite B site

Christopher D. Evans,[a] Simon A. Kondrat,*[a] Paul J. Smith,[a] Troy D. Manning,[b] Peter J. Miedziak,[a] Gemma L. Brett,[a] Robert D. Armstrong,[a] Jonathan K. Bartley,[a] Stuart H. Taylor,[a] Matthew J. Rosseinsky[b] and Graham J. Hutchings[b]

Received 20th November 2015, Accepted 4th December 2015
DOI: 10.1039/c5fd00187k

Gold and gold alloys, in the form of supported nanoparticles, have been shown over the last three decades to be highly effective oxidation catalysts. Mixed metal oxide perovskites, with their high structural tolerance, are ideal for investigating how changes in the chemical composition of supports affect the catalysts' properties, while retaining similar surface areas, morphologies and metal co-ordinations. However, a significant disadvantage of using perovskites as supports is their high crystallinity and small surface area. We report the use of a supercritical carbon dioxide anti-solvent precipitation methodology to prepare large surface area lanthanum based perovskites, making the deposition of 1 wt% AuPt nanoparticles feasible. These catalysts were used for the selective oxidation of glycerol. By changing the elemental composition of the perovskite B site, we dramatically altered the reaction pathway between a sequential oxidation route to glyceric or tartronic acid and a dehydration reaction pathway to lactic acid. Selectivity profiles were correlated to reported oxygen adsorption capacities of the perovskite supports and also to changes in the AuPt nanoparticle morphologies. Extended time on line analysis using the best oxidation catalyst (AuPt/$LaMnO_3$) produced an exceptionally high tartronic acid yield. $LaMnO_3$ produced from alternative preparation methods was found to have lower activities, but gave comparable selectivity profiles to that produced using the supercritical carbon dioxide anti-solvent precipitation methodology.

[a]*Cardiff Catalysis Institute, School of Chemistry, Cardiff University, Main Building, Park Place, Cardiff, CF10 3AT, UK. E-mail: kondratsa@cardiff.ac.uk*
[b]*Department of Chemistry, University of Liverpool, Crown Street, Liverpool, L69 7ZD, UK*

Introduction

The oxidation of alcohols provides a route to carboxylic acids, which are components in many chemical syntheses, including those in the fine chemical and pharmaceutical industries. To achieve this transformation in the liquid phase, low pressures, low temperatures and the use of molecular oxygen as the oxidant are industrially and environmentally advantageous. The oxidation of glycerol, in particular, has attracted significant attention due to its high functionality and its availability from the *trans*-esterification of triglycerides, as a by-product of the biodiesel manufacturing process.[1]

Glycerol can be oxidised with heterogeneous catalysts to produce a range of molecules with applications in polymers, building, cosmetics, food additives, and organic syntheses.[2] Gold nanoparticles have been shown to be active for the oxidation of glycerol in the presence of a base, such as sodium hydroxide.[3,4] In these cases, the highest selectivity was to the C3 oxidation product, glyceric acid. Following this work, investigations into the precious metal present on the catalyst were undertaken. It was found that a synergistic effect was in operation when gold was alloyed with another metal, such as palladium[5] or platinum.[6]

The reaction mechanism (shown in Scheme 1) contains multiple steps with a variety of different possible products. The initial step of the oxidation of glycerol is the formation of glyceraldehyde, which is in equilibrium with dihydroxyacetone. In the presence of a catalyst and base, under oxidising conditions, glyceraldehyde has been shown to rapidly oxidise to glyceric acid, which can then be oxidised further.[7] Recently, further attention has been given to the transformation of glycerol to lactic acid, under oxidative conditions.[8,9] Lactic acid has many uses in the food industry and also it can be polymerised to poly-lactic acid, a biodegradable material.[10] This reaction pathway proceeds *via* the dehydration of

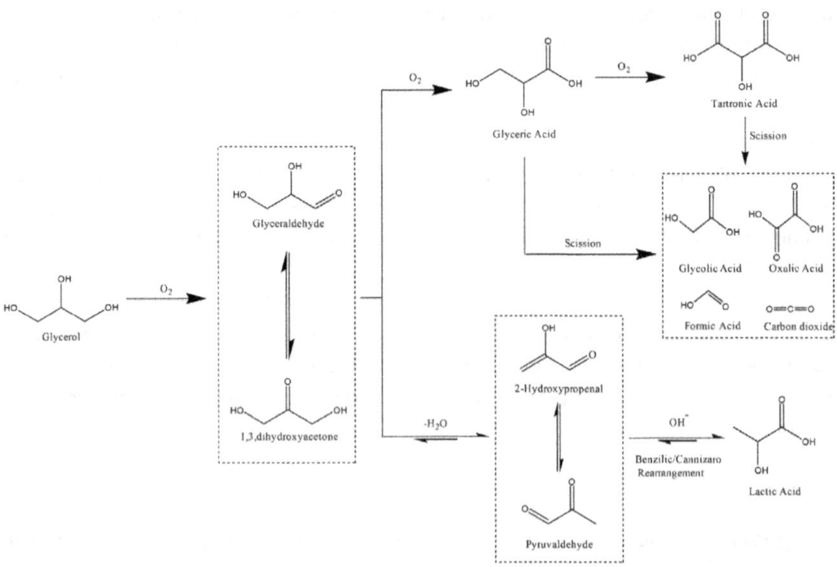

Scheme 1 Possible reaction pathways for glycerol oxidation.

glyceraldehyde or dihydroxyacetone to form pyruvaldehyde, which then rearranges into lactic acid.

The effect of a support on the oxidation of glycerol under basic conditions has been studied in detail.[11–13] Carbon supports have been shown to be more active than titania and iron oxide supports.[12] A study with Au/NiO and Au/$NiO_{1-x}(TiO_2)_x$ showed a very high activity with the NiO support, but a poor selectivity to any particular product.[14] Monometallic Au, Pd and Pt supported on activated carbon have been shown to be active for glycerol oxidation under base free conditions.[15] Further studies have shown that TiO_2,[16] $MgAl_2O_4$ and H-mordenite supported gold catalysts have activity for glycerol oxidation under base free conditions. Villa *et al.* studied the effect of the acid and base properties of a support on the activity and selectivity of Au catalysts for the base free oxidation of glycerol.[17] The study found that basic supports resulted in a high activity, but with the production of a large number of C_1 and C_2 scission products, while acid supports had a lower activity but with improved selectivity towards glyceraldehyde.

Evidently, the support structure has a significant impact on the activity and selectivity of Au and Au alloy catalysts for the oxidation of glycerol. Further study, by systematically altering a property of the support, would be desirable. Metal oxide and mixed metal oxide supports offer a huge range of different metal cations to change the properties, such as the acidity/basicity, metal-support interaction or oxygen adsorption capacity. A key issue with respect to such a study is that in many cases changing the metal cation results in a change in the support structure. This, in turn, results in significant variation in the surface area, surface species and morphologies.

Perovskites have the general formula ABO_3, where cation A is larger than cation B. An interesting aspect of these structures is the fact that the cations, A and B, can be varied, and in so doing, the intrinsic properties of these different perovskites can be tuned to achieve the most desirable characteristics, without affecting the crystal structure of the compound.[18] This would allow for a systematic study of the effect of different transition metal B sites on the activity and selectivity of a liquid phase oxidation reaction, such as glycerol oxidation. Studies on lanthanum based perovskites for the oxidation of propane and iso-butene have shown that the activity is highly dependent on the choice of B site element (Cr, Mn, Fe, Co or Ni).[19] Due to the isostructural nature of these $LaBO_3$ compounds, correlations could be drawn between the activity and B site electronic configuration. Unfortunately, traditionally prepared perovskite structure materials have been synthesised by precipitation methods, which yield small surface area powders in the range of 1–15 $m^2\ g^{-1}$ (ref. 20 and 21), which, in general, are not ideal for supporting metal nanoparticles.

The preparation of transition metal oxide[22] and mixed metal oxide catalysts[23] by a supercritical anti-solvent (SAS) process has been shown to produce large surface area, high activity materials for oxidation reactions. Materials prepared by this method have also been demonstrated to be excellent catalyst supports for precious metal nanoparticles for reactions such as benzyl alcohol oxidation and the direct synthesis of hydrogen peroxide.[24,25] Utilising this preparation methodology would allow for the preparation of large surface area perovskites. In this work, we investigate the use of perovskite supported precious metal alloy nanoparticles in order to tune the selectivity of the glycerol oxidation reaction.

Reaction conditions that facilitate the broadest number of reaction products, including lactic acid, were chosen.

Experimental

Perovskite support preparation

A range of $LaBO_3$ (B denotes Cr, Mn, Fe, Co or Ni) perovskites were prepared using the supercritical anti-solvent (SAS) precipitation method. A brief summary of the preparation method is given below, with a more detailed experimental method reported elsewhere.[22] Lanthanum(III) acetylacetonate hydrate (4 mg ml^{-1}) and one of the B element acetate salts (concentration varied to give the La : B molar ratios shown in Table 1) (Sigma Aldrich ≥99% Puriss) were dissolved in methanol (reagent grade, Fisher Scientific). SAS experiments were performed using apparatus manufactured by Separex. A technical diagram of the SAS apparatus is shown in Fig. 1. CO_2 (BOC) was pumped through the system (held at 130 bar, 40 °C) *via* the outer part of a co-axial nozzle at a rate of 12 kg h^{-1}. The metal salt solution was co-currently pumped through the inner nozzle using an Agilent HPLC pump at a rate of 4 ml min^{-1}. The resulting precipitate was recovered on a stainless steel frit, while the CO_2–solvent mixture passed down stream, where the pressure was decreased to separate the solvent and CO_2. The precipitation vessel has an internal volume of 1 L. Precipitation was carried out for 120 min followed by a purge of the system with CO_2 for 30 min under 130 bar and 40 °C. The system was then depressurised and the dry powder collected. The SAS precipitates were then calcined at 750 °C (with a ramp rate of 2 °C min^{-1}) for 4 h to produce the perovskite materials.

In addition to the SAS preparation method, $LaMnO_3$ was synthesised by mechanochemical milling of the individual oxides and flame spray pyrolysis. The milling procedure used a planetary ball mill (Retsch PM100). La_2O_3 and Mn_2O_3

Table 1 Metal salts used for the SAS precipitations and the physical properties of the resultant perovskites

Sample	B metal salt	Precursor solution La : B molar ratio	Precipitated La : B molar ratio[a]	Phase composition from XRD	Crystallite size (nm)	Surface area (g m^{-2})
$LaCrO_3$	Chromium(III) acetate[b]	1 : 1	1 : 1.06	$LaCrO_3$, trace La_2CrO_6[c]	6	52
$LaMnO_3$	Manganese(II) acetate tetrahydrate	1 : 1.2	1 : 1.03	$LaMnO_3$ (100%)	18	32
$LaFeO_3$	Iron(II) acetate	1 : 1.4	1 : 0.99	$LaFeO_3$ (85%), La_2O_3 (9%), Fe_2O_3 (6%)	21	26
$LaCoO_3$	Cobalt(II) acetate tetrahydrate	1 : 1.1	1 : 1.05	$LaCoO_3$ (90%), Co_3O_4 (10%)	22	22
$LaNiO_3$	Nickel(II) acetate tetrahydrate	1 : 1.1	1 : 1.07	$LaNiO_3$ (90%), La_2O_3 (8%), NiO (2%)	15	36

[a] Ratios calculated by MPAES. [b] Chromium acetate prepared in house from potassium chromate and sodium acetate. [c] Phase composition of La_2CrO_6 could not be quantified.

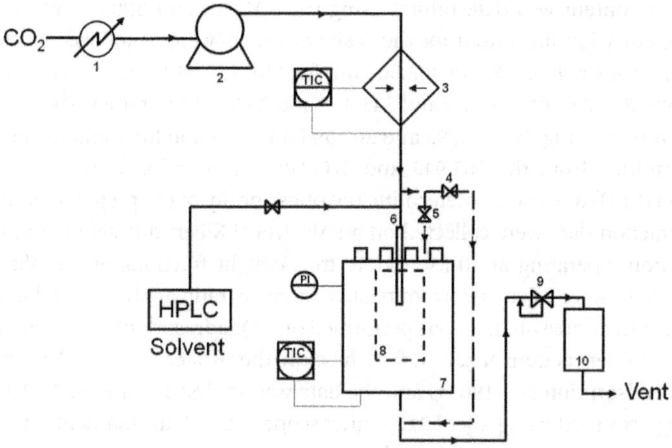

Fig. 1 Technical diagram of the Separex SAS equipment. (1) Chiller; (2) liquid pump; (3) heat exchanger; (4) and (5) by-pass valves, (6) co-axial nozzle for CO_2 and metal salt solution delivery; (7) precipitation vessel; (8) sample recovery vessel; (9) back pressure regulator and (10) separation vessel.

were added to a ZrO_2 milling vessel with six 15 mm ZrO_2 balls before being ground at 700 rpm for 16 h. The resulting dry powder was recovered and calcined in static air at 700 °C for 4 h.

Flame pyrolysis was performed using custom built equipment. Aqueous La/B nitrate solutions (0.1 M) were sprayed at a rate of 0.5 ml min^{-1} *via* a Sonozap ultrasonic nebuliser (2.8 W, 130 kHz) into a horizontally aligned propane (0.5 L min^{-1}) and oxygen (1.4 L min^{-1}) flame (0.082″ diameter stainless steel nozzle). Gas flows were controlled using mass flow controllers. The resulting powder was collected on a water cooled quartz plate 10 cm from the nozzle tip. A typical collection time was 10 minutes.

Addition of gold/platinum nanoparticles to the perovskite supports

Aqueous solutions of $HAuCl_4$ (Johnson Matthey) and H_2PtCl_6 (Johnson Matthey) were prepared at the desired concentrations. Polyvinyl alcohol (PVA, 1 wt% aqueous solution, Aldrich, MW = 10 kDa) was freshly prepared and used as the stabilizer. $NaBH_4$ (Sigma Aldrich, 0.1 M aqueous solution) was also freshly prepared and used as the reducing agent. To an aqueous mixture of $HAuCl_4$ and H_2PtCl_6 of the desired concentration (1 : 1 metal weight ratio, 1 wt% total metal in the final catalyst), the PVA solution was added (PVA/(Au + Pt) (w/w) = 0.65) with vigorous stirring for 2 min. $NaBH_4$ was then added rapidly such that the $NaBH_4$: total metal ratio (mol : mol) was 7.5. After 1 h of stirring, the mixture was filtered, washed with distilled water and dried at 120 °C for 16 h.

Catalyst characterisation

The La : B site elemental ratios of the SAS precipitated perovskites were determined by microwave plasma atomic emission spectroscopy (MP-AES) using an Agilent 4100 instrument. The precipitates were dissolved in aqua regia solutions and the La content was determined using the 394.910 and 398.852 nm emission lines. The emission lines used for the B site elements were as follows: 357.688 and 425.433 nm for Cr, 403.076 and 403.307 nm for Mn, 259.940 and 371.993 nm for Fe, 340.512 and 345.351 nm for Co, and 341.476 and 352.454 nm for Ni. The Au content was determined using the 242.795 and 267.595 nm emission lines and the Pt content was determined from the 265.945 and 270.240 nm emission lines. Powder X-ray diffraction (XRD) was used to determine the phase purity of the prepared perovskites. X-ray diffraction data were collected on a PANalytical X'Pert diffractometer, with Cu $K_{\alpha 1}$ radiation, operating at 40 kV and 40 mA. Weight fractions of the phases and crystallite sizes were calculated from relative intensity ratio analysis and the Scherrer equation. Surface area analysis was performed on a Quadrasorb BET. The catalyst was pre-treated under vacuum at 250 °C for 2 h before the surface area was determined by 5 point N_2 adsorption at −196 °C and the data was analysed using the BET method. TEM was performed using a Jeol 2100 microscope with a LaB_6 filament operating at 200 kV. Samples were prepared by dispersing the powder catalyst in ethanol and dropping the suspension onto a lacey carbon film over a 300 mesh copper grid. XPS was performed using a Kratos Axis Ultra DLD system with a monochromatic Al K_α X-ray source operating at 120 W. Data was collected in the hybrid mode of operation, using a combination of magnetic and electrostatic lenses, at pass energies of 40 and 160 eV for high resolution and survey spectra, respectively. NH_3-Temperature programmed desorption (TPD) was carried out using a Quantachrome Industries ChemBET TPR/TPD chemisorption analyser, fitted with a thermal conductivity detector (TCD). 100 mg of sample was pre-treated for 1 h at 130 °C (15 °C min^{-1}) in He (80 ml min^{-1}). Ammonia was adsorbed at room temperature for 20 min to ensure saturation. Physisorbed ammonia was then removed at 100 °C (1 h, 15 °C min^{-1}) in He (80 ml min^{-1}). Chemisorbed ammonia was subsequently desorbed by heating to 800 °C (15 °C min^{-1}) in a flow of He (80 ml min^{-1}) with the desorption monitored using a TCD, a current of 180 mV, and an attenuation of 1.

Glycerol oxidation testing and product analysis

Catalyst testing was performed using a 50 ml Radleys glass reactor. The aqueous glycerol (or glyceric acid) solution (0.3 M, containing NaOH (NaOH/glycerol ratio = 4, mol/mol)) was added into the reactor. The reactor was then heated to 80 °C prior to being purged three times with oxygen. Following this, the desired amount of catalyst (glycerol/metal ratio = 1000, mol/mol) was suspended in the solution and the reactor was heated to 100 °C. The system was then pressurised to 3 bar O_2 and the reaction mixture was stirred at 900 rpm. After the stated reaction time, the reactor vessel was cooled to room temperature and the reaction mixture was diluted by a factor of 10 before being analysed by HPLC (Agilent 1260 infinity HPLC) equipped with ultraviolet and refractive index detectors and a Metacarb 67H column (held at 50 °C). The eluent was an aqueous solution of H_3PO_4 (0.01 M), used at a flow rate of 0.8 ml min^{-1}. Quantification of the reactants consumed and products generated was determined by an external calibration method. The reaction effluent was analysed for the following products: glyceric acid, tartronic acid, oxalic acid, glycolic acid, formic acid, acetic acid and lactic acid.

To study the reusability of the catalyst, four sequential reactions were performed with the catalyst washed with water, filtered and oven dried (120 °C, 16 h) between reactions.

Results

Properties of the SAS precipitated perovskites

As observed in Table 1, the precipitation of near stoichiometric La and B elements from the SAS precipitations was achieved for all the different perovskites, which is an important factor when producing perovskite materials with high phase purity. However, in some cases, this required an excess of the B site metal salt in the metal salt solution, to prevent excess La in the final precipitate. Non-stoichiometric precipitation was due to the different precipitation yields of the individual metal acetate salts, dictated primarily by the solubility of the salts in supercritical CO_2–methanol under the conditions used. It is envisaged that the yields from the SAS process could be altered to give precipitate ratios closer to 1 : 1 from initial 1 : 1 starting solutions by varying the pressure, solvent : CO_2 ratio and also the solution injection geometry.[26] Given that the purpose of using the SAS technique was to access novel morphologies and large surface areas for catalyst discovery and not catalyst production, the intensive research required to optimise each perovskite preparation was not considered essential and alteration of the initial starting salt ratios was a simpler way of precipitating materials with the correct stoichiometry.

Thermogravimetric analysis (TGA) (Fig. 2) showed multiple stage mass losses for all of the SAS precipitated materials. Previously, we have shown that the SAS precipitation of Ce, Mn, Fe, Co and Ni acetate salts resulted in an acetate salt being retained.[22,25] However, the local coordination geometry around the metal is altered and the sample no longer displays long range order, according to XRD

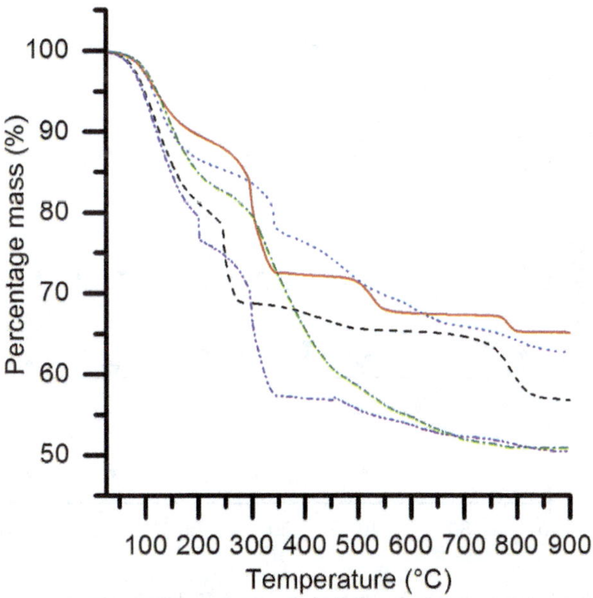

Fig. 2 Thermogravimetric analysis of the SAS precipitated materials, where the B site element is: Cr (red solid); Mn (dashed black); Fe (blue dotted); Co (green dashed-dotted); and Ni (purple dashed-dotted). The experimental conditions used were: 10 °C min^{-1} ramp rate, 50 ml min^{-1} flow rate of air and 15–25 mg of sample.

analysis. This results in their thermal decomposition at temperatures below 400 °C to form their corresponding oxides. It is, therefore, likely that the mass losses observed up to *ca.* 450 °C are indicative of the decomposition of acetate species, with higher temperature mass losses being associated with the transitions between various metal oxide and mixed metal oxide phases. The number of ternary oxide permutations is dependent on the ability of the B site element to adopt different valence states. An example of an extremely complicated system is the La–Cr–O system, where a number of ternary oxide phases are possible ($LaCrO_4$, $La_2Cr_3O_{12}$, La_2CrO_6 and $LaCrO_3$) depending on the phase composition and temperature.[27] These phases decompose at specific temperatures to produce $LaCrO_3$ and O_2. Potentially, the mass losses centred at 520 °C and 785 °C represent phase decomposition of $La_2Cr_3O_{12}$ and La_2CrO_6, respectively. Detailed *in situ* XRD analysis could provide more information on these intermediate states, although this is not included in the scope of the current work. The SAS precipitated materials were calcined at 750 °C, as TGA showed the final mass loss event had started before this temperature.

XRD analysis (Fig. 3 and Table 1) of the calcined materials showed that perovskite phases were dominant. Discernible amounts of by-product phases that contributed to 10 to 15 wt% of the sample were observed in the $LaFeO_3$, $LaCoO_3$ and $LaNiO_3$ samples. The by-phases were the single component La and B site oxides, which could be minimised by higher temperature calcination at the expense of the catalyst surface area. Considering the relatively low composition of

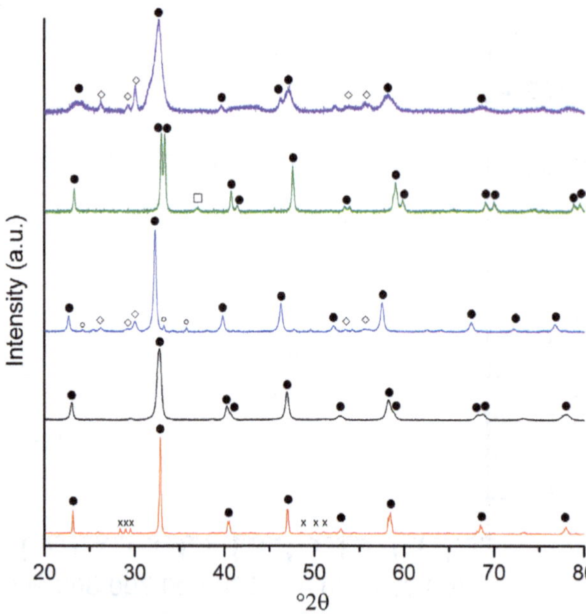

Fig. 3 Powder X-ray diffraction patterns of the SAS La : B precipitates after 750 °C calcination. In ascending order, the B site samples are: Cr (red); Mn (black); Fe (blue); Co (green); and Ni (purple). Phases: ● perovskite phases (for simplicity, the rhombohedral, orthorhombic and cubic phases are not differentiated); ○ Fe_2O_3; ◇ La_2O_3; □ Co_3O_4; and X La_2CrO_6.

the by-product phases, it was concluded that the loss of surface area from a higher temperature calcination would be counterproductive. The other by-product phase of note was a small amount of La_2CrO_6 in the $LaCrO_3$ sample, which could not be quantified by the relative intensity ratio (RIR) method due to the limited data available for this metastable phase. However, the area of the principle reflection at 28.13° 2θ was substantially smaller than that of the $LaCrO_3$ phase, suggesting that this Cr(VI) phase was only present in trace quantities.

The crystallite sizes of the SAS precipitated perovskites were calculated from the Scherrer equation, and were found to be between 6 and 22 nm (Table 1). Relative to perovskites prepared by more conventional methods, the crystallite sizes observed from the SAS precipitations were relatively small. This can be attributed to the expected small primary particle size of the SAS precipitate and the relatively low calcination temperature used, which was made possible by the high degree of A and B element mixing afforded by the SAS technique. The small particle size of the SAS precipitated perovskites resulted in surface areas in the region of 22–52 $m^2\ g^{-1}$, which are larger than the 1–15 $m^2\ g^{-1}$ areas found for perovskites prepared by more conventional techniques.[20,21] The combination of a large surface area and small crystallite size was envisaged to provide a suitable number of surface sites for the anchoring of metal nanoparticles.

The sol immobilisation technique, using PVA as the protecting ligand, was used to deposit 1 wt% AuPt (1 : 1 molar ratio) nanoparticles onto the perovskite materials. For all supports, the desired metal content and Au : Pt ratio was deposited (calculated from MP-AES data shown in Table 2). Representative TEM images, with corresponding particle size distributions, are shown in Fig. 4 and 5. In all cases, a small mean particle size of *ca.* 2 nm with a standard deviation of *ca.* 1 nm was observed. The slight size variation in the metal nanoparticle size between the different perovskite supported catalysts was found to have no strong correlation with either the surface area or the B site element. The observed particle sizes are comparable to those reported for catalysts prepared with more conventionally used supports, such as TiO_2,[28] which might be expected as the SAS precipitated perovskites had comparable surface areas to their 50–60 $m^2\ g^{-1}$. The ability of the anti-solvent precipitation methodology to prepare perovskites with sufficient surface areas to successfully support 1 wt% AuPt led us to investigate these materials as catalysts for a liquid phase oxidation reaction.

The effect of the B site in the AuPt/perovskite catalysts for the glycerol oxidation reaction was investigated. The conversion profiles are shown in Fig. 6 and the turnover frequencies (TOFs) ($mol_{glycerol\ converted}\ mol_{AuPt}^{-1}\ h^{-1}$) are given in Table 3. It can be seen that all the catalysts had similar initial rates, with the TOF of the

Table 2 Au and Pt surface and bulk composition of the AuPt/perovskite catalysts from MP-AES and XPS analysis

Support	Au and Pt content (wt%)	Bulk Au/Pt ratio from MP-AES	Surface Au/Pt ratio from XPS
$LaCrO_3$	0.55 (Au), 0.55 (Pt)	1.0	0.6
$LaMnO_3$	0.47 (Au), 0.46 (Pt)	1.0	1.4
$LaFeO_3$	0.46 (Au), 0.47 (Pt)	1.0	1.0
$LaCoO_3$	0.55 (Au), 0.54 (Pt)	1.0	0.6
$LaNiO_3$	0.50 (Au), 0.48 (Pt)	1.1	0.9

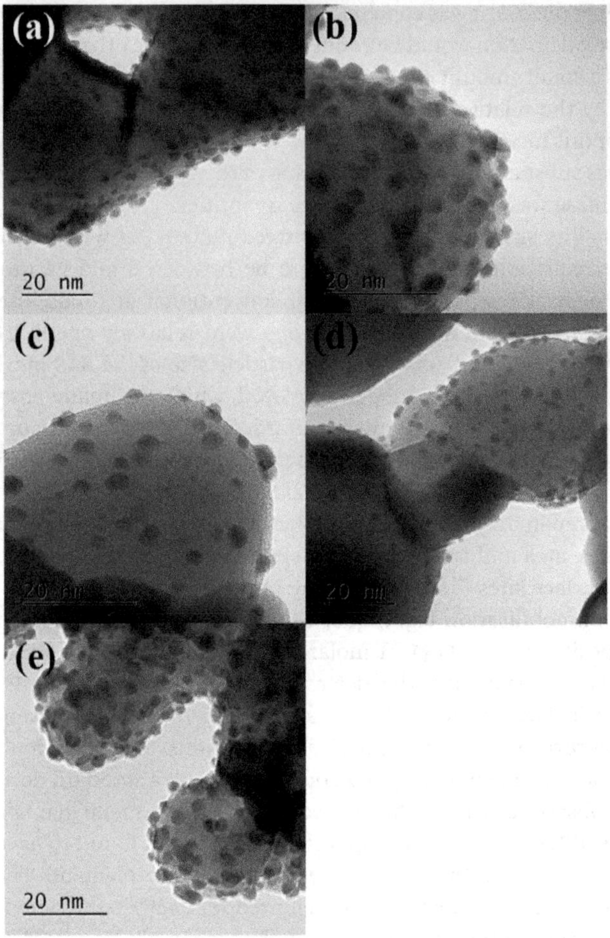

Fig. 4 Representative transmission electron micrographs of AuPt supported on the different SAS prepared $LaBO_3$ perovskites. (a) AuPt/$LaCrO_3$; (b) AuPt/$LaMnO_3$; (c) AuPt/$LaFeO_3$; (d) AuPt/$LaCoO_3$; (e) AuPt/$LaNiO_3$.

AuPt/$LaCrO_3$ and AuPt/$LaNiO_3$ catalysts being slightly higher at 620 h^{-1} and 560 h^{-1}, respectively, compared with the other catalysts, which had TOFs of 440–460 h^{-1}. No correlation between the TOF and the AuPt nanoparticle size was observed, although this was expected as the variance in TOFs and particle sizes was small. The TOFs observed for the AuPt/perovskite catalysts were found to be comparable to the *ca.* 500 h^{-1} observed by Shen *et al.* when using a 1 wt% AuPt/TiO_2 catalyst under similar reaction conditions (90 °C with a base : substrate ratio of 4 : 1).[9]

The product selectivities with different AuPt/perovskite catalysts are shown in Fig. 7. The activities of the different catalysts were very similar, whereas, the product distributions were markedly different. The $LaMnO_3$ supported catalyst was found to favour C_3 oxidation products, with a high selectivity towards glyceric acid, which at high conversions further oxidised to tartronic acid. Selectivities to C–C scission products (oxalic acid, glycolic acid, formic acid and CO_2) and lactic

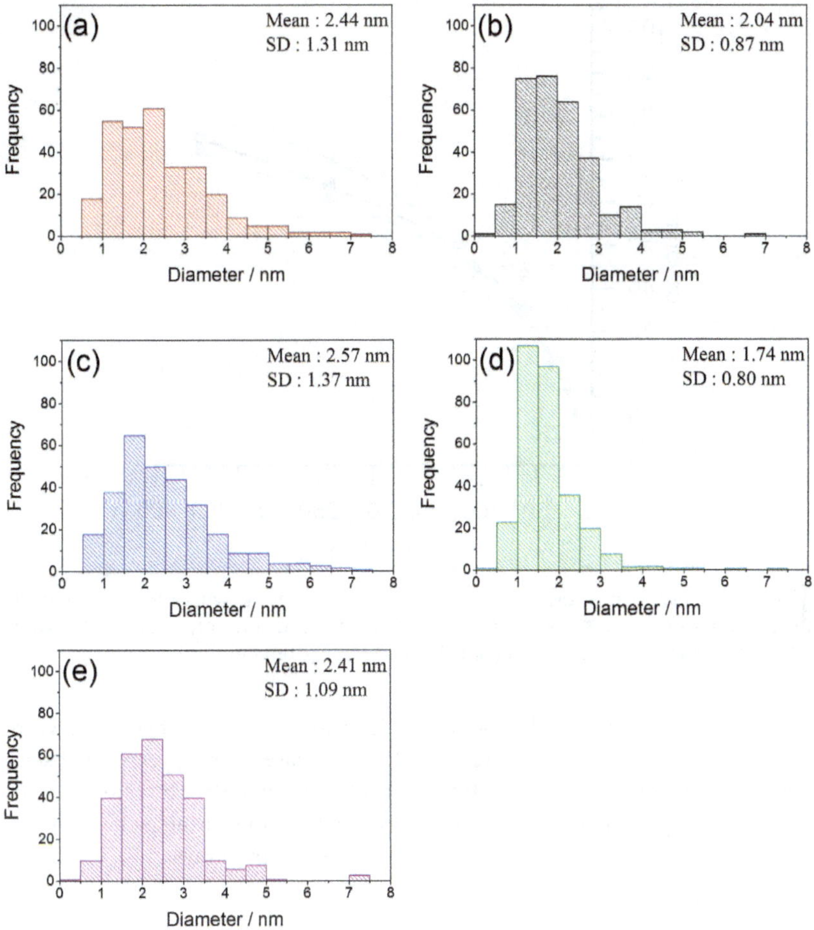

Fig. 5 Particle size distribution histograms of AuPt supported on the different SAS prepared $LaBO_3$ perovskites. (a) AuPt/$LaCrO_3$; (b) AuPt/$LaMnO_3$; (c) AuPt/$LaFeO_3$; (d) AuPt/$LaCoO_3$; (e) AuPt/$LaNiO_3$.

acid were low and consistent across the range of conversions observed for the $LaMnO_3$ supported catalyst. This is an interesting result as the reaction conditions used are reported to enhance the dehydration and re-arrangement of glyceraldehyde to lactic acid. Under these relatively high temperatures and high base concentrations, AuPt nanoparticles supported on CeO_2 or TiO_2 have been reported to give lactic acid selectivities between 60 and 80%.[8,9] Under milder conditions (for example temperatures of *ca.* 60 °C and a base : glycerol ratio of 2 : 1), where the rate of the lactic acid pathway is subdued, the selectivity profile seen for the AuPt/$LaMnO_3$ catalyst is frequently observed.[29] It is apparent that employing the $LaMnO_3$ support switches off the lactic acid pathway and promotes the oxidation pathway.

The selectivity profiles for the $LaCoO_3$ and $LaNiO_3$ supported catalysts are similar, with a moderate selectivity towards glyceric acid, a relatively high C–C scission selectivity and a lactic acid selectivity of *ca.* 30%. It was noted that for the

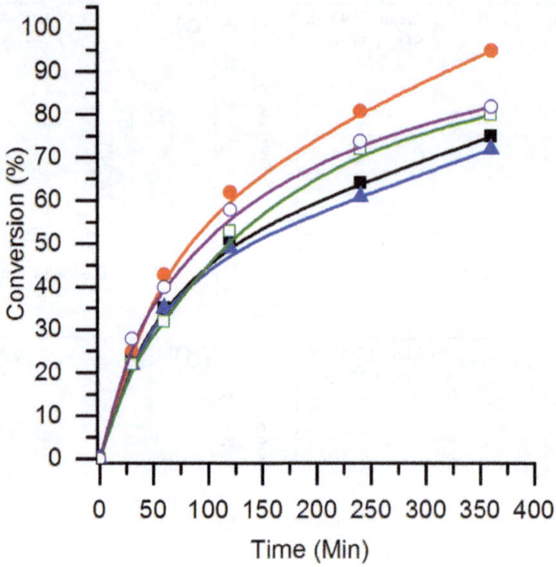

Fig. 6 The conversion of glycerol with AuPt/$LaBO_3$ catalysts, where the B sites of the supports are: Cr (●); Mn (■); Fe (▲); Co (□); and Ni (○). Conditions: glycerol 0.3 M in water, 4 : 1 NaOH : glycerol, glycerol : metal = 1000, 3 bar O_2, temperature = 100 °C.

AuPt/$LaNiO_3$ catalyst, the glyceric acid selectivity decreased when the glycerol conversion increased from 28% to 82%. This decrease in the glyceric acid selectivity did not correspond to a further oxidation to tartronic acid, but was accompanied by an increase in lactic acid formation, indicating a change in the prevalence of the oxidation and dehydration reaction pathways.

Table 3 Glycerol oxidation using AuPt supported on the perovskite and single oxide materials[a]

		Selectivity[c] (%)			
Catalyst	TOF[b] (h^{-1})	Glyc[d]	Tar[d]	C–C scission[d]	Lac[d]
Perovskites					
AuPt/$LaCrO_3$	620	5	7	2	86
AuPt/$LaMnO_3$	460	70	17	8	5
AuPt/$LaFeO_3$	440	10	2	19	69
AuPt/$LaCoO_3$	440	43	9	24	24
AuPt/$LaNiO_3$	560	30	3	26	41
Corresponding single oxides					
AuPt/MnO_2	560	33	3	33	31
AuPt/Fe_2O_3	240	19	2	58	21
AuPt/Co_3O_4	180	23	4	24	49
AuPt/NiO	700	30	10	39	21

[a] Reaction conditions: glycerol 0.3 M in water, 4 : 1 NaOH : glycerol, glycerol : metal = 1000, 3 bar O_2, 100 °C. [b] TOF calculated at 30 min, moles of glycerol converted/moles of metal per h. [c] Selectivity calculated after 6 h of reaction. [d] Key: Glyc = glyceric acid, Tar = tartronic acid, C–C scission = C_1 and C_2 products, and Lac = lactic acid.

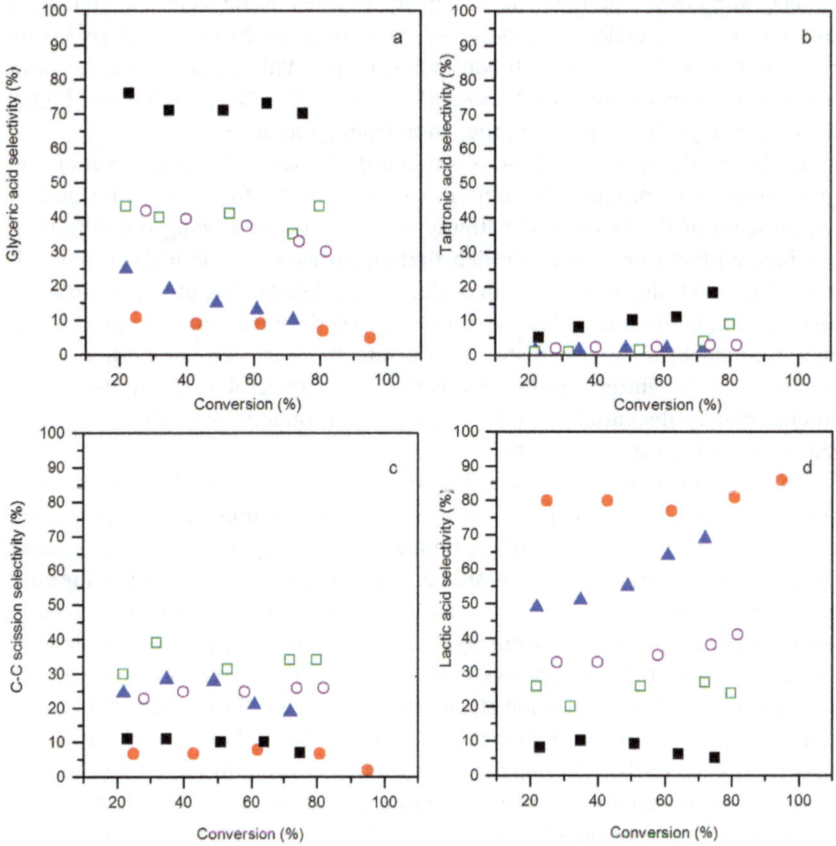

Fig. 7 Conversion–selectivity plots: (a) glyceric acid selectivity; (b) tartronic acid selectivity; (c) C–C scission selectivity; and (d) lactic acid selectivity, from the glycerol oxidation reaction using the AuPt/$LaBO_3$ catalysts, where the B sites of the supports are: Cr (●); Mn (■); Fe (▲); Co (□); and Ni (○). Conditions: glycerol 0.3 M in water, 4 : 1 NaOH : glycerol, glycerol : metal = 1000, 3 bar O_2, temperature = 100 °C.

The selectivity profile of AuPt/$LaFeO_3$ also changed with respect to glycerol conversion. At low conversions, the AuPt/$LaFeO_3$ catalyst had a glyceric acid selectivity of 25%, a scission product selectivity of 24.5% and 49% selectivity towards lactic acid. As the reaction progressed and the glycerol conversion increased, the selectivity towards glyceric acid decreased dramatically. As observed with the AuPt/$LaNiO_3$ catalyst, the change in the selectivity profile was the result of a shift towards lactic acid production, with the selectivity to this product increasing from 49% to 69% over the reaction period. At 69% lactic acid selectivity, the AuPt/$LaFeO_3$ catalyst is similar to previously reported catalysts, under comparable conditions.[8] The shift towards lactic acid formation was far more significant in the $LaFeO_3$ supported catalyst than any of the other perovskite supported catalysts. This may suggest that the oxidation sites on this catalyst are not stable or are blocked by reaction intermediates.

The highest lactic acid yield was observed with the AuPt/$LaCrO_3$ catalyst, with 86% selectivity towards this product at 95% glycerol conversion. Unlike the

$LaFeO_3$ supported catalyst, the selectivity towards lactic acid was relatively insensitive to conversion, with only a slight increase in the lactic acid selectivity from 80% to 86% over the full conversion range. With a TOF of 620 h^{-1} and a selectivity towards lactic acid above 85%, the AuPt/$LaCrO_3$ catalyst is a highly effective catalyst for lactic acid production from glycerol.

Evidently, the variation of the perovskite B site had a dramatic effect on the glycerol oxidation product distribution. The choice of Mn for the B site resulted in suppression of the lactic acid pathway with glyceric acid being the dominant product. With Cr or Fe B sites, the dehydration pathway to lactic acid is promoted, and a Ni or Co B site produces both oxidation and dehydration products. Both the lactic acid pathway and oxidation pathway to glyceric or tartronic acid proceed *via* the glyceraldehyde intermediate. As the rate limiting step for both reaction pathways is the initial proton abstraction from glycerol to form the alkoxy intermediate, substantially different product distributions were observed along-side very similar catalyst activities.

The most obvious reason for this variation in activity was that the different perovskite supports altered the metal support interaction of the AuPt nano-particles. This could result in different nanoparticle morphologies, such as phase separated Au and Pt, random alloy AuPt or core shell morphologies. Given the poor Z contrast between Au and Pt, stands for scanning transmission electron microscopy (STEM) and X-ray energy dispersive spectroscopy (XEDS) analysis of the AuPt nanoparticle morphology is challenging. Modern XEDS detectors, with greatly improved X-ray collection efficiencies, could allow for element mapping of AuPt nanoparticles at low enough energies to limit beam damage and present an opportunity for further development of the current work. In an attempt to gain some initial understanding of the environment and structure of the metal nano-particles, a combination of conventional TEM, reported earlier in Fig. 4, to investigate the support–particle size dependency, and XPS, to probe the metal oxidation state and relative surface composition, was performed. As mentioned previously, no strong variance in the metal nanoparticle size was observed by TEM, discounting the most obvious effect of metal dispersion on product selectivity. XPS analysis of the Au and Pt 4f levels (Fig. 8 and Table 2) showed that both the Au and Pt were in the metallic oxidation state. As seen in Table 2, the surface ratio of the two metals was noted to vary with the different B sites with a minimum Au : Pt ratio of 0.6 for the Cr and Co containing perovskites and a maximum of 1.4 for the Mn perovskite. Given that the bulk Au : Pt ratios, as determined by MP-AES, were found to be unity for all of the catalysts, the changes in the surface ratio are indicative of changes in the metal nanoparticle structure. This could potentially be due to changes in the composition of the alloy nanoparticles, the presence of monometallic phases or the formation of core–shell structures. Previous studies in which both core–shell and random alloy AuPt nanoparticles were prepared by sol immobilisation showed that the morphologies were well represented by the observed XPS metal ratios (*i.e.* a Au : Pt ratio of 1 : 1 for the random alloy or 0.75 : 1 for the core–shell).[30] The exact nature of the nanoparticle structure could be investigated further by STEM-XEDS analysis, although this would be challenging.

Under milder reaction conditions, at lower base concentrations and temper-atures, where the oxidation pathway is prevalent, monometallic Au catalysts have been found to promote the selective oxidation of glycerol to glyceric acid, while Pt and Pd have been shown to promote over-oxidation to form C–C scission

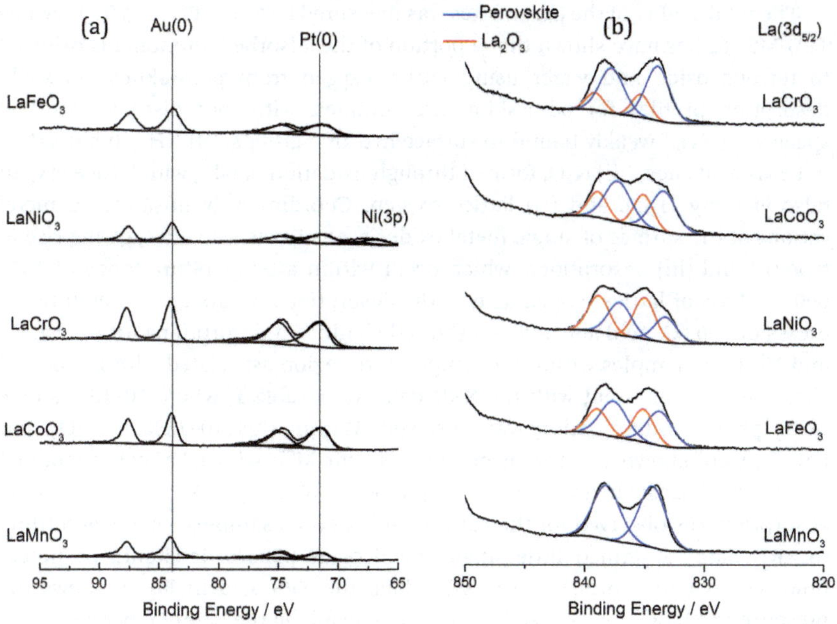

Fig. 8 XPS analysis of the AuPt/$LaBO_3$ catalysts. (a) Au (4f) and Pt (4f) spectra of the catalysts. (b) La ($3d_{5/2}$) spectra.

products.[29] The AuPt/$LaMnO_3$ catalyst with the Au surface enrichment was found to be the most selective towards glyceric acid production, while the Pt enriched surface of the AuPt/$LaCrO_3$ catalyst selectively produced lactic acid. No strong trend between the Pt surface concentration and C–C scission products was observed. The hypothesis that the Au surface concentration correlates with the promotion of the oxidation pathway does, however, appear to be contradicted by literature reports of the glycerol to lactic acid reaction with supported AuPt catalysts with Au : Pt ratios between 3 : 1 and 1 : 3, which showed that these catalysts had very little variation in activity or selectivity towards lactic acid. Monometallic Au and Pt catalysts did show some slight variation in lactic acid yields (Au = 46% lactic acid yield and Pt = 56% lactic acid yield),[8] although this difference is far less significant compared to the difference between the AuPt/$LaMnO_3$ and AuPt/$LaCrO_3$ catalysts (AuPt/$LaMnO_3$ = 4% lactic acid yield and AuPt/$LaCrO_3$ = 82% lactic acid yield).

Consideration of the properties of the perovskite supports and their potential to alter the reaction pathway must also be considered. XPS analysis (Fig. 8b) showed a discernible surface contribution from La species with an alternative binding energy to that of the perovskite that could be assigned to La_2O_3 or $La(OH)_3$. It was found that the materials produced with the Fe, Co and Ni B sites had a higher contribution of this phase, which correlates with the increased La_2O_3 from XRD analysis. A loose correlation between the increased selectivity towards C–C scission and the content of phase separated La_2O_3 and B site oxide was observed, although no correlation with lactic acid or glyceric acid production can be found.

The total acidity of the perovskites was measured by NH_3-TPD (Fig. 9). Previous TPD-MS studies have shown that a portion of the adsorbed ammonia is oxidised to nitrous oxide and water using lattice oxygen from perovskites. As such, desorption profiles for perovskites are complex, with four distinct desorbed species: (i) NH_4^+ weakly bound to surface hydroxyl groups, (ii) NH_3 chemisorbed to Lewis acid sites, (iii) N_2O, formed through oxidation of NH_3 with lattice oxygen (also yielding H_2O), and (iv) lattice oxygen. Coordinatively unsaturated metal cations at the surface of single metal oxides act as Lewis acid sites, giving rise to type (ii) and (iii) desorptions, which occur within a temperature range of 150–600 °C. Loss of lattice oxygen as type (iv) desorptions occurs at temperatures in excess of 500 °C. Acid site analysis showed significant desorptions for the Fe, Co and Ni B site samples within the temperature region associated with Lewis acid sites. This is consistent with the XRD data from Table 1, where 10–15% single oxide phase contamination was observed. Within this 10–15%, $LaNiO_3$ and $LaFeO_3$ were shown to contain mainly trivalent M^{n+}, whilst $LaCoO_3$ contained Co_3O_4, which exists in a normal spinel phase as $Co^{II}Co^{III}{}_2O_3$. Very little ammonia adsorption was observed for the $LaCrO_3$ and $LaMnO_3$ samples, which, according to XRD data, contained little single metal oxide phases. It should be noted, however, that in contrast to the XRD data, the XPS data in Fig. 8 shows the presence of a La_2O_3 phase, itself a Lewis acid, in all but the $LaMnO_3$ perovskite. In this way, the La_2O_3 : perovskite ratio followed the order: $LaNiO_3 > LaFeO_3 > LaCoO_3 > LaCrO_3$. $LaNiO_3$, $LaFeO_3$ and, to a lesser extent, $LaMnO_3$ exhibited a high temperature desorption at *ca.* 650–700 °C, which might be attributed to the loss of highly mobile lattice oxygen. These data suggest that the more Lewis acidic single oxide phases are responsible for the loss of controlled selectivity in the AuPt/$LaFeO_3$, AuPt/$LaCoO_3$ and AuPt/$LaNiO_3$ catalysts and the formation of C–C scission products.

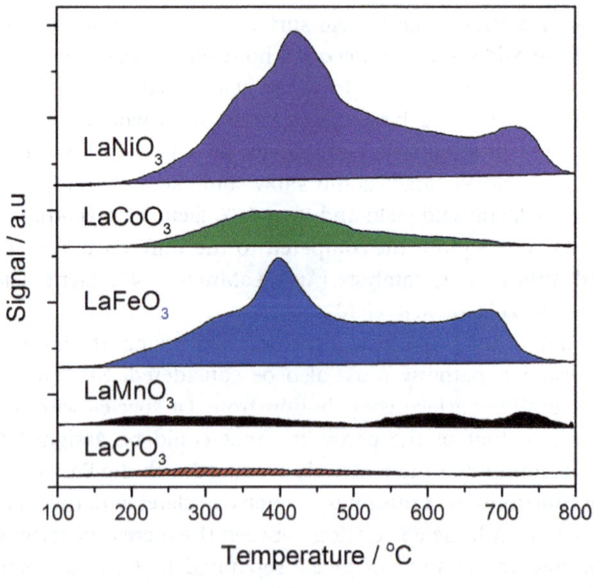

Fig. 9 Temperature programmed desorption of ammonia analysis of the SAS prepared $LaBO_3$ supports, where B is: Cr (red solid); Mn (black); Fe (blue); Co (green); or Ni (purple).

To test the hypothesis that the single metal oxide phases were responsible for the differences in product selectivity, AuPt was deposited on MnO_2, Fe_2O_3, Co_3O_4 and NiO supports prepared by the SAS process (calcined at 750 °C) and tested as catalysts for the glycerol oxidation reaction (Table 3). Unlike the corresponding perovskite supported catalysts, the single oxide supported catalysts had a significant range of TOFs from 180 to 700 h^{-1}, with the TOF for the Ni and Mn single oxides being higher than their corresponding perovskite catalysts. It is important to note that all of the catalysts were less selective than the AuPt/perovskites, with the highest selectivity towards any product being 49% lactic acid with the AuPt/Co_3O_4 catalyst. Specifically, the high selectivity towards glyceric acid with the AuPt/$LaMnO_3$ catalyst (69%) was not replicated with the AuPt/MnO_2 catalyst (33% glyceric acid selectivity) and the high lactic acid selectivity with the AuPt/$LaFeO_3$ catalyst was not replicated compared to the AuPt/Fe_2O_3 catalyst (69% *vs.* 21%). As a point of comparison, Villa *et al.* studied Au/NiO and Au/NiO–TiO_2 (NiO impregnated onto TiO_2) for glycerol oxidation. Similar to the current study, the NiO supported catalyst was found to be highly active but with a broad range of products. Dilution of the NiO sites by supporting onto TiO_2 resulted in a significant decrease in the activity and an improved selectivity towards glyceric acid.[14]

Perovskites have been extensively researched as catalysts for the deep oxidation of alkanes, alkenes and CO.[19,31] A strong correlation between the activity and O_2 coverage profiles has been reported, with perovskites with good O_2 adsorption capacities being more active. Tejuca and co-workers investigated the chemisorption of O_2 and isobutene on a range of $LaBO_3$ catalysts with the same range of B sites used in this study (*i.e.* Cr, Mn, Fe, Co and Ni).[19] The adsorption profiles of O_2 on the $LaBO_3$ clean surfaces from this earlier study have been plotted against the glycerol oxidation selectivity profiles of the various B sites (at 6 h reaction time) used in AuPt/$LaBO_3$, as shown in Fig. 10. If C–C scission products are assumed to be produced from an oxidation process, the sum of the oxidation

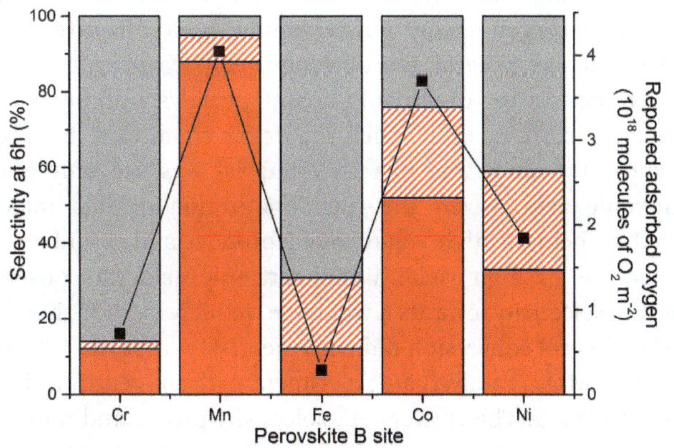

Fig. 10 Selectivity profiles of the AuPt/$LaBO_3$ catalysts compared to reported oxygen adsorption values for the relevant perovskite phases.[19] Sum of the C_3 oxidation products (glyceric and tartronic acid) shown in solid red. C–C scission products (sum of glycolic acid, oxalic acid, formic acid and CO_2) shown in dashed red. Lactic acid selectivity shown by the solid grey bar. (■) read from the second *y* axis represents the reported oxygen adsorption of the perovskite.

pathway products correlates well with the reported oxygen adsorption capacities. The most selective catalyst towards the oxidation pathway used the $LaMnO_3$ support, which had the best oxygen adsorption capacity. The $LaCrO_3$ and $LaFeO_3$ supports with poor oxygen adsorption characteristics were found to give catalysts that favour the production of lactic acid, which is formed from an initial oxidation followed by dehydration to pyruvaldehyde and rearrangement. The $LaCoO_3$ and $LaNiO_3$ supported catalysts were found to produce both oxidation and dehydration products that correspond to the intermediate oxygen adsorption capacities. An interesting point highlighted in Tejuca and co-workers' paper was that the O_2 capacities were enhanced on the addition of the iso-butane substrate for all B site perovskites, but disproportionately across the period (*i.e.* the effect was more significant for Co than Mn B sites).[19] Further studies with more ideal surfaces could probe the effect of glycerol adsorption on influencing oxygen adsorption.

A question raised from observing this range of product distributions when choosing different perovskite supports is to ask whether the support itself is active. Glycerol oxidation testing with the perovskite supports gave conversions of less than 1% after 30 min time on line, demonstrating that the metal oxide system had a very low activity for alcohol oxidation under these conditions. Isotopic labelling and computational studies by Davis and co-workers demonstrated that, on supported Au and Pt catalysts, the role of O_2 in glycerol oxidation was indirect, with oxygen regenerating the active surface hydroxyl species instead of being directly incorporated into the acid products.[32] Given the reaction temperature, it is unsurprising that the perovskites themselves were not active. A possible hypothesis is that the ability of the support to adsorb oxygen would affect the ability of the active site to regenerate. Catalysts with poor oxygen adsorption capacities perform the initial oxidation of glycerol to glyceraldehyde but the rate of the second oxidation process to glyceric acid is slower than that of the dehydration pathway to produce lactic acid. The $LaMnO_3$ support with its higher oxygen adsorption capacity facilitates the regeneration of the oxidation site, allowing for glyceraldehyde oxidation to glyceric acid. Alternatively, the change in B site electronic configuration from Cr (d^3) to Ni (d^7), which is reflected in each $LaBO_3$ perovskite's oxygen adsorption capability, could affect the prevalence of surface species such as glycerol, reaction intermediates or hydroxyl species.

One applied observation of the AuPt/$LaMnO_3$ catalyst's ability to promote the selective oxidation pathway under such aggressive conditions (high temperature and base : substrate ratio) was that this catalyst was the only one tested to produce tartronic acid. Given the functionalization of this molecule, it is acknowledged as being a high value added product and would be a desirable molecule to make in high yield. While tartronic acid was observed at low conversions, the selectivity towards it was noted to increase at higher conversions. Given that the glycerol conversion did not reach 100% over the 6 h reaction time using the AuPt/$LaMnO_3$ catalyst, an experiment with the reaction time extended to 24 h was performed. The conversion, selectivity profile and molar concentrations with respect to reaction time are shown in Fig. 11. 100% conversion was observed after 10 h time on line, at which point the selectivity towards glyceric acid was 66% with a tartronic acid selectivity of 22%. This represents a slight increase in the tartronic acid selectivity from the 18% observed at 6 h reaction time. After all the glycerol had been converted, it can be seen from the molar concentration plot that glyceric acid is being converted to tartronic acid. Under

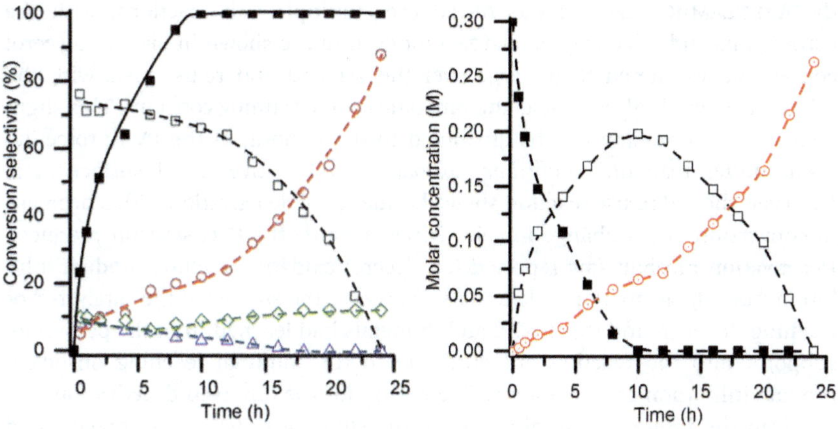

Fig. 11 Extended glycerol oxidation reaction time using the AuPt/$LaMnO_3$ catalyst. (Left) Time on line conversion and selectivity (■ glycerol conversion, selectivity towards: □ glyceric acid, ○ tartronic acid, ◇ C–C scission, and △ lactic acid). (Right) Time on line molar concentration plot (■ glycerol, □ glyceric acid and ○ tartronic acid). Conditions: glycerol 0.3 M in water, 4 : 1 NaOH : glycerol, glycerol : metal = 1000, 3 bar O_2, temperature = 100 °C.

these reaction conditions, it was clear that the sequential oxidation of glycerol to tartronic acid, *via* glyceric acid, takes place. After 24 h reaction time, all of the glyceric acid had been converted to give a final selectivity towards tartronic acid of 88%, with C–C scission accounting for the remaining products. A yield of 88% tartronic acid in the context of the general literature is exceptionally high and can be attributed to the high base : substrate ratio (4 : 1) and temperature (100 °C) combined with suppression of the dehydration route to lactic acid and excessive C–C scission over the AuPt/$LaMnO_3$ catalyst.

An important consideration in the application of heterogeneous catalysts is their reusability and resistance to leaching. Effluent analysis from reactions using AuPt/$LaMnO_3$ and AuPt/$LaFeO_3$ by MP-AES is shown in Table 4. These two catalysts were chosen for their different selectivity profiles and also, in the case of the $LaFeO_3$ sample, due to the notable changes in selectivity during the reaction. With respect to the B site leaching, less than 2% of the possible metal was found in either reaction effluent. This level of leaching would not contribute to the reaction, as shown by the absence of activity for the perovskite catalysts tested without AuPt. Slightly higher levels of La leaching were determined (between 2 and 6% of the total La), although, again, this had little effect on the reaction. The re-use of

Table 4 Effluent analysis following the glycerol oxidation reaction[a]

Catalyst	La in effluent (ppm)	B element in effluent (ppm)	% La leaching	% B site leaching
AuPt/$LaMnO_3$	58	20	2.1	1.1
AuPt/$LaFeO_3$	168	36	6.0	1.9

[a] Reaction conditions: glycerol 0.3 M in water, 4 : 1 NaOH : glycerol, glycerol : metal = 1000, 3 bar O_2, 100 °C, 6 h.

the $AuPt/LaMnO_3$ catalyst was tested over multiple 6 h reactions, with the conversions and selectivity breakdown for each re-use shown in Fig. 12. Glycerol conversion was noted to increase over the 1st and 2nd re-use tests with the selectivity towards glyceric acid and tartronic acid remaining constant. This slight increase in conversion can be attributed to the removal of the PVA protecting agent under reaction conditions, exposing more active metal surface area. However, the 3rd re-use reaction showed evidence of deactivation with a decrease in conversion and a change in selectivities towards the C–C scission products. Deactivation mechanisms reported for glycerol oxidation include product inhibition from the formation of ketonic species over the surface of the catalysts[33] or leaching. It was found that no Au and Pt metals had leached from the perovskite supports into the reaction effluent. Due to the minimal leaching observed, product inhibition could be a possible reason for the observed deactivation.

Given the interesting findings from this study of utilising SAS precipitated perovskites as supports for liquid phase oxidation, the viability of perovskites prepared by other routes should be investigated. It is well known that the precursor salts, residual impurities and preparation technique can influence the properties of metal oxide systems. Therefore, the $LaMnO_3$ support was prepared by two alternative routes, a mechanochemical synthesis from the single metal oxides and flame pyrolysis of the metal nitrate solutions. It was considered that the mechanochemical preparation route would result in minimal impurities with no precipitation agent or organic species that could leave residues. Flame pyrolysis was anticipated to produce highly crystalline materials with a large geometric surface area and, with further development, could produce materials with comparable surface areas to the SAS precipitation method. The surface areas and crystallite sizes of the prepared $LaMnO_3$ supports are shown in Table 5. The small surface area and high crystallite size of the $LaMnO_3$ sample prepared by the milling of the single oxides are typical of perovskite materials. The surface area of the material produced by flame pyrolysis was found to be larger than that

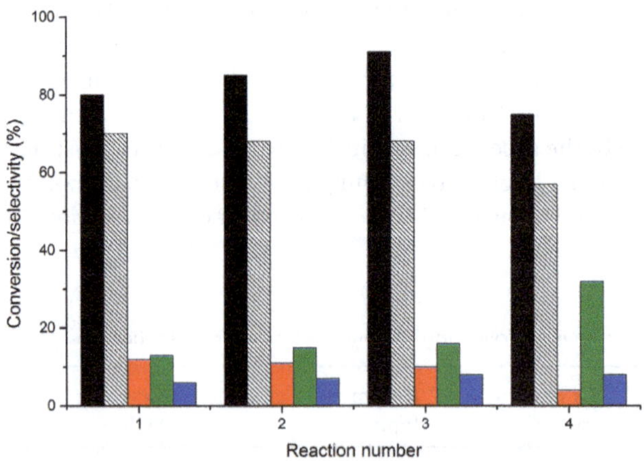

Fig. 12 Re-use testing of the $AuPt/LaMnO_3$ catalyst. Black fill = glycerol conversion; black dashed = glyceric acid selectivity, red fill = tartronic acid selectivity; green fill = C–C scission selectivity and blue fill = lactic acid selectivity. Conditions: glycerol 0.3 M in water, 4 : 1 NaOH : glycerol, glycerol : metal = 1000, 3 bar O_2, temperature = 100 °C.

Table 5 Comparison of the physical properties of $LaMnO_3$ prepared by different synthetic routes

Preparation method	$LaMnO_3$ phase purity[a] (%)	Crystallite size[a] (nm)	Surface area ($m^2 g^{-1}$)
SAS precipitation	100	18	32
Flame spray pyrolysis	90	40	15
Mechanochemical	95	25	1

[a] Determined from XRD analysis.

produced by the mechanochemical preparation route, at 15 $m^2 g^{-1}$, but not as large as the 32 $m^2 g^{-1}$ observed for the SAS prepared sample. It should be noted that both the mechanochemical and flame pyrolysis routes were not optimised and that significant work could be undertaken to improve the surface areas and reduce the crystallite sizes.

After the addition of the 1 wt% AuPt, the $LaMnO_3$ catalysts prepared by mechanochemical synthesis and flame pyrolysis were tested under the same glycerol oxidation reaction conditions as those used for the SAS catalysts (Fig. 13). It was evident that the conversion with these alternative supports was lower than that with the SAS prepared material and corresponded to the surface area of the supports. The reduced surface area would result in fewer sites for the AuPt

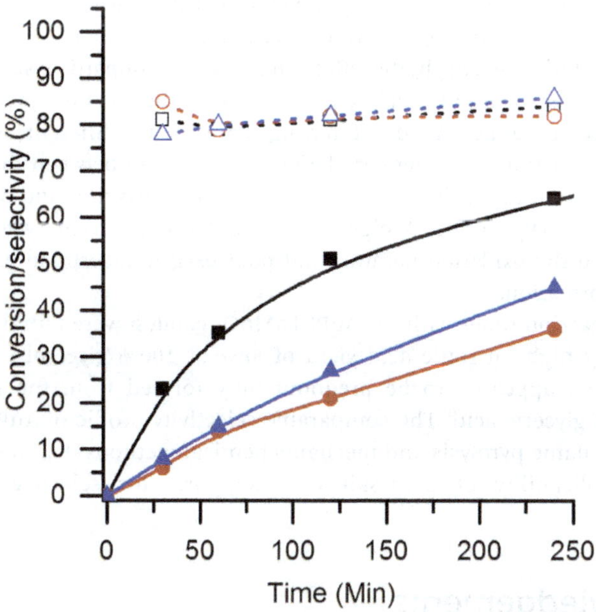

Fig. 13 Conversion and C_3 oxidation product selectivity of $AuPt/LaMnO_3$ using different perovskite preparation routes. Conversions are represented with solid fills (■ = SAS, ▲ = flame pyrolysis and ● = mechanochemical) and selectivities with the corresponding open symbols.

nanoparticles, reducing the dispersion of the metals and, hence, lowering the conversions. Despite the reduced activity, a key observation was that all the catalysts tested had the same selectivity towards the C_3 oxidation products (glyceric and tartronic acid). From an applied perspective, this is a key finding as it illustrates that the phenomenon of $LaMnO_3$ supports inhibiting lactic acid production is not unique to any specific preparation technique. Provided a $LaMnO_3$ support with sufficient surface area can be produced, a catalyst with strong selective oxidation potential can be synthesised.

Conclusion

The SAS precipitation methodology has been successfully utilised as a method of catalyst discovery. More specifically, it has allowed access to a class of materials, namely perovskites, by providing sufficient surface areas to allow them to be used as supports for liquid phase glycerol oxidation. A range of $LaBO_3$ supports (where B is Cr, Mn, Fe, Co or Ni) was prepared with surface areas between 22 and 52 $m^2\ g^{-1}$. This allowed for the successful addition of 1 wt% AuPt nanoparticles to make active glycerol oxidation catalysts. Under the catalytic reaction conditions chosen, the selectivity of the reaction could be tuned between the production of glyceric/tartronic acid, through a sequential oxidation pathway, and the production of lactic acid, through an initial oxidation and dehydration pathway. The factor changing the reaction pathway was the choice of element in the perovskite B site, with Cr and Fe resulting in a high selectivity towards lactic acid, Mn being highly selective towards sequential oxidation products, and Co or Ni giving a range of products. The use of the single metal oxides as supports resulted in a loss of control of the product selectivity. Without the AuPt nanoparticles, the perovskite supports were inactive for the reaction. The choice of B site was found to have a negligible effect on the AuPt nanoparticle size but it did significantly alter the Au : Pt surface ratio, although there is little precedence in the literature for either Au or Pt having significantly different selectivities towards this reaction. A strong correlation was observed between the reported oxygen adsorption capacities of the different B site perovskites and the glycerol oxidation selectivity profile. A high oxygen adsorption capacity was found to correlate with the oxidation pathway and poor oxygen adsorption capacity for lactic acid formation.

Longer reaction times with the AuPt/$LaMnO_3$ catalyst were found to give an exceptionally high tartronic acid yield of 88% at 100% glycerol conversion. Tartronic acid appeared to be predominantly formed from the sequential oxidation of glyceric acid. The comparable selectivity profile of AuPt/$LaMnO_3$ prepared by flame pyrolysis and mechanochemical methods demonstrates the general applicability of perovskites as supports for selective oxidation reactions.

Acknowledgements

The UK Catalysis Hub is kindly thanked for resources and support provided *via* our membership of the UK Catalysis Hub consortium and funded by EPSRC (EP/K014854/1). We thank the Research Complex at Harwell for use of the TEM.

References

1 A. Behr, J. Eilting, K. Irawadi, J. Leschinski and F. Lindner, *Green Chem.*, 2008, **10**, 13–30.

2 C.-H. Zhou, J. N. Beltramini, Y.-X. Fan and G. Q. Lu, *Chem. Soc. Rev.*, 2008, **37**, 527–549.

3 L. Prati and M. Rossi, *J. Catal.*, 1998, **176**, 552–560.

4 S. Carrettin, P. McMorn, P. Johnston, K. Griffin and G. J. Hutchings, *Chem. Commun.*, 2002, 696–697.

5 N. Dimitratos, F. Porta and L. Prati, *Appl. Catal., A*, 2005, **291**, 210–214.

6 C. L. Bianchi, P. Canton, N. Dimitratos, F. Porta and L. Prati, *Catal. Today*, 2005, **102–103**, 203–212.

7 F. Porta and L. Prati, *J. Catal.*, 2004, **224**, 397–403.

8 P. Lakshmanan, P. P. Upare, N.-T. Le, Y. K. Hwang, D. W. Hwang, U. H. Lee, H. R. Kim and J.-S. Chang, *Appl. Catal., A*, 2013, **468**, 260–268.

9 Y. Shen, S. Zhang, H. Li, Y. Ren and H. Liu, *Chem.–Eur. J.*, 2010, **16**, 7368–7371.

10 P. P. Upare, Y. K. Hwang, J. S. Chang and D. W. Hwang, *Ind. Eng. Chem. Res.*, 2012, **51**, 4837–4842.

11 S. Carrettin, P. McMorn, P. Johnston, K. Griffin, C. J. Kiely and G. J. Hutchings, *Phys. Chem. Chem. Phys.*, 2003, **5**, 1329–1336.

12 N. Dimitratos, J. A. Lopez-Sanchez, J. M. Anthonykutty, G. Brett, A. F. Carley, R. C. Tiruvalam, A. A. Herzing, C. J. Kiely, D. W. Knight and G. J. Hutchings, *Phys. Chem. Chem. Phys.*, 2009, **11**, 4952–4961.

13 A. Tsuji, K. T. V. Rao, S. Nishimura, A. Takagaki and K. Ebitani, *ChemSusChem*, 2011, **4**, 542–548.

14 A. Villa, G. M. Veith, D. Ferri, A. Weidenkaff, K. A. Perry, S. Campisi and L. Prati, *Catal. Sci. Technol.*, 2013, **3**, 394–399.

15 A. Villa, G. M. Veith and L. Prati, *Angew. Chem., Int. Ed.*, 2010, **49**, 4499–4502.

16 S. A. Kondrat, P. J. Miedziak, M. Douthwaite, G. L. Brett, T. E. Davies, D. J. Morgan, J. K. Edwards, D. W. Knight, C. J. Kiely, S. H. Taylor and G. J. Hutchings, *ChemSusChem*, 2014, 7, 1326–1334.

17 A. Villa, S. Campisi, K. M. H. Mohammed, N. Dimitratos, F. Vindigni, M. Manzoli, W. Jones, M. Bowker, G. J. Hutchings and L. Prati, *Catal. Sci. Technol.*, 2015, **5**, 1126–1132.

18 M. A. Peña and J. L. G. Fierro, *Chem. Rev.*, 2001, **101**, 1981–2018.

19 G. Kremenic, J. M. L. Nieto, J. M. D. Tascon and L. G. Tejuca, *J. Chem. Soc., Faraday Trans. 1*, 1985, **81**, 939–949.

20 Y. Teraoka, S. Nanri, I. Moriguchi, S. Kagawa, K. Shimanoe and N. Yamazoe, *Chem. Lett.*, 2000, 1202–1203, DOI: 10.1246/cl.2000.1202.

21 D. W. Johnson, Jr, P. K. Gallagher, F. Schrey and W. W. Rhodes, *Am. Ceram. Soc. Bull.*, 1976, **55**, 520–523.

22 R. P. Marin, S. A. Kondrat, R. K. Pinnell, T. E. Davies, S. Golunski, J. K. Bartley, G. J. Hutchings and S. H. Taylor, *Appl. Catal., B*, 2013, **140–141**, 671–679.

23 Z.-R. Tang, C. D. Jones, J. K. W. Aldridge, T. E. Davies, J. K. Bartley, A. F. Carley, S. H. Taylor, M. Allix, C. Dickinson, M. J. Rosseinsky, J. B. Claridge, Z. Xu, M. J. Crudace and G. J. Hutchings, *ChemCatChem*, 2009, **1**, 247–251.

24 R. P. Marin, S. Ishikawa, H. Bahruji, G. Shaw, S. A. Kondrat, P. J. Miedziak, D. J. Morgan, S. H. Taylor, J. K. Bartley, J. K. Edwards, M. Bowker, W. Ueda and G. J. Hutchings, *Appl. Catal., A*, 2015, **504**, 62–73.

25 P. J. Miedziak, Z. Tang, T. E. Davies, D. I. Enache, J. K. Bartley, A. F. Carley, A. A. Herzing, C. J. Kiely, S. H. Taylor and G. J. Hutchings, *J. Mater. Chem.*, 2009, **19**, 8619–8627.

26 E. Reverchon, I. De Marco and E. Torino, *J. Supercrit. Fluids*, 2007, **43**, 126–138.

27 T. J. Kallarackel, S. Gupta and P. Singh, *J. Am. Ceram. Soc.*, 2013, **96**, 3933–3938.

28 J. A. Lopez-Sanchez, N. Dimitratos, P. Miedziak, E. Ntainjua, J. K. Edwards, D. Morgan, A. F. Carley, R. Tiruvalam, C. J. Kiely and G. J. Hutchings, *Phys. Chem. Chem. Phys.*, 2008, **10**, 1921–1930.

29 A. Villa, N. Dimitratos, C. E. Chan-Thaw, C. Hammond, L. Prati and G. J. Hutchings, *Acc. Chem. Res.*, 2015, **48**, 1403–1412.

30 V. Peneau, Q. He, G. Shaw, S. A. Kondrat, T. E. Davies, P. Miedziak, M. Forde, N. Dimitratos, C. J. Kiely and G. J. Hutchings, *Phys. Chem. Chem. Phys.*, 2013, **15**, 10636–10644.

31 L. G. Tejuca, J. L. G. Fierro and J. M. D. Tascon, *Adv. Catal.*, 1989, **36**, 237–328.

32 B. N. Zope, D. D. Hibbitts, M. Neurock and R. J. Davis, *Science*, 2010, **330**, 74–78.

33 B. N. Zope and R. J. Davis, *Green Chem.*, 2011, **13**, 3484–3491.

Faraday Discussions

PAPER

Selective hydrogenation of halogenated arenes using porous manganese oxide (OMS-2) and platinum supported OMS-2 catalysts†

Iain J. McManus, Helen Daly, Haresh G. Manyar, S. F. Rebecca Taylor, Jillian M. Thompson* and Christopher Hardacre*

Received 16th December 2015, Accepted 21st December 2015
DOI: 10.1039/c5fd00227c

Porous manganese oxide (OMS-2) and platinum supported on OMS-2 catalysts have been shown to facilitate the hydrogenation of the nitro group in chloronitrobenzene to give chloroaniline with no dehalogenation. Complete conversion was obtained within 2 h at 25 °C and, although the rate of reaction increased with increasing temperature up to 100 °C, the selectivity to chloroaniline remained at 99.0%. Use of Pd/OMS-2 or Pt/Al_2O_3 resulted in significant dechlorination even at 25 °C and 2 bar hydrogen pressure giving a selectivity to chloroaniline of 34.5% and 77.8%, respectively, at complete conversion. This demonstrates the potential of using platinum group metal free catalysts for the selective hydrogenation of halogenated aromatics. Two pathways were observed for the analogous nitrobenzene hydrogenation depending on the catalyst used. The hydrogenation of nitrobenzene was found to follow a direct pathway to aniline and nitrosobenzene over Pd/OMS-2 in contrast to the OMS and Pt/OMS-2 catalysts which resulted in formation of nitrosobenzene, azoxybenzene and azobenzene/hydrazobenzene intermediates before complete conversion to aniline. These results indicate that for Pt/OMS-2 the hydrogenation proceeds predominantly over the support with the metal acting to dissociate hydrogen. In the case of Pd/OMS-2 both the hydrogenation and hydrogen adsorption occur on the metal sites.

Introduction

Halogenated anilines are used in the manufacture of polymers, pharmaceuticals, dyes and agrochemicals[1,2] and are most easily prepared by the hydrogenation of the corresponding halonitrobenzene. High selectivity for the hydrogenation reaction over any dehalogenation is required to optimise yield as well as minimise poisoning of the metal catalyst by strong adsorption of the halogen.

CenTACat, School of Chemistry and Chemical Engineering, Queen's University, Belfast, David Keir Building, Stranmillis Road, Belfast, BT9 5AG, UK. E-mail: jillian.thompson@qub.ac.uk; c.hardacre@qub.ac.uk

† Electronic supplementary information (ESI) available. See DOI: 10.1039/c5fd00227c

High selectivities to haloanilines have been observed over a range of heterogeneous catalysts including ruthenium,[3] nickel,[4] platinum[5] and a range of transition metal sulphides.[6] In many cases the source of hydrogen is from a stoichiometric reducing agent such as $NaBH_4$;[3,4,6] however, the use of hydrogen gas is preferred to minimise the waste formed. It is generally accepted that platinum is the preferred metal for hydrogenation, giving minimum amounts of dehalogenation with palladium enhancing the dehalogenation reaction leading to increased side products. This is highlighted by the observation that palladium is the catalyst of choice in dehalogenation for pollution control,[7,8] although it should be noted that homogeneous catalysts based on iron,[9] nickel[10,11] and rhodium[12] are also reported to be highly active for dehalogenation reactions. Despite the propensity of palladium to dehalogenate aromatics, a recent study has highlighted the importance of the size of palladium nanoparticles for the selective hydrogenation of halonitrobenzenes over Pd/C. Using hydrogen gas, high conversion and greater than 99% selectivity to chloroaniline were obtained in the absence of a solvent on a 2000 L pilot plant scale over 30 runs.[7]

Dehalogenation can, therefore, occur to varying degrees over different catalysts; however, the overall selectivity of the hydrogenation reaction depends on a number of other factors such as the halogen involved. For example, iodo-substituted aromatics are more easily dehalogenated than the corresponding bromo and chloro molecules and the fluoro molecules are the most stable.[13] This has been attributed to a combination of the C–X bond strength, with C–F > C–Cl > C–Br > C–I, as well as the adsorption strength of the different substrates on the catalyst.[14,15] In addition, severe reaction conditions such as high temperatures or pressures can cause dehalogenation and complete hydrogenation of the aromatic ring. For example, Wang *et al.* showed that for the gas phase hydrogenation of *para*-chloronitrobenzene, the product selectivity was sensitive to the H_2 to substrate ratio with a range of products from *para*-chloroaniline to cyclohexane being formed.[16]

The present study reports on the hydrogenation of halogenated and non-halogenated nitrobenzene over a series of manganese oxide supported catalysts. Despite many reports examining nitrobenzene hydrogenation, the mechanism by which the reaction occurs is still not well understood. The first reported mechanism for this reaction was by Haber in 1898 and involved hydrogenation of nitrobenzene (**1**) to nitrosobenzene (**2**) which could then further hydrogenate to form phenyl hydroxylamine (**3**) or react with either phenyl hydroxylamine or aniline to form azoxybenzene (**4**) or azobenzene (**5**), respectively.[17] The azoxybenzene could also be further hydrogenated to azobenzene and on to hydroazobenzene (**6**) and aniline (**7**). Meanwhile the phenyl hydroxylamine is further hydrogenated to aniline. This process is illustrated in Scheme 1. However, Gelder *et al.* used kinetic isotope effects and the relative rates of reaction between nitrobenzene and nitrosobenzene to show that the nitrosobenzene species could not be the main intermediate in the reaction mechanism over Pd/C.[18] They proposed that a PhNOH intermediate was formed which is hydrogenated to phenylhydroxylamine. In this case, the hydrogen involved results in a kinetic isotope effect however, in the nitrosobenzene reaction, the intermediate is dimerised to give azoxybenzene where no hydrogen is involved and no kinetic isotope effect is observed. A similar idea was proposed by Richner *et al.* who used infra-red spectroscopic analysis of the species adsorbed on the catalyst as well as

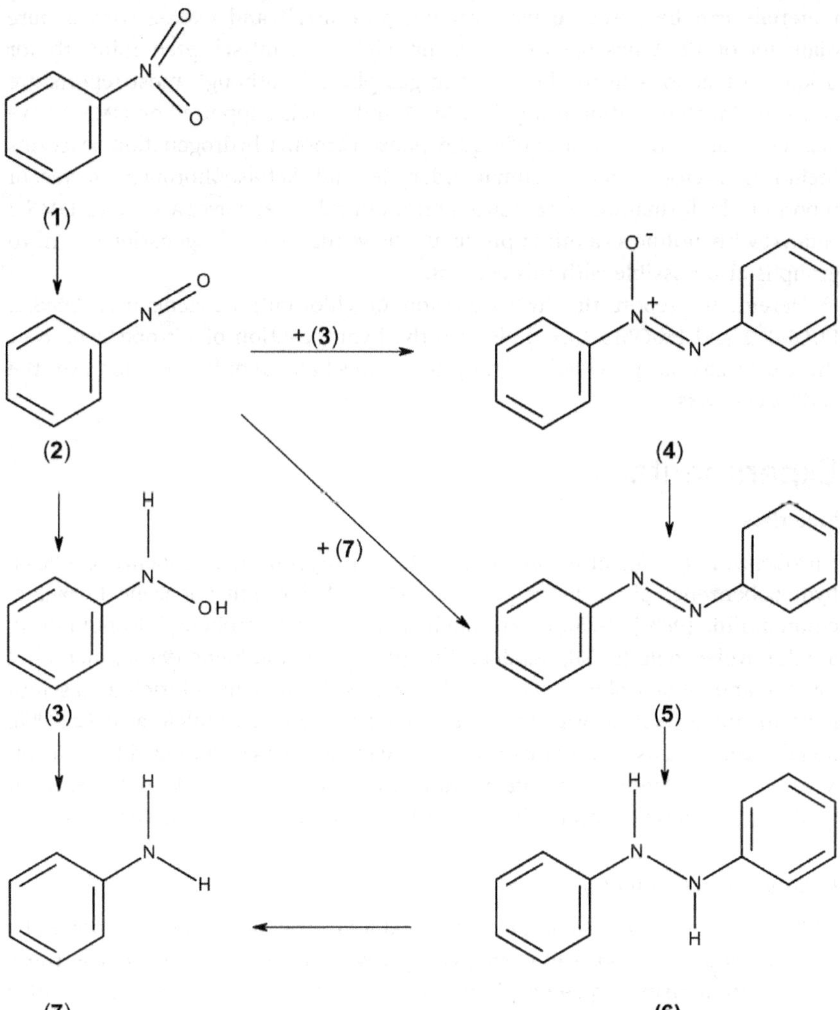

Scheme 1 Mechanism of hydrogenation of nitrobenzene as proposed by Haber.

in the liquid phase to propose the direct hydrogenation mechanism through nitrosobenzene and phenylhydroxylamine over Au/TiO_2. Therein, the hydrogenation of nitrosobenzene was so fast that it was not observed either in the liquid phase or on the catalyst surface;[19] however, adsorbed phenylhydroxylamine was proposed as an adsorbed species. Hydrogenation of either nitrosobenzene or phenyl hydroxylamine formed azoxybenzene and with nitrosobenzene poisoning the catalyst the subsequent reaction pathway depended on the route taken to form azoxybenzene. Currently, as the reaction mechanism is still unclear, it cannot be said whether this reaction occurs *via* different routes over different metals, supports or under different experimental conditions.

Octahedral molecular sieves (OMS) are porous materials based on linked MnO_6 octahedra which utilise the redox behaviour of manganese for use in a variety of applications including catalysis, adsorption[20] and in batteries.[21] These materials can be prepared with varying pore size[22] and OMS-2 with a pore diameter of 4.6 Å has been used extensively as a catalyst[23] predominantly for oxidation reactions in the liquid[24] and gas phase.[25] Although most reports are concerned with oxidation catalysis, OMS-2 and metals supported on OMS-2 have also been used in a number of liquid phase carbonyl hydrogenation reactions including acetophenone,[26] cinnamaldehyde and ketoisophorone.[27] A recent report on the formation of azobenzene from nitrobenzene over a 15% Ag–OMS-2 catalyst with aniline as a minor product[28] shows that the hydrogenation of a nitro group is also possible with this catalyst.

Herein, we report the hydrogenation of chloronitrobenzene over OMS-2, Pt/OMS-2 and Pd/OMS-2. A study into the hydrogenation of nitrobenzene over these catalysts has provided insight into the mechanism of the reaction over the OMS-2 catalysts.

Experimental

Materials

Nitrobenzene (99%), nitrosobenzene (97%), azoxybenzene, azobenzene (98%), hydrazobenzene, phenylhydroxylamine (97%), 4-chloronitrobenzene (99%), 4-chloroaniline (98%), 4-chlorobenzaldehyde (97%), 4-chlorobenzylalcohol (99%), 4-iodonitrobenzene (98%), 4-iodoaniline (98%), 4-iodotoluene (99%), methanol (99.9%) and 1-butanol (99.7%) were all obtained from Sigma Aldrich. Potassium permanganate (≥99%) was obtained from Fluka as was maleic acid (≥98%). Metal precursors used were platinum nitrate (assay 15.14%) supplied by Johnson Matthey and palladium(II)nitrate hydrate procured from Sigma Aldrich. Research grade hydrogen was supplied by BOC. All chemicals were used as supplied.

Catalyst preparation

OMS-2 was synthesized *via* a sol–gel method following the procedure reported by Genuino *et al.*[25] A 5 wt% Pt/OMS-2 catalyst was prepared by incipient wetness impregnation using aqueous platinum nitrate as the metal precursor. After impregnation, the material was dried at 120 °C for 12 h followed by calcination at 450 °C in air for 4 h. The 5 wt% Pd/OMS-2 catalyst was prepared by wet impregnation using palladium nitrate in methanol as the metal precursor at room temperature under agitation for 24 h. The catalyst was dried by removing methanol *in vacuo* and then the material was calcined in air at 450 °C for 4 h. All catalysts were ground and sieved to a less than 45 μm particle size for use in liquid phase hydrogenation reactions. 5 wt% Pt/Al_2O_3 (code 125) was obtained from Johnson Matthey and used as received.

Catalyst characterization

X-ray diffraction measurements were performed using Cu Kα radiation (1.5405 Å) on a PANalytical X'PERT PRO MPD diffractometer in reflection geometry using a NaI scintillation counter, a curved graphite crystal monochromator and a nickel filter. The scattered intensities were collected from 5° to 90° (2θ) by scanning at 0.017° (2θ) steps with a counting time of 0.5 s at each step.

The surface area, pore volume and average pore diameter were measured by N_2 adsorption–desorption isotherms at 77 K using Micromeritics ASAP 2010. The pore size was calculated in the adsorption branch of the isotherms using the Barrett–Joyner–Halenda (BJH) method and the surface area was calculated using the Brunauer–Emmett–Teller (BET) method.

The metal content of samples of OMS-2 and metal supported on OMS-2 was determined by ICP-OES using a PE Optima spectrometer.

The transmission electron microscopy (TEM) studies were performed on a JEOL 2010-FEG instrument, at 100 kV. Platinum particle size distributions were obtained by counting up to 100 particles.

XPS spectra were obtained using a Kratos AXIS Ultra DLD XPS spectrometer with monochromated Al K_α X-rays and a hemispherical analyser with a pass energy of 160 eV. The powdered samples were mounted on conducting copper tape and the binding energies were normalized to the C 1s signal from adventitious carbon at 284.6 eV. Background subtraction was performed using a Shirley background and CasaXPS.[29]

Typical reaction sequence

All reactions were performed in a Hazard Evaluation Laboratory (HEL) autoclave reactor incorporating a gas inducing impeller, baffles and online hydrogen consumption monitoring. The reactor was heated by an internal heating coil and the temperature was monitored by a thermocouple. Hydrogen pressure was maintained throughout the reaction by addition through a mass flow controller.

The reactor was charged with a catalyst (0.03 g) and methanol (30 cm^3) and the reactor was sealed and purged with hydrogen before heating to 60 °C while stirring at 800 rpm. Upon reaching 60 °C the catalyst was reduced by stirring at 1000 rpm for 1 h under 1 bar H_2. The substrate (0.002 mol), 1-butanol (0.4 cm^3) as the internal standard and fresh solvent (30 cm^3) were then added to the reactor which was sealed, purged with hydrogen and heated to the reaction temperature, with stirring at 800 rpm. When the required temperature was reached the stirring was stopped and the reactor pressurised to the required pressure for the reaction which was started by commencing the stirring at 1400 rpm. Reaction samples (2 cm^3) were removed over time, passed through a syringe filter to remove the catalyst and analyzed by GC using a Perkin-Elmer Clarus 500 GC equipped with a FID and an Agilent HP-5 column. For the chloronitrobenzene and halotoluene reactions the products were identified by GC/MS analysis and those for the nitrobenzene reactions were compared with known standards. It was not possible to separate the pure samples of azobenzene and hydrazobenzene using this method on the GC.

Results and discussion

Catalyst characterization

N_2 adsorption–desorption isotherms of OMS-2, Pt/OMS-2 and Pd/OMS-2 catalysts showed a type IV behaviour with a hysteresis loop of type H3, which is characteristic of aggregates of particles with rod-like morphology with tunnel or slit type pores.[30] The BET surface area for OMS-2 was 59.0 $m^2\ g^{-1}$ and on addition of metal, the surface area of the catalyst decreased to 43.2 $m^2\ g^{-1}$ for Pt/OMS-2 and 43.9 $m^2\ g^{-1}$ for Pd/OMS-2. The surface textural properties of the catalysts used are shown in Table S1.†

The XRD patterns obtained for Pt/OMS-2 and Pd/OMS-2 only show peaks assigned to the OMS-2 support (Fig. S1†). The diffractograms show that the material was a pure cryptomelane phase with no impurities of other manganese oxide forms. No peaks due to Pd or Pt were observed by XRD suggesting Pd and Pt particles of less than 4 nm in diameter were present. This is consistent with the TEM characterisation of the Pt/OMS-2 catalyst (Fig. S2†) which shows uniformly distributed spherical Pt nanoparticles with a mean diameter of ∼ 2 nm and the OMS-2 molecular sieves having a rod like morphology.

Hydrogenation of chloronitrobenzene and halogenated reagents

It is desirable for the halogen atom to be retained in the product following hydrogenation giving a high selectivity to halogenated products as well as minimising catalyst poisoning from strong adsorption of halogens on the active metal sites. Fig. 1 shows the selectivity to halogenated products for the hydrogenation of chloronitrobenzene over OMS-2, Pt/OMS-2 and Pd/OMS-2. The reaction over OMS-2 formed chloroaniline as the major product with selectivities of >96% over the temperature range of 25–100 °C. Similarly, high selectivities to halogenated products were observed over Pt/OMS-2 while for a commercial 5% Pt/Al_2O_3 catalyst, the selectivity to halogenated products was reduced to 89% for a reaction at 25 °C and further decreased with increasing temperature. At elevated reaction temperatures over OMS-2 and Pt/OMS-2, no change in selectivity was observed with selectivities to halogenated products of 99% observed at 100 °C over both catalysts.

In contrast, significant dehalogenation of chloronitrobenzene was observed over Pd/OMS-2 where 62% selectivity to all halogenated products was observed again with little variation on increasing temperature. At low chloronitrobenzene conversions (<20%) the selectivity was the highest to nitrobenzene (38.6%) and aniline (37.8%) with a selectivity to chloroaniline of only 8.5%. Therefore, the initial reaction occurring on Pd/OMS-2 is rapid dehalogenation and subsequent hydrogenation of nitrobenzene to aniline. As the reaction proceeded to complete conversion of chloronitrobenzene, the selectivity to chloroaniline increased to 30.3% while the selectivity to aniline decreased to 30.4% as a result of reduction of the aromatic ring. The increase in chloroaniline selectivity suggests that the

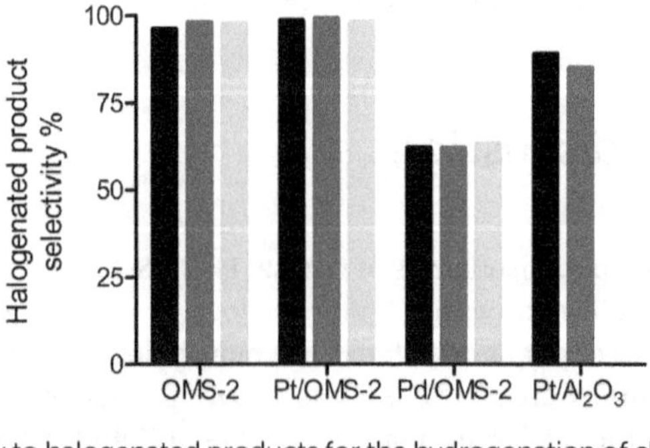

Fig. 1 % selectivity to halogenated products for the hydrogenation of chloronitrobenzene over OMS-2, Pt/OMS-2, Pd/OMS-2 and Pt/Al_2O_3 at 2 bar H_2 and reaction temperatures of 25 °C (black), 50 °C (mid grey) and 100 °C (light grey).

catalyst is either modified by the presence of chlorine or the strong adsorption of the products on the surface which alters the adsorption geometry of chloronitrobenzene. Alternatively, this adsorption could result in a poisoning of the metal which forces the hydrogenation to occur on the OMS-2 support leading to lower hydro-dehalogenation occurring. The further hydrogenation of aniline through to cyclohexane has been previously reported for hydrogenation of chloronitrobenzene over Au and Pd supported on alumina catalysts[16] and selectivity to these products is observed to increase as aniline reacts further over Pd/OMS-2.

To determine the extent to which the halogen is stable under hydrogenation conditions, a range of substrates were tested under different reaction conditions. It is known that iodo-based substrates are less stable to dehalogenation than the analogous chloro molecules, therefore, iodonitrobenzene was tested over OMS-2 under the same reaction conditions as chloronitrobenzene. Here the selectivity to halogenated products decreased to 62% as expected from the relative carbon–halogen bond strengths. Fig. 2 shows that, at low conversions, there was an initially rapid dehalogenation to produce up to 77% selectivity to nitrobenzene. However, as the conversion increased there was no further dehalogenation and the selective hydrogenation of iodonitrobenzene occurred to give a selectivity of 38.5% iodoaniline. It is proposed that iodide formed from dehalogenation adsorbed strongly on the sites responsible for dehalogenation thereby poisoning them to further reaction or modifies the adsorption geometry for the substrate promoting nitro group reduction *vs.* hydro-dehalogenation. In contrast, Fig. 3 shows that the extent of dehalogenation of iodotoluene was again high over the palladium catalyst whilst, even with this easily dehalogenated substrate at elevated temperatures of 100 °C, there was negligible reaction over either OMS-2 or Pt/OMS-2 with only 3.4% conversion observed after 3 h.

To exemplify the system further, hydrogenation of a halogenated carbonyl compound was tested using chlorobenzaldehyde. The results given in Fig. S3† show that for OMS-2 and Pt/OMS-2 there was less than 1% dehalogenation with the major product being chlorobenzyl alcohol. However, when the reaction was carried out over Pd/OMS-2, a rapid reaction occurred to form chlorobenzyl alcohol (58.9% selectivity to chlorobenzyl alcohol at 10% conversion of chlorobenzaldehyde) with the major products at the complete conversion of

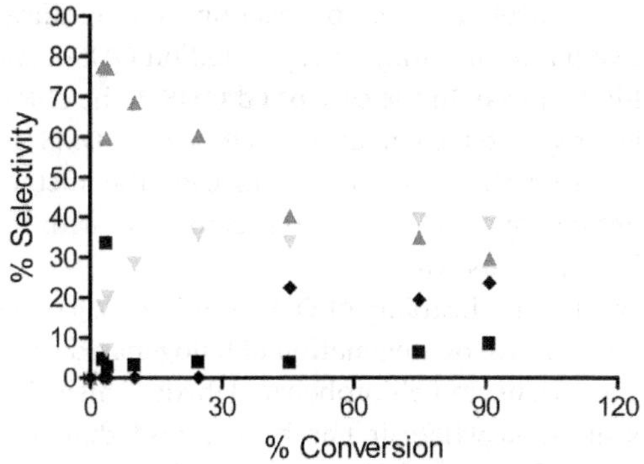

Fig. 2 % selectivity as a function of % conversion for hydrogenation of iodo-nitrobenzene (0.002 mol) to nitrobenzene over OMS-2 at 2 bar H_2 and 25 °C in 60 ml methanol; nitrobenzene (▲), aniline (■), iodoaniline (▼) and coupling products (◆).

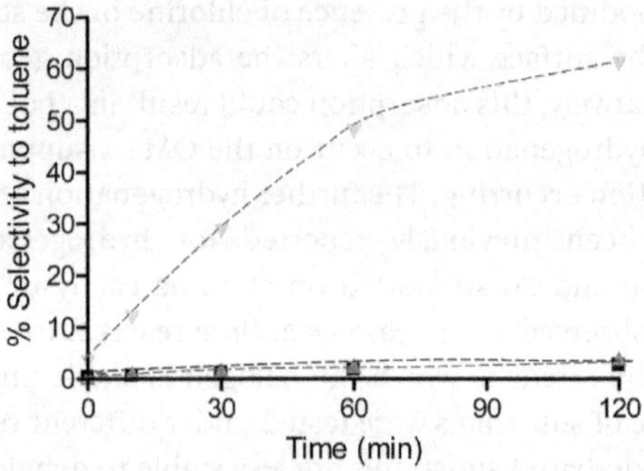

Fig. 3 % selectivity to toluene for reaction of iodotoluene over OMS-2 (■), Pt/OMS-2 (▲) and Pd/OMS-2 (▼). Reactions were carried out using 0.1 g catalyst and 0.002 mol iodotoluene at 2 bar H_2 and 100 °C in 60 cm^3 methanol.

chlorobenzaldehyde found to be dechlorinated benzaldehyde (50.0%) and toluene (38.9%). In the case of Pt/Al_2O_3, the reaction profile was initially very different to the reaction over any of the OMS-2 supported catalysts. In the case of the alumina supported catalyst, half of the substrate formed an acetal with the methanol solvent before the first sample could be taken. However, there was no dehalogenation with the acetal and chlorobenzaldehyde with both undergoing hydrogenation to a chlorobenzyl alcohol and a small amount of di-chlorobenzyl ether (3%), which was probably formed through the reaction between chlorobenzaldehyde and chlorobenzyl alcohol.

In general, the selectivity observed over OMS-2 and Pt/OMS-2 is the same for all reagents although the rate of reaction is typically faster over the platinum containing catalyst. The rate of reaction is much faster over Pd/OMS-2 with, in general, significantly different selectivities being observed compared with that of OMS-2 and Pt/OMS-2 which is mainly associated with dehalogenation reactions. The similarity in the selectivity for OMS-2 and Pt/OMS-2 indicates that the reaction over the Pt containing catalyst may be occurring over the support rather than the metal. The faster rate of reaction is found due to the more efficient dissociation of hydrogen over the metal which can then spillover onto the support, as reported previously for the hydrogenation of unsaturated aldehydes and ketones.[30] A comparison of the rates of reaction and significant differences in selectivity of the two platinum catalysts supported on OMS-2 and alumina is also consistent with this proposal. In the case of Pd/OMS-2, the chloronitrobenzene is converted significantly faster than over Pt/OMS-2 or OMS-2 with a different selectivity implying that the reaction occurs over the metal rather than the support. The higher rate of dehalogenation is consistent with the reported ability of palladium to facilitate this reaction.

This study shows the applicability of OMS-2 either on its own or as a support for platinum for the selective hydrogenation of halogenated aromatic compounds such as chloronitrobenzene and chlorobenzaldehyde. This is in contrast to other supports such as alumina where it has been found that little hydrogenation occurs in the absence of a metal.[31] Although the presence of platinum allows for a faster reaction, the addition of higher catalyst loading of inexpensive OMS-2 would be preferable.

XPS of catalysts after reaction: hydrogenation of chloronitrobenzene

XPS spectra of OMS-2, Pt/OMS-2 and Pd/OMS-2 fresh and used catalysts are shown in Fig. 4. The Mn $2p_{3/2}$ photoelectron peak is positioned at 641.7 eV which can be assigned to the Mn^{4+} state (Fig. 4a). The O 1s photoelectron peak was composed of two features (O_α and O_β) indicating the presence of oxygen species in a mixed valence state (Fig. 4b). The peak positioned at a binding energy of 529.2 eV is characteristic of lattice oxygen (O^{2-}) and the peak positioned at 532.2 eV could be assigned to the surface oxygen ions with lower valence or in the form of hydroxyl, as shown in ref. 32 and 33. The corresponding concentration of O_α calculated from the relative peak areas [$O_\alpha \times 100/(O_\alpha + O_\beta)$] indicated that OMS-2 has a ~16% lattice oxygen concentration which can be depleted and replenished with ease; however, the lattice oxygen concentration increases in the presence of metal to ~33% for Pd/OMS-2 and ~47% for Pt/OMS-2. Fig. 4c shows the peak fit of Pt 4f photoelectron spectral lines. In the case of the fresh Pt/OMS-2 catalyst, the $4f_{7/2}$ peak positioned at the binding energy of 74.8 eV was assigned to PtO_2. In the case of the used Pt/OMS-2 catalyst, the $4f_{7/2}$ feature is shifted to a lower binding energy with 2 components fitted at 71.9 and 73.6 eV. The $4f_{7/2}$ peak positioned at the binding energy of 71.9 eV could be assigned to PtO and the $4f_{7/2}$ peak positioned at the binding energy of 73.6 eV could be assigned to PtO sites with high oxygen concentration.[34] In fresh Pt/OMS-2, Pt metal is present as PtO_2 on the catalyst surface resulting from calcination in air, but under the experimental conditions at 373 K in the presence of H_2 1 MPa, Pt is reduced to Pt^0 metal. In our previous work, we have shown that *in situ* pre-reduction of a catalyst in the liquid phase is a very effective technique to reduce Pt to its metallic state as evident from the Pt L_{III} edge HERFD-XANES spectrum of the catalyst.[35] Fig. 4d shows the peak fit of Pd 3d spectral lines. The $3d_{5/2}$ peak, positioned at a binding energy of 337.7 eV was assigned to PdO, while a small component at the binding energy of 335.3 eV can be assigned to the Pd^0 state. In the case of the used Pd/OMS-2 catalyst, a Pd $3d_{5/2}$ peak was observed in a similar position to that of the fresh Pd/OMS-2 catalyst, positioned at a binding energy of 337.9 eV, but with significantly enhanced peak

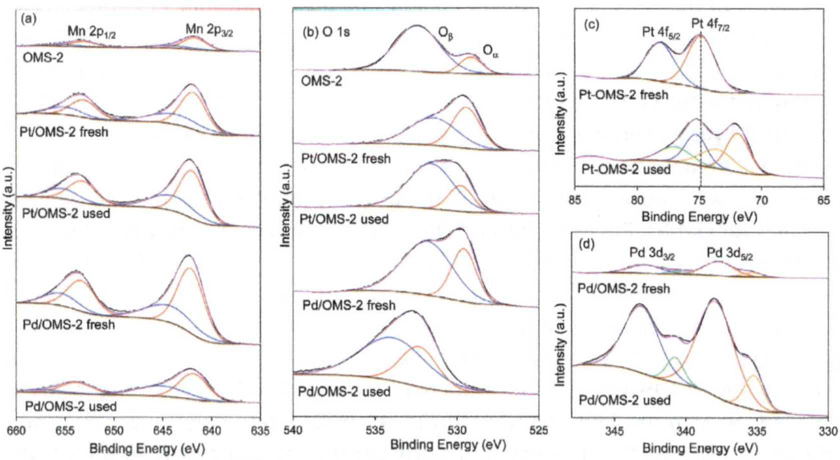

Fig. 4 XPS of fresh and used Pt and Pd/OMS-2 catalysts for the hydrogenation of chloronitrobenzene at 25 °C and 2 bar H_2.

areas indicating a higher surface coverage of Pd nanoparticles, possibly due to redispersion of the Pd metal on the catalyst support. Increased Pd nanoparticle dispersion is also indicated from the decrease in the peak areas of Mn $2p_{3/2}$ features in the used Pd/OMS-2 catalysts in comparison to the fresh Pd/OMS-2 catalyst (Fig. 4a). This has been observed previously for supported Au catalysts on exposure to halogenated molecules.[36] Therein, Au catalysts supported on a range of carbon and oxide supports showed the formation of dimers and trimers of gold stabilised by the halogen from large metal nanoparticles.

Mechanistic study

As the hydrogenation of chloronitrobenzene over OMS-2 and Pt/OMS-2 resulted in a large number of intermediate products which were not available commercially, the mechanistic study of this reaction was carried out using nitrobenzene. This was considered appropriate as negligible dehalogenation occurred over these catalysts meaning the halogen did not influence surface adsorption or catalysis.

Hydrogenation of nitrobenzene

The reaction profiles for the hydrogenation of nitrobenzene over OMS-2, Pt/OMS-2 and Pd/OMS-2 at 25 °C and 2 bar H_2 are shown in Fig. 5. The reaction rate is significantly faster over Pd/OMS-2 compared to Pt/OMS-2 and OMS-2 and, as found in the hydrogenation of chloronitrobenzene, the rate of reaction over Pt/OMS-2 is greater than the OMS-2 support. The major product for all catalysts is aniline. Over Pd/OMS-2, the only additional product identified in the liquid phase is nitrosobenzene. In contrast, for reactions over OMS-2 and Pt/OMS-2, coupling products were also observed. In each of these catalysts similar reaction profiles were found with aniline as well as azoxybenzene, and azo/hydrazobenzene and nitrosobenzene being formed. The similarity in the reaction profiles and selectivities for Pt/OMS-2 and OMS-2 again suggests that, as for hydrogenation of the halogenated reagents, the reaction occurs predominantly over OMS-2 while for the Pd catalyst, the reaction occurs predominantly on the metal. This switch in the site for hydrogenation has been suggested to be due to the relative adsorption strengths of the substrate and hydrogen on the metals and the support.[37]

The differing reaction rates and product distribution suggest that the hydrogenation of nitrobenzene could be occurring through different pathways on Pd compared to the OMS-2 and Pt/OMS-2 catalysts. The two pathways are thought to be direct hydrogenation or hydrogenation *via* the formation of coupling products with the predominant route changing with the extent of conversion. At low conversions of nitrobenzene (~10%), the only products observed in the reaction profiles for OMS-2, Pt/OMS-2 and Pd/OMS-2 were nitrosobenzene and aniline. No coupling products were observed at this conversion suggesting that at a high initial surface coverage of nitrobenzene, the direct hydrogenation of nitrobenzene through nitrosobenzene and phenylhydroxylamine to aniline is the major pathway. The selectivity to aniline is higher for Pd at 85.6% and comparable over OMS-2 and Pt/OMS-2 at 59.0% and 63.5%, respectively, suggesting a faster conversion of nitrosobenzene/phenylhydroxylamine intermediates to aniline over the Pd catalyst. As the conversion of nitrobenzene increased to 40%, the aniline selectivity over Pd/OMS-2 increased to 88.3%. In contrast, over Pt/OMS-2 and OMS-2, the coupled product azoxybenzene was formed with azo/hydrazobenzene

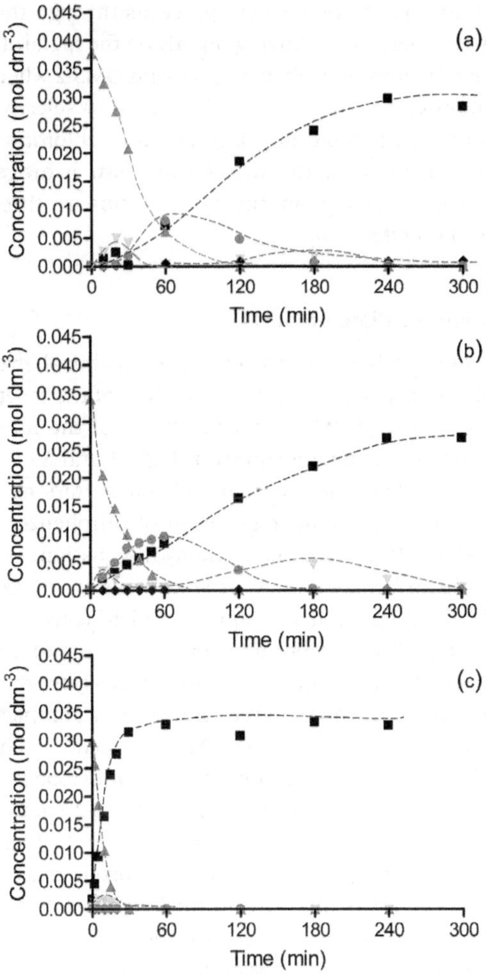

Fig. 5 Reaction profiles for (a) OMS-2, (b) Pt/OMS-2 and (c) Pd/OMS-2 for hydrogenation of nitrobenzene. Reactions were carried out using 0.03 g catalyst and 0.002 mol nitrobenzene at 2 bar H_2 and 25 °C in 60 cm^3 methanol. Nitrobenzene (▲), nitrosobenzene (▼), aniline (■), azoxybenzene (●) and azo/hydrazobenzene (◆).

also being observed for the support alone, giving selectivities to aniline of only 19.7% and 19.1%, respectively.

The formation of coupling products occurs by condensation of either two nitrosobenzene molecules or phenylhydroxylamine and nitrosobenzene. These processes are known to be dependent on the concentration of these species on the catalyst surface with higher concentrations enhancing the coupling reaction. The higher selectivity to nitrosobenzene and coupled products over OMS-2 suggests that this hydrogenation pathway to aniline is occurring on OMS-2. The selectivity at low nitrobenzene conversions suggests that direct hydrogenation can occur over all the catalysts but that the slower rate over OMS-2 and Pt/OMS-2 increases the surface concentration of intermediates, for example nitrosobenzene and phenylhydroxylamine, which can then go through the coupling hydrogenation route. Over palladium the reaction always proceeds through the direct hydrogenation route due to the reaction occurring rapidly on the metal. This is in contrast to the hydrogenation of chloronitrobenzene over Pd/OMS-2 where some coupling products were also observed, albeit at small levels (~7% selectivity). In this case, as discussed previously, the rapid dehalogenation of chloronitrobenzene could result in chloride poisoning of the metal sites with hydrogenation, without dehalogenation, then occurring on the OMS-2 support through a coupling pathway as observed on OMS-2 alone.

Hydrogenation of intermediate products

Reaction of the intermediates nitrosobenzene, hydrazobenzene and azoxybenzene was studied to probe the rate and selectivity of the reactions over OMS-2 (Fig. 6). The reaction over Pt/OMS-2 and Pd/OMS-2 of the intermediates nitrosobenzene and hydrazobenzene are shown in Fig. S4 and S5,† respectively. The hydrogenation of nitrosobenzene, where initial concentrations are much higher than those observed during the hydrogenation of nitrobenzene, shows similar trends to that observed during nitrobenzene hydrogenation. Despite the higher concentration of nitrosobenzene, very little conversion to azoxybenzene was observed over Pd/OMS-2 but rather a rapid complete conversion to aniline. In contrast, the reaction of nitrosobenzene over OMS-2 and Pt/OMS-2 initially showed a fast reaction with high selectivity to azoxybenzene; however, the hydrogenation of nitrosobenzene slowed greatly and complete conversion of nitrosobenzene was not observed for Pt/OMS-2 or OMS-2. The main product initially was azoxybenzene which then formed azo/hydrazobenzene together with a slow formation of aniline. As expected, the Pt/OMS-2 catalyst showed faster hydrogenation of all intermediate products than the OMS-2 support. The main difference between the nitrobenzene and nitrosobenzene reaction profiles is the increased amount of coupled products formed in the latter case, due to the increased concentration of nitrosobenzene on the surface. Reaction of hydrazobenzene, while selectively only forming aniline, was slow and all catalysts were showed to deactivate (Fig. 6, S4 and S5†). This is in line with the slow formation of aniline through the coupled products route observed during hydrogenation of nitrobenzene and nitrosobenzene over OMS-2 and Pt/OMS-2 discussed previously.

To further probe the reaction mechanism, reactions were carried out at higher temperatures. As previously described, at low temperatures, the reaction selectivity at low conversion of nitrobenzene (~10%) followed direct hydrogenation to aniline with only aniline and nitrosobenzene products observed in the liquid phase. At higher conversions of nitrobenzene, the nitrosobenzene concentration increased and, due to the slow hydrogenation of this intermediate and/or poisoning of the catalyst, coupling occurs leading to the much slower hydrogenation route to aniline.

The rate of nitrobenzene hydrogenation over OMS-2 increases with increasing reaction temperature and interestingly, a change in the selectivity is also observed (Fig. 7 and S6†). At temperatures between 25 and 50 °C there is an initial formation of nitrosobenzene and a concomitant increase in coupled products as the concentration of nitrosobenzene increases. As the temperature of the reaction over OMS-2 is increased to 65 °C, there is significantly less azoxybenzene formed

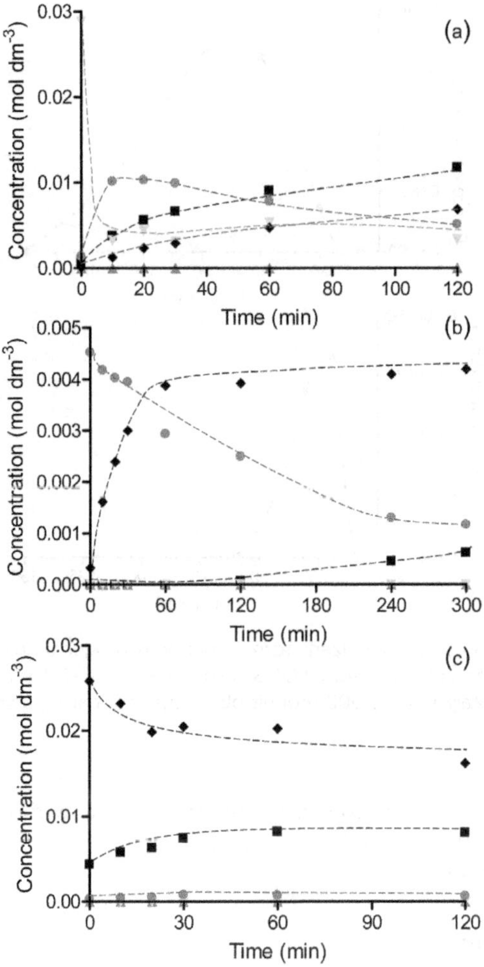

Fig. 6 Hydrogenation of reaction intermediates (a) nitrosobenzene, (b) azoxybenzene and (c) hydrazobenzene over OMS-2. Reaction of nitrosobenzene was carried out using 0.03 g catalyst and 0.002 mol nitrosobenzene at 2 bar H_2 and 25 °C in 60 cm^3 methanol. Reaction of azoxybenzene was carried out using 0.01 g catalyst and 0.0003 mol nitrobenzene at 2 bar H_2 and 25 °C in 60 cm^3 methanol. Reaction of hydrazobenzene was carried out using 0.03 g catalyst and 0.002 mol hydrazobenzene at 2 bar H_2 and 40 °C in 60 cm^3 methanol. Nitrosobenzene (▼), azo/hydrazobenzene (◆), azoxybenzene (●) and aniline (■).

(Fig. 7). Increasing the reaction temperature results in a shift in the reaction pathway favouring the direct hydrogenation pathway over OMS-2. This is likely due to weaker adsorption of nitrosobenzene/an increased rate of hydrogenation of nitrosobenzene and hence lower surface concentration and reduced propensity to undergo coupling.

For reactions over Pt/OMS-2, the same trends are observed as for OMS-2 with the enhancement in direct hydrogenation which occurs at a lower temperature than for OMS-2, at 50 °C rather than 65 °C. The hydrogenation of nitrobenzene over Pd/OMS-2 at 65 °C compared to 25 °C shows a small increase in the reaction

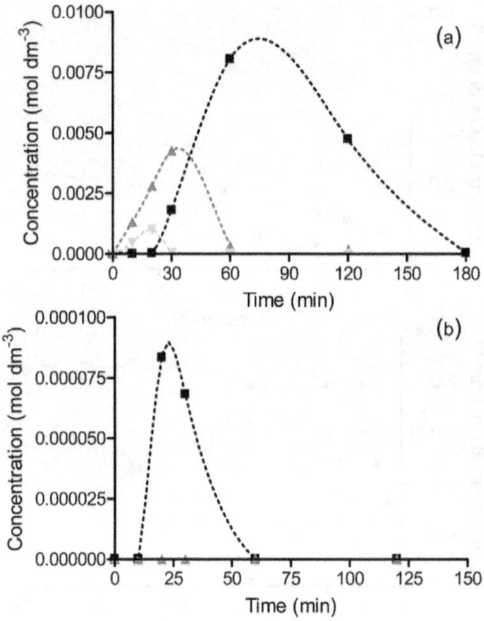

Fig. 7 Concentration of azoxybenzene formed during hydrogenation of nitrobenzene at 25 °C (■), 50 °C (▲) and 65 °C (▼) for (a) OMS-2 and (b) Pd/OMS-2. Reactions were carried out using 0.03 g catalyst and 0.002 mol nitrobenzene at 2 bar H_2 and 25 °C in 60 cm^3 methanol.

rate but no change in the product distribution with only nitrosobenzene and aniline formed with lower selectivity to nitrosobenzene with the increase in reaction temperature.

Conclusions

This study has shown the inherent hydrogenation activity of OMS-2 both on its own as well as a support for platinum group metals. In the case of OMS-2, little hydro-dehalogenation was found, providing a route for non-precious metal based catalytic materials. The rate of the reaction was enhanced by the incorporation of platinum on the surface which provided a more facile pathway for dissociative H_2 adsorption with little change in the selectivity. In contrast, with the use of palladium supported on OMS-2, the reaction proceeds on the metal leading to significant dehalogenated products. The mechanism has been further probed using nitrobenzene as the exemplar molecule. In this case, over Pd/OMS-2 the direct hydrogenation pathway to aniline was preferred with few coupled products being formed. In contrast, over OMS-2 and Pt/OMS-2, both the direct and indirect (*via* the coupled products) pathways were found again consistent with the role of OMS-2 as the hydrogenation site in these catalysts.

Acknowledgements

We acknowledge EPSRC for funding as part of the CASTech grant (EP/G011397/1) and the Department of Employment and Learning for a studentship (IM).

Supporting data are openly available on Queen's University, Belfast Research Portal http://pure.qub.ac.uk/portal/en/datasets.

References

1 H. Blaser, H. Steiner and M. Studer, *ChemCatChem*, 2009, **1**, 210–221.

2 V. Kratky, M. Kralik, M. Mecarova, M. Stolcova, L. Zalibera and M. Hronec, *Appl. Catal., A*, 2002, **235**, 225–231.

3 M. Liu, W. Yu and H. Liu, *J. Mol. Catal. A: Chem.*, 1999, **138**, 295–303.

4 I. Pogorelic, M. Filipan-Litvic, S. Merkas, G. Ljubic, I. Cepanec and M. Litvic, *J. Mol. Catal. A: Chem.*, 2007, **274**, 202–207.

5 S. Ichikawa, T. Seki and T. Ikariyaa, *Adv. Synth. Catal.*, 2014, **356**, 2643–2652.

6 S. Piña Jr, D. M. Cedillo, C. Tamez, N. Izquierdo, J. G. Parsons and J. J. Gutierrez, *Tetrahedron Lett.*, 2014, **55**, 5468–5470.

7 J. Lyu, J. Wang, C. Lu, L. Ma, Q. Zhang, X. He and X. Li, *J. Phys. Chem. C*, 2014, **118**, 2594–2601.

8 M. A. Keane, *ChemCatChem*, 2011, **3**, 800–821.

9 K. C. Christoforidis, E. Serestatidou, M. Louloudi, I. K. Konstantinou, E. R. Milaeva and Y. Deligiannakis, *Appl. Catal., B*, 2011, **101**, 417–424.

10 C. M. King, R. B. King, N. K. Bhattacharyya and M. G. Newton, *J. Organomet. Chem.*, 2000, **600**, 63–70.

11 A. Scrivanti, B. Vicentini, V. Beghetto, G. Chessa and U. Matteoli, *Inorg. Chem. Commun.*, 1998, **1**, 246–248.

12 D. T. Ferrughelli and l. T. Horvath, *J. Chem. Soc., Chem. Commun.*, 1992, 806–807.

13 H. Min, S. Lee, M. Park, J. Hwang, H. M. Jung and S. Lee, *J. Organomet. Chem.*, 2014, **755**, 7–11.

14 R. Baumgartner, G. K. Stieger and K. McNeill, *Environ. Sci. Technol.*, 2013, **47**, 6545–6553.

15 Y. Ukisu and T. Miyadera, *J. Mol. Catal. A: Chem.*, 1997, **125**, 135–142.

16 X. Wang, N. Perret and M. A. Keane, *Chem. Eng. J.*, 2012, **210**, 103–113.

17 F. Haber, *Z. Elektrochem. Angew. Phys. Chem.*, 1898, **22**, 506.

18 E. A. Gelder, S. D. Jackson and C. M. Lok, *Chem. Commun.*, 2005, 522–524.

19 G. Richner, J. A. van Bokhoven, Y. M. Neuhold, M. Makosch and K. Hungerbühler, *Phys. Chem. Chem. Phys.*, 2011, **13**, 12463–12471.

20 P. H. Ho, S. C. Lee, J. Kim, D. Lee and H. C. Woo, *Fuel Process. Technol.*, 2015, **131**, 238–246.

21 W. K. Pang, V. K. Peterson, N. Sharma, C. Zhang and Z. Guo, *J. Phys. Chem. C*, 2014, **118**, 3976–3983.

22 N. Duan, S. L. Suib and C. O'Young, *J. Chem. Soc., Chem. Commun.*, 1995, 1367–1368.

23 C.-H. Chen and S. L. Suib, *J. Chin. Chem. Soc.*, 2012, **59**, 465–472.

24 V. D. Makwana, Y.-C. Son, A. R. Howell and S. L. Suib, *J. Catal.*, 2002, **210**, 46–52.

25 H. C. Genuino, S. Dharmarathna, E. C. Njagi, M. C. Mei and S. L. Suib, *J. Phys. Chem. C*, 2012, **116**, 12066–12078.

26 G. D. Yadav and R. K. Mewada, *Catal. Today*, 2012, **198**, 330–337.

27 H. G. Manyar, B. Yang, H. Daly, H. Moor, S. McMonagle, Y. Tao, G. D. Yadav, A. Goguet, P. Hu and C. Hardacre, *ChemCatChem*, 2013, **5**, 506–512.

28 G. D. Yadav and R. K. Mewada, *Chem. Eng. J.*, 2013, **221**, 500–511.
29 D. A. Shirley, *Phys. Rev. B: Solid State*, 1972, **5**, 4709–4714.
30 K. S. W. Sing, D. H. Everett, R. A. W. Haul, L. Moscou, R. A. Pierotti, J. Rouquerol and T. Siemieniewska, *Pure Appl. Chem.*, 1985, **57**, 603–619.
31 M. Arai, A. Obata, Y. Nishiyama and M. Arai, *J. Catal.*, 1997, **166**, 115–117.
32 R. Wang and J. Li, *Environ. Sci. Technol.*, 2010, **44**, 4282–4287.
33 L. S. Wei, J. H. Li and X. F. Tang, *Catal. Lett.*, 2009, **127**, 107–112.
34 C. R. Parkinson, M. Walker and C. F. McConville, *Surf. Sci.*, 2003, **545**, 19–33.
35 J. Sa, C. Kartusch, M. Makosch, C. Paun, J. A. van Bokhoven, E. Kleymenov, J. Szlachetko, M. Nachtegaal, H. G. Manyar and C. Hardacre, *Chem. Commun.*, 2011, **47**, 6590–6592.
36 J. Sa, S. F. R. Taylor, H. Daly, A. Goguet, R. Tiruvalam, Q. He, C. J. Kiely, G. J. Hutchings and C. Hardacre, *ACS Catal.*, 2012, **2**, 552–560; J. Sa, A. Goguet, S. F. R. Taylor, R. Tiruvalam, C. J. Kiely, M. Nachtegaal, G. J. Hutchings and C. Hardacre, *Angew. Chem., Int. Ed.*, 2011, **50**, 8912–8916.
37 H. G. Manyar, R. Morgan, K. Morgan, B. Yang, P. Hu, J. Szlachetko, J. Sa and C. Hardacre, *Catal. Sci. Technol.*, 2013, **3**, 1497–1500.

Faraday Discussions

PAPER

The development of a new generation of methyl chloride synthesis catalyst

Alastair R. McInroy,[a] John M. Winfield,[a] Chris C. Dudman,[b] Peter Jones[b] and David Lennon*[a]

Received 30th November 2015, Accepted 8th December 2015

DOI: 10.1039/c5fd00202h

In previous work by the authors, aspects of the surface chemistry connected with methyl chloride synthesis over an η-alumina catalyst have been examined. This communication considers a role for Group 1 metal salts to modify the catalytic performance of the well characterised η-alumina catalyst. Firstly, based on a previously postulated mechanism for the reaction of methanol on η-alumina, a mechanism for methyl chloride synthesis over the η-alumina catalyst is proposed. Secondly, the validity of the new mechanism is tested by observing how the (i) type and (ii) loading of the Group 1 metal salt may perturb methyl chloride selectivity. The outcomes of these measurements are rationalised with reference to the postulated mechanism. Overall, this study represents an example of how a proposed reaction mechanism has been used to inform and guide a catalyst development strategy for a large-scale industrial process.

1. Introduction

Methyl chloride is an important industrial product with a global capacity of *ca.* 900 ktonne. It is a valuable starting material in the production of higher chlorinated products, silicones, and methyl cellulose. It also has applications as a methylating agent.[1–3] One method for the large scale manufacture of methyl chloride is performed *via* the hydrochlorination of methanol over a solid acid catalyst at elevated temperatures, eqn (1). Selectivity to methyl chloride is reduced by the reaction shown in eqn (2), where the alcohol is converted to dimethyl ether. The reactions shown in eqn (1) and (2) are both reversible, so at any given feed composition and temperature there is an equilibrium product composition. An improved methyl chloride synthesis catalyst will catalyse reaction (1) more effectively than reaction (2). A catalyst that has received considerable application is γ-alumina.[4,5] Hydrochlorinations using a $ZnCl_2/Al_2O_3$ catalyst also show selectivity to CH_3Cl.[6]

[a]*School of Chemistry, Joseph Black Building, University of Glasgow, Glasgow, G12 8QQ, UK. E-mail: David.Lennon@Glasgow.ac.uk; Tel: +44-(0)-141-330-4372*

[b]*Inovyn, South Parade, Runcorn, Cheshire, WA7 4JE, UK*

$$CH_3OH + HCl \rightarrow CH_3Cl + H_2O \quad (1)$$

$$2CH_3OH \rightarrow CH_3OCH_3 + H_2O \quad (2)$$

In 2000 ICI (UK) filed a patent demonstrating improved yields of CH_3Cl when an η-alumina catalyst was doped with a Group 1 metal salt.[7] In order to improve the understanding of why a doped η-alumina catalyst should convey favourable catalytic performance when applied to synthesis, we examined aspects of synthesis-related surface chemistry over an un-doped commercial grade η-alumina catalyst.[8] Lundie *et al.* probed the surface acidity of η-alumina, identifying and characterising a distribution of active sites. Four different sites were identified: (i) strong, (ii) medium-strong, (iii) medium-weak and (iv) weak Lewis acid sites.[9] That work paved the way for investigations of the adsorption of the reagents (CH_3OH and HCl) on this surface. The interaction of methanol on η-alumina was probed using inelastic neutron scattering (INS) in order to determine the nature and form of the adsorption complex.[10] Further studies of methanol on η-alumina using a combination of infrared spectroscopy and mass spectrometry led to the proposal of a mechanism for the formation of the dimethyl ether by-product.[11] Crucially, the mechanism made a case for site-selective chemistry. Next, from studies of the adsorption of HCl over η-alumina, it was possible to discern how dissociative adsorption of HCl connected with hydroxyl group formation and hydroxyl exchange reactions.[12]

Progressing to the bimolecular reaction over the un-doped η-alumina catalyst, temperature-programmed reaction measurements showed how modest CH_3OCH_3 production tracks the dominant formation of methyl chloride. A coincidence of reaction profiles was attributed to both product and by-product formation processes involving a common intermediate, namely methoxy groups residing in adsorption sites of similar Lewis acidity.[13] Two reviews of these investigations of η-alumina catalysts have recently appeared: Parker and Lennon have described how INS has been used to probe aspects of methyl chloride synthesis related chemistry over η-alumina,[14] and Lennon and Winfield have reviewed adsorption and reactions over η-alumina as part of a wider examination of heterogeneous catalysts used for large-scale syntheses of selected chlorohydrocarbons and fluorohydrocarbons.[15]

Other work carried out elsewhere has produced a series of papers describing various aspects of related chemistry and chemical engineering. Representative studies include the use of micro-reactors[16,17] and the use of $ZnCl_2$ modified zeolite and alumina based catalysts.[18,19] The latter work illustrates how the addition of a zinc salt can modify the acid properties of an γ-alumina catalyst to beneficially influence product yields.

As stated above, we have previously postulated a mechanism for the reaction of methanol over an η-alumina catalyst including site selectivity.[11] Linking this postulate with catalyst characterisation studies,[9] the authors deduce that it should be possible to extend this model to account for actual CH_3Cl production over this substrate. Such a scheme has direct relevance to the industrial operation. Thus, in this paper, we propose a mechanism for the site-selective formation of methyl chloride over an un-doped η-alumina catalyst, Fig. 1.

Following our observations reported in the previous work on η-alumina,[9–13] Fig. 1 shows methyl chloride to be formed from the reaction of adsorbed methoxy

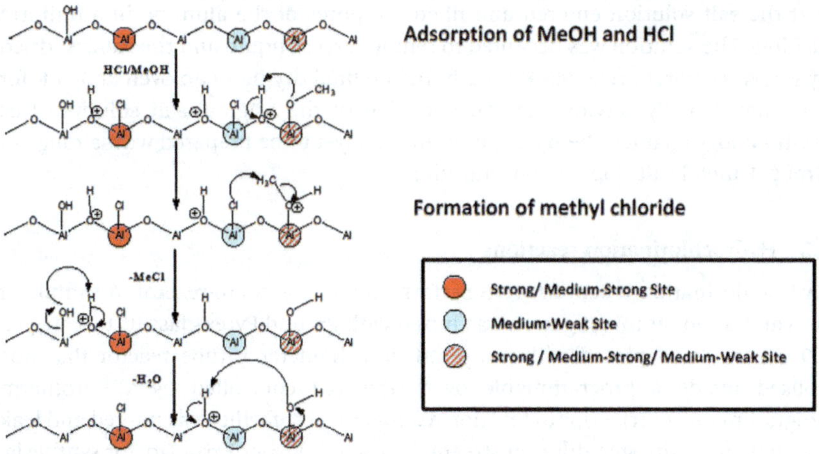

Fig. 1 A schematic representation of the site-selective formation of methyl chloride over η-alumina. The red circles represent a strong/medium-strong Lewis acid site; the blue circles represent a medium-weak Lewis acid site; the red/blue shaded circles represent a strong/medium-strong/medium-weak Lewis acid site. The definition and form of these sites is described in ref. 9.

species and adsorbed chloride. Importantly for methyl chloride production, the chlorine species must be adsorbed on the medium-weak site, as previous temperature-programmed measurements indicate that chloride adsorbed on strong and medium-strong sites is too strongly adsorbed to engage in methyl chloride production.[12,13] In contrast, the methoxy species may be adsorbed on any of the three strongest Lewis acid sites. Following deductions on the principal step for dimethyl ether production over this catalyst,[11] the key step for methyl chloride formation is believed to be concerted C–Cl formation and C–O breakage. The mechanism is consistent with reports on the process of the reverse reaction by Yates and co-workers.[5]

In recognition of the earlier patent report on the benefits of Group 1 metal dopants for η-alumina,[7] the validity of the site-selective chemistry indicated in Fig. 1 is tested by examining how certain additives affect catalytic performance. Given that Li and Na cations may intercalate into the alumina lattice, we have examined (i) the effect of modifying a commercial grade η-alumina with K and Cs salts and (ii) the effectiveness of these new materials for methyl chloride synthesis.

2. Experimental

2.1 Catalyst preparation

The η-alumina catalyst was supplied by Ineos Chlor (Ineos Chlor catalyst ref: 25867);[20] this is the same transition alumina examined in the previous studies and that is characterised in ref. 9. The doping of the alumina catalyst with K and Cs salts was performed by an impregnation method. An aqueous solution of CsCl (Aldrich, purity 99.9%) or KCl (Aldrich, purity 99.999%) of pre-calculated concentration was added to a glass flask *via* a two-way adaptor and dropping funnel whilst under vacuum ($<10^{-2}$ torr). The use of an evacuated vessel ensured that the salt solution entered and filled the pores of the alumina in a uniform fashion. The solution was decanted to remove excess liquid and the catalyst dried by rotary evaporation at 343 K for 2 h, before final drying in an oven at 353 K for a further 2 h. By varying the concentration of the original salt solution, this methodology enabled the base η-alumina catalyst to be prepared with a range of Group 1 metal salt dopant concentrations.

2.2 Hydrochlorination reactions

Hydrochlorination reactions were performed using a micro-reactor. A sample of the catalyst under investigation was mixed with ground Pyrex glass (0.1 g catalyst, 0.9 g Pyrex) and placed within a $\frac{1}{4}$" od mm diameter U-tube reactor that was housed inside a programmable oven (Pye 104 controlled by a Eurotherm programmable unit). Prior to reaction testing, the test facility was purged and leak tested using a nitrogen dilution stream. Methanol (Fisher Extra Dry for synthesis, <0.01% water) was delivered to the reactor by means of a HPLC pump (Gilson) operating with a 0–5 ml min^{-1} pump head. The methanol and anhydrous HCl (Air Products Electronic Grade 99.999% v/v purity) flows were set to 0.045 ml min^{-1} (measured flow 0.040 ml min^{-1} at 20 °C) and 25.9 ml min^{-1} at 20 °C and 1 atmosphere respectively, giving a HCl: methanol v/v ratio of 1.1 : 1. Gas flows were controlled using Brookes Mass Flow Controllers. Products were analysed by in-line gas-liquid chromatography (HP6850, FID, 50 m CP wax 52 column 530 μm diameter 2 μm film thickness). Samples of the reactor exit stream were analysed at 15 min intervals using the GLC. Data were collected using a Perkin Elmer Nelson 900 interface and analysed by Perkin Elmer Turbochrom data handling software. Temperature programmed reaction (TPR) measurements commenced at 443 K, where a temperature ramp of 10 K h^{-1} was applied up to 623 K. A Grant Squirrel data logger recorded reactor temperature and mass flow controller flow rates.

3. Results

3.1 Temperature profiles and activation energies: η-alumina

Fig. 2 presents the temperature-programmed reaction profiles for methyl chloride and dimethyl ether production over the temperature range 453–613 K using the unmodified η-alumina catalyst. Methyl chloride production displays an almost sigmoidal profile, attaining a maximum rate at approximately 600 K. In contrast, the CH_3OCH_3 rate reaches a maximum at ~565 K; thereafter its rate of formation diminishes quite sharply on increasing temperature. With reference to the different intensity ranges used in Fig. 2, formation of CH_3Cl dwarfs the production of CH_3OCH_3; at 560 K and based on relative peak areas, the exit stream comprises 95.3% CH_3Cl and 4.7% CH_3OCH_3. No other hydrocarbons or chlorohydrocarbons were detected in the analysis.

The data from Fig. 2 in the range 487–541 K can be used to generate Arrhenius plots that correspond to constrained methanol conversions of ~20%, Fig. 3. Linearity for both methyl chloride and dimethyl ether production is maintained over this temperature range and yields apparent activation energies of 88.3 (±0.9) and 94.8 (±1.1) kJ mol^{-1} respectively, with the error representing the standard deviation from three replicate runs. The E_a value for dimethyl ether production is comparable with that reported by de Vore *et al.* (98 kJ mol^{-1}) for its formation over

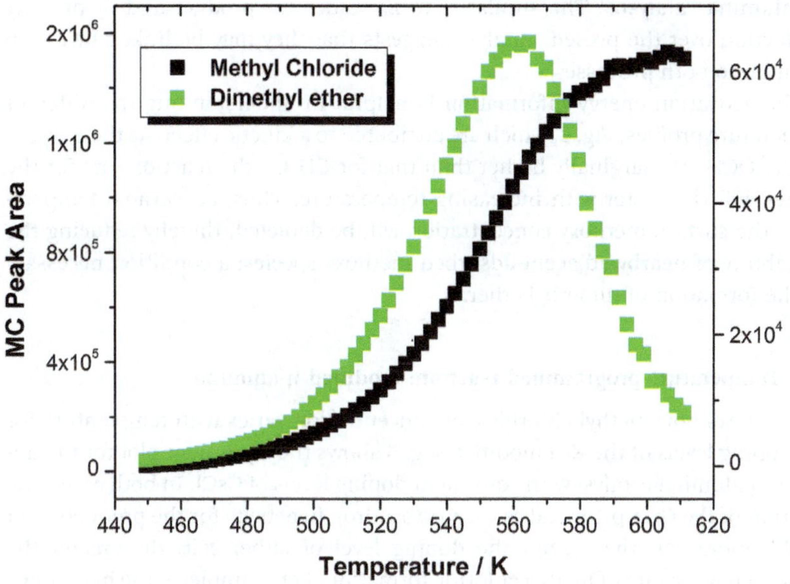

Fig. 2 Temperature-programmed reaction profile for methyl chloride (black squares, scaled according to the left hand side *Y* axis) and dimethyl ether (green squares, scaled according to the right hand side *Y* axis) production over the un-doped η-alumina catalyst. Note the different scales of the two *Y* axes.

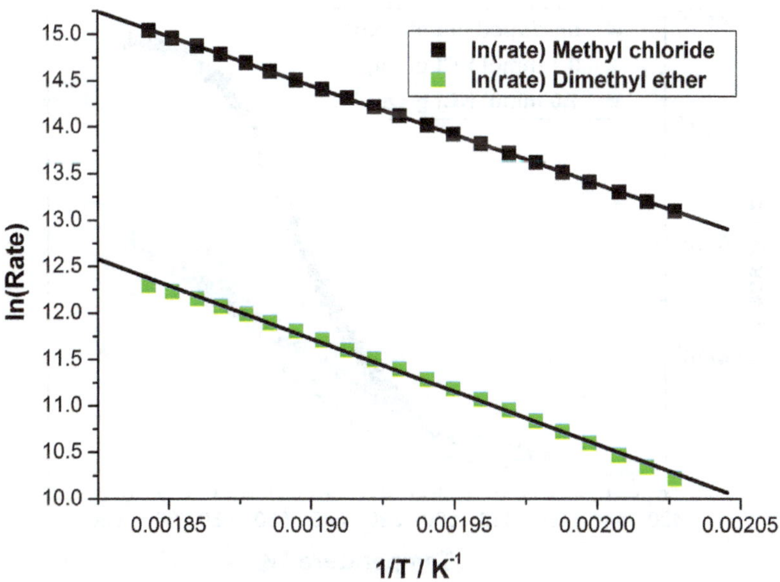

Fig. 3 Arrhenius plots for methyl chloride (black squares) and dimethyl ether (green squares) production over the un-doped η-alumina catalyst.

a γ-alumina catalyst.[21] The similarity in E_a values for product and by-product production over the present catalyst suggests that they may be linked by a step common to both processes.

The activation energy information is helpful in understanding the different temperature profiles, Fig. 2, which are attributed to a kinetic effect. As the E_a value for CH_3OCH_3 is marginally higher than that for CH_3Cl, the reaction rate for the former will rise faster with increasing temperature. Thus, at elevated temperatures, the surface methoxy concentration will be depleted, thereby reducing the probability of nearby/adjacent adsorbed methoxy species; a condition necessary for the formation of dimethyl ether.

3.2 Temperature-programmed reaction: modified η-alumina

Fig. 4 shows how methyl chloride exit concentration varies with temperature for two doping levels of the KCl modifier. Fig. 5 shows the equivalent plot for the un-doped η-alumina catalyst with equivalent doping levels of CsCl. In both cases, the addition of the Group 1 metal salt leads to a drop in activity for the production of CH_3Cl; moreover, the greater the doping level of either salt, the greater the reduction in the rate of methyl chloride formation. For example, at the higher KCl loading, activity with respect to CH_3Cl formation is decreased by 46.3% at 560 K.

Fig. 6 shows how dimethyl ether production varies with respect to temperature for two doping levels of KCl modifier; Fig. 7 shows the equivalent plot for CsCl. Here the effect is more dramatic than that observed for the formation of CH_3Cl (Fig. 4 and 5), where a Group 1 metal salt loading of 1.0 mmol $g_{(cat)}^{-1}$ dramatically reduces the formation of CH_3OCH_3; particularly so with the Cs salt (Fig. 7).

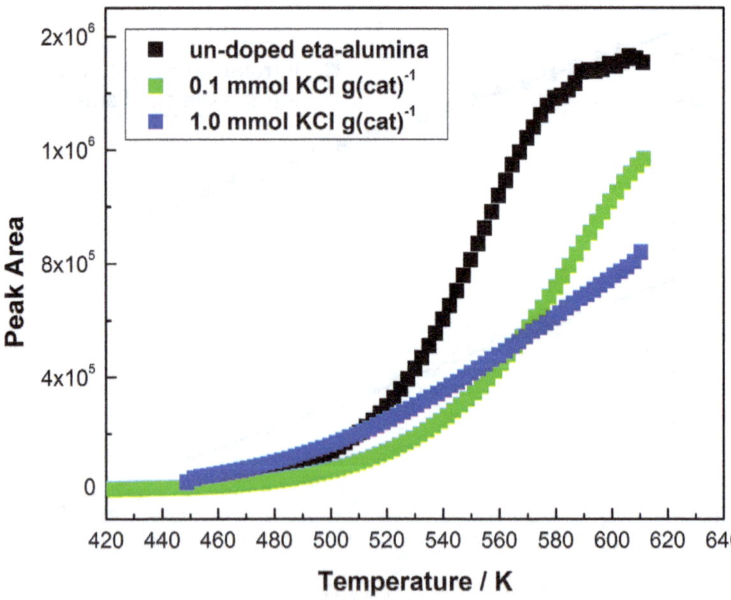

Fig. 4 Plot of methyl chloride concentration as a function of reaction temperature for (a) the un-doped η-alumina catalyst (black squares), (b) η-alumina catalyst + 0.1 mmol KCl $g_{(cat)}^{-1}$ (green squares), and (c) η-alumina catalyst + 1.0 mmol KCl $g_{(cat)}^{-1}$ (blue squares).

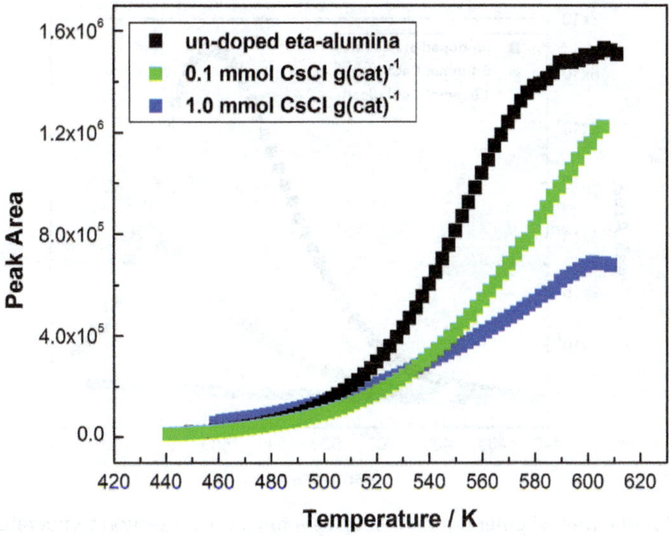

Fig. 5 Plot of methyl chloride concentration as a function of reaction temperature for (a) the un-doped η-alumina catalyst (black squares), (b) η-alumina catalyst + 0.1 mmol CsCl $g_{(cat)}^{-1}$ (green squares), and (c) η-alumina catalyst + 1.0 mmol CsCl $g_{(cat)}^{-1}$ (blue squares).

Specifically, at 560 K and with reference to dimethyl ether production over the plain η-alumina catalyst, Fig. 6 shows the dimethyl ether concentration to be reduced by 92.1% at a loading of 1.0 mmol KCl $g_{(cat)}^{-1}$, whilst Fig. 7 shows the

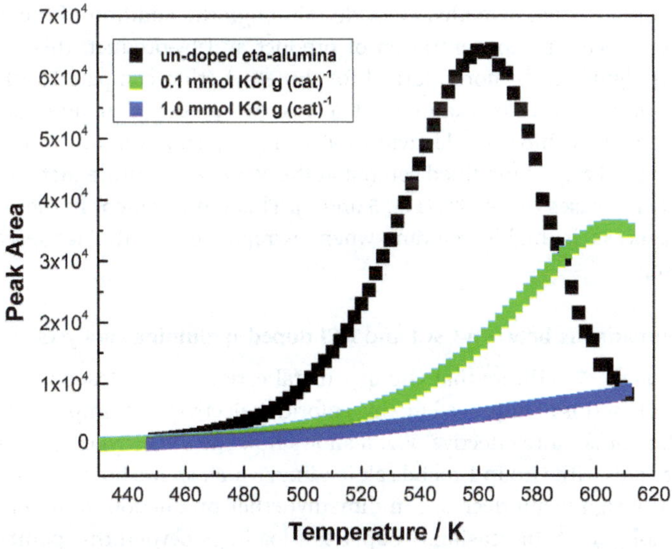

Fig. 6 Plot of dimethyl ether concentration as a function of reaction temperature for (a) the un-doped η-alumina catalyst (black squares), (b) η-alumina catalyst + 0.1 mmol KCl $g_{(cat)}^{-1}$ (green squares), and (c) η-alumina catalyst + 1.0 mmol KCl $g_{(cat)}^{-1}$ (blue squares).

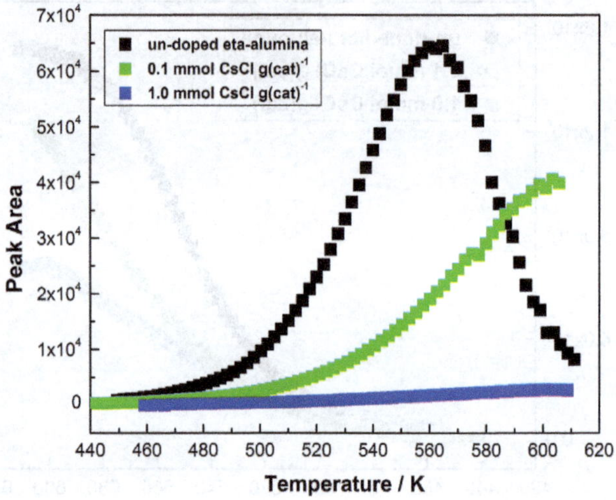

Fig. 7 Plot of dimethyl ether concentration as a function of reaction temperature for (a) the un-doped η-alumina catalyst (black squares), (b) η-alumina catalyst + 0.1 mmol CsCl $g_{(cat)}{}^{-1}$ (green squares), and (c) η-alumina catalyst + 1.0 mmol CsCl $g_{(cat)}{}^{-1}$ (blue squares).

dimethyl ether concentration to be reduced by 97.2% at a loading of 1.0 mmol CsCl $g_{(cat)}{}^{-1}$.

Methyl chloride selectivity, $S_{(MC)}$, is defined according to eqn (3),

$$\text{Methyl chloride selectivity } (\%) = \frac{[\text{methyl chloride}]}{[\text{methyl chloride}] + [\text{dimethyl ether}]} \times \frac{100}{1} \quad (3)$$

Collectively, Fig. 4–7 indicate that the modified alumina catalysts exhibit an improved selectivity for methyl chloride. Although the addition of the chemical modifier reduces the concentration of product and by-product, this decline in activity is significantly more marked for dimethyl ether compared with methyl chloride production. For example, at 573 K the un-doped η-alumina catalyst displays a methyl chloride selectivity of 95% (Fig. 2). However, for the case of the 1.0 mmol CsCl $g_{(cat)}{}^{-1}$ modified catalyst at the same temperature methyl chloride selectivity increases to >99.5% (Fig. 5 and 7). This improvement in the selectivity of the process is highly relevant when extrapolated to the large-scale unit operation.

3.3 Comparisons between CsCl and KCl doped η-alumina catalysts

Whereas Fig. 4–7 indicate that Group 1 metal cation doping leads to favourable outcomes in terms of improved methyl chloride selectivity, it is important to know which dopant is more effective. Fig. 8 shows how the dimethyl ether production rate correlates with Group 1 metal salt loading, which shows both sets of samples to exhibit a significant decrease in dimethyl ether production up to loadings of ~0.3 mmol $g_{(cat)}{}^{-1}$. Increasing Group 1 salt loadings beyond this point leads to little change in by-product concentration. Moreover, it is evident from Fig. 8 that dimethyl ether levels are noticeably higher for KCl doping compared with their CsCl counterparts. Indeed, at the highest modifier concentration studied

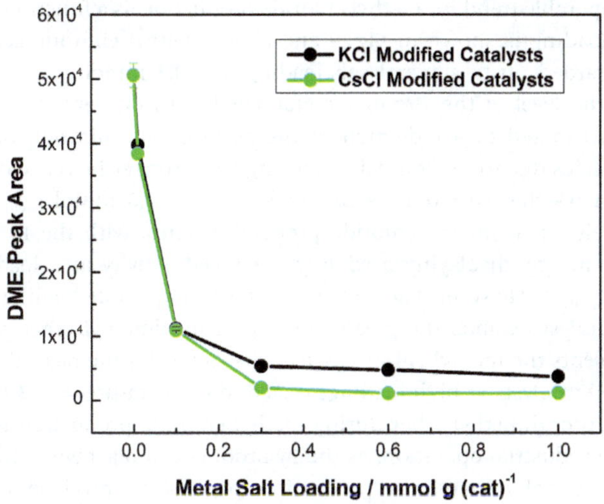

Fig. 8 Dimethyl ether production at 543 K as a function of alkali metal salt loading: (a) KCl modified η-alumina modified catalysts (black circles) and (b) CsCl modified η-alumina modified catalysts (green circles). The error bars represent the range observed for duplicate measurements.

(1.0 mmol $g_{(cat)}{}^{-1}$), dimethyl ether production on the CsCl doped catalyst is one third of that produced over the KCl doped alumina. This action is attributed to Cs^+ possessing a larger ionic radius than K^+ ($Cs^+ = 0.169$ nm, $K^+ = 0.133$ nm)[22] and thereby more efficiently blocking Lewis acid sites.

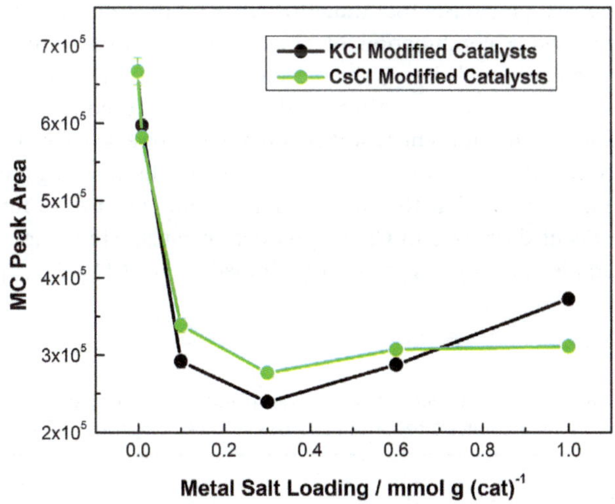

Fig. 9 Methyl chloride production at 543 K as a function of alkali metal salt loading: (a) KCl modified η-alumina modified catalysts (black circles) and (b) CsCl modified η-alumina modified catalysts (green circles). The error bars represent the range observed for duplicate measurements.

The comparable trend for methyl chloride production as a function of Group 1 metal salt loading is shown in Fig. 9 and shows methyl chloride activity to be abruptly retarded up to a metal salt loading of ~0.3 mmol $g_{(cat)}^{-1}$; thereafter increasing the level of the Group 1 metal salt is less influential. The reduced activity is attributed to a reduction in the surface concentration of adsorbed methoxy species due to the ion pair removing the stronger Lewis acid sites.

Fig. 9 shows that Group 1 metal salt loadings ≥0.3 mmol $g_{(cat)}^{-1}$ lead to approximately equal methyl chloride production rates, with the KCl modified alumina seemingly displaying a slightly enhanced activity at a loading of 1.0 mmol KCl $g_{(cat)}^{-1}$. However, Fig. 8 shows that, at comparable loadings, the CsCl modified catalyst exhibits the greater reduction in dimethyl ether production. Table 1 presents the methyl chloride selectivity values for the base alumina plus the modified catalysts at high loadings (reaction temperature = 563 K).

Formation of dimethyl ether during methyl chloride production is problematical in the industrial operation, as the by-product can react with chlorine elsewhere in the chloromethanes production process to produce undesirable chlorinated hydrocarbon derivatives. This means that, from the industrial perspective, enhanced methyl chloride selectivity is the primary objective. Therefore, with reference to Table 1, the CsCl doped catalyst exhibits the more desirable performance: $S_{(MC)} = 99.6\%$. Often in heterogeneous catalysis enhanced selectivity for a specific product is achieved at the cost of activity. That principle is indeed evident here; Fig. 5 shows that doping of the base alumina catalyst with CsCl at a loading of 1.0 mmol CsCl $g_{(cat)}^{-1}$ reduces the methyl chloride production rate at 560 K by 57%. However, this residual activity is economically acceptable and the 5.1% improvement in methyl chloride selectivity indicated in Table 1 minimises significantly purification issues at the end of the commercial process.

4. Discussion

We have proposed previously that dimethyl ether production over an η-alumina catalyst occurs *via* a combination of chemisorbed methoxy units occupying medium-strong Lewis acid sites.[11] The mechanism for methyl chloride formation presented in Fig. 1 proposes that methyl chloride production occurs *via* a combination of chloride chemisorbed on medium-weak sites reacting with methoxy species bound at a combination of strong, medium-strong and medium-weak sites. Fig. 9 shows that Group 1 metal salt loadings of ≥0.3 mmol $g_{(cat)}^{-1}$ lead to significant decreases in CH_3Cl production rates, whilst Fig. 8 indicates that comparable loadings significantly impede the CH_3OCH_3 production pathway.

Table 1 Methyl chloride selectivity values for the η-alumina catalyst and the high loading KCl and CsCl modified catalysts recorded at a reaction temperature of 563 K

Catalyst	Methyl chloride selectivity/%
η-Al_2O_3	94.5
η-Al_2O_3 + 1.0 mmol KCl $g_{(cat)}^{-1}$	99.0
η-Al_2O_3 + 1.0 mmol CsCl $g_{(cat)}^{-1}$	99.6

In order to understand better the site-selective chemistry connected with methyl chloride production, it is informative to review the mechanism proposed for CH_3OCH_3 production.[11] One way to eliminate essentially all dimethyl ether formation would be for the Group 1 metal salt to "cap" the strong and medium-strong Lewis acid sites. Adopting this assumption, the profile observed in Fig. 8 is interpreted to indicate that loadings of $\geq$0.3 mmol $g_{(cat)}^{-1}$ lead to the strong and medium-strong Lewis acid sites being deactivated by association with a Group 1 metal salt ion pair. Fig. 10 presents a variant of the original η-alumina/methanol mechanism (*i.e.* no HCl involvement) proposed in ref. 11 but, crucially, the area of the schematic diagram bounded by the solid black line signifies the "capping" of the strong and medium-strong Lewis acid sites. The unavailability of these sites to the incident methanol molecules prevents formation of CH_3OCH_3. This proposal is entirely consistent with the observations presented in Fig. 6–8. It is also noted that the capping of the strong Lewis acid sites will prevent the formation of surface formate species *via* the evolution of CH_4(g) and H_2(g), as previously reported for the un-doped η-alumina catalyst.[11]

The logic applied above for by-product formation can be readily extended to the industrially relevant formation of methyl chloride over η-alumina catalysts and is represented schematically in Fig. 11, which is a variant of Fig. 1. The shaded area of Fig. 11 indicates the inaccessibility of the strong and medium-strong Lewis acid sites that occurs at a Group 1 metal loading of 1.0 mmol $g_{(cat)}^{-1}$. This leaves the medium-weak Lewis acid site as the active site: both methoxy and Cl chemisorb at this class of site and combine to produce CH_3Cl selectively ($S_{(MC)}$ = 99.6%). This proposal is consistent with the profiles depicted in Fig. 5 and 7. Importantly, the last step shown in Fig. 11, which results in the evolution of water,

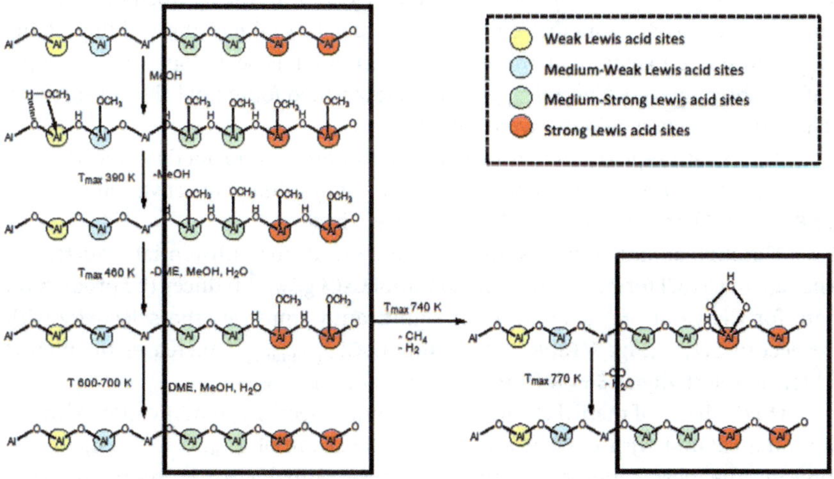

Fig. 10 Proposed methanol transformations during temperature programming of an η-alumina catalyst doped with 1.0 mmol $g_{(cat)}^{-1}$ of a Group 1 metal salt. The areas bounded by the solid black line represent chemistry no longer accessible due to the Group 1 metal "capping" the strong and medium-strong Lewis acid sites. This mechanism is a variation on that proposed for the reaction of methanol on η-alumina,[11] that is reproduced with permission from Lundie *et al.*, *J. Phys. Chem. B*, 2005, **109**, 11592. Copyright 2005 American Chemical Society.

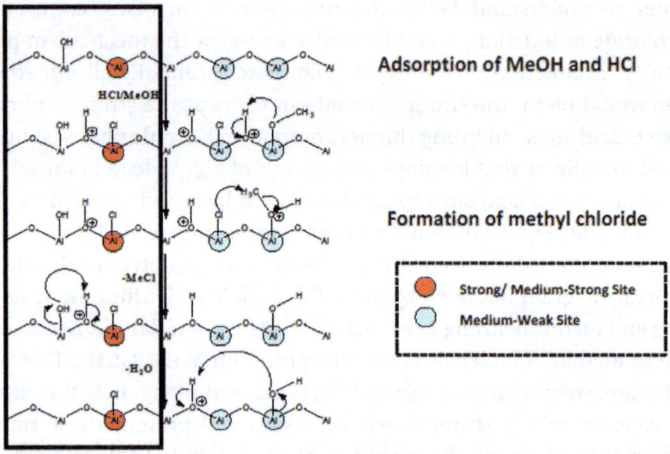

Fig. 11 Proposed reaction mechanism for methyl chloride production over an η-alumina catalyst doped with 1.0 mmol $g_{(cat)}^{-1}$ of an alkali metal salt. The area bounded by the solid black line represents chemistry no longer accessible due to the alkali metal "capping" of the strong and medium-strong Lewis acid sites.

constitutes an active site regeneration step that allows the catalytic cycle to continue; this is a pre-requisite for steady-state operation.

5. Conclusions

A previously reported mechanism for the reaction of methanol on an η-alumina catalyst[11] has been extended and adapted to account for the hydrochlorination of methanol to produce methyl chloride. In order to evaluate the viability of the newly proposed mechanism, the η-alumina catalyst has been doped with Group 1 salts in the range 0–1.0 mmol $g_{(cat)}^{-1}$ and re-examined for methyl chloride activity. The following conclusions have been drawn.

• Temperature-programmed reaction measurements for methyl chloride and dimethyl ether over the un-doped η-alumina catalyst indicate activation energies of 88.3 (±0.9) and 94.8 (±1.1) kJ mol^{-1} respectively.

• Although doping of the η-alumina catalyst leads to a substantial reduction in the rate of CH_3Cl formation (Fig. 5, 1.0 mmol CsCl $g_{(cat)}^{-1}$ reduces the production rate for CH_3Cl at 560 K by 57%), it significantly improves the selectivity with respect to CH_3Cl (Fig. 7/Table 1: 1.0 mmol CsCl $g_{(alumina)}^{-1}$ increases the methyl chloride selectivity at 563 K from 94.5% to 99.6%).

• On the basis of methyl chloride activity and selectivity, modification of the η-alumina catalyst with CsCl at a loading of ≥0.3 mmol $g_{(cat)}^{-1}$ is considered to present the best catalyst formulation for sustained CH_3Cl production with minimal formation of CH_3OCH_3 where kinetic control of the product composition is possible.

• With reference to the modified reaction mechanism, this improvement in product selectivity is attributed to Group 1 salt ion pairs "capping" the strong and medium-strong Lewis acid sites of the η-alumina catalyst,[9] with methyl chloride production occurring solely on medium-weak Lewis acid sites.

Acknowledgements

Ineos Chlor and the EPSRC are thanked for project support. The EPSRC are thanked for the provision of a postgraduate studentship (ARM).

References

1 K. Weissermel and H.-J. Arpe, in *Industrial Organic Chemistry*, VCH Wiley, 3rd edn, 1997, p. 72.
2 V. A. Makhlin and S. I. Ivanov, *Kinet. Catal.*, 1997, **38**, 865.
3 A. M. Becerra, A. E. Castro Luna, D. E. Ardissone and M. I. Ponzi, *Ind. Eng. Chem. Res.*, 1992, **31**, 1040.
4 J. E. Crowell, T. P. Beebe Jr. and J. T. Yates Jr., *J. Chem. Phys.*, 1987, **6**, 3668.
5 T. P. Beebe Jr., J. E. Crowell and J. T. Yates Jr., *J. Phys. Chem.*, 1988, **92**, 1296.
6 E. B. Svetlanov and R. M. Flid, *Russ. J. Phys. Chem.*, 1966, **40**, 1638.
7 C. J. Mitchell and P. Jones, Patent PCT Int. Appl. WO 2000076658, Imperial Chemical Industries Plc, UK, 2000.
8 In January 2001 ICI Chemicals & Polymers Ltd. divested into Ineos Chlor Ltd.
9 D. T. Lundie, A. R. McInroy, R. Marshall, J. M. Winfield, P. Jones, C. C. Dudman, S. F. Parker, C. Mitchell and D. Lennon, *J. Phys. Chem. B*, 2005, **109**, 11592.
10 A. R. McInroy, D. T. Lundie, J. M. Winfield, C. Dudman, P. Jones, S. F. Parker, J. W. Taylor and D. Lennon, *Phys. Chem. Chem. Phys.*, 2005, **7**, 3093.
11 A. R. McInroy, D. T. Lundie, J. M. Winfield, C. Dudman, P. Jones and D. Lennon, *Langmuir*, 2005, **21**, 11092.
12 A. R. McInroy, D. T. Lundie, J. M. Winfield, C. C. Dudman, P. Jones, S. F. Parker and D. Lennon, *Catal. Today*, 2006, **114**, 403.
13 A. R. McInroy, D. T. Lundie, J. M. Winfield, P. Jones, C. C. Dudman and D. Lennon, *Appl. Catal., B*, 2007, **70**, 606.
14 S. F. Parker and D. Lennon, *Acc. Chem. Res.*, 2014, **47**, 1220.
15 J. M. Winfield and D. Lennon, *New Materials for Catalytic Applications*, ed. E. Kemnitz and V. Parvulescu, Elsevier, Amsterdam, 2016, ch. 7, ISBN 9780444635877.
16 S. A. Schmidt, N. Kumar, B. Zhang, K. Eränen, D. Y. Murzin and T. Salmi, *Ind. Eng. Chem. Res.*, 2012, **51**, 4545.
17 S. A. Schmidt, N. Kumar, A. Reinsdorf, K. Eränen, J. Wärnå, D. Y. Murzin and T. Salmi, *Chem. Eng. Sci.*, 2013, **95**, 232.
18 S. A. Schmidt, N. Kumar, A. Shchukarev, K. Eränen, J.-P. Mikkola, D. Y. Murzin and T. Salmi, *Appl. Catal., A*, 2013, **468**, 120.
19 S. A. Schmidt, M. Peurla, N. Kumar, K. Eränen, D. Y. Murzin and T. Salmi, *Appl. Catal., A*, 2015, **490**, 117.
20 In July 2015 Ineos Chlor and Solvay formed a joint venture to produce a new commercial organisation named Inovyn.
21 P. A. Clayborne, T. C. Nelson and T. C. de Vore, *Appl. Catal., A*, 2004, **257**, 225.
22 J. G. Stark and H. G. Wallace, *Chemistry Data Book*, John Murray, London, 1980.

Faraday Discussions

PAPER

A transition metal oxofluoride offering advantages in electrocatalysis and potential use in applications†

H. Svengren,*[a] N. Torapava,[b] I. Athanassiadis,[a] S. I. Ali[a] and M. Johnsson[a]

Received 16th November 2015, Accepted 2nd December 2015
DOI: 10.1039/c5fd00169b

The recently described solid solution $(Co,Ni,Mn)_3Sb_4O_6F_6$ has proved stable and efficient as a catalyst for electrocatalytic water oxidation. The end component $Co_3Sb_4O_6F_6$ was found to be most efficient, maintaining a current density of $j = 10$ mA cm^{-2} at an overpotential of 443 mV with good capability. At this current density, O_2 and H_2 were produced in the ratio 1 : 2 without loss of faradaic current against a Pt-cathode. A morphological change in the crystallite surface was observed after 0.5 h, however, even after 64.5 h, the overall shape and size of the small crystallites were unaffected and the electrolyte contained only 0.02 at% Co. It was also possible to conclude from *in situ* EXAFS measurements that the coordination around Co did not change. The oxofluorides express both hydrophilic and hydrophobic surface sites, incorporate a flexible metalloid element and offer the possibility of a mechanism that differs from other inorganic catalytic pathways previously described.

Introduction

The environmental consequence of human consumption of non-renewable energy is rapidly increasing the demand for advancements in efficient utilization of sustainable energy and storage. Efficient storage is, today, mostly focusing on batteries and reduced fuels, such as hydrogen or hydrocarbons. Rechargeable batteries have undergone tremendous development during recent years and we are continuously searching for new routes to increase battery lifetimes, increase charge–discharge efficiencies and improve energy densities.[1] Research has, accordingly, led to new and intriguing discoveries in electrode materials, including a new class of fluorine-based materials as high capacity electrodes. Fluorine-based materials have historically been discarded as electrode materials

[a]*Department of Materials and Environmental Chemistry, Stockholm University, SE-106 91 Stockholm, Sweden. E-mail: henrik.svengren@mmk.su.se; Fax: +46-8-152187; Tel: +46-8-162386*

[b]*Max IV Laboratory, Lund University, SE-221 00 Lund, Sweden*

† Electronic supplementary information (ESI) available: Table of amounts used in the experiments, SEM micrographs, images of the analytical set up, ECSA plot, EXAFS data graphs, and PXRD patterns. See DOI: 10.1039/c5fd00169b

due to their insulating properties; therefore this class of compound has, to a great extent, been overlooked in the search for efficient electrodes. Sufficient conductivity can be achieved for practical applications by the addition of a conductive agent, such as carbon, into the electrode paste, which has opened up a way for finding new electrode materials. The perovskite-type 3d-transition metal trifluorides, MeF_3, are high capacity ion exchange cathode materials.[2] Transition metal oxofluorides, typically in the family of MeOF, have proven to be high capacity active cathodes for Li-ion batteries.[3] Research in the area of hydrogen fuel generation through water splitting has focused on the oxygen evolution reaction (OER), which constitutes a bottleneck in the overall reaction.[4] Efforts have been made to find ways to lower the activation energy by a catalytically functional electrode. Noble metals, transition metals, and their salts are well known water oxidation catalysts (WOCs) and are also implemented in traditional industrial hydrogen processing.[5] It has been shown how Co^{II} and Ni^{II} from aqueous solutions form catalysts during *in situ* electrolysis with phosphate and borate ions.[6] Oxides[7] containing first row transition metals, such as Fe, Mn, Co, and Ni, or polyoxymetalates[8] and hydroxides[9] have been identified as active catalysts. Recent work proposes that in metal oxides it is a metal hydroxide in the form of an oxyhydroxide that constitutes the active site.[7a,10] Changes in the morphology after exfoliation of oxyhydroxides, generating layered double hydroxide (LDH) nanosheets with a hierarchical structure, increase the active surface area, enhancing gas release and consequently improving the catalytic performance.[11] An outstanding example of a catalytic system optimized for water oxidation is the core cluster [$CaMn_4O_5$] in the oxygen evolving complex of Nature's photosystem II (PSII). The cluster exhibits five oxo-bridges linking the metals to form a distorted cubane structure with an additional Mn site.[12] Reactant H_2O is channelled through a protein environment to the inorganic core complex to preferred oxygen sites in the mechanistic cycle. One oxygen atom is more weakly bonded and may adopt a lower negative charge than a valence of −II, and hence it be important in the mechanism of dioxygen formation.[12] Inspired by this crystal structure, oxides with oxygen coordination in a cubic arrangement around the Mn or Co atoms, with or without incorporated redox-inactive metals such as Ca, La, and Li, have been studied extensively.[13] These systems show a high activity, involving alternation of the metal valence electrons to a higher state in oxo-bridging to $-OH_2$, –OH, –O, and –O–O species during the catalytic cycle.[6a,7a,13a,14] A special class of compound is the transition metal–metalloid–oxofluorides, MeMiOF. Our group has synthesized and characterized several such compounds and among them is the solid solution $(Co_xNi_{1-x})_3Sb_4O_6F_6$ (ref. 15) and $(Co_{0.86}Mn_{0.14})_3Sb_4O_6F_6$, the latter presented as a novel compound in this work. The compound exhibits a cubic arrangement of the active transition metal that oxo-bridges to the metalloid Sb with possible catalytic participation. Recently, we demonstrated that relatively hard (270 HV0.1, Mohs 3.5–4) and fairly large crystals (0.3–0.5 mm) can easily be synthesized, see Fig. 1.[16] In this study, we present the electrochemical characterization of $(Co_xNi_{1-x})_3Sb_4O_6F_6$ ($x = 1, 0.75, 0.5, 0.25, 0$) and the novel compound $(Co_yMn_{1-y})_3Sb_4O_6F_6$ ($y = 0.86$). The study includes linear sweep voltammetry (LSV) polarization curves, chronopotentiometry (CP) at a current density of 10 mA cm^{-2}, Tafel slopes, kinetics, relative comparisons of the electrochemically active surface area (ECSA), faradaic efficiencies of O_2 and H_2 production, long term durability with metal leaching analysis, SEM-EDS and X-ray

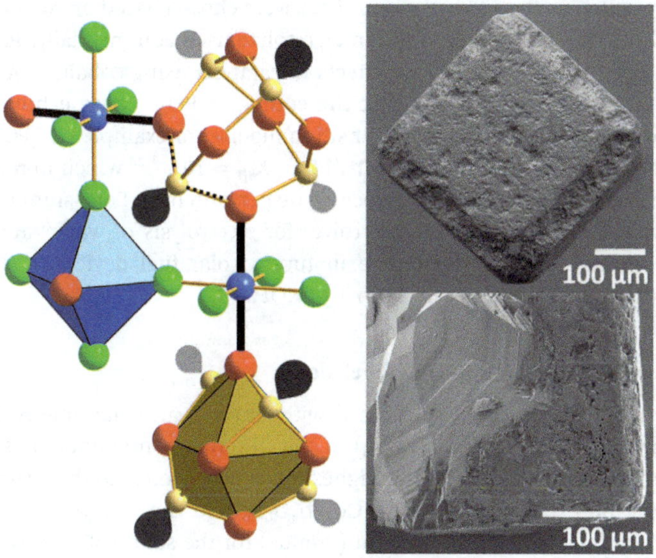

Fig. 1 The crystal structure (left) of $Co_3Sb_4O_6F_6$ is made up of slightly distorted $[CoO_2F_4]$ octahedra that are corner sharing *via* F atoms. The O atoms connect further to $[Sb_4O_6]$ cages, highlighting that there are only two possible locations for oxo-bridging for each cobalt atom. The O–Co–O–Sb–O–Co–O oxo-bridge is illustrated with black dashed bonds. Co = blue, O = red, Sb = yellow with its lone pair in gray, and F = green. A single crystal (right) – the fractured surface reveals the dense bulk.

absorption fine structure (EXAFS) analyses. As the MeMiOF materials exhibit both hydrophilic and hydrophobic sites on the interface, we further discuss the catalytic mechanism.

Results and discussion

Electrochemical evaluations of heterogeneous catalysts in functionalized working electrodes have many independent factors to account for, as described by McCrory *et al.*[24] In this study, efforts were dedicated to formulate a systematic and repeatable method for fabricating working electrodes. The aim of this effort is to minimize variation in the electrocatalytic performance originating from electrode inhomogeneity with respect to particle distribution, catalyst density and working electrode area. Thin film application of a paste with optimal consistency proved to be the best method for the oxofluoride compounds. The electrodes were formulated in this way to yield repeatable electrodes for electrochemical evaluation within this study and were not optimized for benchmarking *vs.* electrodes manufactured in other ways, such as layered double hydroxides grown directly on the current support.[11a] We therefore think that it is not appropriate to directly compare figures of merit, such as overpotentials, with those obtained from electrodes assembled in other ways. As a general summary, the study from McCrory *et al.* found that many promising WOCs had overpotentials between 330 and 500 mV at a current density of $j = 10$ mA cm^{-2}.[24b] The electrochemical analysis of the different compounds was repeated with three different electrodes and is presented as the average. Electrolytes were chosen based on two criteria: (i) a neutral pH stable buffer, and (ii) an electrolyte used commercially. Phosphate buffer at pH 7 has shown positive effects in catalysis using cobalt.[6,16] A negative effect could arise from the fact that the conjugate PO_4^{3-} base exhibits strong coordination and low solubility to transition metals, for example $Co_3(PO_4)_2$ $K_{sp} = 10^{-35}$,[25] $Ni_3(PO_4)_2$ $K_{sp} = 10^{-32}$ and $Mn_3(PO_4)_2$ $K_{sp} = 10^{-24}$,[26] which may spoil the native crystals or decrease the electrocatalytic performance. Potassium hydroxide is already a commercially used electrolyte for electrolysis of water[5] and can be expected to be a realistic candidate in future solar fuel devices.[24] The alkali electrolyte was chosen to be pH 14 in 1 M KOH.

Linear scan voltammograms and Tafel slopes

The electrochemical activity, evaluated with linear scan voltammetry compensating for the internal resistance (R_i) of the different compounds, is shown in Fig. 2. The compound yielding the highest current response to the overpotential scan is the pure cobalt compound $Co_3Sb_4O_6F_6$ and the catalytic activity then decreases with decreasing Co-content (x-value) for the solid solution series. The ranking order of the compounds from high to low was found to be: $x = 1 > y = 0.86 > x = 0.75 >$ CoO/Co_3O_4 $> x = 0.5 > x = 0.25 > x = 0$. The control electrode, not containing catalyst, showed an insignificant response. The two most active compounds were chosen for comparing Tafel plots in the potential range where the low current response starts to increase. From the Tafel slopes, the overpotential required for a current density $j > 0$ is a minimum of 340 mV for $Co_3Sb_4O_6F_6$ and 359 mV for $(Co_{0.86}Mn_{0.14})_3Sb_4O_6F_6$. Interestingly, the very first potential scan on a pristine electrode shows a current response around 1.5–1.6 V_{RHE} in proportion to the cobalt content and for the following scans this characteristic disappears. The Co-oxide electrode did not show this characteristic.

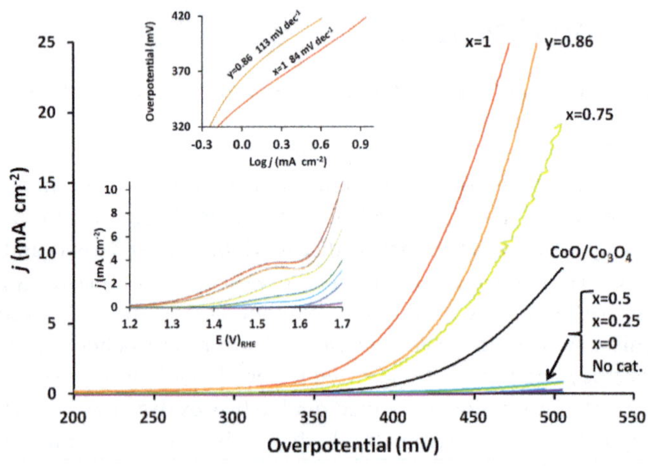

Fig. 2 R_i-compensated polarization curves from preconditioned electrodes that have reached a stable state at a linear scan rate of 2 mV s^{-1}. The upper inset shows the Tafel plots for the two most active phases: $Co_3Sb_4O_6F_6$ at 84 mV dec^{-1} and $(Co_{0.86}Mn_{0.14})_3Sb_4O_6F_6$ at 113 mV dec^{-1}. The lower inset shows redox waves deduced for $Co^{II} \rightarrow Co^{III}$, seen only at the very first scan on a pristine electrode. The electrolyte was 1 M KOH.

An explanation for this could be that the pristine crystal surface exposing Co^{II} is oxidized to higher state Co^{III} for the first oxidation scan, creating a redox-wave and henceforth an active surface in equilibrium with its surrounding. Co-oxide did not show this characteristic behavior, the compound was established to be $Co^{II}O/Co_3{}^{III}O_4$ in a ratio that is already a redox-stable phase. This behavior shows how important it is to precondition the electrodes to a stable state prior to characterization.

Chronopotentiometry and overpotential

The current density (j) output for a 10% efficient solar water splitting device is generally expected to be $j = 10$ mA cm^{-2}.[24] In this study, we have chosen the required overpotential (η) at this current density for assessment of the electrocatalytic activity in water oxidation. The overpotential was determined in chronopotentiometric mode with an initial applied current density of $j = 1$ mA cm^{-2} for 5 min, followed by an increase to $j = 10$ mA cm^{-2} for 10 minutes. The results are presented as an average from three experiments. The overpotential for the respective compound is presented in Fig. 3 in two different electrolytes with different pH values: 0.1 mol dm^{-3} phosphate buffer pH 7 and 1.0 mol dm^{-3} potassium hydroxide pH 14. The large difference in η between the two electrolytes can be attributed to both a large difference in conductivity (11.5 mS for 0.1 M phosphate buffer and 240 mS for 1 M KOH) and different mechanistic properties in the presence of $PO_4{}^{3-}$ ions and different pH conditions. The lowest overpotential, $\eta = 443$ mV, required for $j = 10$ mA cm^{-2} is assigned to the compound with $x = 1$. Interestingly, the phase with $x = 0$ requires a lower η than that with $x = 0.5$ and $x = 0.75$ in phosphate buffer. This is the only deviation from the general trend where a higher fraction of cobalt gives a proportionally higher activity. Graphical visualization of the data from the chronopotentiometry in 1 M KOH can be seen in Fig. 4 and the inset shows a single experiment for the most electrocatalytic compound, $x = 1$, which was continued for 5 hours, demonstrating the

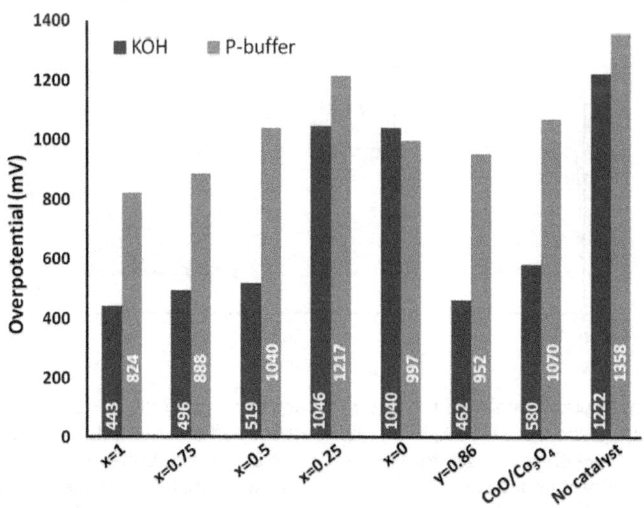

Fig. 3 Overpotentials, η, required at the current density of $j = 10$ mA cm^{-2}.

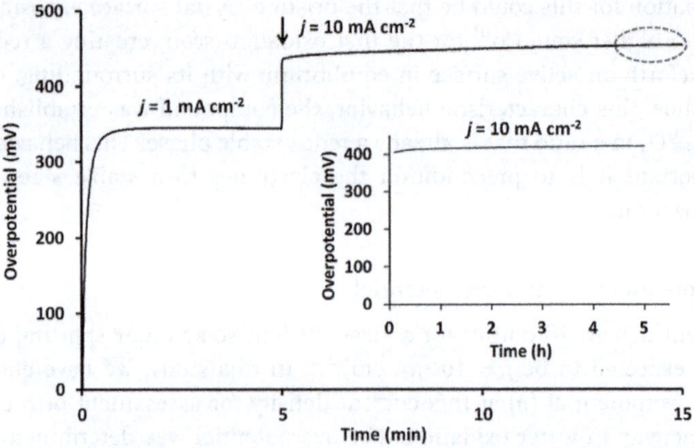

Fig. 4 Chronopotentiometry to evaluate the overpotential, η, at $j = 10$ mA cm^{-2} for $x = 1$ in 1 M KOH. The dotted ellipse highlights the data range used for the calculation of the average overpotential from three of the electrodes presented in Fig. 3. The inset shows the stability over 5 hours.

long term stability with an increase of only ~10 mV. All the compounds show the same type of kinetics in KOH, comparable to the one in Fig. 4, however, in phosphate buffer, the curves show different kinetics, as seen in Fig. 5. The required potential for $j = 1$ mA cm^{-2} does not incite other redox events other than the expected slow oxidation of water. When the current density applies to $j = 10$ mA cm^{-2}, the required potential increases and redox events are seen.

The two most active compounds, $x = 1$ and $y = 0.86$, show a current flow requiring a lower initial potential that ceases after approximately 3 min and the

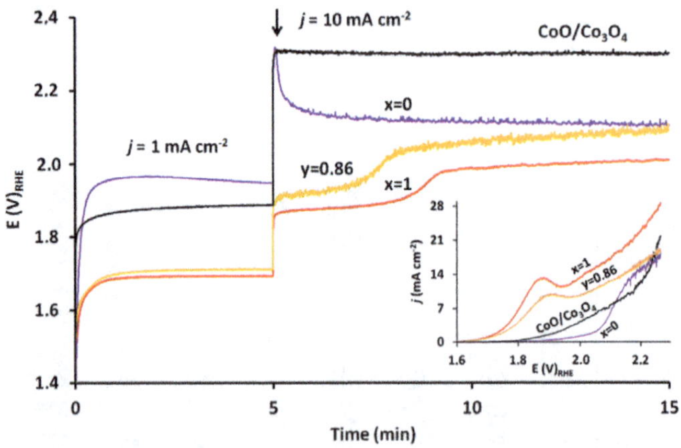

Fig. 5 Chronopotentiometry of the two most active compounds, $x = 1$ and $y = 0.86$, in relation to $x = 0$ and CoO/Co_3O_4 in 0.1 M phosphate buffer at pH 7, showing (1) the trend in the redox wave characteristic for $Co^{II} \rightarrow Co^{III}$ in phosphate buffer, (2) activation of Ni, and (3) insignificant changes in the CoO/Co_3O_4 powder that has reached a stable state. The inset shows the same trends from the R_i-compensated LSV (scan rate 2 mV s^{-1}) using preconditioned electrodes that have reached a stable state.

potential is increased to a higher level. Correspondingly, a redox peak can be seen in the inset voltammogram. This is most likely attributed to a flow of charge originating from $Co^{II} \rightarrow Co^{III}$. Different from the redox event assumed to take place in the very first LSV in KOH seen in Fig. 2, the behavior seen in Fig. 5 takes place on a preconditioned electrode that has reached a stable state prior to analysis and the redox event takes place repeatedly. It is therefore tempting to conclude that the PO_4^{3-} ion is involved in the reaction. Such an event could be that when the potential allows Co^{III} species, less coordination to PO_4^{3-} occurs, but when the potential decreases, $Co^{III} \rightarrow Co^{II}$ formation is pushed to the right to form $Co^{II}{}_3PO_4$ due to a low value of K_{sp}. Nickel shows a different behavior. It appears to require a higher potential to activate catalysis, and the initiated current flow requires a lower potential thereafter. This event is reflected in the inset voltammogram, where a higher potential is needed to induce the current flow, which then increases relatively fast. A conclusion on the difference between the two electrolytes is that the phosphate buffer appears to have more interactions on a mechanistic level for the oxofluorides than KOH, and this would possibly obstruct the catalytic performance.

Electrochemically active surface area

The electrochemically active surface area (ECSA) was estimated by analyzing the capacitive current associated with the double layer charging from the scan rate dependence of R_i-compensated cyclic voltammograms (CV) in the non-faradaic region. The true value of the ECSA in cm^2 is obtained from the ratio of the double layer capacitance C_{dl} (mF cm^{-2}) to the specific capacitance C_s (mF cm^{-2}) of the sample. The C_s values are unknown for the present samples and their determination is beyond the scope of this work. We therefore approximate the ECSA from the double layer capacitance per electrode geometric surface area C_{dl} (mF cm^{-2}) to offer a relative comparison of the surface activity for the two most active compounds, $x = 1$ and $y = 0.86$, in relation to $x = 0$ and CoO/Co_3O_4, as presented in Fig. 6. The difference in C_{dl} is relatively large between the two most active compounds, which are close in Co-content, possibly the 14% Mn on the metal site may have altered the capacitance of the compound. This relatively large difference is not seen in the activity results from LSV in Fig. 2 or CP in Fig. 3, where the electrochemical character is more similar. It was expected that the CoO/Co_3O_4 electrode should show a relatively large value of C_{dl} due to the large difference in the particle sizes – 0.1–50 μm for the crushed oxofluorides and 0.1–1 μm for the Co-oxide powder. It can be concluded from LSV, CP, CV and ECSA investigations that the compounds $Co_3Sb_4O_6F_6$ and $(Co_{0.86}Mn_{0.14})_3Sb_4O_6F_6$ exhibit a higher electrocatalytic activity than the electrode made from CoO/Co_3O_4, which served as a benchmark for the electrodes. However, taking into account the complexity of fabricating electrodes to a very similar molar ratio of Co per surface area, it is difficult to determine exactly how much higher the activity is for the two most active oxofluoride compounds compared to the oxide electrode.

Leaching

A true catalyst should, per definition, not be consumed during the catalytic reaction and should only act to lower the activation energy of the reaction catalyzed. One way to evaluate this is long term experiments studying the phase

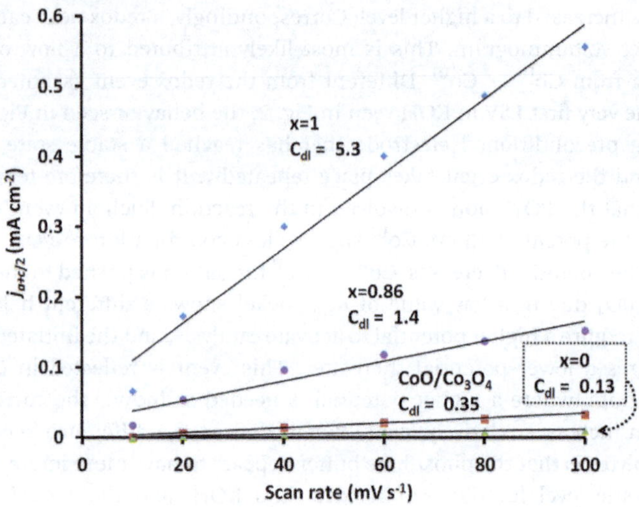

Fig. 6 Electrochemically active surface area (ECSA) approximated from the double layer capacitance C_{dl} (mF cm^{-2}) for relative comparison of the surface activities of the two most active compounds, $x = 1$ and $y = 0.86$, in relation to $x = 0$ and CoO/Co_3O_4 in 1 M KOH. The data were obtained from R_i-compensated voltammograms.

stability and leaching. The most active catalyst, $x = 1$, was chosen for studies at an applied current density of $j = 10$ mA cm^{-2} over 64.5 h in 1 M KOH electrolyte. The electrolyte was thereafter analyzed using ICP-MS, which showed a low leakage of 0.02 at%. The same $x = 1$ phase, this time with 10 mg native crystals, was placed in 1 M KOH electrolyte without an applied potential for 20 h and was then removed for single crystal XRD analysis to investigate the phase stability in the strong alkaline environment. This experiment did not show any phase transformation. We conclude from the two experiments that the crystals under *in situ* conditions show good stability with low leakage and the 1 M KOH electrolyte does not provoke any phase change. Furthermore, we measured the Co-content in an electrode used in the experiments. A 1 cm^2 area of an $x = 1$ electrode was first heated for 30 min in concentrated HNO_3 at 120 °C to dissolve the metal and the solution was subsequently analyzed with ICP-MS. The total cobalt content was determined to be 14.9 μmol, which is 85.5% of the calculated amount.

Kinetics

The kinetics of the catalysis in the studied oxidation reaction $2H_2O \rightarrow O_2 + 4H^+ + 4e^-$ on the anode containing $Co_3Sb_4O_6F_6$ required an overpotential of $\eta = 443$ mV at a steady current density of $j = 10$ mA cm^{-2}, obtained by CP, presented in Fig. 4. This reaction also involves electron conductivity in the catalytic interface and mass transport across the electric double layer between the crystal surface and the electrolyte. Assuming that the cell circuits at faradaic current, the same is valid for the reaction $2H^+ + 2e^- \rightarrow H_2$ at the Pt cathode. The temperature dependent kinetics for the experimental conditions adopted in this study at an applied oxidation overpotential of $\eta_{OX} = 443$ mV were estimated. A water bath surrounding the cell was used to let the temperature be varied three times with fast, medium and slow temperature changes between 16 and 44 °C in the electrolyte. Assuming faradaic efficiency in the reaction $2H_2O \rightarrow O_2 + 4H^+ + 4e^-$, mol O_2 equivalents was calculated from the recorded current density j/nF ($n = 4e^-$) and plotted against the temperature as $1/T$ (K) using the Arrhenius equation $k = be^{-E/RT}$ and the results are presented in Fig. 7. According to the calculated rate constant k (Faraday/4 [mol O_2 equivalents] cm^{-2} s^{-1}), an approximately 20 degrees increase in temperature will double the O_2 yield in the presented temperature range. The mass activity with respect to [O_2 eq.] related to k was calculated at three different temperatures, 16 °C, 25 °C and 44 °C, to be 6.5, 8.8 and 15.8 nmol [O_2 eq.] per mg $x = 1$, respectively, in the present electrode formulation. The inset in Fig. 7 shows the data sets from the three different temperature changes in an Arrhenius plot.

Faradaic efficiency

The claim of having a compound catalyzing the reaction $2H_2O \rightarrow O_2 + 4H^+ + 4e^-$ should be proved by gas evolution and the faradaic efficiency should be determined. This can be done by measuring the oxygen evolution. However, electrolysis of water, according to $2H_2O \rightarrow O_2 + 2H_2$, by using an electrocatalytic counter electrode large enough not to bottleneck the overall electron flow, consequently rising the overpotential, should also be established by the simultaneous detection of H_2 gas. To evaluate this, a gas tight electrochemical single chamber cell was connected to a mass spectrometer and employed with a working electrode of $Co_3Sb_4O_6F_6$ having a geometric surface area of 14.9 cm^2, a reference electrode, double Pt mesh as a counter electrode and 1 M KOH electrolyte. The set up and results are presented in Fig. 8.

The faradaic efficiency at an applied current density of $j = 1$ mA cm^{-2} for the first 5 min revealed that oxygen evolved as expected but hydrogen did not evolve sufficiently. This leads to the conclusion that the cell potential at $j = 1$ mA cm^{-2}

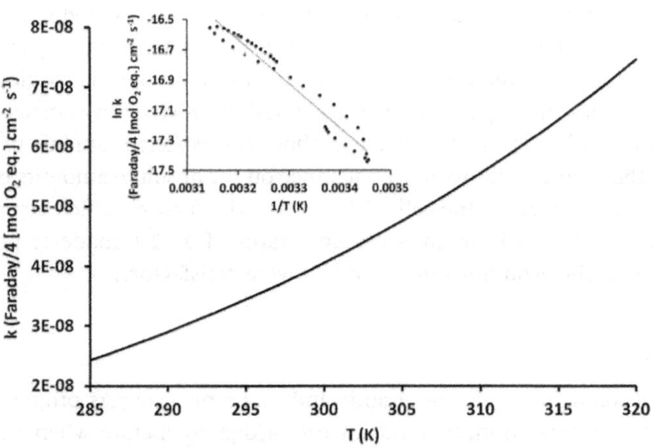

Fig. 7 Kinetic behavior of $Co_3Sb_4O_6F_6$ in 1 M KOH showing the temperature dependence of the rate constant k at an applied overpotential of $\eta = 443$ mV. The inset shows the Arrhenius plot combined from three data sets and the corresponding regression line from which k was calculated.

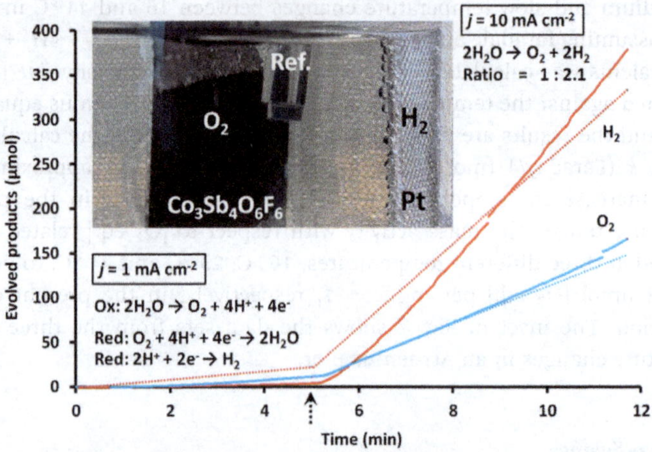

Fig. 8 Faradaic efficiency of $Co_3Sb_4O_6F_6$, plotted as evolved products (smooth lines) at an applied current density of $j = 1$ mA cm^{-2} for the first 5 min followed by an increase to $j = 10$ mA cm^{-2} (indicated by an arrow) in relation to the expected production (dotted lines). The ratio $O_2 : H_2$ was determined to be 1 : 2.1. The faradaic efficiency was measured to be 107% for O_2 and 114% for H_2; an experimental tolerance on the absolute amounts of gaseous and aqueous products in the cell should be applied.

was too small for reduction of $2H^+ + 2e^- \rightarrow H_2$, highlighting the importance of performing faradaic efficiency studies at a high enough current to make the oxidation potential valid for faradaic splitting of water. Another redox reaction must balance the current flow; the reaction that is most likely in a single cell chamber is the reverse recombination reaction: $O_2 + 4H^+ + 4e^- \rightarrow 2H_2O$. After 5 min, the current density was increased to $j = 10$ mA cm^{-2} and the required oxidation potential was measured to be 1.728 V_{RHE}. At this current density, the cell potential was enough to evolve both O_2 and H_2 gas in a 1 : 2.1 proportion at an oxidation overpotential of $\eta = 498$ mV. The slightly higher overpotential in this experiment compared to the CP results at $\eta = 443$ mV presented in Fig. 3 could result in errors when scaling the working electrode by 15 times in geometric size and the fact that the reference electrode was not in an optimal position due to practical reasons in the gas tight cell. The faradaic efficiency was measured to be 107% for O_2 and 114% for H_2, which is above the expected faradaic production, however, the experiment involves a tolerance on the absolute amounts of gaseous and aqueous products in the cell. The fact that the faradaic efficiency was larger than expected but still at an acceptable ratio of 1 : 2.1 made it possible to conclude that the evolutions of H_2 and O_2 were satisfactory.

Mechanism

Several mechanistic studies of organic and inorganic catalysts propose that two active metal atoms connected *via* an oxo-bridge co-operate when bonding to water. Reaction intermediates like –OH_2, –OH, –O, –O–O and finally molecular oxygen form during the catalytic cycle, which involves changes in the valence states of the active metals; in the case of Co, it means a change between +III and +IV.[7a,14a] In the present oxofluoride compounds, the metal (Co, Ni, Mn) is octahedrally coordinated to four F and two O atoms, see Fig. 1. The oxygen atoms bridge further to Sb and are potential sites for the catalytic reactions. If those oxygen atoms play a role during the catalytic reaction, it means that they are involved in the formation of new bonds to form reactant intermediates, which would break the oxo-bridges in between Co and Sb. Such a process would thus destroy the crystal structure. However, this does not match with the fact that the crystals perform well as catalysts both chemically[16] and electrochemically for a long time period. The electrocatalytic activity shows a fairly stable oxidation potential over 5 h at an applied current density of $j = 10$ mA cm^{-2}, as seen in Fig. 4. The cell potential decreased slightly over a period of 64.5 h, showing a low leakage of 0.02 at% Co in the alkaline electrolyte. The electrodes were analysed with SEM-EDS, and using high magnification, the surface morphology of the crystallites was seen to change from being smooth in the electrode before electrolysis to rough after just 0.5 h. This rough surface morphology seems then to not change with time and is similar for the electrode used for 64.5 h, see Fig. 9. The EDS results show that all ingoing elements are present in comparable ratios both in the outer thin edge and in the centre of a crystal, see Fig. 10 and Table 1. This result supports the fact that the crystal and its outer layer, which seems to be amorphous, have the same composition. Also, the EXAFS experiments, obtained at a current density of $j = 1$ mA cm^{-2}, see Fig. 11, did show that the crystalline phase persists after the applied potential, supporting the findings of the stable crystallite shape seen in the SEM images.

The EXAFS spectra reflected, most likely, the vast bulk crystal in the 200 μm thick, 1 cm^2 large electrode, rather than the thin electrochemically active interface, as we could not detect any difference in the oxidation state of either cobalt or nickel when the potential was on or off. A mechanism where both Co and Sb are

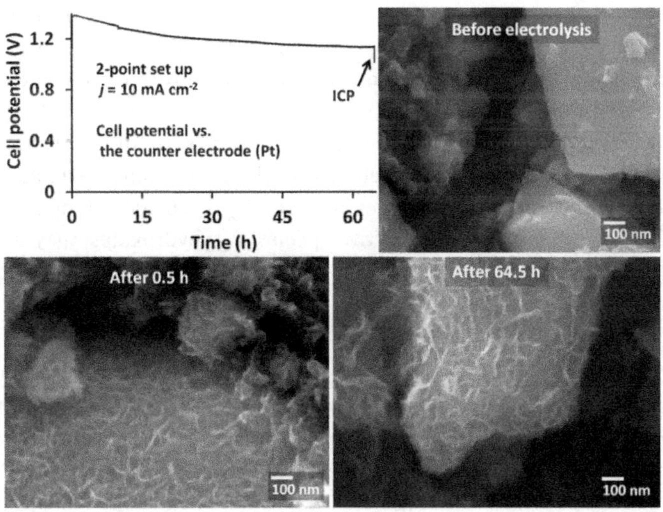

Fig. 9 The cell potential over 64.5 h, after which the electrode was withdrawn from the electrolyte for ICP analysis and SEM studies. The image before electrolysis shows a homogenous smooth crystallite surface, after 0.5 h of electrolysis a morphological change in the surface is seen, and after 64.5 h of electrolysis the morphology appears the same.

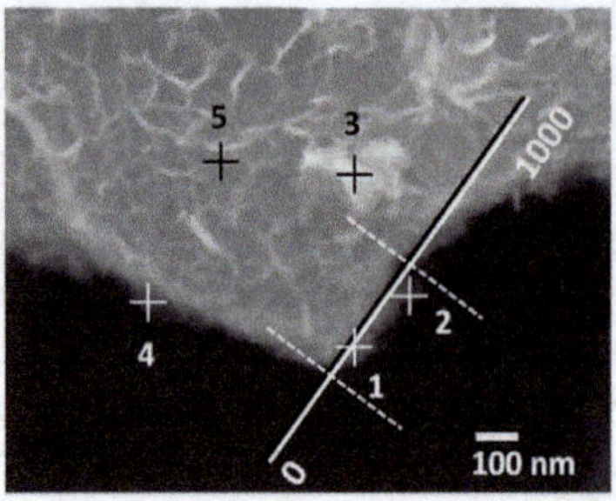

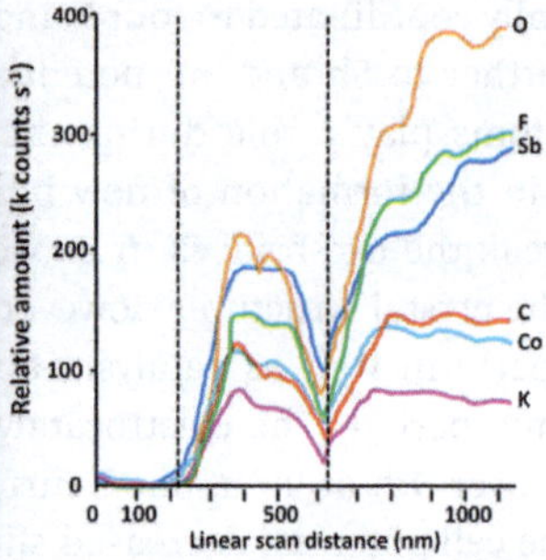

Fig. 10 SEM-EDS analysis of a $Co_3Sb_4O_6F_6$ crystallite on the electrode surface after 64.5 h of use with an applied current density of $j = 10$ mA cm^{-2}. The change in the surface morphology is seen in the picture. The EDS linear scan spectra show that all the elements are represented in the thin surface. The results were also observed (Table 1) with point spectra analysis at markers 1–5.

involved appears as an alternative when considering the electrocatalytic performance and the results from EXAFS, SEM, and EDS. When Co^{II} oxidises to Co^{III} or Co^{IV}, it forms a stronger bond to oxygen in the bridge over to Sb^{III}. This Sb–O bond may then weaken, allowing the coordination around Sb^{III} to change. Sb^{III} is flexible and can adopt both trigonal pyramidal coordination (three bonds), like in the present oxofluoride compounds, and see-saw coordination (four bonds). If a forth weak bond is formed from Sb to $-OH_2$, it will lead to further weakening of the –Co–O–Sb–O– oxo-bridge, resulting in longer Sb–O bond distances. The weaker Sb–O bonds may be important for the dioxygen formation process, and hydroxide and intermediate oxygen species can thus be stabilized. Formation of molecular oxygen from water is a process that is not thermodynamically favored and a gaseous phase would be less energy demanding if a non-polar surface is accessible in the highly polar electrolyte. Hence, F could also passively mediate the reaction by lowering the formation energy of the dioxygen molecule. Fluorine may also play a role in directing water molecules to the oxo-bridges. In this study there is thus evidence for the oxofluoride phase persisting in shape, activity and composition with a morphological change in the surface and there are no indications that the compound collapses to form *e.g.* Co-oxide nanoparticles. The possibility that other catalytically active species, *e.g.* hydroxides, may form from the mechanism outlined here would require less change from the initial structure.

Table 1 SEM-EDS elemental composition (at%) of sample points 1–5 shown in Fig. 10

Point	C	O	F	K	Co	Sb	Ratio Co : Sb
1	34	27	17	2	13	7	1 : 1.7
2	27	16	16	3	22	17	1 : 1.3
3	27	29	28	1	9	6	1 : 1.6
4	39	15	14	2	17	12	1 : 1.4
5	17	28	37	1	8	8	1 : 1.0
Std. dev.	8.5	7.2	9.9	0.9	5.7	4.5	—

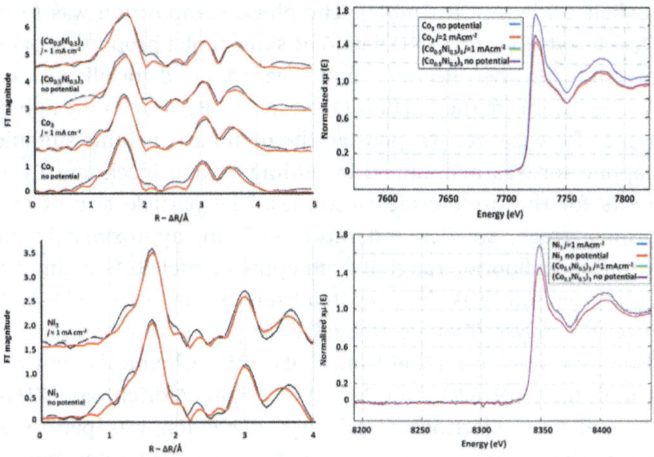

Fig. 11 *In situ* measured EXAFS (left) and XANES (right) of Co and Ni K edges, respectively, from $x = 1$, $x = 0.5$ and $x = 0$ with and without an applied current of 1 mA cm^{-2} in 0.1 mol dm^{-3} phosphate buffer pH 7. The EXAFS plot shows Fourier transforms with the calculated model (red line) and the experimental data (black line). FT data are offset by +2.

Experimental

The hydrothermal synthesis of compounds within the solid solution $(Co_xNi_{1-x})_3Sb_4O_6F_6$ ($x = 1, 0.75, 0.5, 0.25, 0$) was carried out according to Hu *et al.*[15] Also, the isostructural compound $(Co_yMn_{1-y})_3Sb_4O_6F_6$ ($y = 0.86$) was synthesized under similar conditions, starting with the off-stoichiometric mixture of 4 mmol CoF_2 (STREM Chemicals), 1 mmol MnF_2 (Sigma-Aldrich) and 1 mmol Sb_2O_3 (Sigma-Aldrich) together with 2 ml of deionized water. Crystals were manually separated from the synthesis product for further investigations. Single crystal X-ray diffraction (XRD) experiments were carried out by using a Bruker D8 Venture diffractometer equipped with a PHOTON 100 CMOS detector. Data integration, including the application of a correction for the oblique incidence, was done with software package SAINT.[17] An absorption correction was applied by the computer program SADABS.[18] The structure solution was done by Superflip.[19] Structure refinements were carried out by using JANA2006.[20] Powder X-ray data (PXRD) were collected on a PANalytical X'Pert PRO powder X-ray diffractometer in Bragg–Brentano geometry with Cu-Kα radiation. Scanning electron microscopy (SEM) studies were carried out by using a JEOL JSB-7000F equipped with an energy-dispersive spectrometer (EDS). The working electrodes were prepared by mixing the powder of each corresponding crystal phase in the molar proportion of 889 μmol (approx. 250 mg, 75 wt%) of transition metal content with 333.13 mg (approx. 10 wt%) of carbon black (Imerys, Super C65), 50 mg (approx. 15 wt%) of polyvinylidene fluoride (Arkema, Kynar FLEX 2801) and 0.8 ml of *N*-methyl-2-pyrrolidone (VWR/Prolabo). The suspension was ground in a mortar to form a homogenous paste. It was then applied with a doctor blade technique using a film applicator of 16.5 mm width and 200 μm film thickness on carbon fibre paper (Freudenberg, GDL-H23) and dried at 60 °C for three hours. The thin film was typically 10 cm long and was cut to form electrodes with a catalytic film area of 1 cm^2. An electrode for benchmarking was prepared from commercially available cobalt oxide (ABCR GmbH). The phase composition was found to be $CoO/Co_3O_4 = 0.54/0.46$ from PXRD data. The same molar proportion of transition metal (Co + Ni + Mn) per electrode unit area was used for all the oxofluoride electrodes, comparable to the CoO/Co_3O_4 electrode. Because of the lighter weight in the volume of the CoO/Co_3O_4 powder, the total paste volume suitable for the thin film applicator was maintained by adding carbon black and PVDF in the same ratio as for the oxofluoride electrodes. The particle size of CoO/Co_3O_4 appeared, on average, smaller and more uniform, approximately 0.1–1 μm, compared to the oxofluoride ranging from approximately 0.1–50 μm. Serving as a reference, an electrode was also prepared from carbon black and PVDF without the presence of catalyst. Two electrolytes were prepared: potassium hydroxide solution 1.00 mol dm^{-3} pH 13.99 from KOH (VWR Chemicals) and phosphate buffer 0.1 mol dm^{-3} pH 6.91 from Na_2HPO_4 (Sigma-Aldrich) and KH_2PO_4 (J.T. Baker Chemical Co.). Electrochemical characterization was performed using a potentiostat (BioLogic, model SP-50) with a standard three point set up: a working electrode (WE), a counter electrode (CE) of Pt-mesh and an Ag/AgCl reference electrode (ALS Co.), except in the experiment lasting for 64.5 h and the *in situ* EXAFS experiment, where a two point set up was used composed of a working electrode and a counter electrode (Pt-mesh) with the cell voltage referring to the counter electrode. The electrodes were placed in the centre of a single chamber open cell containing 35 ml of electrolyte and stirred with a magnet at 2500 rpm placed under the WE. Prior to investigation, the electrodes were preconditioned to reach a stable state using cyclic voltammetry (CV) until negligible change in the current response was observed, 15 cycles between 0–1.6 $V_{Ag/AgCl}$ in phosphate buffer and 6 cycles between 0–0.7 $V_{Ag/AgCl}$ at a scan rate of 20 mV s^{-1}. The ohmic drop was measured using the current interrupt (CI) method and the internal resistance (R_i) was subsequently corrected for by employing BioLogic software for the following potentiostatic techniques: R_i-compensated linear sweep voltammetry (LSV) at a scan rate of 2 mV s^{-1} used for the Tafel plot and R_i-compensated CV estimating the electrochemically active surface area (ECSA). The ECSA was analyzed using CV in the non-faradaic window of 0.05–0.15 $V_{Ag/AgCl}$ (1 M KOH) by measuring the capacitive current originating from the double layer charging depending on the different scan rates 10, 20, 40, 60, 80, and 100 mV s^{-1}. Potential scans were held at each potential vertex for 10 s before the subsequent scan. The double layer capacitance (C_{dl}) was calculated by plotting Δj ($j_a - j_c$) against the scan rate where the slope of the fitted linear regression corresponds to $2C_{dl}$. Chronopotentiometry (CP) was performed at a current density (j) of 1 mA cm^{-2} and 10 mA cm^{-2}. The potential scale has been calculated *versus* the reversible hydrogen electrode (RHE) using the Nernst equation $E_{RHE} = E_{Ag/AgCl} + 0.059\ pH + 0.209$, where the reference electrode filled with NaCl electrolyte at 3 mol dm^{-3} has a potential offset of +0.209 V_{SHE}. The overpotential was calculated as $\eta\ (V) = E_{RHE} - 1.23$ V. In the faradaic efficiency experiment, a mass spectrometer (Pfeiffer, model Thermostar GSD320-QMG220) was used to quantify the evolved amounts, n (moles), of molecular H_2 and O_2 during an applied current density of 1 mA cm^{-2} for 300 s followed by an increase to 10 mA cm^{-2} for 402 s, interrupted due to rising pressure. Calculations were made according to faradaic efficiency $= n_{gas}/Q(nF)^{-1}$, where Q (C) is the charge passed, n is the number of electrons, and n_{gas} is the amount of gas (mol O_2 or H_2) in the reaction. The reactions were performed in a 440 cm^3 gas tight single chamber containing 185 ml KOH electrolyte at room temperature with a working electrode exposing 14.9 cm^2 of catalytic film and magnetic stirring at 1000 rpm. The EXAFS data of nickel (Ni) and cobalt (Co) K edges were recorded at the wiggler beamline I811 at the Max IV Laboratory, Lund University, Lund, Sweden. The facility operated at 1.5 GeV with a maximum current of 250 mA. The EXAFS station was equipped with a Si [111] double crystal monochromator. Working electrodes of $(Co_xNi_{1-x})_3Sb_4O_6F_6$, ($x = 1, 0.5, 0$) were cut to an area of 1 cm^2 and placed in a 1 mm thick plastic frame covered with Kapton film, constituting an electrochemical cell with a volume of 0.23 cm^3. Phosphate buffer of 1.0 mol dm^{-3} at pH 7 was used as the electrolyte. The cell was connected in a two point set up. The experiments were performed in the transmission and fluorescence mode using gas filled ion chambers. Higher order harmonics were reduced by detuning the second monochromator crystal to 40% of the maximum intensity for both Ni and Co at the end of the scan. The energy scale of the X-ray absorption spectra was calibrated by assigning the first inflection point of the K edges of Ni to 8331.49 eV and of Co to 7708.78 eV.[21] The EXAFSPAK program package was used for data treatment.[22] For each sample, 1–2 scans were averaged after energy calibration by means of the EXAFSPAK program package. The experimental spectra were Fourier-transformed using the k-window over the k-space ranging from 2 to 14 Å^{-1}. The theoretical phase shift and backscattering amplitudes were obtained using the known crystal structures of $Co_3Sb_4O_6F_6$ and $Ni_3Sb_4O_6F_6$. The calculated EXAFS structure parameters were refined using FEFF ver. 7.02 based on models considering only single scattering contributions.[23] The distances, R, and the Debye–Waller factors, σ^2, were refined simultaneously. The amplitude reduction factors, $S_0{}^2$, and the shift in threshold energy, ΔE, were varied as a global parameter in the fit procedure. Determination of the Co-levels in the KOH electrolyte while using an electrode based on oxofluorides with $x = 1$ was carried out by ICP-MS analysis (Thermo X-Series 2). The sample from the electrolyte was acidified with konc. HNO_3 and the electrode sample, having a catalytic area of 1 cm^2, was heated to120 °C for 30 min, diluted, and quantified against internal standards.

Conclusions

The recently described oxofluoride solid solution $(Co_xNi_{1-x})_3Sb_4O_6F_6$ and the novel compound $(Co_yMn_{1-y})_3Sb_4O_6F_6$ ($y = 0.86$) were characterized with electrochemical methods for water oxidation catalysis. Efforts were focused on forming a repeatable method for electrode formulation. Two electrolytes were studied, 1 M KOH (pH 14) and 0.1 M phosphate buffer (pH 7). An electrode with a similar formulation and with a comparable molar ratio of Co to the oxofluorides was prepared as a benchmark from cobalt oxide powder with a composition of 54 mol% CoO and 46 mol% Co_3O_4. Linear scan voltammetry reveals a ranking order so that the activity decreases with decreasing amounts of Co for both electrolytes. A higher current response to potential was seen in the alkaline electrolyte, and Tafel plots, comparable for the two most active compounds, were calculated at 84 mV dec^{-1} for $x = 1$ and 113 mV dec^{-1} for $y = 0.86$. The characteristic redox wave $Co^{II} \rightarrow Co^{III}$ was seen only for the very first scan in the alkaline electrolyte, but for the phosphate buffer, the redox wave remained. This observation is likely to depend on the low solubility product of $Co_3{}^{II}(PO_4)_2$ (K_{sp} 10^{-35}) forcing most Co^{III} or Co^{IV} to go back to the low oxidation state when the potential decreases. Overpotentials were determined for all compounds by chronopotentiometry at the current density of $j = 10$ mA cm^{-2}. In general, the overpotential was lower for the alkaline electrolyte than for the phosphate electrolyte, however, one should take into account the difference in conductivity between 1 M KOH at 240 mS and the phosphate buffer at 11.55 mS. In the alkaline electrolyte, the two most active compounds operated at overpotentials of 443 mV for $x = 1$ and 462 mV for $y = 0.86$, in comparison with 580 mV for CoO/Co_3O_4. The overpotential for $x = 1$ was maintained over 5 h with a small increase of ~10 mV. In a longer experiment lasting for 64.5 h using a two point set up measuring the cell potential referring to the counter Pt-electrode, the cell potential did not increase. After 64.5 h, the electrolyte was analysed with ICP showing a leakage of only 0.02 at%. As an indication, the study by McCrory *et al.* concluded that several promising WOCs showed an overpotential in the range 330–500 mV, and that the complexity in different electrode compositions has to be considered.[24b] Hence, no direct comparison with other types of electrodes was made in this study for the benefit of internal ranking. The electrochemically active surface area was approximated to C_{dl} and the data showed a similar trend to the activity with $x = 1$ being highest at 5.3 mF cm^{-2}, then $y = 0.86$ at 1.4 mF cm^{-2} and then CoO/Co_3O_4 at 0.35 mF cm^{-2}. At an applied overpotential of 443 mV for $x = 1$, using chronoamperometry, there was a temperature dependence of the output current roughly equivalent to doubling the O_2 yield for a 20 °C temperature increase. The faradaic efficiency from mass spectrometry at an applied current density of $j = 10$ mA cm^{-2} was determined to be an $O_2 : H_2$ ratio of 1 : 2.1 with no faradaic current loss. Investigations by SEM show that even small crystallites survive in terms of their shape and size after 64.5 h of electrolysis at $j = 10$ mA cm^{-2}, but the surface shows a change in morphology. SEM-EDS and EXAFS show that the surface interface still contains all the elements of the parent phase and the metal coordination does not change after the potential was applied. Speculating that there could be diversity in the mechanistic pathways for different catalysts to obtain an oxygen evolution reaction, Sb could play an active role in the present oxofluorides if changing from trigonal pyramidal to see-saw coordination with weaker bonds to oxygen, consequently having longer bond distances. This could be important for O_2 formation in a –O–Co–O–Sb–O– cluster, coordinating reactant H_2O and intermediate species during the reaction. However, such a transformation would require flexibility in the crystalline phase and possibly the observed morphological surface change could be an amorphous active interface. As a final remark, the present oxofluoride shows good stability and is a good performing WOC. Further studies of the mechanism will be carried out.

Acknowledgements

Pär Hjelmquist is acknowledged for performing the ICP-MS analysis. Kjell Jansson is acknowledged for assistance with the SEM imaging. The Swedish Research Council is acknowledged for financial support.

References

1 J. Goodenough, *Energy Environ. Sci.*, 2014, 7, 14.

2 (*a*) M. Nishijima, I. Gocheva, S. Okada, T. Doi, J. Yamaki and T. Nishida, *J. Power Sources*, **150**, A1318; (*b*) F. Badway, F. Cosandey, N. Pereira and G. Amatucci, *J. Electrochem. Soc.*, 2003, **110**, 6474–6502.

3 (*a*) I. Gocheva, I. Tanaka, T. Doi, S. Okada and J. Yamaki, *Electrochem. Commun.*, 2009, **11**, 1583; (*b*) N. Pereira, F. Badway, M. Wartelsky, S. Gunn and G. Amatucci, *J. Electrochem. Soc.*, 2009, **156**, A407; (*c*) M. Reddy, S. Madhavi, G. Subba Rao and B. Chowdari, *J. Power Sources*, 2006, **162**, 1312.

4 (*a*) N. Lewis and G. Nocera, *Proc. Natl. Acad. Sci. U. S. A.*, 2006, **103**, 15729; (*b*) T. Cook, D. Dogutan, S. Reece, Y. Surendranath, T. Teets and D. Nocera, *Chem. Rev.*, 2010, **110**, 6474; (*c*) M. Walter, E. Warren, J. McKone, S. Boettcher, Q. Mi, E. Santori and N. Lewis, *Chem. Rev.*, 2010, **110**, 6446.

5 B. Tilak, P. Lu, J. Colman and S. Srinivasan, in *Comprehensive Treatise of Electrochemistry*, ed. J. Bockris, B. Conway, E. Yeager and R. White, Plenum, New York, 1981, vol. 2, pp. 1–97.

6 (*a*) Y. Surendranath, M. Kanan and D. Nocera, *J. Am. Chem. Soc.*, 2010, **132**, 16501–16509; (*b*) M. Kanan and D. Nocera, *Science*, 2008, **321**, 1072; (*c*) Y. Surendranath, M. Dinca and D. Nocera, *J. Am. Chem. Soc.*, 2009, **131**, 2615; (*d*) M. Kanan, Y. Surendranath and D. Nocera, *Chem. Soc. Rev.*, 2009, **38**, 109; (*e*) A. Esswein, Y. Surendranath, S. Reece and D. Nocera, *Energy Environ. Sci.*, 2011, **4**, 499–504; (*f*) M. Dinca, Y. Surendranath and D. Nocera, *Proc. Natl. Acad. Sci. U. S. A.*, 2010, **107**, 10337; (*g*) M. Kanan, J. Yano, Y. Surendranath, M. Dinca, V. Ychandra and D. Nocera, *J. Am. Chem. Soc.*, 2010, **132**, 13692–13701.

7 (*a*) J. Gerken, J. McAlpin, J. Chen, M. Rigsby, W. Casey, R. Britt and S. Stahl, *J. Am. Chem. Soc.*, 2011, **133**, 14431; (*b*) L. Bloor, P. Molina, M. Symes and L. Cronin, *J. Am. Chem. Soc.*, 2014, **136**, 3304; (*c*) Y. Gorlin and T. Jamillo, *J. Am. Chem. Soc.*, 2010, **132**, 13612; (*d*) Q. Yin, J. Tan, C. Besson, Y. Geletii, D. Musaev, A. Kutznetsov, Z. Luo, K. Hardcastle and C. Hill, *Science*, 2010, **328**, 342; (*e*) M. Wiechen, I. Zaharieva, H. Dau and K. Kurz, *Chem. Sci.*, 2012, **3**, 2330; (*f*) I. Zaharieva, P. Chernev, M. Risch, K. Klingan, M. Kohlhoff, A. Fischer and H. Dau, *Energy Environ. Sci.*, 2012, **5**, 7081; (*g*) M. Lyons and M. Brandon, *Int. J. Electrochem. Sci.*, 2008, **132**, 1386; (*h*) M. Lyons and M. Brandon, *Int. J. Electrochem. Sci.*, 2008, **132**, 1425; (*i*) A. Singh, S. L. Chang, R. K. Hocking, U. Bach and L. Spiccia, *Energy Environ. Sci.*, 2013, **6**, 579; (*j*) J. Rosen, G. Hutchings and F. Jiao, *J. Am. Chem. Soc.*, 2013, **135**, 4516; (*k*) J. Kirner, J. Stracke, B. Gregg and R. Finke, *ACS Appl. Mater. Interfaces*, 2014, **6**, 13367; (*l*) J. McAlpin, T. Stich, H. William and R. Britt, *Coord. Chem. Rev.*, 2012, **256**, 2445; (*m*) A. Han, H. Wu, Z. Sun, H. Jia and P. Du, *Phys. Chem. Chem. Phys.*, 2013, **15**, 12534–12538; (*n*) N. Chou, P. Ross, A. Bell and T. Tilley, *ChemSusChem*, 2011, **4**, 1566; (*o*) C.-C. Yang, T. Eggenhuisen, M. Wolters, A. Agiral, H. Frei, P. Jongh, K. Jong and G. Mul, *ChemCatChem*, 2013, **5**, 550; (*p*) M. Bajdich, M. Garcia-Mota, A. Vojvodic, J. Nørskov and A. Bell, *J. Am. Chem. Soc.*, 2013, **135**, 13521.

8 (*a*) J. Soriano-Lopez, S. Goberna-Ferron, L. Vigara, J. Carbo, J. Poblet and J. Galan-Mascaros, *Inorg. Chem.*, 2013, **52**, 4753; (*b*) J. Stracke and R. Finke, *ACS Catal.*, 2013, **3**, 1209; (*c*) J. Stracke and R. Finke, *ACS Catal.*, 2014, **4**, 909; (*d*) J. Stracke and R. Finke, *J. Am. Chem. Soc.*, 2011, **133**, 14872.

9 (*a*) G. Elizarova, G. Zhidomirov and V. Parmon, *Catal. Today*, 2000, **58**, 71–88; (*b*) X. Wang, H. Luo, H. Yang, P. Sebastian and S. Gamboa, *Int. J. Hydrogen Energy*, 2004, **29**, 967–972.

10 (*a*) L. Trotochaud, J. Ranney, K. Williams and S. Boettcher, *J. Am. Chem. Soc.*, 2012, **134**, 17253; (*b*) M. Gong, Y. Li, H. Wang, Y. Liang, J. Wu, J. Zhou, J. Wang,

T. Regier, F. Wei and H. Dai, *J. Am. Chem. Soc.*, 2013, **135**, 8452; (*c*) R. Doyle, I. Godwin, M. Brandon and M. Lyons, *Phys. Chem. Chem. Phys.*, 2013, **15**, 13737; (*d*) I. Godwin and M. Lyons, *Electrochem. Commun.*, 2013, **32**, 39; (*e*) K. Bediako, B. Lassalle-Kaiser, Y. Surendranath, J. Yano, V. Yachandra and D. Nocera, *J. Am. Chem. Soc.*, 2012, **134**, 6801; (*f*) Y. Surendranath, D. Lutterman, Y. Liu and D. Nocera, *J. Am. Chem. Soc.*, 2012, **134**, 6326.

11 (*a*) H. Liang, F. Meng, M. Cabán-Acevedo, L. Li, A. Forticaux, L. Xiu, Z. Wang and S. Jin, *Nano Lett.*, 2015, **15**, 1421; (*b*) F. Song and X. Hu, *Nat. Commun.*, 2014, **5**, 4477.

12 Y. Umena, K. Kawakami, J.-R. Shen and N. Kamiya, *Nature*, 2011, **473**, 55.

13 (*a*) U. Maitra, B. Naidu, A. Govindaraj and C. Rao, *Proc. Natl. Acad. Sci. U. S. A.*, 2013, **110**, 11704; (*b*) N. McCool, D. Robinson, J. Sheats and G. Dismukes, *J. Am. Chem. Soc.*, 2011, **133**, 11446; (*c*) M. Symes, D. Lutterman, T. Teets, B. Anderson, J. Breen and D. Nocera, *ChemSusChem*, 2013, **6**, 65; (*d*) E. Tsui, R. Tran, J. Yano and T. Agapie, *Nat. Chem.*, 2013, **5**, 293.

14 (*a*) M. Zhang, M. Respinis and H. Frei, *Nat. Chem.*, 2014, **6**, 362; (*b*) J. McAlpin, Y. Surendranath, M. Dinca, T. Stich, S. Stoian, W. Casey, D. Nocera and D. Britt, *J. Am. Chem. Soc.*, 2010, **132**, 6882; (*c*) A. Indra, P. Menezes, I. Zaharieva, E. Baktash, J. Pfrommer, M. Schwartze, H. Dau and M. Driess, *Angew. Chem., Int. Ed.*, 2013, **52**, 13206.

15 S. Hu, M. Johnsson, P. Lemmens, D. Schmid, D. Menzel, J. Tapp and A. Möller, *Chem. Mater.*, 2014, **26**, 3631.

16 H. Svengren, S. Hu, I. Athanassiadis, T. Laine and M. Johnsson, *Chem.–Eur. J.*, 2015, **21**, 12991.

17 *Bruker manual*, AXS Inc., Madison, 2012.

18 G. Sheldrick, *SADABS 2008/1*, Univ. of Göttingen, Göttingen, 2008.

19 V. Petricek, M. Dusek and L. Palatinus, *Kristallogr.*, 2014, **229**, 345.

20 L. Palatinus and G. Chapius, *J. Appl. Crystallogr.*, 2007, **40**, 786.

21 A. Thompson, D. Attwood, E. Gullikson, M. Howells, K. Kim, J. Kirz, J. Kortright, I. Lindau, P. Pianatta, A. Robinson, J. Scofield, J. Underwood, D. Vaughan, G. Williams and H. Winick. *X-ray Data Booklet*, Berkeley, California, 2001.

22 G. George and I. Pickering, *EXAFSPAK – A Suite of Computer Programs for Analysis of X-ray Absorption Spectra*, SSRL, Stanford, 1993.

23 (*a*) A. Filliponi, *J. Phys.: Condens. Matter*, 1994, **6**, 8415; (*b*) A. Filipponi, A. Di Cicco and C. Natoli, *Phys. Rev. B: Condens. Matter Mater. Phys.*, 1995, **52**, 15122.

24 (*a*) C. McCrory, S. Jung, J. Peters and T. Jaramillo, *J. Am. Chem. Soc.*, 2013, **135**, 16977; (*b*) C. McCrory, S. Jung, I. Ferrer, S. Chatman, J. Peters and T. Jaramillo, *J. Am. Chem. Soc.*, 2015, **137**, 4347.

25 P. Kelter, M. Mosher and A. Scott, *Chemistry the Practical Science*, Houghton Mifflin Company, Boston, 2009, p. A15.

26 D. Nordstrom, L. Plummer, D. Langmuir, E. Busenberg, H. May, B. Jones and D. Parkhurst, *WATEQ4F thermodynamic database*, www.brr.cr.usgs.gov/projects/GWC_coupled/phreeqc; W. Ball and D. Nordstrom, *Chemical Modeling of Aqueous Systems II, A.C.S. Symp. Ser. 416*, ed. D. C. Melchior and R. L. Basett, Am. Chem. Soc., Washington, 1990, pp. 398–413; *WATEQ4F, User's manual with revised thermodynamic data base and test cases for calculating speciation of major, trace and redox elements in natural waters*, U.S. Geological Survey Open-File Report 90-129, p. 185.

Faraday Discussions

PAPER

Acetylene hydrogenation over structured Au–Pd catalysts†

Alan J. McCue,*[a] Richard T. Baker[b] and James A. Anderson*[a]

Received 23rd November 2015, Accepted 24th November 2015
DOI: 10.1039/c5fd00188a

AuPd nanoparticles were prepared following a methodology designed to produce core–shell structures (an Au core and a Pd shell). Characterisation suggested that slow addition of the shell metal favoured deposition onto the pre-formed core, whereas more rapid addition favoured the formation of a monometallic Pd phase in addition to some nanoparticles with the core–shell morphology. When used for the selective hydrogenation of acetylene, samples that possessed monometallic Pd particles favoured over-hydrogenation to form ethane. A sample prepared by the slow addition of a small amount of Pd resulted in the formation of a core–shell structure but with an incomplete Pd shell layer. This material exhibited a completely different product selectivity with ethylene and oligomers forming as the major products as opposed to ethane. The improved performance was thought to be as a result of the absence of Pd particles, which are capable of forming a Pd-hydride phase, with enhanced oligomer selectivity associated with reaction on uncovered Au atoms.

Introduction

Olefin manufacture *via* naphtha cracking yields alkene streams containing small quantities of alkyne impurities, which require removal to avoid complications with downstream polymerisation steps. The preferred industrial route is *via* selective hydrogenation of the alkyne since this yields additional alkene.[1–3] This is generally accomplished over a bimetallic catalyst with Pd acting as the active metal and Ag as a structural modifier.[4] Achieving high selectivity at close to complete alkyne conversion remains difficult as the alkene has unhindered access to the catalyst's surface so can readily adsorb and react to give undesired alkane. CO is often co-fed as a transient poison to compete with the alkene for adsorption sites, although the amount added must be carefully regulated depending on the catalyst activity.[2] This process has been widely studied since the pioneering works

[a]*Surface Chemistry and Catalysis Group, Materials and Chemical Engineering, School of Engineering, University of Aberdeen, Aberdeen, AB24 3UE, UK. E-mail: a.mccue@abdn.ac.uk; j.anderson@abdn.ac.uk; Fax: +44 (0)1224 272901; Tel: +44 (0)1224 272838*

[b]*School of Chemistry, University of St Andrews, St Andrews, KY16 9ST, UK*

† Electronic supplementary information (ESI) available. See DOI: 10.1039/c5fd00188a

of Bond and Wells,[5] however, recent advances in experimental techniques and computational methodologies have resulted in an increased understanding of the factors that promote over-hydrogenation. Over a monometallic Pd catalyst, it is now thought that the interplay between Pd hydride and carbide phases governs alkene selectivity.[6–15] It has also been shown that alloying Pd with Ag minimises hydride formation resulting in increased alkene selectivity.[16]

A number of strategies have been explored in recent years to improve selectivity over Pd catalysts,[17–19] or even circumvent the need for Pd.[20–22] For example, the selectivity of Pd catalysts can be significantly improved by adsorption of organic modifiers[23–27] onto the metal surface, which is thought to create sites that favour acetylene adsorption while hindering adsorption of ethylene. Whilst industrial catalysts typically use Ag as a structural modifier, a number of other metals have been explored.[28–30] In general, improved alkene selectivity occurs when hydride formation is minimised, although a growing number of reports suggest that Pd atom isolation is beneficial.[31–33] A number of alternative metals, such as Cu,[34,35] Ag[36,37] and Au,[38] offer inherently better selectivity than Pd but require operation at relatively high temperatures (*ca.* 523 K) since hydrogen dissociation is restricted by an activation energy barrier. Such temperatures significantly exceed those used by current industrial reactors (approx. 313–413 K), which may limit potential implementation. To overcome this limitation, attempts have been made to increase the activity of Cu catalysts by adding small quantities of Pd to act/create sites that permit hydrogen dissociation at low temperatures (*ca.* 373 K).[39–46] Recent literature reports hint that a similar approach may be achieved by the addition of Pd to Ag[47] and Au[48] catalysts. Metal-free ceria has also been shown to be extremely selective for acetylene and methylacetylene hydrogenation but it requires operation at high temperatures (>473 K).[49] This can be overcome by doping with Ga, which appears to create sites that allow for hydrogen dissociation at lower temperatures,[50] although no data based on operation at elevated pressure is available.

The ability to synthesise bimetallic nanoparticles has advanced rapidly in recent years with significant attention devoted to AuPd nanoparticles.[51,52] These materials often possess unique structures/orders on the nanoscale (*i.e.* the manner in which the two metals interact is controlled). For example Ma *et al.*[53] compared the performance of AuPd octahedra with nanoflowers (organised agglomerates of the octahedra) and identified that the nanoflower samples exhibited superior performance for the selective hydrogenation of acetylene. One class of bimetallic nanomaterial that attracts particular attention is core–shell nanoparticles,[54,55] where one metal forms a shell over the second metal. In principle, when the shell is relatively thin, the electronic properties of the shell metal may be perturbed, which offers exciting possibilities with regards to adsorption and catalytic behaviour.[56–58] In-spite of the obvious appeal, synthesis methods rarely seem to produce only core–shell nanoparticles. Instead, a mixture of core–shell, bimetallic and monometallic particles are often formed, making evaluation of the catalytic properties of individual species challenging.[59] Additional complications arise from the synthesis methods employed with stabilising agents, such as polyvinylpyrrolidone, necessary to avoid nanoparticle agglomeration. If the polymer is not removed, it can influence the catalytic performance, although sometimes for the better.[60,61] Various removal methods have been explored including ozone/UV radiation[62] and solvent treatments,[63] however moderate to high temperature calcination still remains the most common method. This may not be appropriate for core–shell structures since the two metal components are likely to mix, resulting in the loss of the unique structure.[64]

A recent report by Wilson *et al.*[64] described a synthesis method where core–shell AuPd structured nanoparticles could be synthesised with different Pd shell thicknesses, with EDX-STEM and FTIR spectroscopy of adsorbed CO results consistent with this assumption. In this study, the same preparation methodology is further explored by adding different amounts of Pd to pre-formed Au nanoparticles. The results indicate that the nanoparticle structure and catalytic performance varies considerably depending on the Pd loading.

Experimental

Catalyst preparation

Nanoparticles with assumed Au cores and Pd shells of different thicknesses were prepared by following the procedure of Wilson *et al.*[64] with minor modifications. This method has been shown to result in core–shell nanoparticles with the Pd shell thickness controllable to some extent by adding different amounts of Pd. Briefly, Au nanoparticles were synthesised by the reduction of an aqueous solution of $HAuCl_4$ with pre-chilled $NaBH_4$ to yield a red sol. Hydroquinone was subsequently added to the mixture before an aqueous solution of Na_2PdCl_4 was added at a constant rate *via* a syringe pump (Table 1) with the intention of depositing Pd onto the pre-formed Au nanoparticles. Following the complete addition of the Pd solution, the reaction was quenched by adding aqueous HCl and the nanoparticles were stabilised with polyvinylpyrrolidone (PVP). The nanoparticles were immobilised onto a support by adding an appropriate portion of the metal solution to an aqueous slurry of TiO_2 (P25, Evonik, acidified with H_2SO_4) to give a 1 wt% loading. Finally, the solid was filtered, washed thoroughly with ultrapure water, dried at 60 °C and stored in this state prior to use. Following this methodology, three samples were prepared with different amounts of Pd and Au (Table 1). The samples are denoted as *x*Pd(Au) with '*x*' indicating the nominal Pd : Au atomic ratio. Significantly, the samples were not subjected to high temperature calcination, which may change the surface composition by allowing

Table 1 Chemical characteristics of the catalysts

Sample	Pd addition rate [μmol h^{-1}]	Nominal composition [wt%]		Nominal Pd : Au atomic ratio	Actual Pd : Au atomic ratio[a]	Average particle size[b] [nm]	CO uptake[c] [μmol g^{-1}]
		Pd	Au				
Pd	—	1.00	—	—	—	—	3.7
0.2Pd(Au)	5	0.10	0.91	0.2	0.26	3.5	2.2
1.5Pd(Au)	36	0.45	0.55	1.5	0.80	3.6	6.6
3.3Pd(Au)	57	0.65	0.37	3.2	1.10	3.7	5.3
Au	—	—	1.00	—	—	—	Negligible

[a] Determined by EDS using TiO_2 supported samples. [b] Determined by TEM using TiO_2 supported samples. [c] Determined by CO pulse chemisorption.

for metal mixing or preferential segregation.[59,64] This does mean, however, that the samples will have retained some PVP on the surface. In order to investigate the significance of this, the samples were pre-treated in hot water and water/UV-light, which have been shown to be effective in removing PVP to some extent.[62,63] Monometallic samples (1 wt% Pd or Au) were also prepared by the reduction of the appropriate metal with $NaBH_4$ followed by immobilisation onto the same TiO_2 support for comparison purposes.

Catalyst testing

Catalysts were tested for activity/selectivity in the gas phase hydrogenation of acetylene using a Microactivity reference reactor (PID Eng & Tech, supplied by Micromeritics). Approximately 200 mg of sample was diluted with 800 mg of SiC and supported in a 9 mm i.d. stainless steel reactor. Prior to reaction, the samples were reduced in 30% H_2/N_2 for 1 h at 373 K. The relatively low reduction temperature was deliberately chosen to avoid perturbation of the surface composition. The reactions were typically performed at 1 bar using a mixture of 1.2% acetylene/2.2% hydrogen/balance N_2 to give a H_2 : acetylene ratio of 1.8 : 1 and a space velocity of 24 000 h^{-1}. The main reaction parameter varied was the temperature (323–363 K) with 5 h time on stream permitted under each condition to ensure catalyst stabilisation. One run was conducted with a 3.6 : 1 H_2 : acetylene ratio. Analysis of the effluent gas was performed using a PE Clarus 580 GC fitted with an FID and a 30 m × 0.53 mm elite alumina capillary column. The acetylene conversion was calculated as the amount reacted divided by the amount introduced. The selectivity to ethylene and ethane was calculated as the amount formed divided by the amount of acetylene reacted, and the selectivity to oligomers was calculated based on a carbon balance. Under these conditions, full acetylene conversion was generally achieved, which makes the extraction of kinetic parameters inappropriate. These conditions allow for the assessment of ethylene selectivity under challenging conditions, which mirror industrial requirements (<5 ppm acetylene in effluent gas).

Characterisation

TEM images were obtained using a JEOL JEM 2011 instrument operating with a LaB_6 filament at an accelerating voltage of 200 kV and equipped with an Oxford Instruments EDS spectrometer. The images were captured using a Gatan CCD camera and analysed using Digital Micrograph 3.4.4 software. This software was also used to prepare Digital Diffraction Patterns (DDPs) by applying Fourier transforms to specific regions of the TEM images. The TEM instrument was also used to obtain EDS spectra and to record two-dimensional elemental distribution maps. For TEM examination, the samples were deposited from a suspension onto holey carbon-coated Cu grids. TiO_2-supported samples were deposited from a suspension in acetone. Unsupported nanoparticles were prepared as described earlier, but were additionally subjected to centrifugation and washed with ultrapure water before being suspended in aqueous media and deposited onto grids. A few images were obtained from the grids prepared from the sample suspensions prior to centrifugation and washing, but these proved unstable in the microscope because of the effect of dissolved HCl on the grids and the formation of salt crystals. In order to avoid the effects of these impurities (*e.g.* overlap of the

Cl and Pd spectra) and compositional segregation effects during centrifugation, EDS spectra of the supported nanoparticle samples were used to calculate the Pd : Au ratios presented below.

Temperature programmed reduction (TPR) experiments were conducted on a TPDRO 1100 instrument with a TCD detector using a 5% H_2/N_2 mixture. Profiles were collected over a temperature range of 313–373 K using a heating rate of 5 K min^{-1} and signal output normalised per gram of sample. CO pulse chemisorption experiments were carried out on the same instrument to evaluate CO uptake. Samples were reduced at 373 K for 1 h before being cooled to 308 K. Diluted CO (19% in He) was pulsed (0.285 ml loop volume) over the sample until saturation was attained.

FTIR spectroscopy of adsorbed CO was performed using a PE Spectrum 100 spectrometer with the sample presented as a 16 mm diameter self-supporting disc. The disc was suspended in a quartz holder and held in a vacuum system, which permitted *in situ* evacuation and gas manipulation. The sample was initially outgassed before being reduced in 50 ml min^{-1} H_2 at 373 K for 1 h (note: the same pretreatment temperature as used for the catalytic measurements). The sample was cooled and the system evacuated to *ca.* 4×10^{-5} mbar. An initial spectrum (25 scans, 4 cm^{-1} resolution) was collected prior to the stepwise introduction of increasing CO overpressures (0.1–40 Torr) at beam temperature. The results are presented as difference spectra using a scan before exposure to CO as the background. The spectra collected following this approach generally exhibited weak absorption bands (likely due to the presence of PVP on the surface blocking sites) but the bands due to CO observed subsequently reflect the surface sites present available during the catalytic measurements.

Results

Synthesis and characterisation

The synthesis of bimetallic AuPd nanoparticles is known to be complex with the composition often varying depending on the metal particle size when both metals are introduced simultaneously.[51,52] There are numerous groups who have reported preparation methods that produce nanoparticles with gold cores and palladium shells,[56–59,64] however, these often form in addition to monometallic particles and bimetallic particles where the two metals are mixed/alloyed. In this work, the synthesis method reported by Wilson *et al.*[64] was adopted, which appeared to produce core–shell nanoparticles based on the characterisation presented.[64] In order to synthesise samples, three equivalent Au colloidal solutions were prepared before different amounts of Na_2PdCl_4 was added at different rates (Table 1), followed by deposition onto a TiO_2 support. A material, designated as 0.2Pd(Au), was prepared by adding a small amount of Pd (0.1 wt% once supported) at a very slow addition rate (5 μmol h^{-1}). Under these conditions, the concentration of total palladium in solution is expected to be very low, meaning that the formation of monometallic Pd particles is unlikely (*i.e.* Pd is more likely to deposit onto the existing Au particles). The Pd : Au ratio estimated by EDS for this sample was 0.26, which correlates well with the nominal ratio of 0.20 (Table 1). Interestingly, Wilson *et al.* determined a Pd : Au ratio of 0.23 by ICP-MS for an equivalent material and highlighted that such low Pd loadings are likely to result in less than monolayer coverage of the Au core.[64] Dynamic pulse chemisorption of

CO was useful for gaining an understanding of the number of Pd surface sites because, under these conditions, CO adsorption on a monometallic Au sample was negligible (Table 1). CO uptake for this sample was 2.2 μmol g^{-1} (Table 1), lower than expected based on the amount of Pd added (CO : Pd ratio = 0.2), which implies that at least a fraction of Pd is inaccessible due to coverage by PVP.

A second sample, denoted as 1.5Pd(Au), was prepared by adding a larger amount of Pd (0.45 wt% once supported) at an addition rate of 36 μmol Pd h^{-1}. The Pd : Au ratio determined by EDS for this sample was 0.80, which falls below the nominal ratio (1.5) and may be as a result of incomplete reduction of the Pd precursor and subsequent loss during the filtration and washing steps. CO chemisorption resulted in 6.6 μmol g^{-1} uptake and a CO : Pd ratio of 0.2 assuming the nominal Pd loading. 1.5Pd(Au) has a higher Pd loading than 0.2Pd(Au) but an equivalent CO : Pd ratio, and this is tentatively taken as evidence of multilayer deposition on the Au core (*i.e.* not all of the Pd present are surface sites). However, given that Pd was added more rapidly, the concentration of the Pd species in solution during the synthesis is expected to be higher, which enhances the probability of forming monometallic Pd clusters/nanoparticles. A third sample, 3.3Pd(Au), was prepared and had the highest nominal Pd loading (0.65 wt % once supported) and Pd addition rate (57 μmol g^{-1}). Characterisation of this sample (presented later) suggests that some nanoparticles may be 'Pd rich' or even exclusively contain Pd. The Pd : Au ratio obtained by EDS is lower than expected (1.1 *vs.* 3.2 – see Table 1) and, again, incomplete reduction of Pd is thought to be the cause (note: Wilson *et al.* also reported lower Pd : Au ratios than the nominal ratios).[64] CO uptake for this sample was 5.3 μmol g^{-1}, which corresponds to a CO : Pd ratio of only 0.09 (based on the nominal Pd loading). This implies that not all of the Pd atoms can be surface atoms. This is consistent with both multilayer formation around the Au cores and the presence of monometallic Pd particles.

Following the deposition of the nanoparticles onto TiO_2, they were thoroughly washed with water to remove as much of the PVP stabilising agent as possible. DRIFTS spectra of the supported nanoparticles following drying confirm the presence of PVP on the surface (bands at 2960, 2924, 2880 and 2868 cm^{-1} due to the CH_3 asymmetric, CH_2 asymmetric, CH_3 symmetric and CH_2 symmetric stretching modes, respectively). Whilst it is desirable to remove the PVP to enable the relationship between the exposed sites and catalyst performance to be determined, it did not appear to have a significant effect on the catalytic measurements (see later). Wilson *et al.*[64] treated samples post synthetically at 573 K to remove the remaining PVP but this (intentionally) resulted in the two metals mixing. Given that the aim of this work was to assess the impact of the Pd shell thickness on the catalytic performance, treatments that induce metal mixing are undesirable. The effectiveness of a lower temperature calcination step (373–473 K) was evaluated, although no evidence of PVP decomposition/loss was apparent, as assessed by DRIFTS. Instead, alternative methods were explored for comparison purposes. The 'hot-water' method reported by Lopez-Sanchez *et al.*,[63] which is effective for the partial dissolution of polyvinylalcohol, was tested along with irradiation of the samples dispersed in water with UV light. Both methods appeared to be successful in exposing additional metal sites, although the intensity of the C–H stretching modes did not decrease significantly, suggesting some PVP was retained.

One method that can be used to evaluate whether monometallic Pd particles exist within a sample is temperature programmed reduction (TPR). Exposure of Pd to hydrogen at room temperature leads to the formation of a hydride phase, which decomposes above ambient temperature producing a characteristic negative peak.[14] TPR profiles generated by heating various samples in 5% H_2/N_2 at 5 K min^{-1} from 313 to 373 K are shown in Fig. 1 (note: the data is presented as detector signal *vs.* time to improve clarity). Monometallic Pd prepared by the reduction of Na_2PdCl_4 shows a negative feature at 343 K, which is attributed to the decomposition of β-Pd-hydride. The equivalent profiles for 0.2Pd(Au), 1.5Pd(Au) and 3.3Pd(Au) show no negative features. Instead, a very gentle increase in the detector signal suggests a small degree of reduction as the temperature approaches 373 K. As such, the ability to form a hydride phase, as judged by TPR alone, cannot be used to rationalise the differences in product selectivity (see later). This may be because the technique is not sufficiently sensitive or the materials do not possess large enough clusters of Pd to form a hydride phase. Indeed, it is known that hydride formation tends to zero as Pd dispersion increases to 100%.[65]

TEM imaging

Fig. 2 presents TEM images taken at a range of magnifications of the unsupported 3.3Pd(Au) nanoparticles deposited onto the TEM grid directly from their native suspension. Despite the adverse effects of the other constituents of this suspension (see Experimental section), these images clearly show the highly dispersed nature of the nanoparticles, their approximately spherical morphology and their uniform diameter, of around 3–4 nm. Crystal planes and atomic columns are evident in the higher magnification images and a DDP was obtained from the

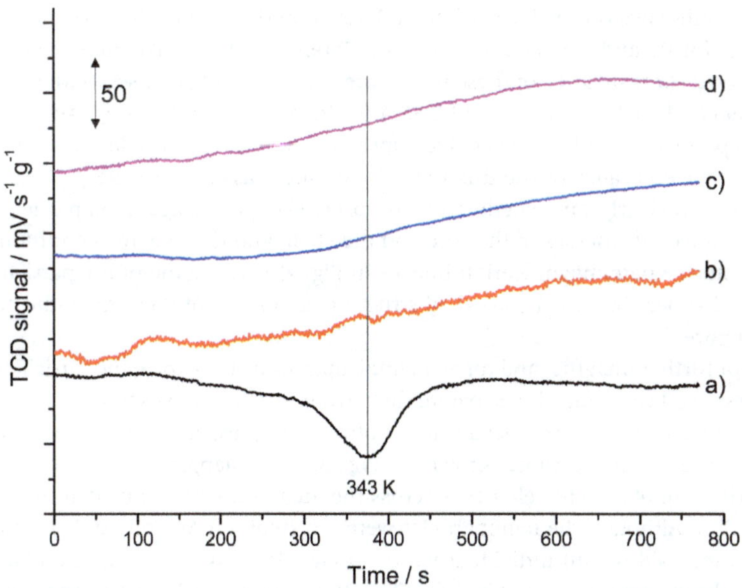

Fig. 1 TPR profiles for the samples heated from 313 to 373 K at 5 K min^{-1}. (a) Pd, (b) 0.2Pd(Au), (c) 1.5 Pd(Au) and (d) 3.3 Pd(Au).

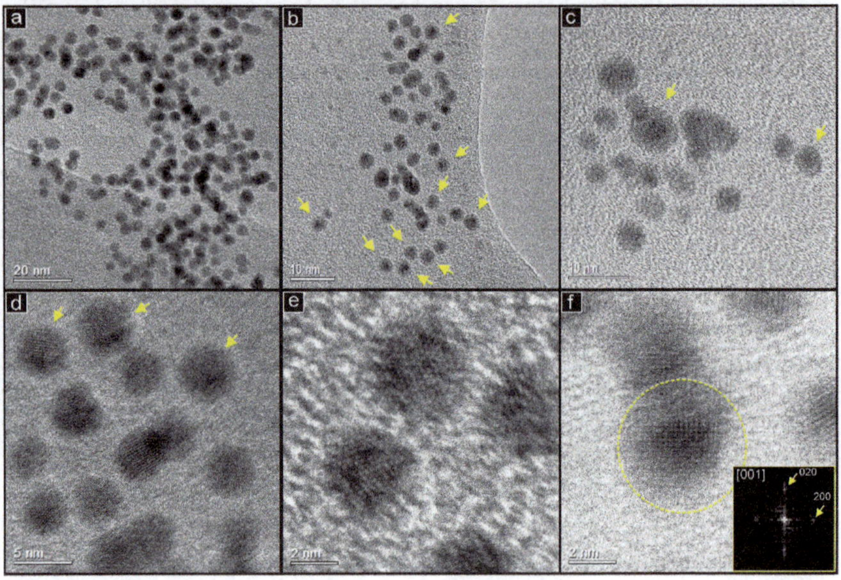

Fig. 2 TEM images of the unsupported 3.3Pd(Au) nanoparticles at a range of magnifications. The DDP of the nanoparticle circled in image (f) is shown in the inset. Particles with apparent core–shell morphology are indicated. See the text for further details of all the TEM images.

nanoparticle circled in Fig. 1(f). This was indexed to the face centred cubic crystal structure (space group 225, $Fm\bar{3}m$) expected for both Pd and Au, viewed along the [001] zone axis. Because of the relatively low spatial resolution of DDPs obtained from such small particles, it is unlikely that Pd and Au could be distinguished with confidence on the basis of the difference in their unit cell dimensions (a = 3.89 Å for Pd and a = 4.08 Å for Au, a difference of 4.5%). The most commonly observed spots in DDPs of these metals are the {200} and {111} sets with spacings of 1.95 and 2.04 Å and 2.25 and 2.35 Å for Pd and Au, respectively. Au–Pd core–shell particles would be expected to appear as a light ring with a dark centre in the TEM image because of the difference in atomic mass between Pd (106) and Au (197). How clearly, and whether, this is observed depends, again, on particle size, the relative thicknesses of the core and the shell, and the imaging conditions. In the intermediate magnification images in Fig. 2(b)–(d), some of the particles do indeed appear to have a core–shell structure, and some of these are indicated in the figure.

For further imaging, and for elemental analysis by EDS, the nanoparticles were washed and centrifuged to separate them from their native suspension before re-suspending them in deionised water. In these samples, clusters of nanoparticles are typical, such as those shown in Fig. 3. EDS mapping allows the spatial distribution of selected elements across the area of interest to be determined in the TEM. Although the nanoparticles were too small for Au and Pd distributions to be mapped within individual nanoparticles, Pd and Au maps were obtained from the clusters shown in Fig. 3 for the 3.3Pd(Au) sample. Both Au and Pd extend throughout the structures seen in the TEM images and, importantly, the

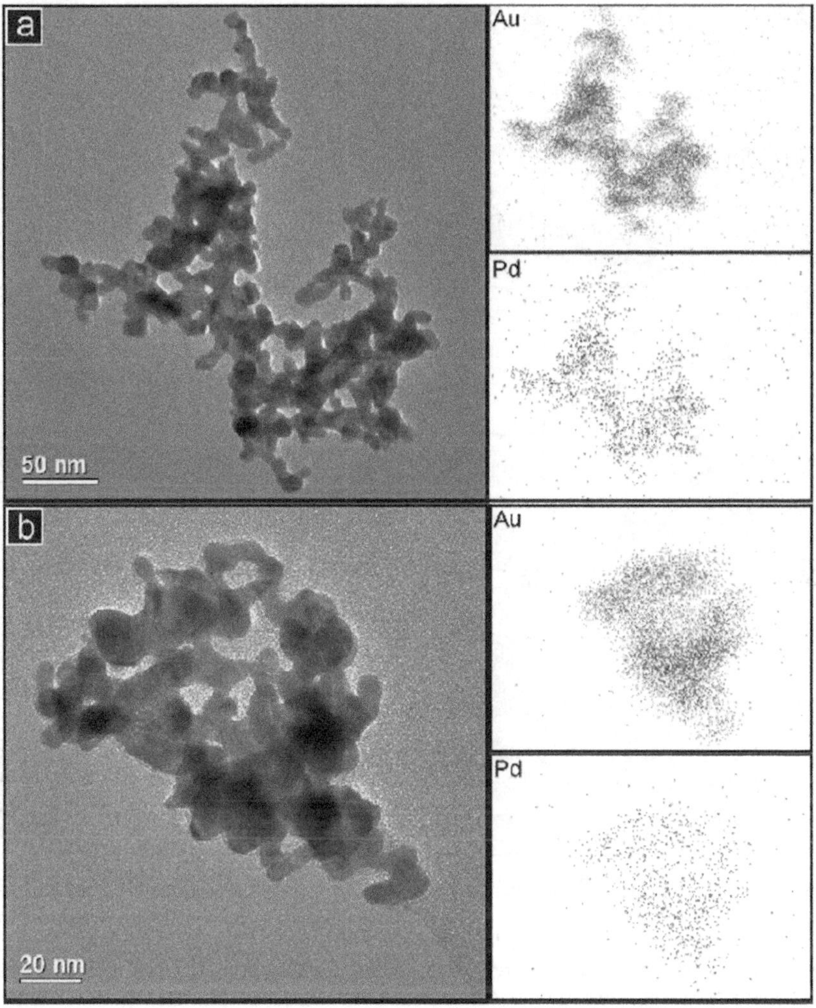

Fig. 3 TEM Images of two groups of unsupported 3.3Pd(Au) nanoparticles (after centrifugation) along with the corresponding EDS elemental maps for Au and Pd.

distribution of Pd closely matches that of Au. This indicates an intimate mixing of these two elements. It follows that at least most nanoparticles contain both elements. To investigate this further, ED spectra were obtained from further clusters and also from points within clusters, as presented in Fig. 4. For 0.2Pd(Au), whole clusters as well as points within clusters were seen to have a consistently high Au content and low Pd content (Fig. 4(a) and (b)). This was also the case for the majority of clusters of 1.5Pd(Au) and 3.3Pd(Au) – as shown in Fig. 4(c) and (f) and for the two nanoparticles shown in Fig. 4(g). However, a minority of the particles analysed were almost pure Pd. This is the case within the circled area of Fig. 4(d) and in the expanded view in Fig. 4(e) for 1.5Pd(Au), as well as for the nanoparticles of 3.3Pd(Au) seen in Fig. 4(h). The image and DDP in Fig. 4(i) confirm the expected crystallography of these particles.

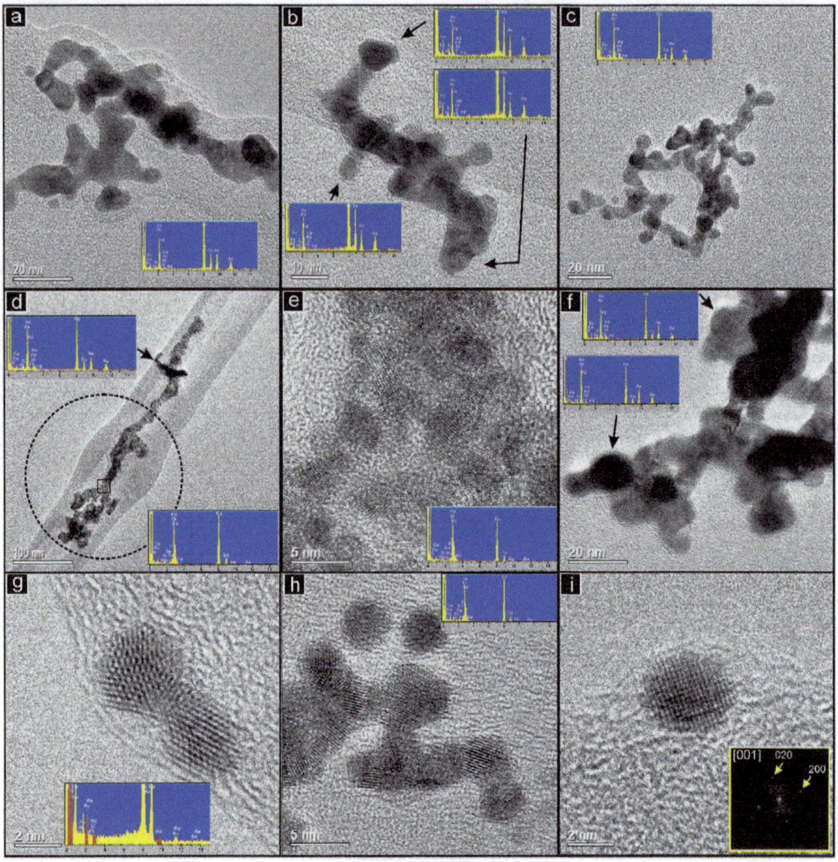

Fig. 4 TEM images of unsupported nanoparticles (after centrifugation) with corresponding EDS spectra. The spectra are of the whole field of view, the circled area in (d) or the regions indicated with arrows. Images (a) and (b) show 0.2Pd(Au) nanoparticles, (c) to (e) show 1.5Pd(Au) nanoparticles and (f) to (i) show 3.3Pd(Au) nanoparticles. The DDP of the nanoparticle in image (i) is shown in the inset. Image (e) is an expanded view of the small square area in image (d).

Fig. 5–7 present, respectively, images of the 0.2, 1.5 and 3.3Pd(Au) nanoparticles supported on titania, which were used as catalysts in this work. The nanoparticle size distributions of these samples were determined from the TEM images and are presented in Fig. 8. These were very similar for all three samples, each of which showed an average particle size of 3.5 to 3.7 nm. The high dispersion of these nanoparticles across the support is shown for 0.2Pd(Au) in Fig. 5(a). In Fig. 5(b), a metal particle is viewed in profile and the crystal structure of the nanoparticle and the support each afford complete DDPs. The support can be indexed to the anatase phase of titania viewed along the [100] (or [010]) direction (the presence of the 002 spots – normally systematically absent – suggests either a defective structure or some double diffraction) and the additional spots in DDP(ii) can be assigned to the metal viewed along its [011] zone axis. As is typical for these samples, the crystal lattices of the metal and the support are not aligned. This is consistent with the

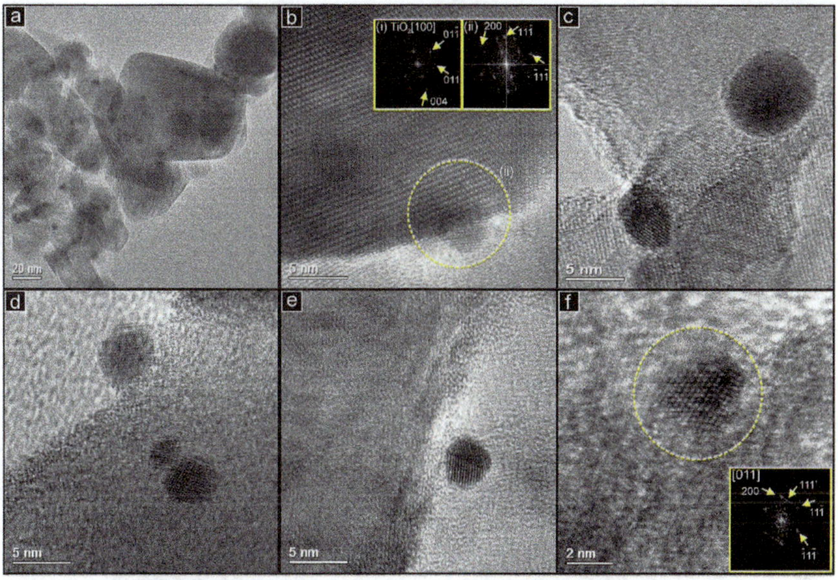

Fig. 5 TEM images of 0.2Pd(Au) nanoparticles supported on TiO_2 at a range of magnifications. Image (b) includes the DDPs of (i) the support, which is indexed to anatase, and (ii) the metal nanoparticle and support (circled). The DDP of the twinned nanoparticle (circled) in image (f) shows a full diffraction pattern relating to the larger crystal domain (LHS) and additional spots (′) from the smaller domain (RHS).

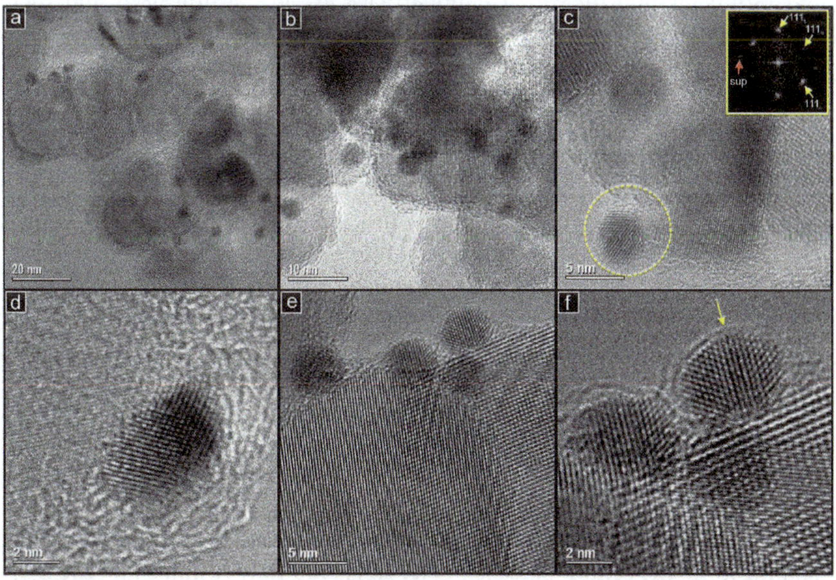

Fig. 6 TEM images of 1.5Pd(Au) nanoparticles supported on TiO_2 at a range of magnifications. The DDP of the nanoparticle circled in image (c) is shown in the inset. Spots due to two crystal domains in the nanoparticle (t = top, b = bottom, c = common to both domains) and the oxide support (sup) are indicated. Image (f) is an expanded view from (e) in which the twin boundary of one nanoparticle is indicated.

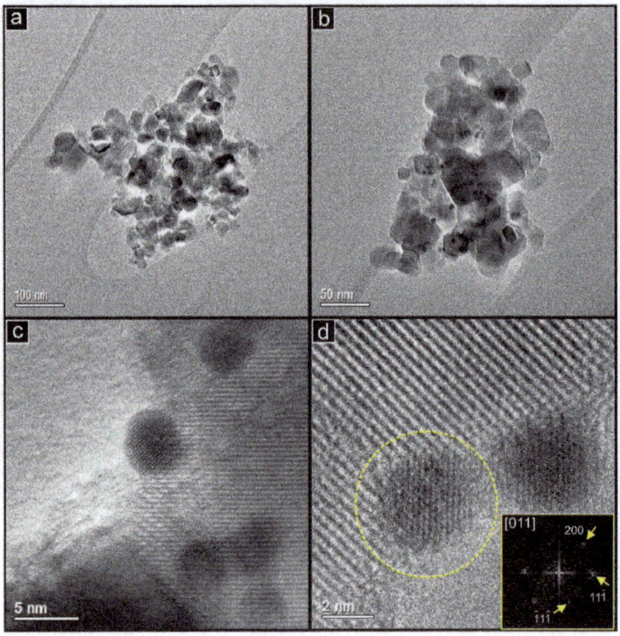

Fig. 7 TEM images of 3.3Pd(Au) nanoparticles supported on TiO_2 at a range of magnifications. The DDP of the nanoparticle circled in image (d) is shown in the inset.

preparation method, in which nanoparticles are first synthesised and then deposited onto the support. The majority of the nanoparticles are also polycrystalline, exhibiting multiple crystal domains. This is the case for the 7 nm particle in Fig. 5(c) and for the singly-twinned particles in Fig. 5(e) and (f). The DDP of the latter contains a full diffraction pattern for the larger domain (on the left) plus additional spots for the smaller domain (right). It was very rare to observe core–shell structures in the nanoparticles supported on titania. One example is the particle shown in profile in Fig. 5(d), which has a dark centre and lighter region around it, both clearly facetted. Although more speculative, it is also possible that the nanoparticle in image (e) has a very thin, light shell, of maybe only one atomic layer. The images in Fig. 6 show similar results for 1.5Pd(Au). The metal nanoparticles are well dispersed across the surface of the support particles. At high resolution, the majority appear to be polycrystalline – as in Fig. 6(c), (e) and (f) – and not crystallographically aligned with the support, as shown in the DDP for image (c). Image (d) appears to show two overlapping nanoparticles and in image (f) the very clear twin boundary is indicated. Finally, for 3.3Pd(Au), the low magnification images in Fig. 7 again show the dispersion of the metal nanoparticles across the support. At high magnification, the central particle in Fig. 7(c) is multiply-twinned and the DDP of the circled particle in Fig. 7(d) can be indexed to Au or Pd viewed along the [011] direction. No features were seen in the images of the titania-supported samples of either 1.5Pd(Au) or 3.3Pd(Au) that could be identified as core–shell structures.

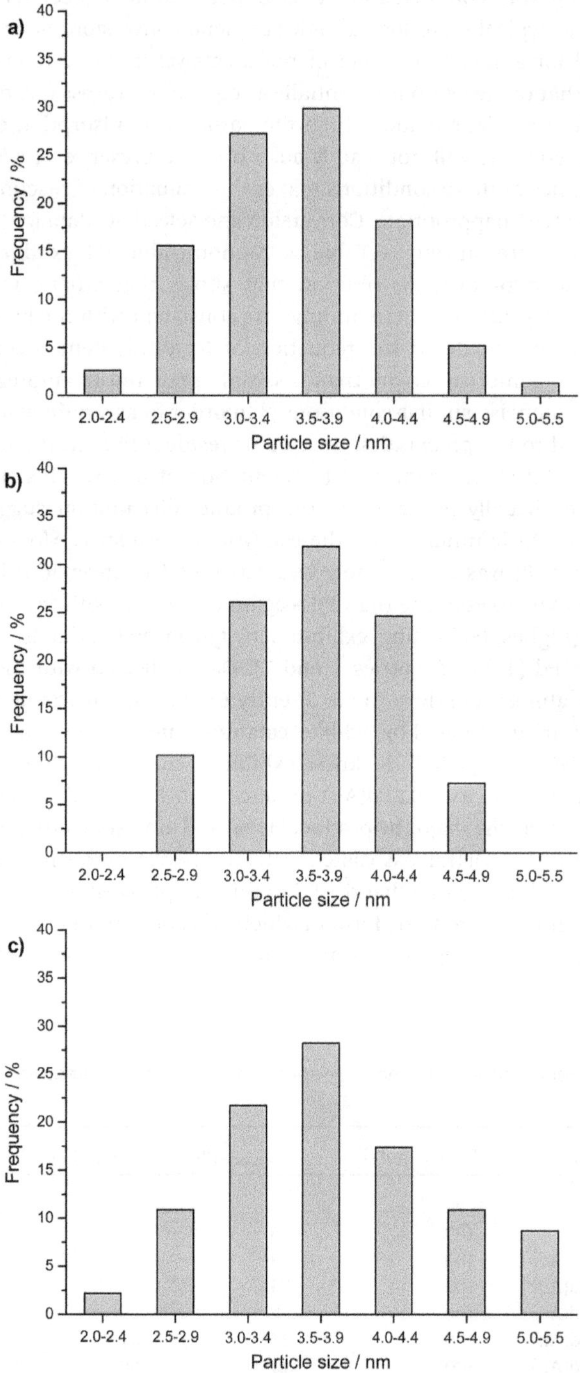

Fig. 8 Metal nanoparticle size distributions determined by TEM for (a) 0.2Pd(Au), (b) 1.5Pd(Au) and (c) 3.3Pd(Au).

Catalyst testing

Catalyst testing was conducted in a fixed bed stainless steel reactor under conditions that typically produce a high acetylene conversion. Such conditions are beneficial for assessing whether or not a catalyst is prone to over-hydrogenation given that the reaction is essentially a sequential process (*i.e.* the enthalpy of adsorption of ethylene is lower than the enthalpy of adsorption of acetylene meaning that ethylene will not readily adsorb in the presence of acetylene). Of course, the choice of these conditions makes the evaluation of reaction rates and kinetic parameters inappropriate. Conversion and selectivity data for the samples tested at 363 K are shown in Table 2. Monometallic Pd exhibits complete conversion and a product distribution that shows that ethane is the major product (Table 2 – entry 1). These findings are consistent with the literature,[8–10,15] although a sample made by the reduction of Na_2PdCl_4 demonstrated higher ethylene and oligomer selectivity than a sample prepared by impregnation but tested under otherwise similar conditions.[39] In order to ascertain whether or not this was related to the preparation method or residual PVP on the nanoparticle surface, this catalyst was subjected to calcination at 673 K. This thermal pre-treatment had virtually no effect on the product distribution suggesting that surface PVP has little influence on the catalytic performance. Monometallic Au (Table 2 – entry 8) was considerably less active (21% conversion) but showed a higher selectivity to ethylene than Pd – again consistent with the literature.[38]

3.3Pd(Au) (highest Pd loading) exhibits activity/conversion that is equivalent to monometallic Pd (Table 2, entries 1 and 7). Data collected with the 1.5Pd(Au) sample looks almost identical (Table 2, entry 6). As such, it would appear that these samples are dominated by 'Pd-like' catalysis. Interestingly, the sample with the smallest Pd loading/shell thickness exhibits a completely different catalytic performance. In this case, 0.2Pd(Au) produces ethylene as the major product (53%), oligomers as the major by-product (41%) and almost no ethane (6%). High activity at only 363 K is a trait associated with Pd, whereas high ethylene selectivity is certainly not. This suggests that the Pd in this sample is either electronically or geometrically different from the form in which Pd is present in the other samples studied in this work. One point of significance is that the oligomer selectivity for

Table 2 Acetylene conversion and ethylene/ethane/oligomers selectivity over various catalysts at 363 K

Entry	Sample	Conversion [%]	$S_{ethylene}$ [%]	S_{ethane} [%]	$S_{oligomers}$ [%]
1	Pd	100	23	55	22
2	Pd[a]	100	25	51	24
3	0.2Pd(Au)	100	53	6	41
4	0.2Pd(Au)[b]	100	53	7	40
5	0.2Pd(Au)[c]	100	52	9	39
6	1.5Pd(Au)	100	27	51	22
7	3.3Pd(Au)	100	25	50	25
8	Au	21	62	21	17

[a] Sample calcined at 673 K. [b] Sample was pre-treated by heating in water at 363 K to remove PVP. [c] Sample was pre-treated in water whilst exposed to UV light to remove PVP.

0.2Pd(Au) is abnormally high with respect to Pd. In order to determine whether this was associated with residual PVP on the surface, different pre-treatments were explored with an objective to remove the stabiliser. Treatments in hot water or with UV irradiation (Table 2, entries 4 and 5, respectively) did not appear to significantly affect the oligomer (or any other product) selectivity. From this it is inferred that a high oligomer selectivity is not associated with the presence of residual PVP but is more likely associated with some other function related to the exposed metal sites – most likely the presence of Au surface atoms.

Time on stream data for the 0.2Pd(Au) sample using 1.8 equivalents of H_2 relative to acetylene (the same conditions as used in Table 2) and with a larger 3.6 fold excess of hydrogen is presented in Fig. 9. A number of informative points are apparent. Firstly, using 1.8 equivalents of H_2, the selectivity is remarkably stable across a range of temperatures with very little tendency to form ethane. For example, the ethane selectivity is only increased from 3 to 6% as the temperature was raised from 323 to 363 K (even though full acetylene conversion is readily achieved at low temperature). As the temperature is increased, the most significant change to the selectivity is actually an increase in the ethylene selectivity (44 to 53%) at the expense of oligomer formation (53 to 41%). This suggests that oligomer formation is associated with hydrogen availability. Secondly, at low temperatures (323 and 333 K), 0.2Pd(Au) is prone to deactivation with a notable decrease in acetylene conversion over 5 h time on stream. This is attributed to the high oligomer selectivity, which will eventually lead to a reduction in the available adsorption sites. Operation at higher temperatures results in the partial recovery of the activity, although most likely masks any further deactivation. Lastly, whilst operation of 0.2Pd(Au) with a slight excess of H_2 relative to acetylene limits over-hydrogenation, the use of a larger H_2 excess does result in increased ethane formation, although largely as a result of decreased oligomer formation (Fig. 9(b)). Under these conditions, deactivation at 323 K is still apparent, although to a lesser extent, again suggesting a correlation between hydrogen availability and oligomer formation. No evidence of deactivation was observed for the 1.5Pd(Au) and 3.3Pd(Au) samples (see ESI Fig. S2 and S3,† respectively).

FTIR spectroscopy of adsorbed CO

FTIR spectroscopy using CO as a probe molecule is a useful technique for evaluating the nature of surface sites on a supported metal particle since band positions are sensitive to both the type of adsorption site (*i.e.* type of metal) and the mode of adsorption (*e.g.* linear *vs.* bridge bound CO). FTIR spectra of the samples that had been reduced at 373 K and exposed to CO at beam temperature are shown in Fig. 10. Following reduction, the 0.2Pd(Au) sample exhibits bands at 2071 and 2040 cm^{-1}, as well as a contribution from gas phase CO centred at 2143 cm^{-1} (Fig. 10(a) and (b)). The band at 2071 cm^{-1} is likely associated with CO adsorbed linearly on Pd corner atoms,[66] although the band position is slightly lower than typically expected, suggesting that the Pd atoms are electronically perturbed. The broad shoulder at *ca.* 2040 cm^{-1} occurs at a frequency seldom observed, although a band at 2050 cm^{-1} has previously been assigned to CO linearly adsorbed on edge sites.[66] Interestingly, there is almost no band intensity in the region where bridge bound CO would be expected (1950–1850 cm^{-1}). Instead, there is only a faint broad band centred at 1920 cm^{-1}, although the

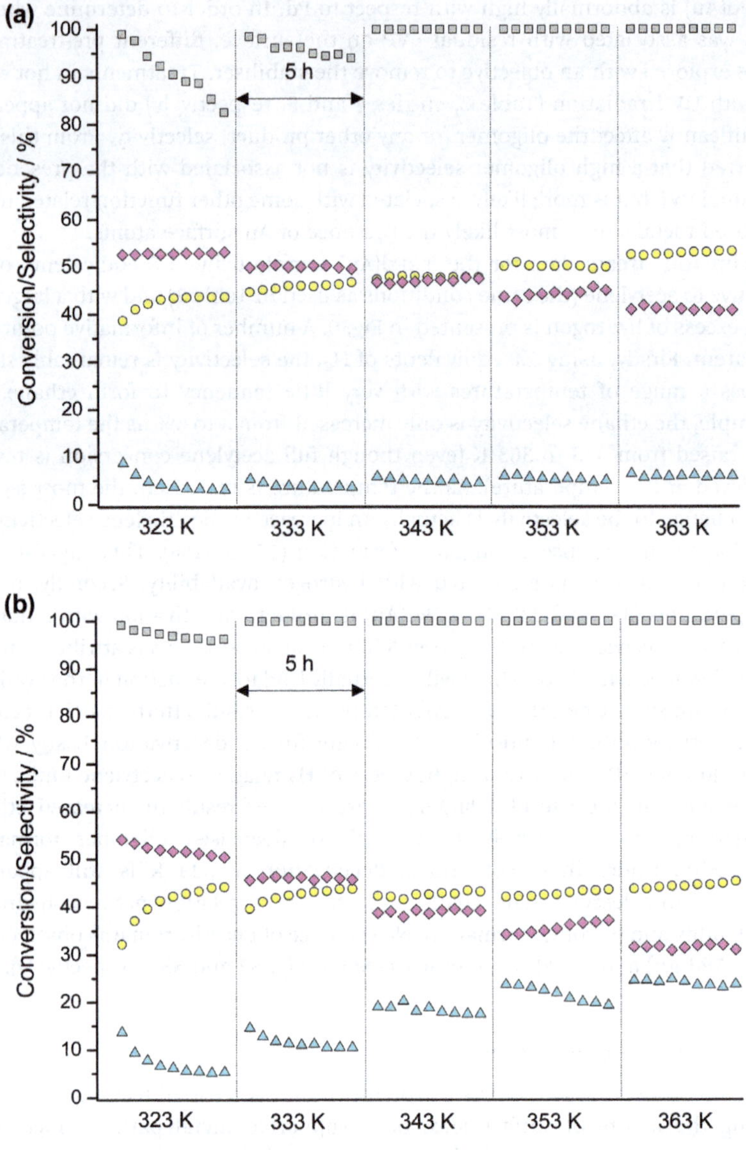

Fig. 9 Time on stream data (5 h periods) for 0.2Pd(Au) at temperatures between 323 and 363 K using H_2 : acetylene ratios of (a) 1.8 : 1 and (b) 3.6 : 1. Acetylene conversion (grey squares), ethylene selectivity (yellow circles), ethane selectivity (cyan triangles) and oligomer selectivity (pink diamonds).

intensity is only marginally above the instrumental signal to noise in that region. This suggests that very few sites exist with neighbouring Pd atoms, or that these sites are not available for CO adsorption after reduction at 373 K. Similarly, there is no evidence of a band at *ca.* 2100 cm^{-1} which could be assigned to CO adsorbed linearly on low coordinated Au atoms.[67,68] Again, this implies that these sites do not exist (*i.e.* Pd preferentially occupies low coordination sites) or are not available for adsorption.

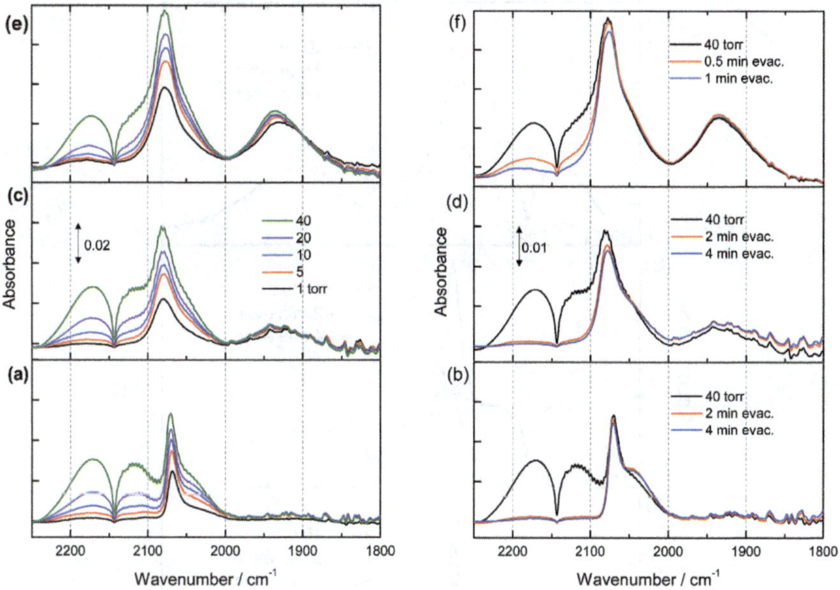

Fig. 10 FTIR spectra of the samples after reduction at 373 K with exposure to CO at room temperature. (a) 0.2Pd(Au), (b) 0.2Pd(Au) after evacuation, (c) 1.5Pd(Au), (d) 1.5Pd(Au) after evacuation, (e) 3.3Pd(Au) and (f) 3.3Pd(Au) after evacuation.

In order to ascertain the nature of the sites that were blocked by PVP, spectra of adsorbed CO were collected following a UV irradiation/water pre-treatment to remove as much PVP as possible. These are presented in Fig. 11. Following such a treatment, the 0.2Pd(Au) sample (Fig. 11(a)) presents bands at 2074 and 2037 cm^{-1}, which are assigned as before. A much more intense band associated with CO bridging two Pd atoms is apparent at 1970 cm^{-1} (bridging on edge sites), which suggests that PVP does influence the surface site availability. However, this appears to have little effect on the catalytic performance (compare Table 2, entries 3 and 5). A further peak is apparent at 2110 cm^{-1} and this is attributed to CO linearly adsorbed on Au atoms, which correlates well with the vibrational frequency calculated by Zhu *et al.*[69] for CO adsorbed on Au-edge atoms (2112 cm^{-1}) using DFT. The appearance of a band associated with CO adsorption on Au is consistent with sub-monolayer coverage of the Au cores with Pd. As a result, it is thought that 0.2Pd(Au) will possess a number of different surface sites, including Pd atoms with only Au neighbours and Pd atoms with one Pd neighbour and further Au neighbours. Importantly, there is no evidence of 3-fold hollow Pd sites, which suggests the absence of larger Pd ensembles.

The spectra of 1.5Pd(Au) after reduction at 373 K (Fig. 10(c) and (d)) show similarities to those collected for the 0.2Pd(Au) sample. Absorption bands are observed at 2070 and 2043 cm^{-1}, which correspond to CO adsorbed linearly on Pd atoms, as before. However, a far more intense band related to bridge bound CO on Pd is apparent and is centred at 1936 cm^{-1}. The spectra of this sample following UV/water treatment (Fig. 11(b)) are only subtly different but highlight two key points. Firstly, no band is observed above 2100 cm^{-1}, suggesting the absence of exposed Au atoms. This may be because the Au core is completely covered by Pd

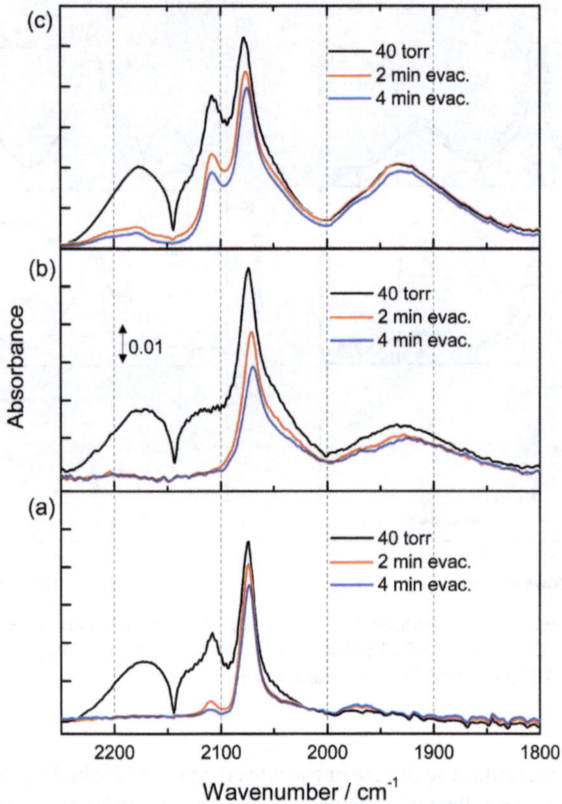

Fig. 11 FTIR spectra of the samples exposed to UV irradiation after reduction at 373 K with exposure to CO at room temperature and subsequent evacuation. (a) 0.2Pd(Au), (b) 1.5Pd(Au) and (c) 3.3Pd(Au).

atoms (*i.e.* a true core–shell structure has been formed). Secondly, the region where the bridge bound CO is seen is more complex. The centre of the band shifts from 1924 to 1932 cm^{-1} as CO overpressure/coverage is increased. This suggests a degree of dipolar coupling, which implies that Pd–Pd dimer sites must be in close proximity to one another. A second feature begins to become resolved at 1969 cm^{-1}, which is likely associated with bridge bound CO on Pd edge sites.[66] Finally, the band tails considerably towards a lower wavenumber, which hints that some 3-fold hollow adsorption sites exist. This implies that larger ensembles of Pd exist in this sample as compared with the 0.2Pd(Au) sample. This would correlate with the deposition of more than 1 layer of Pd on the Au core or with the formation of some monometallic Pd particles.

Following reduction, the 3.3Pd(Au) sample (Fig. 10(e) and (f)) displays the same bands as observed for the 1.5Pd(Au) sample (2077, 2044 and 1936 cm^{-1}). This would, in principle, be consistent with a thicker Pd shell grown around a gold core. However, the spectra collected after a treatment intended to remove PVP (Fig. 11(c)) suggest differently, given that a prominent band was observed at 2108 cm^{-1}. This band is attributed to CO adsorbed on Au-edge atoms and highlights that Au atoms definitely contribute to the surface sites. Given that this band is considerably more intense than for the 0.2Pd(Au) sample, it is thought that more of these sites exist for the 3.3Pd(Au) sample. In order for this to be the case, less of the Au core must be covered by Pd. As a result, it is highly likely that monometallic Pd particles exist within this sample. Lastly, the band associated with bridge bound CO on Pd edge sites is more pronounced for this sample (1975 cm^{-1}).

Discussion

0.2Pd(Au) sample

Detailed characterisation of the 0.2Pd(Au) material suggests that this sample is best described as an incomplete monolayer of Pd sitting on an Au core. The proposed structure is presented in Fig. 12. This structure appears to be close to the intended design and was likely possible due to the exceptionally slow Pd addition rate (5 μmol h^{-1}), which should minimise the Pd solution concentration during the synthesis – disfavouring the formation of monometallic Pd particles whilst favouring deposition of Pd onto pre-existing Au cores. Indeed, FTIR spectroscopy confirms the absence of monometallic Pd particles since there is no evidence of three-fold hollow adsorption sites (Fig. 10(a) and (b) and 11(a)). There are only weak bands associated with bridge bound CO (more evident after partial removal of PVP – Fig. 11(a)). Interestingly, the removal of PVP did appear to uncover Pd edge sites (1970 cm^{-1}), although these sites appeared to have little effect on the catalytic performance (Table 2 – entries 3, 4 and 5). The appearance of an absorption band associated with CO adsorbed on low coordinated Au sites[67,68] after PVP removal is consistent with incomplete coverage of the Au core (Fig. 11(a)). Combined TEM-EDS analysis of unsupported 0.2Pd(Au) nanoparticles tends to suggest that both metals were present together but with a consistently high Au content and low Pd content, suggesting good sample uniformity (Fig. 4(a) and (b)). Immobilisation of the 0.2Pd(Au) nanoparticles onto TiO_2 is not thought to have disrupted the structure with evidence of a faint core–shell structure in a small minority of particles and no evidence of crystal lattice alignment between the nanoparticles and the support (*i.e.* no surface wetting) – see Fig. 5.

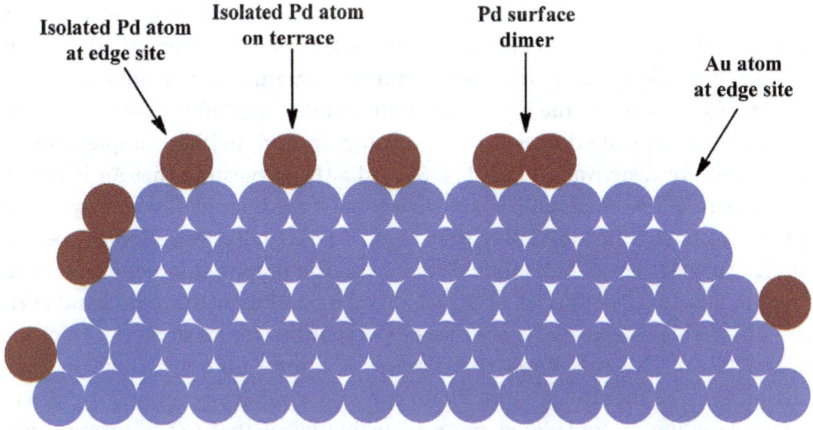

Fig. 12 Proposed structure of 0.2Pd(Au) after removal of PVP.

Catalytic data collected for the selective hydrogenation of acetylene shows that the 0.2Pd(Au) sample produces ethylene and oligomers as the major reaction products (Table 2 and Fig. 9). This holds true, irrespective of whether an attempt was made to remove residual surface PVP or not (Table 2 – entries 3, 4 and 5). The catalyst also shows remarkably low ethane selectivity even at temperatures well in excess of that necessary to fully react acetylene (*i.e.* under conditions where ethylene is free to adsorb and react). This is in contrast to a monometallic Pd catalyst, which produces ethane as the major product (Table 2 – entry 1). In principle, two explanations can be used to rationalise the low ethane selectivity of 0.2Pd(Au). Firstly, any Pd present is highly dispersed, meaning that formation of the β-Pd-hydride phase is avoided.[65] Given that this phase is generally associated with ethane formation, its absence would account for the decreased ethane formation/increased ethylene selectivity.[6–16] Secondly, since the amount of Pd present is insufficient to cover the Au core, it is possible that Au acts as an additional reaction site. This could be as a result of direct hydrogenation on Au sites or *via* Pd assisted hydrogen dissociation (*i.e.* H_2 dissociation on Pd and spillover onto Au), as has been reported for CuPd catalysts[39–42] (and possibly for AgPd[47] and AuPd[48]). Regardless of the mechanism, the participation of Au would explain the relatively high oligomer selectivity and deactivation at lower temperatures since both are characteristic traits of Au during alkyne hydrogenation,[38] especially at the low temperatures employed in this study. The involvement of Au as a reaction site may also contribute to the low ethane selectivity given that ethylene adsorption on Au is an activated process (*i.e.* subject to an energy barrier).[38] These results also suggest that the interplay between Pd and Au may be further modified by changing the Pd surface coverage and may warrant further investigation. The low ethane selectivity is most likely associated with the combined effects of high Pd dispersion and the role of Au as a reaction site. However, given that the activity of 0.2Pd(Au) is far higher than that of monometallic Au, it would suggest that direct hydrogenation on Au is the least likely process to occur.

1.5Pd(Au) and 3.3Pd(Au) samples

Both the 1.5Pd(Au) and 3.3Pd(Au) samples show similar catalytic performances with product distributions that favour ethane (Table 2 – entries 6 and 7, respectively) and closely mirror that of monometallic Pd (Table 2 – entry 1). That being the case, it is highly likely that both samples contain monometallic Pd nanoparticles (formed due to the higher Pd addition rate – see Table 1) that act as the dominant reaction site during catalytic testing. Indeed, neither sample displays any evidence of deactivation (ESI Fig. S2 and S3†), suggesting that Au is not an active site in either material (*i.e.* no surface Au sites exist or the activity of any exposed Au is significantly lower than that of Pd). Characterisation of the two samples suggests small but subtle differences. The proposed structures are presented in Fig. 13. FTIR spectra of CO adsorbed on 1.5Pd(Au) show no band at *ca.* 2100 cm^{-1}, which suggests the absence of low coordinated Au sites (Fig. 10(c) and (d), and 11(b)). This may imply that exposed Au atoms do not occupy low coordination sites, or that the pre-formed Au cores are completely covered in Pd. The later explanation is considered more plausible given that CO adsorption was apparent on Au for both the 0.2Pd(Au) and 3.3Pd(Au) samples (Fig. 11). That being

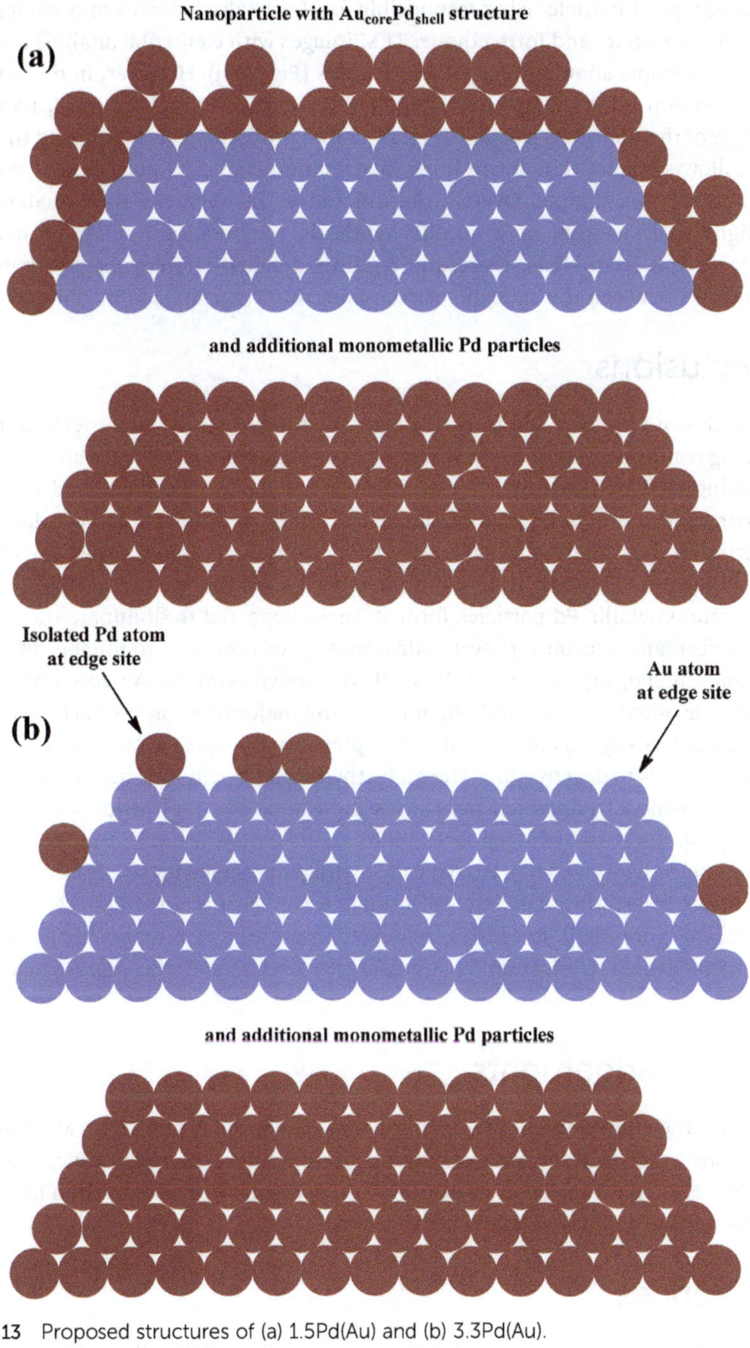

Fig. 13 Proposed structures of (a) 1.5Pd(Au) and (b) 3.3Pd(Au).

the case, it is thought that some nanoparticles in the 1.5Pd(Au) sample will possess a core–shell structure. In addition, TEM-EDS analysis showed that some Pd rich regions are present (Fig. 4(d) and (e)). As such, the 1.5Pd(Au) sample probably contains core–shell nanoparticles and monometallic Pd particles, with the latter type of particle being responsible for the catalytic activity and tendency to over-hydrogenate and form ethane. TEM images with elemental analysis of the 3.3Pd(Au) sample show similar Pd rich regions (Fig. 4(h)). However, in this case, CO adsorption on Au is apparent (Fig. 11(c)), suggesting that, at most, partial coverage of the Au core is possible (Fig. 13). More pronounced bridging and three-fold hollow sites are thought to imply that monometallic Pd particles are more common for this sample. Overall, the differences between the three materials highlight that, in-spite of a similar synthesis methodology, different nano-structures can form. This, therefore, requires detailed characterisation before useful links between the catalytic performance and structure can be obtained.

Conclusions

It was shown that the addition of Pd to pre-formed Au nanoparticles under reducing conditions can selectively result in the deposition of Pd onto an Au core, producing nanoparticles with a core–shell morphology. The likelihood of this occurring was found to depend on the rate of the Pd addition with slow addition favouring deposition onto Au, whereas faster addition allowed for the formation of monometallic Pd particles in addition to core–shell nanoparticles. In samples where monometallic Pd particles formed, these appeared to dominate the catalytic performance, favouring over-hydrogenation of acetylene to ethane. In one material – 0.2Pd(Au) – a partial Pd shell was grown onto the Au core and this material favoured ethylene and oligomers as the major reaction products. A high oligomer selectivity was attributed to the presence of exposed Au atoms, which ultimately leads to deactivation. The high ethylene selectivity was attributed to the high dispersion of Pd, which minimises the extent of β-Pd-hydride formation, although spillover of hydrogen from Pd to Au sites may also result in increased ethylene selectivity given that Au typically hydrogenates alkenes to a lesser extent. Overall, an interesting catalytic performance can be obtained by the careful synthesis of core–shell structures, however, the synthesis remains challenging and detailed characterisation is necessary to understand the nanostructures formed.

Acknowledgements

We thank the University of Aberdeen for financial support and Dr K. McManus (University of Aberdeen) for performing preliminary experiments with these samples. Electron microscopy and EDS were performed by RTB at the Electron Microscopy Facility, University of St Andrews.

References

1 A. Borodziński and G. C. Bond, *Catal. Rev.: Sci. Eng.*, 2006, **48**, 91–144.
2 A. Borodziński and G. C. Bond, *Catal. Rev.: Sci. Eng.*, 2008, **50**, 379–469.
3 S. A. Nikolaev, I. L. N. Zanaveskin, V. V. Smirnov, V. A. Averyanov and K. I. Zanaveskin, *Russ. Chem. Rev.*, 2009, **78**, 231–247.
4 D. B. Tiedtke, T. T. P. Cheung, J. Leger, S. A. Zisman, J. J. Bergmeister and G. A. Delzer, in *13th Ethlyene Producers Conference*, 2001, vol. 10, p. 1.
5 G. C. Bond and P. B. Wells, *Adv. Catal.*, 1964, **15**, 91–226.

6 D. Teschner, J. Borsodi, A. Wootsch, Z. Révay, M. Hävecker, A. Knop-Gericke, S. D. Jackson and R. Schlögl, *Science*, 2008, **320**, 86–89.
7 D. Teschner, J. Borsodi, Z. Kis, L. Szentmiklósi, Z. Révay, A. Knop-Gericke, R. Schlögl, D. Torres and P. Sautet, *J. Phys. Chem. C*, 2010, **114**, 2293–2299.
8 M. Armbrüster, M. Behrens, F. Cinquini, K. Föttinger, Y. Grin, A. Haghofer, B. Klötzer, A. Knop-Gericke, H. Lorenz, A. Ota, S. Penner, J. Prinz, C. Rameshan, Z. Révay, D. Rosenthal, G. Rupprechter, P. Sautet, R. Schlögl, L. Shao, L. Szentmiklósi, D. Teschner, D. Torres, R. Wagner, R. Widmer and G. Wowsnick, *ChemCatChem*, 2012, **4**, 1048–1063.
9 M. García-Mota, B. Bridier, J. Pérez-Ramírez and N. López, *J. Catal.*, 2010, **273**, 92–102.
10 B. Bridier, N. Lopez and J. Pérez-Ramírez, *Dalton Trans.*, 2010, **39**, 8412–8419.
11 W. Ludwig, A. Savara, K. H. Dostert and S. Schauermann, *J. Catal.*, 2011, **284**, 148–156.
12 W. Ludwig, A. Savara, R. J. Madix, S. Schauermann and H. J. Freund, *J. Phys. Chem. C*, 2012, **116**, 3539–3544.
13 S. Schauermann, N. Nilius, S. Shaikhutdinov and H. J. Freund, *Acc. Chem. Res.*, 2013, **46**, 1673–1681.
14 J. Sá, G. D. Arteaga, R. A. Daley, J. Bernardi and J. A. Anderson, *J. Phys. Chem. B*, 2006, **110**, 17090–17095.
15 B. Yang, R. Burch, C. Hardacre, G. Headdock and P. Hu, *J. Catal.*, 2013, **305**, 264–276.
16 N. A. Khan, S. Shaikhutdinov and H. J. Freund, *Catal. Lett.*, 2006, **108**, 159–164.
17 A. J. McCue and J. A. Anderson, *Front. Chem. Sci. Eng.*, 2015, **9**, 142–153.
18 A. D. Benavidez, P. D. Burton, J. L. Nogales, A. R. Jenkins, S. a. Ivanov, J. T. Miller, A. M. Karim and A. K. Datye, *Appl. Catal., A*, 2014, **482**, 108–115.
19 Y. K. Gulyaeva, V. V. Kaichev, V. I. Zaikovskii, E. V. Kovalyov, A. P. Suknev and B. S. Bal'zhinimaev, *Catal. Today*, 2015, **245**, 139–146.
20 S. A. Nikolaev and V. V. Smirnov, *Catal. Today*, 2009, **147**, S336–S341.
21 S. A. Nikolaev, V. V. Smirnov, A. Y. Vasil'kov and V. L. Podshibikhin, *Kinet. Catal.*, 2010, **51**, 375–379.
22 B. Bridier and J. Pérez-Ramírez, *J. Am. Chem. Soc.*, 2010, **132**, 4321–4327.
23 F. M. McKenna, R. P. K. Wells and J. A. Anderson, *Chem. Commun.*, 2011, **47**, 2351–2353.
24 A. J. McCue, F.-M. McKenna and J. A. Anderson, *Catal. Sci. Technol.*, 2015, **5**, 2449–2459.
25 F. M. McKenna and J. A. Anderson, *J. Catal.*, 2011, **281**, 231–240.
26 F. M. McKenna, L. Mantarosie, R. P. K. Wells, C. Hardacre and J. a. Anderson, *Catal. Sci. Technol.*, 2012, **2**, 632–638.
27 A. J. McCue and J. A. Anderson, *Catal. Sci. Technol.*, 2014, **4**, 272–294.
28 F. Studt, F. Abild-Pedersen, T. Bligaard, R. Z. Sørensen, C. H. Christensen and J. K. Nørskov, *Angew. Chem., Int. Ed.*, 2008, **47**, 9299–9302.
29 S. K. Kim, J. H. Lee, I. Y. Ahn, W. J. Kim and S. H. Moon, *Appl. Catal., A*, 2011, **401**, 12–19.
30 M. Friedrich, S. A. Villaseca, L. Szentmiklósi, D. Teschner and M. Armbrüster, *Materials*, 2013, **6**, 2958–2977.
31 J. Osswald, R. Giedigkeit, R. E. Jentoft, M. Armbrüster, F. Girgsdies, K. Kovnir, R. Thorsten, Y. Grin and R. Schlögl, *J. Catal.*, 2008, **258**, 210–218.

32 J. Osswald, K. Kovnir, M. Armbrüster, R. Giedigkeit, R. E. Jentoft, U. Wild, Y. Grin and R. Schlögl, *J. Catal.*, 2008, **258**, 219–227.
33 G. Vilé, D. Albani, M. Nachtegaal, Z. Chen, D. Dontsova, M. Antonietti, N. López and J. Pérez-Ramírez, *Angew. Chem., Int. Ed.*, 2015, **54**, 11265–11269.
34 B. Bridier, N. López and J. Pérez-Ramírez, *J. Catal.*, 2010, **269**, 80–92.
35 B. Bridier, M. a G. Hevia, N. López and J. Pérez-Ramírez, *J. Catal.*, 2011, **278**, 167–172.
36 G. Vilé, D. Baudouin, I. N. Remediakis, C. Copéret, N. López and J. Pérez-Ramírez, *ChemCatChem*, 2013, **5**, 3750–3759.
37 E. Oakton, G. Vilé, D. S. Levine, E. Zocher, D. Baudouin, J. Pérez-Ramírez and C. Copéret, *Dalton Trans.*, 2014, **43**, 15138–15142.
38 Y. Segura, N. López and J. Pérez-Ramírez, *J. Catal.*, 2007, **247**, 383–386.
39 A. J. McCue, C. J. McRitchie, A. M. Shepherd and J. A. Anderson, *J. Catal.*, 2014, **319**, 127–135.
40 A. J. McCue, A. M. Shepherd and J. A. Anderson, *Catal. Sci. Technol.*, 2015, **5**, 2880–2890.
41 A. J. McCue and J. A. Anderson, *J. Catal.*, 2015, **329**, 538–546.
42 A. J. McCue, A. Gibson and J. A. Anderson, *Chem. Eng. J.*, 2016, **285**, 384–391.
43 H. L. Tierney, A. E. Baber, J. R. Kitchin and E. C. H. Sykes, *Phys. Rev. Lett.*, 2009, **103**, 246102.
44 G. Kyriakou, M. B. Boucher, A. D. Jewell, E. ALewis, T. J. Lawton, A. E. Baber, H. L. Tierney, M. Flytzani-stephanopoulos and E. C. H. Sykes, *Science*, 2012, **335**, 1209–1212.
45 M. B. Boucher, B. Zugic, G. Cladaras, J. Kammert, M. D. Marcinkowski, T. J. Lawton, E. C. H. Sykes and M. Flytzani-Stephanopoulos, *Phys. Chem. Chem. Phys.*, 2013, **15**, 12187–12196.
46 Q. Fu and Y. Luo, *ACS Catal.*, 2013, **3**, 1245–1252.
47 G. Pei, X. Liu, A. Wang, A. F. Lee, M. A. Isaacs, L. Li, X. Pan, X. Yang, X. Wang, Z. Tai, K. Wilson and T. Zhang, *ACS Catal.*, 2015, **5**, 3717–3725.
48 G. X. Pei, X. Y. Liu, A. Wang, L. Li, Y. Huang, T. Zhang, J. W. Lee, B. W. L. Jang and C.-Y. Mou, *New J. Chem.*, 2014, **38**, 2043–2051.
49 G. Vilé, B. Bridier, J. Wichert and J. Pérez-Ramírez, *Angew. Chem., Int. Ed.*, 2012, **51**, 8620–8623.
50 G. Vilé, P. Dähler, J. Vecchietti, M. Baltanás, S. Collins, M. Calatayud, A. Bonivardi and J. Pérez-Ramírez, *J. Catal.*, 2015, **324**, 69–78.
51 G. J. Hutchings and C. J. Kiely, *Acc. Chem. Res.*, 2013, **46**, 1759–1772.
52 P. Paalanen, B. M. Weckhuysen and M. Sankar, *Catal. Sci. Technol.*, 2013, **3**, 2869–2880.
53 C. Ma, Y. Du, J. Feng, X. Cao, J. Yang and D. Li, *J. Catal.*, 2014, **317**, 263–271.
54 R. G. Chaudhuri and S. Paria, *Chem. Rev.*, 2012, **112**, 2373–2433.
55 Q. Zhang, I. Lee, J. B. Joo, F. Zaera and Y. Yin, *Acc. Chem. Res.*, 2013, **46**, 1816–1824.
56 Y. Mizukoshi, K. Sato, J. Kugai, T. A. Yamamoto, T. J. Konno and N. Masahashi, *J. Exp. Nanosci.*, 2015, **10**, 235–247.
57 H. Wang, C. Wang, H. Yan, H. Yi and J. Lu, *J. Catal.*, 2015, **324**, 59–68.
58 W. Jones, R. Su, P. P. Wells, Y. Shen, N. Dimitratos, M. Bowker, D. Morgan, B. B. Iversen, A. Chutia, F. Besenbacher and G. Hutchings, *Phys. Chem. Chem. Phys.*, 2014, **16**, 26638–26644.

59 R. Tiruvalam, J. Pritchard, N. Dimitratos, J. Lopez-Sanchez, J. Edwards, A. Carley, G. Hutchings and C. Kiely, *Faraday Discuss.*, 2011, **152**, 63–86.
60 P. T. Witte, P. H. Berben, S. Boland, E. H. Boymans, D. Vogt, J. W. Geus and J. G. Donkervoort, *Top. Catal.*, 2012, **55**, 505–511.
61 P. T. Witte, S. Boland, F. Kirby, R. van Maanen, B. F. Bleeker, D. a M. de Winter, J. a. Post, J. W. Geus and P. H. Berben, *ChemCatChem*, 2013, **5**, 582–587.
62 M. Crespo-Quesada, J. M. Andanson, A. Yarulin, B. Lim, Y. Xia and L. Kiwi-Minsker, *Langmuir*, 2011, **27**, 7909–7916.
63 J. A. Lopez-Sanchez, N. Dimitratos, C. Hammond, G. L. Brett, L. Kesavan, S. White, P. Miedziak, R. Tiruvalam, R. L. Jenkins, A. F. Carley, D. Knight, C. J. Kiely and G. J. Hutchings, *Nat. Chem.*, 2011, **3**, 551–556.
64 A. R. Wilson, K. Sun, M. Chi, R. M. White, J. M. Lebeau, H. H. Lamb and B. J. Wiley, *J. Phys. Chem. C*, 2013, **117**, 17557–17566.
65 M. Boudart and H. S. Hwang, *J. Catal.*, 1975, **39**, 44–52.
66 T. Lear, R. Marshall, J. A. Lopez-Sanchez, S. D. Jackson, T. M. Klapötke, M. Bäumer, G. Rupprechter, H.-J. Freund and D. Lennon, *J. Chem. Phys.*, 2005, **123**, 174706.
67 R. Meyer, C. Lemire, S. K. Shaikhutdinov and H.-J. Freund, *Gold Bull.*, 2004, **37**, 72–124.
68 Y. Zhu, J. He, C. Shang, X. Miao, J. Huang, Z. Liu, H. Chen and Y. Han, *J. Am. Chem. Soc.*, 2014, **136**, 12746–12752.
69 B. Zhu, G. Thrimurthulu, L. Delannoy, C. Louis, C. Mottet, J. Creuze, B. Legrand and H. Guesmi, *J. Catal.*, 2013, **308**, 272–281.

Faraday Discussions

PAPER

Ammonia decomposition catalysis using lithium–calcium imide†

Joshua W. Makepeace,[ab] Hazel M. A. Hunter,[b] Thomas J. Wood,[b] Ronald I. Smith,[b] Claire A. Murray[c] and William I. F. David*[ab]

Received 18th November 2015, Accepted 24th November 2015
DOI: 10.1039/c5fd00179j

Lithium–calcium imide is explored as a catalyst for the decomposition of ammonia. It shows the highest ammonia decomposition activity yet reported for a pure light metal amide or imide, comparable to lithium imide–amide at high temperature, with superior conversion observed at lower temperatures. Importantly, the post-reaction mass recovery of lithium–calcium imide is almost complete, indicating that it may be easier to contain than the other amide–imide catalysts reported to date. The basis of this improved recovery is that the catalyst is, at least partially, solid across the temperature range studied under ammonia flow. However, lithium–calcium imide itself is only stable at low and high temperatures under ammonia, with *in situ* powder diffraction showing the decomposition of the catalyst to lithium amide–imide and calcium imide at intermediate temperatures of 200–460 °C.

Introduction

The catalytic decomposition of ammonia to release its stored hydrogen is an important step in enabling the use of ammonia as a potential replacement for fossil fuels in technologies ranging from transportation to seasonal energy storage. The partial decomposition of ammonia when used in an internal combustion engine enables more favourable combustion conditions to be used,[1] while low-temperature fuel cells require the complete decomposition of ammonia to hydrogen and nitrogen:

$$2NH_3 \rightarrow N_2 + 3H_2 \quad (1)$$

Using current catalyst technology, the kinetics of ammonia decomposition are slow and require very high operating temperatures that, in turn, compromise the

[a]*Inorganic Chemistry Laboratory, University of Oxford, South Parks Road, Oxford, OX1 3QR, UK*
[b]*ISIS Facility, STFC Rutherford Appleton Laboratory, Harwell Campus, Didcot, OX11 0QX, UK. E-mail: bill.david@stfc.ac.uk*
[c]*Diamond Light Source Ltd, Rutherford Appleton Laboratory, Harwell Campus, Didcot, OX11 0DE, UK*
† Electronic supplementary information (ESI) available. See DOI: 10.1039/c5fd00179j

efficiency of the ammonia-based energy cycle.[2] The development of affordable catalysts with higher activity can therefore represent a significant step in facilitating the widespread use of ammonia as a sustainable fuel.

Currently, the most active catalysts for ammonia decomposition involve transition metals[3] and use a range of support architectures (such as porous carbon, metal oxides and carbon nanotubes[4]) along with alkali- and alkaline earth-metal salts[5,6] to enhance activity. Among the metals studied, ruthenium, nickel and iron have been the most widely investigated. Ruthenium is generally considered to be the most active metal per mole while iron-based catalysts are favourable per unit mass since their cost advantage over ruthenium allows the use of higher metal loadings.

Light metal amides and imides have recently emerged as a new class of catalyst for the decomposition of ammonia. While the precise mechanism of their catalytic action is not yet fully elucidated, sodium amide ($NaNH_2$),[7] lithium imide (Li_2NH)[8] and composites of lithium imide with transition metals and their nitrides[9,10] show high ammonia decomposition activity. There are indications from isotope exchange measurements that these materials have a more complex interaction with ammonia than that of simple surface catalysis and that there are bulk interactions between the catalysts and the ammonia to which they are exposed.[8,11] The active forms of the catalysts are dependent on the precise ammonia decomposition conditions (ammonia flow rate, temperature). For the sodium amide system a mixture of sodium and sodium amide is observed[7] while a non-stoichiometric lithium imide ($Li_{1+x}NH_{2-x}$, $0 \leq x \leq 1$) is seen in the lithium-based system.[8] Composites of lithium imide with metal nitrides, which show an enhanced activity, have been proposed to form a ternary nitride during ammonia decomposition.[9]

There are many avenues for the further development of these catalysts in order to provide a low-cost and effective method of releasing hydrogen from ammonia. One challenge encountered with the sodium/sodium amide system is that both components of the system are liquid at the temperature of operation; indeed, sodium itself has a vapour pressure in the order of 25 mbar at 600 °C,[12] leading to issues associated with catalyst containment within the heated reaction zone. While a reflux-type arrangement may help contain a liquid catalyst, it is anticipated this would reduce the efficiency of the reaction and may introduce overly complex engineering requirements.

Stoichiometric lithium imide is solid across the temperature range in which the ammonia decomposition reaction occurs. However, the reaction with ammonia to form a non-stoichiometric phase

$$Li_2NH + NH_3 \rightarrow 2LiNH_2 \quad (2)$$

$$xLiNH_2 + (1 - x)Li_2NH \rightarrow Li_{2-x}NH_{1+x} \quad (3)$$

introduces a lithium amide-like character to the catalyst structure. Lithium amide melts at around 370 °C. In an *in situ* neutron powder diffraction experiment, it was observed that a stoichiometry value close to $Li_{1.5}ND_{1.5}$ was the approximate limit of maintaining a solid phase at 500 °C under very low ammonia flows (of around 5 standard cubic centimetres per minute (sccm)).[8] This evidence indicates that the ammonia decomposition reaction conditions must be maintained within specific limits if the catalyst is to be kept in the solid state during the reaction.

A logical progression from this work is to examine related materials which may be easier to maintain as a solid. In particular, imide materials have much higher melting points than their corresponding amides, and so an imide which forms at low temperature would seem to be an ideal candidate for a practical ammonia decomposition catalyst. Taking the example of the lithium amide–lithium hydride hydrogen storage system,[13] reductions in the hydrogen desorption temperature (*i.e.* a lower temperature of imide formation) were achieved by using ternary systems combining lithium amide with magnesium hydride[14,15] or calcium hydride.[16,17] Of particular interest to this study is that a 2 : 1 mixture of lithium amide with the relevant hydride results in the formation of a ternary imide:

$$2LiNH_2 + MgH_2 \rightarrow Li_2Mg(NH)_2 + 2H_2 \quad (4)$$

$$2LiNH_2 + CaH_2 \rightarrow Li_2Ca(NH)_2 + 2H_2 \quad (5)$$

The lower temperatures of these reactions compared with the lithium-only analogue raise the prospect that these imides may be more stable in the presence of ammonia than lithium imide alone. The elevated pressures required to hydrogenate these ternary imides relative to lithium imide may indicate that this hypothesis has merit; however, there are no studies to date that examine their interaction with ammonia. The aim of this work is to examine the ammonia decomposition properties of these two ternary imides with a focus on whether they offer enhanced practicality compared to lithium imide.

Experimental

Sample preparation

All sample preparation and handling was undertaken inside an argon-filled glove box (Unilab, mBraun, O_2 and H_2O < 0.1 ppm). Samples of lithium–calcium imide were prepared by the reaction of lithium amide ($LiNH_2$, 98%, Sigma Aldrich) and calcium hydride (CaH_2, 98%, Sigma Aldrich). The powdered reactants were mixed in a 2 : 1 molar ratio by hand using an agate mortar and pestle for 5 minutes per gram of reactant mixture. 2 g of the resultant mixture was loaded into a 125 mL tungsten carbide (WC) milling jar filled with nine 10 mm tungsten carbide milling balls. The sample was milled under argon in the WC jar for four hours at 400 rpm in a Retsch PM100 planetary mill, in 15 minute segments with two minute rest periods between each segment. The direction of rotation was reversed after each segment. The ball-milled mixture was then heated in 0.25 g portions to affect the reaction to form lithium–calcium imide (eqn (5)).

The portion was placed in a cylindrical alumina crucible housed inside a quartz tube. The sample was put under flowing argon gas (99.998%, BOC), placed inside a horizontal tube furnace (Carbolite MTF 12/38/400) and heated to 380 °C at a ramp rate of 2 °C min^{-1} and held at that temperature for 12 hours. Powder X-ray diffraction (XRD) of the samples, performed using a PANalytical X'Pert diffractometer using copper $K\alpha_1$ radiation, confirmed that the synthesised lithium–calcium imide was at least 90% pure as assessed by Rietveld analysis of the powder pattern with the TOPAS Academic v5.0 package (see ESI†).

Lithium–magnesium imide was synthesised in an analogous fashion to lithium–calcium imide, using magnesium hydride (MgH_2, 98%, Aldrich) in place of calcium hydride.

A sample of deuterated lithium–calcium imide ($Li_2Ca(ND)_2$) was prepared for the neutron powder diffraction measurement. The synthesis method was as described above, using the deuterated isotopomers of the starting materials. Deuterated lithium amide ($LiND_2$) was prepared by the reaction of lithium nitride (Li_3N, 98%, Sigma Aldrich) with deuterated ammonia (ND_3, 99%, 99 atom% D, Aldrich):

$$Li_3N + 2ND_3 \rightarrow 3LiND_2 \tag{6}$$

A hand-ground 1 g sample of lithium nitride was placed in a cylindrical stainless steel reactor, sealed and attached to the deuterated ammonia supply. The system was evacuated and then filled with 3 bar of ammonia, heated at 5 °C min^{-1} to 300 °C and held at that temperature for 3 hours. The resultant powder was confirmed to be deuterated lithium amide using XRD and Raman spectroscopy (Bruker Senterra).

Calcium deuteride (CaD_2) was prepared by the reaction of *ca.* 1 mm pieces of calcium metal (99.5%, Alfa Aesar) with deuterium gas (D_2, 99.8 atom% D, SIP Analytical). The calcium pieces were used as-received and were placed into a cylindrical stainless steel reactor, sealed and attached to the deuterium supply. The system was evacuated and then filled with 10 bar of deuterium. The sample was subsequently heated at 5 °C min^{-1} to 315 °C and held at that temperature overnight. The composition of the resultant off-white powder was confirmed as calcium deuteride using XRD.

The purity of the deuterated lithium–calcium imide synthesised from the reaction of these starting materials was confirmed to be in excess of 90% using XRD (see ESI†).

Ammonia decomposition testing

The ammonia decomposition testing used an identical setup to that described previously.[7,8] Catalyst powders were loaded into a 46.9 cm^3 cylindrical nickel-plated 316 stainless steel reactor with a gas flow input along a pipe through the centre of the cylinder, entering 1 cm from the bottom of the reactor (see ESI†). Out flow was through the lid of the reactor. Gas flow of argon or ammonia (NH_3, 99.999%, SIP Analytical) was controlled using a mass flow controller (HFC-302, Teledyne Hastings Instruments) housed in a custom-designed gas panel. The reactor was heated using a vertical tube furnace (Severn Thermal Solutions). The composition of the gas leaving the reactor was analysed using a Quantitative Gas Analyser mass spectrometry system (HPR20 QIC R&D, Hiden Analytical), run in multiple ion detection mode with a Faraday detector. Mass-to-charge ratios (m/z) corresponding to hydrogen ($m/z = 2$), ammonia ($m/z = 17$), nitrogen ($m/z = 28$) and argon ($m/z = 40$) were monitored. In a typical ammonia decomposition experiment, ammonia was flowed at 60 sccm through the system and the reactor was heated in steps between 250 °C and 600 °C, with 3 hour dwell periods at each temperature set point.

Ammonia decomposition percentages were calculated using an in-house computer program which removes the background signal for each species and calculates the ammonia decomposition percentage based on the average value over the dwell period at each temperature set point.

Synchrotron X-ray powder diffraction

Two synchrotron X-ray powder diffraction experiments were performed using the I11 beam line at the Diamond Light Source, United Kingdom. Diffraction data were collected in 2 minute intervals using the multi-analyser crystal (MAC) detectors and an X-ray wavelength of 0.826209 Å was used.

For each experiment, a sample of lithium–calcium imide was placed in the centre of a 0.9 mm i.d. single crystal sapphire capillary, which was open at both ends. The powder sample was held in position by a plug of quartz wool on either side. The capillary was loaded into a flow-through gas cell and mounted on the beam line (see ESI†). After leak testing with helium gas, the sample was aligned and a hot air blower was positioned to heat the sample. The sample was rocked between −5 and +15° to minimise preferred orientation effects during data collection.

In the first experiment, the sample of lithium–calcium imide was heated from room temperature to 550 °C at 5 °C min^{-1} under a constant flow of 1 sccm of helium gas. The sample was held at 550 °C for half an hour and then cooled to room temperature at 10 °C min^{-1}. In the second experiment, the sample was heated under a 1 sccm flow of ammonia, initially to 350 °C at 5 °C min^{-1}, with a one hour dwell at that temperature. Subsequently, the sample was heated in 50 °C steps at 5 °C min^{-1} with one hour dwells at each temperature, up to a maximum temperature of 600 °C. Finally, the sample was cooled at 10 °C min^{-1}.

Differential scanning calorimetry–thermogravimetric analysis

A Netzsch STA 441 F1 Jupiter was used for simultaneous differential scanning calorimetry and thermogravimetric analysis of lithium–calcium imide samples. 8.7 mg samples of the lithium–calcium imide were sealed inside 20 μL aluminium pans. The lid of each pan was pierced as it was loaded into the instrument. A purge flow of nitrogen gas was used during sample loading to protect the sample from moisture in the atmosphere. Once loaded, the sample was heated from 25 °C to 575 °C at 5 °C min^{-1} under a 20 sccm flow of either nitrogen or ammonia gas. The sample mass and relative heat flow to the sample pan were recorded throughout the experiment.

Neutron powder diffraction

In situ time-of-flight neutron powder diffraction data for lithium–calcium imide were collected on the POLARIS instrument at the ISIS Neutron and Muon Source in the United Kingdom. The experimental setup for the *in situ* experiment was very similar to that described previously when studying lithium imide.[8] The sample of deuterated lithium–calcium imide (1.9 g) was loaded into a stainless steel flow-through cell (see ESI†) which was placed into a standard furnace housed in the POLARIS sample tank. Neutron diffraction data were collected in 100 s intervals throughout, and the furnace was set to the desired temperature (or temperature ramp rate). The sample cell was attached to the same gas delivery panel, as used for ammonia decomposition testing, equipped with argon, ammonia and deuterated ammonia input gases. Outgas composition and ammonia decomposition percentages were calculated as described for the ammonia decomposition experiments. Given that deuterated species were used in the neutron diffraction experiment, the following mass-to-charge (m/z) fragments were analysed in addition to the standard fragments: ND_3 ($m/z = 20$), ND_2H ($m/z = 19$), NDH_2 ($m/z = 18$), D_2 ($m/z = 4$) and HD ($m/z = 3$), with the partially-deuterated species (and NH_3/H_2) included for monitoring during the H–D exchange experiment. Additionally a reference neutron powder diffraction dataset was collected of the as-prepared sample in a sealed 6 mm thin-walled vanadium can over 2 hours at room temperature.

Results and discussion

The variable-temperature ammonia conversion percentages of lithium–calcium imide and lithium–magnesium imide are shown in Fig. 1, with datasets for lithium amide and that of the blank reactor shown for comparison.

Below 430 °C, both lithium–calcium imide and lithium–magnesium imide show superior conversion activity to lithium imide–amide and very similar activity to each other. Above this temperature, the activity of lithium–magnesium imide is significantly lower than both lithium imide–amide and lithium–calcium imide. In contrast, above 430 °C the ammonia decomposition activity of lithium–calcium imide is very similar to that of lithium imide–amide. Overall, the variable-temperature ammonia conversion percentage profile of lithium–calcium imide in this system is the highest observed to date out of the pure metal amide or metal imide systems studied.

In the published reports of the ammonia decomposition activity of light metal amides and imides there has been some disagreement as to how active the

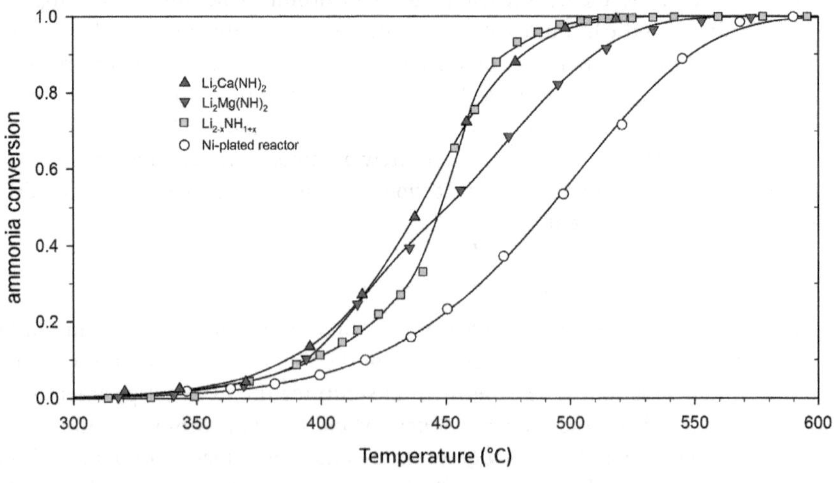

Fig. 1 Ammonia conversion fraction as a function of temperature for a number of metal amide/imide catalysts, along with the nickel-plated blank reactor. Measurement conditions were 0.5 g of catalyst and an ammonia flow rate of 60 sccm. The lines indicate the fits to the data.

materials are on their own. It has been reported that composites of lithium amide–imide with transition metal nitrides display synergistic ammonia decomposition activity.[9,10] These results, measured for intimate mixtures, were used to infer that measurements in stainless steel reactors, such as those in this study and our previously-reported work, were not measuring the isolated activity of the amide/imide. Measurements of the activity of lithium amide–imide in quartz reactors show very low ammonia decomposition activity. Of course, it is difficult to isolate the effect of the reactor from the catalyst. In most cases, quartz is an inert material which is a good choice for reactors for ammonia decomposition. However, in our laboratory tests we have established that lithium amide and sodium amide react strongly with quartz at elevated temperature (see ESI†), forming oxide and unknown phases that are most likely metal silicates. This may go some way to explain the very low ammonia decomposition activity observed for lithium and sodium amides when tested in quartz reactors. It is for this reason that this work continues to use metal reactors for measuring ammonia decomposition activity. The interaction between metals/metal nitrides and these catalysts is of great interest, but experimental results to date suggest that the amide materials are actively decomposing ammonia, rather than playing a simple promotion role (as is the case with other group I and II metal salts[18]).[11] Indeed, a recent study on the interaction of ruthenium with lithium amide–imide suggested that the enhanced ammonia decomposition activity from the composite material may in fact be due to the metal influencing the decomposition pathway of lithium amide.[19]

To provide a quantitative measure of the data presented in Fig. 1, the ammonia decomposition data were fitted by a least-squares procedure to one- or two-process sigmoidal functions of the form:

$$C_\mathrm{i} = 1 - \exp(-\exp(A/R - E_\mathrm{A}/RT)) \tag{7}$$

where C_i is the conversion (1-ammonia fraction), A is a constant, R is the gas constant, E_A is an activation energy and T is the temperature (in Kelvin). Two-process functions were fitted to a sum of two equations shown in eqn (7), with different refined A and E_A values for each, with the total function given by:

$$C_\mathrm{total} = 1 - [fC_1 + (1 - f)C_2] \tag{8}$$

where f is the fractional contribution of the first sigmoidal function and $1 - f$ is the fractional contribution of the second function. The outputs of the fitting process are summarised in Table 1.

Table 1 Parameters from least-squares optimisation of sigmoidal functions fitted to variable-temperature ammonia conversion data. Uncertainties are given in brackets.

Species	A_1	E_{A1}	A_2	E_{A2}	f
$Li_2Ca(NH)_2$	0.16(2)	120(10)	0.28(3)	200(30)	0.70(4)
$Li_2Mg(NH)_2$	0.15(1)	112(11)	0.61(4)	421(30)	0.85(4)
$Li_{1+x}NH_{2-x}$	0.165(3)	120(4)	0.67(2)	490(17)	0.573(5)
Blank	0.141(9)	110(7)			

As outlined previously, one of the challenges with the use of light metal amides and imides for ammonia decomposition is containing the catalyst melt within the reactor hot zone. It is known[8] that lithium imide–amide can be kept solid under controlled conditions. When it is heated under argon to the temperature of high ammonia conversion, lithium imide will remain solid, and so mass recovery from the reactor is generally in excess of 80%. In contrast, if it is heated under ammonia, the mass recovery is generally less than 30% due to the formation and subsequent melting of lithium amide.[8] The mass recovery of the lithium–calcium imide sample after the ammonia decomposition reaction shown in Fig. 1 was almost complete (98%), whereas the lithium–magnesium imide recovery was less than 5%. The combination of high activity and high mass recovery—which implies ease of containment—makes lithium–calcium imide worthy of further investigation as an attractive ammonia decomposition catalyst. Since lithium–magnesium imide showed neither of these attributes, it was not investigated further in this study.

The high mass recovery obtained for lithium–calcium imide is consistent with the catalyst remaining solid throughout the range of conditions used in the ammonia decomposition experiment. However, a post-reaction X-ray powder diffraction pattern, shown in Fig. 2, indicates a more complex behaviour. Specifically, a mixture of lithium amide and calcium imide was identified, with no lithium–calcium imide observed in the diffraction pattern.

In order to provide more detailed information on this transformation and to gain an understanding of the active form of the catalyst and its practical limitations, *in situ* diffraction experiments were undertaken. In the first experiment using synchrotron X-ray powder diffraction, lithium–calcium imide was heated under a 1 sccm flow of helium gas to examine whether simple heating could cause the material to separate into lithium imide and calcium imide. The published crystal structure of lithium–calcium imide, Fig. 3, is a layered structure which can be described as alternating layers of lithium imide and calcium imide.[20] As such, it is possible that the separation of those layers into the constituent imide phases may not require a large input of energy. After separation, lithium imide would readily react with ammonia to form lithium amide, consistent with the powder diffraction pattern obtained for the post-reaction sample.

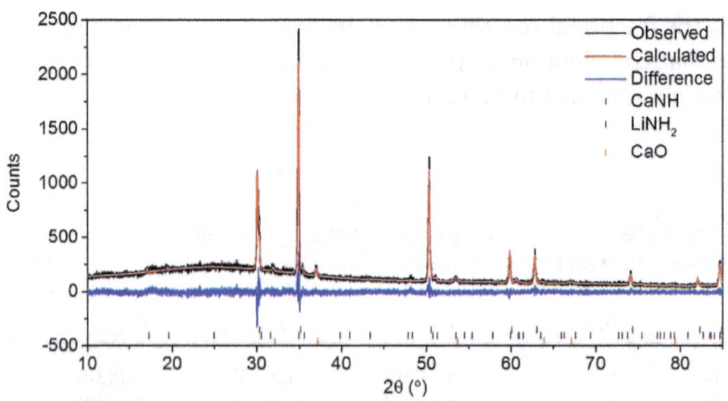

Fig. 2 X-ray powder diffraction pattern of the lithium–calcium imide catalyst after the ammonia decomposition reaction.

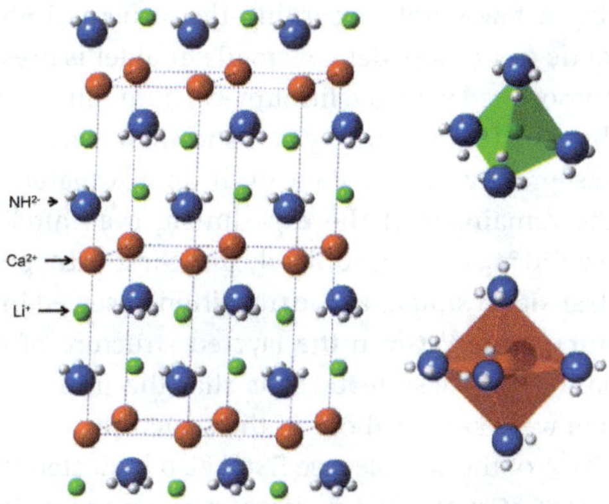

Fig. 3 Crystal structure of lithium–calcium imide, with lithium atoms shown in green, calcium in red, nitrogen in blue and hydrogen in grey. The local geometries of the $Li(NH)_4$ tetrahedron and $Ca(NH)_6$ octahedron are shown to the right.

A contour plot of the diffraction data for the heating of lithium–calcium imide between room temperature and 550 °C under helium is shown in Fig. 4. What is immediately apparent from these data is that lithium–calcium imide is present over the entire temperature range and does not transform into lithium imide and calcium imide. Therefore it is reasonable to conclude that the decomposition of the catalyst is due to its interaction with ammonia.

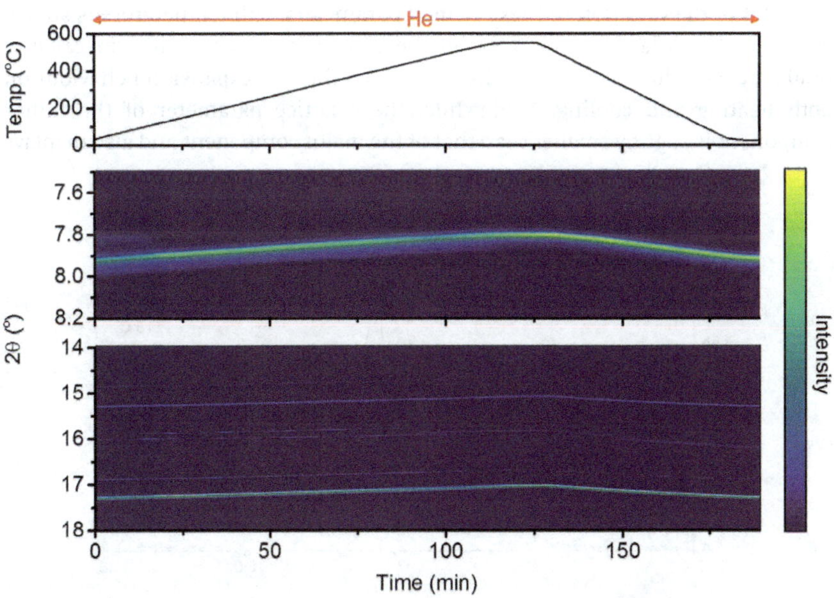

Fig. 4 Contour plot of sections of the *in situ* synchrotron X-ray powder diffraction data for lithium–calcium imide. The sample was heated at 5 °C min^{-1} under 1 sccm flowing helium gas.

It is worth noting a few points regarding the structural characterisation of lithium–calcium imide from these data. A small shoulder is present on a number of the Bragg peaks associated with the lithium–calcium imide phase. An example of this feature as observed for the (001) peak is shown in Fig. 5. As can be seen, the feature at 7.84° was present in the data until approximately 120 °C and not observed during the remainder of the experiment, even after the sample was cooled back to below 120 °C. It is thus unlikely that these extra peaks are related to an order–disorder transition similar to the transition observed in lithium imide at a slightly lower temperature.[21] Given the layered structure of the material, the most likely explanation for these features is that the initial sample possessed stacking faults which were annealed out of the structure.

Combined DSC-TGA of the sample (see ESI†) also indicated the presence of an endothermic event at 86 °C which is consistent with an order–disorder transition of the lithium imide impurity in the sample. Annealing of stacking faults in the sample would be expected to be exothermic, but no thermal events were observed around the temperature of the loss of the low-angle shoulder. The DSC-TGA data also show no significant mass loss, consistent with the stability of the catalyst over the temperature range.

Further peak asymmetry is observed in the data comprising a small shoulder at higher 2-theta values, see Fig. 5. This feature was particularly prevalent in the [00*l*] reflections and persisted throughout the experiment. This may be indicative of a degree of compositional variation or disorder in the lithium–calcium imide structure—for example a degree of non-stoichiometry could be envisaged similar to that observed in lithium imide.[22–24] This peak asymmetry was modelled in the Rietveld analysis of the diffraction data as a second component to the lithium–calcium imide phase, one with a very similar a lattice parameter to the main phase and a smaller (approx. 0.5%) c lattice parameter.

A plot of the variation in these refined parameters with temperature is shown in Fig. 6. The a lattice parameters of the two components of lithium–calcium imide are close in magnitude and show similar thermal expansion behaviour on both heating and cooling. Meanwhile, the c lattice parameter of the minor component is slightly smaller than that of the major component and also displays

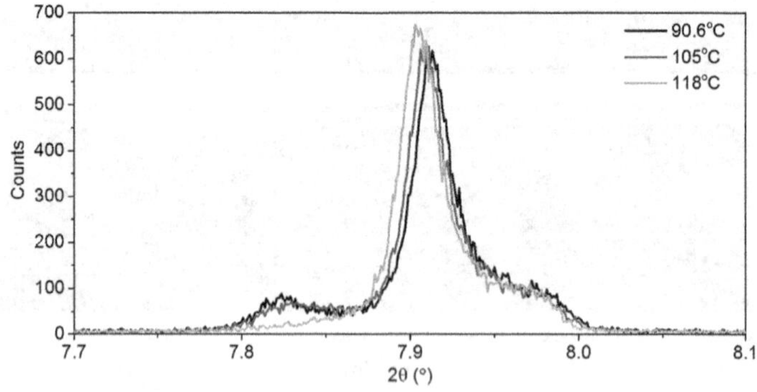

Fig. 5 The synchrotron X-ray powder diffraction (001) peak of lithium–calcium imide, showing the disappearance of the low-angle shoulder with heating under 1 sccm flowing helium, and the presence of high-angle asymmetry.

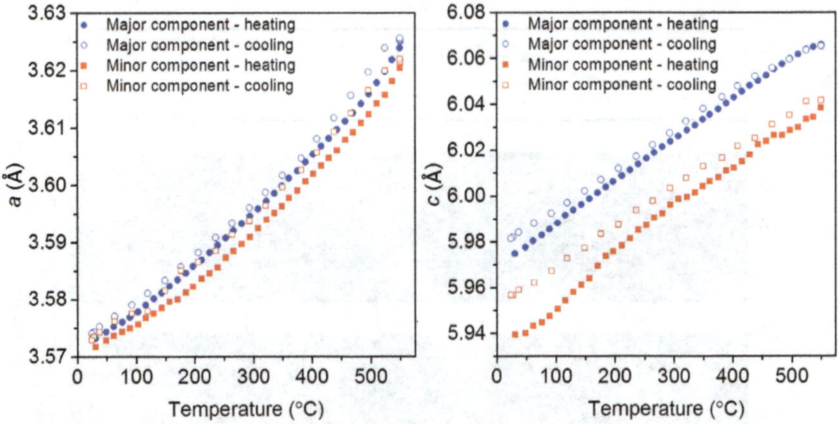

Fig. 6 Lattice parameters *a* and *c* for the two lithium–calcium imide components included in the refinement of the synchrotron diffraction data for the heating of lithium–calcium imide under flowing helium gas. Errors in the lattice parameter values are $1–2 \times 10^{-4}$ Å.

differing thermal expansion behaviour. Upon heating, the major component displays almost linear thermal expansion behaviour, with a subtle change in the slope of the expansion at 120 °C, consistent with the disappearance of the low-angle asymmetry attributed to stacking faults in the structure. The minor component exhibits relatively steep thermal expansion behaviour below 180 °C during heating. Above this temperature, the rate of thermal expansion decreases to a similar rate to the major component. On cooling, both components display linear thermal contraction. The irreversible steep thermal expansion of the *c* axis of the minor component towards the value of the major component is consistent with the hypothesis that this asymmetry reflects compositional disorder in the structure which partially orders as the sample was heated at the beginning of the experiment. This partial ordering results in a narrowing of the Bragg peak, which produces the shift in the *c* parameter of the minor component.

Having established that, under an inert gas atmosphere, lithium–calcium imide is stable in the temperature range used for ammonia decomposition, a second experiment was conducted with the sample heated under ammonia flow. The experimental results are summarised in Fig. 7, showing contour plots of sections of the diffraction data. These data indicate that, unlike under helium gas flow, the catalyst is not stable under ammonia flow. This is consistent with the *ex situ* diffraction data which indicated that lithium amide and calcium imide were formed during the ammonia decomposition reaction.

The *in situ* data show that the interaction of lithium–calcium imide with ammonia is more complex than a simple separation into lithium amide and calcium imide as may be surmised from the *ex situ* analysis. The data from the experiment were analysed using the Rietveld method, and the resultant molar percentages extracted for each phase included in the refinement process are shown in Fig. 8. The analysis of these data is complicated by overlapping Bragg peaks for lithium amide-imide, calcium imide and lithium–calcium imide, so

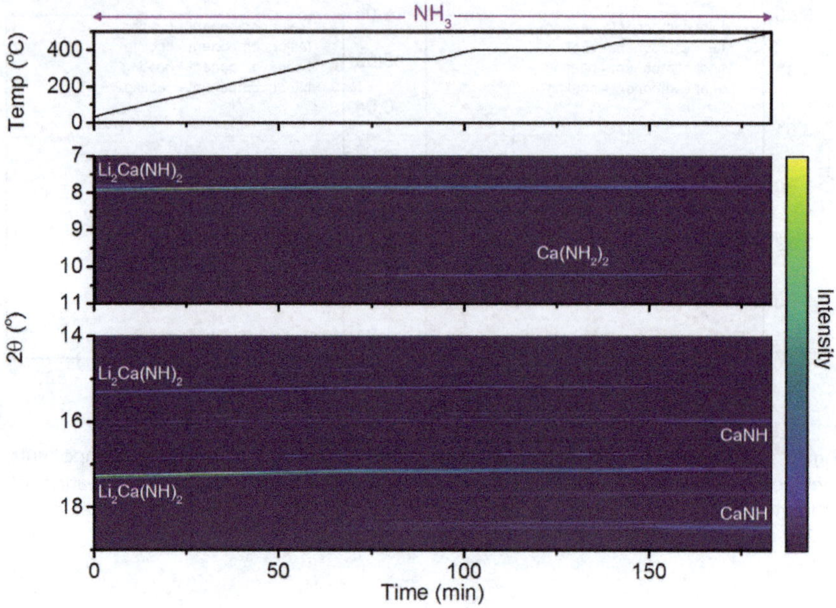

Fig. 7 Selected contour plots of synchrotron X-ray powder diffraction data for lithium–calcium imide heated under 1 sccm flowing ammonia gas.

a manual refinement of every fifth dataset was chosen instead of a batch refinement process as used in our previous *in situ* studies.[8]

Below 200 °C, the most significant change in the sample is the loss of the lithium imide impurity, which is most likely to have formed lithium amide by reaction with ammonia. Above 200 °C, the lithium–calcium imide begins to diminish, with calcium imide and lithium oxide phases growing in to replace it. Since lithium amide was not detected in the diffraction data at this stage, and the temperature is below its melting point, it was presumed that all of the lithium amide was oxidised by reaction with the sapphire capillary. Although sapphire has proven to be reasonably inert for the study of lithium amide systems, it clearly will react with lithium amide under certain conditions. Above 370 °C, no lithium amide formed from the decomposition of lithium–calcium imide will be observed, due to it being molten, and this should be taken into account when interpreting the data in Fig. 8. However, it would appear that the lithium oxide phase may well be a good proxy for the amount of lithium amide formed from the decomposition of lithium–calcium imide; the molar ratio of lithium oxide to calcium imide at the end of the experiment is approximately 1 : 1, in keeping with the lithium–calcium ratio in the starting material. As the sample was cooled at the end of the experiment, a small amount of lithium amide was observed in the diffraction data, which is consistent with the proposed decomposition mechanism from lithium–calcium imide to calcium imide and lithium imide.

During the first dwell in temperature at 350 °C, a new Bragg peak is observed to form at 10.2°; this angular position is not indicative of the structures of lithium amide or calcium imide, the expected decomposition products. Rather, it matches the expected position of the (011) Bragg peak of calcium amide

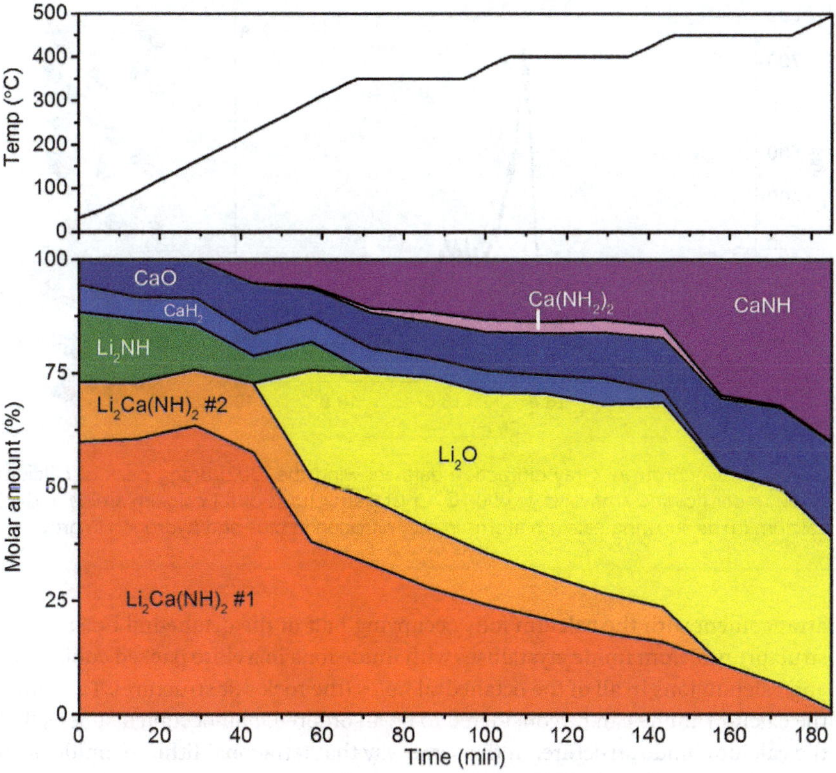

Fig. 8 Molar percentages of the crystalline phases present during exposure of lithium–calcium imide to ammonia, as extracted from Rietveld analysis of every fifth synchrotron X-ray powder diffraction data set. $Li_2Ca(NH)_2$ #1 is the major component of the lithium–calcium imide peak; $Li_2Ca(NH)_2$ #2 is the minor component.

($Ca(NH_2)_2$). The presence of calcium amide, even in the small quantity observed in this experiment, was unexpected, as the synthesis of calcium amide has typically required reactive ball-milling of calcium hydride under elevated ammonia pressure[25,26] or reaction between calcium metal and liquid ammonia over long timescales.[27] These data suggest that calcium amide can be formed from calcium imide under more moderate conditions than from calcium hydride or calcium metal.

Although the calcium imide molar amount in Fig. 8 is presented as a single phase, close examination of the diffraction data toward the end of the experiment (*i.e.* at the highest temperatures studied) reveals complex lineshapes in the Bragg peaks associated with calcium imide, Fig. 9a. In the Rietveld analysis of these data, these peak shapes were fitted using three calcium imide components. The presence of these complex lineshapes and their evolution in the context of the presence of ammonia may be indicative of stoichiometry variation associated with a non-stoichiometric calcium imide phase.

Examining the crystal structures of calcium amide (as reported by Senker *et al.*[28]) and calcium imide, Fig. 9b and c, it can be seen that calcium amide adopts a tetragonal structure where the amide groups form a pseudo-cubic close packed

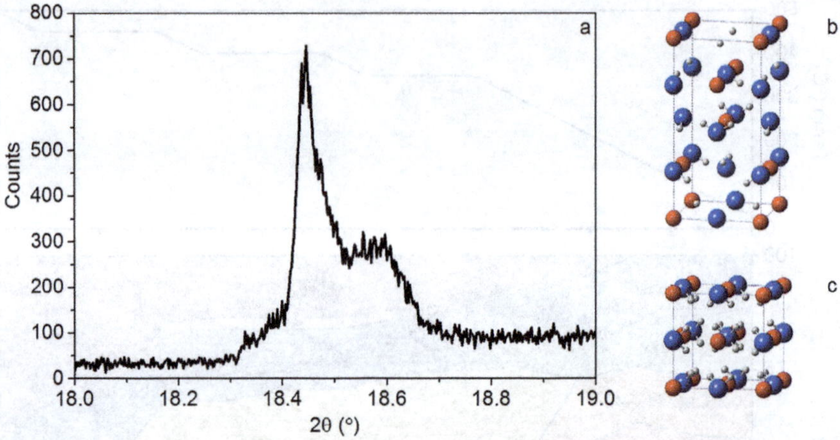

Fig. 9 (a) Synchrotron X-ray diffraction data showing the (002) Bragg peak of calcium imide under flowing ammonia at 450 °C. Crystal structures of (b) calcium amide and (c) calcium imide showing calcium atoms in red, nitrogen in blue and hydrogen in grey.

arrangement with the calcium ions occupying half of the octahedral holes in the structure. Calcium imide crystallises with imide ions in a close packed cubic array and calcium ions in all of the octahedral holes (the rock salt structure). Therefore, the calcium amide can be considered to be an ordered, half-occupied supercell of the calcium imide structure, in the same way that tetragonal lithium amide is an ordered supercell of the fluorite structure of lithium imide, where lithium ions occupy the tetrahedral, rather than the octahedral, sites in the close-packed cubic lattice. Given that this topological similarity facilitates the formation of non-stoichiometric phases by Frenkel defect migration in the lithium amide–imide system, it is reasonable to suggest that the same mechanism may be possible in the case of calcium amide–imide. Further investigation would be required to confirm this hypothesis. The extensive oxidation of the sample observed above 450 °C precluded the investigation of the behaviour of the system at higher temperatures.

The behaviour of the lithium–calcium imide catalyst in the synchrotron diffraction experiment was mirrored in a combined DSC-TGA experiment, the results of which are plotted in Fig. 10. The same endothermic event at 86 °C, as was observed under helium flow, was seen under ammonia, consistent with the lithium imide order–disorder transition. The sample began gaining mass at around 200 °C, consistent with the diffraction data and was correlated with a broad exothermic event in the data. The peak of this exothermic event coincides with a sharp endothermic event at 367 °C. This is consistent with the absorption of ammonia by lithium imide to form lithium amide (or lithium amide-like non-stoichiometric species), which melts at around 370 °C.

The maximum mass increase of 2.5 mg, reached at 400 °C, represents a 31% gain over the starting sample mass of 8.7 mg. Considering the reaction of lithium imide to lithium amide, the release of the maximum amount of lithium imide from the structure (*i.e.* full decomposition to lithium imide and calcium imide) and its subsequent conversion to lithium amide would result in a mass increase

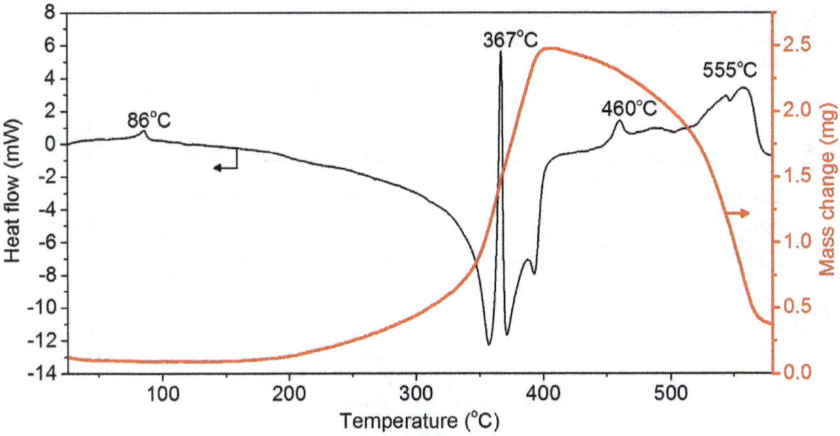

Fig. 10 Combined DSC-TGA experiment on lithium–calcium imide under 20 sccm flowing ammonia gas.

of 20%. Therefore, the larger mass increase observed in the DSC-TGA experiment is consistent with the additional formation of calcium amide and/or non-stoichiometric calcium imide, as observed in the diffraction experiment.

Above 400 °C, the mass decreases until the maximum temperature of the experiment, 575 °C. There are at least two endothermic events during the mass loss. Most of the mass loss occurs at temperatures which were not accessed in the synchrotron experiment and so cannot be related to those data. However, the mass appears to be returning to the initial mass value, which either suggests the formation of a mixture of lithium imide and calcium imide, the decomposed catalyst, or potentially the reformation of lithium–calcium imide.

The behaviour of the lithium–calcium imide catalyst under ammonia was investigated further in an *in situ* neutron powder diffraction experiment. The use of neutrons allowed for a stainless steel reactor to be used, removing the complication of the sample oxidising and thus allowing the higher temperature behaviour to be studied. A contour plot of the diffraction data from the 90° detector bank of the POLARIS instrument is shown in Fig. 11, along with the sample temperature, gas composition and flow, and the ammonia conversion fraction, where applicable.

Initially, the sample of deuterated lithium–calcium imide was heated to 550 °C under flowing argon in order to examine whether exposing the sample to ammonia above the temperature where mass uptake was favourable would stabilise the lithium–calcium imide structure. It can be seen from the diffraction data that this is the case: when deuterated ammonia was added after five hours, the sample was essentially unchanged, remaining as lithium calcium imide. As such, the decomposition of the catalyst can be avoided in the same way as the conversion of lithium imide into lithium amide.[8] A relatively high ammonia conversion fraction of 0.6 was observed at this temperature, though the reactor geometry in the neutron diffraction experiment is not optimised to give the highest ammonia decomposition performance.

The lattice parameters of the lithium–calcium imide extracted from Rietveld analysis of the diffraction data during this period show that the exposure of the

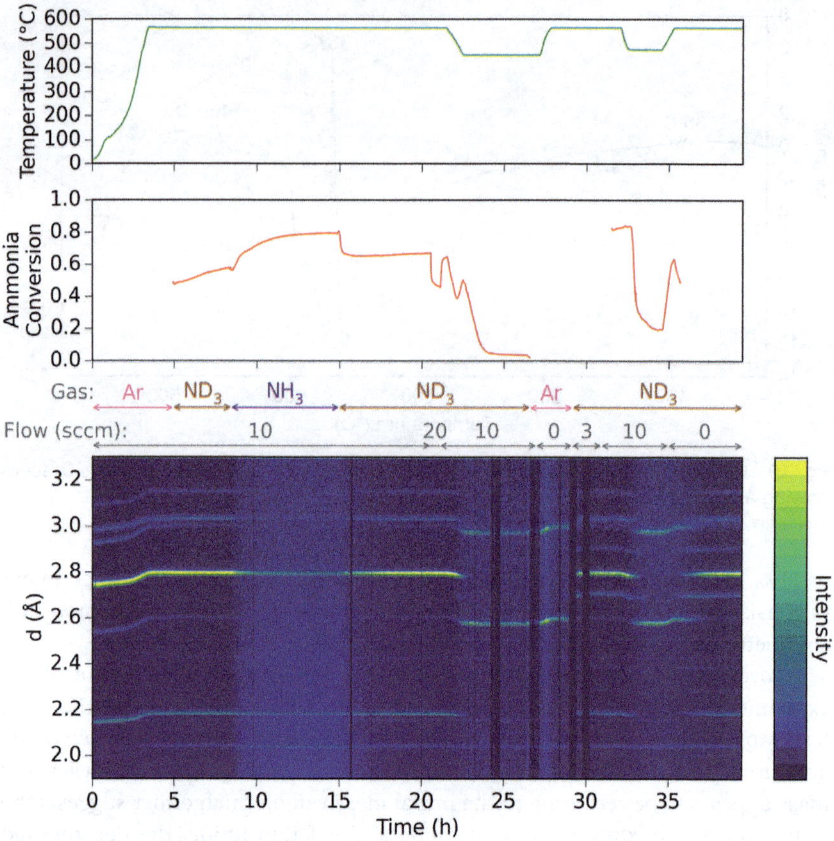

Fig. 11 Neutron powder diffraction experiment on deuterated lithium–calcium imide, showing (top) the sample temperature, (middle) the ammonia conversion fraction and (bottom) a contour plot of the diffraction data, with gas composition and flow indicated above. Gaps in the contour plot data represent periods where the ISIS synchrotron was not operational.

sample to ammonia resulted in a very slight (0.05%) decrease in the *a* parameter and a larger increase (0.1%) in the *c* parameter (see ESI†). Based on our analysis of lithium imide–amide,[8,23] these changes in the lattice parameters may indicate a degree of non-stoichiometry which can be accommodated in the lithium–calcium imide structure without causing the decomposition of the phase. More detailed analysis would be required to establish whether this was unambiguously the case.

Having observed that lithium–calcium imide can be stabilised under ammonia at high temperature, an isotope exchange experiment was then conducted. Approximately 8 hours into the experiment, the gas flow over the sample was switched from ND_3 to NH_3. At this point, and for the duration of the exchange, an increased background signal can be observed in the contour plot of the diffraction data along with a weaker signal from the sample, see Fig. 11. This is consistent with the incorporation of hydrogen into the sample, as hydrogen has a very large incoherent neutron scattering cross section,[29] which manifests as higher noise in diffraction data. In order to quantify the isotope exchange reaction, Rietveld analysis was performed which allowed the occupancy of the deuterium position in the lithium–calcium imide structure to vary completely between fully occupied by deuterium and fully occupied by hydrogen. The results of this analysis are shown in Fig. 12.

It can be seen that there is exchange of deuterium for hydrogen in the structure, with the deuteration fraction exhibiting an exponential decay, with a time constant of 74(2) minutes, after the NH_3 was introduced, down to a minimum of around 0.2. After approximately 14 hours experimental run time the flow was returned to ND_3, and the deuteration fraction returned to 1.0 with a more rapid time constant of 20(2) min. These data indicate that there is bulk interaction of the lithium–calcium imide sample with the ammonia flowing over and being decomposed by it. This bulk interaction shows how substantially the behaviour of amide- and imide-based ammonia decomposition catalysts differs from traditional surface catalysts.

Fig. 13 presents the mass spectrometry data collected during the isotope exchange experiment. These data indicate that the ammonia decomposition rate is higher when the sample is not deuterated. This can be seen most clearly by examination of the nitrogen trace, which increases when the NH_3 is added, and decreases when the gas flow was switched back to ND_3. This variation gives some indication that there may be a kinetic isotope effect in this reaction, which would indicate that the rate determining step of the ammonia decomposition reaction using lithium–calcium imide involves a bond which contains H/D. A more detailed study of this effect, such as has been performed for sodium amide,[11] would be needed to confirm that this is the case.

Following the isotope exchange reaction, the sample was cooled to 440 °C, whereupon the lithium–calcium imide phase disappeared and was replaced by calcium imide and, what is presumed to be, molten lithium amide. The absorption of ammonia associated with this transformation is observed in the apparent spike in the ammonia conversion fraction at approximately 23 hours. This result was consistent with the synchrotron and DSC-TGA data. At this point in the experiment, a blockage in the pipework formed which is most likely associated

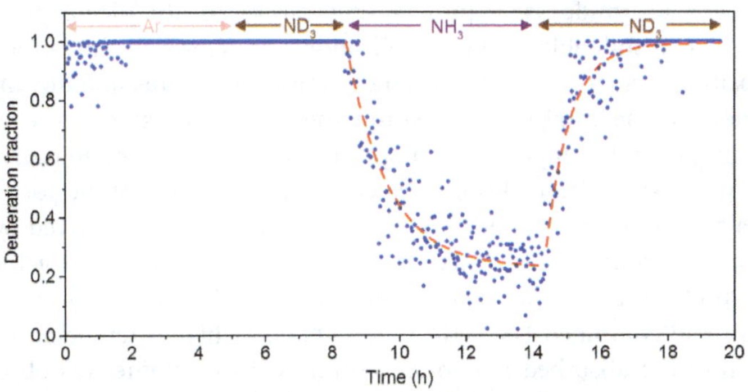

Fig. 12 Isotope exchange experiment during neutron powder diffraction of lithium–calcium imide, showing the deuteration fraction of the sample over time, with the gas species which the sample was exposed to indicated at the top of the figure. Red dashed lines are fits of the deuteration fraction data to simple exponential functions.

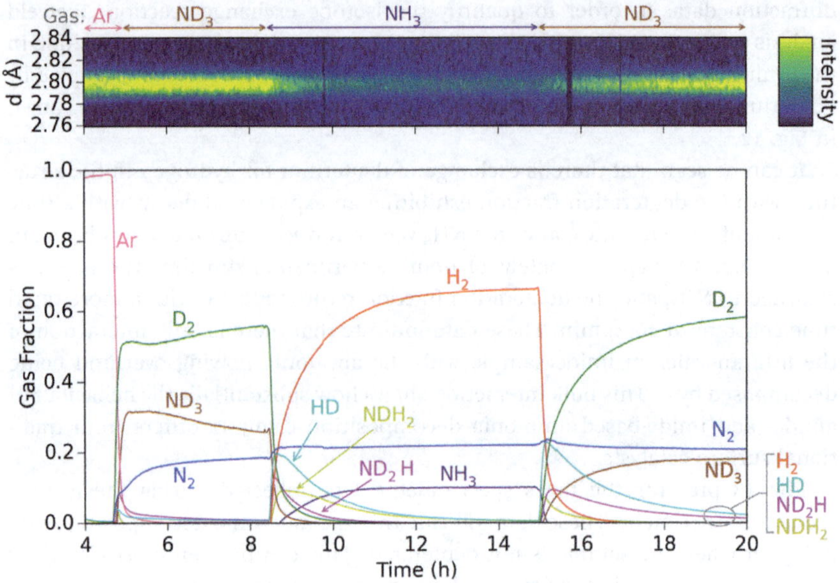

Fig. 13 Mass spectroscopy data during the isotope exchange reaction, with the neutron diffraction (011) Bragg peak of lithium–calcium imide shown for reference.

with the formation of molten lithium amide. By increasing the temperature of the sample to 550 °C and flowing argon across it, solid lithium imide was formed, which removed the blockage and allowed for the resumption of flowing ND_3. During this period, lithium–calcium imide was observed to reappear in the diffraction data. The reversibility of this transformation was confirmed by once again lowering and then raising the temperature of the sample, which resulted in the decomposition and reformation of the lithium–calcium imide catalyst before the system blocked again.

Conclusion

Lithium–calcium imide is a promising catalyst for the decomposition of ammonia for sustainable energy applications. It shows a similar ammonia decomposition activity to lithium imide–amide and substantially improved containment of the catalyst mass within the reactor, making it the best-performing ammonia decomposition catalyst in this family studied to date.

The layered structure of lithium–calcium imide is susceptible to attack by ammonia to form lithium amide and calcium imide, as was observed by *in situ* diffraction experiments. Evidence for the formation of non-stoichiometric calcium imide and calcium amide was observed in the synchrotron X-ray diffraction and combined DSC-TGA experiments. At higher temperatures, after desorption of the absorbed ammonia, lithium–calcium imide was observed to reform. One way to consider this behaviour is that the ternary imide behaves in a very similar manner to the combination of its two imide components: the ammonia absorption behaviours of lithium and calcium imide are superimposed on the behaviour of the ternary imide. However, the ternary structure does appear to raise the temperature at which the lithium imide layer reacts with ammonia compared with lithium imide alone.

While the presence of calcium imide and the reformation of lithium–calcium imide at high temperatures ensure that at least part of the catalyst remains solid over a wide range of ammonia decomposition reaction conditions, the formation of lithium amide is not ideal for the containment of the catalyst in a heated reactor. Furthermore, lithium amide will react with a number of common support materials, making the task of increasing the catalyst surface area more difficult. In the design of future metal amide and imide catalysts, one of the goals must be to find a catalyst which is solid and stable across the range of ammonia decomposition reaction conditions. It is possible that ternary imide structures, with two integrated metal ions, may provide a route to achieving this goal.

Financial support

This work was supported by an STFC Innovations Proof of Concept Award (Phase 3 POCF1213-14) and an STFC CLASP Award (ST/L006278/1).

Acknowledgements

The ISIS Neutron and Muon Source and the Diamond Light Source Ltd are acknowledged for the provision of experimental time. Chris Goodway, Paul McIntyre, Adam Sears, Mark Kibble and James Taylor are thanked for technical support. Thanks are given also to Beth Evans, Kate Ronayne and Liam Brennan for project management. JWM thanks St John's College, Oxford for financial support.

References

1 A. J. Reiter and S.-C. Kong, *Energy Fuels*, 2008, **22**, 2963–2971.
2 G. Thomas and G. Parks, *Potential Roles of Ammonia in a Hydrogen Economy*, U.S. Department of Energy, Washington, 2006, pp. 1–23.
3 S. F. Yin, B. Q. Xu, X. P. Zhou and C. T. Au, *Appl. Catal., A*, 2004, **277**, 1–9.
4 S. Yin, B. Xu, C. Ng and C. Au, *Appl. Catal., B*, 2004, **48**, 237–241.
5 Z. Kowalczyk, S. Jodzis, W. Raróg, J. Zieliński and J. Pielaszek, *Appl. Catal., A*, 1998, **173**, 153–160.
6 W. Raróg, Z. Kowalczyk, J. Sentek, D. Składanowski and J. Zieliński, *Catal. Lett.*, 2000, **68**, 163–168.
7 W. I. F. David, J. W. Makepeace, S. K. Callear, H. M. A. Hunter, J. D. Taylor, T. J. Wood and M. O. Jones, *J. Am. Chem. Soc.*, 2014, **136**, 13082–13085.
8 J. W. Makepeace, T. J. Wood, H. M. A. Hunter, M. O. Jones and W. I. F. David, *Chem. Sci.*, 2015, **6**, 3805–3815.
9 J. Guo, P. Wang, G. Wu, A. Wu, D. Hu, Z. Xiong, J. Wang, P. Yu, F. Chang, Z. Chen and P. Chen, *Angew. Chem., Int. Ed.*, 2015, **54**, 2950–2954.
10 J. Guo, F. Chang, P. Wang, D. Hu, P. Yu, G. Wu, Z. Xiong and P. Chen, *ACS Catal.*, 2015, **5**, 2708–2713.
11 T. J. Wood, J. W. Makepeace, H. M. A. Hunter, M. O. Jones and W. I. F. David, *Phys. Chem. Chem. Phys.*, 2015, **17**, 22999–23006.

12 H. N. Gilbert, N. D. Scott, W. F. Zimmerli and V. L. Hansley, *Ind. Eng. Chem.*, 1933, **25**, 735–741.
13 P. Chen, Z. Xiong, J. Luo, J. Lin and K. L. Tan, *Nature*, 2002, **420**, 302–304.
14 H. Leng, T. Ichikawa and H. Fujii, *J. Phys. Chem. B*, 2006, **110**, 12964–12968.
15 Y. Nakamori, G. Kitahara, K. Miwa, N. Ohba, T. Noritake, S. Towata and S. Orimo, *J. Alloys Compd.*, 2005, **404–406**, 396–398.
16 K. Tokoyoda, S. Hino, T. Ichikawa, K. Okamoto and H. Fujii, *J. Alloys Compd.*, 2007, **439**, 337–341.
17 H. Chu, Z. Xiong, G. Wu, T. He, C. Wu and P. Chen, *Int. J. Hydrogen Energy*, 2010, **35**, 8317–8321.
18 D. Szmigiel, H. Bielawa, M. Kurtz, O. Hinrichsen, M. Muhler, W. Raróg, S. Jodzis, Z. Kowalczyk, L. Znak and J. Zieliński, *J. Catal.*, 2002, **205**, 205–212.
19 J. Guo, Z. Chen, A. Wu, F. Chang, P. Wang, D. Hu, G. Wu, Z. Xiong, P. Yu and P. Chen, *Chem. Commun.*, 2015, **51**, 15161–15164.
20 H. Wu, *J. Am. Chem. Soc.*, 2008, **130**, 6515–6522.
21 M. P. Balogh, C. Y. Jones, J. F. Herbst, L. G. Hector and M. Kundrat, *J. Alloys Compd.*, 2006, **420**, 326–336.
22 W. I. F. David, M. O. Jones, D. H. Gregory, C. M. Jewell, S. R. Johnson, A. Walton and P. P. Edwards, *J. Am. Chem. Soc.*, 2007, **129**, 1594–1601.
23 J. W. Makepeace, M. O. Jones, S. K. Callear, P. P. Edwards and W. I. F. David, *Phys. Chem. Chem. Phys.*, 2014, **16**, 4061–4070.
24 D. J. Bull, E. Weidner, I. L. Shabalin, M. T. F. Telling, C. M. Jewell, D. H. Gregory and D. K. Ross, *Phys. Chem. Chem. Phys.*, 2010, **12**, 2089–2097.
25 S. Hino, T. Ichikawa, H. Leng and H. Fujii, *J. Alloys Compd.*, 2005, **398**, 62–66.
26 H. Y. Leng, T. Ichikawa, S. Hino, N. Hanada, S. Isobe and H. Fujii, *J. Power Sources*, 2006, **156**, 166–170.
27 J. Senker, H. Jacobs, M. Müller, W. Press, H. M. Mayer and R. M. Ibberson, *Z. Anorg. Allg. Chem.*, 1999, **625**, 2025–2032.
28 J. Senker, H. Jacobs, M. Müller, W. Press, H. M. Mayer and R. M. Ibberson, *J. Phys. Chem. B*, 1998, **102**, 931–940.
29 National Institute of Standards and Technology, Neutron scattering lengths and cross sections, http://www.ncnr.nist.gov/resources/n-lengths/, accessed November 16, 2015.

Faraday Discussions

PAPER

Novel bifunctional catalysts based on crystalline multi-oxide matrices containing iron ions for CO_2 hydrogenation to liquid fuels and chemicals

N. Utsis, R. Vidruk-Nehemya, M. V. Landau* and M. Herskowitz

Received 16th November 2015, Accepted 24th November 2015
DOI: 10.1039/c5fd00174a

Seven solid mono-, bi- and tri-metallic oxide matrices where Fe(2+,3+) ions are distributed in different chemical/spatial environments were synthesized and characterized by XRD, N_2-adsorption and EDAX methods. After basification with potassium, all matrices were activated by carburization or reduction–carburization under conditions selected based on the TPC/TPR spectra, tailoring the carburization extent of iron. The performances of the activated Fe-based catalysts with respect to CO_2 conversion and C_{5+} selectivity were measured in a fixed-bed reactor under standard conditions in transient and continuous operation modes in units containing one or three reactors in series with water separations between the reactors. The catalysts were characterized by XRD, N_2-adsorption, HRTEM-EELS and XPS before and after steady-state operation in the reactors. It was found that the rate of CO_2 conversion is not limited by thermodynamic equilibrium but is strongly restricted by water inhibition and it depends on the nature of the Fe-oxide precursor. The ratio between the FTS and RWGS rates, which determines the C_{5+} hydrocarbons productivity, is strongly affected by the nature of the Fe-oxide matrix. The catalysts derived from the Fe–Al–O spinel and Fe–Ba–hexaaluminate precursors displayed the best balance of the two functions R_{FTS}/R_{RWGS} = 0.77–0.78. They were followed by magnetite, CuFe–delafossite, K–ferrite, Fe–La–hexaaluminate and LaFe–perovskite with a gradual lowering of R_{FTS}/R_{RWGS} from 0.60 to 0.15 and a gradual decrease in the C_{5+} productivity. The active sites that enhance the RWGS reaction are located on the surface of the Fe-oxide phases, while the FTS and methanation reactions occur on the surface of the Fe-carbide phases.

Introduction

Carbon dioxide hydrogenation is a viable catalytic process that converts the greenhouse gas and hydrogen potentially produced from water into renewable

Chemical Engineering Department, Blechner Center for Industrial Catalysis and Process Development, Ben-Gurion University of the Negev, Beer-Sheva, 84105, Israel. E-mail: mlandau@bgu.ac.il; Fax: +972 8 6479427; Tel: +972 8 6472141

chemicals and fuels. A variety of catalysts used in this process have been studied, most of them based on iron.[1–10] The catalytic process of hydrogenating carbon dioxide into hydrocarbon fuels C_{5+} (CDHF) consists of the reverse water gas shift (RWGS) reaction followed by the Fischer–Tropsch synthesis (FTS). The challenge is to produce a catalytic material that properly balances the two catalytic functions, thus achieving a high activity, selectivity and stability. The catalysts published in the open literature and patents have only partially achieved the desired performance, displaying a relatively low activity and producing mostly lower hydrocarbons. The other issue in this complex process is the reversibility of the RWGS reaction and the strong inhibition of both reaction rates by the water product, which limits the conversion under the process conditions. Most studies report CO_2 conversions up to 40%.

Recently, the Blechner Center at BGU has published a study[11] of CDHF conducted on two Fe-based catalysts prepared according to procedures described in the literature and on a novel Fe–Al spinel catalyst. The CO_2 conversion measured in one packed-bed reactor was limited to about 60% because of the excessive amount of water produced in this process. Therefore, increasing the conversion requires removal of water. Switching to a system of three packed-bed reactors in series with interim removal of water and condensed higher hydrocarbons increased the CO_2 conversion to as much as 89%. The novel pure-spinel catalyst displayed a significantly higher activity and selectivity than the other Fe catalysts. This process produced a product called green feed, similar in its composition to the product of the high-temperature Fe-based Fischer–Tropsch process from syngas. A follow-up study[12] showed that potassium is a key promoter that improves the reaction rates of RWGS and FTS and increases the selectivity to higher hydrocarbons while producing mostly olefins. It changed the texture of the catalyst, stabilized the Fe–Al–O spinel thus preventing decomposition into Fe_3O_4 and Al_2O_3, increased the content of Fe_5C_2 while shifting Fe in the oxide and carbide phases to a more reduced state, increased the relative exposure of carbide iron on the catalyst surface, and increased the CO_2 adsorption and adsorption strength. A detailed kinetic model of the RWGS, FTS and methanation reactions was developed for the Fe spinel catalyst based on extensive experimental data measured over a range of operating conditions. Significant oligomerization activity of the catalyst was found. Testing the pelletized catalyst with CO_2, CO and H_2 mixtures over a range of operating conditions demonstrated its high productivity to higher hydrocarbons. The composition of the liquid (C_{5+}) was found to be a function of the potassium content and the composition of the feedstock.

The location and control of the surface active sites catalyzing the RWGS, methanation and FTS reactions in bifunctional Fe-based catalysts determine their performance. The active sites could belong to either the Fe-oxide or carbide phases or both, where the ratio of the two active sites is critical. Fe-based catalysts after activation and self-organization under reaction conditions contain two phases, nanoparticles of magnetite Fe_3O_4 and FeC_x carbides, among which the most active and abundant is Haegg carbide Fe_5C_2.[13–15] The relative amounts of the Fe-oxide and carbide phases are governed by the thermodynamic equilibrium of Fe-oxide carburization/oxidation, depending on the temperature and partial pressures of CO, H_2, H_2O at the steady state.[14,16]

The Fe-oxide phase as well as reduced metallic iron are not active in FTS and the formation of hydrocarbons starts only after partial carburization, as shown in the transient kinetics studies.[6,8,17] Actually, the iron oxide phase catalyzes the side WGS reaction yielding CO_2.[18] However, since RWGS is a critical step in CO_2 hydrogenation, the iron oxide phase fulfills this function. The FTS takes place on the active sites located on the surface of the Fe-carbide phase.[6] The spatial distribution of the Fe-oxide and carbide phases under the process conditions is essential. A model of Fe-catalyst particles was proposed based on X-ray absorption spectroscopy data,[17] where the Fe-oxide or carbide core is covered by a layer of a dynamic FeC_x carbide phase rich with vacancies and the catalytic performance was assumed to be independent of the carbide or oxide nature of the particle cores. It supposed the existence of only one type of active Fe-phase in the steady state catalyst, whose catalytic performance may be controlled only by changing the reaction conditions and by the addition of electronic promoters. However, further investigations implementing high resolution electron microscopy combined with EELS revealed that at the steady state Fe-catalysts[11,19,20] consist of separate nanoparticles of Fe-oxide and carbide phases embedded in or mixed with graphitic carbon. In these materials, "clean" or original surfaces of oxide and carbide phases are exposed to the gas phase containing the reactants and products. Therefore, Fe-oxide and carbide nanoparticles located in close proximity provide the active sites for RWGS and FTS.

Adjusting the contributions of the Fe-oxide and carbide surfaces in the catalytic material affects the relative performance of the two catalytic functions. The potassium promoter shifts the Fe-oxide carburization equilibrium to the carbide phases.[12] Insertion of Al atoms into the Fe_3O_4 spinel by isomorphous substitution of Fe(3+) ions produces an opposite effect.[11] It may be expected that the composition and crystal structure of a solid oxide matrix containing iron ions as a component would have a significant impact on the composition of its multiphase carburized product obtained at steady state operation of CO_2 hydrogenation after proper activation. Some hardly reducible Fe-matrices may be carburized to a controlled extent under much more severe operating conditions. This is a potentially new tool for balancing the catalyst functions by determining the carbide/oxide ratio at the activation step.

In the present study, a number of different crystalline oxide materials containing Fe-ions in their lattice were prepared and tested. Activation and measurement of the RWGS, methanation and FTS rates in CO_2 hydrogenation revealed the effect of structure and composition of the Fe-oxide precursor on the degree of balancing the two catalyst functions and performance.

Experimental

Catalyst preparation

Fe–Ba- and Fe–La-hexaaluminates were prepared by a hot co-precipitation strategy.[21] A clear solution was obtained by dissolving $Ba(NO_3)_2$ (Alfa Aesar) or $La(NO_3)_3 \cdot xH_2O$ (Aldrich) and $Fe(NO_3)_3 \cdot 9H_2O$ (Fisher Chemical) in hot distilled water. It was acidified to pH $\sim$ 1 with diluted HNO_3 (Gadot) followed by the addition of $Al(NO_3)_3 \cdot 9H_2O$ (Riedel de Haën). This solution was poured under vigorous stirring into a large excess of $(NH_4)_2CO_3$ solution heated at 60 °C, followed by precipitation of mixed hydroxides at pH 7.5–8.0. The slurry was aged at

60 °C for 3 h, filtered, washed with distilled water to remove nitrates, and dried in air overnight at 110 °C. The dried precursors were calcined at 450 °C for 2 h (heating rate of 2 °C min^{-1}) and at 1200 °C (2 °C min^{-1}) for 3 h.

Potassium ferrite was prepared by a sol–gel method.[22] A 1 M aqueous solution of KNO_3 (Aldrich) and a 2 M solution of ferric nitrate ($Fe(NO_3)_3 \cdot 9H_2O$ (Fisher Chemical)) were mixed and to this mixture was added a 2 M solution of citric acid along with 7 ml of ethylene glycol. The obtained solution was preheated under stirring at 80–90 °C for the period of time required for the formation of the sol, followed by its gelation and drying. The obtained powder was washed with ethanol, dried under vacuum overnight at 60 °C and calcined in a nitrogen atmosphere at 850 °C (heating rate of 1 °C min^{-1}) for 4 h.

Copper delafossite was prepared by a hydrothermal synthesis strategy.[23] 0.01 mol of $Fe(NO_3)_3 \cdot 9H_2O$ (Fisher Chemical) and 0.01 mol of $Cu(NO_3)_2 \cdot 3H_2O$ (Aldrich) were dissolved in 80 ml of distilled water to form a clear solution. Then, 0.2 mol of NaOH (Biolab) was added into the solution with stirring to get a colloidal suspension. 1 ml of propionaldehyde was added to the suspension as a reducing agent, followed by hydrothermal treatment in a Teflon-lined autoclave at 180 °C for 48 h. The solid product was collected by filtration, washed with distilled water, and dried in air at room temperature.

Fe–La–perovskite was prepared by a combustion synthesis[24] using glycine as the fuel. Equal amounts of moles of iron nitrate ($Fe(NO_3)_3 \cdot 9H_2O$ (Fisher Chemical)) and lanthanum nitrate ($La(NO_3)_3 \cdot 6H_2O$ (Merck)) (8.86×10^{-3} mol) were dissolved separately in 15 ml of water by stirring for 15 min at r.t. After this, the two solutions were mixed together and stirred for an additional 30 min. 0.048 mol of glycine (Sigma Aldrich) was added and the solution was mixed for another 30 min. Then, the solution was evaporated under stirring at 80 °C until gelation occurred. Combustion was carried out in air at 200 °C (heating rate of 10 °C min^{-1}) for 2 h and then under air flow at 600 °C for 4 h (heating rate of 5 °C min^{-1}, flow rate of 10 L h^{-1} g^{-1}).

Fe–Al–O spinel $Fe(2+)[Fe(3+)_{0.5}Al_{0.5}]_2O_4$ was prepared by co-precipitation.[11] 21.0 g of $Al(NO_3)_3 \cdot 9H_20$ (Riedel de Haën) and 72.3 g of $Fe(NO_3)_3 \cdot 9H_2O$ (Fisher Chemical) were dissolved in distilled water (20 and 20 ml, respectively) and mixed together. Then, a 5% solution of NH_4OH (Biolab) was gradually added at room temperature under stirring, up to pH 8.0. The solid precipitated from the solution was separated by filtration, washed with distilled water, dried in air at 110 °C for 24 h, and calcined at 450 °C in air for 6 h.

The powder of nanostructured magnetite with a crystal size of 3–4 nm and a surface area of 250 m^2 g^{-1} was purchased from MACH I Inc.

All materials besides K–ferrite were modified by the addition of potassium as a promoter. This was inserted (an amount of 4 wt%) to the as-prepared samples by incipient wetness impregnation with an aqueous solution of K_2CO_3 (Sigma Aldrich), followed by drying in air at 110 °C overnight. In the case of the Fe–Al–O spinel, potassium was inserted after drying and before calcination at 450 °C.

Catalyst characterization

The elemental compositions of the catalytic materials were measured by the EDAX method (Quanta-200, SEM-EDAX, FEI Co. instrument). The surface area, pore volume, and pore size were calculated from N_2 adsorption–desorption isotherms by using conventional Brunauer–Emmett–Teller (BET) and Barrett–Joyner–Halenda (BJH) methods. The samples were outgassed under vacuum for 2 h at 250 °C, and isotherms were obtained on a NOVA3200e (Quantachrome) instrument at the temperature of liquid nitrogen. The thermo-programmed carburization (TPC) measurements were performed by using a Chemisorption Analyzer AutoChem II 2920 instrument, Micrometrics Co. TPC spectra were recorded in the temperature interval 35–900 °C in 100 ml min^{-1} flow of 10% CO/10% H_2/Ar with a heating rate of 5 °C min^{-1}.

Conventional wide-angle XRD patterns were obtained with a Phillips 1050/70 powder diffractometer fitted with a graphite monochromator at 40 kV and 28 mA. Software developed by Crystal Logic was used. The phase identification was performed by using a SBDE ZDS computer search/match program coupled with the ICDD. The relative contents of the iron oxide and carbide phases were obtained using Rietveld refinement of the XRD profile implementing the DBWS-9807 program. The average crystal size of the identified phases was estimated based on the characteristic peaks broadening using the Scherrer equation.

XPS data were collected using X-ray photoelectron spectrometer ESCALAB 250 ultrahigh vacuum (1×10^{-9} bar) apparatus with an Al K_α X-ray source and a monochromator. The spectral components of the C and Fe signals were found by fitting a sum of the single component lines to the experimental data by means of a non-linear least-square curve fitting. The EX05 argon gun system performed controlled removal of surface layers, allowing depth profile analyses of elements. The morphology and in-depth analysis of individual nanoparticles were investigated using a field emission analytical transmission electron microscope (JEOL JEM-2100F) operated at an accelerating voltage of 200 kV and equipped with a Gatan imaging filter (GIF Quantum)/EELS system.[11] The obtained elemental distribution maps were further combined in one RGB image, marking Fe with red, C with green, O with blue and Al with yellow. This yielded multicomponent element distribution maps for the nanoparticles observed in the regular HRTEM images. The samples for HRTEM were prepared by depositing a drop of an ethanol suspension of the solid catalyst on a silica-coated copper grid (silicone monoxide substrates on 300 mesh copper grids, Electron Microscopy Sciences Co.) that formed a contrast oxide background for carbon-rich particles.

Testing of the catalyst performance

All catalysts were tested at 20 bar, 320 °C, a molar H_2/CO_2 ratio of 3.0 and a weight hourly space velocity (WHSV) of 3 g CO_2 per g_{cat} per h in the one fixed-bed reactor configuration described elsewhere.[11,12] Some experiments were conducted in a three-reactor system.[11,12]

The hydrocarbon liquid mixture and oxygenates from the aqueous products were separated and analyzed by GC-MS (Agilent Technologies 6890N network GC system equipped with a 5973 Network mass-selective detector). CO_2, H_2, CO and C_1–C_6 hydrocarbons were analyzed on-line using a GC instrument, Agilent Technologies, model 7890A equipped with 5 valves/7 columns (PLOT and packed)/2 TCD/FID detectors. The organic content in the aqueous product was measured by TOC analysis (instrument TOC-V_{CPN} Shimadzu Co.).

The CO_2 conversion was calculated according to the equation:

$$X_{CO_2} = \frac{F_{i,CO_2} - F_{o,CO_2}}{F_{i,CO_2}} 100\%$$

where F_{i,CO_2} and F_{o,CO_2} are the inlet and outlet CO_2 molar flow rates, respectively. The mass selectivity of the products was calculated as the mass rate of carbon in the product, W_{C_j}, divided by the mass rate of carbon reacted, as shown below:

$$S_i = \frac{W_{C_j}}{12(F_{i,CO_2} - F_{o,CO_2})} 100\%$$

The productivity of C_{5+} hydrocarbons was calculated as the mass rate of the products, $W_{C_{5+}}$, at the reactor outlet divided by the mass of the catalyst, M_{cat}:

$$P_{C_{5+}} = \frac{W_{C_{5+}}}{M_{cat}}$$

The rates of the RWGS, FTS and methanation (M) reactions were calculated according to the equation:[12]

$$\frac{d\omega_i}{d\left(\frac{1}{WHSV}\right)} = R_i \times \omega_{CO_2,in} \times MW_i$$

where ω_i is the mass fraction of component i, WHSV is the weight hourly space velocity of CO_2, R_i is the rate of change in the reaction where component i is a reactant/product, $\omega_{CO_2,in}$ is the mass fraction of CO_2 at the inlet, and MW_i is the molecular weight of component i. The calculation of the reaction rate was based on the rate balance assuming that the productions of methane and C_{2+} hydrocarbons are two parallel processes preceded by the formation of CO according to the RWGS reaction (Scheme 1).

Results and discussion

Characteristics of the Fe-oxide matrices as catalyst precursors

The recorded X-ray diffraction patterns (Fig. 1A) and the results of the elemental analyses of the as-synthesized oxide matrices including Fe-ions in their lattices allowed us to identify their structures. The purchased Fe-oxide consisted of the pure magnetite Fe_3O_4 phase (Fig. 1A-1; ICDD Card # 19-629) with an average crystal size of 3 nm. The Fe–Al–O material represented a pure spinel phase (Fig. 1A-6, ICDD Card # 34-192) with partial isomorphous substitution of the

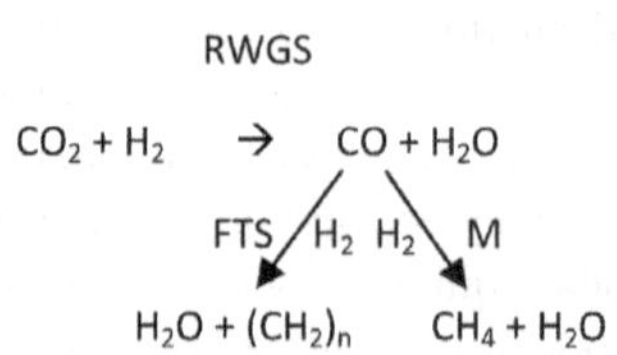

Scheme 1

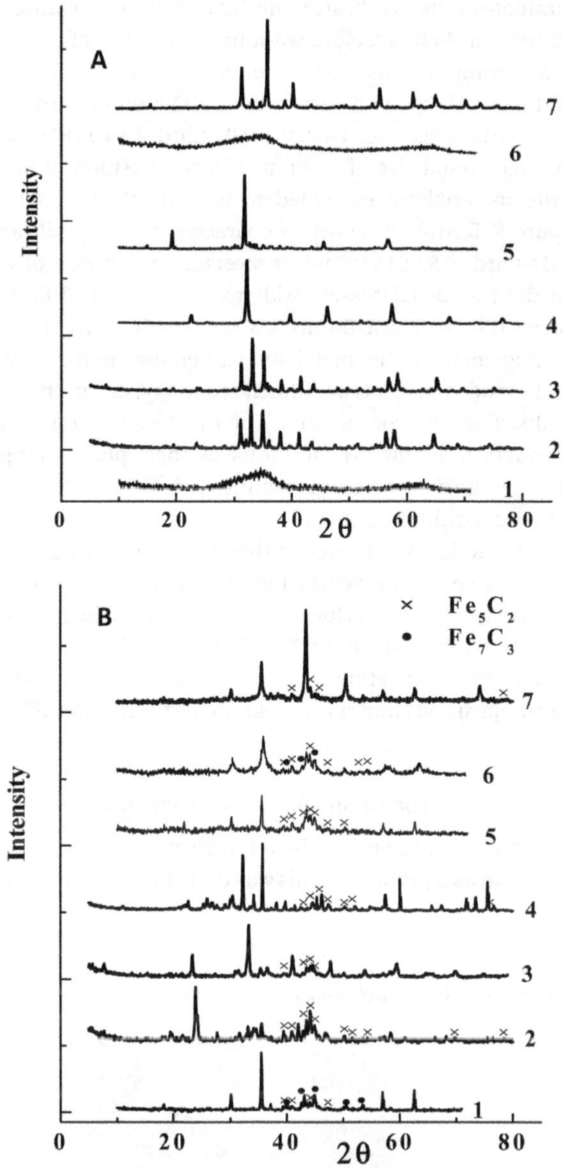

Fig. 1 X-ray diffraction patterns of the synthesized oxide matrices containing Fe-ions: (A) as-synthesized; (B) after activation and testing in CDHF. (1) Magnetite; (2) Fe–Ba-hexaaluminate; (3) Fe–La-hexaaluminate; (4) Fe–La-perovskite; (5) K–ferrite; (6) Fe–Al-spinel; (7) Fe–Cu-delafossite.

Fe(3+) ions for Al(3+), shifting the X-ray diffraction peak positions to an extent corresponding to formula $Fe(2+)[Fe(3+)_{0.5}Al_{0.5}]_2O_4$.[11,25] The chemical composition of this material corresponded to a Fe/Al ratio of 2 and its average crystal size was 2 nm. The Fe–La-perovskite demonstrated an X-ray diffraction pattern corresponding to a pure $LaFeO_3$ phase (Fig. 1A-4; ICDD Card # 37-1493) with a Fe/La ratio of 1 and an average crystal size of 40 nm. The Fe–Ba– and

Fe–La–hexaaluminates were synthesized under conditions yielding the maximal content of iron ions in their structure without segregation of separate Fe-oxide. Their chemical compositions corresponded, respectively, to formulae $BaFe_7Al_5O_{19}$ and $LaFe_6Al_6O_{19}$, and their X-ray diffraction patterns corresponded to hexaaluminate structures (Fig. 1A-2,3; ICDD Card # 84-1787) with a purity > 97% and an average crystal size of >50 nm. The composition of the synthesized potassium ferrite material corresponded to formula $KFeO_2$ containing Fe(3+) ions. It was a pure K–ferrite phase with a characteristic X-ray diffraction pattern (Fig. 1A-5; ICDD Card # 83-2153) and an average crystal size of >50 nm. The composition of the Fe–Cu–delafossite oxide material matched formula $CuFeO_2$, being a pure phase (Fig. 1A-7; ICDD Card # 39-246) with an average crystal size of >50 nm. The arrangements of the metal and oxygen ions in the crystal lattices of the synthesized Fe-oxide materials are illustrated in Fig. 2. Significant differences in the spatial distribution (coordination) of the Fe-ions and their elemental environment renders different organizations of their phase composition and particle morphology during activation treatment in the H_2/CO atmosphere and under the reaction conditions of CDHF.

The texture characteristics of the as-synthesized oxide matrices containing Fe-ions are listed in Table 1. The values for the surface area, pore volume and diameter varied over a wide range. However, these features do not determine their catalytic properties in CDHF since the Fe-oxide materials serve as precursors of the catalysts that after activation and self-organization display completely different phase compositions and texture parameters (see Table 3 in Section 3.3).

Selection of the activation conditions for the Fe-oxide matrices

The as-synthesized Fe-oxide matrices before activation and testing were doped with 4 wt% potassium as a promoter. This was done to adjust the selectivity of the

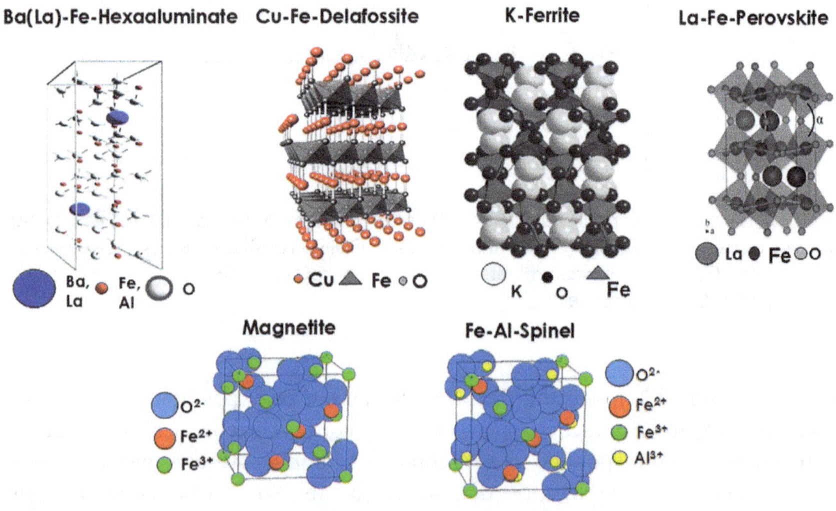

Fig. 2 Arrangements of the metal and oxygen ions in the lattices of the synthesize Fe-oxide matrices.

Table 1 Texture parameters of the as-synthesized Fe-oxide matrices

Fe-oxide matrix	Surface area, $m^2\ g^{-1}$	Pore volume, $cm^3\ g^{-1}$	Aver. pore diam., nm
Magnetite	250	1.40	8
Fe–Al–O spinel	140	0.57	3
Fe–Ba–hexaaluminate	6	0.03	32
Fe–La–hexaaluminate	7	0.03	13
Fe–Cu–delafossite	4	0.01	19
K–ferrite	5	0.02	20
Fe–La–perovskite	18	0.17	39

Fe-carbide phases produced during the material activation, depressing the production of methane and shifting the hydrocarbon distribution towards heavier higher hydrocarbons.[12] The selection of the activation conditions for the K-modified Fe oxide matrices was done based on the results of TPC experiments. According to the data, all the synthesized materials may be divided into two groups, consuming CO at low and high temperatures. The typical TPC spectra for these two groups – Fe–Al–O spinel and Fe–La–hexaaluminate, are shown in Fig. 3. The temperatures corresponding to the beginning of the CO consumption and the positions of the TPC peaks in the spectra are listed in Table 2.

The principal difference here is that the materials belonging to the first group (K/magnetite, K/Fe–Al–O-spinel, K/Fe–Cu–delafossite and K–ferrite) may be efficiently carburized at the relatively low temperature of 300 °C in a H_2/CO flow, yielding Fe-carbide phases without accumulation of graphitic carbon formed by the Boudouard reaction. These materials were activated by carburization before testing in CDHF with a CO/H_2 stream of 30 : 30 : 150 ml $min^{-1} \times g_{cat}{}^{-1}$ of CO, H_2, and He at 300 °C for 3 h.[11] For the second group of materials (K/Fe–Ba–hexaaluminate, K/Fe–La–hexaaluminate and K/Fe–La–perovskite), carburization does not proceed at temperatures < 400 °C. However, at higher temperatures, carbon dioxide is converted to bulk carbon, embedding the catalyst particles and blocking their pores and surface. For such materials, a different activation strategy was implemented – reduction in H_2 flow at 900 °C for 2 h based on TPR data (not shown) in order to convert some of the iron ions to the metallic state, followed by low-temperature carburization.

Effect of the Fe-matrix structure on its performance in CDHF after activation/self-organization

All the K-doped Fe-oxide matrices after appropriate activation were tested in CDHF in a fixed-bed reactor under the standard conditions of 320 °C, 20 bar, a molar H_2/CO_2 ratio of 3.0 and a $WHSV_{CO_2}$ of 3 h^{-1}. Each test was conducted for 150 h. During the initial 100 h on stream, the materials underwent self-organization of their chemical and phase composition and morphology. Over this period the system reached steady-state. The data collected during the last 50 h on stream characterized the catalyst's performance.

The spent catalysts after testing were treated in the reactor with He flow at 300 °C for 6 h to eliminate residual heavy hydrocarbons, then passivated in 1% O_2/Ar flow at room temperature for 1 h, followed by treatment in pure CO_2 flow at room

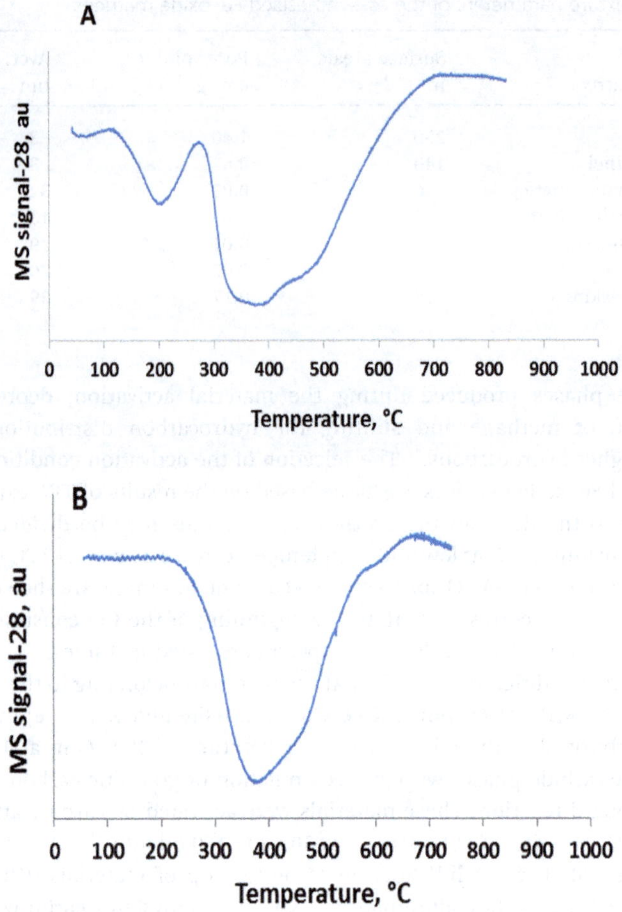

Fig. 3 TP-carburization spectra recorded with K-doped Fe–Al–O spinel (A) and Fe–La-hexaaluminate (B).

temperature for 1 h. After discharging from the reactor their phase composition and texture were characterized by X-ray diffraction and N_2-adsorption–desorption methods. The characterization results are presented in Table 3. All the spent

Table 2 Temperatures of the initial CO consumption and the peak positions in the TPC spectra recorded with the K-doped Fe-oxide materials

Samples	Starting temperature of CO consumption, °C	TPC peak positions, °C
K/magnetite	230	260, 420, 560
K/Fe-Al-O-spinel	150	230, 400, 490
K/Fe-Ba-hexaaluminate	380	530, 730
K/Fe-La-hexaaluminate	370	500, 720
K/Fe-Cu-delafossite	200	300, 380
K-ferrite	210	300, 360
K/Fe-La-perovskite	340	600

materials showed multiphase compositions, including Fe-carbide Fe_5C_2 or Fe_5C_2–Fe_7C_3 and residual oxide phases (which were the same as in the as-prepared materials or the products of their decomposition during activation and self-organization). The catalysts after reaction obtained from K-doped $BaFe_7Al_5O_{19}$, $LaFe_6Al_6O_{19}$, $LaFeO_3$, $KFeO_2$ and $CuFeO_2$ matrices contained only Haegg carbide Fe_5C_2 in amounts of 17–38 wt%. The magnetite and Fe–Al–O spinel produced a mixture of two carbides, where Fe_5C_2 was more abundant.

The phase compositions of the residual oxide phases, excluding magnetite, were evidence for the partial or complete transformations of the Fe-oxide precursors to other oxide phases in addition to carburization of the Fe-ions included in the parent crystal lattices. 31–45 wt% of unconverted parent Fe-hexaaluminates and perovskite phases with the same Fe content were found in the catalysts after reaction. Carburization of Fe-ions to carbide phases in those hexaaluminate and perovskite materials was accompanied by their decomposition forming $BaCO_3$, Al_2O_3, $LaAlO_3$, La-oxide-carbonate and La-oxides, which did not contain iron, thus expected to be inactive in CDHF. K/Fe–Cu–delafossite

Table 3 Phase composition and texture characteristics of the spent catalysts

	Phase composition			Texture parameters		
Fe-oxide matrix	Phase	Wt%	Particle size, nm	Surface area, $m^2 g^{-1}$	Pore volume, $cm^3 g^{-1}$	Average pore diameter, nm
K/Fe–Ba–HAl	$BaFe_7Al_5O_{19}$	45	45	10	0.025	20
	$BaCO_3$	16	45			
	Fe_5C_2	30	45			
	Al_2O_3	9	2.5			
K/Fe–La–HAl	$LaFe_6Al_6O_{19}$	44	50	3	0.005	13
	$LaAlO_3$	17	50			
	Fe	3	45			
	Fe_5C_2	26	40			
	Al_2O_3	10	45			
K/Fe–La–perovskite	$LaFeO_3$	31	40	2	0.005	32
	FeO	27	50			
	$La_2O_2(CO_3)$	19	50			
	La_2O_3	6	45			
	Fe_5C_2	17	15			
K–ferrite	Fe_3O_4	25	45	3	0.018	24
	Fe_5C_2	49	40			
	Amorphous	26	N.D.			
K/Fe–Cu–delafossite	Fe_3O_4	40	31	3	0.0136	15
	Cu	40	32			
	Fe_5C_2	35	28			
	CuO	20	9			
K/magnetite	Fe_3O_4	39	50	9	0.67	7
	Fe_5C_2	38	40			
	Fe_7C_3	23	45			
Fe–Al–O spinel	$Fe(Fe_{0.34}Al_{0.66})_2O_4$	55	10	60	0.62	8
	Fe_5C_2	36	35			
	Fe_7C_3	9	35			

besides Fe_5C_2 is converted to metallic Cu (40 wt%) magnetite and Cu-oxide. Potassium ferrite, in parallel with Fe carburization, was converted to magnetite and an unidentified amorphous phase probably rich in potassium. The Fe–Al–O spinel during carburization and self-organization under CDHF conditions lost some of its Fe-ions to form carbides, while the residual oxide phase became a Fe–Al–O spinel enriched with aluminum. Only the magnetite phase demonstrated a simple conversion of some of the oxide precursor to Fe-carbides, leaving the unconverted part in the parent form.

The catalysts derived from the Fe-oxide matrices represent assemblies of contacting Fe-oxide and Fe-carbide nanoparticles diluted by graphitic carbon. The space distributions of these nanoparticles in the materials obtained from Fe–Al–O spinel and Fe–Ba–hexaaluminate are illustrated in the HRTEM-EELS images shown in Fig. 4. It may be expected that iron species in ionic (oxide) and near-metallic (carbide) states existing at the surface of nanoparticles of a similar size but with different chemical compositions (Fig. 4) may catalyse different steps of the CDHF process. The close assembling of these nanoparticles should favour surface diffusion or spillover of reaction intermediates (*i.e.* CO molecules or activated hydrogen atoms) from one type of active phase to another. No indication of the formation of a Fe-oxide layer at the surface of carbide nanoparticles or a carbide layer on the surface of oxide nanoparticles was found. The surfaces of the Fe-oxide and Fe-carbide phases were directly exposed to the gas phase, indicating the possibility of two types of active sites existing at the surface of bifunctional Fe-catalysts under the conditions of the CDHF process. Also, the total surface areas of the catalysts after reaction and their pore dimensions varied over a wide range among the seven catalysts (Table 3), *e.g.* the surface area ranged from 2 to 60 $m^2\ g^{-1}$, which may have an impact on the catalyst performance.

The range of the RWGS, FTS and methanation reaction rates depends on the balance between the different catalytic functions, which may vary extensively, demonstrating important trends not investigated in previous studies. The RWGS, FTS and methanation rates calculated from experimental data and the C_{5+} hydrocarbons productivity are plotted in Fig. 5 to reflect the gradual increase in the C_{5+} hydrocarbons productivity. The RWGS rates in this sequence increased hyperbolically. The FTS rates also increased gradually, which follows, as expected, the increase in the C_{5+} productivity. However, the FTS rates demonstrated a parabolic trend, so that the difference between the RWGS and FTS reaction rates decreased progressively with increasing C_{5+} productivity. The methanation rates measured for all catalysts were similar and relatively low.

This indicates that the nature of the oxide catalyst precursor – the solid crystalline matrix containing Fe-ions with different valence states and environments – strongly affects the balance between the RWGS and FTS steps. The results shown in Fig. 5 are evidence that in all the tested materials the catalytic functions responsible for RWGS and FTS are not balanced. The hydrocarbons productivity is limited by the rate of the FTS reaction. The ratio of R_{FTS}/R_{RWGS} reached a value of about 0.8 with the highest-performance catalysts produced from the Fe–Al–O spinel and Fe–Ba–hexaaluminate precursors, followed by magnetite, Fe–Cu-delafossite, K-ferrite, Fe–La–hexaaluminate and Fe–La–perovskite with a gradual lowering of R_{FTS}/R_{RWGS} from 0.60 to 0.15. A similar conclusion about limiting the rate of the CDHF tandem reaction by the FTS rate was recently seen in ref. 29

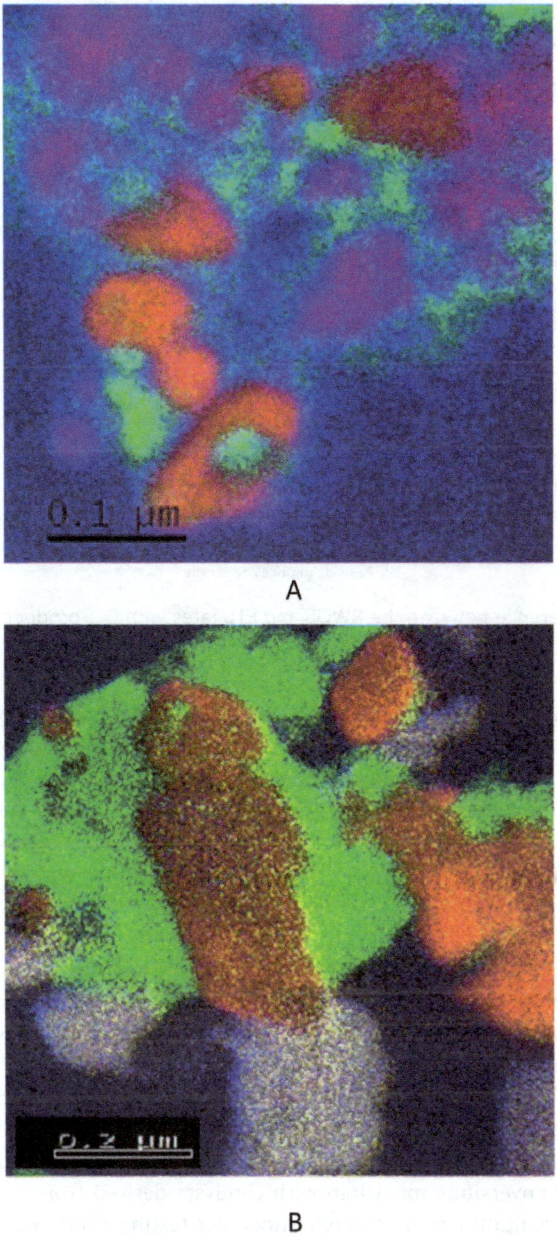

Fig. 4 TEM-EELS images of the steady state catalysts after testing derived from the Fe–Al–O-spinel (A) and Fe–Ba–hexaaluminate (B) matrices: red-orange – Fe-carbide; violet – Fe-oxide, blue-gray – Fe–Ba–hexaaluminate, green – carbon.

based on calculations of the RWGS, FTS and M reaction rate constants for the CDHF process conducted with a K–Mn–Fe/Al_2O_3 catalyst.

The ratio $\psi = R_{FTS}/R_{RWGS}$ reflects the balancing of the RWGS and FTS functions and may be defined as "extent of functions balancing". According to Scheme 1,

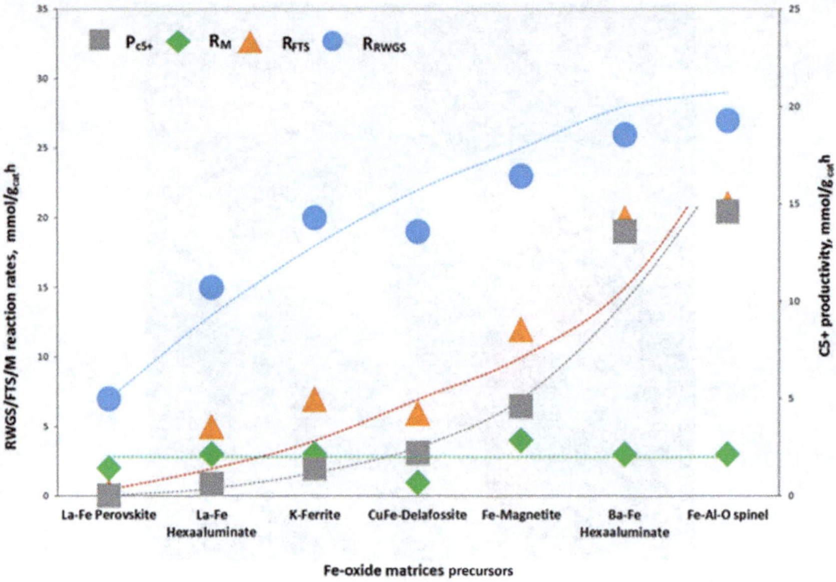

Fig. 5 The difference between the RWGS and FTS rates with C_{5+} productivity.

a more general definition of the balancing extent of the two catalyst functions should relate to both routes of CO transformation: FTS and M, producing components of different fuels: methane, LPG analogue C_2–C_4, and liquid hydrocarbons for transportation fuels. In the general case, the "extent of catalyst functions balancing" reflecting the productivity of hydrocarbon fuels should be defined as: $\psi = (R_{FTS} + R_M)/R_{RWGS}$. This value may also be simply estimated based on the measured selectivity of the CDHF process to CO (S_{CO}, %): $\psi = 100 - S_{CO}$ (%).

The weak effect of the catalyst composition and texture on the RWGS reaction rates for the most active materials on the right side of Fig. 5 poses a question: if this reaction as well as the whole CDHF tandem are thermodynamically limited, then that does not allow measurement of the intrinsic catalytic activities under selected conditions. According to our estimations, the equilibrium conversion of CO_2 in the RWGS reaction conducted in a closed system at 320 °C is ~25%. Consumption of CO in consecutive FTS and M reactions shifts the equilibrium so that the CO_2 conversions measured with catalysts derived from Fe–Al–O spinel and Fe–Ba–hexaaluminate precursors under our testing conditions became 38–40%. Since the CO hydrogenation to hydrocarbons is not limited thermodynamically at 320 °C (ref. 26), then the CO_2 conversion in the tandem CDHF reaction should also not be restricted by thermodynamic equilibrium.

Conversion values lower than that permitted by chemical equilibrium may be caused by a low catalyst activity and/or strong inhibition of tandem reactions by water evolved in both of them.[7,12,27] In order to verify this, tests with catalysts derived from Fe–Al–O spinel and Fe–Ba–hexaaluminate precursors were conducted under the same conditions but in a three reactor unit. Here, the catalysts were divided into three equal parts, loaded into the three cascade reactors, and during testing, water and heavy hydrocarbons were separated from the reagents/products stream between the 1st and 2nd and 2nd and 3rd reactors.[11] The results in Table 4 indicate, as expected, that the performance in the three reactors improves and the CO_2 conversion (68–71%) became significantly higher. This means that the rate of the tandem CDHF reaction under selected testing conditions is not limited by thermodynamics or catalyst activity, but is strongly restricted by water inhibition. This may explain the weak dependence of the CO_2 conversion rate on the chemical nature and texture of the catalytic materials.

Contribution of the Fe-oxide and Fe-carbide phases in the steady-state catalysts to the different steps of the CO_2 hydrogenation reaction

A series of transient kinetic experiments aimed at determining the function of the two main active sites was conducted. The consumption of CO for carburization of the Fe-oxide precursors as well as oxidation of the pure Fe-carbide phases with CO_2 and water complicate the calibration of the testing unit and interpretation of such data.[8,9,17] The Fe-oxide catalysts that require high temperatures for carburization (Table 2) were used to determine the time needed for establishing steady state after the He flow is switched to CO_2/H_2. Also, their testing without activation yields only the catalytic performance of the Fe-oxide surface for comparison with the activated catalysts containing Fe-oxide and carbide phases.

The results of the four transient experiments are presented in Fig. 6. The experiment conducted with inactivated K/Fe–Ba–hexaaluminate (Fig. 6A) under standard conditions indicated stable CO_2 conversion after 30–50 min. A CO_2 conversion of 23% with CO as the only product was measured, very close to the RWGS equilibrium conversion. This means that the surface of the oxide K/Fe–Ba–hexaaluminate particles contains only active sites for the RWGS reaction.

The consumption of CO evolved according to the RWGS reaction at the surface of the inactivated K/Fe_3O_4 nanocrystals, which masks the CO evolution at the beginning of run (Fig. 6B). After saturating the surface with CO, its production increased with time on stream. In the case of K/Fe–Ba–hexaaluminate, the CO_2 conversion on magnetite corresponds to the RWGS equilibrium value. Traces of hydrocarbons appear after 60–80 min and may be attributed to the formation of the 4 wt% Fe_5C_2 phase detected in the catalyst. Similar results were obtained in the transient experiment carried out with inactivated K/Fe–Al–O spinel. Therefore, Fe-containing oxide components of steady state catalysts derived from different Fe-oxide precursors and stabilized under CDHF conditions efficiently catalyse only the RWGS reaction.

Table 4 Results of the comparative testing of the catalysts derived from Fe–Al–O spinel and Fe–Ba–hexaaluminate precursors – 320 °C, 20 bar, $H_2/CO_2 = 3$, $WHSV_{CO_2} = 3\ h^{-1}$ in one- and three-reactor configurations

	CO_2 conversion, %	R_{FTS}/R_{RWGS}	Productivity C_{5+}, mmol C g_{cat}^{-1} h^{-1}
Fe-Al–O spinel 1 reactor	40	0.77	6.8
Fe-Al–O spinel 3 reactors	68	—	17
Fe–Ba–hexaaluminate 1 reactor	38	0.78	7.3
Fe–Ba–hexaaluminate 3 reactors	71	—	19

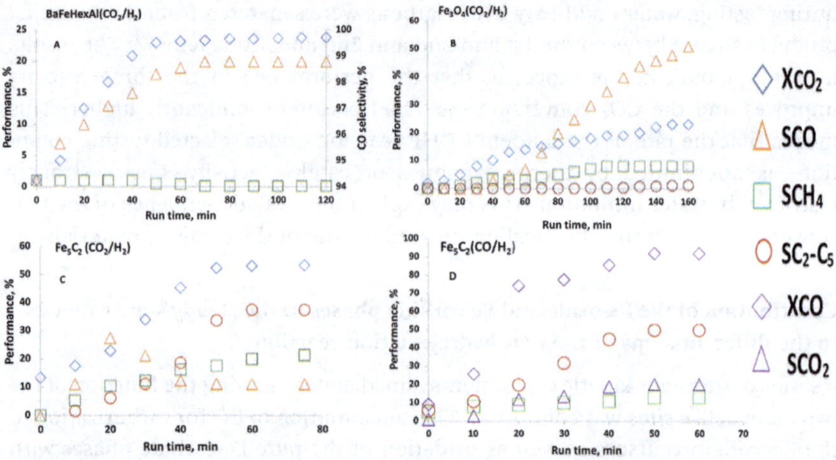

Fig. 6 Results of the transient experiments of CO_2 and CO hydrogenation conducted with pure Fe-oxide and Fe-carbide materials: (A) K/Fe–Ba–hexaaluminate (CO_2/H_2); (B) K/Fe_3O_4 (CO_2/H_2); (C) K/Fe_5C_2 (CO_2/H_2); (D) K/Fe_5C_2 (CO/H_2).

CO_2 hydrogenation with Fe_5C_2 carbide synthesized by the carburization of nanocrystalline K/magnetite at 300 °C yielded, after 50 min, a full range of products (Fig. 6C). The initial high production of CO decreasing after 50–60 min is not a result of the RWGS reaction on the pure Fe-carbide surface. It is probably a result of the oxidation of the Fe-carbide phase with CO_2: $FeC_x + CO_2 \rightarrow C + CO + Fe_xO_y$.[13] This produces a Fe-oxide phase, initially active in the RWGS reaction, which increases with time due to additional oxidation of the Fe-carbide phase with water: $FeC_x + H_2O \rightarrow C + H_2 + Fe_xO_y$.[13]

In the transient CO hydrogenation with the pure K/Fe_5C_2 material, a high ~90% conversion of CO was achieved after 50 min with products of the WGS reaction opposite to those of the RWGS reaction, appearing in small amounts after only 20–30 min when the Fe-carbide was partially oxidized to Fe_3O_4 by water evolved in the FTS. A relatively high methanation selectivity was measured in the transient experiments with the Fe-carbide catalyst.

The two types of iron species, oxide and carbide, have different energies of the $Fe_{3/2}$ core electrons.[12] The peaks in the XPS spectra at lower binding energies around 707.1–707.5 eV correspond to the carbides with a near-metallic state of iron. The peaks at higher binding energies of 710.4–711.5 eV belong to the iron in the oxide phases, representing a superposition of signals contributed to by Fe(2+) and Fe(3+) ions. The XPS spectra were recorded for the spent catalysts derived from the above listed Fe-oxide matrices after argon sputtering for periods of 10–90 s (required for cleaning the surface from adsorbed molecules and the oxide layer formed during catalyst passivation). The relative contributions of the Fe_{oxide} and $Fe_{carbide}$ surface species may be clearly distinguished in the XPS spectra. Fig. 7 presents the spectra recorded with the catalysts at steady state derived from three Fe-oxide matrices after Ar-sputtering for 30 s. The contribution of the $Fe_{carbide}$ species to the total peak area in the spectra gradually increased from K/Fe–La-perovskite to K-ferrite, in agreement with the results from the XRD phase analysis and the similar particle size of the oxide and carbide phases (Table 3, Fig. 4).

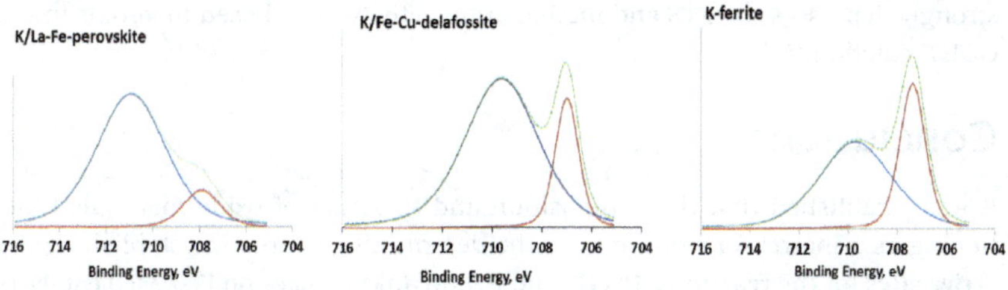

Fig. 7 XPS spectra recorded for the $Fe_{3/2}$ core with the steady state catalysts derived from the K/Fe–La–perovskite, K/Fe–Cu–delafossite and K–ferrite precursors after testing in the CDHF reaction.

The ratios of the integral intensities of the signals obtained at binding energies of 707.1–707.5 eV and 710.4–711.5 eV, corresponding to the relative surface concentrations of the $Fe_{carbide}/Fe_{oxide}$ species, were calculated as their average values obtained at different sputtering times. The plot of the ratio of reaction rates as a function of the ratio of the surface concentrations of $Fe_{carbide}/Fe_{oxide}$ in Fig. 8 displays a linear dependency for the six different Fe-matrix precursors containing 4 wt% potassium. This confirms that the two types of active sites responsible for the formation of CO (RWGS) and hydrocarbons (FTS–M) are surface carbide and oxide species, respectively. The K–ferrite catalyst displays the highest exposure of surface Fe-carbide species ($Fe_{carbide}/Fe_{oxide}$ = 0.53), which does not fit the linear plot. Indeed, this catalyst should be considered as a special case since, according to the precursor stoichiometry after activation and self-organization, it contained about 30 wt% potassium. Potassium as a promoter in high amounts

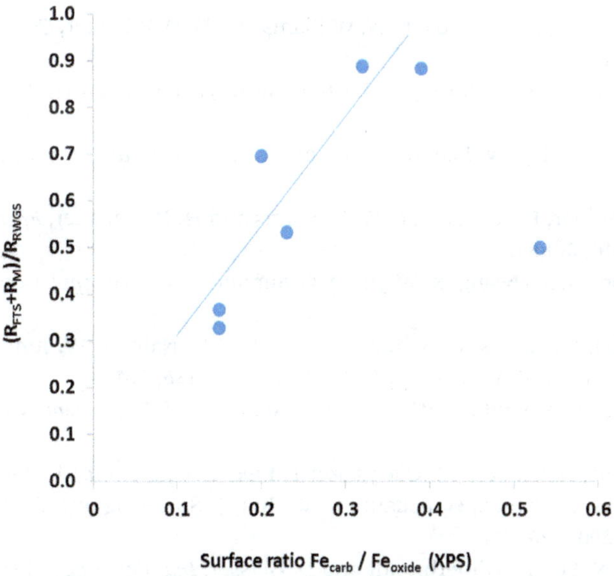

Fig. 8 Correlation between the relative surface concentrations of the Fe-species of the carbide and oxide phases and the ratio of the FTS–M and RWGS reaction rates.

strongly depresses the FTS and methanation activity of Fe-based materials in the CDHF reaction.[12,28]

Conclusions

It was established that the composition and structure of oxide materials functioning as a matrix for iron ions are the determining factors in establishing the active sites for the reactions, RWGS and FTS/M, taking place on Fe-based catalysts in the CDHF process. The results of this study indicate that the relative content of the Fe-oxide and Fe-carbide phases in catalysts working at steady state and the exposure of Fe-oxide and carbide species on the catalyst surface are determined by the state of the iron in the oxide precursor. The catalytic materials represent assemblies of nanoparticles of Fe-carbide and Fe-oxide diluted with carbon. They expose the surface of the carbide or oxide phases to reacting molecules in the gas phase. This explains the strong effect of the catalyst phase composition on balancing the two catalyst functions – RWGS catalysed by oxide (ionic) iron species and FTS/M catalysed by iron in the near-metallic state at the surface of the carbide phases. Catalysts derived from Fe–Al–O spinel and Fe–Ba–hexaaluminate materials yield a high productivity of C_{5+} hydrocarbons.

Acknowledgements

This work was supported by the Israel Science Foundation (grant No. 582/13). The authors are grateful to Dr A. Erenburg, Dr N. Froumin and Dr V. Ezersky for the characterization of the catalytic materials by XRD, XPS and TEM-EELS methods, respectively.

Notes and references

1 R. W. Dorner, D. R. Hardy, F. W. Williams and H. D. Willauer, *Catal. Commun.*, 2010, **11**, 816.
2 J. S. Kim, S. Lee, S. B. Lee, M. J. Choi and K. W. Lee, *Catal. Today*, 2006, **115**, 228.
3 P. S. S. Prasad, J. W. Bae, K.-W. Jun and K. W. Lee, *Catal. Surv. Asia*, 2008, **12**, 170.
4 R. W. Dorner, D. R. Hardy, F. W. Williams and H. D. Willauer, *Energy Environ. Sci.*, 2010, **3**, 884.
5 J. S. Hong, J. S. Hwang, K. W. Jun, J. C. Sur and K. W. Lee, *Appl. Catal., A*, 2001, **218**, 53.
6 T. Riedel, M. Claeys, H. Schulz, G. Schaub, S. S. Nam, K. W. Jun, M. J. Choi, G. Kishan and K. W. Lee, *Appl. Catal., A*, 1999, **186**, 201.
7 T. Riedel, G. Schaub, K. W. Jun and K. W. Lee, *Ind. Eng. Chem. Res.*, 2001, **40**, 1355.
8 H. Schulz, G. Schaub, M. Claeys and T. Riedel, *Appl. Catal., A*, 1999, **186**, 215.
9 T. Riedel, H. Schulz, G. Schaub, K. W. Jun, J. S. Hwang and K. W. Lee, *Top. Catal.*, 2003, **26**, 41.
10 Y. Yao, X. Liu, D. Hilderbrandt and D. Glasser, *Ind. Eng. Chem. Res.*, 2011, **50**, 11002.
11 M. V. Landau, M. Herskowitz and R. Vidruk, *ChemSusChem*, 2014, **7**, 785.

12 N. Meiri, Y. Dinburg, M. Amoyal, V. Koukouliev, R. Vidruk Nehemya, M. V. Landau and M. Herskowitz, *Faraday Discuss.*, 2015, **183**, 197.

13 E. de Smit and B. M. Weckhuysen, *Chem. Soc. Rev.*, 2008, **37**, 2758.

14 E. de Smit, F. Cinquini, A. M. Beale, O. V. Safonova, W. van Beek, P. Sautet and B. M. Weckhuysen, *J. Am. Chem. Soc.*, 2010, **132**, 14928.

15 B. H. Davis, *Catal. Today*, 2009, **141**, 25.

16 P. E. Højlund-Nielsen and J. Bøgild-Hansen, *J. Mol. Catal.*, 1982, **17**, 183.

17 S. Li, W. Ding, G. D. Meitzner and E. Iglesia, *J. Phys. Chem. B*, 2002, **106**, 85.

18 A. N. Pour and M. R. Housaindokht, *Prog. React. Kinet. Mech.*, 2014, **39**, 137.

19 M. Ding, Y. Yang, B. Wu, T. Wang, L. Ma, H. Xiang and Y. Li, *J. Mol. Catal. A: Chem.*, 2011, **351**, 165.

20 X. Cui, J. Xu, C. Zhang, Y. Yang, P. Gao, B. Wu and Y. Li, *J. Catal.*, 2011, **282**, 35.

21 Y. Zhang, X. Wang, Y. Zhu, B. Hou, X. Yang, X. Liu, J. Wang, J. Li and T. Zhang, *J. Phys. Chem. C*, 2014, **118**, 1999.

22 L. Khana and N. K. Verma, *Mater. Sci. Eng., B*, 2013, **178**, 1230.

23 Q. Qiu, M. Liu, K. Sunada, M. Miyauchi and K. Hashimoto, *Chem. Commun.*, 2012, **48**, 7365.

24 J. Faye, E. Guélou, J. Barrault, J. M. Tatibouët and S. Valange, *Top. Catal.*, 2009, **52**, 1211.

25 J. W. Halloran and H. K. Bowen, *J. Am. Ceram. Soc.*, 1980, **63**, 58.

26 L. Torrente-Murciano, D. Mattia, M. D. Jones and P. K. Plucinski, *J. CO_2 Util.*, 2014, **6**, 34.

27 M.-D. Lee, J.-F. Lee and C.-S. Chang, *J. Chem. Eng. Jpn.*, 1990, **23**, 130.

28 X. Ma, D. Sun, Y. Qiwen, F. Weiyong and D. Fang, *J. Nat. Gas Chem.*, 2009, **18**, 354.

29 H. D. Willauer, R. Ananth, M. T. Olsen, D. M. Drab, D. R. Hardy and F. W. Williams, *J. CO_2 Util.*, 2013, **3–4**, 56.

Faraday Discussions

DISCUSSIONS

Bridging model and real catalysts: general discussion

Charles Campbell, Michael Bowker, Michail Stamatakis, Graham Hutchings, Philip Davies, James Earley, Mark Howard, Bruce Garrett, Femi Oloye, Elad Gross, Andrzej Kotarba, Miron V. Landau, Itai Panas, James Anderson, Bert Weckhuysen, Wilke Dononelli, Sarwat Iqbal, Burcin Temel, Ewa Nowicka, Albert Bruix, Christian Reece, Jacob Moulijn, Bruce Gates, Richard Catlow, Natasa Novak Tusar, David Willock, Daniel Wotton, Francisco Ivars Barcelo, Richard Palmer, Robert Madix, Cynthia Friend, Avelino Corma, Hans-Joachim Freund, Gareth Whiting and Joachim Sauer

DOI: 10.1039/C6FD90017H

Charles Campbell opened the discussion of the paper by Hans-Joachim Freund: If you have a 3D gold particle and it spreads out to be a 2D particle when you adsorb CO_2, it must gain energy stability. Did you estimate the energy change of the overall process to do that?

Hans-Joachim Freund responded: I note that the particles are two-dimensional without CO_2 exposure and oxalate formation, driven by the electron transfer from the Ag substrate to the Au nanoparticle. As for the energetics, we know that CO_2 desorbs when we heat to 340 K. Hannu Häkkinen has performed DFT calculations for the formation of oxalate from CO_2.[1]

1 F. Calaza, C. Stiehler, Y. Fujimori, M. Sterrer, S. Beeg, M. Ruiz-Oses, N. Nilius, M. Heyde, T. Parviainen, K. Honkala, H. Häkkinen and H.-J. Freund, *Angew. Chem., Int. Ed.*, 2015, **54**, 12484.

Philip Davies remarked: I understand from your paper that the gold particles grow in a 2D form only where there are electrons available from the metallic layer underneath the oxide. In a real catalyst could that negative charge be supplied from a defect in the oxide? If that were the case, presumably oxidation of the surface could then change the growth of the gold to a 3D mode? Finally, do you have any information about the differences in the shape of the 7 and 12 nm silver particles?

Hans-Joachim Freund replied: This is an important question in the context of the present study. The growth and morphology of the Au particles change as the thickness of the MgO is varied. For *ca.* 10 monolayers (ML) of MgO, the particles adopt a 3D morphology, while for 3ML of MgO they are 2-dimensional.[1] The reasons are connected with the polaronic response of ultrathin oxide films, as discussed by Pacchioni *et al.*[2,3] We demonstrated that the concept based on the action of the metal substrate in the film studies reported may be transferred to bulk oxides by doping the oxide[4] with suitable dopants, which act as a source for electrons. In that sense, reduced oxides with oxygen vacancies may act in the same way provided the energetics are appropriate. The second question is referring to the 7 and 12 nm Pd particles, I suppose. The particle shape is similar and they are basically truncated cubo-octaedra similar to Hansen *et al.*[5]

1 M. Sterrer, T. Risse, M. Heyde, H.-P. Rust and H.-J. Freund, *Phys. Rev. Lett.*, 2007, **98**, 206103.
2 G. Pacchioni, L. Giordano and M. Baistrocchi, *Phys. Rev. Lett.*, 2005, **94**, 226104.
3 D. Ricci, A. Bongiorno, G. Pacchioni and U. Landman, *Phys. Rev. Lett.*, 2007, **97**, 036106.
4 X. Shao, S. Prada, L. Giordano, G. Pacchioni, N. Nilius and H.-J. Freund, *Angew. Chem., Int. Ed.*, 2011, **50**, 11525.
5 K. Højrup Hansen, T. Worren, S. Stempel, E. Lægsgaard, M. Bäumer, H.-J. Freund, F. Besenbacher and I. Stensgaard, *Phys. Rev. Lett.*, 1999, **83**, 4120.

Andrzej Kotarba said: I was intrigued by the electron transfer through an electrical insulating MgO film. How effective can the electron transfer from the Ag substrate to Au cluster be through the magnesia film? How does it depend on the film's thickness? Can you observe it experimentally? By STM? What is the explanation for the more effective electron tunnelling to 2D clusters than to 3D? How does the electron feel the geometrical difference between these two cluster geometries through the insulating film?

Ewa Nowicka added: Does the thickness of the thin metal oxide affect the electron transfer from Ag to Au?

Hans-Joachim Freund responded: We have studied the electron transfer through a MgO film as a function of thickness.[1] Close to 10 monolayers, the MgO film behaves in a similar way to a bulk material, which shuts off the electron transfer because of the missing necessary response of the oxide film.[2,3] The explanation is given in the cited papers.

1 M. Sterrer, T. Risse, M. Heyde, H.-P. Rust and H.-J. Freund, *Phys. Rev. Lett.*, 2007, **98**, 206103.
2 G. Pacchioni, L. Giordano and M. Baistrocchi, *Phys. Rev. Lett.*, 2005, **94**, 226104.
3 D. Ricci, A. Bongiorno, G. Pacchioni and U. Landman, *Phys. Rev. Lett.*, 2007, **97**, 036106.

Albert Bruix said: From a theoretical modelling point of view, one always tries to minimize the thickness of the slab representing an extended surface in order to reduce the computational cost of the calculations. Thin oxide films are also routinely grown on metals as model catalysts of more bulk-like crystalline surfaces. However, the results presented in this paper (DOI: 10.1039/C5FD00143A) indicate that the properties of very thin films may significantly differ from surfaces of more bulk-like crystals, especially when in contact with an underlying metal. This raises concerns about how representative such thin slab models are for bulk-like crystal surfaces. At what thickness does a MgO film stop allowing charge transfer through it? Are there other oxide supports where this happens and where such thickness (or thinness) effects should be considered?

Hans-Joachim Freund answered: In general, it depends on the system in question. Some systems approach the bulk properties faster than others. CaO, for example, behaves in a similar way to MgO. TiO_2 appears to be different.

Graham Hutchings asked: Presumably you could use gold or other metals as an electron source as the core of the nanoparticle? What if you put Pd into the Au nanoparticle, presumably this could also form a pancake-shaped surface structure, but would there be any structure in this? Would the Pd segregate to the periphery? This could be interesting as you might be able to probe the hydrogenation of acrolein further.

Hans-Joachim Freund replied: This is an important point. We should try to extend the ideas tested for Au to Au–Pd alloys. I could easily imagine that segregation might happen. This has to be investigated further.

Bruce Gates said: Can you build defects into the MgO? What do you think you might learn from samples that incorporated controlled defects?

Hans-Joachim Freund answered: Defects may be built in by a number of methods. 1) Electron bombardment to create oxygen defects, *e.g.* color centers. 2) Doping. This will lead to specific modifications of the materials chemistry, *e.g. via* electron transfer.

Mark Howard commented: You described the initial electron transfer to form a 'solvated' carboxylate dimer ion as 'downhill' by 1.6 eV. A second electron transfer then leads to the formation of surface oxalate. Are further reactions of the oxalate energetically feasible, or is it too stable?

Hans-Joachim Freund responded: We are in the process of studying the reactivity of the oxalate at the rim in detail using a number of reactants, such as hydrogen, water, and ammonia. The results are not yet available. What is perhaps interesting to note, is that the oxalate decomposes into two CO_2 molecules and not into the generally more stable carbonate by releasing CO.[1]

1 F. Calaza, C. Stiehler, Y. Fujimori, M. Sterrer, S. Beeg, M. Ruiz-Oses, N. Nilius, M. Heyde, T. Parviainen, K. Honkala, H. Häkkinen and H.-J. Freund, *Angew. Chem., Int. Ed.*, 2015, **54**, 12484 (see supplementary information).

Avelino Corma asked: What about using a semiconductor, generating your negative charges and then reducing your CO_2?

Hans-Joachim Freund replied: I guess this could be possible if the electronic conditions are chosen appropriately.

Michael Bowker remarked: We might expect that for the 'pancake', gold would reflect the oscillating potential of the oxide to which it is directly attached. Have you got any evidence from STM for such an oscillation of the electron density?

Hans-Joachim Freund responded: You are right in noting that scanning probe techniques work best on flat samples, and have difficulties imaging three-dimensional objects. However, depending on how well the tip is prepared and how large the distance is between nanoparticles, it is possible to not only image the top facet, but, as has been shown in a paper on Pd/alumina nanoparticles,[1] the side facets may also be imaged.

1 K. Højrup Hansen, T. Worren, S. Stempel, E. Lægsgaard, M. Bäumer, H.-J. Freund, F. Besenbacher and I. Stensgaard, *Phys. Rev. Lett.*, 1999, **83**, 4120.

Elad Gross enquired: Can the structural reconstruction of the supported Au nanoparticles from 3D to 2D be reversed once CO^2 is removed from the Au surface?

Hans-Joachim Freund replied: If we do not go beyond 500 K, then the removal of the formed oxalate occurs through decomposition into CO^2 without changing the morphology of the particles. If temperatures beyond this are used, then morphological changes will occur.

Bruce Gates opened the discussion of the paper by Albert Bruix: Can you extend your method to obtain a support that is closer to that of a real (industrial) catalyst? If you go to the more complex structure of a real catalyst, is there any technique that characterises the reactivity of those edge sites in a more complex environment?

Albert Bruix replied: From the computational modelling point of view, a straightforward next step would be to replace the Au(111) slab with another periodic slab of γ-alumina, which is the support used in the "real" industrial catalyst.

The same DFT-based methods and strategies used for the Au(111)-supported system can also be routinely applied to systems with an alumina support. A difficult but interesting question to address for such models is if a MoS_2 particle would lie flat down or stand up on an oxide, and perhaps if there is a size-dependence for such behaviour. This is of pivotal importance since steric effects, which are strong for flat-lying MoS_2, might be reduced for vertical MoS_2.

From the experimental point of view, one faces the well-known pressure, temperature, and materials gap with respect to the "real" industrial case.

The materials gap is difficult to overcome, since using alumina supports makes it more challenging to synthesise the MoS_2 nanoparticles with well-defined structures and, in addition, the insulating characteristic of alumina does not allow us to use STM to study the atomic and electronic structure of the model systems. However, other imaging techniques, such as atomic force microscopy (AFM) and transmission electron microscopy (TEM), can be used to probe such structures instead. Actually, good models for the γ-alumina surface have been prepared recently,[1] which may allow characterization of nanoparticles, including MoS_2, on such surfaces, using atom-resolved AFM. The pressure gap can be partially overcome by *in situ* spectroscopic experiments at elevated reactant (*e.g.* H_2) pressures, which bring the model catalyst closer to *in operando* conditions. Such deviations from the typical UHV conditions allow us to explore how the spectroscopic fingerprints (from *e.g.* XPS, XRD, or IR spectra) of the different sites evolve, but under high temperatures relevant for catalysis it is difficult to image the catalyst structure with atomic sensitivity with any available method.

1 T. N. Jensen, K. Meinander, S. Helveg, A. S. Foster, S. Kulju, T. Musso and J. V. Lauritsen, *Phys. Rev. Lett.*, 2014, **113**, 106103.

Richard Palmer remarked: I understand the "molybdenum" edges of your triangular clusters are terminated with extra sulfur atoms. We have made more or less stoichiometric MoS_2 clusters where we do not supply extra sulfur.[1] Can you please elaborate on the role that the extra sulfur plays in the properties of the "molybdenum" edges and what properties we might expect without it?

1 M. J. Cuddy, K. P. Arkill, Z. W. Wang, H.-P. Komsa, A. V. Krasheninnikov and R. E. Palmer, *Nanoscale*, 2014, **6**, 12463.

Albert Bruix responded: It should be noted that MoS_2 nanoparticles terminated with Mo-edges and 100% S coverage are over-stoichiometric (*i.e.* $Mo_{21}S_{66}$, $Mo_{36}S_{104}$, and $Mo_{55}S_{150}$, for sizes $n = 6$, 8, and 10, respectively). In turn, the same nanoparticles with 50% S coverage are also over-stoichiometric, but closer to the MoS_2 ratio (*i.e.* $Mo_{21}S_{48}$, $Mo_{36}S_{80}$, and $Mo_{55}S_{120}$, for sizes $n = 6$, 8, and 10, respectively). The relative stability of the different edge terminations of MoS_2 has been extensively studied by means of DFT calculations and it is well known that both the "Mo-edges" and "S-edges" of MoS_2 nanoparticles are S-terminated under any range of T and p where the nanoparticles are stable.[1] Under reducing conditions, edges with lower S content (*e.g.* with 50% S) become stable, but the Mo-terminated edges (*i.e.* 0% S) are very unstable and never appear. This means that under very reducing environments the MoS_2 nanoparticles would partially or completely decompose into metallic Mo instead of exposing Mo-terminated edges. This indicates that the MoS_x clusters you have made are also S-terminated and over-stoichiometric, although obviously for the larger clusters with larger basal planes the ratio is closer to the MoS_2 case. In fact, S-terminated nanoparticle models (with 100% S) have also been used in ref. 2 to calculate the edge energy of the MoS_2 flakes.

With respect to the role of the extra sulfur, the edges of MoS_2 nanoparticles usually constitute the active sites of MoS_2-based catalysts and, therefore, their sulfur coverage determines, to a large extent, the catalytic activity. In fact, the 50% S Mo-edges have recently been recognised as the relevant edge-phase in industrial-style catalysts.[3] We have also given a detailed account of how different properties change with edge structure and composition (see DOI: 10.1039/C5FD00203F and ref. 4). In particular, we describe the changes in the metallic character of the edges upon reduction and how this affects the XPS fingerprint of the edge sites. We also report on the S-vacancy formation energy and H adsorption energies at nanoparticles with different edges, showing that these relevant descriptors for HDS and HER activity strongly depend on both the edge termination direction and the amount of sulfur at the edge.

1 M. V. Bollinger, K. W. Jacobsen and J. K. Nørskov, *Phys. Rev. B*, 2003, **67**, 085410.
2 M. J. Cuddy, K. P. Arkill, Z. W. Wang, H.-P. Komsa, A. V. Krasheninnikov and R. E. Palmer, *Nanoscale*, 2014, **6**, 12463.

3 L. P. Hansen, Q. M. Ramasse, C. Kisielowski, M. Brorson, E. Johnson, H. Topsøe and S. Helveg, *Angew. Chem., Int. Ed.*, 2011, **50**, 10153.
4 A. Bruix, H. G. Füchtbauer, A. K. Tuxen, A. S. Walton, M. Andersen, S. Porsgaard, F. Besenbacher, B. Hammer and J. V. Lauritsen, *ACS Nano*, 2015, **9**, 9322.

Michael Bowker asked: Why choose gold as the support for the MoS_x particles? And what effect does the Au have on the electronic charge distribution compared with a bare slab?

Albert Bruix replied: There are several reasons to choose Au as the support material for studying the properties of MoS_2 nanoparticles in model systems. The preferential deposition of metallic Mo on the elbows of the herringbone reconstruction of Au(111) facilitates the synthesis of well-dispersed and relatively similar-sized MoS_2 nanoparticles upon exposure to H_2S. Furthermore, the interaction between Au(111) and MoS_2 is also sufficiently weak to prevent strong epitaxial effects during growth of the MoS_2 particles/layers. At the same time, such interaction is sufficiently strong as to prevent nanoparticle stacking, even at a relatively high coverage. Alternatively, growth on less (more) strongly interacting substrates, such as graphite (TiO_2), leads to stacking (epitaxial growth). With respect to the MoS_2–Au interaction, the main effect between Au(111) and the most abundant type of nanoparticle, the completely sulfided Mo-edge, takes place at the interface between the basal plane (the central region) of the nanoparticle and the metal surface. Charge redistribution therein gives rise to charge transfer from Au(111) to the MoS_2 nanoparticle and to the formation of an interface dipole.[1] The interaction between single-layer MoS_2 and Au(111) also leads to hybridisation between the Au and MoS_2 bands and to a band gap renormalization for MoS_2.[2] Recently reported adsorption energies on supported stripe models[3] also suggest that such charge transfer might also change the reactivity of the edges. Furthermore, very recent results, which will be presented in an upcoming report,[4] indicate that the interaction between completely sulphided S-edges and the Au(111) surface is quite strong, which greatly stabilizes nanoparticles with S-edges.

1 M. Farmanbar and G. Brocks, *Phys. Rev. B*, 2015, **91**, 161304.
2 A. Bruix, J. A. Miwa, N. Hauptmann, D. Wegner, S. Ulstrup, S. S. Grønborg, C. E. Sanders, M. Dendzik, A. Grubišić Čabo, M. Bianchi, J. V. Lauritsen, A. A. Khajetoorians, B. Hammer and P. Hofmann, *Phys. Rev. B*, 2016, **93**, 165422.
3 C. Tsai, F. Abild-Pedersen and J. K. Nørskov, *Nano Lett.*, 2014, **14**, 1381.
4 A. Bruix and B. Hammer, manuscript in preparation.

Michail Stamatakis commented: With extended structures of MoS_2, there is a transition in which Mo changes its coordination from octahedral (1T phase) to trigonal prismatic (2H phase). You showed results on the 2H phase for your nanoparticles, but have you considered simulating the 1T phase as well? If so, what is the energy of the transition between the two phases, and what is known about the properties of the 1T nanoparticles?

Albert Bruix responded: The 2H to 1T phase transition is usually induced by electron-doping extended MoS_2 structures. This follows from the energy of the conduction band minima, which is lower for the 1T phase (which is metallic) than for the 2H phase (which is semi-conducting). In addition to the MoS_2 nanoparticles of the 2H phase, we have also considered 1T nanoparticles of different sizes and sulfur-contents, although these results are still unpublished.[1] Our results indicate that the stable nanoparticles of the different phases (and terminations) contain different numbers of sulfur atoms. Therefore, the transition energies between 2H (Mo-edge and S-edge) and 1T nanoparticles depend on the chemical potential of S. In particular, at high chemical potentials (oxidizing, S abundant conditions), 2H nanoparticles are significantly more stable, with transition energies of around 10 eV (for the $n = 8$ particle, for example). In turn, at a subsequently lower chemical potential of S, the transition energies become smaller and, at very low potentials (reducing, sulfur deficient conditions), the 1T nanoparticles are even more stable. This is consistent with the induction of phase transitions by doping with electrons, which in our computational models is achieved by reduction (sulfur removal). Interestingly, these 1T nanoparticles are metallic just like the extended 1T single layer, which results in very low calculated Mo3d core-level-shifts close to those of metallic Mo. To the best of my knowledge, these nanoparticles have not been identified in experiments, but the extended single layers of the 1T phase have been shown to be active for the hydrogen evolution reaction.

1 Manuscript in preparation.

Burcin Temel said: In the presence of Au under the MoS_2 clusters, even though the energy levels change in terms of H_2 binding, you still see the same pathways and mechanistic schemes regardless of whether or not you have gold for the hydrotreating activity. The hydrogenation pathway is still on the moly edge. The support effect may ultimately change the edge reactivities for the hydrogen evolution reaction but the hydrotreating reaction mechanisms remain unaffected, as also seen in previous STM, DFT and IR studies.

Albert Bruix responded: To the best of my knowledge, the effect of the underlying metallic support on the hydrotreating mechanism has not been addressed by means of DFT calculations; virtually all studies consider free-standing stripe or nanoparticle models. One very recent article[1] addresses the effect of Au on the edge reactivity using stripe models. It reveals that hydrogen adsorption and, by extension, the HER activity, are significantly affected by the presence of the metallic support. This suggests that Au(111) could have an effect on the HDS activity of the MoS_2 edges as well.

In turn, STM studies of the MoS_2/Au(111) model catalysts have extensively and successfully characterised the structure of un-promoted and promoted MoS_2 nanoparticles prepared under different conditions. This technique also allows visualisation of the bonding modes of different molecules with the two types of particle edges, confirming that these are the active sites. However, information regarding the effect of the Au(111) substrate on the edge activity is not available from STM experiments.

1 C. Tsai, F. Abild-Pedersen and J. K. Nørskov, *Nano Lett.*, 2014, **14**, 1381.

Miron V. Landau commented: It is well established that in sulphided Co(Ni) Mo/Al_2O_3 catalysts, the MoS_2 phase exists in the form of curved single long slabs standing vertically at the alumina surface due to the interaction of its reactive edge plane with the reactive moieties at the surface of the support. In your experiments, using non-reactive gold plates (not nanoparticles) as a support for single MoS_2 slabs brings them to their flat orientation since the contacting of two non-reactive surfaces – gold and the basal plane of the MoS_2 slab – reduces the system's surface in a maximal way. Is the orientation of the MoS_2 slabs at the support's surface essential for their performance in hydrodesulfurization? The novel catalysts may be prepared on supports with non-reactive surfaces, depositing pre-prepared small MoS_2 slabs. Is this a real challenge in the preparation of MoS_2-based catalysts?

Albert Bruix replied: It is indeed known that Co and Ni-promoted MoS_2 forms extended and curved layers on alumina, which erratically align (not just vertically). However, the structure of such Co(Ni)-MoS_2/Al_2O_3 catalysts is actually still unclear, mainly because the techniques used to probe these systems lack single-atom sensitivity. The edge stoichiometry and structure of industrial-style catalysts can instead be studied for flat-lying nanocrystals on a carbon support.[1] The edge-structure of carbon and alumina-supported nanocrystals is expected to be the same, but the different alignment is likely to affect the hydrodesulfurisation activity. In particular, for flat-lying MoS_2 nanoparticles, some bonding modes of large S-containing molecules (see edge adsorption *vs.* brim adsorption in ref. 2) may be sterically hindered by the interaction with the support, independent of it being Au(111) or carbon. In fact, the main challenge in HDS catalysts consists of removing S heteroatoms from large organic molecules, such as 4,6-dimethyldibenzothiophene, which contain the last remaining ppm of sulfur in ultra-low sulfur diesel (ULSD). Thus, from a synthetic point of view, one of the goals is to prepare catalysts with maximal exposure of edges and where neighbouring layers are far enough away to avoid steric repulsions that make adsorption of such large molecules unfavourable.

1 Y. Zhu, Q. M. Ramasse, M. Brorson, P. G. Moses, L. P. Hansen, C. F. Kisielowski and S. Helveg, *Angew. Chem., Int. Ed.*, 2014, **53**, 10723.
2 S. Rangarajan and M. Mavrikakis, *ACS Catal.*, 2016, **6**, 2904.

Cynthia Friend opened the discussion of the paper by Elad Gross: The technique you describe is very powerful. Is it possible to adapt it to a laboratory scale or is a synchrotron radiation source required?

Elad Gross responded: It is possible to adapt the near-field IR technique to a laboratory scale by using an IR laser as a source (see, for example, ref. 1). However, the broadband emission of a synchrotron source provides a wide range near-field IR spectrum, while only a limited part of the IR spectrum can be detected with an IR laser source (according to the laser's emission range).

1 B. Pollard, E. A. Muller, K. Hinrichs and M. B. Raschke, *Nat. Commun.*, 2014, **5**, 3587.

Graham Hutchings said: You are using a Pt nanoparticle that is 100 nm in diameter. For most catalytic reactions this would be inactive. If you could use a smaller nanoparticle then the ratio of the edge sites (which you indicate are active) to the bulk sites will increase and hence your sensitivity can increase. Is this something you have plans for?

Elad Gross responded: In this project we studied the reactivity of molecules on the surface of Pt nanoparticles in the size range of ~100 nm, which are larger than the industrial catalytic particles that are in the size range of 1–10 nm. This compromise was made due to the limited spatial resolution in IR nano-spectroscopy measurements, which is determined by the tip diameter (~20 nm). By studying larger nanoparticles as a model system, we gained the capabilities to differentiate between the chemical reactivity of surface atoms that reside on the center of the nanoparticle *versus* atoms that reside on the perimeter of the particle. This capability could not be gained while studying smaller, industrial-like catalytic nanoparticles. An IR nanospectroscopy signal from molecules that are anchored to the surface of small nanoparticles (<10 nm) can be detected, however, it will provide only an averaged IR signal from all the surface adsorbed molecules with no capability to distinguish between the different adsorption sites.

Bert Weckhuysen commented: This is a very impressive study that shows what is currently possible when you make use of tunable IR synchrotron radiation and couple it to an AFM setup to investigate a chemical reaction. Clearly, this is pushing the spatial resolution down to the nanometer regime and puts it in the field of nano-spectroscopy, similar to what one can do with Raman, *i.e.* TERS (Tip Enhanced Raman Spectroscopy). I am wondering what the time resolution is of your measurements? Assume when you would be willing to scan over a 100 nm × 100 nm domain. What would be the time resolution to monitor such a catalytic process with the proposed IR technique? What about the sensitivity of the method as you use a special molecule attached to the metal nanoparticle? What other molecules could you use with this methodology? How does this methodology compare with the TERS approach, which has already been applied to photo-catalytic systems?

Elad Gross answered: 1. Time resolution: Each single point measurement in this method takes a few minutes. Therefore, it is not possible to directly detect a chemical reaction on surfaces while using synchrotron IR and AFM. Since acquiring a single point measurement requires between 1 and 20 min, we are currently limited to line scans at the synchrotron. Scanning an area of 100nm × 100nm will require a few shifts and also thermal drift will start to become a real problem while scanning large areas. There are other approaches that use femtosecond IR lasers coupled with AFM, by which both high temporal and high spatial resolution can be achieved. However, in this case, the IR spectrum is limited to the IR frequencies of the laser source and only a fraction of the IR spectrum can be studied while using this approach.[1]

2. Sensitivity: The method can detect various molecular vibrations on surfaces. It is easier to detect highly IR-absorbing bonds, such as carbonyl and O–H bonds, but the detection of C–H bonds is also feasible.

3. Tip enhanced IR *versus* TERS: Tip enhanced IR is complementary to TERS. There are a few advantages for coupling IR with AFM: (1) The cross section for IR absorption is much higher than that of Raman and, as a consequence, a lower flux of photons is required in this measurement. (2) Unlike Raman, IR absorption is a linear phenomenon and therefore the locally enhanced IR signal is less dependent on the structure of the AFM tip. The main drawbacks of tip enhanced

IR are the need for a synchrotron source for broadband IR measurements and the currently limited temporal resolution.

1 V. Kravtsov, R. Ulbricht, J. M. Atkin and M. B. Raschke, *Nat. Nanotechnol.*, 2016, **11**, 459.

James Anderson said: My question relates to the sensitivity of your experiment. Given the signal to noise ratio in the spectra you display for carbonyl (C=O) and O–H stretching modes, do you envisage, given the much lower absorption coefficients, that you would be able to detect a feature due to *e.g.* a C–H or C–C stretching mode?

Elad Gross responded: The relatively high cross-section for IR absorption in C–O and O–H stretching modes makes it easier to study changes in these stretches with our approach. Detection of C–H and C–C stretches will be more challenging, though feasible while using a metal surface (*e.g.* Au) that will induce improved local AFM-enhanced IR signals.

Philip Davies asked: I believe the approach that you describe could be a very important tool in catalysis and thin film research but I wonder how the sensitivity of the infrared spectroscopy differs between the oxide surface and the metal nanoparticle? I note that the XP spectra of the N(1s) region very clearly show more than one nitrogen state. It is unlikely, in my view, that these differences arise from the changes to the oxidation state of the OH group since that is some distance from the nitrogen in the N-heterocyclic carbenes. I wonder whether the XPS is detecting adsorbates on the oxide surface, which is not showing up in the infrared spectra because of a lack of sensitivity to vibrations on that surface. Have you looked at the adsorption of the N-heterocyclic carbene on clean silica surfaces with XPS?

Elad Gross replied: 1. When comparing the IR sensitivity to adsorbates on the metal and the oxide surfaces, the IR sensitivity is higher on the metal surface.

2. Both AFM (Fig. 6 in DOI: 10.1039/C5FD00194C) and XPS measurements did not detect any N(1s) signal (which is induced by N-heterocyclic carbenes) on the bare Si surface. These results indicate that there is no adsorption of N-heterocyclic carbenes on the bare Si.

3. The width of the N(1s) XPS signal in our results is similar to that in previously published XPS data.[1] Different Au adsorption sites might change the N–Au interaction and therefore increase the XPS peak width. In order to check this effect, additional experiments are planned in which we will study the XPS signal of carbenes that will be attached to the surface of well-defined Au nanocrystals.

1 C. M. Crudden, J. H. Horton, I. I. Ebralidze, O. V. Zenkina, A. B. McLean, B. Drevniok, Z. She, H.-B. Kraatz, N. J. Mosey, T. Seki, E. C. Keske, J. D. Leake, A. Rousina-Webb and G. Wu, *Nat. Chem.*, 2014, **6**, 409.

Michael Bowker enquired: You heat the Si crystal oxide layer to a very high temperature, is it in air or a vacuum? This is important because you are close to the temperature at which the Si oxide grows in air, which will probably be accelerated by the presence of Pt. Silica grows up and can encapsulate the nanoparticles.[1]

1 M. Bowker, J. J. Crouch, A. F. Carley, P. R. Davies, D. J. Morgan, G. Lalev, S. Dimov and D.-T. Pham, *J. Phys. Chem. C*, 2013, **117**, 21577.

Elad Gross responded: Our samples were prepared by annealing at 550 °C in air. No major changes were observed in the Pt XPS signal following this treatment. However, when the annealing temperature was increased to 800 °C, we detected a decrease in the Pt XPS signal, which indeed can be correlated to coating of the Si supported Pt nanoparticles with a thin silica layer.

Richard Palmer said: Regarding the general issue of experimental developments for imaging catalysts, it is of course true that all experimental techniques have their limitations. In the field of atomic-resolution electron microscopy, the need to employ the thin samples required by transmission electron microscopes is one such limitation. A powerful development for the future would be the widespread availability of aberration-corrected electron microscopy in reflection mode, "super-SEM", if you like, which would allow catalyst particles on solid supports to be characterised to atomic resolution.

Elad Gross answered: It should be noted that HR-TEM tomography measurements can detect the 3D structure and composition of small (~100–200 nm) mesoporous crystals and detect the structural properties of nano-sized catalytic particles that reside within the pores with a resolution of few nm. Therefore, this method can detect the properties of real-life catalysts.

Itai Panas asked: We learnt about possible different surface chemical bonds to a Pt nanoparticle if the chemisorption is centered on a facet or if it occurs at an edge of the nanoparticle. As evidence for this, spatially resolved IR spectral fingerprints for the two situations were presented. Is the alternative interpretation, that the IR spectra are orientation dependent and that the character of the actual surface chemical bond is instead the same for the two sites, also possible?

Elad Gross replied: Differences in the IR signals from the edge and the centre of the nanoparticle were only obtained once the sample was exposed to mild oxidising or reducing conditions. The IR signals of molecules that were anchored to the edge and centre of the nanoparticle were similar following exposure of the sample to harsher oxidising or reducing conditions. These results indicate that there is no major impact of the anchored molecules' orientation on their absorption properties and the changes in the IR signals at different surface locations were induced by dissimilarities in the chemical properties of surface anchored molecules.

Francisco Ivars Barcelo communicated: Taking into account that the Pt nanoparticles, scanned by the IR nanospectroscopy technique, are deposited on a planar crystalline surface, is it not the angle of incidence, formed between the long-axis of the AFM tip and the scanned surface, limiting the observed absorption intensity, according to the regular selection rules for dipole excitation of vibrational modes by infrared light on reflective surfaces? If so, how do you deal with the AFM tip scanning angle and the location of the IR detector?

Elad Gross communicated in response: The AFM tip polarizes the electric field along the long axis of the tip. Therefore, in this measurement we are primarily sensitive to vibrational modes that have a dipole moment parallel to the tip. In structured systems, such as highly ordered self-assembled monolayers on single crystal surfaces, vibrations with a dipole moment that is parallel to the surface will not be detected (see, for example, ref. 1). However, in amorphous samples or in samples in which the molecules are not well-ordered on the surface, the different vibrations will be detected.

In the presented research (DOI: 10.1039/C5FD00194C), the nanoparticles are not in their equilibrium state and, as detected by AFM measurements, they do not have highly ordered facets (a Wulff-like structure). Therefore, it can be concluded that in a single point measurement (an area of 20nm × 20nm) the imaged surface is not constructed from a single highly-ordered crystal and the surface-anchored carbenes are not well-ordered on the metal surface. As a consequence, the different vibrations of surface-anchored molecules will be detected by IR measurements.

1 I. Amenabar, S. Poly, W. Nuansing, E. H. Hubrich, A. A. Govyadinov, F. Huth, R. Krutokhvostov, L. Zhang, M. Knez, J. Heberle, A. M. Bittner and R. Hillenbrand, *Nat. Commun.*, 2013, **4**, 2890.

Wilke Dononelli opened a general discussion: In the literature, the size and the shape of nanoparticles are often discussed. Also, the talks in this session showed the influence of the size and the shape of nanoparticles on their general or catalytic capabilities. What is the influence of the distance between the nanoparticles? Do you see any effects in your systems?

Michael Bowker replied: The distance between particles can have important consequences in catalysis. Too few particles and the activity may not be optimised. Too many has a number of consequences. There are spillover and reverse spillover processes where communication between the metal and the support affects activity. If the support area becomes too small, then deactivation may occur. However, also one of the most important causes of plant closure is sintering. Here, the active surface area is lost due to a decline in the number of the smallest particles, leading to an increased maximum in the particle size distribution, and a reduction in the integral in that distribution (that is, the total number of particles per unit area of support). This can happen right at the beginning of the plant operation (due to coalescence of too proximate particles), while longer term sintering is due to surface-mediated Ostwald ripening, occurring by interparticle atomic diffusion across the support.

Elad Gross answered: It was recently demonstrated by the Somorjai group[1] that hydrogen atoms dissociated on Pt nanoparticles may be transferred to neighboring Co nanoparticles *via* long-distance (~100 nm) hydrogen atom spillover. These measurements revealed that different metallic nanoparticles do not have to be in high proximity in order to activate cascade reactions. However, if the lifetime of the intermediate is short, then the distance between the nanoparticles should be minimized (until a specific distance in which coalescence between neighboring nanoparticles might occur).

1 S. K. Beaumont, S. Alayoglu, C. Specht, W. D. Michalak, V. V. Pushkarev, J. Guo, N. Kruse and G. A. Somorjai, *J. Am. Chem. Soc.*, 2014, **136**, 9898.

Hans-Joachim Freund responded: The distribution of nanoparticles can play an important role in a number of aspects, for example in the delivery of educts and the diffusion of products due to the presence of a collection zone around a particle, or with respect to sintering through Ostwald ripening. We have not studied these aspects for the present system in a systematic way, but with respect to other systems studied, we do not expect the first aspect to play a very important role. There is no considerable sintering in the studied temperature range.

Albert Bruix replied: The distance between neighbouring MoS_2 nanoparticles is quite relevant to the hydrodesulfurization activity, since the edges of flat-lying or stacked MoS_2 layers should be far enough apart to avoid steric repulsions that make adsorption of large S-containing molecules unfavourable. In fact, as mentioned previously, the main challenge in HDS catalysts consists of removing S heteroatoms from large organic molecules, such as 4,6-dimethyldibenzothiophene, which contain the last remaining ppm of sulfur in ultra-low sulfur diesel (ULSD).

Jacob Moulijn continued the general discussion: The shell (molybdenum oxide)–core (hematite) systems that you describe are interesting model catalysts. Could they be transformed into industrial catalysts? Because only the outer surface will be covered, I assume that it is not possible to do this by using a porous support as is often done in heterogeneous catalysis. Could you envisage other techniques for catalyst production, such as layered manufacturing methods (for instance 3D-printing)?

Michael Bowker replied: I think impregnation into the oxide powder could be used, as we have done, followed by the usual processing steps (drying, adding binder, pelleting, calcination). Alternatively, maybe impregnation into the prepared (calcined) pellets could be achieved, although homogeneity here would probably be a problem. The fundamental catalyst particles are not porous, but are solid iron oxide nanoparticles of ~20–50 nm, depending on the preparation method, and so impregnation, combined with appropriate calcination conditions, coats all of the surface of the material. Regarding scale-up, this is not something we generally deal with in academia and at present we certainly have not scaled-up in the sense of making pellets in significant number. All I can say is that the type of method described above is, I believe, how the current industrial catalysts are made – by converting powder into pellets. As described elsewhere, I would envisage that in the future web-based methods will be used to make such catalysts, and could avoid using aqueous streams, thereby reducing pollution burdens and cost burdens from contaminated aqueous effluent.

Graham Hutchings commented: All three speakers in this session have used model materials in which the nanoparticles or the nanoclusters have a very strong interaction with the support. However, it is known that the support can influence the nanoparticle immensely. For example, supported nanoparticles prepared from the same colloidal sol have very little interaction with the support if supported on carbon and show no faceting, whereas on titania the nanoparticles are faceted but weakly bound and on zinc oxide they are faceted and strongly interact with the support.[1] Is it possible that there are techniques that can probe such materials?

1 J. Pritchard, L. Kesavan, M. Piccinini, Q. A. He, R. Tiruvalam, N. Dimitratos, J. A. Lopez-Sanchez, A. F. Carley, J. K. Edwards, C. J. Kiely and G. J. Hutchings, *Langmuir*, 2010, **26**, 16568.

Hans-Joachim Freund responded: There tends to be a moderate interaction between the tips and substrates for scanning tunnelling microscopy, so that very weakly bound species may be particularly strongly influenced. There are microscopes that allow us to probe the tunnelling current or the tip surface force (in atomic force microscopy) at the same point on the substrate with a possibly smaller influence of the latter on the surface species (see *e.g.* ref. 1).

1 H.-P. Rust, M. Heyde and H.-J. Freund, *Rev. Sci. Instr.*, 2006, 77, 043710.

Elad Gross answered: Tip dependent measurements, such as AFM and STM, mainly probe nanoparticles that are strongly bound to the surface, otherwise the nanoparticles will drift on the surface due to the scanning tip. Light tapping mode AFM measurements can be used to minimize the tip–surface interactions and to detect weakly surface-bound nanoparticles. However, while coupling AFM and IR spectroscopy, the longer distance between the AFM tip and the surface in light tapping measurements will reduce the IR signal that is induced by molecules that are adsorbed on the supported nanoparticles.

Albert Bruix responded: Please see my previous responses to the questions regarding the effect of the Au substrate on the structure and properties of MoS_2 catalysts. I should also add that STM measurements allow characterisation of the structural differences between MoS_2 particles on different supports, as long as the latter has a well-defined structure and is not insulating. When insulating supports, such as γ-alumina, are used, other techniques with atomic sensitivity, such as AFM, can be used instead.

Bruce Garrett opened the discussion of the paper by Cynthia Friend: You showed convincingly that van der Waals' interactions are critically important for controlling the selectivity on gold nanostructures. How general is this finding? Does it apply to oxide surfaces?

Cynthia Friend answered: As discussed in our paper (DOI: 10.1039/C5FD00192G), the hierarchy for oxide surfaces is different than for coinage metals, although there is only limited data. Since oxides are less polarizable and have very different bonding, it is likely that both the primary bonding to the surface and the van der Waals' interactions will be different. This is an interesting topic for further study.

Philip Davies commented: This paper (DOI: 10.1039/C5FD00192G) provides a very thought provoking insight into the surface reaction steps involved in the oxidation of alcohols at gold surfaces with possibly wide ranging implications.

The ester product measured depends on two kinetically controlled steps: the reaction of the alcohol with a surface oxygen and a β-hydride elimination, but as I understand the paper, your proposal is that despite the higher acidity of the fluorinated molecule and the lower probability of it undergoing β-hydride elimination, it is the stability of the alkoxy intermediate on the surface that controls the relative ratio of fluorinated molecules in the ester product. Whilst this trend is consistent with the DFT calculations, I am surprised by the size of the effect and I wonder whether there could be some other effect going on. In a study of the analogous fluorinated and non-fluorinated amines reacting with oxygen on a copper surface, we saw a change in reaction mechanism due to the basicity of the reactant affecting the structure of the surface intermediate.[1] Can you rule out structural effects, such as a change in the islanding of the alkoxy intermediates?

1 P. R. Davies, D. Edwards and M. Parsons, *Surf. Sci.*, 2007, **601**, 3253.

Cynthia Friend replied: We have not yet studied islanding of the trifluorethoxy. Unpublished studies just completed in our lab for ethoxy and methoxy on Au(110) show that there is no islanding. Your point is a good one, though, in that we do observe islanding of acetate on Au(110). This work is under review.

Christian Reece asked: Which method of dispersion correction do you use in your method? Have you considered the Grimme's D3 dispersion correction, which seems to decrease the degree of over prediction of binding energies?

Cynthia Friend answered: We used the DFT+vdWsurf method33, an approach to model adsorbates on surfaces, which goes beyond the pairwise approximation by including the collective response of the substrate electrons in the calculation of the adsorbate–substrate van der Waals' interactions. This method combines the Tkatchenko–Scheffler (TS) DFT+vdW method[1] for intermolecular interactions with the Lifshitz–Zaremba–Kohn (LZK) theory.[2,3]

1 A. Tkatchenko, M. Scheffler, *Phys. Rev. Lett.*, 2009, **102**, 073005.
2 E. M. Lifshitz, *Sov. Phys. JETP.*, 1956, **2**, 73.
3 E. Zaremba, W. Kohn, *Phys. Rev. B*, 1976, **13**, 2270.

Charles Campbell commented: In our TPD studies of alkanes on MgO(100), Pt(111) and graphite(0001), we find that the heats of adsorption increase by a few kJ mol^{-1} for each CH_2 group you add to the alkane due to van der Waals' interactions. What about the CF_3 group you added at the end of your alkoxys. Do you know what contribution that has to the adsorption energy? I am interested in the stability difference predicted from the experimental ratio of populations, as well as the DFT energy estimation.

Cynthia Friend responded: As you noted, the addition of addition CH_2 groups increases the interaction energy with the surface. In the case of alkoxides bound to Au(111), the increase is on the order of 0.1 eV per methylene group. The CF_3 group decreases the interaction energy with the Au(111) surface in the case of trifluorethanol. The reason for this is that there is Pauli repulsion between the electron rich trifluoro group and the surface, which leads to a different geometry.

Christian Reece said: Including the van der Waals' dispersion correction always over predicts binding energies. Is it possible that you are seeing a trend in your dispersion correction functional rather than a trend in the chemistry? In other words, is the dispersion correction method saturating the chemical interactions giving false trends? This is a common problem with dynamics in which if your thermo/barostat is too strong, the chemistry you see is a function of your thermo/barostat rather than the "real" chemistry.

Cynthia Friend answered: Without the van der Waals' correction, the binding energies for the homologous series of molecules – alkoxides bound to Au(111) – are essentially the same. When the van der Waals' correction is added, the trends are consistent with the experimental measurements. While the calculations may not be quantitatively accurate, they do reproduce experiments, indicating that the trends are due to the van der Waals' interactions of the alkyl groups with the surface.

Itai Panas commented: In second-order Møller–Plesset perturbation theory (MP2), the dispersion interaction between two subsystems emerges as coupled virtual transition moments between occupied and unoccupied orbitals on either subsystem, *i.e.* coupled virtual single-excitations. The interaction is composed of an attractive and a repulsive contribution. The attractive part has the leading term being of the order $(R^{-3})^2$ due to its dipole–dipole transition moment origin. It is strongly attractive at short inter-system distances. The repulsive part has a Pauli exclusion principle type origin. It states that the virtual excitations on either subsystem suffer from the overlap between unoccupied orbitals on the one subsystem with the occupied orbitals of the other. Effectively, this boils down to inter-subsystem distance dependent polarisabilities of either subsystem. If this is not treated properly, *e.g.* if only inter-subsystem Pauli repulsion between occupied Kohn–Sham states is included, erroneous conclusions may be drawn. A sensitivity test of the claimed detailed impact of the dispersion interaction on the results would be to employ the structures obtained by dispersion free DFT and subsequently augment these calculations by the dispersion contribution while not allowing for the structures to relax further at this step. Would this be possible in this case?

Cynthia Friend answered: This is an interesting suggestion for further study.

Michael Bowker queried: Methoxy is less stable than ethoxy; as well as van der Waals' effects can this be to do with stabilisation of the alkoxy anion by inductive effects *via* the hydrocarbon chain that decreases with increasing chain length?

Cynthia Friend responded: Given that the primary bonding *via* the O in the alkoxides is essentially the same for all chain lengths studied, it is unlikely that an inductive effect can explain the result.

David Willock commented: The data you present in Table 1 (DOI: 10.1039/C5FD00192G) show nicely that the van der Waals' energy is an important factor in determining the binding of the oxy anions to the Au surface. However, the data also show that there are some electronic effects that should be remembered. For example, CH_3CH_2O and CF_3CH_2O each have van der Waals' contributions that differ by only 0.02 eV but their calculated PBE binding energies differ by 0.27 eV. This is probably due to the electron withdrawing nature of the CF_3 group. The overall binding strength of the anions to the surface should be considered as a combination of the two effects. The PBE data is useful as it suggests that the fluorinated reactant should not be as effective in coupling to the aldehyde, which is the rate determining step in Fig. 3 (DOI: 10.1039/C5FD00192G).

Cynthia Friend responded: We observed that the CF_3 group decreases the binding energy of the fluorinated ethoxide. It is also correct that fluorination affects the rate determining step for aldehyde formation and coupling. This effect is described in the paper (DOI: 10.1039/C5FD00192G).

Bruce Gates asked: On the basis of your results, can you anticipate the form of the kinetics of your catalytic reaction, accounting for competitive adsorption? Can you anticipate selectivity effects? Can you resolve the intrinsic reactivities of the various surface species?

Cynthia Friend replied: Yes, we are able to anticipate selectivity to some extent in the coupling based on the relative concentrations of the species that are dictated by the competition for binding of the different alkoxides. There is a second effect in the selectivity and that is the rate of β-hydrogen elimination from the alkoxide to form the respective alkoxide. The rate of H elimination also depends on the chain length and F substitution. As described in the paper (DOI: 10.1039/C5FD00192G), the rate of β-hydrogen elimination is retarded by fluorine substitution based on experimental coupling reactions. We are in the process of studying this effect using theoretical methods.

Charles Campbell said: Since the adsorbed alkoxy ratio is directly reflected in the product selectivities, as you showed so beautifully in your paper (DOI: 10.1039/C5FD00192G), this implies that their coupling (or the coupling of some species whose concentrations are proportional to the alkoxy concentrations) is the rate determining step. Is that the case? If so, how?

Cynthia Friend responded: Once the alkoxides are bound to the surface, the most difficult step that affects the rate is elimination of H to yield the respective aldehydes from the alkoxides on the surface. Thus, the concentration of the alkoxides is pivotal. One other factor that we are examining is whether the rate of aldehyde production also depends on the nature of the alkoxide. Formaldehyde formation from methoxy is a slower process than, *e.g.* acetaldehyde from ethoxy.

Fluorination of the alkoxide also affects the rate of hydrogen elimination. As described in our paper (DOI: 10.1039/C5FD00192G), β-H elimination from the ethoxy is significantly faster than from 2,2,2-trifluoroethoxy based on the product distribution measured.

Graham Hutchings commented: Following on from the discussion raised by Professor Campbell on competitive adsorption, there is one molecule that is missing from the hierarchy list in Fig. 1 (DOI: 10.1039/C5FD00192G), namely water. Where does water fit on this scale and does it influence the reaction? Perhaps co-feeding water could be instructive?

Cynthia Friend responded: Water is very weakly bound on Au based on experimental studies. In the calculations, we use the water formation as our reference state so that it is indirectly included there. The overall reactions used in the DFT calculations are of the form:

ROH (g) + OH (ads) → RO (ads) + H_2O (gas).

Michael Bowker asked: Do you think the nanogold is essentially gold(111)?

Cynthia Friend answered: The nanoporous Au is clearly mainly (111), however, there is a significant fraction of under coordinated sites. The orientation is clear based on atomically-resolved TEM.

Daniel Wotton communicated: I was wondering whether you could be sure that the reaction involving trifluoroethanol and ethanol on the catalyst surface still proceeded with an initial hydrogen scission reaction on the ethanol in an S_N1 fashion or whether it would be possible for electrons from the trifluoroethanol to first attach to the carbon resulting in the hydrogen being kicked out similar to an S_N2 type mechanism? Would it be possible to buy deuterated ethanol and use the kinetic isotope effect to determine this, where a change in the rate of the reaction can confirm a more S_N1 or S_N2 type relationship?

Cynthia Friend communicated in reply: We did not probe for an S_N1 *vs.* S_N2 mechanism.

Bruce Gates opened the discussion of the paper by Gareth Whiting: What you observed with the clay binder was quite different from what you observed with the other binders. Can you explain the differences? Under your conditions, were there phase changes in the clay binder, was it vitrified? What happens to it under the severe conditions of the regenerator? With these changes, is the clay a realistic binder?

Gareth Whiting answered: It is true that the clay binder displayed a far greater effect on the catalyst reactivity compared to SiO_2, Al_2O_3 and the SiO_2:Al_2O_3 mixed binder. This was due to the fact that we calcined all samples at 600 °C in order to convert the NH_4-ZSM-5 to the protonated-ZSM-5 form, and each sample was treated in the same way in order to make a fair comparison. However, we found that upon this high temperature heat treatment, the kaolinite converts to metakaolin *via* dehydroxylation at around 550–650 °C. The synthesis of spray dried FCC particles actually involves the mixing of hydrated and non-hydrated forms (*i.e.* metakaolin) of kaolin, but given the temperature of the synthesis (~700 °C), the overriding phase should then be metakaolin. From what I know from the literature and patents, this phase actually benefits the FCC catalysts, offering greater physical integrity and mechanical strength, and lowering the surface area (a high matrix surface area is known to increase coke production) of the FCC particles. However, in our case, the conversion of kaolinite to metakaolin after calcination is apparently showing that it indeed does impart high mechanical strength, but due to the fact that it is in an over-exaggerated concentration (80 wt.%), it is actually leading to the limited reactivity of the pellet.

Sarwat Iqbal asked: Are there any other techniques you could use to study these materials? You have selected a 30 : 70 ratio of zeolite : binder, is there a particular reason to choose this ratio? What would happen if you increase or decrease the amount of binder? The treatment temperature for the preparation of these bound materials was always 600 °C, what is the reason for selecting this temperature? Do you expect an effect on the acidity migration if you increase the temperature?

Gareth Whiting replied: Regarding the use of additional techniques, indeed there are quite a few that we hope to use to help characterize these complex materials in the future. For example, we have access to an iLEM (integrated Light-Electron Microscope), which works by using a laser of a specific wavelength to excite fluorescent molecules within the catalyst material (*i.e.* after the reaction – thiophene oligomerization), and then, within the same setup, TEM (Transition Electron Microscopy) can be used to overlay the structural location with these chemical species in the sample. Hence, this will be a nice way to reinforce our results provided so far from the UV-Vis spectrometer and the confocal fluorescence microscopes. We also envisage using STXM (Scanning Transmission X-ray Microscopy) to obtain high spatial resolution maps at the zeolite–binder interface within these materials and correlating them with chemical species present after reaction. In reference to your second question, we chose the ratio of 30 : 70 (zeolite : binder) in order to observe any binder effects with enough intensity to confidently assign them. In previous work,[1] we used a lower amount of binder (80 : 20), and we did observe slight differences compared to using pure zeolite, but not to the same extent as when we used 80% binder and 20% zeolite. The treatment at 600 °C was used in order to fully calcine the zeolite, which was already in the NH_4 form, which will then be converted to the protonated form, hence forming Brønsted acid sites. We can imagine that upon increasing the treatment temperature, Al migration from the Al_2O_3 binder (in the ZSM-5:Al_2O_3 bound pellet) to the zeolite interface will certainly increase, as the species will be far more mobile than at lower temperatures.

1 G. T. Whiting, F. Meirer, M. M. Mertens, A.-J. Bons, B. M. Weiss, P. A. Stevens, E. de Smit and B. M. Weckhuysen, *ChemCatChem*, 2015, 7, 1312.

James Earley said: All of the techniques that you have shown analyse the surface of the tablet. Did you do any post reaction analysis of the inside of the zeolite/binder tablet? If so, did you do any analysis of how the product distribution changed through the tablet? Have you looked at how the binder selection changes the activity of the catalyst in an industrially relevant reaction? With this you could see which binder gives you the most favourable properties for the reaction.

Gareth Whiting responded: Indeed, you are correct in stating that the techniques used so far are used to probe the near surface of the zeolite–binder materials. We have performed some post-reaction analysis of similar materials (zeolite:binder-bound extrudates) (not published in this paper) by cutting the cylindrical samples with a scalpel after the thiophene oligomerization reaction in order to visualize the aspects you suggested. Although the cut is quite uneven using a scalpel (*i.e.* not close to a perfectly flat surface), we were able see that the product distribution does indeed change from the outside to the inside of the extrudate. There is also a significant heterogeneous nature in the distribution of the zeolite crystals from the outside to the inside of the cylindrical extrudate, which could play a part in influencing the product distribution. However, further work will be focused on understanding this. Concerning the use of an industrially relevant reaction, we have not yet performed these reactions, but this is a good point and we will pursue this in the future.

Femi Oloye commented: I observed that the type of acid site of interest in this research is the Brønsted acid site. I also noted that you used ammonia TPD for the determination of the strength of the acidity in the catalyst. However, I am a bit worried that considering the clay/binder in the catalyst, the Brønsted acid site may not be the major acid site in the material. It would be wise to determine the nature/type of acid site that is present in the catalyst. Furthermore, I suggest that pyridine or other probes, which could be used to differentiate between the acid sites, should be used.

Gareth Whiting answered: This is a good question, and although the assignment of desorption peaks in ammonia TPD is tentative when suggesting either Brønsted or Lewis acid sites in the literature, there is enough confidence to say that the peak of desorption at higher temperatures (~400 °C) is assigned to stronger acid sites (*i.e.* Brønsted acid sites) and those at lower temperature (*i.e.* 200 °C) are due to weaker acid sites. In our case, the use of ammonia TPD confirms the presence of both strong and weak acid sites, but it also confirms the presence of moderate acid sites in the Al_2O_3 bound samples, suggesting Al migration to create additional acid sites, which is what we expected.

The use of pyridine IR is in our future plans and is a good suggestion as it will give more detail into the type of acid sites present, and we will also use larger probe molecules, such as collidine, in order to probe acid sites at the zeolite–binder interface (*i.e.* not inside the zeolite pores).

Mark Howard asked: Do you have any insights into the mechanism of the aluminium migration in your systems, or indeed more generally? Do you believe that this is a surface migration, through the bulk, or could it even be in the vapour phase?

Gareth Whiting replied: This is a really interesting question and it is something that researchers and myself are still wondering about. As far as I know, the mechanism of aluminium migration is unknown in the literature, but I could imagine that it migrates in an aluminium hydroxide form from the near surface of the Al_2O_3 binder to the interface of the zeolite–Al_2O_3 or the zeolite surface (filling defects in the framework). These are, of course, postulations, but the mobility of these species could be enhanced by temperature and solvent effects during the preparation of the catalyst bodies. However, as this is still an over-riding question in the field of industrial zeolite-bound catalysts, we plan to purse this further using model catalyst systems together with solid state Al NMR spectroscopy.

Natasa Novak Tusar enquired: If silica is added to the ZSM-5, does this not just dilute the amount of aluminium in the zeolite? What was the amount of aluminium in the ZSM-5?

Gareth Whiting responded: I am not sure I understand the question correctly, but it is true that by binding the ZSM-5 in SiO_2, the relative concentration of aluminium compared to silicon is diluted in the overall sample compared to pure ZSM-5 alone. However, the amount of aluminium in the ZSM-5 in both the SiO_2 bound and pure zeolite catalysts will be the same, as the framework of the zeolite will most likely remain unchanged when bound with SiO_2, although it has been known that Si from the SiO_2 binder can fill voids/defects in the zeolite. The $SiO_2 : Al_2O_3$ ratio was 80.

Cynthia Friend opened the discussion of the paper by Emma Gibson: Is it possible for you to generalize your results to predict which other oxides will remain on the surface and also which will yield formaldehyde? Putting it another way, can you explain why molybdenum oxide is used over other materials?

Michael Bowker replied: I think the basic rule is that materials with high oxidation state components will give good yields if they are present at the surface. It also appears that, at least for these materials, high oxidation state elements segregate to the surface. Then, the efficiency probably also depends on energy level differences between fully oxidised and the next lower oxidation state, and differences in the reducibility of terminal and bridged oxygen species.

Richard Catlow asked: What do we know about the surface chemistry of MoO_3 and the contrast between the surface structure of this oxide and that of Fe_2O_3? Is there a great preponderance of terminal oxygen atoms on the former?

Michael Bowker answered: A significant difference is evident in the Raman spectra – the terminal stretch is present at 990 cm^{-1} for MoO_3 but it is absent from the Fe_2O_3 spectra. Furthermore, formate (the combustion intermediate) is easily formed on Fe_2O_3, as shown in our paper by *in situ* DRIFTS (DOI: 10.1039/C5FD00225G), but is quenched when hematite is covered with MoO_3, which produces only the methoxy (the selective intermediate for formaldehyde).

Itai Panas enquired: Was any significant evaporation of MoO_3 (g) or $MoO_2(OH)_2$ (g) from the MoO_3 monolayers coating observed at the temperature of the working catalyst? Such evaporation would obviously display irreversible long-term catalyst deactivation.

Michael Bowker replied: Please see our responses to the questions on paper DOI: 10.1039/C5FD00225G. In our experiments, the timescale was too short to see any Mo loss, and we had no evidence of loss of performance on our timescale

Graham Hutchings commented: In this discussion so far we have discussed the surface of molydena catalysts but what about iron oxide? Everybody uses iron molybdate as a catalyst for methanol oxidation, what is unique about this combination? Does vanadium molybdate work? What about nickel? Cobalt?

Michael Bowker answered: There appears to be a number of factors for having Fe present.

1) It appears to make a support of iron molybdate, which anchors the active Mo-only layer at the surface, perhaps in an epitaxial relationship.

2) We have evidence that it encourages anion mobility in the subsurface layers compared to bulk MoO_3.

3) It enables the formation of larger surface area Mo catalysts than MoO_3 alone.

4) (More speculatively), it anchors the Mo phase more strongly than other formulations, which reduces volatilisation.

Regarding V – yes it works quite well, but with a worse yield than with Mo. Ni and Co, being in low oxidation states, are combustors.

Robert Madix opened a general discussion: I think it is appropriate at this time and in this setting that I offer some perspective to the discussion, particularly with regards to the importance of surface science studies for the progress of catalysis. I appreciate your patience with me in doing so. It occurred to me sitting in the audience today that it has been nearly 50 years since I first set out to study reactions on well-defined single crystal surfaces of metals. What motivated me to do so was the fact that most discussions of catalytic reactions and mechanisms that I heard as a graduate student were not even based on either the surface structure or composition of the active material. Obviously, this was glaringly insufficient for a scientific discourse. Furthermore, reaction intermediates were suggested in an *ad hoc* fashion without substantial proof of their existence, and values of rate constants for surface reactions were actually non-existent, partly because the intermediates themselves were not identified and, certainly, their surface concentrations were unknown. Fundamentally, there was a dire need to study defined reactions on surfaces for which the surface composition structure was controlled and well-defined. How else could the field progress on a firm foundation? The coincident development of the tools of ultrahigh vacuum physics and the progress in the synthesis of high purity single crystal materials offered the possibility of controlling these features at precisely that time.

The intent in my laboratory was, from the outset (and continues to be), to study reactions and concepts of fundamental importance in heterogeneous catalysis. Let me provide a few examples derived from the work in my laboratory. I was surprised to learn that there was no knowledge of either the adsorption probability or desorption rate constant of any alkane on platinum – the activation of which lies at the heart of the reforming process. Poisoning of metallic catalysts by simple elements, such as C, S or O, was not understood in any detail. Selective partial oxidation was poorly understood. A better understanding could lead to improved or new processes.

At the time, the decomposition of formic acid was indeed commonly used as a test reaction for catalysis by metals. Much attention focused on dehydrogenation *versus* dehydration and its causes. In fact, the results of such studies produced among the first of the now well-known volcano curves – as was presented as a testimony to the principle of Sabatier. In fact, this example is cited in Wikipedia! There were mysteries governing even this simple reaction, however. One was the phenomenon of the so-called "catalytic superactivity" metallic nickel for this reaction, which was a phenomenon whereby order of magnitude increases in the rate were observed suddenly under certain reaction conditions. What value did one then attribute to this reaction on Ni? Using systematic studies of this reaction on Ni single crystals prepared in various ways – atomically clean, carbided and graphitized – we showed conclusively that superactive nickel was simply clean nickel. There was no "pressure gap" anomaly about the result (unless you consider higher pressure to mean impurities in the reactants); the result was clear. To summarize briefly, in the next decade or so, we examined the nature of catalytic poisoning on Ni, Pd and Pt, delineating the effect of partial selective poisoning that clearly had implications for catalytic operation under normal conditions. Not long thereafter, we developed the modern mechanism for methanol partial oxidation to formaldehyde over metallic silver catalysts and developed a set of principles that govern a wide variety of oxygen-assisted reactions over group 1B metals that is still of general conceptual utility today. The point is, a lot of understanding emerged that was highly relevant to actual catalytic processes.

I have been puzzled for quite some time over the muted response that these developments received from the catalysis community in general. In general, studies of this type, as they emerged around the World, became pigeonholed as "surface science" and their impact isolated by conceptual misconstructions, such as the "pressure gap". There were, in fact, some attempts in industry to capitalize on some of the concepts developed, but, of course, these were not public. The fact is, these studies appeared to have very little resonance among the research community in catalysis. If there was general knowledge of this activity in the community of researchers in catalysis (then separate from its surface science component), it may have been met with the questions "What do we do with this information?" "Where do we take it from here?" I believe the recent advances in quantum theory and its potential for advancing both concepts and quantitative understanding in surface reactions and catalysis have brought a turning point to the general importance of such studies of surface reactions under well-defined conditions. This interplay has received considerable discussion at this Faraday Discussion, as it should. The fact is that theory requires measurables on well-defined systems with which to benchmark its accuracy and drive further development and progress. Clearly, theoretical calculations utilize very well-defined structures and local compositions. But without data to which to compare the results, they are as "rarefied" (if not more) as the surface science type experiments. It is essential that experiments under well-defined conditions be done in parallel with advances in theory in order to validate the computational developments.

There is not a lack of surface chemistry to be explored. For example, metal surfaces are not static under reaction conditions; they are quite dynamic. On Ni, for example, adsorbed H facilitates Ni atom migration *via* the formation of NiH_x complexes, which may themselves facilitate hydrogenation of unsaturated hydrocarbons. Condensed two-dimensional surface oxides of varying sizes form under net oxidizing conditions on noble metals; how do these affect oxidation reactions? Subsurface interstitials or vacancies in metal oxides change the rate constants for surface reactions or the electron distribution on small metal clusters at the surface; how do we predict or take advantage of these effects? Furthermore, catalysis is about kinetics. Intelligent approaches to the accurate measurement and systematic trends for surface reaction kinetics are needed. The current problems may be more complex, but they can be well addressed with modern methodology that combines theory and surface science with the development and understanding of catalytic processes. Interested young scientists and academic departments or research institutes should not hesitate to engage in such surface science research. Often, theorists say (quite reasonably) that they are examining "trends". Experimentalists should also be examining these trends, and, in fact, the exploration of some of them. Methods have evolved recently, for example, for creating large surfaces with defined compositional variation over the surface that can be locally probed for chemical reactivity in order to short-circuit the requirement for synthesizing many different samples. In fact, the theorists need guidance from well-defined experiments, because there is a limit to the parameter space they can probe. Yes, there may come the time when theoretical computations are sufficiently rapid and robust to lead the way to catalyst development, but currently well-executed surface science and theory need each other. Fundamentally, a sound understanding is always of use when appropriately applied. The need for creative studies of catalytically relevant systems on well-defined surfaces has not diminished; indeed, it is greater than ever before.

Joachim Sauer answered: I could not agree more with your statement that a close dialogue is needed between theory and experiment when working on the same surface or catalysis problem. I also fully agree that benchmarks are needed for which advanced computational methods have been used and agreement has been reached with reliable experiments on well-characterized samples.

There is a (slowly) increasing number of such studies. From our own work on zeolite catalysis, I can name a study on enthalpies and entropies of adsorption of small hydrocarbons[1] and a study on pre-exponentials and energy barriers for the methylation of alkenes.[2]

1 G. Piccini, M. Alessio, J. Sauer, Y. Zhi, Y. Liu, R. Kolvenbach, A. Jentys and J. A. Lercher, *J. Phys. Chem. C*, 2015, **119**, 6128.
2 G. Piccini, M. Alessio and J. Sauer, *Angew. Chem., Int. Ed.*, 2016, **55**, 5235.

Avelino Corma commented: The quality of the metal, the trends, and the results you get from modern catalytic systems are inspiring for me, as are isotopic labelling, smaller particles, well-defined single isolated sites, and discussions with people working in organocatalysis. Hans-Joachim Freund showed that we can change the states, but we have to know which ones and maximise them.

Jacob Moulijn said: The remark was made that elucidation of the mechanisms in heterogeneous catalysis is a very different topic than in homogeneous catalysis. In some cases this will be true, but in general it is wise to consider the possibility that the mechanisms are probably the same: why not help elucidate the mechanism in a heterogeneous catalysis case by studying the homogeneous catalysis analogue? A good example is the metathesis of alkenes. In our earlier work in the study of heterogeneous systems,[1] we came to the wrong (cyclobutane) mechanism, while a little bit later, in studies dealing with homogeneous catalysis, the correct mechanism (carbene complexes converting in four membered-metallocycles) was found relatively quickly. There was little doubt that in homogeneous and heterogeneous catalysis, the metathesis mechanism was the same and, as a consequence, the idea that the metallocycle was the key intermediate also in heterogeneous metathesis was quickly generally accepted. This similarity is not valid in all cases, for instance in more complex cases (such as specific planes of nanoparticles carrying the active site(s), structured catalyst particles, *etc.*) there is no simple homogeneous analogue.

1 J. C. Mol and J. A. Moulijn, Catalytic Metathesis of Alkenes, *Catalysis*, Springer, Berlin Heidelberg, 1987, pp. 69–129.

Cynthia Friend answered: I agree that guiding principles can be derived from homogeneous systems. As suggested by Muetterties and others,[1] there are many analogies in bonding between soluble clusters and surfaces. On the other hand, the specific structures and electronic properties of extended materials are different, which means that there are and will be differences.

It is interesting to note that there are many similarities in our work on metallic Au *vs.* the work described by Avelino Corma in the introductory lecture. The hierarchy of binding seemed qualitatively similar, based on his presentation.

1 E. L. Muetterties, T. N. Rhodin, E. Band, C. F. Brucker and W. R. Pretzer, *Chem. Rev.*, 1979, **79**, 91.

Itai Panas commented: Intuitive conceptual contact between homogeneous, supported nanoparticles and heterogeneous catalysis – heterogeneous catalysis of late transition metal single crystal surfaces utilizes two quasi-separable polarisabilities of the substrate. One is related to the broad conduction band built mainly by overlapping (n + 1)s atomic states. The second is associated with the more narrow band formed mainly from nd dominated atomic states. An incoming molecule upon approach must first decide whether it is an electrophile or a nucleophile and if the gains forming the corresponding surface chemical bond are greater than the costs associated with its formation. Let's assume here for simplicity that the adsorbate is unable to distinguish between cleave planes. To form a local bond to the surface, the local electron density needs to be either accumulated (electrophile) or pushed away (nucleophile). In both cases, however, the result is a nucleophile, which on penetration of the (n + 1)s shell experiences a less screened atomic core. In addition, Pauli repulsion between the incoming lone pairs and the quasi-atomic localized nd orbitals of the polarisable nd ionic core causes the latter states to split/broaden. The degree of splitting depends on the overlap with the incoming lone pair and occupation of the nd orbitals. A closer proximity can be afforded if the nd shell has half or unoccupied orbitals compared to if all these orbitals are filled, with the overall Coulomb attraction being decided by the effective core radius. All the latter contributions are decisive for catalytic selectivity among transition metals. Alloying, *e.g.* Au by Pd or Pt, allows for the emergence of local catalytic "hot spots" owing to the greater on-site polarisability of the d9 state, offering locally enhanced stability compared to the d10 (Au) of the host matrix. While complex, these are easily incorporated into a homogeneous catalysis conceptual framework. However, what truly distinguishes a metal from a supported ionic transition metal catalyst is the cost for attracting/displacing quasi-free electrons. This particle-in-a-box type kinetic energy cost for localization, reflecting local electron density modulation, is due to the local nature of the adsorbate, and includes *e.g.* quantum size effects. The fruitfulness in the ability to attenuate the latter energy component is understood to be partly reflected in the emergence of supported nanoparticle based transition metal catalysis.

Faraday Discussions

PAPER

Concluding remarks: progress toward the design of solid catalysts

Bruce C. Gates

Received 13th May 2016, Accepted 17th May 2016
DOI: 10.1039/c6fd00134c

The 2016 Faraday Discussion on the topic "Designing New Heterogeneous Catalysts" brought together a group of scientists and engineers to address forefront topics in catalysis and the challenge of catalyst design—which is daunting because of the intrinsic non-uniformity of the surfaces of catalytic materials. "Catalyst design" has taken on a pragmatic meaning which implies the discovery of new and better catalysts on the basis of fundamental understanding of the catalyst structure and performance. The presentations and discussion at the meeting illustrate the rapid progress in this understanding linked with improvements in spectroscopy, microscopy, theory, and catalyst performance testing. The following text includes a statement of recurrent themes in the discussion and examples of forefront science that evidences progress toward catalyst design.

Catalysis and catalyst design

Catalysis is the key to controlling chemical change, in processes ranging from the biological to the technological. It is used to make products including chemicals, fuels, materials, food, beverages, and personal care products, and together these have a value of roughly 5–10 trillion dollars (US) per year worldwide. Catalysis is also essential for the removal of environmental pollutants, such as those generated in motor vehicles and fossil fuel-fired power plants. Thus, the science underlying catalytic technology is essential.

Catalysis science is also challenging, because almost all large-scale industrial catalysts are solids. These work at their surfaces—and these surfaces are notoriously non-uniform in both composition and structure, often being substantially different from simple terminations of the bulk material—and they undergo changes when exposed to reactants.

The 2016 Faraday Discussion "Designing New Heterogeneous Catalysts" attracted a full house of researchers and practitioners, eliciting novel contributions and stimulating discussion that brought forth a diversity of ideas about

Department of Chemical Engineering, University of California, Davis, Davis CA, USA. E-mail: bcgates@ucdavis.edu

what the terms "catalyst design" and "heterogeneous" mean; we address these points below.

Avelino Corma opened the discussion (DOI: 10.1039/C6FD00066E), introducing the subject with a few history lessons and a linked set of case studies recounting a family of new catalysts based on insights guided by the emerging understanding of supported metal catalysts. This introductory lecture provided a basis for assessing the premise of catalyst design and the discussants' efforts to home in on the meaning of this term.

The history of catalyst discovery reflects the complexity of the typical catalyst surface and the reactions that occur on it as well as the enormous array of materials that can be considered candidate catalysts. The original approach to catalyst discovery was trial and error experimentation, and this is linked to the alchemists' search for the philosopher's stone and their early discoveries of "contacts" which today we would call solid catalysts. The work of Fritz Haber, Carl Bosch, and Alwin Mittasch more than a hundred years ago stands as the monument to the trial-and-error approach to catalyst discovery. Their work at BASF involved the systematic testing of some 20 000 materials as catalysts for the synthesis of ammonia from H_2 and N_2, with the resulting iron-containing catalyst being only a little different from today's technological ammonia synthesis catalysts. These pioneers worked from the emerging principles of chemical equilibrium, chemical kinetics, and process engineering, guiding them to optimal testing. They realized the need to evaluate candidate catalysts in reactors and under conditions that mimic a potential process—thus, they used high pressures (to favor the equilibrium for ammonia formation), recycled unconverted H_2 and N_2 in a continuous process for efficient use of these raw materials, and tested the catalysts in operation long-term to evaluate their lifetimes. They found a highly active and stable catalyst, a porous iron-containing mineral with a high internal surface area, and the high catalytic activity allowed operation at relatively low temperatures, favoring the equilibrium of the exothermic ammonia synthesis reaction, maximizing the conversion, minimizing the recycling of unconverted reactants, and pointing toward optimization of the process.

The trial-and-error approach is still widely practiced in technology, and it will not go away soon, because the catalyst design challenge is enormous. However, catalyst testing technology has advanced markedly since the Haber–Bosch–Mittasch discoveries, and advances in catalyst testing methodology keep improving, with many highly automated rapid-throughput test systems now in operation worldwide, many of them operating on quite a small scale.

Most important for this essay, and reflecting the give and take of the Faraday Discussion, the science of surface catalysis has advanced markedly since Haber's day. A historically important example illustrating how fundamental understanding guided catalyst discovery was introduced by Corma to start the discussion (DOI: 10.1039/C6FD00066E). The prototypical process is petroleum naphtha reforming, which emerged more than 50 years ago in work by Vladimir Haensel and coworkers at UOP.[1] The major goal was to increase the octane numbers of hydrocarbons in fuels, such as petrol (gasoline). We can describe the reasoning that led to a breakthrough catalyst and a novel process as emerging from the inventors' recognition of catalyst functions—they realized that hydrogenation and dehydrogenation reactions are essential in the naphtha reforming process and are catalyzed by the surfaces of metals (*e.g.* nickel, palladium, platinum), and that reactions that lead to octane number increases result from increased branching of hydrocarbons such as olefins—which are known to be catalyzed by strong acids, which protonate the olefins to give reactive intermediates (carbocations) that readily undergo skeletal isomerization.

A practical goal was thus to increase the branching of paraffins (the principal components of naphtha) to increase their octane numbers, and a route to this branching involves (1) a catalyst that dehydrogenates the paraffin on a metal surface, to give olefins, which are (2) isomerized by the acid catalyst and then (3) rehydrogenated on the metal surface. Seeing the value of having a high catalyst surface area per unit volume for high reaction rates per unit volume, Haensel and coworkers saw the value of having their metal dispersed finely as small particles (nanoparticles) to give a high metal surface area and placing these particles directly on a porous acidic support, so that the products of the initial metal-catalyzed dehydrogenation immediately migrate to the nearby acidic sites, where isomerization occurs, allowing the ensuing branched product, in turn, to move quickly to the metal for the final step of rehydrogenation.

Sterba and Haensel[1] selected transition aluminas as their catalyst supports, because these are inexpensive and can be prepared as porous materials with high internal surface areas. Knowing that the acid had to be strong and that aluminum chloride (in the presence of water) is a strong Brønsted acid, they treated their support to convert it to a surface chloride to increase its acid strength and its catalytic activity for isomerization. The resulting catalyst consisted of metal nanoparticles on the support, and it is called a bifunctional catalyst because the metal function gives the olefin from the paraffin, the acid function gives the branched olefin from the unbranched olefin, and the metal function gives the branched (high-octane-number) paraffin from the branched olefin. Because the equilibrium limitation allows the formation of only a little olefin under the processing conditions, the olefin has to be transported efficiently to the metal to drain it away and speed up the overall reaction, and so the two functions have to be located close to each other on the catalyst surface and the metal particles need to be small.

The optimal metal was found to be platinum, and catalyst performance testing showed it to offer not only the expected hydrogenation/dehydrogenation activity but also high activity for dehydrocyclization reactions to give aromatic compounds such as benzene and toluene formed from *n*-hexane and *n*-heptane, respectively. These aromatic compounds have high octane numbers and were considered good fuel components (but they are minimized in applications today because they are carcinogenic).

The bifunctional catalyst was discovered substantially on the basis of early catalyst design principles—an understanding of what combination of catalyst functions was needed and what kinds of materials provided them. Subsequent work in various industrial organizations, again mostly done by trial-and-error, led to better catalysts that contain not just platinum in the supported particles, but a combination of platinum with another metal, rhenium or tin.

Today, researchers seeking better catalysts have far better tools for the characterization of catalysts and surface reactions than were available half a century ago. Corma illustrated the thought processes that led his research group to new, markedly improved catalysts for the selective reduction of nitroaromatic compounds, as described in the following section. His methodology leads us toward an understanding of how to approach the idea of catalyst design in a realistic and nuanced way.

Corma's case studies of supported metal catalysts for the selective reduction of nitroaromatic compounds: toward a working definition of catalyst design

Only rather a little of what was presented at the Faraday Discussion could be described as direct first-principles design of catalysts. Nonetheless, the theme of catalyst design was widely evident implicitly, and the discussion on more than one occasion got to the point of addressing how the term "catalyst design" should be interpreted when we are far from having the capability of using first principles to determine the catalyst structure needed for a given conversion—and then to synthesize the catalyst.

The case studies that provide our foundation are the work of Serna and Corma, and a recently published perspective[2] by these authors provides a summary of nearly a decade's work. Our purpose here is to highlight the thought processes of these researchers, link them to the aforementioned work by Haensel, and use them as a basis for a functional interpretation of "catalyst design".

Recognizing the potential value of catalysts to selectively hydrogenate nitro groups in compounds with other groups that react with H_2, such as olefinic groups, Serna and Corma investigated supported gold catalysts. They worked from the discovery that gold nanoclusters supported on TiO_2 are selective catalysts for the hydrogenation of nitrobenzene to give aniline and for the hydrogenation of nitrobenzenes substituted with groups such as C=C to give anilines substituted with these groups.

Part of the foundation of the work of Serna and Corma was the measurement of product distributions and the inference of reaction networks in their conversions. They also used infrared spectroscopy to identify surface reaction intermediates, including nitrosobenzene, which could participate in condensation reactions and give products other than the desired substituted anilines. The authors also worked from a hypothesis advanced by several authors working with supported gold catalysts—the gold-support interface can play an important role in the catalysis. Serna and Corma used density functional theory to show that the nitrobenzene hydrogenation reaction is favored at this gold–TiO_2 interface rather than on just the gold surface, and their infrared spectra showed that the interaction of nitrobenzene with the gold–TiO_2 interface is strong and hinders the adsorption of other molecules incorporating C=C bonds, in contrast to the situation when the support is SiO_2.

These results pointed to the benefit of small gold particles on TiO_2, maximizing the gold-support interface. The measurement of the kinetics of the hydrogenation reaction with such catalysts further showed that the gold particle size influenced the rate and led the authors to the further inference that the rate depended on the density of low-coordinated gold sites—which they quantified by using infrared spectroscopy of CO molecules as probes of the surface. Measuring the activities of the catalysts for HD exchange in the reaction of H_2 with D_2, they found (consistent with their kinetics of the hydrogenation of nitrobenzene) that the activity for H_2 dissociation increased with the number of low-coordinated gold sites, and their kinetics of the nitrobenzene conversion pointed to H_2 dissociation as the rate-determining step.

Thus, Serna and Corma inferred that they could further increase the activity of their catalyst by boosting the rate of H_2 dissociation. Realizing that platinum is highly active for this reaction, they modified their catalyst by incorporating platinum—choosing a platinum content small enough (50–2000 ppm) to increase the rate of the desired reaction without substantially lowering the selectivity. They also observed that just platinum (without gold) supported on TiO_2 at such a low loading was not active, and so inferred that their platinum-containing gold catalyst was bifunctional (this reasoning connects their work to that of Haensel, mentioned above).

The authors further evaluated their improved catalyst in experiments with liquid-phase reactants under conditions (with and without solvent) that might be representative of potential applications. They found that it worked for more than about 1350 turnovers without substantial deactivation.

Further work by these authors led to a new generation of catalyst for the selective hydrogenation of substituted nitrobenzenes. Serna and Corma reasoned how to transform highly active but usually unselective metals exemplified by platinum into highly selective catalysts. Testing supported platinum particles (now at a higher loading) as catalysts, probing their surfaces with CO and using infrared spectroscopy, they found that the selectivity for the reduction of nitro groups benefited from minimizing the number of terrace sites on the platinum particles—achieved by minimizing the platinum particle size. A similar benefit was observed by the adsorption of thiol ligands or the partial decoration of the platinum surface by TiO_x species from the support to cover these terrace sites (the coverage by fragments of the support is linked to the idea of so-called strong metal-support interactions (SMSI)). Other species were used comparably to decorate the metal surface, and the ideas were extended to metals other than platinum.

The foregoing summary is simplified but sufficient to illustrate how a family of increasingly selective and active catalysts resulted from the following:

- Elucidation of the catalyst performance (reaction network elucidation; reaction kinetics determination, including elucidation of the activity for H_2 activation with the simple test reaction $H_2 + D_2 \rightarrow 2HD$);
- Elucidation of the role of a support as not just an inert platform but rather a catalyst component that works in concert with the metal particles on it;
- Extension of the idea of the support's role to recognize that support-derived species, among others, present as metal surface modifiers, work in similar ways;
- Use of CO as a probe molecule to identify metal surface sites;
- Use of theory in combination with experiments;
- Catalyst testing under potentially applicable conditions to demonstrate satisfactory behavior in long-term operation.

The idea of catalyst functions (metal, support, second metal) enters into the reasoning and the terminology used by Serna and Corma. However, an essential point for us is that their depth of reasoning and access to data providing fundamental insight into the workings of the catalysts go far beyond those illustrated by Haensel's work, reflecting marked advances in the tools for understanding catalysts that have emerged in the intervening decades.

The work of Serna and Corma set the table for this Faraday Discussion because it is exemplary in its thoroughness and integration of numerous techniques and thought processes. The inferences in their reasoning, leading them to better and better catalysts (including new ones that we have not addressed here), and their assembly of the information illustrate how we should think of catalyst design today. A fundamental understanding guided these researchers to hypotheses that they tested with a battery of complementary methods and then—step by step—to the synthesis of a family of improved catalysts.

A key result—quite generally valid—is that as Serna and Corma made better catalysts, they made catalysts that were more complex in composition and structure. Oversimplifying some, we can say that the recognition of and the incorporation of more functions led to better catalysts. Again, there is a parallel with the Haensel example of naphtha reforming. This pattern is the rule in industrial catalysis: for the most part, as catalysts are improved, they become more complex in composition and structure.

Thus, it is the fundamental understanding of catalysts and their functions, understanding the roles of the various catalyst components, that provides a path toward new and better catalysts. We could call this statement a functional definition of catalyst design. A number of contributions to this Faraday Discussion illustrate the point, and we present some examples below to illustrate it, as well as some recurring themes. It is worth keeping in mind that the discussion was based on single original research contributions and did not represent longer-term advances, such as those summarized in the preceding paragraphs.

Recurring themes of the discussion

A number of themes recurred in the discussion, including the following:

- Catalyst preparation, especially of catalysts incorporating regular structures;
- Multitechnique catalyst characterization, by experiments and theory;
- Determination of crystalline structures in catalysts;
- Characterization of catalysts by methods that have their roots in ultrahigh vacuum single-crystal surface science;
- Investigation of atomic architectures of catalysts, with a focus on active sites;
- Investigation of supported metal particle catalysts, with some focus on bimetallic particles with core–shell structures;
- Attention to defect sites on surfaces and interactions of supported species such as metal clusters with these sites.

Preparation of supported metal catalysts with beams of metal clusters

A novel cluster beam route for the preparation of model supported metal catalysts was shown to be applicable to the production of much larger amounts (grams per hour) of catalysts than had been available by comparable methods, and even powder samples could be formed (DOI: 10.1039/C5FD00178A). This new methodology opens the door to large-scale preparation of supported metal catalysts, even bimetallics, without the complications of metal salts or the expense of organometallics, and it might be expected to provide a boost to research in this important area. However, the methods do not lend themselves to the preparation of metals supported uniformly in porous supports.

Synthesis and characterization of crystalline porous catalysts

Catalyst preparations focused on materials with regular porous structures, including metal nodes connected by organic linkers in metal organic frameworks (MOFs) (DOI: 10.1039/C5FD00198F) and Keggin ions with inorganic linkers creating matrices for redox catalysis (DOI: 10.1039/C6FD00006A). These frameworks consist of cationic and anionic components, and the composition of each can be tuned to incorporate various catalytic species (the materials can be catalyst supports). The inorganic materials in this class offer better prospects in terms of stability than MOFs, but recent work has shown that some MOFs are stable at temperatures up to about 650 K, and MOFs are drawing rapidly increasing attention as catalysts with tailorable properties. The opportunities for the design of both physical properties (such as porosity) and chemical properties by the choice of the node and linker components explain the rapidly growing interest in these crystalline porous materials as catalysts and catalyst supports.

Catalyst synthesis with guidance by theory and spectroscopy

Many groups now use theoretical tools, and collaborations between experimentalists and theorists have become the norm. It is too early for theory to predict activation barriers of reactions and rates of reactions that are accurate within better than a couple of orders of magnitude, but the quantitative limitations of theory are not the essential point: theory is widely accepted for the guidance it provides in uncovering trends and in providing frameworks for researchers to think about how catalysts work and to try to make better ones. Thus, theory plays a central role in catalyst design as we understand the term and seems destined to play a greater role as the computations get faster and the theories become more accurate.

However, when catalysts are as structurally complex as the typical industrial solid catalysts, experiments still provide more useful guidance than theory. For example, manganese tungstate catalysts for the selective oxidation of short-chain alkanes prepared by Li *et al.* (DOI: 10.1039/C5FD00191A) were intended to have a balance of acid–base and redox sites. The preparation of these catalysts was guided by Raman spectra recorded during the synthesis, as well as high-resolution transmission electron microscopy and field emission scanning electron microscopy images, X-ray diffraction data, X-ray photoelectron spectra, and porosity data characterizing the resulting materials. The data allowed some rationalization of the synthesis chemistry and some hypotheses about the nature of the catalytic sites, but the complexity of the materials and the presence of defect sites, notwithstanding the presence of some pure phases, limited the understanding of the surface species to hypotheses that are no doubt simplified. This work demonstrates the magnitude of the challenge of catalyst design even when powerful arrays of techniques are available to characterize the syntheses and the properties of the resulting materials.

Catalyst characterization by X-ray diffraction, spectroscopy, and imaging

The presentations included many characterization results determined by X-ray diffraction crystallography, pointing to the fact that many of the reported catalysts incorporated (portions of) crystalline material. Commonly reported surface characterization methods were X-ray photoelectron spectroscopy and the workhorse infrared spectroscopy, continuing long-established trends. Raman spectroscopy was also much in evidence, suggesting that this technique is on the upswing in catalysis research.

Imaging methods, some of them quite novel in catalysis research (DOI: 10.1039/C5FD00200A, DOI: 10.1039/C5FD00210A), were also used widely in the work reported at the Faraday Discussion. The work included the characterization of zeolite-containing cracking catalysts investigated on the scale of microns. Important components of industrial catalysts for hydrocarbon cracking in the manufacture of petrol incorporate binders and amorphous components in addition to zeolite crystallites, and the results presented at the Faraday Discussion demonstrate their high degrees of heterogeneity—on the scale of microns—perhaps a surprising result in view of the enormous scale of the application of cracking catalysts.

High-resolution electron microscopy was also widely applied in the work reported at the discussion. This has become a workhorse technique and is evidently widely accessible. It was frequently used to provide evidence of bulk phases present in catalysts, but rarely to provide evidence of individual atoms in catalysts, although the recent literature indicates that numerous groups now have the capability of imaging individual heavy metal atoms on supports consisting of light atoms—because the contrast is maximized when the atomic numbers of the supported atoms and those of the support are far different from each other. This capability has triggered a surge of research in single-site supported metal catalysts, commented on further below.

Catalyst characterization by techniques with a foundation in ultrahigh vacuum single-crystal surface science

Another theme in the discussion was the application of experimental methods that have emerged from the foundation of classic ultrahigh vacuum surface science. It was evident from the discussion how much this science has evolved since its beginnings in the investigations of single crystals of metal more than half a century ago. Researchers now routinely investigate structurally complex materials such as metal particles on thin oxide films mounted on single metal crystals, which are good electrical conductors and facilitate experimentation with UHV spectroscopies. Experiments done with such samples are providing insight into how the underlying support (substrate) can influence the electronic properties of the metal particle and thus its reactivity and catalytic character.

Thus, O'Brien *et al.* (DOI: 10.1039/C5FD00143A) prepared thin films of MgO(111) on Ag(100) crystals and formed nanoparticles of gold on the MgO surface. The samples were investigated under UHV conditions with low-energy electron diffraction, X-ray photoelectron spectroscopy, infrared spectroscopy, and scanning tunneling microscopy. The authors found that charge transfer through the MgO film markedly affected the chemistry of the gold and species adsorbed on it. When the gold was present as small, essentially two-dimensional clusters, CO_2 adsorbed at the gold edges (the metal-support interface—see the summary above of the Serna–Corma work), reacting to form oxalate species. However, when the temperature of the sample was raised to 500 K, the gold morphology changed markedly, resulting in three-dimensional particles, which were not reactive with

CO_2. Electron transfer from the underlying silver was inferred to have facilitated the morphological change and thus the marked change in reactivity of the gold. These results indicate a strong influence of the support that is important in catalysis broadly, and well-defined model catalysts were needed to elucidate the phenomena.

A link between the UHV surface science measurements and those made with high-area porous catalysts was also illustrated by the work of Wang *et al.* (DOI: 10.1039/C5FD00161G). These investigators carried out kinetics experiments with a TAP (temporal analysis of products) reactor to infer that atomic oxygen is an intermediate in the catalytic oxidation of methanol to formaldehyde (but not in the oxidation of CO) taking place on porous catalysts consisting of silver and gold. The same research group had earlier found evidence of such intermediates on single-crystal gold by using classic UHV surface science methodology.[3] The link between the flow reactor experiments with a porous catalyst and the results of the UHV experiments with single crystals is noteworthy—and uncommon—such connections are not general because the reactive surface species that are important in catalysis are often not stable enough to survive under UHV conditions.

Both the O'Brien and Wang examples illustrate how single-crystal surface science has evolved since its inception and how the methods of that science continue to provide fundamental insights into the workings of more complex (heterogeneous) catalysts.

Catalytic sites and their characterization by theory

A number of contributions were also directed to the dominant class of crystalline porous catalysts, zeolites. With their regular structures, these materials attracted the attention of theorists, extending a tradition of investigations of zeolite deprotonation energies, with evidence of how the deprotonation energy varies from a two- to a three-dimensional material (DOI: 10.1039/C5FD00207A) and how framework topology and Brønsted acid site location affect the key steps initiating processes such as the industrially important conversion of methanol to olefins (DOI: 10.1039/C6FD00010J).

Some work involved calculations based on theory to elucidate the role of individual catalytic sites on surfaces, which were, however, represented in a simplified way as periodic slabs without taking into account the electrolyte present during catalysis (DOI: 10.1039/C5FD00213C). The catalyst sites incorporated reactive oxo groups with various degrees of coordination by cobalt cations corresponding to three different terminations of Co_3O_4. These were investigated for catalysis of the oxygen evolution reaction whereby two molecules of water are oxidized to give O_2 with the generation of electrons for the photocatalytic or electrocatalytic production of transportation fuels by the reduction of water or carbon dioxide. The calculated results led Plaisance *et al.* (DOI: 10.1039/C5FD00213C) to a statement of design principles for the optimal catalyst for the oxygen evolution reaction at a given overpotential (this was shown to be the rate determining step). This was a timely topic for the Faraday Discussion because it deals with specific, realistically represented catalytic sites, an important reaction, and an application of theory that leads to a clearly articulated statement of a design principle for the reaction investigated.

Confronting structures of highly heterogeneous catalysts

Except for the relatively regular crystalline catalysts exemplified by zeolites, MOFs, and linked Keggin ions, almost all of the catalysts reported in the discussion were heterogeneous to a substantial degree, and some were highly heterogeneous. Attempts to resolve the heterogeneity involved laborious characterization by multiple complementary techniques. The approach is illustrated by the work of Maganas *et al.* (DOI: 10.1039/C5FD00193E), whose supported catalysts (represented in a simplified way as VO_x/SBA-15) were characterized by Raman and X-ray absorption spectroscopies and theory, and by comparisons with molecular catalysts. The results led to hypotheses about the key surface species; the work illustrates the magnitude of the challenge of the structural elucidation of complex solid catalysts.

Themes seldom stated in the Faraday Discussion

Amorphous catalysts

Many of the catalysts prepared in the work described at this discussion are rather complex materials, but, remarkably, most of them are, to a degree, crystalline, notwithstanding the importance of amorphous materials in practical catalysis—exemplified by the transition aluminas, such as those investigated by Haensel. However, understanding of the details of the structures of amorphous solids still severely lags behind the understanding of the details of the structures of crystalline materials. Thus, the emphasis on crystalline materials in the work presented at the Faraday Discussion reflects a tendency of researchers to go where the established techniques allow progress, rather than to attack the most challenging catalysts and reactions.

Industrial catalysts and catalysis

Except for work with technological catalysts for hydrocarbon cracking, there was scant mention of industrial catalysts, consistent with the reality that the design of such catalysts is, for the most part, not quite realistic. The work on the cracking catalysts made clear that they are highly heterogeneous and raised questions about whether preparations to decrease the degree of heterogeneity—likely requiring greater expense—might bring practical benefits. There is a lack of information to address this question.

There was correspondingly a lack of input into the meeting regarding catalyst deactivation; often catalyst stability is crucial to economic application, and it is not surprising that this topic (of much proprietary industrial research) was largely missing. The whole notion of designing catalysts for increased stability is largely untouched.

Catalysts for new and emerging processes

There was little said at the discussion regarding processes for the conversion of natural gas, although these have grown markedly in importance recently with the advent of low-cost natural gas in some parts of the world, especially North America. This inexpensive gas has abruptly changed the emphases in catalytic practice and catalysis research in ways that point to opportunities for future Faraday Discussions.

Catalytic biomass conversion was also largely missing from this discussion. Growth of this field has slowed after a massive surge, as the field is faced with challenging economics and an uncertain future in terms of the commitment of public subsidies to make processes realistic. It is another highly active area of catalysis research and worthy of attention in future discussions, and the question looms about whether we may envision a sustainable future fuels industry without massive applications of biofuels.

Characterization of catalysts in the working state and novel applications of catalytic reactors

Characterization of catalysts while they are functioning provides the best opportunities to distinguish the surface species that are engaged in catalysis rather than being bystanders, although the heterogeneity of surfaces and the possible minority status of the catalytically relevant species mean that even this depth of characterization is often insufficient. The catalysis research community has made great strides in adapting spectroscopic and other measurement techniques to catalysts in reactive atmospheres and in operation, often in flow reactors. Examples of the spectroscopies used for working catalysts are infrared, Raman, ultraviolet-visible, electron paramagnetic resonance, X-ray absorption, and Mössbauer.

Catalysts to which the term "design" might most accurately apply

The complexity and heterogeneity of most solid catalysts make the challenge of true design daunting, but there are catalysts that are simple enough in structure that the idea of design takes on a significant meaning. Catalysts in this class that are being intensely investigated today are so-called supported "single-site" catalysts.[4,5] These typically consist of isolated metal cations on the surfaces of oxide, zeolite, and MOF supports, and the recent advent of atomic-resolution scanning transmission electron microscopy has led to reports from numerous groups showing these cations, especially when the metals have high atomic numbers and the supports consist of atoms with low atomic numbers, so that the contrast between the supported metals and the support is maximized. An example of catalysts in this class are the cobalt oxide catalysts described in the discussion by Plaisance *et al.* (DOI: 10.1039/C5FD00213C), but the cobalt is too low in atomic number to facilitate atomic-resolution imaging.

One can formulate a list of design variables for single-metal–atom complexes on supports that are crystalline (*e.g.* zeolites and MOFs). These include the metal, the metal nuclearity (number of metal atoms in the catalytic sites), the support, and other ligands on the metal.[5] Such catalysts can be synthesized by reactions of organometallic precursors with support surfaces, and when the chemistry is precise, the result is essentially molecular species bonded to the support by metal–oxygen bonds. Such catalysts when synthesized precisely are much less heterogeneous than the typical solid catalyst and provide quite some legitimacy to the notion of catalyst design.

Conclusions

"Heterogeneous" is a term that describes essentially all solid catalysts, and the degrees of non-uniformity are often pronounced and provide daunting challenges to detailed structure characterization, let alone design from first principles. Consequently, the term "catalyst design" has taken on a pragmatic meaning which implies the discovery of new and better catalysts on the basis of fundamental understanding of the catalyst structure and performance, often including a recognition of specific catalyst components and structural features that perform specific functions, such as bonding to one reactant molecule strongly to the exclusion of other potential reactant molecules and bonding of reactant molecules to more than one feature of a surface, as well as recognition of which surface compositions (*e.g.* of one metal *vs.* another) provide optimal activity of selectivity for a given reaction. The fundamental understanding of catalysis has advanced rapidly as catalyst characterization methods have improved, and especially significant are experiments with complementary spectroscopy, microscopy, and catalyst testing linked with theory; particular benefits are associated with spectroscopic investigations of catalyst surfaces in the functioning state and with density functional theory representing surface catalytic sites that are identified spectroscopically and have a relatively high degree of uniformity. It is these scientific advances that have gradually made the notion of catalyst design more and more a reality.

Acknowledgements

The author is grateful for support from the U. S. Department of Energy, Office of Science, Basic Energy Sciences, Grant DE-FG02-04ER15513.

References

1 M. J. Sterba and V. Haensel, *Ind. Eng. Chem. Prod. Res. Dev.*, 1976, **15**, 2–17.
2 P. Serna and A. Corma, *ACS Catal.*, 2015, **5**, 7114–7121.
3 K. J. Stowers, R. J. Madix and C. M. Friend, *J. Catal.*, 2013, **308**, 131–141.
4 M. Flytzani-Stephanopoulos and B. C. Gates, *Annu. Rev. Chem. Biomol. Eng.*, 2012, **3**, 545–574.
5 P. Serna and B. C. Gates, *Acc. Chem. Res.*, 2014, **47**, 2612–2620.

Poster titles

Towards the development of new-generation ammonia synthesis catalysts operating *via* Mars-van Krevelen type mechanisms, **S. Laassiri, C. D. Zeinalipour-Yazdi, R. A. Catlow and J. S. J. Hargreaves**, *University of Glasgow, UK*

Hierarchical ZSM-5: promising solid acid catalysts for *para*-selective tertiary-butylation of phenol, **R. K. Parsapur and P. Selvam**, *Indian Institute of Technology-Madras, India*

Designing structured AuPd nanoparticles by sol immobilsation and heat treatment for improved photocatalytic hydrogen production reactions, **W. Jones and M. Bowker**, *Cardiff University, UK*

A novel acid-mediated synthesis of ordered mesoporous aluminosilicate, H-AlSBA-15, **N. V. Krishna and P. Selvam**, *Indian Institute of Technology-Madras, India*

Molecular insights into alumina-supported Ziegler-Natta catalysts, **A. Piovano, T. Thushara, E. Morra, M. Chiesa, S. Bordiga and E. Groppo**, *Università degli Studi di Torino, Italy*

MnS-1 nanocrystals with manganese local environment on demand, **D. Maučec, M. Mazaj, I. Arčon, M. Baghbanzadeh, O. Kappe, R. Gläser, N. Zabukovec Logar and N. Novak Tusar**, *National Institute of Chemistry, Slovenia*

Hydrogenolysis of glycerol over copper and rhenium catalysts supported on zinc-oxide, **M. Shozi and H. Friedrich**, *University of KwaZulu-Natal, South Africa*

Utilising metal–metal spillover for selective alkyne hydrogenation in CuPd/Al_2O_3 bimetallic catalysts, **A. McCue**, *University of Aberdeen, UK*

A step in the production of sustainable cellulose-derived biodiesel, **L. M. Mullins, C. Ryan and J. A. Sullivan**, *University College Dublin, Ireland*

Preparation, characterization and application of Fe_3O_4/PEG nanoparticles functionalized with ferrous oxalate or ferric citrate for removal of bisphenol A, **M. Neamtu, C. Nadejde, V.-D. Hodoroaba, R. J. Schneider, G. Ababei and U. Panne**, *"Al. I. Cuza" University of Iasi, Romania*

Au nanoparticle–polydopamine–reduced graphene oxide ternary heterogeneous system for selective C–H bondoxidation in benzylic hydrocarbons under mild conditions, **B. Majumdar, T. Bhattacharya and T. Sarma**, *Indian Institute of Technology Indore, India*

Dehydration of alcohols over heteropoly acid catalysts in the gas phase, **W. Alharbi and I. Kozhevnikov**, *University of Liverpool, UK*

Selective alkylation of benzene with propane over bifunctional Pt–heteropoly acid catalyst, **A. Alotaibi, E. F. Kozhevnikova and I. Kozhevnikov**, *University of Liverpool, UK*

Atrazine degradation through PEI–copper nanoparticles immobilized on montmorillonite and sand, **K. Sethu, I. Dror and B. Berkowitz**, *Weizmann Institute of Science, Israel*

Hydrodeoxygenation of biomass-derived ketones over bifunctional metal-acid catalysts in the gas phase, **K. H. Alharbi, M. A. Alotaibi, E. F. Kozhevnikova and I. Kozhevnikov**, *University of Liverpool, UK*

The surface–stress dependence of the H binding position onPt(111), **I. Shuttleworth,** *Nottingham Trent University, UK*

New heterogeneous catalyst systems for organic synthesis using tubular continuous flow reactors, **C. Hornung, *et al.***, *CSIRO, Australia*

Alkylaminocyclophosphazene-functionalized silica as solid support for polyoxometalate-catalyzed biphasic oxidation with hydrogen peroxide, **M. Craven, E. F. Kozhevnikova, C. Kunstmann-Olsen, A. Steiner, I. V. Kozhevnikov**, *University of Liverpool, UK*

Sulfonated graphene oxide and derivatives as solid catalysts for the dehydration of glucose to levulinic acid, **N. Said, N. Theyssen and W. Leitner**, *Max-Planck Institut für Kohlenforschung, Germany*

Functionalization of graphene for use in biosensors and gold catalysis, **D. Wotton, D. Morgan, P. R. Davies and O. Guy**, *Cardiff University, UK*

Cinnamaldehyde hydrogenation with bimetallic AuPd catalysts, **S. Cattaneo, N. Dimitratos, S. J. Freakley, M. Sankar and G. J. Hutchings**, *Cardiff University, UK*

A copper chloride catalyst for phosgene synthesis: surface science of Cu(100)/O(a)/HCl, **P. Davies, H. Altass, S. Guan, D. J. Willock, G. Rossi and D. Lennon**, *Cardiff University, UK*

The effect of Al distribution in H-ZSM-5 framework on alkene oligomerization, **E. Tabor, M. Bernauer, V. Pashkova, J. Dědeček, Z. Sobalík and B. Wichterlová**, *J. Heyrovský Institute of Physical Chemistry of the ASCR, v. v. i., Czech Republic*

Al-rich beta zeolites. Al organization in the framework and catalytic performance, **P. Sazama, E. Tabor, P. Klein, B. Wichterlova, G. Sadovska, V. Pashkova, S. Sklenak and J. Dedecek**, *J. Heyrovský Institute of Physical Chemistry of the ASCR, Czech Republic*

The role of Mo loadings on the properties of zirconia based catalytic materials, **F. Oloye, J. A. Anderson and A. J. McCue**, *University of Aberdeen, UK*

The effect of doping and laser irradiation on the photocatalytic water splitting over TiO_2 nanoparticles, **R. Fiorenza, S. Scire, G. Compagnini, L. D'Urso, M. E. Fragalà and S. Filice**, *University of Catania, Italy*

Bioethanol dehydrogenation over copper catalysts supported on functionalized-graphenes and a large surface area graphite, **M. V. Morales, E. Asedegbega-Nieto, B. Bachiller-Baeza and A. Guerrero-Ruiz**, *Universidad Nacional de Educación a Distancia (UNED), Spain*

One-step colloidal Cu/ZnO nanoparticles for the liquid phase direct synthesis of DME, **A. García-Trenco, E. White, M. S. P. Shaffer and C. K. Williams**, *Imperial College London, UK*

Cobalt oxides deposited by various methods on stainless steel sieves and their activity in the ethanol total oxidation, **F. Kovanda, K. Jirátová, J. Balabánová and P. Kšírová**, *University of Chemistry and Technology, Prague, Czech Republic*

The influence of storage solvent on the hydrogen content and morphological properties of Raney Ni, **J. A. Mendes-Burak, B. Spliethoff and R. Rinaldi**, *Max Planck Institute, Germany*

Theoretical study of confinement and Si/Al ratio effects on the pyridine adsorption in faujasite, **M. Castellà-Ventura and E. Kassab**, *Laboratoirede Chimie Théorique – CNRS/UPMC, France*

Electrochemical modification of diamond electrodes with core–shell nanoparticles for emerging applications, **C. Mavrokefalos, M. Hasan, W. Khunsin, S. Maier, J. Rohan, R. Compton and J. Foord**, *University of Oxford, UK*

Phenanthrene hydrocracking using a Cr doped NiMo/Al_2O_3 catalyst, **C. Heide and T. Reina**, *Imperial College London (Chemical Engineering), UK*

Study of the reversible formation of PdH and its impact on the control of partial hydrogenation of butadiene selectivity, **E. Asedegbega Nieto, A. Iglesias Juez, B. Bachiller-Baeza, M. Fernandez Garcia, I. Rodriguez Ramos, A. Guerrero Ruiz**, *UNED, Spain*

Pt-segregation on CeO_2 particles: characterization and kinetic studies, **F. Pilger, A. Testino and C. Ludwig**, *Paul Scherrer Institute (PSI)/École Polytechnique Fédérale de Lausanne (EPFL), Switzerland*

Computational investigation of the Brønsted acidity in three- and two-dimensional zeolites, **V. T. Ho and P. Nachtigall**, *Charles University in Prague, Czech Republic*

Methanol oxidation on nanoporous gold, **W. Dononelli, L. Moskaleva and T. Klüner**, *Carl-von-Ossietzky Universität Oldenburg, Germany*

Trends in ethylene hydrogenation over transition metal surfaces, **C. Heard, A. Posada Borbón and H. Grönbeck**, *Chalmers University of Technology, Sweden*

Preparation of Au-$Mn_{3-x}Fe_xO_4$/SiO_2 nanocatalysts through an *in-situ* post-treatment strategy, **M. Colombo, C. Wu, R. Brescia, M. Prato and L. Manna**, *Istituto Italiano di Tecnologia, Italy*

Two-dimensional NMR relaxometry as a non-invasive tool for comparing adsorption strengths in silica-supported catalysts, **N. Robinson, S. Jenkins and C. D'Agostino**, *University of Cambridge, UK*

Stabilizer-free gold and gold–palladium catalysts prepared by colloidal methods for alcohol oxidation, **L. Abis, S. J. Freakley, N. Dimitratos, M. Sankar and G. J. Hutchings**, *Indian Institute of Technology, Delhi, India*

Insights into the ring-opening and decarboxylation of biomass derived cyclic esters, **S. Gupta, Md. I. Alam, N. Sinha and M. A. Haider**, *Cardiff Catalysis Institute/Cardiff University, UK*

Development of sulfided catalysts based on micro/mesoporous materials for middle distillates hydrotreating, **E. Naranov, O. Golubev, A. Sadovnikov, A. Maximov and E. Karakhanov**, *Lomonosov Moscow State University, Russian Federation*

Determination of exposed Au sites in 1% Au/TiO_2 synthesised by deposition, precipitation and sol immobilisation methods, **E. Nowicka, D. J. Morgan, Simon A. Kondrat and G. J. Hutchings**, *Cardiff University, UK*

Ab initio simulations of alcohol dehydrogenation on Au–Pd bimetallic nanocatalysts, **L. M. Molina, A. Benito and J. A. Alonso**, *Universidad de Valladolid, Spain*

Molybdenum carbide as a potential substitute for noble metals in olefins hydrogenation, **K. Sebakhy, G. Vitale and P. Pereira-Almao**, *University of Calgary, Canada*

Composite β-$AgVO_3$@$V_{1.6}{}^{5+}$ $V_{0.4}{}^{4+}O_{4.8}$ hydrogels and xerogels for catalytic applications, **E. Serrano Larrea, R. Fernández de Luis, M. Iglesias and M. I. Arriortua**, *Universidad del País Vasco/Euskal Herriko Unibertsitatea, Spain*

Muon spin rotation (μSR) investigation on TiO_2 single crystals – a new probe to the defect in oxide, **H. Ariga, K. Shimomura, A. Pant, E. Torikai, K. Nagamine, S. Takakusagi and K. Asakura**, *Hokkaido University, Japan*

Bimetallic Zr(Ti) and Zr(Ce)-MOFs for Lewis acid catalysis, **M. Iglesias, A. M. Rasero-Almansa and F. Sánchez**, *Inst. Ciencia de Materiales de Madrid, Spain*

Selective Raney-Ni-catalyzed hydrodeoxygenation of phenol to benzene, **G. Calvaruso and R. Rinaldi**, *Max-Planck-Institute für kohlenforschung, Germany*

High-purity hydrogen production by sorption-enhanced steam glycerol reforming (SESGR) over a metal-stabilized CaO–nickel hybrid sorbent–catalyst, **M. C. Iliuta, M. Shokrollahi Yancheshmeh and H. Radfarnia**, *Laval University, Canada*

Nanosized cobalt oxides over aluminum monoliths for VOC oxidation, **K. Jirátová, F. Kovanda, J. Klempa and J. Balabanova**, *Institute of Chemical Process Fundamentals of the CAS, Czech Republic*

Synthesis of hierarchical Sn-containing zeolite beta for the catalytic valorisation of bulky molecules and continuous operation, **K. Yakabi, A. Al-Nayili and C. Hammond**, *Cardiff University, UK*

Hydrogenation reactions for green chemsitry: levulinic acid to gamma-valerolactone, **C. Reece, D. Willock, D. Jones, S. Iqbal, S. Ishikawa, W. Ueda and G. J. Hutchings**, *Cardiff University, UK*

HCN synthesis from CH_4 and NH_3 with ternary alloys, **B. Temel and P. E. Højlund Nielsen**, *Haldor Topsoe A/S, Denmark*

Ruthenium nanoparticles supported on carbon – an active catalyst for the hydrogenation of lactic acid to form 1,2-propane diol, **D. Jones, S. Iqbal, D. Schonmakers, L. Lu, S. Kondrat, P. Wells, E. Gibson, D. Morgan, C. Kiley and G. J. Hutchings**, *Cardiff University, UK*

Decarboxylative fluorination of carboxylic acids with heterogeneous catalysts, **G. Tarantino and C. Hammond**, *Cardiff University, UK*

TiO_2 –SiO_2 films from organic-free colloidal TiO_2 anatase nanoparticles as a photocatalyst for removal of VOCs from indoor air, **A. Šuligoj, U. Lavrenčič Štangar and N. Novak Tušar**, *National Institute of Chemistry, Slovenia*

Conversion of glucose into gluconic acid under base free reaction conditions, **S. Iqbal, Y. Cao, X. Liu, P. Miedziak, J. Edwards, R. Armstrong, D. Morgan, J. Wang and G. J. Hutchings**, *Cardiff University, UK*

Designing an effective low-temperature soot oxidation catalyst based on the cryptomelane matrix, **T. Jakubek, M. Kolodziej, P. Stelmachowski and A. Kotarba**, *Materials and Surface Chemistry group, Jagiellonian University, Poland*

Catalytic conversion of CO_2 by magnetite (Fe_3O_4), **D. Santos-Carballal, A. Roldan and N. H. de Leeuw**, *Cardiff University, UK*

Transformations of aromatic alcohols catalysed by a molybdate modified metal–organic framework, **R. Smith and L. Brammer**, *University of Sheffield, UK*

Production of silver clusters for catalysis from a carbon dioxide matrix in the matrix assembly cluster source (MACS), **R.Griffin, V. Oiko, T. Mathieu, L. Cao and R. Palmer**, *University of Birmingham, UK*

Population and hierarchy of active gold species in iron oxide catalysts for low temperature CO oxidation, **S. Freakley, Q. He, J. Edwards, A. Carley, A. Borisevich, C. Kiely, G. J. Hutchings and M. Haruta**, *Cardiff Catalysis Institute, UK*

Competitive hydrogenation catalysed by ruthenium nanoparticles in carbon nanoreactors, **M. Aygun, C. T. Stoppiello, M. Lebedeva, T. W. Chamberlain, M. D. C. Gimenez-Lopez and A. N. Khlobystov**, *University of Nottingham, UK*

Synthesis of nanosized Pt–Co alloy catalysts by thermal decomposition of molecular heterobimetallic precursors, **S. Linke, R. Naumann d'Alnoncourt, F. Rosowski and M. Driess**, *Technische Universität Berlin, Germany*

Creating selective sites for the partial oxidation of *n*-butane by atomic layer deposition of PO_x on the surface of V_2O_5, **V. Strempel, R. Naumann d'Alnoncourt, M. Driess and F. Rosowski**, *Technische Universität Berlin, Germany*

Multilayer stacks of supported size-selected clusters as a novel means of catalyst production, **N. Jian, K. Bauer and R. Palmer**, *University of Birmingham, UK*

Carbon nanodots enhanced C_3N_4 for photocatalystic CO_2 conversion, **X. Liu, M. Bayazit and C. Jiang**, *University College London (UCL), UK*

First-principles insights into the Cu(I) oxide-catalyzed conversion of CO_2 to fuels and chemicals, **A. Mishra and N. de Leeuw**, *University College London, UK*

Do state-of-the-art computational models explain why RuO_2 is more active than IrO_2 for the oxygen evolution reaction?, **L. Briquet, M. Sarwar, J. Mugo, G. Jones and F. Calle-Vallejo,** *Johnson Matthey, UK*

Highly active magnetic catalysts for efficient degradation of bisphenol A from aqueous media, **C. Nadejde, M. Neamtu, V.-D. Hodoroaba, R. J. Schneider, A. Paul, G. Ababei and U. Panne,** *'Alexandru Ioan Cuza' University, Romania*

Stability of nickel supported over CeO_2 catalysts in aqueous phase reforming, **C. Á. Jarauta Córdoba, L. García, J. Ruíz, M. Oliva and J. Arauzo,** *Univeridad de Zaragoza, Spain*

Mesoporous ZSM-5 zeolites hosting Fe-species as effective catalysts in NH_3-SCR of NO and NH_3-SCO reactions, **K. Brylewska, K. Tarach, M. Rutkowska, M. Jabłońska, M. Choi, L. Chmielarz and K. Góra-Marek,** *AGH University of Science and Technology in Kraków, Poland*

Novel microwave-assisted synthesis of magnetic palladium catalysts for the selective hydrogenation of cinnamaldehyde to hydrocinnamaldehyde, **A. M. Raspolli Galletti, C. Antonetti, E. Bertolucci, M. Marracci and B. Tellini,** *University of Pisa, Italy*

Mesoporous ferrierite: influence of structural, textural and acidic parameters on catalytic performance in ethanol dehydration, **K. Brylewska, K. Tarach, U. Filek, W. Mozgawa, Z. Olejniczak and K. Góra-Marek,** *AGH University of Science and Technology in Kraków, Poland*

Advances in time-resolved spectroscopy: New cell design for combined, *in situ* XAFS/DRIFTS, **E. Dann, E. Gibson, P. Collier, A. Goguet, C. Hardacre and P. Wells,** *University College London, UK*

Novel magnetic mesoporous silica nanospheres (MMSN) for ring opening polymerization, **P. Cruz, Y. Pérez Cortés, I. del Hierro Morales and M. Fajardo,** *Universidad Rey Juan Carlos, Spain*

The regeneration of deactivated heterogeneous catalysts by Fenton and non-Fenton H_2O_2-based oxidation routes, **D. Wigger, H. Musch, L. Falco, J. Socci, K. Gora-Marek and I. Melián-Cabrera,** *Aston University, UK*

Monitoring solid-state ion-exchange of copper into zeolites at low temperatures facilitated by ammonia, **R. R. Tiruvalam, S. Shwan, M. Skoglundh, L. F. Lundegaard, T. V. W. Jansens, A. Carlsson and P. N. R. Vennestrøm,** *Haldor Topsoe A/S, Denmark*

Tuning the surface termination of Fe_3O_4(001) thin films on Pt(001), **E. Davis, K. Zhang, Y. Cui, H. Kuhlenbeck, S. Shaikhutdinov and H.-J. Freund,** *Fritz Haber Institute of the Max Planck Society, Germany*

Influence of transport limitation on NO_x adsorption in dual-layer automotive catalysts, **M. Vaclavik, V. Novak, P. Koci, G. Gregory and D. Thompsett,** *University of Chemistry and Technology Prague, Czech Republic*

Selective ethanol dehydrogenation on nanoporous CuNi catalysts, **N. Janvelyan, J. Shan, T. Egle, J. Ye, J. Biener, M. Flytzani-Stephanopoulos and C. M. Friend,** *Harvard University, USA*

Synthesis and characterization of cobalt–iron supported catalyst, **A. Ismail, B. Liu, M. Casavola and F. de Groot**, *Utrecht University, Netherlands*

Unravelling the complexity of CO oxidation catalysis on Au nanoclusters, **N. Nikbin, N. Austin, M. Christiansen, D. G. Vlachos, M. Stamatakis and G. Mpourmpakis**, *University College London, UK*

A first-principles investigation of CO_2 interaction with nano-cluster modified TiO_2 rutile(110) and anatase(101) surfaces, **M. Fronzi and M. Nolan**, *Tyndall National Institute, University College Cork, Ireland*

Investigation of MOF-253$\cdot x$$PdCl_2$ derivatives as catalysts for CO_2 hydrogenation, **A. R. Auxilio, D. F. Kennedy, M. Marshall, M. Fulcher, J. Taghavimoghaddam and A. L. Chaffee**, *Monash University, Australia*

Magnetite iron oxide as a low temperature catalyst to activate CO_2 and water, **F. Ivars, F. Mirabella, P. Dementyev, K.-H. Dostert, S. Schauermannand H.-J. Freund**, *Fritz-Haber-Institutder Max-Planck-Gesellschaft, Germany*

The Skinner Prize for the best poster was jointly awarded to Mr Ross Griffin of University of Birmingham, UK, for his poster on production of silver clusters for catalysis from a carbon dioxide matrix in the matrix assembly cluster source (MACS) and Mr Neil Robinson of University of Cambridge, UK, for his poster on two-dimensional NMR relaxometry as a non-invasive tool for comparing adsorption strengths in silica-supported catalysts.

List of participants

Miss Laura Abis, *Cardiff Catalysis Institute/Cardiff University*, *United Kingdom*
Miss Einas Abood, *Newcastle University*, *United Kingdom*
Dr Shakeel Ahmed, *King Fahd University of Petroleum and Minerals*, *Saudi Arabia*
Mrs Walaa Alharbi, *University of Liverpool*, *United Kingdom*
Mrs Khadijah Alharbi, *University of Liverpool*, *United Kingdom*
Dr Mohammad Ashraf Ali, *King Fahd University of Petroleum and Minerals*, *Saudi Arabia*
Mr Abdullah Alotaibi, *University of Liverpool*, *United Kingdom*
Professor James Anderson, *University of Aberdeen*, *United Kingdom*
Dr Hiroko Ariga, *Hokkaido University*, *Japan*
Dr Esther Asedegbega Nieto, *UNED*, *Spain*
Miss Mehtap Aygun, *University of Nottingham*, *United Kingdom*
Professor Donald Bethell, *University of Liverpool*, *United Kingdom*
Professor Michael Bowker, *Cardiff University*, *United Kingdom*
Dr Ludovic Briquet, *Johnson Matthey Technology Centre*, *United Kingdom*
Dr Albert Bruix, *Aarhus University - iNANO & Department of Physics and Astronomy*, *Denmark*
Ms Kamila Brylewska, *Faculty of Materials Science and Ceramics, AGH University of Science and Technology in Kraków*, *Poland*
Mr Jorge Burak, *Max Planck Institute*, *Germany*
Professor Robbie Burch, *Queen's University Belfast*, *United Kingdom*
Mr Gaetano Calvaruso, *Max-Planck-Institute für Kohlenforschung*, *Germany*
Prof Dr Charles Campbell, *University of Washington*, *USA*
Dr Martine Castellà-Ventura, *Laboratoire de Chimie Théorique - CNRS/UPMC*, *France*
Professor Richard Catlow, *University College London*, *United Kingdom*
Mr Stefano Cattaneo, *Cardiff University*, *United Kingdom*
Professor Alan Chaffee, *Monash University*, *Australia*
Professor Dr Michael Claeys, *University of Cape Town*, *South Africa*
Dr Paul Collier, *Johnson Matthey*, *United Kingdom*
Dr Gillian Collins, *Tyndall National Institute, University College Cork*, *Ireland*
Miss Nicola Collis, *Johnson Matthey Technology Centre*, *United Kingdom*
Mr Massimo Colombo, *Istituto Italiano di Tecnologia*, *Italy*
Professor Dr Avelino Corma, *Instituto de Tecnología Química UPV-CSIC*, *Spain*
Mr Michael Craven, *University of Liverpool*, *United Kingdom*
Miss Paula Cruz, *Universidad Rey Juan Carlos*, *Spain*
Miss Ellie Dann, *University College London*, *United Kingdom*
Professor William David, *STFC*, *United Kingdom*
Professor Philip Davies, *Cardiff University*, *United Kingdom*
Mr Earl Davis, *Fritz Haber Institute of the Max Planck Society*, *Germany*
Mr Giampiero De Notarpietro, *Italcementi S.p.A.*, *Italy*
Dr Jiří Dědeček, *J. Heyrovský Institute of Physical Chemistry of the ASCR*, *Czech Republic*

Mrs Frederique Desmedt, *Solvay, Belgium*
Professor Domingo Dominguez, *University of Santiago de Compostela, Spain*
Mr Wilke Dononelli, *Carl von Ossietzky Universität Oldenburg, Germany*
Mr John Mark Douthwaite, *Cardiff University, United Kingdom*
Dr James Earley, *Johnson Matthey, United Kingdom*
Dr Mark Feaviour, *Johnson Matthey, United Kingdom*
Mr Roberto Fiorenza, *University of Catania, Department of Chemical Sciences, Sicily, Italy*
Dr Nico Fischer, *University of Cape Town, South Africa*
Dr Janet Fisher, *Johnson Matthey, United Kingdom*
Dr Simon Freakley, *Cardiff Catalysis Institute, United Kingdom*
Professor Hans-Joachim Freund, *Fritz-Haber-Institut der Max-Planck-Gesellschaft, Germany*
Professor Cynthia Friend, *Harvard University, USA*
Dr Marco Fronzi, *Tyndall National Institute, University College Cork, Ireland*
Dr Andrés García-Trenco, *Imperial College London, United Kingdom*
Dr Bruce Garrett, *Pacific Northwest National Laboratory, USA*
Professor Bruce Gates, *University of California, USA*
Dr Emma Gibson, *University College London, United Kingdom*
Mr Ross Griffin, *University of Birmingham, United Kingdom*
Dr Alex Grigoropoulos, *University of Liverpool, United Kingdom*
Dr Elad Gross, *Institute of Chemistry, The Hebrew University of Jerusalem, Israel*
Mr Shaoliang Guan, *Cardiff University, United Kingdom*
Miss Shelaka Gupta, *Indian Institute of Technology, Delhi, India*
Miss Alice Halman, *University of Central Lancashire, United Kingdom*
Mr Z. Hao, *Delft University of Technology, Netherlands*
Dr Christopher Heard, *Chalmers University of Technology, Sweden*
Miss Chiara Heide, *Imperial College London (Chemical Engineering), United Kingdom*
Prof Dr Moti Herskowitz, *Ben-Gurion University of the Negev, Israel*
Mr Viet Thang Ho, *Faculty of Science, Charles University in Prague, Czech Republic*
Dr Christian Hornung, *CSIRO, Australia*
Dr Mark Howard, *Cardiff University, United Kingdom*
Dr Hazel Hunter, *STFC, United Kingdom*
Professor Graham Hutchings, *Cardiff University, United Kingdom*
Dr Marta Iglesias, *Instituto Ciencia de Materiales de Madrid, Spain*
Professor Maria C. Iliuta, *Laval University, Canada*
Dr Sarwat Iqbal, *Cardiff University, United Kingdom*
Dr Francisco Ivars Barcelo, *Fritz Haber Institut der Max Planck Gesellschaft, Germany*
Mr Eoin Jackman, *Cardiff University, United Kingdom*
Mr Tomasz Jakubek, *Faculty of Chemistry, Jagiellonian University, Poland*
Miss Nare Janvelyan, *Harvard University, USA*
Mrs Clara Á. Jarauta Córdoba, *Universidad de Zaragoza, Spain*
Mr Nan Jian, *University of Birmingham, United Kingdom*
Dr Květa Jirátová, *Institute of Chemical Process Fundamentals of the CAS, Prague, Czech Republic*
Dr Medlock Jonathan, *DSM Nutritional Products Ltd., Switzerland*
Mr Wilm Jones, *Cardiff University, United Kingdom*
Mr Daniel Jones, *Cardiff University, United Kingdom*
Mrs Meltem Karaismailoglu, *Yildiz Technical University, Turkey*
Dr Emile Kassab, *Laboratoire de Chimie Théorique - CNRS/UPMC, France*
Dr Simon Kondrat, *Cardiff Catalysis Institute, United Kingdom*

Prof Dr Andrzej Kotarba, *Materials and Surface Chemistry group, Jagiellonian University, Poland*
Professor František Kovanda, *University of Chemistry and Technology, Prague, Czech Republic*
Dr Said Lassiri, *University of Glasgow, United Kingdom*
Mr Michael Lamb, *Durham University, United Kingdom*
Prof Dr Miron V. Landau, *Ben-Gurion University of the Negev, Israel*
Dr Philip Landon, *Sasol Technology (UK) Ltd, United Kingdom*
Professor David Lennon, *University of Glasgow, United Kingdom*
Dr Felix Leyssner, *SPECS Surface Nano Analysis GmbH, Germany*
Professor Jyh Shing Lin, *Department of Chemistry, Tamkang University, Chinese Taipei*
Ms Stephanie Linke, *BasCat, UniCat BASF JointLab, Technische Universität Berlin, Germany*
Prof Dr Robert Madix, *Harvard University, USA*
Mr Biju Majumdar, *Indian Institute of Technology, Indore, India*
Dr Josh Makepeace, *University of Oxford, United Kingdom*
Miss Grazia Malta, *Cardiff University, United Kingdom*
Dr Dimitrios Manganas, *Max Planck Institut for Chemical Energy Conversion, Germany*
Mr Haresh Manyar, *Queens University Belfast, United Kingdom*
Mr Christos Mavrokefalos, *University of Oxford, United Kingdom*
Dr Alan McCue, *University of Aberdeen, United Kingdom*
Dr Ignacio V Melian Cabrera, *Aston University, United Kingdom*
Dr Nathalie Meyer, *Huntsman Polyurethanes, Belgium*
Miss Amy Miller, *University of Cambridge Chemistry Department, United Kingdom*
Dr Abhishek Mishra, *University College London, United Kingdom*
Dr Luis M. Molina, *Universidad de Valladolid, Spain*
Ms María Virtudes Morales, *Universidad Nacional de Educación a Distancia (UNED), Spain*
Mr Raphaell Moreira, *Freie Universität Berlin, Germany*
Emeritus Professor Jacob Moulijn, *Delft University of Technology, Netherlands*
Ms Lisa M. Mullins, *UCD School of Chemistry and Chemical Biology, Belfield, Dublin 4, Ireland*
Dr Claudia Nadejde, *Interdisciplinary Research Department – Field Science, 'Alexandru Ioan Cuza' University, Romania*
Mr Evgeny Naranov, *Lomonosov Moscow State University, Russian Federation*
Dr Raoul Naumann d'Alnoncourt, *Technische Universität Berlin, BasCat, Germany*
Dr Mariana Neamtu, *Alexandru Ioan Cuza University, Iași, Romania*
Prof Dr Natasa Novak Tusar, *National Institute of Chemistry, Slovenia*
Dr Ewa Nowicka, *Cardiff Catalysis Institute, United Kingdom*
Mr Alex O'Malley, *University College London, United Kingdom*
Mr Femi Oloye, *University of Aberdeen, United Kingdom*
Professor Richard Palmer, *University of Birmingham, United Kingdom*
Prof Dr Itai Panas, *Chalmers University of Technology, Sweden*
Mr Luke Parker, *Cardiff University, United Kingdom*
Mr Frank Pilger, *Paul Scherrer Institut (PSI) / École Polytechnique Fédérale de Lausanne (EPFL), Switzerland*
Mr Alessandro Piovano, *Università degli Studi di Torino, Italy*
Dr Craig Plaisance, *Technishe Universität München, Germany*
Miss Antonia María Rasero Almansa, *Instituto de Ciencia de Materiales de Madrid, CSIC, Spain*

Prof Dr Anna Maria, Raspolli Galletti, *Department of Chemistry, University of Pisa, Italy*
Mr Christian Reece, *Cardiff University, United Kingdom*
Miss Nia Richards, *Cardiff University, Cardiff Catalysis Insitute, United Kingdom*
Mr Neil Robinson, *Department of Chemical Engineering and Biotechnology, University of Cambridge, United Kingdom*
Professor Notker Rösch, *Technische Universität München, Germany*
Dr Alberto Roldan, *Cardiff University, United Kingdom*
Ms Nesrine Said, *Max-Planck-Institut für Kohlenforschung, Germany*
Dr David Santos-Carballal, *Cardiff University, United Kingdom*
Prof Dr Joachim Sauer, *Humboldt University Berlin, Germany*
Mr Khaled Sebakhy, *University of Calgary, Canada*
Professor Parasuraman Selvam, *Indian Institute of Technology, Madras, India*
Dr Edurne Serrano Larrea, *Universidad del País Vasco/Euskal Herriko Unibertsitatea, Spain*
Dr Kalidhasan Sethu, *Weizmann institute of Science, Israel*
Mr Parag Shah, *Cardiff University, United Kingdom*
Dr Mzamo Shozi, *University of KwaZulu-Natal, South Africa*
Dr Ian Shuttleworth, *Nottingham Trent University, United Kingdom*
Ms Rebecca Smith, *University of Sheffield, United Kingdom*
Mr Joseph Socci, *Aston University, United Kingdom*
Dr Michail Stamatakis, *Department of Chemical Engineering, University College London, United Kingdom*
Dr Jakub Staszak-Jirkovský, *Machavert Pharmaceuticals, USA*
Ms Verena Strempel, *BasCat- UniCat BASF JointLab, Technische Universität Berlin, Germany*
Mr Henrik Svengren, *Department of Materials and Environmental Chemistry, Stockholm University, Sweden*
Dr Edyta Tabor, *J. Heyrovský Institute of Physical Chemistry of the ASCR, Czech Republic*
Dr Yasuyuki Takamitsu, *Tosoh corporation, Japan*
Dr Junwang Tang, *University College London, United Kingdom*
Miss Giulia Tarantino, *Cardiff University, United Kingdom*
Dr Burcin Temel, *Haldor Topsoe A/S, Denmark*
Dr David Thompsett, *Johnson Matthey Technology Centre, United Kingdom*
Mr Ram Tiruvalam, *Haldor Topsoe A/S, Denmark*
Dr Annette Trunschke, *Fritz-Haber-Institut der Max-Planck-Gesellschaft, Germany*
Professor Wataru Ueda, *Kanagawa University, Japan*
Professor Nisar Ullah, *King Fahd University of Petroleum and Minerals, Saudi Arabia*
Mrs Natalie Utsis, *Ben-Gurion University of the Negev, Israel*
Mr Marek Václavík, *University of Chemistry and Technology, Prague, Czech Republic*
Prof Dr Rutger van Santen, *Eindhoven University of Technology, Netherlands*
Prof Dr Bert Weckhuysen, *Utrecht University, Netherlands*
Dr Peter Wells, *University College London, UK Catalysis Hub, Research Complex at Harwell, United Kingdom*
Dr Gareth Whiting, *Utrecht University, Netherlands*
Dr David Willock, *Cardiff Catalysis Institute, United Kingdom*
Dr Tom Wood, *STFC, United Kingdom*
Mr Daniel Wotton, *Cardiff University, United Kingdom*
Mr Chunzheng Wu, *Istituto Italiano di Tecnologia, Italy*
Miss Keiko Yakabi, *Cardiff University, United Kingdom*
Dr Andraž Šuligoj, *National Institute of Chemistry, Slovenia*

